INTERNATIONAL SERIES OF MONOGRAPHS ON PHYSICS

International Series of Monographs on Physics

147. A. Yaouanc, P. Dalmas de Réotier: *Muon spin rotation, relaxation, and resonance*
146. B. McCoy: *Advanced statistical mechanics*
145. M. Bordag, G.L. Klimchitskaya, U. Mohideen, V.M. Mostepanenko: *Advances in the Casimir effect*
144. T.R. Field: *Electromagnetic scattering from random media*
143. W. Götze: *Complex dynamics of glass-forming liquids—a mode-coupling theory*
142. V.M. Agranovich: *Excitations in organic solids*
141. W.T. Grandy: *Entropy and the time evolution of macroscopic systems*
140. M. Alcubierre: *Introduction to* $3+1$ *numerical relativity*
139. A. L. Ivanov, S. G. Tikhodeev: *Problems of condensed matter physics—quantum coherence phenomena in electron-hole and coupled matter-light systems*
138. I. M. Vardavas, F. W. Taylor: *Radiation and climate*
137. A. F. Borghesani: *Ions and electrons in liquid helium*
136. C. Kiefer: *Quantum gravity, Second edition*
135. V. Fortov, I. Iakubov, A. Khrapak: *Physics of strongly coupled plasma*
134. G. Fredrickson: *The equilibrium theory of inhomogeneous polymers*
133. H. Suhl: *Relaxation processes in micromagnetics*
132. J. Terning: *Modern supersymmetry*
131. M. Mariño: *Chern-Simons theory, matrix models, and topological strings*
130. V. Gantmakher: *Electrons and disorder in solids*
129. W. Barford: *Electronic and optical properties of conjugated polymers*
128. R. E. Raab, O. L. de Lange: *Multipole theory in electromagnetism*
127. A. Larkin, A. Varlamov: *Theory of fluctuations in superconductors*
126. P. Goldbart, N. Goldenfeld, D. Sherrington: *Stealing the gold*
125. S. Atzeni, J. Meyer-ter-Vehn: *The physics of inertial fusion*
123. T. Fujimoto: *Plasma spectroscopy*
122. K. Fujikawa, H. Suzuki: *Path integrals and quantum anomalies*
121. T. Giamarchi: *Quantum physics in one dimension*
120. M. Warner, E. Terentjev: *Liquid crystal elastomers*
119. L. Jacak, P. Sitko, K. Wieczorek, A. Wojs: *Quantum Hall systems*
118. J. Wesson: *Tokamaks, Third edition*
117. G. Volovik: *The Universe in a helium droplet*
116. L. Pitaevskii, S. Stringari: *Bose-Einstein condensation*
115. G. Dissertori, I.G. Knowles, M. Schmelling: *Quantum chromodynamics*
114. B. DeWitt: *The global approach to quantum field theory*
113. J. Zinn-Justin: *Quantum field theory and critical phenomena, Fourth edition*
112. R.M. Mazo: *Brownian motion—fluctuations, dynamics, and applications*
111. H. Nishimori: *Statistical physics of spin glasses and information processing—an introduction*
110. N.B. Kopnin: *Theory of nonequilibrium superconductivity*
109. A. Aharoni: *Introduction to the theory of ferromagnetism, Second edition*
108. R. Dobbs: *Helium three*
107. R. Wigmans: *Calorimetry*
106. J. Kübler: *Theory of itinerant electron magnetism*
105. Y. Kuramoto, Y. Kitaoka: *Dynamics of heavy electrons*
104. D. Bardin, G. Passarino: *The Standard Model in the making*
103. G.C. Branco, L. Lavoura, J.P. Silva: *CP Violation*
102. T.C. Choy: *Effective medium theory*
101. H. Araki: *Mathematical theory of quantum fields*
100. L. M. Pismen: *Vortices in nonlinear fields*
99. L. Mestel: *Stellar magnetism*
98. K. H. Bennemann: *Nonlinear optics in metals*
94. S. Chikazumi: *Physics of ferromagnetism*
91. R. A. Bertlmann: *Anomalies in quantum field theory*
90. P. K. Gosh: *Ion traps*
87. P. S. Joshi: *Global aspects in gravitation and cosmology*
86. E. R. Pike, S. Sarkar: *The quantum theory of radiation*
83. P. G. de Gennes, J. Prost: *The physics of liquid crystals*
73. M. Doi, S. F. Edwards: *The theory of polymer dynamics*
69. S. Chandrasekhar: *The mathematical theory of black holes*
51. C. Møller: *The theory of relativity*
46. H. E. Stanley: *Introduction to phase transitions and critical phenomena*
32. A. Abragam: *Principles of nuclear magnetism*
27. P. A. M. Dirac: *Principles of quantum mechanics*
23. R. E. Peierls: *Quantum theory of solids*

Muon Spin Rotation, Relaxation, and Resonance

Applications to Condensed Matter

Alain Yaouanc
Commissariat à l'Energie Atomique, Grenoble, France

Pierre Dalmas de Réotier
Commissariat à l'Energie Atomique, Grenoble, France

OXFORD
UNIVERSITY PRESS

OXFORD
UNIVERSITY PRESS

Great Clarendon Street, Oxford OX2 6DP

Oxford University Press is a department of the University of Oxford.
It furthers the University's objective of excellence in research, scholarship,
and education by publishing worldwide in

Oxford New York

Auckland Cape Town Dar es Salaam Hong Kong Karachi
Kuala Lumpur Madrid Melbourne Mexico City Nairobi
New Delhi Shanghai Taipei Toronto

With offices in

Argentina Austria Brazil Chile Czech Republic France Greece
Guatemala Hungary Italy Japan Poland Portugal Singapore
South Korea Switzerland Thailand Turkey Ukraine Vietnam

Published in the United States
by Oxford University Press Inc., New York

First published 2011

British Library Cataloguing in Publication Data
Data available

Library of Congress Cataloging in Publication Data

Data available

Typeset by SPI Publisher Services, Pondicherry, India
Printed in Great Britain
on acid-free paper by
CPI Antony Rowe, Chippenham, Wiltshire

ISBN 978–0–19–959647–8

1 3 5 7 9 10 8 6 4 2

Contents

Preface

Muon spin rotation, relaxation, and resonance techniques, the so-called μSR techniques, provide information at an atomic level on the chemical and physical properties of matter. Their utilisation has enormously increased during the past decades, rendering *Muon Spin Rotation Spectroscopy: Principles and Applications in Solid State Physics* by Schenck (Adam Hilger, Bristol, 1985) incomplete. The more recent textbook *Solid State Phenomena As Seen by Muons, Protons, and Excited Nuclei* by Karlsson (Clarendon Press, Oxford, 1995) nicely presents some of the most striking results obtained by μSR in various subfields of solid state physics. However, the methodological and technical aspects are usually left aside. Smilga and Belousov have published a remarkable monograph on *The Muon Method in Science* (Nova Science, Commack, NY, 1994) in which the quantum mechanical description of the evolution of the muon polarisation in matter is given in great detail. This book reflects the interest of the Russian school which has always been more on the chemical and physical properties of muon and muonium in semiconductors and dielectrics and in conventional metals rather than on various aspects of magnetism and superconductivity. Nagamine has published *Introductory Muon Science* (Cambridge University Press, Cambridge, 2003), which provides an overall picture of muon-related scientific research, i.e. (i) application to muon catalysed nuclear fusion, (ii) investigation at the microscopic level of the magnetic properties of new materials and biomolecules, and (iii) radiography on a large scale. As the title of Nagamine's book suggests, the investigation of condensed matter is only discussed at an introductory level.

The present textbook focuses on the use of the muon as a probe of condensed matter, i.e. we discuss the μSR techniques. Our aim is to give a comprehensive and unified description of these techniques. While it is mostly intended for PhD students or researchers wishing to learn about the insights μSR can provide into condensed matter science, chemical physics, and materials science, it should also be accessible to undergraduate students. Our presentation is probably biased by our own interest in magnetism and superconductivity topics which nowadays represent more than a half of the activity of the μSR community. In particular, we try to describe the compounds under study in terms of conventional concepts such as the generalised susceptibility for magnetic materials. Mainly because of our ignorance, the study of liquids and purely chemical processes is only discussed in the last section of the chapter on muonium centres in materials. General references to the chemistry literature can be found in the books or review papers of Walker (1983), Chappert and Grynszpan (1984), Cox (1987), Roduner (1988), Smilga and Belousov (1994), and Reid and Roduner (2000) listed in the reference list of Chapter 1. The proceedings of the latest μSR conferences (edited by Heffner and collaborators, Blundell and collaborators, and Kojima and Higemoto, see the reference list in Chapter 1) give a feeling of the recent developments.

This book is divided into three parts. In the first part (Chapters 1 and 2) we briefly discuss some typical results of the application of μSR to condensed matter science and explain the basic principles involved. In Chapter 1, after an introduction to the μSR techniques, the reader is taught μSR through simple examples taken from the literature. Chapter 2 gives a description of the available muon beams and spectrometers and explain how to extract the physical quantities of interest, namely the muon polarisation functions, from the positron counts. The second part (Chapters 3, 4, 5, 6, 7, and 8) is the bulk of the monograph. It presents a comprehensive discussion of the polarisation functions. Many details are given with the hope that they do not overshadow the physics. In the third part (Chapters 9, 10, 11, 12), we analyse selected examples taken from fields for which the μSR techniques have provided key information: diffusion properties of light interstitials, magnetism, superconductivity, and muonium centres in materials. Note that the division between magnetism and superconductivity may seem somewhat arbitrary because of their intertwined nature in new materials, such as cuprates or iron-based superconductors. The superconductivity chapter only discusses properties unique to this state, such as vortex matter. Our purpose has not been to write a comprehensive review for each of these fields, but rather to present simple typical examples of the application of μSR. Each of the four chapters starts with a discussion of the physical properties probed by μSR. In addition, a list of relevant review articles and books is provided. To round up our discussion of the μSR techniques, we have written an epilogue which focuses on possible technical developments.

Finally, there are eight appendices at the end of this textbook. The first deals with the μSR data analysis. The appendix that follows lists the most commonly used polarisation functions. Then, mathematical concepts and formulae are recalled in Appendix C. The next three appendices summarise physical concepts and their consequences. They are intensively employed in the main body of this book. We finish with two appendices providing nuclear and electronic data in table format.

The International System of Units (SI) is used throughout the text, and fundamental constants like the Boltzmann and Dirac constants are retained in all the expressions.

The Commissariat à l'Energie Atomique in Grenoble has provided us with a stimulating environment in which to work. Interactions with numerous colleagues have shaped many aspects of our thinking.

We are indebted to Paul C. M. Gubbens for enlighting discussions over the years and his comments on the content of the book. The numerous single crystals that he has been able to obtain has allowed us to refine our insight into the μSR response in magnetic and superconducting compounds. The authors express their sincere thanks to Erik B. Karlsson and Douglas E. MacLaughlin, who have read the entire manuscript and whose insightful questions have immensely improved the text. We are grateful to Steve F. J. Cox, who, in addition to comments on the manuscript, has provided us with key notions on the muonium formation mechanisms and the properties of muonium centres in materials. Obviously, all the defects and errors which remain in this book are ours. All the errors that we discover or are communicated to us by our readers will be posted at http://ukcatalogue.oup.com/product/9780199596478.do .

List of important symbols

Symbol	Definition
a_0	Initial muon asymmetry.
$a_{\mathrm{Mu},0}$	Free muonium ground-state radius.
$\mathbf{B}_0$	Spontaneous magnetic field vector at the muon site.
$B_{\mathrm{c}1}$	Lower critical field for a type II superconductor.
$B_{\mathrm{c}2}$	Upper critical field for a type II superconductor.
$\mathbf{B}_{\mathrm{dem}}$	Demagnetisation field vector. This is the dipole field which arises from the magnetic moments of the sample and it is computed in the continuum approximation.
$\mathbf{B}_{\mathrm{dip}}$	Dipole magnetic field vector operator at the muon site arising from the dipole interaction between the muon magnetic moment and the localised magnetic moments of the whole sample.
$\mathbf{B}'_{\mathrm{dip}}$	Dipole magnetic field vector at the muon site due to the magnetic moments inside the Lorentz sphere.
$\mathbf{B}_{\mathrm{ext}}$	Vector representing the external applied magnetic field.
$\mathbf{B}_{\mathrm{Fermi}}$	Magnetic field vector operator at the muon site arising from the Fermi contact interaction between the muon magnetic moment and the conduction electron magnetisation density at the muon site.
$\mathbf{B}_{\mathrm{hyp}}$	Hyperfine magnetic field vector operator at the muon site arising from the short-range interaction between the muon and localised magnetic moments.
$\mathbf{B}_{\mathrm{loc}}$	Local magnetic field vector at the muon site; it is an operator in a quantum mechanical context.
$\mathbf{B}_{\mathrm{Lor}}$	Lorentz field vector: it is the opposite of the dipole magnetic field vector at the muon site due to the magnetic moments within the Lorentz sphere computed in the continuum approximation.
c	Speed of light in vacuum.
c_{imp}	Fractional concentration of impurities or defects.
$\mathbf{d}$	Vector specifying the position of an atom in a unit cell.
$D_{\mathrm{v},\mathbf{B}_{\mathrm{ext}}}(\mathbf{B}_{\mathrm{loc}})$	Distribution function of vector $\mathbf{B}_{\mathrm{loc}}$, measured in an external field $\mathbf{B}_{\mathrm{ext}}$. Sometimes written as $D_{\mathrm{v},\mathbf{B}_{\mathrm{ext}}}(B_{\mathrm{loc}})$ if the distribution is isotropic. If $\mathbf{B}_{\mathrm{ext}} = 0$, the subscript $\mathbf{B}_{\mathrm{ext}}$ is generally omitted.

Symbol	Definition
$D_{\mathrm{c},\mathbf{B}_{\mathrm{ext}}}(B^{\alpha}_{\mathrm{loc}})$	Distribution function of the α component of $\mathbf{B}_{\mathrm{loc}}$, measured in an external field $\mathbf{B}_{\mathrm{ext}}$. If $\mathbf{B}_{\mathrm{ext}}=0$, the subscript $\mathbf{B}_{\mathrm{ext}}$ is generally omitted.
$D_{\mathrm{m},\mathbf{B}_{\mathrm{ext}}}(B_{\mathrm{loc}})$	Distribution function of the modulus B_{loc}, measured in an external field $\mathbf{B}_{\mathrm{ext}}$. If $\mathbf{B}_{\mathrm{ext}}=0$, the subscript $\mathbf{B}_{\mathrm{ext}}$ is generally omitted.
D_{diff}	Diffusion constant, i.e. diffusivity.
$\mathcal{D}$	Constant used in several expressions for the relaxation rates; $\mathcal{D} \equiv (\mu_0/4\pi)^2 \gamma_\mu^2 (g\mu_{\mathrm{B}})^2 / v_{\mathrm{c}}$.
e	Proton electric charge.
$F^{\alpha\beta}(\mathbf{q},\omega)$	$\alpha\beta$ component of the spectral-weight tensor.
g	Spectroscopic splitting factor, e.g. g_{e} for a free electron and the Landé factor for a free rare-earth ion.
h	Planck constant.
$\hbar \equiv h/(2\pi)$	Reduced Planck constant or Dirac constant.
$\mathcal{H}_{\mathrm{dip}}$	Dipole Hamiltonian operator accounting for the dipole interaction between the muon magnetic moment and the localised magnetic moments of the system.
$\mathcal{H}_{\mathrm{sys}}$	Hamiltonian operator for the system under study, including the possible effects of the perturbation due to the muon.
$\mathcal{H}_{\mathrm{tot}}$	Hamiltonian operator modelling the muon–system ensemble.
$\mathcal{H}_{\mathrm{Q},i}$	Quadrupole Hamiltonian operator for nucleus i.
$\mathcal{H}_{\mathrm{Z},\mu}$	Zeeman Hamiltonian operator for the muon spin.
$\mathcal{H}_{\mathrm{Z,sys}}$	Zeeman Hamiltonian operator for the system under study.
$\mathbf{i}$	Direct lattice vector.
$\mathbf{I}_i$	Nuclear spin vector operator of nucleus at site i.
$\mathcal{I}$	Unit operator.
$\mathbf{J}_i$	Total angular momentum vector operator, sometimes referred to as spin, of atom at site i.
$\mathbf{k}$	Propagation vector of a magnetic structure.
$\mathbf{K}$	Reciprocal lattice vector.
k_{B}	Boltzmann constant.
K'_{dip}	Frequency shift associated to the dipole field from magnetic moments inside the Lorentz sphere.
K_{exp}	Experimental frequency shift.
K_μ	Intrinsic muon Knight shift.
ℓ_{mfp}	Electronic mean free path.
$\mathbf{m}_i$	Magnetic moment vector of atom at site i; it is an operator in a quantum mechanical context.
$\mathbf{M}$	Magnetisation vector of modulus M.
M_{e}	Electron rest mass.
M_μ	Muon rest mass.
M_{p}	Proton rest mass.

Symbol	Definition
N	Depending on the context, N is either the number of magnetic atoms (or nuclei) in the sample or the number of flux lines in the case of a superconducting sample.
$n_{\rm c}$	Number of unit cells in the sample.
$n_{\rm d}$	Number of magnetic ions in the unit cell.
$N_{\pm}$	Positron count in the forward (+) or backward (−) direction.
$P_X(t)$	Transverse polarisation function.
$P_Z(t)$	Longitudinal polarisation function.
$P_X^{\rm stat}(t)$	Static transverse polarisation function.
$P_Z^{\rm stat}(t)$	Static longitudinal polarisation function.
$P_X^{\rm exp}(t)$	Experimentally measured transverse polarisation function.
$P_Z^{\rm exp}(t)$	Experimentally measured longitudinal polarisation function.
$\mathbf{q}$	Depending on the context, either a wave vector in the first Brillouin zone or a wave vector in the reciprocal space.
q_i	Electric field gradient at site i.
Q_i	Quadrupole moment of nucleus at site i.
$\mathbf{r}_0$	Vector specifying the muon site.
$\mathbf{r}_{i+d}$	Vector radius from the muon to atom at site $i + d$.
$\mathbf{S}_\mu$	Muon spin vector; it is an operator in a quantum mechanical context. When there is no risk of confusion, we take $\mathbf{S}_\mu$ as the initial polarisation vector of the muon beam which is proportional to $\mathbf{S}_\mu(t = 0)$.
t	Time as in $P_\alpha(t)$.
T	Temperature of the compound under study.
$T_{\rm c}$	Critical superconducting or magnetic phase transition temperature.
$T_{\rm C}$	Curie temperature.
$T_{\rm N}$	Néel temperature.
V	Volume of the sample.
v_0	Volume per magnetic ion.
$v_{\rm c}$	Volume of the magnetic unit cell.
$v_{\rm c}^*$	Volume of the first Brillouin zone.
$\{X, Y, Z\}$	Orthogonal reference frame attached to the laboratory with the Z axis parallel to $\mathbf{B}_{\rm ext}$ or the muon initial polarisation if $\mathbf{B}_{\rm ext} = 0$.
$\{x, y, z\}$	Orthogonal reference frame attached to the axes of the crystal under study.
$\alpha_{\rm ani}$	Uniaxial anisotropy parameter of $D_{{\rm v},\mathbf{B}_{\rm ext}}(\mathbf{B}_{\rm loc})$.
$\alpha_{\rm bal}$	Instrumental balance parameter.
β	Depending on the context, β is either the thermal parameter $(k_{\rm B}T)^{-1}$ or the critical exponent for the magnetisation.
$\gamma_{\rm e}$	Gyromagnetic ratio of the free electron.
γ_i	Gyromagnetic ratio of nucleus or atom at site i.

Symbol	Definition
$\gamma_{\rm p}$	Proton gyromagnetic ratio.
γ_μ	Muon gyromagnetic ratio.
$\Gamma_{\rm Mu}$	Muonium hopping or tunnelling rate between two sites.
$\Gamma_{\rm t}$	Generic term: either $\Gamma_{\rm Mu}$ or Γ_μ.
Γ_μ	Muon hopping or tunnelling rate between two sites.
$\Gamma(\mathbf{q})$	Decay rate of a magnetic fluctuation or excitation of wave vector $\mathbf{q}$.
$\delta_{i,j}$	Kronecker symbol; $\delta_{i,j} = 1$ if $i = j$, 0 otherwise. The indices i and j can refer to vectors.
$\Delta_{\rm G,TF}$, $\Delta_{\rm G,ZF}$	Standard deviation of a component of a Gaussian field distribution; $\Delta_{\rm G,TF}$ ($\Delta_{\rm G,ZF}$) refers to the transverse- (zero- or longitudinal-) field geometry. Its square $\Delta^2_{\rm G,TF}$ ($\Delta^2_{\rm G,ZF}$) is the variance of the distribution.
$\Delta_{\rm GbG}$	Standard deviation of the distribution of $\Delta_{\rm G,TF}$ or $\Delta_{\rm G,ZF}$ centred at Δ_0.
$\Delta_{\rm L,TF}$	Half-width at half-maximum of a Lorentzian field-component distribution.
$\Delta_{\rm L,ZF}$	Parameter characterising an isotropic zero-field distribution $D_{\rm v}(B_{\rm loc})$ in the case of diluted magnetic impurity. The half-width at half-maximum of $D_{\rm v}(B_{\rm loc})$ is equal to $(2^{1/2} - 1)^{1/2}\Delta_{\rm L,ZF}$.
$\Delta_{Z,\rm v}$	Standard deviation of the distribution of the field Z component arising from the flux-line lattice of a superconductor.
$\Delta_{Z,\rm L}$	Value of $\Delta_{Z,\rm v}$ according to the London model.
ε_0	Permittivity of free space.
η_i	Electric field gradient asymmetry parameter at site i.
ϑ^α	Skewness of $D_{\rm c,\mathbf{B}_{\rm ext}}(B^\alpha_{\rm loc})$.
$\kappa_{\rm m}$	Inverse of $\xi_{\rm m}$.
$\kappa_{\rm GL}$	Ginzburg–Landau parameter; $\kappa_{\rm GL} = \lambda_{\rm L}/\xi_{\rm GL}$.
$\lambda_{\rm L}$	London penetration depth.
λ_X	Exponential decay rate of $P_X(t)$. It is denoted by $1/T_2$ in the nuclear magnetic resonance literature.
λ_Z	Exponential relaxation rate of $P_Z(t)$. It is denoted by $1/T_1$ in the nuclear magnetic resonance literature.
μ_0	Permeability of free space.
$\mu_{\rm B}$	Electronic Bohr magneton.
$\mu_{\rm n}$	Nuclear magneton.
ν_0	Free muonium hyperfine frequency.
$\nu_\mu \equiv \omega_\mu/(2\pi)$	Muon precession frequency.
$\nu_{\rm c} \equiv 1/\tau_{\rm c}$	Magnetic correlation frequency.
ξ_0	Pippard–BCS coherence length.
$\xi_{\rm GL}$	Ginzburg–Landau coherence length.
$\xi_{\rm m}$	Magnetic correlation length to characterise either the disorder or the critical dynamics.

Symbol	Definition
$\boldsymbol{\sigma}$	Pauli vector operator; $\boldsymbol{\sigma} = 2\mathbf{S}_\mu$ when dealing with the muon.
$\tau_c \equiv 1/\nu_c$	Magnetic correlation time.
τ_μ	Muon lifetime.
Φ_0	Magnetic flux quantum.
χ_0	Uniform magnetic susceptibility; $\chi_0 \equiv \chi(\mathbf{q} = \mathbf{0})$.
$\chi(\mathbf{q})$	Wave vector-dependent static susceptibility; $\chi(\mathbf{q}) \equiv \chi(\mathbf{q}, \omega = 0)$.
$\chi(\mathbf{q}, \omega)$	Generalised magnetic susceptibility (complex number).
ω	Angular frequency. In Fourier transforms, it is the t-conjugate variable.
ω_μ	Muon angular precession frequency. Depending on the context $\omega_\mu = \gamma_\mu B_{\rm loc}$, $\omega_\mu = \gamma_\mu B_0$, or $\omega_\mu = \gamma_\mu B_{\rm ext}$.
$\mathcal{R}e\{A\}$, $\mathcal{I}m\{A\}$	Real and imaginary parts of complex number or function A
$A^\dagger$	Hermitian conjugate of operator A.
$\langle A \rangle$	Thermal average of A.
$\overline{A}$	Ensemble average of A.
$\mathrm{Tr}\{A\}$	Trace of operator A: $\sum_m \langle m\|A\|m \rangle$; the sum is over all the basis states of the Hilbert space in which A is defined.
$[A, B]$	Commutator of operators A and B: $[A, B] = AB - BA$.
$\{AB\}$	Symmetrised product of operators A and B: $\{AB\} = (AB + BA)/2$.
$\boldsymbol{A}$	Denotes a tensor. Vectors which are tensors of rank one are denoted as $\mathbf{A}$.
$\hat{\mathbf{A}}$	Unit vector collinear to $\mathbf{A}$; $\hat{\mathbf{A}} = \mathbf{A}/A$.
$\mathbf{A} \cdot \mathbf{B}$	Scalar product of vectors or vector operators $\mathbf{A}$ and $\mathbf{B}$.
$\mathbf{A} \times \mathbf{B}$	Vector product of vectors or vector operators $\mathbf{A}$ and $\mathbf{B}$.

Numerical values of fundamental constants

All the numerical values given here have been taken from the website http://physics.nist.gov/cuu/Constants/index.html as of October 2009, except for four of them. The muon lifetime τ_μ is from Amsler *et al.* (2008),[1] and γ_μ, $a_{\rm Mu,0}$, and ν_0 are derived using Eqs. 1.4, 8.2, 8.3, and 8.6. Note that with our convention the gyromagnetic ratio of the free electron is negative. No uncertainty is reported for the different values either because some of the constants are taken as definitions in the International System of Units or because the standard deviations do not affect the five or six digits that we reproduce here.

Quantity	Symbol	SI value
Muonium ground-state radius	$a_{\rm Mu,0}$	$5.31736\times10^{-11}\,{\rm m}$
Speed of light in vacuum	c	$2.99792\times10^{8}\,{\rm m\,s^{-1}}$
Proton electric charge	e	$1.60219\ \times10^{-19}\,{\rm C}$
Spectroscopic splitting factor of the free electron	$g_{\rm e}$	2.00232
Planck constant	h	$6.62607\times10^{-34}\,{\rm J\,s}$
Reduced Planck constant	$\hbar\equiv\frac{h}{2\pi}$	$1.05457\times10^{-34}\,{\rm J\,s}$
Boltzmann constant	$k_{\rm B}$	$1.38065\times10^{-23}\,{\rm J\,K^{-1}}$
Electron rest mass	$M_{\rm e}$	$9.10938\times10^{-31}\,{\rm kg}$
Proton rest mass	$M_{\rm p}$	$1.67261\times10^{-27}\,{\rm kg}$
Muon rest mass	M_μ	$1.88353\times10^{-28}\,{\rm kg}$
		$2.06768\times10^{2}\,{\rm M_e}$
		$1.12609\times10^{-1}\,{\rm M_p}$
		$1.05658\times10^{2}\,{\rm MeV\,c^{-2}}$
Gyromagnetic ratio of the free electron	$\gamma_{\rm e}$	$-1.76086\times10^{11}\,{\rm s^{-1}\,T^{-1}}$
Proton gyromagnetic ratio	$\gamma_{\rm p}$	$2.67522\times10^{8}\,{\rm rad\,s^{-1}\,T^{-1}}$
Positive muon gyromagnetic ratio	γ_μ	$8.51616\times10^{8}\,{\rm rad\,s^{-1}\,T^{-1}}$
Permittivity of free space	ε_0	$8.85419\times10^{-12}\,{\rm F\,m^{-1}}$
Permeability of free space	μ_0	$4\pi\times10^{-7}\,{\rm N\,A^{-2}}$
Electronic Bohr magneton	$\mu_{\rm B}$	$9.27401\times10^{-24}\,{\rm J\,T^{-1}}$
Nuclear magneton	$\mu_{\rm n}$	$5.05078\times10^{-27}\,{\rm J\,T^{-1}}$
Positive muon magnetic moment	$\mathfrak{m}_\mu$	$4.49045\times10^{-26}\,{\rm J\,T^{-1}}$
Free muonium hyperfine frequency	ν_0	$4.4642\times10^{9}\,{\rm Hz}$
Muon lifetime	τ_μ	$2.1970\times10^{-6}\,{\rm s}$
Magnetic flux quantum	Φ_0	$2.06783\times10^{-15}\,{\rm T\,m^2}$

[1] Amsler C. *et al.* (Particle Data Group) (2008) *Physics Letters* **B 667**, 1–1340.

Part I

Basic principles

1
A short survey of μSR research

The principles on which the muon spin techniques rest were established experimentally by Garwin, Lederman, and Weinrich (Garwin *et al.* 1957), following the proposal by Lee and Yang (1956) that the long-held principle of invariance under space reflection (parity) was violated by the weak interaction responsible for the decay of the pion and muon. The muon results from the pion decay. It is remarkable that already in 1957, Garwin and collaborators noted that “it seems possible that polarized positive and negative muons will become a powerful tool for exploring magnetic fields in nuclei, atoms, and interatomic regions.” We hope that this monograph will convince the reader that this prediction has been realised for condensed matter.

Two muon species exist in nature and they differ by the sign of their electric charge. Except in Section 4.6, which specifically deals with the investigation of condensed matter with negative muons, this textbook will consider only the positive muon. Since there is no risk of confusion, the positive muon will simply be referred to as the muon.

The muon spin techniques are used for the physical and chemical investigation of condensed-matter systems. They are usually denoted collectively as the μSR techniques. The acronym stands for muon spin rotation, relaxation, or resonance. It is a method for studying the structure and dynamics of matter on the atomic scale. It involves implanting spin-polarised muons into a sample and observing the effects of the local environment on the behaviour of the muon spins.

Primarily, a muon acts as a sensitive magnetometer: it has a spin one-half and a magnetic moment of $\mathfrak{m}_\mu = 8.8906\,\mu_{\rm n}$ ($\mu_{\rm n} = 5.05078\times10^{-24}\,{\rm J\,T^{-1}}$ is the nuclear magneton). Once inside a material, it probes the effects of local fields at its site, and its precession or resonance signal provides a measurement of these fields. Implantation of spin-polarised muons and subsequent measurements of the time evolution of the polarisation enables the distribution and fluctuations of the internal fields to be examined.

The implanted muons can also tell us about the behaviour of hydrogen in materials: a muon has the same charge as a proton and about one-ninth of its mass, so it acts like a light proton isotope (although strictly speaking, the muon is not an isotope of the proton since they do not belong to the same class of particles; the muon is a lepton like the electron but unlike the proton which is a hadron). An implanted muon can also pick up an electron to form muonium; the muonium response mimics that of a hydrogen atom.

Muons are unstable particles, decaying with a lifetime of $\tau_\mu = 2.1970\,\mu$s to produce a positron and two neutrinos. The muon spin behaviour inside a material is relayed to us *via* its decay positron. Indeed, this daughter particle is preferentially emitted in the direction of the muon spin at the instant of decay. Detection of the positrons emitted from a sample following implantation of polarised muons therefore reveals the muon spin interactions within the compound.

The μSR techniques have many attractive features which explain their present fast development. Compared to most other microscopic techniques, a μSR experiment is relatively straightforward. The μSR spectroscopy is of general use: the sample does not need to contain specific nuclei. A relatively complicated sample environment can be used as illustrated, for example, by the large number of measurements performed in the 10 mK – 1 K range using dilution refrigerators. At present, measurements have been performed in applied magnetic fields up to ∼8 T and in external pressure up to at least ∼1.5 GPa with the sample temperature far below 1 K. Recording of a μSR spectrum requires in most cases no more than one hour. Therefore, in a reasonable amount of time, it is possible to map out the μSR response to changes in temperature, pressure, or applied magnetic field in a compound.

As does any experimental technique, the μSR method has its own limitations. One of the major drawbacks is the large sample mass needed relative, for example, to samples probed at third generation synchrotron radiation sources. The smallest sample studied to date had a cross section of 6.5 mm^2 and a thickness of 0.07 mm (Sonier *et al.* 2003). For the more widely used beamlines, the so-called surface muon beams (i.e. relatively slow muon beams) and with a conventional setup, 150 mg of material per square centimetre of beam cross section is required to stop the muons. This means that the sample thickness needs not to be much larger than ∼1 mm for organic materials (density $\simeq 1\,\mathrm{g/cm^3}$) and much less (100 to 300 μm) for denser material such as transition metals.

A vast amount of μSR studies have been carried out. A large number of them have been described in previously published general references which are still of great interest: Seeger (1978), Walker (1983), Chappert and Grynszpan (1984), Schenck (1985), Chappert and Yaouanc (1986), Cox (1987), Brewer (1994), Smilga and Belousov (1994), Karlsson (1995), Schatz and Weidinger (1996), Blundell (1999), Kalvius, Noakes, and Hartmann (2001), and Nagamine (2003). A book edited in 1999 by Lee, Kilcoyne, and Cywinski presents many examples explained by experts of the applications of the μSR techniques in physics, chemistry, and material science.

Permanent muon facilities presently exist at four places: (i) the Swiss Muon Source (SμS) of the Paul Scherrer Institute (PSI, Villigen, Switzerland); (ii) the Centre for Molecular and Materials Science (CMMS) of TRIUMF, Vancouver, Canada; TRIUMF is a name formerly abbreviated from TRI-University Meson Facility where the TRI acknowledges the three founding Canadian universities: the universities of British Columbia, Alberta, and Victoria; and (iii) the ISIS pulsed neutron and muon source of the Rutherford Appleton Laboratory (RAL, Didcot, United Kingdom). A fourth facility located at the Japan Proton Accelerator Research Complex (J-PARC, Tokai, Japan) has recently become available, replacing the one at the High Energy Accelerator Research Organisation (KEK, Tsukuba, Japan), which was shut down. The first two (SμS and CMMS) have pseudo-continuous sources, while the others provide pulsed muon beams. The experience gained at ISIS and at SμS has shown that having neutron and muon beams at the same institution is a real advantage as it provides opportunities for the neutron and muon communities to interact.

In this chapter a simplified description of the μSR techniques is given, followed by a section describing a number of examples of the application of the μSR techniques in condensed-matter research. Some of these examples have been selected due to their historical interest and the fact they have provided timely key information to the condensed-matter community.

1.1 Basic principles of the μSR techniques

Here we shall first describe the basic purpose of a μSR experiment and the associated experimental setup. Then we shall present simple examples of the polarisation functions which are the physical quantity of interest in most experiments. While muons are thermalised as free particles in most materials, this is not always the case. In the third part, we shall discuss the muon thermalisation process and explain that muonium can sometimes be formed. The physics of muonium centres is summarised in the last part.

1.1.1 μSR measurements and experimental setups

As already mentioned, in the μSR techniques polarised muons are implanted into a sample where their polarisation evolves under the action of the local magnetic field at the muon sites, $\mathbf{B}_{\rm loc}$, until they decay. While the muon thermalisation process is to be presented later on (see Section 1.1.3), for the time being one must retain that this slowing down does not affect the muon polarisation and that a muon implanted in a metal can be considered as an electronically dressed particle which may be mobile. In semiconductors and molecular compounds, there is usually an appreciable proportion of muons forming muonium atoms.

The purpose of most μSR experiments is to follow the muon polarisation as a function of time, i.e. to perform a so-called time-differential measurement. Because of its positive charge, the muon localises at an interstitial site. Due to the absence of quadrupolar electric moment (spin $\frac{1}{2}$) the muon does not couple to electric field gradients. The decay positron is emitted preferentially along the muon spin direction; by collecting several millions positrons, one can reconstruct the time dependence of the muon polarisation function which, in turn, reflects the spatial and temporal distribution of the magnetic fields at the muon sites. Because of their large kinetic energy ($\approx 30\,\mathrm{MeV}$), the positrons easily escape the sample: they are also weakly absorbed by cryostat or furnace walls so that a complex sample-environment equipment may be used. In summary, the μSR techniques provide a local probe of hyperfine fields, as do nuclear magnetic resonance (NMR), perturbed angular correlations (PAC), and Mössbauer spectroscopy. Therefore the basic physical concepts developed for the latter techniques can be adapted to μSR. One of its unique characteristics is its ability to detect the presence of very small magnetic fields and therefore small magnetic moments. In fact, effects of nuclear magnetic moments, which are $\sim 10^3$ smaller than usual electronic magnetic moments, can be quite easily studied. They correspond to a field of at most a few hundredths of millitesla at the muon site. In the other limit, local magnetic fields at the muon site as high as $8\,\mathrm{T}$ are routinely investigated at CMMS.

Two types of experimental geometries are generally employed; see Fig. 1.1. In the transverse geometry an external magnetic field $\mathbf{B}_{\rm ext}$ is applied perpendicular to the

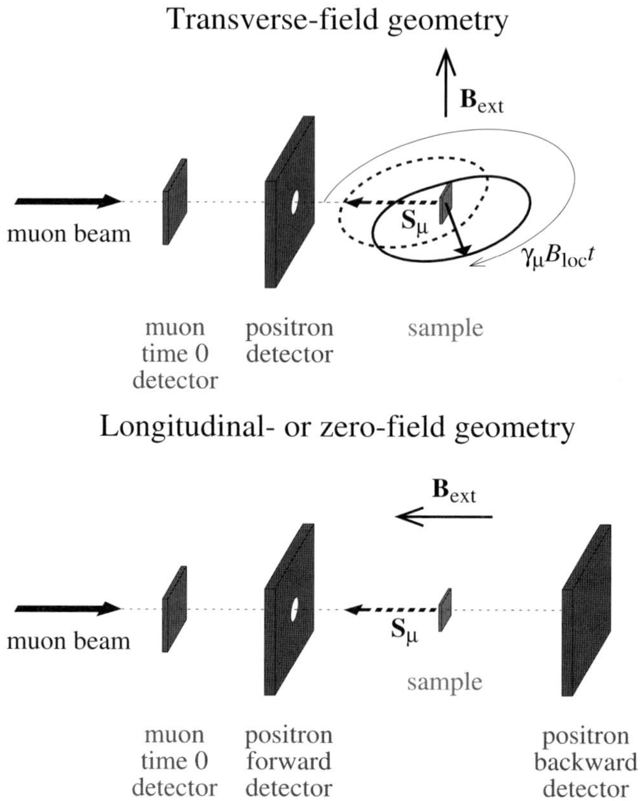

Fig. 1.1 The two types of experimental geometries of common use. The arrows originating from the sample sketch the muon spin direction: the dashed one at time $t = 0$ and the solid one at t. The dashed line drawn for the transverse-field geometry panel represents the probability $W(\theta)$ of positron emission along the muon beam direction when the muon is implanted into the sample. The solid line is for the same probability at a later time t. The muon spin is then along a direction making an angle $\gamma_\mu B_{\rm loc} t$ from the initial muon beam polarisation vector.

initial muon beam polarisation $\mathbf{S}_\mu$. In the longitudinal geometry $\mathbf{B}_{\rm ext}$ is parallel to $\mathbf{S}_\mu$. Zero-field experiments are also common practice: they are considered as a special case of longitudinal-geometry measurements.

In a basic time-differential measurement, an electronic clock is started just when a muon is implanted into the sample. This clock is stopped when the positron produced by the muon is detected. Thus an histogram is built from the positron counts versus the time spent by the muons in the sample. The decay of the average count reflects the finite muon lifetime. The anisotropy of the muon decay is reflected by an excess or a shortage in the actual counts. This deviation from the mean provides the physical quantity of interest, $a_0 P_\alpha(t)$, which is often called asymmetry, since it quantifies the asymmetry of the muon decay. The parameter a_0 is the initial asymmetry and $P_\alpha(t)$ the normalised polarisation function ($|P_\alpha(t)| \leq 1$), which will be usually referred to

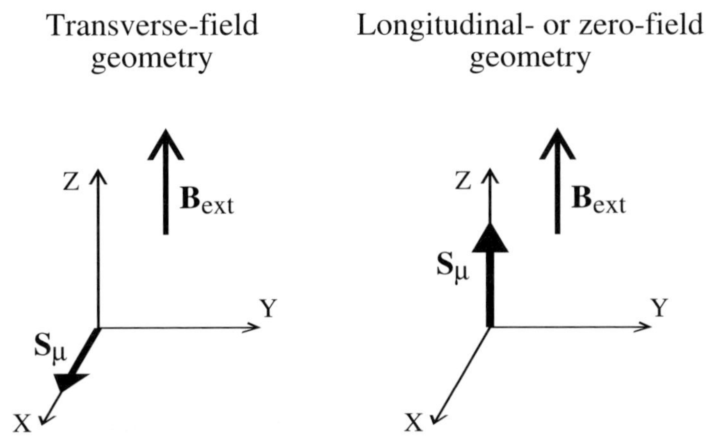

Fig. 1.2 Definition of the axes in the laboratory reference frame used throughout this book. $\{X, Y, Z\}$ is an orthogonal reference frame.

as the polarisation function. The Cartesian label α denotes the direction along which the muon polarisation is measured, i.e. it can be X, Y, or Z. The value of a_0 depends on experimental parameters; typically $a_0 \simeq 0.25$. Here, we will use an orthonormal laboratory reference frame. The vector $\mathbf{B}_{\rm ext}$ determines the direction of the Z axis and $\mathbf{S}_\mu$ is parallel to X and Z in the transverse and longitudinal geometry, respectively; see Fig. 1.2. Whereas in a longitudinal geometry, only $P_Z(t)$ is of interest, in the transverse geometry, $P_X(t)$ and $P_Y(t)$ can be measured.

In the left panel of Fig. 1.3 we display a simulated spectrum typical of a transverse applied field measurement. The amplitude of the observed wiggles may also decay as a function of time because of the potential existence of a distribution of fields at the muon site.

In the right panel we present two simulated spectra typical of the paramagnetic phase of a compound in zero field. The data are taken with the longitudinal geometry. As for the transverse geometry, a spectrum is built up from the positron count versus the time spent by the muons in the sample. We present the counts recorded in the forward and backward detectors. Although no wiggles are observed, the study of the decay of the count rate versus time is interesting. It yields information on the spatial and temporal properties of the field distribution at the muon site.

The positron count in the simplest case is modelled by the formula

$$N(t) = N_0 \exp\left(-t/\tau_\mu\right)\left[1 + a_0 P_\alpha(t)\right], \tag{1.1}$$

where the exponential term accounts for the finite muon lifetime τ_μ and the constant N_0 gives the scale of the positron count.

As already mentioned, the polarisation function is the essential physical quantity measured in most μSR experiments. Section 1.1.2 will be devoted to the description of the functions encountered in simple physical contexts.

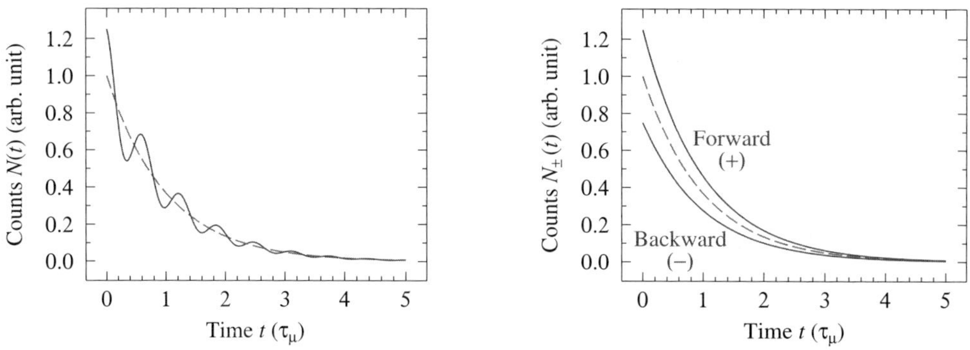

Fig. 1.3 (Left) Simulation of a raw μSR spectrum in the transverse geometry. The full line represents the positron count $N(t)$. A value of 5.3 mT (corresponding to $\omega_\mu \tau_\mu \simeq 10$) has been chosen for the transverse field. (Right) Simulation of raw μSR spectra in the case of the longitudinal geometry. The two full lines represent the counts in the forward and backward detectors. In both plots, an initial asymmetry of $a_0 = 0.25$ has been taken and the dashed lines show the average (isotropic) exponential decay which accounts for the finite muon lifetime.

Before embarking on this description, it is important to mention another type of experiments, the so-called time-integrated measurements, which are not meant to determine the evolution of the muon spin polarisation. Instead, they are mainly intended to investigate the energy-level scheme of a system including a spin species coupled to the muon spin. For definiteness, let us assume a muon spin is interacting *via* the dipole interaction with a proton spin located nearby. This can be generalised to the case of muonium. The spins of the muon, proton, and the rest of the compound form a quantum system characterised by energy levels. As a function of a longitudinal field, two of these levels may become near-degenerate. If the eigenstates of the related levels are built up with an appreciable weight of muon and proton states of opposite magnetic quantum number, an oscillatory inversion of the muon spin results accompanied by a flip of the proton spin. This induces a change in the polarisation of the muon beam, which can be quantified by the ratio

$$\mathcal{R}_{\rm int} = \frac{\sum N_+ - \sum N_-}{\sum N_+ + \sum N_-}, \qquad (1.2)$$

where $\sum N_+$ and $\sum N_-$ are the total number of positrons detected in the forward and backward directions, respectively. Hence, this is not the usual time-differential technique but a time-integrated method. A high counting rate is possible, in particular at a pseudo-continuous source, because the muon beam intensity is usually high at such a source and there is no requirement of having only one muon at a given time in the sample. A characteristic structure appears in $\mathcal{R}_{\rm int}(B_{\rm ext})$ near the resonance field. This technique is called the avoided-level-crossing (ALC) spectroscopy: the two energy levels involved in the resonance do not cross but repel each other because of the residual interactions, namely the dipole interaction in the current case.

1.1.2 Basic μSR polarisation functions

The polarisation functions monitor the properties of $\mathbf{B}_{\text{loc}}$. If all the muon spins precess in the same local field $\mathbf{B}_{\text{loc}}$ oriented at an angle θ from $\mathbf{S}_\mu$, the Larmor equation yields the important result:

$$P_Z(t) = \cos^2\theta + \sin^2\theta\cos(\omega_\mu t). \tag{1.3}$$

In this equation, $\omega_\mu = \gamma_\mu B_{\text{loc}} = 2\pi\nu_\mu B_{\text{loc}}$ where γ_μ is the muon gyromagnetic ratio ($\gamma_\mu = 851.616\,\text{Mrad}\,\text{s}^{-1}\,\text{T}^{-1}$). This ratio is related to the muon magnetic moment $\mathfrak{m}_\mu$ through the relation

$$\mathfrak{m}_\mu = \gamma_\mu\,\hbar S_\mu, \tag{1.4}$$

with $S_\mu = \frac{1}{2}$.

As a simple application of the previous equation, let us consider a magnet in zero external magnetic field. The local field at the muon site is usually not zero in the ordered phase: we denote this spontaneous field by $\mathbf{B}_0$. For a magnet in polycrystalline form, the spatial average of Eq. 1.3 must be performed. If the crystallites in the sample have no preferred orientation, we obtain for the ordered magnetic state

$$P_Z(t) = \frac{1}{3} + \frac{2}{3}\cos(\omega_\mu t). \tag{1.5}$$

The oscillating component reflects the magnetic order in the compound. We assume the presence of only one type of muon localisation site characterised by a single spontaneous local field. If the magnet is not perfectly ordered, i.e. if its correlation length is relatively small, the local field can take a large number of values close to each other. Then the oscillation is strongly damped and can even disappear. If the muon spins precess too quickly relative to the time resolution of the spectrometer, the oscillation will be averaged out to zero and the resulting muon polarisation will be constant at a level of $\frac{1}{3}$. Then, at a magnetic phase transition, if no wiggles are observed in the muon signal, one expects a drop in the effective initial asymmetry, a_{eff}, from $a_{\text{eff}} = a_0$ in the paramagnetic state to $a_0/3$ in the ordered state.

An exotic case is also possible: the magnetic transition of a compound is crossed as it is cooled down, but there is apparently no μSR signature of the crossing. This can be understood if the local field at the muon site vanishes or does not keep a constant value for a sufficiently long time, for the muon spin precession to be observed.

The time evolution of the polarisation of the muons is caused either by a distribution of local fields (dephasing of the muon spins) or by an exchange of energy between the muon spin system and the system under study (relaxation of the muon spins). In this short survey we only discuss the first type of mechanism.

Let us first consider the transverse-field polarisation function for a paramagnet or a diamagnet and assume the field distribution to be static, i.e. the muons do not move in the microsecond time range and the source of the distribution is static as well. The field distribution leads to a dephasing of the muon spin ensemble. If B_{ext} is sufficiently large relative to the field width, $P_X(t)$ only probes the field distribution along $\mathbf{B}_{\text{ext}}$, i.e. along the Z axis, and is given in the simplest case as a Gaussian envelope multiplied by a cosine function:

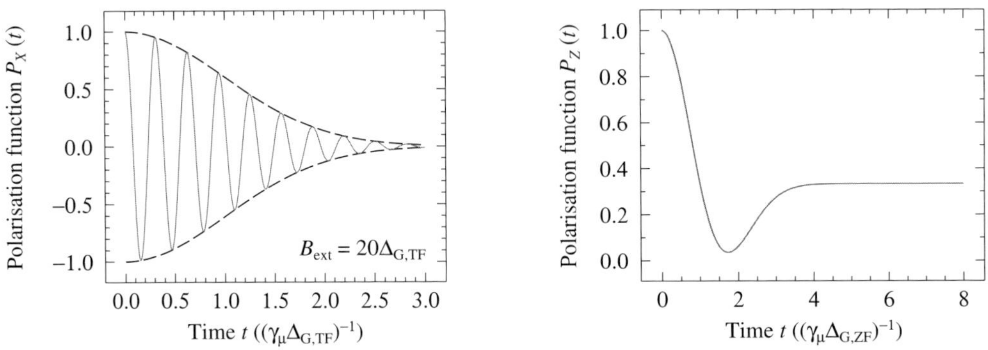

Fig. 1.4 Examples of commonly encountered polarisation functions. (Left) The transverse polarisation function is the product of a Gaussian envelope (shown as dashed lines) by a cosine function. The time scale is given in units of $(\gamma_\mu \Delta_{G,TF})^{-1}$ where $\Delta_{G,TF}$ is the root-mean-square of the Gaussian field distribution. The plot has been drawn for an external field B_{ext} equal to $20\Delta_{G,TF}$. (Right) The Kubo–Toyabe function, which is sometimes observed in zero-field measurements. The time scale is similarly given with respect to the corresponding field distribution root-mean-square.

$$P_X(t) = \exp\left(-\frac{\gamma_\mu^2 \Delta_{G,TF}^2 t^2}{2}\right) \cos(\omega_\mu t), \tag{1.6}$$

where $\omega_\mu = \gamma_\mu B_{ext}$ and $\Delta_{G,TF}^2$ is the variance of the component of the field distribution along Z. An example of this polarisation function is shown in Fig. 1.4.

Usually the field distribution is not static. A useful first approximation is to account for the dynamics with a single magnetic correlation frequency, ν_c. The so-called Abragam function gives a fair description for the polarisation function envelope:

$$P_X(t) = \exp\left\{-\frac{\gamma_\mu^2 \Delta_{G,TF}^2}{\nu_c^2}\left[\exp(-\nu_c t) - 1 + \nu_c t\right]\right\} \cos(\omega_\mu t). \tag{1.7}$$

In the motional-narrowing limit, i.e. when the field dynamics is fast, that is to say $\nu_c t \gg 1$, the envelope becomes an exponential function:

$$P_X(t) = \exp(-\lambda_X t) \cos(\omega_\mu t) \quad \text{with} \quad \lambda_X = \gamma_\mu^2 \Delta_{G,TF}^2 / \nu_c. \tag{1.8}$$

Up to the 1970s, the experimentalists favoured the transverse technique because it can yield useful information even if the muon beam is not of very good quality. Nowadays, in practice this technique is mainly used for three purposes: (i) measurement of frequency shifts, (ii) investigation of phase separation in magnetic systems, and (iii) determination of the physical properties of field distributions in matter. A typical example is the field distribution in the mixed phase of type II superconductors.

The longitudinal geometry is preferred in most experiments since a measurement can be performed even in zero field. This is a definitive advantage of μSR. Obviously, it is of interest to measure both in zero field and in longitudinal fields to get a complete picture of the system under study.

In zero field, an analytical formula is found for $P_Z(t)$ if the field distribution is static, isotropic, Gaussian and no spontaneous magnetic field is present (i.e. outside the ordered magnetic phase in the case of a magnet):

$$P_Z(t) = \frac{1}{3} + \frac{2}{3}(1 - \gamma_\mu^2 \Delta_{\rm G,ZF}^2 t^2) \exp\left(-\frac{\gamma_\mu^2 \Delta_{\rm G,ZF}^2 t^2}{2}\right). \qquad (1.9)$$

This is the famous Kubo–Toyabe function (Kubo and Toyabe 1967), which is drawn in the right panel of Fig. 1.4. It exhibits a dip at $t = \sqrt{3}/(\gamma_\mu \Delta_{\rm G,ZF})$ and levels off to $\frac{1}{3}$ at large times. Relative to the envelope of $P_X(t)$, $P_Z(t)$ is more sensitive to very slow dynamics which is usually not accessible with any other experimental technique. The shape of $P_Z(t)$ is also very sensitive to a small longitudinal external field in the quasi-static limit. Examples of these sensitivities are given in the next section. In the motional narrowing limit, $P_Z(t)$ is an exponential function:

$$P_Z(t) = \exp(-\lambda_Z t) \quad \text{with} \quad \lambda_Z = 2\gamma_\mu^2 \Delta_{\rm G,ZF}^2 / \nu_{\rm c}. \qquad (1.10)$$

At this point where we have discussed the motional narrowing limit which corresponds to fast fluctuating magnetic correlations as well as the quasi-static limit, it is of interest to specify the orders of magnitude. Figure 1.5 shows the range of magnetic correlation times that the μSR techniques are able to probe. In addition, a comparison is made with other experimental techniques. Since it is actually difficult to define precisely the extent of the time windows covered by each of them, some caution must be exercised when considering this graph. In addition it must be realised that a μSR experiment has not the capability of probing the dynamics of a given system over six or seven orders of magnitude in time scale as a quick inspection of Fig. 1.5 could suggest. The actual time range depends on the coupling between the system under investigation and the probe, i.e. the muon spin. The relation between λ_X (λ_Z) and $\nu_{\rm c}$ in Eq. 1.8 (1.10) provides an illustration of the importance of the coupling strength $\Delta_{\rm G,TF}$ ($\Delta_{\rm G,ZF}$) in the transverse (longitudinal) geometry.

1.1.3 Muon thermalisation and muonium formation

The energetic muons from a muon beam are slowed down in the material into which they are implanted in a multiple-stage process. Their initial kinetic energy, ranging from ~4 MeV to ~100 MeV for the most current beams, is initially reduced to a few tens of kiloelectronvolts by inelastic muon scattering which involve the Coulomb interaction. This process mainly implies atomic excitations and ionisations. At lower energies, collective excitations and charge exchange become important. The positive muon can pick up an electron to form a muonium atom. Generally the muon undergoes many cycles of electron capture and subsequent electron loss. The next stage of the thermalisation is constituted by collisions between muonium and atoms. Finally the muonium might dissociate into a free muon.

In metals a bare muon is actually the stable state. A discussion by Cox (2009) argues that muon–electron bound states can be formed, but these states lose their

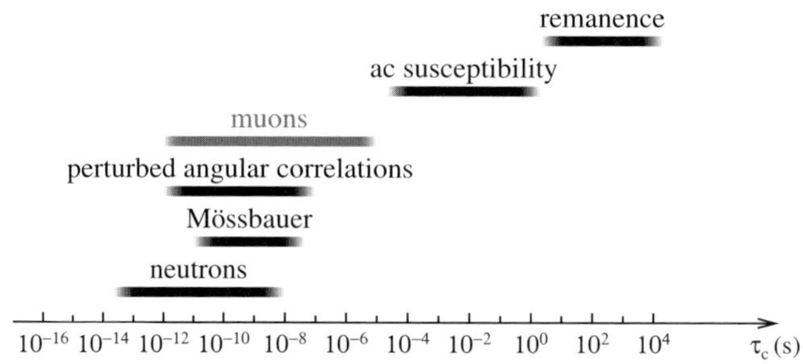

Fig. 1.5 Comparison of the time windows covered by different experimental techniques. Each horizontal line gives the range of the magnetic correlation time $\tau_c \equiv \nu_c^{-1}$ which can be probed. Since the range is not sharply defined the line extremities are blurred.

identity because of the impossibility for the electron to complete its classical orbit. In the case of semiconductors, dielectrics, and molecular compounds, a large fraction of muons are found in a form of a stable bound state with an electron, i.e. a so-called muonium centre is observed. This is understood because the Debye screening length is quite long, and therefore the screening is not impediment to bound-state formation. When implanted in a sufficiently heavily doped semiconductor a muon behaves as a diamagnetic state, although the physics involved is quite complicated, as discussed Section 12.5.

No matter the muon electronic state, the basic principles of the μSR techniques are the same. In the muonium case, the physics may just be more complex because of the number of involved energy levels. Note that a muonium is a paramagnetic centre and a free muon a diamagnetic species.

Because of its electric charge, a diamagnetic muon state may still influence the system under study and *a priori* there is always the possibility that the μSR measurements might not reflect the material intrinsic physical properties.

The μSR techniques give the possibility of studying the response of the system to a perturbation. Typical cases are diamagnetic or paramagnetic materials for which the interest might be the investigation of the local magnetic susceptibility, the muon diffusion or the electronic response of the system. In the latter experiments one can measure the response to the muon charge *via* the local dilation of the lattice or the induced electric field gradient.

One has always to keep in mind that the muon-induced perturbation might be strong enough to affect the physical properties of the system under study, while one is only interested in the pristine compound. A typical albeit rare example is noted at the end of Section 5.3. Supporting the idea that the muon is usually a gentle probe of magnetism is the excellent agreement observed between μSR and NMR in the determination of the phase diagram of the high-T_c superconductor copper oxide family $La_{2-x}Sr_xCuO_4$ (Julien 2003).

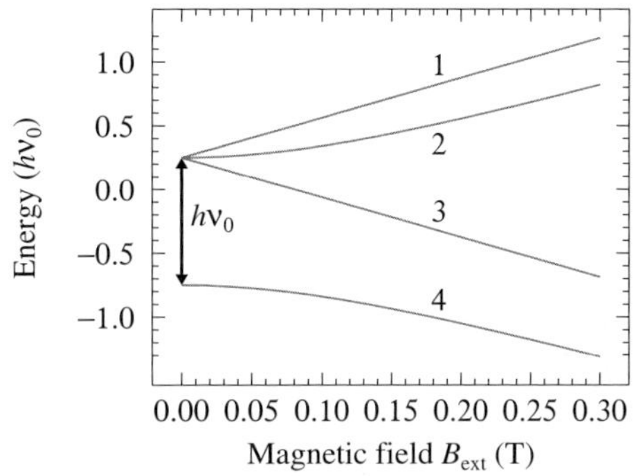

Fig. 1.6 The Breit–Rabi (hyperfine) energy-level diagram appropriate for free muonium (muonium in vacuum) in low field. The hyperfine frequency for free muonium is $\nu_0 = 4.4642\,\mathrm{GHz}$.

1.1.4 Physics of muonium centres

The evolution of the muon spin when the muon forms a muonium atom is governed by a spin Hamiltonian, which comprises at least the Fermi contact interaction between the muon and electron spins and the Zeeman interactions between the external field (if any) and the muon and electron spins, respectively. The dependence of the energy levels of a muonium atom in vacuum on the applied field is represented in the Breit–Rabi diagram presented in Fig. 1.6. Between these four levels, all transitions which involve a flip of the muon spin can *a priori* be observed by μSR as is typical for a genuine resonance technique.

These transitions might be detected in a transverse-field experiment as wiggles which appear at different frequencies. This is because the muons in muonium atoms are implanted in a non-stationary state which is a coherent superposition of spin-up and spin-down Zeeman states. Each and every state which contains one of these Zeeman states is subsequently occupied with a probability that sinusoidally oscillates in time. As for the polarisation of the muon ensemble, its behaviour is equivalent to precession or oscillation at the corresponding frequencies.

Although it has been observed in specialised experimental setups, the zero-field frequency of free or nearly free muonium is too high to be accessible with a conventional experimental setup; the free muonium frequency is indeed $\nu_0 = 4464.2\,\mathrm{MHz}$. However, the low-field "triplet precession" can readily be detected by the transverse-field technique if the field inhomogeneity is not too strong to broaden the precession signal beyond detection. It corresponds to transitions between levels 1 and 2 on the one hand and levels 2 and 3 on the other hand; see Fig. 1.6 for the level labels. Since at low field $\nu_{12} \simeq \nu_{23} \simeq |\gamma_e| B_{ext}/(4\pi)$ with $|\gamma_e|/(4\pi) = 14\,\mathrm{MHz/mT}$ (γ_e being the electron gyromagnetic ratio), the field distribution width should not exceed a few milliteslas. Note that this expression of ν_{12} and ν_{23} is independent of the hyperfine interaction intensity. In intermediate fields, the degeneracy of these transitions is lifted: their splitting is inversely proportional to the hyperfine coupling. This property provides a means for determining the magnitude of the hyperfine interaction, which contains information on the muonium environment. The detection of the pair of frequencies, if they do fall inside the frequency window of the spectrometer, is less problematic than the low-field frequency measurement because the condition on the field width is less stringent.

Longitudinal-field experiments are also useful since they give information on the muonium energy levels even when the field distribution at the muonium site is so large that a coherent precession cannot be detected. As the applied field is increased, the polarisation at the initial time also increases to reach the high-field limit (Paschen–Back limit), where the muon spin is completely decoupled from its environment and the μSR signal is no more depolarised. The analysis of the asymmetry near $t = 0$ as a function of the intensity of the external field also yields the magnitude of the hyperfine coupling.

1.2 A flavour of μSR: examples of experimental results

In the proceedings of the latest international conferences on the μSR techniques edited by Heffner *et al.* (2003), Blundell *et al.* (2006), and Kojima and Higemoto (2009) the studies on various aspects of magnetism account for approximately one-third of the contributions. This is the largest share. Superconductors, semiconductors, applications in chemistry, and technical developments cover about 15 to 20% each. The number of studies on muon diffusion has decreased and is now small.

Surface and interface science is a major topic of modern physics for which, so far, only very few μSR works have been published; an overview is given in a paper of Morenzoni (2009). The reason for this small number of publications is that the first dedicated spectrometer adapted for this type of studies has only been available at SμS since 2007. It requires a low energy muon beam. Examples from this promising field will be presented; see Section 1.2.3. Although we shall not discuss it here, we note the remarkable determination of the spin diffusion length of injected charge carriers within a buried active layer of a functional organic spin-valve device (Drew *et al.* 2009).

An additional experimental technique similar to μSR, β-detected NMR, is now under development at one facility (CMMS). Like μSR with low-energy muons, β-detected NMR is well suited for the study of surfaces and interfaces. It will be introduced in Chapter 4.

The intention with the examples given below is simply to give a feeling of what can be studied. Most of them are of historical interest as well.

1.2.1 Muon and muonium diffusion

Gurevich *et al.* (1972) have reported the first systematic μSR experiment aimed at getting information on the properties of a metallic system. It was the investigation of the muon diffusion in metallic copper.

The measurements were carried out with the transverse-field technique. The left panel of Fig. 1.7 shows the temperature dependence of the damping rate, Λ. The damping of the μSR oscillations is induced by the copper nuclear magnetic moments surrounding the muon. It is known from NMR measurements that these magnetic moments are static in the time scale of the experiment, i.e. the muon lifetime. At 77 K the envelope of the signal approaches a Gaussian law, meaning that the muon is also static. From the level of damping, Gurevich and co-workers deduced $\gamma_\mu \Delta_{\mathrm{G,TF}} = \sqrt{2}\Lambda$

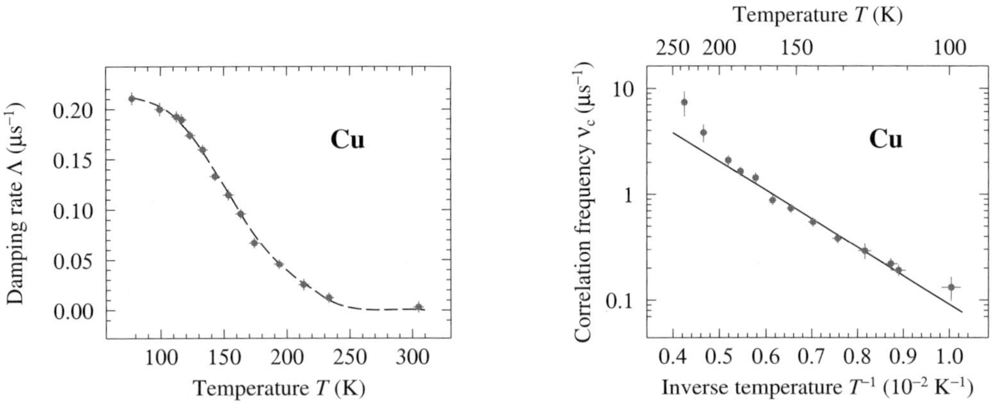

Fig. 1.7 (Left) Temperature dependence of the damping rate Λ measured by the transverse-field technique for a polycrystalline sample of copper. The applied field $B_{\rm ext}$ is equal to 6.2 mT. The damping rate Λ is defined by the authors of this work such that $1/\Lambda$ is the time needed for the envelope of $P_X(t)$ to decrease by a factor exp(1). The dashed line is a guide to the eye. (Right) Dependence of the correlation frequency $\nu_{\rm c}$ as a function of $1/T$ extracted from the experimental spectra using Eq. 1.7. The full line is the result of a fit to an Arrhenius law. The plots are adapted from Gurevich *et al.* (1972).

$= 0.310\,(11)\,\mu{\rm s}^{-1}$, a value in the expected range. Then, fitting the spectra taken above 77 K with Eq. 1.7, they extracted the magnetic correlation frequency $\nu_{\rm c}$ which they took as the muon hopping frequency. The frequency $\nu_{\rm c}$ is plotted versus $1/T$ in the right panel of Fig. 1.7. The full line in this plot is the Arrhenius law prediction:

$$\nu_{\rm c}(T) = \nu_{\rm c}^0 \exp\left(-\frac{E_{\rm ac}}{k_{\rm B}T}\right). \tag{1.11}$$

From more precise measurements, an activation energy $E_{\rm ac}$ corresponding to 551 (15) K and a prefactor $\nu_{\rm c}^0 = 10^{7.46(4)}\,{\rm s}^{-1}$ was obtained for a polycrystalline sample (Grebinnik *et al.* 1976). The relatively small value of $\nu_{\rm c}^0$ was interpreted as a signature of a sub-barrier diffusion process, i.e. a quantum mechanical tunnelling process. The deviation of the data from the Arrhenius law at high temperature was attributed to an over-barrier diffusion mechanism, in which case the muon hops over the energy barrier. Further studies performed by many groups confirmed the interpretation first given by the Russian group.

This study of muon diffusion in copper has been followed by similar investigations in many other metals, alloys, and compounds — see Karlsson (1995). In many cases the muon diffusion is so fast that it can reach impurities and possibly also escape from them during its lifetime. The first example was given for niobium doped with nitrogen (Borghini *et al.* 1978). Figure 1.8 illustrates the effect of impurities on the damping rate Λ defined previously (see caption of Fig. 1.7). The curve is the result of a fit to the trap-limited model described in Section 6.4.2 with each of the two states modellised by Eq. 1.7. Qualitatively, $\Lambda(T)$ is understood as follows. At low temperature the

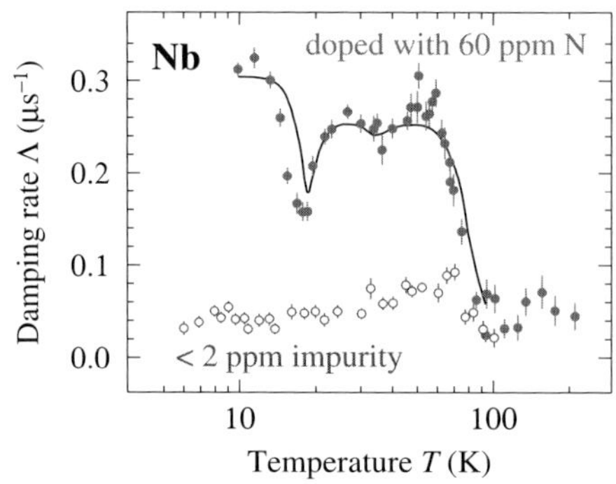

Fig. 1.8 Damping rate Λ as a function of temperature for a niobium sample doped with 60 parts per million (ppm) of nitrogen and for the purest possible niobium sample (i.e. less than 2 ppm impurity at interstitial sites). The curve for the doped sample represents a fit to a trap-limited model. Adapted from Borghini *et al.* (1978) and Karlsson (1995).

muons localise mainly in undisturbed interstitial sites (the concentration of impurities is small). As the temperature goes up, they hop and can reach trap sites induced by the impurities and their strain field. As the temperature is further increased, they escape from the traps and if the temperature is high enough they can revisit them during their lifetime. In agreement with this picture, Λ is much smaller for the purest sample. Interestingly, doping a compound is a method for studying the muon diffusion if it is too fast to be studied in the purest compound.

Thanks mainly to μSR experiments, it has been shown that the diffusion of light interstitial particles in a metal cannot usually be described as a classical over-barrier jump between interstitial sites. The cleanest example of tunnelling is perhaps given by the studies of doped aluminium samples at low temperature (Karlsson 1995). The muon diffusion is very fast in pure aluminium and it can only be studied *via* the investigation of the effect of impurities on the polarisation functions. In Fig. 1.9 is presented the first published proof of the drastic influence of superconductivity on muon diffusion, suggesting the importance of the effect of the electrons in the diffusion mechanism. Note that in the figure we write $P_\alpha^{\rm exp}(t)$ instead of $P_\alpha(t)$. In this way we distinguish a measured from a theoretically motivated polarisation function. The material investigated is aluminium doped with 75 ppm of lithium atoms, which is superconducting in zero field at the measured temperature, 0.20 K. An applied field of 15 mT, which is larger than the critical field, is sufficient to suppress superconductivity.

Not only can muon diffusion be analysed by the μSR techniques. It is also possible to probe the diffusion of muonium centres. An example is given in Fig. 1.10. The muonium tunnelling rate in solid nitrogen (Storchak *et al.* 1994) displays a temperature-independent behaviour between 18 and 32 K and slopes following T^7 and T^{-7} laws at lower and higher temperatures, respectively. The plateau region is understood as indicating a delocalisation of the muonium wavefunction. The behaviour for $32 \lesssim T \lesssim 50$ K is attributed to a two-phonon scattering process which limits muonium diffusion. Below 18 K coherent diffusion is slowed down by the intrinsic disorder introduced by

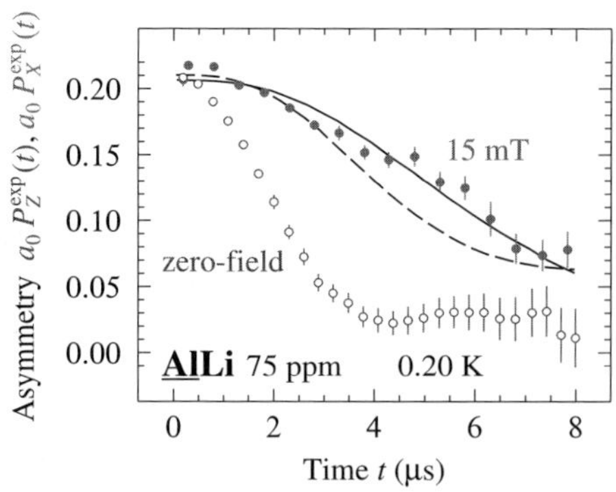

Fig. 1.9 μSR spectra recorded at 0.20 K on a aluminium sample doped with 75 ppm of lithium atoms. The open circles display the zero-field spectrum $a_0 P_Z^{\rm exp}(t)$ and the filled circles the envelope of the transverse polarisation function $a_0 P_X^{\rm exp}(t)$ taken with a field of 15 mT. The solid line results from a fit to the envelope. The dashed line presents the zero-field spectrum expected from the recorded transverse-field data. The observed larger damping for the measured zero-field spectrum shows that the muon diffusion is strongly influenced by superconductivity. This diffusion towards the lithium-induced traps is faster in the superconducting state than in the normal state. Adapted from Hartmann *et al.* (1989).

the anisotropic interaction between the N_2 molecules.

As shown by the data for solid nitrogen, the delocalisation of the muonium wavefunction, i.e. a coherent process, can be observed in a restricted temperature range. In contrast, the muon diffusion seems to occur only through incoherent processes, although indications of muon delocalisation have been found in some metallic compounds for which the distance between the interstitial sites is particularly small; see, e.g. Mulders *et al.* (2003). This difference stems from the difference in the interaction of the muon and muonium with their environment. In crystals such as solid nitrogen

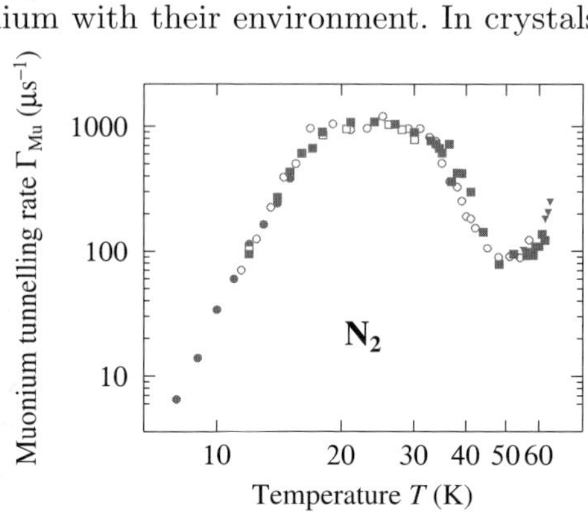

Fig. 1.10 Temperature dependence of the muonium tunnelling rate in solid nitrogen (Storchak *et al.* 1994). Logarithmic scales are used for both axes. The different symbols correspond to longitudinal- or transverse-field measurements in different samples.

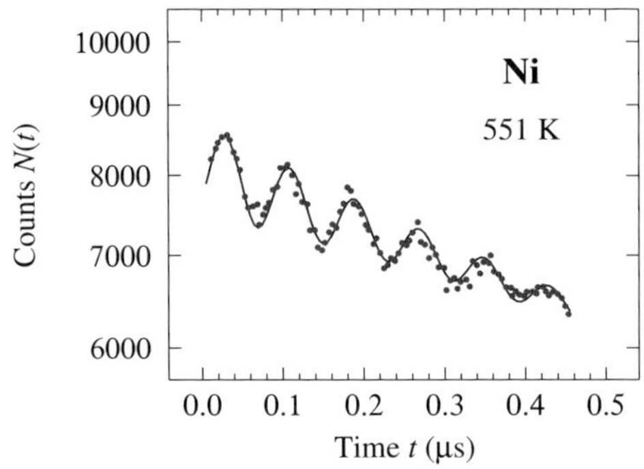

Fig. 1.11 Positron counts, in logarithmic scale, versus time for metallic nickel recorded at 551 K, i.e. in the ferromagnetic state (Foy *et al.* 1973). The solid line is the result of a fit.

the muonium interaction with the crystal is of the van der Waals type, and is therefore weaker than for the muon case, for which it is the screened Coulomb interaction which is important.

1.2.2 Magnetic materials

The mean magnetic moments in a magnetically ordered compound have well-defined directions relative to the crystal axes. In usual compounds, the ordered moments form a periodic magnetic structure. No matter the magnetic structure, we expect in general a spontaneous local magnetic field at the interstitial sites. This field can be measured by detecting the spontaneous (i.e. in the absence of external field) precession of the angular distribution of the decay positrons. This precession was first observed in 1973; see Fig. 1.11. Both the magnitude and the direction of the field relative to an applied field were determined. In the case of nickel, the zero-field measurement probes the sum of the Lorentz and hyperfine fields which are antiparallel. The measurement provides information on the polarisation of the electron density at the muon site which is modified by the presence of the muon.

The μSR techniques, notably the zero-field method, are well known for their high sensitivity to magnetism and their capability to detect magnetic fluctuations from the picosecond to the microsecond time range; see Fig. 1.5. Interestingly, key information can be extracted from compounds which do not display long-range magnetic correlations. We discuss two examples.

Let us first consider the high-T_c superconductor $YBa_2Cu_3O_x$. The very first indication for magnetic order in an unspecified member of the YBaCuO-family was provided by Nishida *et al* (1987a) using the zero-field technique. These authors did not observe a spontaneous precession frequency but rather a relatively sharp drop in the asymmetry as the temperature was lowered below 20 K. The remaining μSR component at low temperature was just a simple relaxing signal. The asymmetry drop occurs because the oscillating component could not be observed; see Eq. 1.5 and comments nearby. Two reasons could be invoked. Either the spontaneous precession frequency was too large to be detected by the spectrometer or the disorder in the compound was so impor-

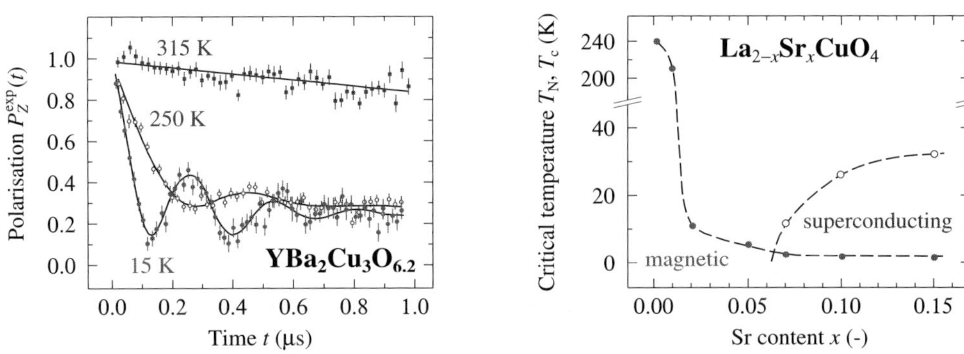

Fig. 1.12 (Left) Zero-field polarisation functions recorded in $YBa_2Cu_3O_{6.2}$ showing spontaneous muon-spin precession at 250 K and below and therefore the presence of a rather long-range magnetic order below that temperature. The solid lines are results of fits. Adapted from Nishida *et al.* (1987b). (Right) Magnetic and superconducting phase diagram of $La_{2-x}Sr_xCuO_4$ as a function of the strontium content parameter x. The magnetic transitions are detected from zero-field muon-spin rotation and relaxation measurements, while the superconducting transitions are determined from macroscopic measurements. The coexistence of internal magnetic fields and superconductivity is evidenced in this study. The dashed lines are guides to the eye. Adapted from Weidinger *et al.* (1989).

tant that the wiggles were strongly damped. Subsequently, the same group (Nishida *et al.* 1987b) proved the existence of a rather long-range order in the tetragonal phase $YBa_2Cu_3O_{6.2}$ through the observation of a spontaneous muon-spin precession at 250 K and below as illustrated in the left panel of Fig. 1.12. Long-range antiferromagnetic order in $YBa_2Cu_3O_{6.2}$ was confirmed by neutron diffraction measurements shortly after the publication of the μSR results. It is not possible from μSR measurements to determine a magnetic structure if its propagation vector is unknown. However, with the knowledge of this vector, combined with the application of space group theory which provides the possible magnetic structures compatible with the crystal symmetry (Bertaut 1968), and the determination of the muon site, magnetic structures can be proposed from zero-field μSR data; see Maeter *et al.* (2009) for a nice example. In addition, ambiguities left by neutron and magnetic X-ray diffraction, such as a phase parameter in the description of a magnetic structure, can possibly be resolved. The nice advantage of μSR is the ease with which a magnetic phase transition is detected, even in compounds with magnetic moments as small as one-hundredth of a Bohr magneton. When small moments are involved, the observation of a magnetic phase transition by neutron diffraction is only possible for single crystal samples. This is usually a difficult task if the wave vector of the magnetic structure is unknown. Thus, good estimates of the magnetic phase transition temperature and magnitude of the magnetic moments are essential. This type of information is exactly what μSR can provide quite easily. This explains the large number of phase diagram determinations carried out with the zero-field μSR technique. The right panel of Fig. 1.12 shows, for example, the phase diagram of a high-T_c superconducting material. While in this study, only the magnetic

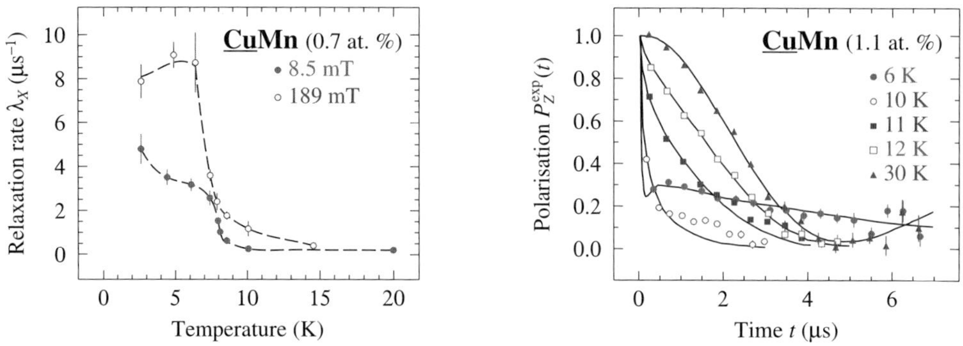

Fig. 1.13 Data on the spin-glass alloys CuMn. (Left) Transverse-field relaxation rate plotted against temperature for two values of the applied magnetic field. The cusp of the susceptibility is at $T_c = 7.7$ K. The data have been recorded with a field-cooling procedure. The solid lines connect points taken under similar conditions. Adapted from Murnick *et al.* (1976). (Right) Zero-field polarisation functions as a function of the temperature. The critical temperature is $T_c = 10.8$ K for that alloy. The solid lines result from fits. Adapted from Uemura *et al.* (1980).

part of the phase diagram is determined from μSR measurements, both the magnetic and superconducting transitions are commonly detected in subsequent μSR studies; see, e.g. Sanna *et al.* (2004).

As a second example of detection and characterisation of a magnetic transition, we discuss the spin-glass case. The first detection of a sharp transition in the magnetic susceptibility at T_c of alloys with a few percent concentration of magnetic impurities of iron in gold at the beginning of the seventies led to an intense activity (Cannella *et al.* 1971). The term glass connotes the absence of long-range spatial spin order. Surprisingly, no evidence was found for a transition in the specific heat. Using the transverse-field technique, Murnick *et al.* (1976) observed a sharp rather than a gradual onset of the transverse relaxation rate at the muon site for CuMn (0.7 at. %); see left panel of Fig. 1.13. A steep slope of the relaxation rate takes place near T_c. At the time of its publication, this work was important since it was indicating that a magnetic transition does occur.

The longitudinal-field technique was introduced by a group of physicists from the University of Tokyo working in Vancouver. One of its first applications was the study of the dynamical character of the spin-glass transition (Uemura *et al.* 1980). Some zero-field polarisation functions are shown in the right panel of Fig. 1.13 for CuMn (1.1 at. %). A rapid change of $P_Z(t)$ is observed near T_c, showing that the spin-glass transition involves a drastic change in the spin dynamics. Note the characteristic tail of the spectrum at 6 K which is a definitive proof of the quasi-static (characteristic time in the microsecond range) nature of the dynamics at that temperature.

A priori, the muon localisation site is unknown. Usually this does not prevent a qualitative analysis of the data. However, it can turn out to be a problem for extracting detailed information from the measurements. Fortunately, in many cases the

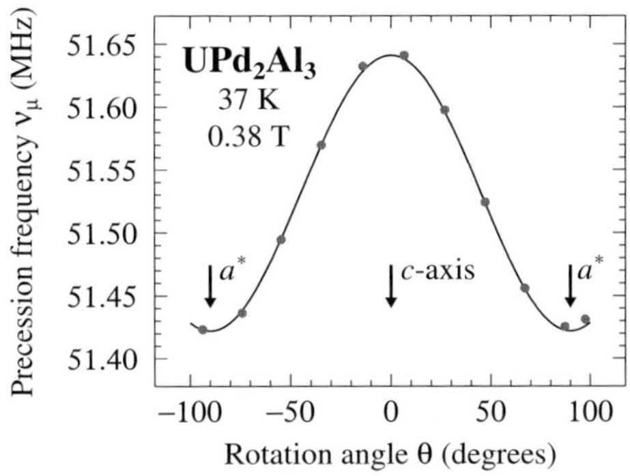

Fig. 1.14 Angular dependence of the muon precession frequency measured for a single crystal sample of the hexagonal compound UPd_2Al_3 in a transverse field of 0.38 T at 37 K. The field is rotated in the $a^* - c$ plane and θ denotes the angle between the field and the c axis. The solid line is the result of a fit. Adapted from Amato (1997).

muon site can be determined experimentally. For example, the angular dependence, with respect to the crystallographic axes, of the frequency in the paramagnetic state of a compound, detected in a transverse-field measurement, provides information on the local symmetry at the muon site. An example in given in Fig. 1.14. The intermetallics UPd_2Al_3 crystallises in an hexagonal structure with space group $P6/mmm$. This compound has a magnetic phase transition to an antiferromagnetic structure at $T_N \simeq 14.5$ K. Its interest stems from the existence of a superconducting phase below $T_c \simeq 1.5$ K which coexists with the magnetic phase. A previous study of that phase had suggested the muon to occupy solely the b site $(0, 0, 1/2)$ (Wyckoff notation). As a matter of fact, the shape of the angular dependence of the frequency shown in Fig. 1.14 is only compatible with this site assignment (Amato 1997).

It must be noted that, in the absence of data such as the one we have just presented, it is always possible to attempt the identification of the muon site from a calculation of the electrostatic potential energy. For that purpose one considers different possible sites, usually interstitial sites of high symmetry except for oxide compounds for which the muon is expected to bind to an oxygen with a bond length of ~1 Å (Holzschuh *et al.* 1983). For insulators, the electrostatic potential for a muon at a given site can be computed using lattice sums, assuming the formal point charges located at the centre of the ionic positions. For metallic compounds the computation is more involved; see, for examples, the work of Teichler (1978) for metallic copper described in Section 5.1.2 and Maeter *et al.* (2009) for iron pnictides. The modified Thomas–Fermi model seems appropriate to account for the screening of the muon charge by the conduction electrons. The self-consistent distribution of the valence electron density is computed from which the electrostatic potential is deduced (Reznik *et al.* 1995). The local minima of this potential at interstitial positions are regarded as possible muon stopping sites.

1.2.3 Superconductors

The study of a superconductor by the μSR techniques always involves the investigation of one of its magnetic properties. This is understandable since these techniques are based on the muon spin which is only sensitive to a magnetic field. Five types of subjects have been touched upon: the characterisation of the vortex matter in the mixed phase, the determination of the superconducting electron density, the measurement of the susceptibility of the Cooper pairs, the search for a spontaneous magnetic field induced by superconductivity, and the determination of the magnetic or superconducting phase diagram. The last topic was shortly mentioned in the previous section. Here, we will present examples for the first two subjects.

There are two types of superconductors: the so-called type I and type II. An applied magnetic field does not penetrate a type I superconductor except for a layer of extremely small thickness. However, most superconductors are of type II. A magnetic field penetrates the bulk of such a superconductor if its intensity is larger than the so-called lower critical field, B_{c1}. Superconductivity is destroyed if the field is larger than the upper critical field, B_{c2}. As for a type I superconductor, if the field intensity applied to a type II superconductor is smaller than B_{c1}, it penetrates only a layer of the material. When $B_{c1} \leq B_{\rm ext} \leq B_{c2}$, the field $B_{\rm ext}$ (the magnetisation is usually so small that it is a good approximation to take the flux density as equal to $B_{\rm ext}$) which enters a type II superconductor is not homogeneously distributed throughout the sample but concentrated in field tubes, the so-called vortices, parallel to the applied field. Each tube contains a unit of magnetic flux screened by circulating currents, which die out within a characteristic distance known as the London penetration depth, $\lambda_{\rm L}$. In the simplest case, these tubes are arranged in a triangular lattice, i.e. they form a flux-line lattice (FLL).

Since the FLL parameter, which scales with $B_{\rm ext}^{-1/2}$, and $\lambda_{\rm L}$ are much larger than the distance between interstitial muon sites and since the muons are assumed to stop randomly among these sites, the field distribution sampled by the transverse-field technique reflects the distribution of the induction $\mathbf{B}$ on a coarse grain nanometre scale arising from the FLL in the bulk of the compound. This is in contrast to magnetic compounds for which it is the local field at the muon site, $\mathbf{B}_{\rm loc}$, which is probed. In fact, one is usually only sensitive to the component of the distribution of $\mathbf{B}$ along the applied field; that is, one measures the distribution $D_{c,\mathbf{B}_{\rm ext}}(B^Z)$. We recall that $\mathbf{B}_{\rm ext}$ is applied along the Z axis. The profile of that distribution is determined with high resolution. μSR supplements small-angle neutron scattering, which has a limited spatial resolution but gives structural information more directly than μSR. NMR probes the FLL field distribution within the skin depth and decoration experiments only near the sample surface. The first measurements of a FLL field distribution are due to Delrieu and Winter (1966) and Redfield (1967). They were performed by NMR.

The FLL field distribution is not symmetric about $B_{\rm ext}$. It exhibits van Hove singularities related to the minimum, maximum, and saddle-point values of the magnetic flux density in the sample. Imperfections of the FLL smear out these singularities. The first convincing visualisation of the FLL field distribution arising from the bulk of a superconducting material as well as its singularities using the transverse-field muon technique was presented by Herlach *et al.* (1990) for niobium. In fact, the distribution

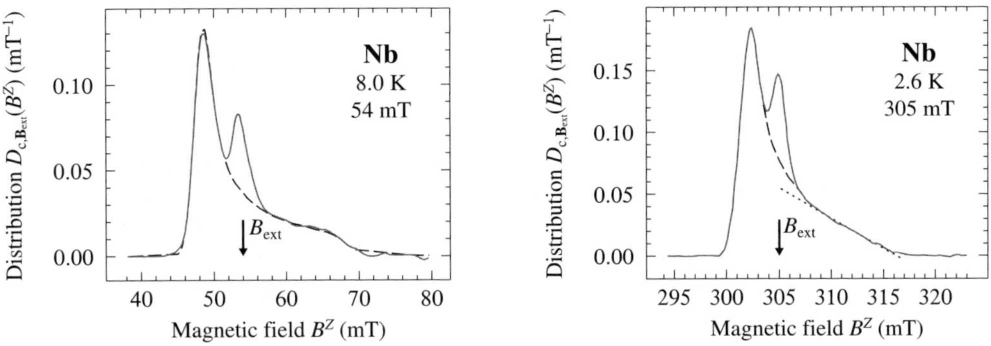

Fig. 1.15 This figure presents in full line the measured field distribution arising from the flux-line lattice in niobium. The dashed line which is mostly hidden by the data line, refers to the result of a theoretical model. The left (right) panel corresponds to a sample temperature and an applied magnetic field of 8.0 K and 54 mT (2.6 K and 305 mT), respectively. The arrow indicates the value of the external field B_{ext}. The peak near B_{ext} originating from muons stopped in the sample holder was excluded from the fit. The dotted line in the right-hand plot emphasises the linear tail at high field. Adapted from Herlach *et al.* (1990).

is not directly measured. It is obtained *via* a Fourier transform of the measured oscillating spectrum. Niobium is a type II superconductor in the clean limit with a critical superconducting temperature $T_c = 9.25$ K. The measurements were carried out in the mixed phase, i.e. $B_{c1} \leq B_{ext} \leq B_{c2}$. The measured field distribution shown in the left panel of Fig. 1.15 corresponds to expectation: the van Hove singularities are observed at $B_{ext} \simeq 46$, 48, and 68 mT. The dashed line is the result of a fit to the microscopic BCS–Gor'kov–Eilenberger (BCS: Bardeen, Cooper, and Schrieffer) theory. In fact, since the temperature at which the μSR spectrum has been recorded is relatively near the superconducting critical temperature, the much simpler Ginzburg–Landau theory provides also a good description. However, Herlach and collaborators have also recorded a spectrum at 2.6 K for a value of the field near B_{c2} (equal to 360 mT at that temperature); see the right panel of Fig. 1.15. A remarkable linear field dependence of the distribution at high field is found. This can only be explained using the BCS–Gor'kov–Eilenberger model, as predicted a long time ago by Delrieu (1972, 1974) for an ideal FLL. This result, which does not seem to be recognised in the literature, is probably the strongest experimental support for the exotic FLL predicted by the microscopic theory at low temperature and for B_{ext} close to B_{c2}.

We have just shown that the transverse-field technique gives the opportunity to unravel the FLL field distribution. This distribution contains information on the lattice itself and the basic physical properties of the superconductor. A single-crystal sample is needed for an optimised experiment. However, if one is ready to lose part of the potential information, a good quality polycrystalline sample is sufficient. As a first step, instead of using an involved model as for niobium in Fig. 1.15, the analysis of the spectra can be done with Eq. 1.6. In the left panel of Fig. 1.16 we give examples for $\Delta_{G,TF}(T)$. As expected the FLL field distribution width vanishes above T_c. The

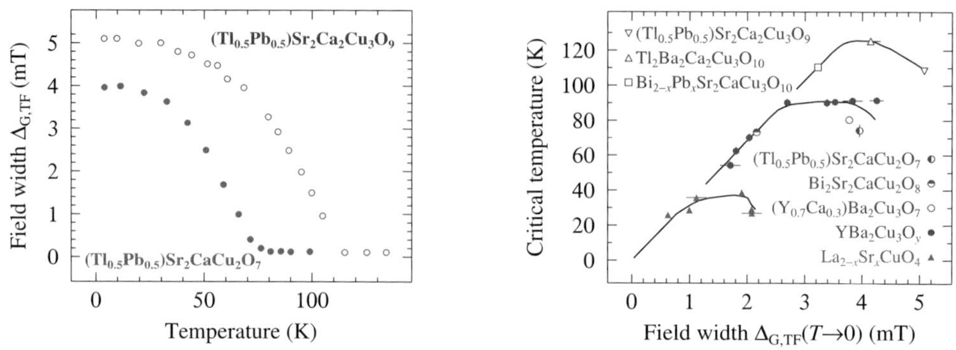

Fig. 1.16 (Left) Examples of the standard deviation, $\Delta_{\mathrm{G,TF}}$, of the FLL field distribution versus temperature for two high-T_{c} superconductors. (Right) First published example of what has been subsequently called Uemura plot. This is a plot of the critical superconducting temperature, as determined from $\Delta_{\mathrm{G,TF}}(T)$, as a function of $\Delta_{\mathrm{G,TF}}(T \to 0)$ for a relatively large number of high-T_{c} superconductors. The linear dependence of T_{c} versus $\Delta_{\mathrm{G,TF}}(T \to 0)$ in the underdoped regime for a given family of material is now called the "Uemura line". Adapted from Uemura *et al.* (1989).

most surprising result is the linear dependence shown in the right panel of the same figure: $T_{\mathrm{c}} \propto \Delta_{\mathrm{G,TF}}(T \to 0)$ if $\Delta_{\mathrm{G,TF}}(T \to 0)$ is not too large for a given family of material. In this linear regime, all the materials fall on the same universal line, the so-called "Uemura line". Since $\Delta_{\mathrm{G,TF}} \propto 1/\lambda_{\mathrm{L}}^2 \propto n_{\mathrm{car}}/m_{\mathrm{car}}$ (superconducting carrier density over effective mass), a linear relation between T_{c} and $n_{\mathrm{car}}/m_{\mathrm{car}}$ is established for increasing carrier doping in underdoped compounds. However, this relation is not obeyed for overdoped materials (Niedermayer *et al.* 1993) and electron-doped systems (Homes *et al.* 1997). A truly universal scaling relation for high-T_{c} materials has been proposed (Homes *et al.* 2004). It generalises the Uemura relation.

Niedermayer *et al.* (1999) have been able to monitor the spatial evolution of the magnetic field distribution as the flux lines emerge through the surface of a superconducting $\mathrm{YBa_2Cu_3O_{7-\delta}}$ film. This was achieved using a low-energy muon beam. Examples of distributions are presented in Fig. 1.17. The two top distributions are computed from spectra recorded using muons having kinetic energies of 29 and 3 keV, respectively, which are to be compared to the 4 MeV of the most common beams. The bottom distribution is measured with a sample consisting of the $\mathrm{YBa_2Cu_3O_{7-\delta}}$ film covered with silver and using again a beam of 3 keV. Whereas the distribution resulting from the bulk of the $\mathrm{YBa_2Cu_3O_{7-\delta}}$ sample (top panel) displays the expected strongly asymmetric field distribution, the distribution taken with the silver-covered superconductor is much more symmetric: both the cusp shift and the tail on the high-field side are drastically reduced. This is understood in the framework of a simple London equation, taking explicitly into account the surface of the $\mathrm{YBa_2Cu_3O_{7-\delta}}$ film. The variation of the magnetic field intensity as a function of depth beneath the surface of a film of the same superconductor in the Meissner state has also been measured using a low-energy muon beam (Jackson *et al.* 2000). The depth of implantation was varied from

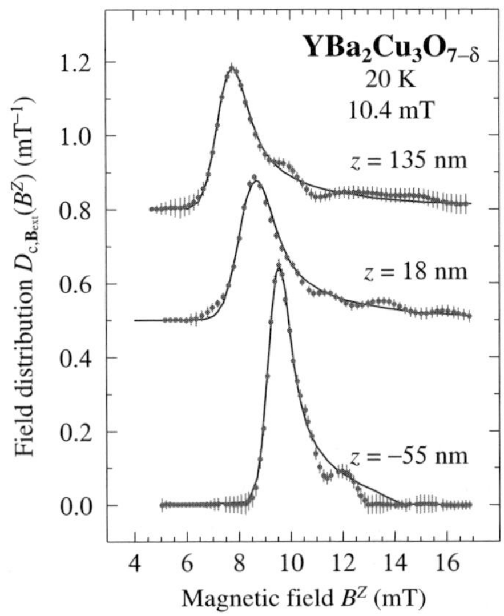

Fig. 1.17 Measured field distributions at $T = 20\,\mathrm{K}$ and $B_{\mathrm{ext}} = 10.4\,\mathrm{mT}$ as a function of the distance z from the surface of a $\mathrm{YBa_2Cu_3O_{7-\delta}}$ film. Since z is taken positive inside the film, a negative value of z means that the muons are stopped in the silver capping layer. The superconductor is in its mixed state with the chosen experimental conditions. The solid lines are the results of fits. For clarity, the distributions and fits for $z = 18$ and $135\,\mathrm{nm}$ are shifted by 0.5 and 0.8, respectively. The data have been recorded with a muon beam of low and tunable energy. Adapted from Niedermayer *et al.* (1999).

20 to 150 nm by tuning the energy of the implanted muons from 3 to 30 keV. These measurements provide a direct confirmation of the validity of the London exponential decay law for a field applied parallel to the surface of a high-T_{c} superconductor compound. In addition the characteristic decay length is found identical to that observed around the vortices in the mixed phase.

It is expected that the London model breaks down in certain superconductors because of the non-local nature of their electrodynamics response. This has been shown for a film of lead maintained in the Meissner state (Suter *et al.* 2004).

1.2.4 Probing matter with muonium

A muonium centre in a solid was observed for the first time in the Soviet Union in crystalline and fused quartz samples as well as in solid carbon dioxide and ice by Myasishcheva *et al.* (1968). In the left panel of Fig. 1.18 we present the recorded spectrum for crystalline quartz. The wiggles are observed in an external transverse field of 0.72 mT. Their precession frequency is too large, by about two orders of magnitude (exactly a factor of 103), to be due to free (diamagnetic) muons. However, they correspond to expectation if assumed to originate from the "triplet precession" of muonium. Because the applied field is very small, the extreme low-field limit of the Breit–Rabi diagram applies and therefore only one muonium frequency is expected. In agreement

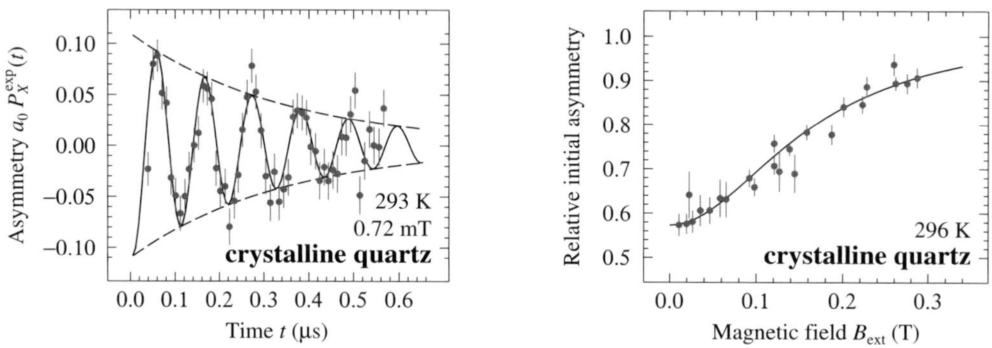

Fig. 1.18 Muonium in crystalline quartz. (Left) The muonium precession spectrum observed in an external transverse field of 0.72 mT at 293 K. The full line is the result of the fit with an exponential envelope which is in turn shown as dashed lines. Adapted from Myasishcheva *et al.* (1968). (Right) The relative initial asymmetry measured at room temperature versus the applied longitudinal-field intensity. The solid line is the result of a fit. In zero field the asymmetry is slightly larger than the theoretical value of 0.5. This may be caused by a small background signal contribution. Adapted from Minaichev *et al.* (1970).

with this interpretation, the detected signal corresponds to half of the available asymmetry, since there is 50% chance for a muonium centre to be formed in the triplet state, the other half of the signal corresponding to muonium in the singlet state.

As explained in Section 1.1.4, the value of the frequency observed in extremely low transverse field does not provide any information on the hyperfine interaction between the muon and electron spins. A repolarisation experiment has the capability of providing such information. It consists in measuring spectra for increasing values of the external longitudinal-field. Specific features in the short time evolution of the muon polarisation may appear: they will not be considered here since they are difficult to interpret. The interest is rather in the amplitude of the extrapolation to time zero of the signal at intermediate time. It is expected to increase with the field as the muon spin is decoupled from its environment. The first convincing experiment was carried out for crystalline quartz. The result is given in the right panel of Fig. 1.18. The measured hyperfine constant is consistent with the known constant for muonium in vacuum. Therefore the interaction of the muonium atom with its environment in quartz is weak.

The repolarisation experiment may seem to be an efficient way of measuring the hyperfine constant. Yet, if the precession frequency(ies) in transverse field is (are) not observed, a certain caution must be exercised in the interpretation of the repolarisation measurement. Field distributions due to randomly oriented magnetic fields and exchange interactions between the muonium and the substance may render the analysis of the data complicated. However, relative to the transverse-field experiment for which the precession frequency may not be seen because of a too large field distribution, as already mentioned, the longitudinal experiment always gives information on the system. In favourable instances, the magnitude of the hyperfine interaction can be

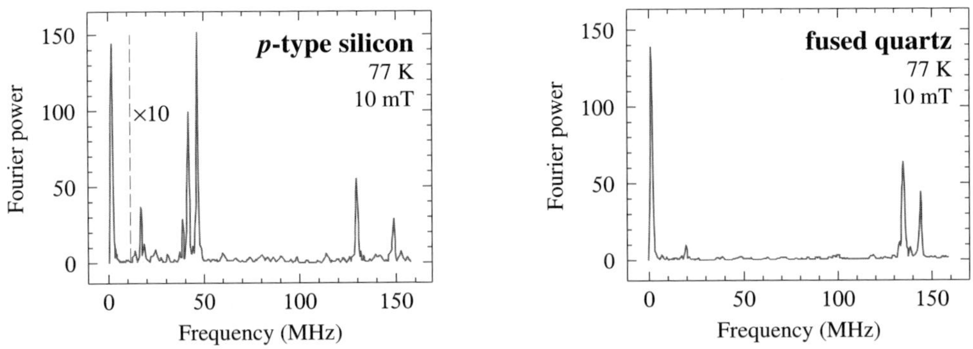

Fig. 1.19 Fourier spectra for p-type doped silicon and fused quartz at 77 K. The measurements were performed in an applied transverse field of 10 mT. For p-type silicon, the vertical scale is expanded by a factor of 10 to the right of the dashed line. The prominent peaks of this graph are (from left to right) the free muon precession signal at 1.36 MHz, a characteristic background signal at 19.2 MHz due to the radio-frequency structure in the proton beam, the two anomalous muonium frequencies at 43.6 ± 2.9 MHz, and the conventional muonium doublet centred at about 139 MHz. Similar lines as for p-type silicon are observed in fused quartz except the anomalous muonium frequencies which are absent. The wider splitting of the conventional muonium doublet in silicon relative to fused quartz is due to the weaker hyperfine coupling. Adapted from Brewer *et al.* (1973).

determined.

Muonium centre physics is not usually as simple as found originally for quartz at low field. In Fig. 1.19 a Fourier spectrum of a transverse-field measurement performed at 77 K for p-type doped silicon is presented and compared with a spectrum for fused quartz, both taken in a field of 10 mT. First, we note that the line corresponding to the free (diamagnetic) muon state is always present in addition to muonium. The exact fraction which remains free depends on the material. In the present example, it is about 20%. Because the applied field is much larger than for the measurement on crystalline quartz in Fig. 1.18, the intermediate field approximation of the Breit–Rabi diagram must be considered. This means that the two transitions of the triplet state are no longer degenerate. They appear as two distinct lines for both p-type silicon and fused quartz. However, the major surprise is the appearance of a couple of extra lines around 43.6 MHz for the silicon sample. These are the signature of a second muonium centre. The complete study shows that it can be modelled with an anisotropic hyperfine interaction of axial symmetry. It is sometimes referred to as anomalous muonium and has been found in numerous semiconductors. Whereas, for a long time, the normal muonium centre has been believed to be formed epithermally prior to thermalisation and to localise at a tetrahedral site of the crystal structure, the origin of the anomalous muonium centre has only been clarified in a second stage. It is now thought to arise from a radiolysis electron binding to a thermalised muon located on a [111] axis between two silicon atoms. Originally this mechanism of formation of the anomalous muonium atom was revealed by Krasnoperov *et al.* (1992) working with superfluid helium.

Neutral muonium centres have been identified in a large number of semiconductors and dielectrics, as well as positive and negative muonium centres. These latter species are diamagnetic, i.e. their electronic shell is closed. Muonium hyperfine couplings have been measured covering the range from some gigahertz (for example, Si) to a few hundreds of kilohertz in II–VI semiconductors.

The μSR techniques have played a pioneering role in the discovery and identification of hydrogen-like states in semiconductors. These states have been detected by the electron-spin-resonance technique only posteriorly to the muon experiments (Gorelkiskii and Nevinnyi 1991).

We now present a last example which belongs to the realm of chemical physics. It is one of the first investigation of free muoniated radicals adsorbed on a surface.

The muoniated cyclohexadienyl radical is detected when benzene is exposed to muons. We recall that a radical is an atom or group of atoms with an unpaired electron. Therefore it is a paramagnetic species. The created muonium centre is characterised by a hyperfine constant about an order of magnitude smaller than for muonium in vacuum. When implanting muons in silica spheres of diameter 7 nm covered with a monolayer of benzene, muoniated radicals are detected by the usual transverse-field technique. This means that the muons migrate to the grain surface and react with benzene in the muon lifetime scale. This opens up new opportunities because it enables the study of radicals which exist for only a short time and whose concentration at any instant therefore tends to be very low. We note that in heterogeneous catalysis, surface-adsorbed molecules are generally the principal reactive species. This explains the interest in such studies.

However, as pointed out by Reid *et al.* (1990), since radical formation is slow there is a dephasing of the spins of the muons forming radicals owing to a significant precession in the precursor state. Thus the signal amplitude in a transverse field is greatly reduced and studying its temperature dependence is barely possible. The avoided-level-crossing technique is *a priori* not as sensitive to dephasing and is therefore expected to provide the proper tool for studying the adsorbed free radicals. This is effectively the case as shown in Fig. 1.20.

Referring to a previous work, the investigators infer the resonance to be the signature of exchange of spin polarisation between the muon and the methylene proton in muoniated cyclohexadienyl radicals. The line shape is Lorentzian at all temperatures, signifying that there is no residual anisotropy in the system C_6H_6Mu-SiO_2 in powder form (Mu stands for muonium). The randomisation of orientation is achieved in a time of 10 ns or less. The resonance width continuously decreases with increasing temperature, in total by a factor of 5 over the whole temperature range. This is due to motional narrowing of the field inhomogeneities induced by the free radical environment. Therefore the data give information on it and on the radical dynamics.

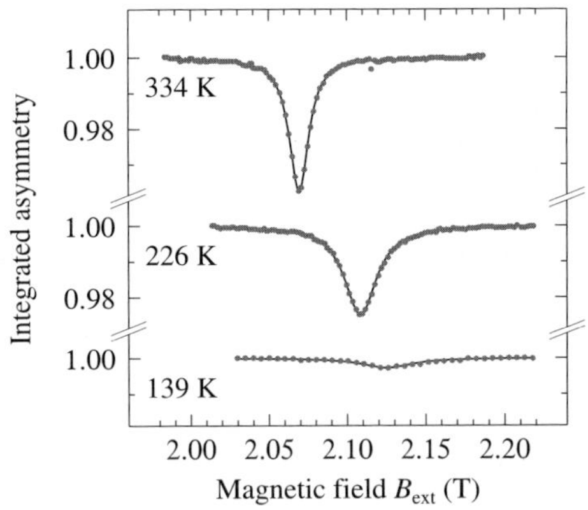

Fig. 1.20 The $\Delta M = 0$ avoided-level-crossing resonances recorded for muons implanted in a monolayer of benzene covering 7-nm SiO_2 grains. The measured signal probes the methylene proton in muoniated cyclohexadienyl radicals. The quantum number $M = m_e + m_\mu + m_p$ is the total spin projection, where m_e, m_μ, and m_p are the magnetic quantum numbers for muonium-electron, muon, and proton, respectively. Each plot gives the normalised time-integrated asymmetry versus the applied magnetic field for a given temperature. The solid curves result from fits assuming Lorentzian resonance shapes. Adapted from Reid *et al.* (1990).

References

Amato A. (1997) *Reviews of Modern Physics* **69**, 1119–1180.

Bertaut E.F. (1968) *Acta Crystallographica* **A 24**, 217–231.

Blundell S.J. (1999) *Contemporary Physics* **40**, 175–192.

Blundell S.J., King P.J.C., and Pratt F.L. (2006) Proceedings of the Tenth International Conference on Muon Spin Rotation, Relaxation and Resonance, *Physica* **B 374-375**.

Borghini M., Niinikoski T.O., Soulié J.C., Hartmann O., Karlsson E., Norlin L.O., Pernestål K., Kehr K.W., Richter D., and Walker E. (1978) *Physical Review Letters* **40**, 1723–1726.

Brewer J.H. (1994) in *Encyclopedia of Applied Physics*, vol. 11, edited by Trigg S.G., Vera E., and Greulich W. (VCH Publishers).

Brewer J.H., Crowe K.M., Gygax F.N., Johnson R.F., Patterson B.D., Fleming D.G., and Schenck A. (1973) *Physical Review Letters* **31**, 143–146.

Cannella V., Mydosh J.A., and Budnick J.I. (1971) *Journal of Applied Physics* **42**, 1689–1690.

Chappert J. and Grynszpan R.I. (1984) editors of *Muons and Pions in Materials Research* (North-Holland, Amsterdam).

Chappert J. and Yaouanc A. (1986) in *Microscopic Methods in Metals*, edited by Gonser U., Topics in Current Physics, vol. 40 (Springer, Berlin).

Cox S.F.J. (1987) *Journal of Physics C: Solid State Physics* **20**, 3187–3319.

Cox S.F.J. (2009) *Reports on Progress in Physics* **72**, 116501 1–130.

Delrieu J.M. (1972) *Journal of Low Temperature Physics* **6**, 197–219.

Delrieu J.M. (1974) PhD Thesis, Université Paris-Sud.

Delrieu J.M. and Winter J.M. (1966) *Solid State Communications* **4**, 545–548.

Drew A.J., Hoppler J., Schulz L., Pratt F.L., Desai P., Shakya P., Kreouzis T., Gillin W.P., Suter A., Morley N.A., Malik V.K., Dubroka A., Kim K.W., Bouyanfif H., Bourqui F., Bernhard C., Scheuermann R., Nieuwenhuys G.J., Prokscha T., and Morenzoni E. (2009) *Nature Materials* **8**, 109–114.

Foy M.L.G., Heiman N., Kossler W.J., and Stronach C.E. (1973) *Physical Review Letters* **30**, 1064–1067.

Garwin R.L., Lederman L.M., and Weinrich M. (1957) *Physical Review* **105**, 1415–1417.

Gorelkiskii Y.V. and Nevinnyi N.N. (1991) *Physica* **B 170**, 155–167.

Grebinnik V.G., Gurevich I.I., Zhukov V.A., Manych A.P., Meleshko E.A., Muratova I.A., Nikol'skiĭ B.A., Selivanov V.I., and Suetin V.A. (1976) *Soviet Physics JETP* **41**, 777–780.

Gurevich I.I., Mel'eshko E.A., Muratova I.A., Nikol'sky B.A., Roganov V.S., Selivanov V.I., and Sokolov B.V. (1972) *Physics Letters* **A 40**, 143–144.

Hartmann O., Karlsson E., Harris S., Wäppling R., and Niinikoski T.O. (1989) *Physics Letters* **A 142**, 504–508.

Heffner R.H., Noakes D.R., and Percival P.W. (2003) Proceedings of the Nineth International Conference on Muon Spin Rotation, Relaxation and Resonance, *Physica* **B 326**.

Herlach D., Majer G., Major J., Rosenkranz J., Schmolz M., Schwarz W., Seeger A., Templ W., Brandt E.H., Essmann U., Fürderer K., and Gladisch M. (1990) *Hyperfine Interactions* **63**, 41–48.

Holzschuh E., Denison A.B., Kündig W., Meier P.F., and Patterson B.D. (1983) *Physical Review* **B 27**, 5294–5307.

Homes C.C., Clayman B.P., Peng J.L., and Greene R.L. (1997) *Physical Review* **B 56**, 5525–5534.

Homes C.C., Dordevic S.V., Strongin M., Bonn D.A., Liang R., Hardy W.N., Korniya S., Ando Y., Yu G., Kaneko N., Zhao X., Greven M., Basov D.N., and Timusk T. (2004) *Nature* **430**, 539–541.

Jackson T.J., Riseman T.M., Forgan E.M., Glückler H., Prokscha T., Morenzoni E., Pleines M., Niedermayer Ch., Schatz G., Luetkens H., and Litterst J. (2000) *Physical Review Letters* **84**, 4958–4961.

Julien M.-H. (2003) *Physica* **B 329-333**, 693–696.

Kalvius G.M., Noakes D.R., and Hartmann O. (2001) in *Handbook on the Physics and Chemistry of Rare Earths*, vol. 32, edited by Gschneidner K.A., Eyring L., and Lander G.H. (Elsevier, Amsterdam).

Karlsson E.B. (1995) *Solid State Phenomena As Seen by Muons, Protons, and Excited Nuclei* (Clarendon Press, Oxford).

Kojima K.M. and Higemoto W. (2009) Proceedings of the Eleventh International Conference on Muon Spin Rotation, Relaxation and Resonance, *Physica* **B 404**, issues 5–7.

Krasnoperov E., Meilikhov E., Abela R., Herlach D., Morenzoni E., Gygax F.N., Schenck A., and Eschenko D. (1992) *Physical Review Letters* **69**, 1560–1563.

Kubo R. and Toyabe T. (1967) in *Magnetic Resonance and Relaxation*, edited by Blinc R. (North-Holland, Amsterdam).

Lee S.L., Kilcoyne S.H., and Cywinski R. (1999) editors of *Muon Science: Muons in Physics, Chemistry, and Materials* (Institute of Physics, Bristol).

Lee T.D. and Yang C.N. (1956) *Physical Review* **104**, 254–258.

Maeter H., Luetkens H., Pashkevich Yu.G., Kwadrin A., Khasanov R., Amato A., Gusev A.A., Lamonova K.V., Chervinskii D.A., Klingeler R., Hess C., Behr G., Büchner B., and Klauss H.-H. (2009) *Physical Review* **B 80**, 094524, 1–19.

Minaichev E.V., Myasishcheva G.G., Obukhov Yu.V., Roganov V.S., and Firsov V.G. (1970) *Soviet Physics JETP* **31**, 849–852.

Morenzoni E. (2009) *Physica* **B 404**, 577–580.

Mulders A.M., Kaiser C.T., Harker S.J., Gubbens P.C.M., Amato A., Gygax F.N., Schenck A., Dalmas de Réotier P., Yaouanc A., Buschow K.H.J., and Menovsky A.A. (2003) *Physical Review* **B 67**, 014303, 1–10.

Murnick D.E., Fiory A.T., and Kossler W.J. (1976) *Physical Review Letters* **36**, 100–104.

Myasishcheva G.G., Obukhov Yu.V., Roganov V.S., and Firsov V.G. (1968) *Soviet Physics JETP* **26**, 298–301.

Nagamine K. (2003) *Introductory Muon Science* (Cambridge University Press, Cambridge).

Niedermayer Ch., Bernhard C., Binninger U., Glückler H., Tallon J.L., Ansaldo E.J., and Budnick J.I. (1993) *Physical Review Letters* **71**, 1764–1767.

Niedermayer Ch., Forgan E.M., Glückler H., Hofer A., Morenzoni E., Pleines M., Prokscha T., Riseman T.M., Birke M., Jackson T.J., Litterst J., Long M.W., Luetkens H., Schatz A., and Schatz G. (1999) *Physical Review Letters* **83**, 3932–3935.

Nishida N., Miyatake H., Shimada D., Hikami S., Torikai E., Nishiyama K., and Nagamine K. (1987a) *Japanese Journal of Applied Physics* **26**, L799–L801.

Nishida N., Miyatake H., Shimada D., Okuma S., Ishikawa M., Takabatake T., Nakazawa Y., Kuno Y., Keitel R., Brewer J.H., Riseman T.M., Williams D.L., Watanabe Y., Yamazaki T., Nishiyama K., Nagamine K., Ansaldo E., and Torikai E. (1987b) *Japanese Journal of Applied Physics* **26**, L1856–L1858.

Redfield A.G. (1967) *Physical Review* **162**, 367–374.

Reid I.D. and Roduner E. (2000) in *Encyclopedia of Spectroscopy and Spectrometry*, edited by Tranter G.E. and Holmes J.L. (Academic Press, San Diego).

Reid I.D., Azuma T., and Roduner E. (1990) *Nature* **345**, 328–330.

Reznik I.M., Vagizov F.G., and Troć R. (1995) *Physical Review* **B 51**, 3013–3020.

Roduner E. (1988) *The Positive Muon As a Probe in Free Radical Chemistry*, edited by Berthier G., Lecture Notes in Chemistry, vol. 49 (Springer, Berlin).

Sanna S., Allodi G., Concas G., Hillier A.D., and De Renzi R. (2004) *Physical Review Letters* **93**, 207001, 1–4.

Schatz G. and Weidinger A. (1996) *Nuclear Condensed Matter Physics: Nuclear Methods and Applications* (Wiley, Chichester).

Schenck A. (1985) *Muon Spin Rotation Spectroscopy* (Adam Hilger, Bristol).

Seeger A. (1978) in *Hydrogen in Metals I*, edited by Alefeld G. and Völkl J., Topics in Current Physics, vol. 28 (Springer, Berlin).

Smilga V.P. and Belousov Yu.M. (1994) *The Muon Method in Science*, Proceedings of the Lebedev Physics Institute, vol. 219 (Nova Science, Commack, NY).
Sonier J.E., Poon K.F., Luke G.M., Kyriakou P., Miller R.I., Liang R., Wiebe C.R., Fournier P., and Greene R.L. (2003) *Physical Review Letters* **91**, 147002, 1–4.
Storchak V., Brewer J.H., Hardy W.N., Kreitzmann S.R., and Morris G.D. (1994) *Physical Review Letters* **72**, 3056–3059.
Suter A., Morenzoni E., Khasanov R., Luetkens H., Prokscha T., and Garifianov N. (2004) *Physical Review Letters* **92**, 087001, 1–4.
Teichler H. (1978) *Physics Letters* **A 67**, 313–315.
Uemura Y.J., Yamazaki T., Hayano R.S., Nakai R., and Huang C.Y. (1980) *Physical Review Letters* **45**, 583–587.
Uemura Y.J., Luke G.M., Sternlieb B.J., Brewer J.H., Carolan J.F., Hardy W.N., Kadono R., Kempton J.R., Kiefl R.F., Kreitzman S.R., Mulhern P., Riseman T.M., Williams D.Ll., Yang B.X., Uchida S., Takagi H., Gopalakrishnan J., Sleight A.W., Subramanian M.A., Chien C.L., Cieplak M.Z., Xiao Gang, Lee V.Y., Statt B.W., Stronach C.E., Kossler W.J., and Yu X.H. (1989) *Physical Review Letters* **62**, 2317–2320.
Walker D.C. (1983) *Muon and Muonium Chemistry*. (Cambridge University Press, Cambridge).
Weidinger A., Niedermayer Ch., Golnik A., Simon R., Recknagel E., Budnick J.I., Chamberland B., and Baines C. (1989) *Physical Review Letters* **62**, 102–105.

2

Recording data and extracting the polarisation functions

The method for recording a μSR spectrum and the principles for its analysis is independent of whether the muon forms a bound state with an electron. Therefore the discussion in this chapter applies equally well for muon and muonium.

After a brief history of the muon, the different μSR setups are considered. We first describe the muon production and decay and then the time structure, energy, and polarisation characteristics of the presently available muon beams. We conclude this chapter by an exposition of the principles of the μSR data treatment.

2.1 A brief history of the muon

The interested reader will benefit from reading the book of Weissenberg (1967) or any introductory textbook on elementary particles.

The positive muon techniques are based on the physical properties of the positive muon which is an elementary particle believed, like the electron, to have no internal structure. The muon was first detected in a cosmic ray experiment by Kunze (1933) and rediscovered later by Anderson and Neddermeyer and by Street and Stevenson. At about the same time Yukawa proposed his theory of the strong nuclear force. The mass of the particle mediating this force, the so-called π meson, was predicted to be in the 0.1 GeV/c^2 range. Unfortunately, it occurs that the muon has about this mass. This led to a confusion still perceptible in the 1980s: in some μSR papers the positive muon was mentioned as a meson instead of a lepton as the electron. The paradox was yet resolved in 1947 by Lattes, Occhialini, and Powell, who proved that the muon is the product of the disintegration of the π meson also called pion. The short lifetime of the pion ($\tau_\pi = 26\,\mathrm{ns}$) explains why it was not detected in the 1930s.

Like the pion, the muon disintegrates spontaneously by β-decay. Here the probability of emission of the positron is larger along the direction of the muon spin than in the opposite direction. This property is due to the non-conservation of the parity in the weak nuclear interaction which drives the decay of the muon (as well as the decay of the pion). If the muon spin direction is known at the time of its birth and measured at its death, the temporal evolution of the muon spin can be used as a tool to measure the magnetic field acting on it at the site where it is implanted in matter.

The development of the muon techniques was first relatively slow because, in the early times, the particle accelerators did not provide sufficiently intense muon beams for systematic studies of condensed matter. The first studies providing key information

Table 2.1 Some characteristics of the positive muon. The data are taken from the numerical values of fundamental constants tabulated at the begining of the book after the list of important symbols. The proton mass at rest is denoted by $M_{\rm p}$. The lifetime is the period after which a population of muons at rest is expected to have decayed by a factor exp(1). The negative muon has the same characteristics except for its electric charge and gyromagnetic ratio which are opposite to their positive muon counterparts. In practice, its lifetime in matter can be reduced drastically relative to the value in vacuum because of its possible capture by a nucleus. This is discussed further in Section 4.6

Quantity	Symbol	Value
Mass at rest ($M_{\rm p}$)	M_μ	1.12609×10^{-1}
Electric charge (C)	e	$+1.60219 \times 10^{-19}$
Spin ($\hbar$)	S_μ	$\frac{1}{2}$
Gyromagnetic ratio (Mrad s^{-1} T^{-1})	γ_μ	851.616
Lifetime (μs)	τ_μ	2.1970

on its physics were published at the beginning of the 1970s. More information on this development is given in Section 1.2.

2.2 μSR experimental setups

The book of Nagamine (2003) provides a useful introduction to the available experimental setups.

2.2.1 Muon production and decay

In this section we give a short description of the muon production and decay. Some properties of the positive muon are listed in Table 2.1. A more detailed discussion can be found in the book by Schenck (1985).

Intense muon beams are commonly produced at intermediate energy particle accelerators in a two-step process. First, a thin target of graphite or beryllium is bombarded with a proton beam, at SμS for example a beam of about 2 milliamperes of protons with kinetic energy 590 MeV, inducing the following reactions between the protons of the beam and the protons and neutrons of the target:

$$\begin{aligned} {\rm p} + {\rm p} &\longrightarrow \pi^+ + {\rm p} + {\rm n}, \\ {\rm p} + {\rm n} &\longrightarrow \pi^+ + {\rm n} + {\rm n}, \end{aligned} \tag{2.1}$$

where p stands for the proton, n for the neutron and π for the pion. For a usual target (thickness of ~1 cm) only a few percent of the protons interact with it. The pion itself is an instable particle and it decays with a lifetime of $\tau_\pi = 26$ ns into a muon and a muon neutrino:

$$\pi^+ \longrightarrow \mu^+ + \nu_\mu. \tag{2.2}$$

This nuclear reaction is at the origin of the first principle on which the μSR techniques are based. For a pion at rest, conservation of linear momentum requires the muon and

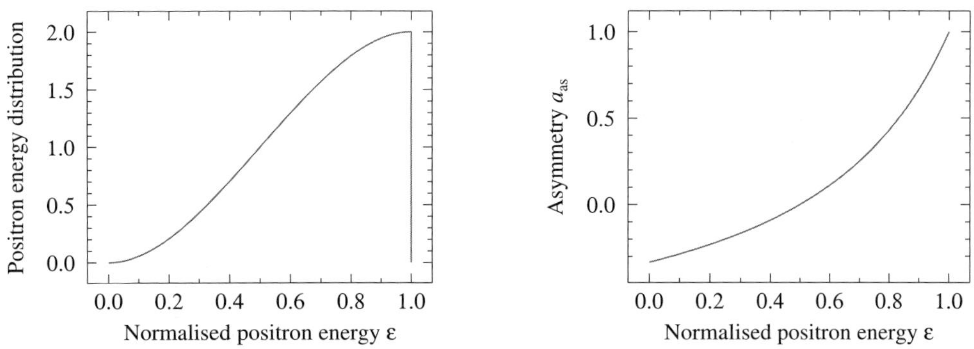

Fig. 2.1 (Left) The energy distribution of the positron as a function of its kinetic energy normalised to $E_{\max}$. The distribution vanishes for $\varepsilon > 1$. (Right) The asymmetry parameter a_{as} as a function of the same variable. The maximum energy $E_{\max}$ is 52.3 MeV.

the neutrino to be emitted in opposite directions. Because of the maximum parity violation in this weak decay process, the neutrino has a negative helicity. This means that its spin is antiparallel to its momentum. The pion has no spin. Thus, because of the principle of angular momentum conservation, the muon has its spin antiparallel to its linear momentum as well. Therefore, if a muon beam can be produced, it will be polarised antiparallel to its linear momentum. So far, the discussion assumed a pion at rest. If the pion moves, the muon will be partly kinematically depolarised.

The muon decays through a weak nuclear process according to

$$\mu^+ \longrightarrow \mathrm{e}^+ + \nu_\mathrm{e} + \bar{\nu}_\mu, \tag{2.3}$$

where ν_e and $\bar{\nu}_\mu$ are the neutrino and antineutrino associated with the positron and muon, respectively. Of these three particles, only the positron is detected. Its energy ranges between 0 and $E_{\max} = 52.3\,\mathrm{MeV}$. The energy distribution of the positrons (Fig. 2.1 left panel) is given by the law $2(3 - 2\varepsilon)\varepsilon^2$, where ε is the positron kinetic energy normalised to $E_{\max}$.

There is a correlation between the direction of emission of the positron and that of the spin of the muon at decay time. This is the second property which enables the muon spin to be used to probe matter. This correlation allows the observation of the direction of the muon spin at the time of decay, by noting in which direction the positron is emitted. Let us define the probability for a positron to be emitted at an angle θ, referred to the direction of the muon spin, as $W(\theta)$. According to the theory of weak interaction,

$$W(\theta) \propto [1 + a_{as}(\varepsilon)\cos\theta]\,. \tag{2.4}$$

Clearly, the positron is most likely emitted along the muon spin axis (when $a_{as}(\varepsilon) > 0$). The asymmetry parameter $a_{as}(\varepsilon)$ varies with the positron kinetic energy as $(2\varepsilon - 1)/(3 - 2\varepsilon)$ as seen in the right panel of Fig. 2.1. According to the plots in the two panels of this figure, the positrons that acquire the maximum allowed kinetic energy have $a_{as} \to 1$ and are therefore the most desirable, but are relatively few in number.

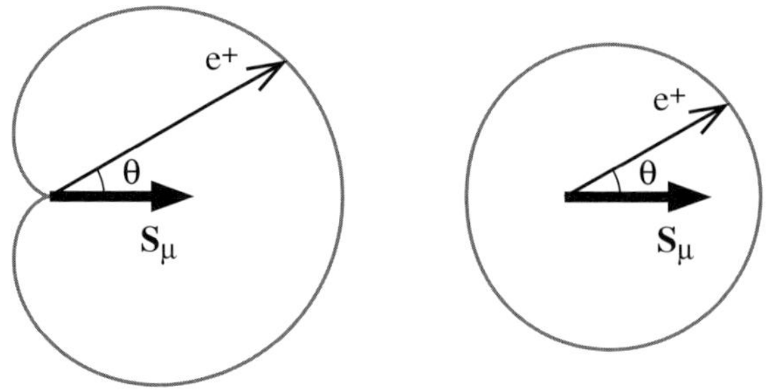

Fig. 2.2 Angular distribution of the positrons resulting from the decay of the muon. The left (right) panel shows the distribution for the maximum energy (average energy) of the positrons. In the drawings we indicate the direction of the muon spin at the time of decay. These curves are cardioids.

Lower energy positrons have lower values of $a_{\rm as}$, and $a_{\rm as}$ is even negative for the low-energy positrons, but these are very few. Integrating over the energy, i.e. when all the positrons are counted, gives $\bar{a}_{\rm as} = \frac{1}{3}$.

Note that the positrons are quite energetic. This means that they can easily escape from the sample and pass through cryostat, furnace or pressure cell walls. This is a definite experimental advantage.

Figure 2.2 presents $W(\theta)$ for two values of $a_{\rm as}$. The left plot shows $W(\theta)$ as a function of θ for $a_{\rm as} = 1$, which corresponds to the highest positron energy. The right plot displays the situation for which $a_{\rm as} = \bar{a}_{\rm as} = \frac{1}{3}$.

2.2.2 Muon beamlines and spectrometers

Here we present some basic facts about muon beams. We start by discussing their time structure. Then we focus on their energy characteristics. This leads us to introducing the three available types of beams which differ by the momentum of the muons. Finally their polarisation properties are discussed.

This presentation, which is only introductive, should be sufficient for understanding how the μSR techniques work and be useful for the selection of the proper type of beam and spectrometer for a given problem. More information is available from the muon facilities. A detailed description of the muon beamlines is given by Eaton and Kilcoyne (1999).

2.2.2.1 Time structure. Here, we assume that only the usual time-differential muon technique is of interest. The time-integrated techniques are discussed in Chapter 4.

The proton beam, which is directed to the target for producing the pions, is provided by an accelerator. The time distribution of the protons in the beam determines the time distribution of the muons in the muon beam. Two types of proton beams exist: pseudo-continuous and pulsed. In the pseudo-continuous case, the proton bunches are separated by a time interval of a few tens of nanoseconds compared to tens of milliseconds in pulsed beams.

In the case of a pseudo-continuous beam, a muon detector is placed in the muon beam close to the sample; see Fig. 1.1. A clock is started when a muon passes through it and is stopped when the positron arising from the muon decay is detected. This constitutes a so-called *event*. A histogram of the number of events versus time is built. Recording a typical histogram of 10^7 events takes less than one hour. For the logic to work as described above, ideally only one muon should be in the sample at a given instant. If a second muon stops before the positron arising from the decay of the first muon is detected, coincidence electronics pauses data-taking to prevent mixing. To avoid as much as possible such an undesirable situation, the intensity of the muon beam is controlled by tuning the elements of the muon beam optics. The decay positron of a given muon is not necessarily detected since, for example, the positron detectors do not cover a 4π steradian solid angle around the sample. Therefore a timeout corresponding to a few muon lifetimes is also set so that one does not have to wait forever after the implantation of a muon before accepting a new incoming muon. In addition, a rather involved logical coincidence electronics is used to limit the background due to other sources of radiation — see Schenck (1985). With such a beam, a muon precession frequency of up to about 600 MHz (this corresponds to a local field of 4.4 T) can be detected with a conventional setup, although with a reduced asymmetry at the highest frequencies. Higher frequencies can be observed with optimised spectrometers. For instance the muonium hyperfine frequency corresponding to the transition between triplet and singlet states (which is as high as 4.464 GHz in zero field) was directly observed in quartz by Holzschuh *et al.* (1981), who used a compact spectrometer and high time resolution detectors. In addition, it is possible with electronics and a veto counter to discriminate between the events arising from muons stopped in the sample and in its environment. The data then reflect the physics of the system under study with no additional signal. Sonier *et al.* (2003) obtained high-quality data for a sample of only 6.5 mm^2 cross section and of 3.74 mg weight.

A disadvantage of a pseudo-continuous beam is the relatively high electronic background which remains despite the coincidence electronics. This complicates the extraction of the polarisation function from the raw data and restricts the observation of slowly damped relaxation signals. A beam-sharing device that allows the extraction of one muon from the beam upon request has been developed (Abela *et al.* 1999). This device drastically decreases the electronic background and allows one to extend the range of measurements of slowly damped signals at a pseudo-continuous beam.

In a pulsed muon beam, a bunch of muons (on the order of thousand muons per bunch) is implanted into the sample at a time taken as zero and a set of detectors is activated by the positrons arising from the muon decays. The number of detectors is much larger than at a continuous source in order to cope with the intense muon pulse. The amount of time to record a spectrum is currently about the same as at a continuous muon beam. Two features of the beam time structure are of importance: the pulse width and the separation between the pulses. The latter is long compared to the muon lifetime. For example, the pulse shape of the muon beam feeding the spectrometers of the ISIS facility at the time of writing is approximately Gaussian with a full-width at half-maximum of about 80 ns and the separation between pulses is 20 ms. This unique time structure, with extremely few background particles appearing between the pulses, enables the detection of very weakly damped signals or of very low precession frequencies. Let us estimate the order of magnitude of the smallest field that can be measured. We have for the muon angular frequency $\omega_\mu = \gamma_\mu B_{\rm loc}$ where γ_μ is the muon gyromagnetic ratio ($\gamma_\mu = 851.616\,{\rm Mrad\,s^{-1}\,T^{-1}}$) and $B_{\rm loc}$ the local field. Since an angle of 0.5 rad is detectable and a μSR measurement at a pulsed source can routinely be carried out up to $15\,\mu$s, $B_{\rm loc}$ as small as 0.04 mT can be detected. To put this field into perspective, it might be produced by a nuclear magnetic moment, which is $\sim 10^3$ times smaller than a usual electronic magnetic moment. Furthermore, the negligible background greatly simplifies the determination of the shape of the polarisation function. On the other hand, the relatively large pulse width limits the time resolution. A muon beam of full-width at half-maximum of 80 ns is unsuitable for investigating systems exhibiting relaxation with exponential rate larger than $\sim 10\,\mu{\rm s}^{-1}$ or $B_{\rm loc}$ exceeding $\sim$50 mT. The spot size of a pulsed beam is also relatively large. However, if the "fly-past" technique can be used (a setup with a minimum amount of material around the sample), samples of $\sim 50\,{\rm mm}^2$ cross section can be reasonably well investigated. In fact this fly-past technique is inspired by the low background setup available at continuous sources, but with no coincidence electronics and veto counters since they cannot be used when more than one muon is present in the sample at a given time.

Independent of the time structure of the muon beam, the implanted muon density is very small; the muon bombardment is relatively gentle. However, because the incident energy of most beams exceeds the ionisation threshold, some radiation damage is induced and therefore the generation of radiolysis products is expected. The various processes by which the muons lose their initial kinetic energy are fast and involve only electrostatic interactions which are not spin-dependent. So we do not expect any loss of polarisation during thermalisation. Thus, at time t_0 after which valuable data are available in the positron count histograms, the polarisation of the muon ensemble is equal to the polarisation of the muon beam when it was created by the decay of the pion. At least, this seems true in non-magnetic metallic samples. The situation is less clear-cut in insulator and molecular materials. Here a certain discrepancy is not uncommon between the intrinsic muon beam polarisation and the measured polarisation at $t = t_0$. In the literature this is referred to as the "missing fraction" and it has always been a matter of discussion. When muonium formation is involved, or is suspected of being involved, the "prompt" depolarisation, which leads to the missing fraction, is now attributed to spin-flip encounters with paramagnetic radiolysis products. A prompt depolarisation can occur in compounds such as magnetic oxides because the fraction of muons which captures electrons to form muonium can be strongly depolarised in a short time interval by the magnetic fields naturally present. The effect of radiolysis electrons in semiconductors is further discussed in Section 12.4.

The time t_0 is on the order of a nanosecond at a continuous muon source. It is on the order of the pulse width at a pulsed source. Of course, no detailed information can be obtained for times shorter than t_0.

2.2.2.2 Energy characteristics. Three types of muon beams are available, for which the muon kinetic energy differs. We first discuss the high momentum beam, then the so-called surface muon beam, and complete our presentation by describing the newly developed low energy muon beam. For each case, the beam properties as well as the muon characteristic stopping range are given.

Historically, high momentum muon beamlines were the first to be developed. In these beamlines muons are obtained from *pions decaying in flight.* A characteristic device for these beamlines is a several-metres-long superconducting solenoid producing a field of a few teslas, called a decay channel, to which pions of selected momentum are directed and decay. This solenoid ensures collection of the emerging muons which fly with helical trajectories to its exit. The muons are then directed onto the sample. Note that muons selected from backward decay of the pions have their spins parallel to their momentum, unlike those resulting from forward decay and also unlike surface muons. Therefore, the muon beam polarisation is not nearly 100% as for a surface beam, but on the order of 70-80%. Such beamlines are appropriate for producing muons with a momentum between 50 and 100 MeV/c. For such a momentum, the stopping range for muons in matter is one to several grams per square centimetre of cross section. Samples with large masses are therefore needed. However, additional material can be introduced into the beam in front of the sample so as to degrade the muon momentum and limit the sample mass. Nowadays these beamlines are essentially used when specific experimental conditions are required, e.g. pressure cells or special sample containers that the muons must penetrate before reaching the sample.

At the beginning of the 1980s it became obvious that the amount of sample required to perform a μSR study was a limiting factor for the increased use of these techniques by the condensed-matter community. Fortunately, in 1976 the concept of a so-called surface muon beam was introduced. It consists of tuning the beamline to select muons decaying from *pions at rest* in the pion production target. Muons decaying from pions at rest have a momentum of 29.79 MeV/c (corresponding to a kinetic energy of 4.1 MeV). Such muons are stopped by a small amount of matter and therefore only those produced at the target surface or close to it can escape and be used for a beam. This justifies the name surface muons. As these muon beams were originally developed in Berkeley by a group from the University of Arizona, they are also sometimes referred to as "Arizona beams".

An additional device, called a separator, is needed in a surface muon beamline in order to remove contaminant particles such as positrons which are massively produced at the pion production target. The operation principle of a separator is similar to that of the spin rotator to be discussed in Section 2.2.2.3. It consists of a magnetic field and an electric field acting on the beam. These fields are oriented perpendicular to each other and also perpendicular to the muon momentum. In addition, their directions and intensities are such that the Lorentz and Coulomb forces acting on the muons are antiparallel. Hence, the total force vanishes and the muon beam is unaffected by the device. However, for particles with the same charge as the muons but different velocities (e.g. the positrons), the Lorentz force is different and the balance with the Coulomb force is not satisfied. Such particles are deviated and not transmitted further down in the beamline. It must yet be noted that the magnetic field interacts with the muon spin, which leads to Larmor precession, whereas the electric field does not. It results in a rotation of the muon spin polarisation away from the direction of the momentum. However, the magnetic field intensity, as well as the time spent by the muons in the separator, is relatively modest. This means that the rotation angle of the muon polarisation is small but not fully negligible, usually on the order of $5°$ to $10°$.

The momentum of the so-called surface muons is still too large for studying thin films, nanomaterials, or multilayers. Indeed the muons are stopped at best in a few hundred micrometres since 100 to 200 mg of material per square centimetre of cross section are required. Slow muons, i.e. with kinetic energy in the kiloelectronvolt range or below, are needed. To produce them, the challenge is to keep a high degree of polarisation in the slowing down process together with a reasonable efficiency. This is achieved using a moderator made of a van der Waals gas (typically Ne, N_2, or Ar) condensed on a cold substrate — see Morenzoni *et al.* (1997) and Morenzoni (1999) for details. Surface muons (4.1 MeV) impinge the substrate which is so thin that it stops no more than half of the muons. They subsequently cross the gas layer of typical thickness 200-300 nm, where they are slowed down through interactions with electrons. This slowing-down process is very effective only in the kiloelectronvolt range and above, and not for slower muons, so that muons with epithermal energy ($\sim$50 eV or less) are released by the moderator. As the process occurs in a relatively short time (a few picoseconds) the muon polarisation is barely affected. The efficiency, of order 10^{-4}, is sufficient to get about 8000 muons per second at the sample allowing for routine experiments. A spectrometer dedicated to studies with low energy muons (LEM) is available at SμS (Paul Scherrer Institute). The possibility to reaccelerate the slow muons provides the opportunity for fine-tuning the stopping range of the muons in the sample from the nanometre range at $\sim$10 eV to the 100-nm region at $\sim$20 keV.

2.2.2.3 Muon spin polarisation. In a surface-muon beam, the muon spin is antiparallel to its linear momentum. However, it can be useful to have the spin at a variable angle from the momentum. In fact, this is possible using a device called a *spin rotator*, which is inserted in the beamline and which is able to turn the muon spin by up to almost $90°$ relative to the muon linear momentum. If it is not installed on all of them, this is mostly because of its cost. A spin rotator is a separator (see Section 2.2.2.2), but with magnetic and electric fields of much higher intensities.

An important application of a spin rotator is for high transverse-field measurements: the external field is then applied along the muon beam linear momentum and the initial muon beam polarisation, $\mathbf{S}_\mu$, is set almost perpendicular to the beam with a spin rotator. With this configuration, the muon trajectories before implantation into the sample are not affected by the external field since no Lorentz force acts on the muons (remember a muon carries a unit electric charge!).

A second advantage of a spin rotator is the possibility of investigating, with a *unique* crystal, the μSR response of a specimen with $\mathbf{S}_\mu$ parallel and perpendicular to a relevant axis of the crystal, respectively. This can be achieved by orienting this crystal axis parallel to the muon beam and recording spectra either using the spin rotator or not. Detectors in adequate directions are of course required.

2.3 Principles of μSR data treatment

Let us consider a simple experimental setup with one positron detector placed in the direction of $\mathbf{S}_\mu$ and a magnetic field applied perpendicular to this direction. This setup is sketched in Fig. 1.1. Since $\mathbf{B}_{\rm ext}$ is perpendicular to $\mathbf{S}_\mu$, we are considering a transverse geometry experiment. We assume that the sample just serves to stop the muons, i.e. it does not influence the muon spins. The time is digitised into $N_{\rm c}$ channels so that a given time t_n is such that $t_n = n\Delta t + t_{\rm offset}$ with Δt being the time resolution, $n = \{0, 1, \ldots, N_{\rm c} - 1\}$ and $t_{\rm offset}$ a time offset.

In general terms, we start from Eq. 2.4 and substitute $a_{\rm as}(\varepsilon)$ by a_0, recognising that the effective initial asymmetry is dependent on the experimental conditions, such as the solid angle subtended by the detector as seen from the sample, or the range of energy of the positrons which are detected. The initial asymmetry a_0 is an experimental parameter, typically $a_0 \sim 0.25$. The positron count at t_n is then written as

$$N(t_n) = N_0 \exp\left(-t_n/\tau_\mu\right)\left[1 + a_0 \cos(\omega_\mu t_n + \varphi)\right]. \tag{2.5}$$

The constant N_0 sets the scale of the number of counts. The exponential function accounts for the finite muon lifetime. According to the Larmor equation (Section 3.1.1), the muon spins precess in the field and therefore the angle θ (Eq. 2.4) is equal to $\theta(t_n) = \gamma_\mu B_{\rm loc} t_n = \gamma_\mu B_{\rm ext} t_n = \omega_\mu t_n$ (we have assumed $B_{\rm loc} = B_{\rm ext}$).

In the forward direction, i.e. in the direction of $\mathbf{S}_\mu$, the phase $\varphi = 0$. We shall label the detector in this direction as $+$. Hence

$$N_+(t_n) = N_0 \exp\left(-t_n/\tau_\mu\right)\left[1 + a_0 \cos(\omega_\mu t_n)\right]. \tag{2.6}$$

We can choose to have the detector at 180° of $\mathbf{S}_\mu$, i.e. in the so-called backward direction denoted as $-$. This means that $\varphi = \pi$ in Eq. 2.5 and therefore

$$N_-(t_n) = N_0 \exp\left(-t_n/\tau_\mu\right)\left[1 - a_0 \cos(\omega_\mu t_n)\right]. \tag{2.7}$$

So far, we do not account for a possible multi-site implantation, muon spin dynamics, and field distribution. This is achieved by the following natural generalisation of the previous two equations:

$$N_\pm(t_n) = N_0 \exp\left(-t_n/\tau_\mu\right)\left[1 \pm a_0 P_X^{\rm exp}(t_n)\right], \tag{2.8}$$

where $P_X^{\rm exp}(t_n)$ is the normalised polarisation function ($P_X^{\rm exp}(t \to 0) \equiv 1$), which contains the physics of the compound under study (see Chapter 3). Unless stated we shall refer to $P_X^{\rm exp}(t_n)$ as simply the polarisation function. As mentioned previously, a fast relaxation or dephasing process may occur between times $t = 0$ and $t_0 \equiv t_{\rm offset}$. Its physical origin is usually different from that governing the polarisation evolution observed for $t \geq t_{\rm offset}$ and described by $P_X^{\rm exp}(t_n)$. It may reflect the fact that a fraction of the muons are in an environment which differs from the others. Since the shape of $P_X^{\rm exp}(t < t_{\rm offset})$ cannot be experimentally determined, it is often difficult to ascertain the origin of the process at play. Anyhow, the presence of this prompt depolarisation gives information which should not be neglected. It can be viewed as a missing fraction of the muon polarisation and is accounted for by a multiplicative coefficient, η_μ, with $0 \leq \eta_\mu \leq 1$ such that

$$N_\pm(t_n) = N_0 \exp(-t_n/\tau_\mu) \left[1 \pm a_0 \eta_\mu P_X^{\rm exp}(t_n)\right], \tag{2.9}$$

where again $P_X^{\rm exp}(t \to 0) \equiv 1$ by definition. The product $a_0\eta_\mu$ is an effective asymmetry. The function $P_X^{\rm exp}(t_n)$ describes the evolution of the polarisation of the muons, excluding the missing fraction.

We now consider an experimental setup with two positron detectors, one in the $\mathbf{S}_\mu$ direction and the second at 180° of $\mathbf{S}_\mu$. We are still considering a transverse geometry experiment. The quantity $a_0\eta_\mu P_X^{\rm exp}(t)$ can be visualised by directly combining the counts of the two detectors:

$$a_0 \eta_\mu P_X^{\rm exp}(t_n) = \frac{N_+(t_n) - N_-(t_n)}{N_+(t_n) + N_-(t_n)}. \tag{2.10}$$

This nice result is valid only if the two detectors are equivalent. In practice, this is rarely the case. As shown in Section A.1, to extract $a_0\eta_\mu P_X^{\rm exp}(t_n)$ from the measurements, the normalised weighted difference of the counts should be considered with the result,

$$a_0 \eta_\mu P_X^{\rm exp}(t_n) = \frac{N_+(t_n) - \alpha_{\rm bal} N_-(t_n)}{N_+(t_n) + \alpha_{\rm bal} N_-(t_n)}, \tag{2.11}$$

where $\alpha_{\rm bal}$ is the so-called balance parameter between the two detectors. Once $\alpha_{\rm bal}$ has been measured, $a_0\eta_\mu P_X^{\rm exp}(t_n)$ can be obtained directly from the counts of the two detectors. There is no need to determine the parameters $N_{0\pm}$ defined in Section A.1.

However, it may be that one must take into account a time-independent background in Eq. 2.9. This means that

$$N_\pm(t_n) = N_0 \{\exp(-t_n/\tau_\mu) \left[1 \pm a_0 \eta_\mu P_X^{\rm exp}(t_n)\right] + b_{\rm elec,\pm}\}. \tag{2.12}$$

In practice, this background exists only at a pseudo-continuous muon beam source. It reflects the difficulty of the data acquisition system to reject events recorded while more than one muon is in the sample. If $b_{\rm elec,\pm}$ is appreciable, $a_0\eta_\mu P_X^{\rm exp}(t_n)$ may not reliably be determined from Eq. 2.11 because a small variation of the electronic background leads to notable differences in the shape of the polarisation function. A possibility is to extract it directly from Eq. 2.12, i.e.

$$\pm a_0 \eta_\mu P_X^{\rm exp}(t_n) = \exp(t_n/\tau_\mu) \left[\frac{N_\pm(t_n)}{N_0} - b_{\rm elec,\pm}\right] - 1. \tag{2.13}$$

With this method, the values of N_0 and $b_{\rm elec,\pm}$ are needed to visualise $a_0\eta_\mu P_X^{\rm exp}(t_n)$. It is also common to build a ratio in the same spirit as given in Eq. 2.11. More details can be found in Section A.1.

Let us now discuss a zero- or longitudinal-field experiment with the two-detector setup; see Fig. 1.1. Obviously,

$$N_\pm(t_n) = N_0 \{\exp(-t_n/\tau_\mu) \left[1 \pm a_0 \eta_\mu P_Z^{\rm exp}(t_n)\right] + b_{\rm elec,\pm}\}. \tag{2.14}$$

As for $a_0\eta_\mu P_X^{\rm exp}(t)$, the product $a_0\eta_\mu P_Z^{\rm exp}(t_n)$ can be derived from the weighted difference only if $b_{\rm elec,\pm}$ is negligible.

Apart from the background of electronic origin already mentioned, other corrections might be necessary for a detailed data treatment. These are the dead-time of the detectors if the counting rate is high and possibly the muon beam shape for measurements performed at a pulsed source. In addition, to cope with the high counting rate at such a source, many detectors (up to a few hundreds) are installed, each of them covering a small solid angle. Then each detector is characterised by its own geometrical phase which is not simply 0 or π, as in the previous discussions. All these points are discussed in detail in Section A.1.

When the muon precession frequency is large, it is convenient to observe the asymmetry spectrum $a_0 \eta_\mu P_\alpha^{\rm exp}(t_n)$ in the rotating reference frame; see Section A.2 for more information. Although the final analysis should only be carried out in the time domain, it is often helpful to inspect the frequency or field distribution associated to $a_0 \eta_\mu P_\alpha^{\rm exp}(t_n)$. This distribution, which is obtained through a time Fourier transform, may help to understand the physics involved. The discrete Fourier transform technique is recalled in Section A.2.1. The more efficient maximum entropy method is exposed in Section A.2.2. The Fourier and the maximum entropy techniques may result in distributions with spurious effects of numerical origin. To deal with them, apodisation of the distributions is needed. This is explained in Section A.2.3.

References

Abela R., Amato A., Baines C., Donath X., Erne R., George D.C., Herlach D., Irminger G., Reid I.D., Renker D., Solt G., Suhi D., Werner M., and Zimmermann U. (1999) *Hyperfine Interactions* **120/121**, 575–578.

Eaton G.H. and Kilcoyne S.H. (1999) in *Muon Science: Muons in Physics, Chemistry, and Materials*, edited by Lee S.L., Kilcoyne S.H., and Cywinski R. (Institute of Physics, Bristol).

Holzschuh E., Kündig W., and Patterson B.D. (1981) *Helvetica Physica Acta* **54**, 552–560.

Kunze P. (1933) *Zeitschrift für Physik A, Hadrons and Nuclei* **83**, 1–18.

Morenzoni E. (1999) in *Muon Science: Muons in Physics, Chemistry, and Materials*, edited by Lee S.L., Kilcoyne S.H., and Cywinski R. (Institute of Physics, Bristol).

Morenzoni E., Prokscha Th., Hofer A., Matthias B., Meyberg M., Wutzke Th., Glückler H., Birke M., Litterst J., Niedermayer Ch., and Schatz G. (1997) *Journal of Applied Physics* **81**, 3340–3347.

Nagamine K. (2003) *Introductory Muon Science* (Cambridge University Press, Cambridge).

Schenck A. (1985) *Muon Spin Rotation Spectroscopy* (Adam Hilger, Bristol).

Sonier J.E., Poon K.F., Luke G.M., Kyriakou P., Miller R.L., Liang R., Wiebe C.R., Fournier P., and Greene R.L. (2003) *Physical Review Letters* **91**, 147002, 1–4.

Weissenberg A.O. (1967) *Muons* (North-Holland, Amsterdam).

Part II

Polarisation functions

3

Basics of muon and muonium polarisation functions

A conventional muon experiment involves the measurement as a function of time of the projection of the muon beam polarisation vector along a given direction specified by the unit vector, $\mathbf{d}_p$, which is the direction of the positron momentum. Note that at $t = 0$ the beam is fully polarised in a direction which may not be parallel to $\mathbf{d}_p$. Since a few millions of decay positrons are recorded in a typical experiment, we must consider the average over all the decay events of the muon spin vector, $\mathbf{S}_\mu(t)$. This average denoted as $\langle \mathbf{S}_\mu(t) \rangle$ is called the polarisation vector. It is usually a thermal average, although an additional ensemble average must be performed for distributed or disordered systems. For simplicity, we only mention the thermal average. Therefore a conventional muon experiment provides access to the scalar product

$$P_{i_\mu,d_p}(t) = \frac{\langle \mathbf{S}_\mu(t) \rangle}{S_\mu(0)} \cdot \mathbf{d}_p, \tag{3.1}$$

where $\mathbf{i}_\mu$ specifies the initial muon beam polarisation vector:

$$\mathbf{i}_\mu = \mathbf{S}_\mu(0)/S_\mu(0). \tag{3.2}$$

We obviously do not have to perform the ensemble average over the initial polarisation $\mathbf{S}_\mu(0)$. The quantity $P_{i_\mu,d_p}(t)$ is called the polarisation, depolarisation, or relaxation function, depending on the physics involved as explained in Chapters 6 and 7. Its value at origin $P_{i_\mu,d_p}(0)$ is equal to the projection along $\mathbf{d}_p$ of $\mathbf{i}_\mu$.

As explained briefly in Section 1.1.1 and in more detail in Section 2.3, most of the experiments are performed with either the longitudinal or transverse geometry. If required, a magnetic field is applied along the Z axis. Then, the polarisation function to be considered is $P_{d_p,d_p}(t)$, which we abbreviate as $P_{d_p}(t)$.

The time evolution of the muon spin is only dictated by the local magnetic field $\mathbf{B}_{\rm loc}(t)$ encountered by the muon during its lifetime. Since the muon is a point-like particle, i.e. it has neither an electric dipole moment nor further multipole terms (quadrupole, octupole, ...), its spin is not influenced by electric fields. However, because the muon has also an electric charge, electric fields exert a force on the particle itself.

We shall first provide in this chapter the basic framework for the muon polarisation function and then consider this function for muonium.

3.1 Muon polarisation function

Two descriptions are possible. A classical treatment of $\mathbf{S}_\mu(t)$ can be used and an average performed for a given field distribution. In this picture, the effect on the muon spin of the possible time dependence of $\mathbf{B}_{\rm loc}$ is described as a stochastic process. Alternatively, $P_{i_\mu,d_p}(t)$ is computed using the Heisenberg (or Schrödinger) equation. In this fully quantum approach, the system under investigation — consisting of the muon and its interaction with the host where it is implanted — is modelled by a Hamiltonian. Whereas the latter method is theoretically more complete (for example it takes into account the energy level degeneracy for a half-integer nuclear spin interacting with an electric field gradient provided the spin is strictly larger than $\frac{1}{2}$; this degeneracy is absent in a purely classical treatment), the former method is usually simple enough that it leads to a useful intuitive picture of the physics involved.

We first describe $P_{i_\mu,d_p}(t)$ with classical mechanics and then turn to quantum mechanics.

3.1.1 Classical treatment

Two complementary methods can be used to derive $P_{i_\mu,d_p}(t)$. In the traditional vectorial method $\mathbf{S}_\mu(t)$ is described and the scalar product $\langle \mathbf{S}_\mu(t)\rangle \cdot \mathbf{d}_p$ is calculated. The second approach, which has mostly been advocated in the Soviet Union (Belousov *et al.* 1979), uses projection operators. It has the advantage of leading to a single compact formula for $P_{i_\mu,d_p}(t)$, which allows one to easily discuss any experimental geometries. We first present the vectorial method.

3.1.1.1 The vectorial method. A basic principle of mechanics states that the time derivative of the angular momentum of a classical object is equal to the sum of the torques applied to that object. Since the muon spin is not sensitive to an electric field, the only torque it senses is due to $\mathbf{B}_{\rm loc}(t)$. This torque is given by the vector product $\mathfrak{m}_\mu \times \mathbf{B}_{\rm loc}(t)$, where $\mathfrak{m}_\mu$ is the muon magnetic moment. Because the angular momentum of the muon is $\hbar\mathbf{S}_\mu$, we have the equation

$$\hbar\frac{\mathrm{d}\mathbf{S}_\mu(t)}{\mathrm{d}t} = \mathfrak{m}_\mu(t) \times \mathbf{B}_{\rm loc}(t), \tag{3.3}$$

or

$$\frac{\mathrm{d}\mathbf{S}_\mu(t)}{\mathrm{d}t} = \gamma_\mu\, \mathbf{S}_\mu(t) \times \mathbf{B}_{\rm loc}(t). \tag{3.4}$$

As for any magnetic moment, the muon gyromagnetic ratio γ_μ is, by definition, the ratio of the magnetic moment over the angular momentum of the muon (here, $\hbar = h/2\pi$, where h is the Planck constant). Equation 3.4 is known as the Larmor equation.

Even without solving Eq. 3.4, we can derive characteristics of $\mathbf{S}_\mu(t)$ by direct inspection of its mathematical structure. First, we note that $\frac{\mathrm{d}\mathbf{S}_\mu(t)}{\mathrm{d}t} \cdot \mathbf{S}_\mu(t) = 0$. This means that the modulus of $\mathbf{S}_\mu$ is a constant of the motion, i.e. it is time independent. Therefore we can safely write $S_\mu(t) = S_\mu(0)$. Secondly, we also note that $\frac{\mathrm{d}\mathbf{S}_\mu(t)}{\mathrm{d}t} \cdot \mathbf{B}_{\rm loc}(t) = 0$, which implies that $\frac{\mathrm{d}\mathbf{S}_\mu(t)}{\mathrm{d}t}$ is perpendicular to $\mathbf{B}_{\rm loc}(t)$.

To proceed further, we shall assume from now on in this chapter that $\mathbf{B}_{\rm loc}$ is time independent. Although this approximation may seem so strong as to jeopardise its

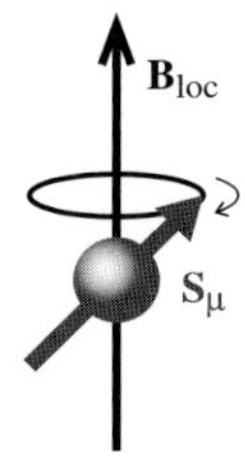

Fig. 3.1 Schematic view of the Larmor precession of spin $\mathbf{S}_\mu$ around $\mathbf{B}_{\rm loc}$.

usefulness, we shall show in Chapter 6 that it is in fact quite useful. In many cases of interest the effect of the dynamical character of $\mathbf{B}_{\rm loc}$ on $\mathbf{S}_\mu$ can approximately be described as a stochastic process using as a basic ingredient the solution of the Larmor equation we are going to derive.

Because $\mathbf{B}_{\rm loc}$ is time independent, we get $\frac{{\rm d}\mathbf{S}_\mu(t)}{{\rm d}t}\cdot\mathbf{B}_{\rm loc} = \frac{{\rm d}[\mathbf{S}_\mu(t)\cdot\mathbf{B}_{\rm loc}]}{{\rm d}t} = 0$. Accounting for the fact that S_μ is a constant of motion, we then deduce that the angle between $\mathbf{S}_\mu(t)$ and $\mathbf{B}_{\rm loc}$ is constant. This analysis shows that if $\mathbf{S}_\mu(0)$ is collinear to $\mathbf{B}_{\rm loc}$, $\mathbf{S}_\mu(t) = \mathbf{S}_\mu(0)$, i.e. $\mathbf{S}_\mu(t)$ is a constant vector.

Now we solve Eq. 3.4. when $\mathbf{S}_\mu(0)$ is not collinear to $\mathbf{B}_{\rm loc}$. We consider a reference frame defined by a set of three orthogonal unit vectors $\{\mathbf{u}, \mathbf{v}, \mathbf{w}\}$ such that $\mathbf{u}$ is parallel to $\mathbf{B}_{\rm loc}$, $\mathbf{v}$ is parallel to $\mathbf{S}_\mu(0) - [\mathbf{S}_\mu(0)\cdot\mathbf{u}]\mathbf{u}$, and $\mathbf{w} = \mathbf{u}\times\mathbf{v}$. The vector set $\{\mathbf{u}, \mathbf{v}, \mathbf{w}\}$ therefore defines a direct orthogonal reference frame. The solution of the Larmor equation then reads

$$\mathbf{S}_\mu(t) = S_\mu^{\parallel}(0)\,\mathbf{u} + S_\mu^{\perp}(0)[\cos(\omega_\mu t)\,\mathbf{v} - \sin(\omega_\mu t)\,\mathbf{w}], \qquad (3.5)$$

where we have defined the angular frequency, $\omega_\mu = \gamma_\mu B_{\rm loc}$. The quantities $S_\mu^{\parallel}(0)$ and $S_\mu^{\perp}(0)$ are the components of the projections of $\mathbf{S}_\mu(0)$ parallel and perpendicular to $\mathbf{B}_{\rm loc}$, respectively. Equation 3.5 shows that the component of $\mathbf{S}_\mu$ collinear to $\mathbf{B}_{\rm loc}$ remains constant, whereas the perpendicular component, $\mathbf{S}_\mu^{\perp}(t)$ (the second term on the right-hand-side of the equation), rotates at the Larmor angular frequency ω_μ. Therefore $\mathbf{S}_\mu(t)$ follows a cone around $\mathbf{B}_{\rm loc}$ (Fig. 3.1). This motion is called *precession.* The precession frequency is $\nu_\mu = \omega_\mu/(2\pi)$. We note that it depends solely on $B_{\rm loc}$ and not on the angle between $\mathbf{B}_{\rm loc}$ and $\mathbf{S}_\mu(0)$. Looking at $\mathbf{B}_{\rm loc}$ head-on, the rotation of $\mathbf{S}_\mu^{\perp}(t)$ is clockwise.

We now discuss the usual experimental geometry for which $\mathbf{d}_p$ is parallel to $\mathbf{i}_\mu$, i.e. $\mathbf{d}_p = \mathbf{S}_\mu(0)/S_\mu(0)$. Since, according to Eq. 3.5,

$$\mathbf{S}_\mu(0) = S_\mu^{\parallel}(0)\,\mathbf{u} + S_\mu^{\perp}(0)\,\mathbf{v}, \qquad (3.6)$$

we get

$$\left[\frac{\mathbf{S}_\mu(t)}{S_\mu(0)}\right]\cdot\mathbf{d}_p = \left[\frac{\mathbf{S}_\mu(t)}{S_\mu(0)}\right]\cdot\left[\frac{\mathbf{S}_\mu(0)}{S_\mu(0)}\right] = \left[\frac{S_\mu^{\parallel}(0)}{S_\mu(0)}\right]^2 + \left[\frac{S_\mu^{\perp}(0)}{S_\mu(0)}\right]^2\cos(\omega_\mu t). \qquad (3.7)$$

Accordingly, if there is a unique local field seen by all the muons, Eq. 3.7 yields the expression of $P_{d_p,d_p}(t)$. It shows that it is possible to determine the modulus of $\mathbf{B}_{\rm loc}$ from the angular frequency of the muon polarisation function and its direction relative to the initial muon beam polarisation by measuring the amplitude of its oscillating component. If $\mathbf{B}_{\rm loc}$ is parallel to the Z axis, the transverse polarisation function is written

$$P_X(t) = \cos(\omega_\mu t), \tag{3.8}$$

and the longitudinal polarisation function $P_Z(t) = 1$ since $\mathbf{S}_\mu^{\parallel}(0) = 0$ and $\mathbf{S}_\mu^{\perp}(0) = 0$ for the transverse and longitudinal geometry, respectively. Let us consider $P_{X,Y}(t)$ that for convenience we abbreviate by $P_Y(t)$, although it is a misnomer. By definition, the initial beam polarisation is along the X axis (see Chapter 2). This means that $S_\mu^{\parallel}(0) = 0$ and $\mathbf{d}_p = \mathbf{w}$ in Eq. 3.5. Then this equation yields

$$P_Y(t) = -\sin(\omega_\mu t) = \cos(\omega_\mu t + \pi/2). \tag{3.9}$$

Therefore, at first glance it does not seem that $P_Y(t)$ contains more physical information than $P_X(t)$ since there is only an angular shift of $\pi/2$ between these two functions. However, it is sometimes necessary to record both $P_X(t)$ and $P_Y(t)$, as for example to display a transverse μSR spectrum in the rotating frame; see Section A.2.

3.1.1.2 The projector operator method. It is possible to express the solution of the Larmor equation given in Eq. 3.5 in terms of the longitudinal operator $\boldsymbol{P}_{\rm L}(\mathbf{u})$ and the two transverse projectors $\boldsymbol{P}_{\rm T_1}(\mathbf{u})$ and $\boldsymbol{P}_{\rm T_2}(\mathbf{u})$ defined in Section C.1. We get

$$\mathbf{S}_\mu(t) = \boldsymbol{M}(t)\, \mathbf{S}_\mu(0), \tag{3.10}$$

with

$$\boldsymbol{M}(t) = \boldsymbol{P}_{\rm L}(\mathbf{u}) + \boldsymbol{P}_{\rm T_1}(\mathbf{u}) \cos(\omega_\mu t) - \boldsymbol{P}_{\rm T_2}(\mathbf{u}) \sin(\omega_\mu t). \tag{3.11}$$

The expression of the Cartesian components of tensor $\boldsymbol{M}(t)$ is easily derived:

$$M^{\alpha\beta}(t) = u^\alpha u^\beta + \left(\delta^{\alpha\beta} - u^\alpha u^\beta\right) \cos(\omega_\mu t) - \sum_\nu \epsilon^{\alpha\nu\beta} u^\nu \sin(\omega_\mu t). \tag{3.12}$$

We have introduced the unit axial tensor $\epsilon^{\alpha\nu\beta}$; see Section C.1. From the definition of $\boldsymbol{M}(t)$ and Eq. 3.1, we derive

$$P_{i_\mu,d_p}(t) = \sum_{\alpha,\beta} \langle M^{\alpha\beta}(t)\rangle\, i_\mu^\beta\, d_p^\alpha. \tag{3.13}$$

We recall that i_μ^β and d_p^β refer to the β component in the laboratory reference frame of the directions of the initial muon beam polarisation and positron detector, respectively.

The tensorial method does not contain new physics relative to the vectorial method, but it is sometimes more practical as clearly shown by the following example. We have

$$P_Y(t) = \langle M^{YX}(t)\rangle = \left\langle \frac{B_{\rm loc}^Y B_{\rm loc}^X}{B_{\rm loc}^2} [1 - \cos(\omega_\mu t)] - \frac{B_{\rm loc}^Z}{B_{\rm loc}} \sin(\omega_\mu t) \right\rangle, \tag{3.14}$$

since $\mathbf{u} = \mathbf{B}_{\rm loc}/B_{\rm loc}$. As expected, this expression is consistent with Eq. 3.9 when $\mathbf{B}_{\rm loc}$ is parallel to Z. In addition, it describes the case for which there is no restriction on the direction of $\mathbf{B}_{\text{loc}}$. The fact that Eq. 3.12 depends on the projection of $\mathbf{u}$ on the three unit vectors $\{\hat{\mathbf{X}}, \hat{\mathbf{Y}}, \hat{\mathbf{Z}}\}$ which form a laboratory reference frame, and not on the three unit vectors $\{\mathbf{u}, \mathbf{v}, \mathbf{w}\}$ as Eq. 3.5, leads to the simple compact expression given in Eq. 3.13.

3.1.2 Quantum treatment

From quantum mechanics we know that only the expectation value of an operator can be measured — see for example Dirac (1958) or Schiff (1968). With the μSR techniques we measure the time evolution of the expectation value of the muon spin operator projected along the positron detector direction. This means that

$$P_{i_\mu,d_p}(t) = \langle\psi|\boldsymbol{\sigma}(t)|\psi\rangle \cdot \mathbf{d}_p, \tag{3.15}$$

where $|\psi\rangle$ is the wavefunction of the muon–system ensemble at initial time and $\boldsymbol{\sigma}$ the Pauli vector operator (see Section C.5). A statistical average is implicit in the expression for $P_{i_\mu,d_p}(t)$. Using the expansion of $|\psi\rangle$ on the Hilbert basis of the muon–system ensemble, $|\psi\rangle = \sum_n C^n|n\rangle$, we write

$$\begin{aligned} P_{i_\mu,d_p}(t) &= \sum_{n,m}(C^m)^* C^n \,\langle m|\boldsymbol{\sigma}(t)|n\rangle \cdot \mathbf{d}_p = \sum_n \langle n|\rho_{\text{tot}}\,\boldsymbol{\sigma}(t)|n\rangle \cdot \mathbf{d}_p \\ &= \text{Tr}\left\{\rho_{\text{tot}}\,\boldsymbol{\sigma}(t)\cdot\mathbf{d}_p\right\}. \end{aligned} \tag{3.16}$$

Here Tr$\{A\}$ stands for the trace of operator A and ρ_{tot} is the density operator of the muon–system ensemble, i.e. $\langle n|\rho_{\text{tot}}|m\rangle = (C^m)^*C^n$. Like any density operator its trace is equal to 1 — see, for example, Messiah (1965). The equation of motion for $\boldsymbol{\sigma}(t)$ is given by the Heisenberg equation:

$$\hbar\frac{\mathrm{d}\boldsymbol{\sigma}(t)}{\mathrm{d}t} = i\left[\mathcal{H}_{\text{tot}}, \boldsymbol{\sigma}(t)\right]. \tag{3.17}$$

Here $P_{i_\mu,d_p}(t)$ is written in the Heisenberg representation, i.e. its time evolution is exclusively ascribed to the Pauli operator representing the muon spin. The Hamiltonian $\mathcal{H}_{\text{tot}}$, which describes the muon–system ensemble, does not depend on time. Also, the density operator does not evolve with time. We factorise ρ_{tot} as $\rho_{\text{tot}} = \rho_\mu\rho_{\text{sys}}$, where ρ_μ and ρ_{sys} are the density operators for the muon and the system, respectively. It should be noticed that the product of ρ_μ and ρ_{sys} is actually a tensor product, but for ease of notation and because there is no risk of confusion, we do not use a special symbol for this product. As usual, ρ_{sys} is the Boltzmann density operator,

$$\rho_{\text{sys}} = \frac{\exp\left(-\beta\mathcal{H}_{\text{sys}}\right)}{\text{Tr}\left\{\exp\left(-\beta\mathcal{H}_{\text{sys}}\right)\right\}}, \tag{3.18}$$

with $\beta = (k_{\text{B}}T)^{-1}$ and $\mathcal{H}_{\text{sys}}$ the Hamiltonian operator for the system, including the possible effects of the perturbation of the muon on the system. Then ρ_μ can be written as

$$\rho_\mu = \frac{1}{2}(\mathcal{I} + \mathbf{i}_\mu\cdot\boldsymbol{\sigma}), \tag{3.19}$$

where $\mathcal{I}$ is the unit operator.

This form of ρ_μ can be understood as follows. Let us first suppose that the muon beam is fully polarised along the Z axis. Then we expect $\langle + | \rho_\mu | + \rangle = 1$ and $\langle - | \rho_\mu | - \rangle = 0$ ($|+\rangle$ and $|-\rangle$ are the two muon Zeeman eigenstates). This is exactly the prediction of Eq. 3.19 since $\mathbf{i}_\mu \cdot \boldsymbol{\sigma} = \sigma^Z$ if $\mathbf{i}_\mu \parallel \mathbf{Z}$. If the beam is polarised along the X axis, we expect

$$(\rho_\mu)_{\mathbf{i}_\mu \parallel \mathbf{X}} = R_Y(\pi/2)\,(\rho_\mu)_{\mathbf{i}_\mu \parallel \mathbf{Z}}\, R_Y(-\pi/2) = R_Y(\pi/2) \left[\frac{1}{2} \left(\mathcal{I} + \sigma^Z \right) \right] R_Y(-\pi/2), \quad (3.20)$$

where $R_Y(\pi/2)$ is the operator associated with a rotation of angle $\pi/2$ about the Y axis (see Section C.4). Because $R_Y(\pi/2)\sigma^Z R_Y(-\pi/2) = \sigma^X$ (see Section C.5), we obtain a result consistent with Eq. 3.19 since $\mathbf{i}_\mu \cdot \boldsymbol{\sigma} = \sigma^X$ for $\mathbf{i}_\mu \parallel \mathbf{X}$. Using the same method for $\mathbf{i}_\mu \parallel \mathbf{Y}$, we derive a formula for the density operator in agreement with Eq. 3.19. Therefore this equation provides a proper general expression for the muon density operator.

From the expression of the density operator and Eq. 3.16 we get the fundamental result

$$P_{i_\mu, d_p}(t) = \frac{1}{2} \mathrm{Tr} \left\{ \rho_{\mathrm{sys}}\, (\mathbf{i}_\mu \cdot \boldsymbol{\sigma})\, [\boldsymbol{\sigma}(t) \cdot \mathbf{d}_p] \right\} = \frac{1}{2} \sum_{\alpha,\beta} \mathrm{Tr} \left\{ \rho_{\mathrm{sys}} \sigma^\beta \sigma^\alpha(t) \right\} i_\mu^\beta d_p^\alpha, \quad (3.21)$$

if $\mathrm{Tr}\left\{\rho_{\mathrm{sys}} \boldsymbol{\sigma}(t) \cdot \mathbf{d}_p\right\} = 0$. This condition is fulfilled since $\mathrm{Tr}\left\{\rho_{\mathrm{sys}} \boldsymbol{\sigma}(t) \cdot \mathbf{d}_p\right\}$ is proportional to the mean value of the muon spin for an initially unpolarised muon beam for which $\rho_\mu = \mathcal{I}/2$. The polarisation which can be induced by an applied magnetic field is too small at usual temperatures to have any effect on the state of polarisation of the muon beam. Indeed, the Zeeman energy of a muon spin submitted to a field of 1 T equals the thermal energy at a temperature of 6.5 mK.

The measurements are usually performed either with the longitudinal or transverse geometry and $P_Z(t)$ or $P_X(t)$ is measured. From the general formula we get

$$P_Z(t) = \frac{1}{2} \mathrm{Tr} \left\{ \rho_{\mathrm{sys}} \sigma^Z \sigma^Z(t) \right\} \quad \text{and} \quad P_X(t) = \frac{1}{2} \mathrm{Tr} \left\{ \rho_{\mathrm{sys}} \sigma^X \sigma^X(t) \right\}. \quad (3.22)$$

The evolution of the Pauli operator is given by the formal solution of the Heisenberg equation (Eq. 3.17):

$$\sigma^\alpha(t) = \exp\left(i \frac{\mathcal{H}_{\mathrm{tot}}\, t}{\hbar} \right) \sigma^\alpha \exp\left(-i \frac{\mathcal{H}_{\mathrm{tot}}\, t}{\hbar} \right). \quad (3.23)$$

We recall that ρ_{sys} is time independent in the Heisenberg representation we are using.

To round up our discussion of the quantum treatment, we now consider the simple model of a muon spin in an applied external magnetic field. The classical treatment leads to the result given in Eqs. 3.10 and 3.11. The Hamiltonian describing the muon–system ensemble in quantum mechanics is simply the muon Zeeman Hamiltonian, $\mathcal{H}_{Z,\mu} = -(\hbar \omega_\mu \mathbf{u} \cdot \boldsymbol{\sigma})/2$. Since we consider here that the muon is isolated, the system density operator is taken proportional to the unit operator, and we have

$$P_Z(t) = \frac{1}{2} \mathrm{Tr} \left\{ \sigma^Z \sigma^Z(t) \right\}, \quad P_X(t) = \frac{1}{2} \mathrm{Tr} \left\{ \sigma^X \sigma^X(t) \right\}, \quad (3.24)$$

with

$$\sigma^\alpha(t) = \exp\left(i\frac{\mathcal{H}_{Z,\mu}\, t}{\hbar}\right)\sigma^\alpha \exp\left(-i\frac{\mathcal{H}_{Z,\mu}\, t}{\hbar}\right) = R^\sigma_u\left(\frac{\omega_\mu\, t}{2}\right)\sigma^\alpha R^\sigma_u\left(-\frac{\omega_\mu\, t}{2}\right). \quad (3.25)$$

Here $R^\sigma_u(\varrho)$ is the operator associated with a rotation of angle ϱ around the u axis for a Pauli operator. Using the expression for this operator given in Section C.5, we derive

$$\begin{aligned}\mathrm{Tr}\left\{\sigma^\beta\sigma^\alpha(t)\right\} &= \mathrm{Tr}\left\{\sigma^\beta\sigma^\alpha\right\}\cos^2\left(\frac{\omega_\mu\, t}{2}\right) + i\,\mathrm{Tr}\left\{\sigma^\beta\sigma^\alpha\sigma^u\right\}\sin\left(\omega_\mu\, t\right) \\ &\quad + \mathrm{Tr}\left\{\sigma^\beta\sigma^u\sigma^\alpha\sigma^u\right\}\sin^2\left(\frac{\omega_\mu\, t}{2}\right). \end{aligned} \quad (3.26)$$

Evaluating the three traces of products of Pauli operators using Eqs. C.23 and C.3, we find $\mathrm{Tr}\left\{\sigma^\beta\sigma^\alpha(t)\right\} = 2M^{\alpha\beta}(t)$, where $\boldsymbol{M}(t)$ is the tensor defined by Eq. 3.11. Therefore the classical and quantum mechanical computations yield the same result for the model considered: the muon spins interact with a unique time-independent magnetic field.

3.2 Muonium polarisation function

Whatever the electronic state of the muon (diamagnetic as considered in the previous section, or bound to an electron to form a muonium and therefore in a paramagnetic state), a muon experiment always consists of measuring the scalar product, $[\langle\mathbf{S}_\mu(t)\rangle/S_\mu(0)]\cdot\mathbf{d}_p$, introduced at the beginning of this chapter. Therefore the results obtained for a diamagnetic muon are still valid provided that the trace operation specified by Eq. 3.16 is performed over the whole system including the muonium Hilbert space and the relevant density operator is used. Here, $\rho_{\rm tot}$ factorises as $\rho_{\rm tot} = \rho_{\rm Mu}\rho_{\rm sys}$, where $\rho_{\rm Mu}$ is the muonium density:

$$\rho_{\rm Mu} = \frac{1}{4}(\mathcal{I} + \mathbf{i}_\mu\cdot\boldsymbol{\sigma}_\mu). \quad (3.27)$$

This is like the expression for ρ_μ except for the prefactor which is $\frac{1}{4}$ rather than $\frac{1}{2}$, reflecting the doubling of the size of the Hilbert space for a muonium atom relative to a diamagnetic muon. Note that we have now added the subscript μ to operator $\boldsymbol{\sigma}$ to emphasise that it refers to the muon and not to its bound electron. Inserting the $\rho_{\rm Mu}$ density operator in Eq. 3.16 we derive

$$P_{i_\mu,d_p}(t) = \frac{1}{4}\mathrm{Tr}\left\{\rho_{\rm sys}\left(\mathbf{i}_\mu\cdot\boldsymbol{\sigma}_\mu\right)\left[\boldsymbol{\sigma}_\mu(t)\cdot\mathbf{d}_p\right]\right\} = \frac{1}{4}\sum_{\alpha,\beta}\mathrm{Tr}\left\{\rho_{\rm sys}\sigma^\beta_\mu\sigma^\alpha_\mu(t)\right\}i^\beta_\mu d^\alpha_p, \quad (3.28)$$

if $\mathrm{Tr}\left\{\rho_{\rm sys}\boldsymbol{\sigma}_\mu(t)\cdot\mathbf{d}_p\right\} = 0$. This hypothesis is valid as already argued when discussing the muon polarisation function. We are usually interested in either $P_Z(t)$ or $P_X(t)$:

$$P_Z(t) = \frac{1}{4}\mathrm{Tr}\left\{\rho_{\rm sys}\sigma^Z_\mu\sigma^Z_\mu(t)\right\}, \quad P_X(t) = \frac{1}{4}\mathrm{Tr}\left\{\rho_{\rm sys}\sigma^X_\mu\sigma^X_\mu(t)\right\}. \quad (3.29)$$

Sometimes the analysis of muonium data requires the consideration of a multi-spin system formed by a muonium atom interacting with a small number of nuclear spins as explained in Section 8.4.1. In this case the temperature can be taken as infinite since the nuclei are not polarised. This results in

$$P_Z(t) = \frac{1}{4(2I+1)^{N_{\rm Mu}}} {\rm Tr}\left\{\sigma_\mu^Z \sigma_\mu^Z(t)\right\}, \quad P_X(t) = \frac{1}{4(2I+1)^{N_{\rm Mu}}} {\rm Tr}\left\{\sigma_\mu^X \sigma_\mu^X(t)\right\}. \quad (3.30)$$

For simplicity, it is assumed that $N_{\rm Mu}$ identical nuclear spins I are interacting with the muonium atom.

Up to this point the evolution of the polarisation functions has been ascribed to the muon spin, see Eq. 3.16, and the Hamiltonian $\mathcal{H}_{\rm tot}$ has been assumed to be time independent. However, it may be convenient to shift the origin of the time dependence of the polarisation functions to the density operator, with the result,

$$P_{i_\mu,d_p}(t) = {\rm Tr}\left\{\rho_{\rm tot}(t)\,\boldsymbol{\sigma}_\mu \cdot \mathbf{d}_p\right\}, \quad (3.31)$$

where

$$i\hbar\frac{{\rm d}\rho_{\rm tot}(t)}{{\rm d}t} = \left[\mathcal{H}_{\rm tot}(t), \rho_{\rm tot}(t)\right]. \quad (3.32)$$

This equation of motion is often referred to as the Liouville equation since it coincides with the equation of motion for the probability-density distribution in phase space in classical mechanics. Note that the Hamiltonian is now time dependent. The two types of formula we have derived for computing the polarisation functions, that is Eqs. 3.16 and 3.31, mirror the two well-known available methods for solving a quantum mechanical problem involving the time variable: the Heisenberg and Schrödinger representations. A detailed derivation can be found in many textbooks, e.g. Abragam (1961), Atherton (1993), Smilga and Belousov (1994), and Slichter (1996).

The Liouville formalism is of particular interest for modelling the effect on the muonium atoms of their environment as shown in Section 8.4.3. In fact, the muon, the muonium-electron, and the cross muon–electron polarisations are intertwined. These polarisation functions are defined as follows:

$$P^\mu_\alpha(t) = {\rm Tr}\{\rho_{\rm tot}(t)\sigma^\alpha_\mu\}, \quad P^{\rm e}_\alpha(t) = {\rm Tr}\{\rho_{\rm tot}(t)\sigma^\alpha_{\rm e}\}, \quad P^{\mu{\rm e}}_{\alpha\beta}(t) = {\rm Tr}\{\rho_{\rm tot}(t)\sigma^\alpha_\mu\sigma^\beta_{\rm e}\}. \quad (3.33)$$

Whereas at $t = 0$ the muonium electron is not polarised (unless the muon binds to an electron of an ordered magnet), i.e. $P^{\rm e}_\alpha(0) = P^{\mu{\rm e}}_{\alpha\beta}(0) = 0$, we have $P^\mu_X(0) = 1$ ($P^\mu_Z(0) = 1$) for a transverse- (longitudinal-)field measurement. These equalities set the initial conditions. However, as time elapses, $P^{\rm e}_\alpha(t)$ and $P^{\mu{\rm e}}_{\alpha\beta}(t)$ acquire finite values. This must be taken into account for getting an expression for $P^\mu_X(t) = P_X(t)$ or $P^\mu_Z(t) = P_Z(t)$ as shown explicitly in Section 8.4.3. While $\rho_{\rm tot}(t)$ depends on the muon and electron degrees of freedom, $\rho_{\rm Mu}$ is only a function of the muon degree of freedom. A natural extension of Eq. 3.27 for $\rho_{\rm tot}(t)$ is therefore

$$\rho_{\rm tot}(t) = \frac{1}{4}\left[\mathcal{I} + \mathbf{P}^\mu(t)\cdot\boldsymbol{\sigma}_\mu + \mathbf{P}^{\rm e}(t)\cdot\boldsymbol{\sigma}_{\rm e} + \sum_{\alpha,\beta} P^{\mu{\rm e}}_{\alpha\beta}(t)\sigma^\alpha_\mu\sigma^\beta_{\rm e}\right]. \quad (3.34)$$

For simplicity of notation, and because there is no risk of confusion, we do not explicitly specify the direct product of muon and electron Pauli operators. It is important to realise that the time dependence sits on the polarisation functions, and not on the Pauli operators.

References

Abragam A. (1961) *The Principles of Nuclear Magnetism* (Clarendon, Oxford).

Atherton N.M. (1993) *Principles of Electron Spin Resonance* (Ellis Horwood and Prentice Hall, New York).

Belousov Yu.M., Gorelkin V.N., Mikaélajan A.L., Miloserdin V.Yu., and Smilga V.P. (1979) *Soviet Physics Uspekhi* **22**, 679–702.

Dirac P.A.M. (1958) *The Principles of Quantum Mechanics*, fourth edition (Oxford University Press, London).

Messiah A. (1965) *Quantum Mechanics*, third reprint (North-Holland, Amsterdam).

Schiff L.I. (1968) *Quantum Mechanics*, third edition (McGraw-Hill, Auckland).

Slichter C.P. (1996) *Principles of Magnetic Resonance*, third enlarged and updated edition (Springer, Berlin).

Smilga V.P. and Belousov Yu.M. (1994) *The Muon Method in Science*, Proceedings of the Lebedev Physics Institute, vol. 219 (Nova Science, Commack, NY).

4 Miscellaneous techniques

In the bulk of this monograph, we focus on the transverse and longitudinal time-differential μSR techniques and their applications to the study of condensed matter. While the vast majority of the μSR studies are performed with these techniques, the reader should be aware of the existence of alternative methods which might be extremely valuable in certain instances. An introduction to them is provided in this chapter.

μSR shares conceptual foundations with the methods of nuclear magnetic resonance (NMR) and electron paramagnetic resonance. Despite this common ground, the transverse- and longitudinal-field μSR techniques have yet more in common with the $\gamma\gamma$ perturbed angular correlation technique — for an introduction, see e.g. Karlsson (1995) — than NMR. The application of radio-frequency (RF) electromagnetic waves to the sample under study is indeed quite rare in μSR, while it is ubiquitous in NMR. However, the use of these waves in μSR can be of great benefit. The first section of this chapter discusses the applications of such RF pulses to μSR.

When trying to understand a physical system at a microscopic level, one of the first question often is: what are its energy levels? To answer it, methods using local probe techniques have been developed. They are based on the investigation of the probe's response to an applied magnetic field which drives the energy levels of the system to resonance at certain values. In the second section of this chapter we present these so-called avoided-level-crossing (ALC) techniques as applied to μSR.

As said above, in this monograph we focus on the time-differential μSR techniques. However, it is not always necessary to measure the time dependence of the polarisation functions since the total number of positron counts in a detector provides in itself an information which is sufficient in certain circumstances. The so-called time-integrated techniques have therefore been developed since the 1970s for specific applications. They are particulary advantageous at pseudo-continuous muon sources. Indeed, they do not suffer from the counting rate limitation inherent to time-differential experiments at this type of sources, which request no more than one muon to be present in the sample at a given time; see Section 2.2.2.1. In addition to a presentation of the time-differential RF and ALC techniques, the first two sections of this chapter provide illustrations of the time-integrated method applications. Basically, a time-integrated measurement can be viewed as a measurement in energy domain rather than in time domain, as for the usual time-differential technique.

The time-integrated technique also provides a means for accurately measuring the intensity of the field at the muon site through a stroboscopic method. This method which can be used for transverse-field experiments with pseudo-continuous muon sources is introduced in Section 4.3.

Finally, three experimental techniques aside from the main stream of μSR research are mentioned: lattice steering, β-detected NMR, and muon spin rotation with negative muons.

Nagamine (2003) discusses ALC spectroscopy, resonance measurements, and experiments with negative muons.

4.1 Radio-frequency μSR spectroscopy

The first RF-μSR experiment dates back to 1958 (Coffin *et al.* 1958). A proof-of-principle experiment displaying spin echoes in transverse geometry was first reported by Kreitzman *et al.* (1988). However, the development of the RF techniques has been rather slow, probably because of the lack of a high-intensity muon pulsed source. Indeed, as shown, e.g. by Shiroka *et al.* (1999) and Cottrell *et al.* (2000), with such a source innovative experiments can be performed. The sample heating induced by the RF power can be minimised by applying brief RF excitation pulses in coincidence with the muon pulses.

We shall first recall the notion of the rotating reference frame before discussing RF-μSR techniques.

4.1.1 Rotating reference frame

We shall have to consider the temporal evolution of the polarisation function, $P_\alpha(t)$, under the combined action of static and RF fields. A convenient method for treating this problem is to examine the behaviour of $P_\alpha(t)$ in the rotating reference frame.

As usual (see Section 1.1.1), we first assume a magnetic field applied along the Z axis of the laboratory reference frame, $\{X, Y, Z\}$. However, instead of labelling that field $\mathbf{B}_{\rm ext}$, in this section we prefer the notation $\mathbf{B}_{\rm stat}$ to stress the static character of the field. The evolution of $P_\alpha(t)$ follows from the Larmor equation; see Section 3.1.1.1. In the longitudinal geometry, the initial muon beam polarisation, $\mathbf{S}_\mu$, is by definition along the Z axis, and therefore parallel to $\mathbf{B}_{\rm stat}$. Then $P_Z(t)$ is constant, i.e. $P_Z(t) = 1$. The physics is more interesting in the transverse geometry. The spin $\mathbf{S}_\mu$ is along the X axis and, according to Eq. 3.8, $P_X(t) = \cos(\gamma_\mu B_{\rm stat} t)$, where γ_μ is the muon gyromagnetic ratio.

We now superimpose an alternating field $\mathbf{B}_{\rm alt}(t)$, applied in a plane perpendicular to the Z axis, say parallel to X. This is a typical NMR arrangement. The effect of this field is most readily analysed by breaking it into two rotating components, one rotating clockwise and the other counter-clockwise:

$$\begin{aligned}\mathbf{B}_{\rm alt}(t) = {} & B_1 \left[\cos(\omega_{\rm RF} t + \varphi)\,\hat{\mathbf{X}} - \sin(\omega_{\rm RF} t + \varphi)\,\hat{\mathbf{Y}}\right] \\ & + B_1 \left[\cos(\omega_{\rm RF} t + \varphi)\,\hat{\mathbf{X}} + \sin(\omega_{\rm RF} t + \varphi)\,\hat{\mathbf{Y}}\right], \end{aligned} \tag{4.1}$$

where $\omega_{\rm RF}$ is the angular frequency of the RF field, φ its phase, and B_1 the amplitude of each of its rotating component. The set $\{\hat{\mathbf{X}}, \hat{\mathbf{Y}}, \hat{\mathbf{Z}}\}$ forms the basis of the laboratory reference frame. Since the two components differ only by the replacement of $\omega_{\rm RF}$ and

φ by $-\omega_{\rm RF}$ and $-\varphi$, we shall only analyse the effect of one component. The total local field acting on the muons is written as

$$\mathbf{B}_{\rm loc} = B_{\rm stat}\hat{\mathbf{Z}} + B_1\left[\cos\left(\omega_{\rm RF}t+\varphi\right)\hat{\mathbf{X}} - \sin\left(\omega_{\rm RF}t+\varphi\right)\hat{\mathbf{Y}}\right]. \qquad (4.2)$$

We have selected the clockwise component of the RF field. We shall need to solve the Larmor equation for $P_\alpha(t)$. It is not practical to do it in the laboratory frame since $\mathbf{B}_{\rm loc}$ is time dependent. So, we resort to the usual trick of writing the equation in the so-called rotating frame in which $\mathbf{B}_{\rm loc}$ is static — Slichter (1996), see also Appendix A where we use a complex number notation. The basis of this frame, $\{\hat{\mathbf{X}}', \hat{\mathbf{Y}}', \hat{\mathbf{Z}}'\}$, is related to $\{\hat{\mathbf{X}}, \hat{\mathbf{Y}}, \hat{\mathbf{Z}}\}$ by

$$\begin{aligned}\hat{\mathbf{X}}' &= \cos\left(\omega_{\rm RF}t\right)\hat{\mathbf{X}} - \sin\left(\omega_{\rm RF}t\right)\hat{\mathbf{Y}},\\ \hat{\mathbf{Y}}' &= \sin\left(\omega_{\rm RF}t\right)\hat{\mathbf{X}} + \cos\left(\omega_{\rm RF}t\right)\hat{\mathbf{Y}},\\ \hat{\mathbf{Z}}' &= \hat{\mathbf{Z}}.\end{aligned} \qquad (4.3)$$

That is, the new frame rotates clockwise around $\mathbf{B}_{\rm stat}$ with the RF angular frequency and hence the RF field appears static. Recalling the Larmor equation (Eq. 3.4) we need to calculate $\frac{{\rm d}\mathbf{S}_\mu(t)}{{\rm d}t}$ using the components of $\mathbf{S}_\mu(t)$ in the rotating reference frame. Taking the derivative of $\mathbf{S}_\mu(t) = S_\mu^{X'}\hat{\mathbf{X}}' + S_\mu^{Y'}\hat{\mathbf{Y}}' + S_\mu^{Z'}\hat{\mathbf{Z}}'$, we get

$$\frac{{\rm d}\mathbf{S}_\mu(t)}{{\rm d}t} = \frac{{\rm d}S_\mu^{X'}}{{\rm d}t}\hat{\mathbf{X}}' + \frac{{\rm d}S_\mu^{Y'}}{{\rm d}t}\hat{\mathbf{Y}}' + \frac{{\rm d}S_\mu^{Z'}}{{\rm d}t}\hat{\mathbf{Z}}' + S_\mu^{X'}\frac{{\rm d}\hat{\mathbf{X}}'}{{\rm d}t} + S_\mu^{Y'}\frac{{\rm d}\hat{\mathbf{Y}}'}{{\rm d}t} + S_\mu^{Z'}\frac{{\rm d}\hat{\mathbf{Z}}'}{{\rm d}t}. \qquad (4.4)$$

Expressing the time derivative of the basis vectors in the rotating frame, it is shown that the last three terms on the right-hand-side sum of Eq. 4.4 reduce to $-\omega_{\rm RF}\hat{\mathbf{Z}}' \times \mathbf{S}_\mu$. Then the Larmor equation reads

$$\frac{{\rm d}S_\mu^{X'}}{{\rm d}t}\hat{\mathbf{X}}' + \frac{{\rm d}S_\mu^{Y'}}{{\rm d}t}\hat{\mathbf{Y}}' + \frac{{\rm d}S_\mu^{Z'}}{{\rm d}t}\hat{\mathbf{Z}}' = \gamma_\mu \mathbf{S}_\mu \times \left(\mathbf{B}_{\rm loc} - \frac{\omega_{\rm RF}}{\gamma_\mu}\hat{\mathbf{Z}}'\right). \qquad (4.5)$$

Hence, when the components of $\mathbf{S}_\mu$ are expressed in the rotating frame, we obtain an equation similar to Eq. 3.4 after substitution of $\mathbf{B}_{\rm loc}$ with the effective field

$$\mathbf{B}_{\rm eff} = \mathbf{B}_{\rm loc} - \frac{\omega_{\rm RF}}{\gamma_\mu}\hat{\mathbf{Z}}' = B_1\left[\cos\varphi\,\hat{\mathbf{X}}' - \sin\varphi\,\hat{\mathbf{Y}}'\right] + \left(B_{\rm stat} - \frac{\omega_{\rm RF}}{\gamma_\mu}\right)\hat{\mathbf{Z}}'. \qquad (4.6)$$

Since $\mathbf{B}_{\rm eff}$ is static in the rotating reference frame, the results given in, e.g., Eq. 3.5 or 3.10 can be used to derive $S_\mu^{X'}(t)$, $S_\mu^{Y'}(t)$, and $S_\mu^{Z'}(t)$.

Finally, had we selected the counter-clockwise instead of the clockwise component, we would have only plus signs in Eq. 4.6. This is a key result. While for the clockwise component of the RF field the component of $\mathbf{B}_{\rm eff}$ along $\hat{\mathbf{Z}}'$ vanishes for a particular $\omega_{\rm RF}$ value, it is finite for the counter-clockwise component.

4.1.2 Time-differential radio-frequency μSR

The polarisation functions $P_\alpha(t)$ are directly measured in the time-differential RF technique. With the help of the rotating reference frame and the results of Sections 3.1.1.1 or 3.1.1.2, we immediately derive the clockwise component of, for example, $P_{Z'}(t)$, the polarisation function in the rotating reference frame. Noting that the Z' and Z axes match,

$$P_Z(t) = \cos^2\vartheta + \sin^2\vartheta\cos(\omega_{\rm eff}t)\,, \tag{4.7}$$

with the definition

$$\omega_{\rm eff} = \gamma_\mu\left[\left(B_{\rm stat} - \frac{\omega_{\rm RF}}{\gamma_\mu}\right)^2 + B_1^2\right]^{1/2} = \left[(\omega_{\rm stat} - \omega_{\rm RF})^2 + \gamma_\mu^2 B_1^2\right]^{1/2}. \tag{4.8}$$

The angle ϑ is defined through the relation

$$\sin^2\vartheta = B_1^2/B_{\rm eff}^2. \tag{4.9}$$

Therefore, scanning the RF frequency, a resonance occurs in the clockwise component; i.e. an exchange of energy between the RF field and the muon spin system takes place when $\omega_{\rm RF} = \omega_{\rm stat}$. At resonance $\sin^2\vartheta = 1$ and therefore $P_Z(t) = \cos(\gamma_\mu B_1 t)$. Note that the counter-clockwise component does not display such an exchange of energy at the resonance field value. We leave the interested reader to derive the expressions for $P_X(t)$ and $P_Y(t)$.

This predicted oscillation for $P_Z(t)$ at resonance has experimentally been checked (Nagamine and Yamazaki 1981). Note that B_1 must be sufficiently intense ($\sim$1 mT) for observing the oscillation of $P_Z(t)$ on the muon time scale. This observation is most easily achieved using a pulsed source since the RF power source can be triggered on the arrival of the muon pulse; i.e. the RF source does not have to continuously work, thus avoiding sample overheating.

Interestingly, Brown *et al.* (1979) used a RF field in transverse-field geometry to observe electron paramagnetic resonance transitions of muonium in quartz which had not previously been observed with the conventional method. This experiment is a double electron–muon resonance experiment since it involves muonium, i.e. a bound state of the muon with an electron. The $P_X(t)$ expression for muonium with an applied RF field has been derived by Estle and Vanderwater (1983).

Let us now consider an application of RF pulses in the longitudinal-field geometry for which $\mathbf{S}_\mu$ is parallel to Z. According to Eq. 4.6, if the resonance condition is fulfilled, i.e. $\omega_{\rm RF} = \omega_{\rm stat}$, the effective field in the rotating frame is perpendicular to the Z axis and its modulus is equal to B_1. If we turn on the RF field for a short time, that is, if we apply a wave train of duration $t_{\rm w}$, $\mathbf{S}_\mu$ will precess through the angle $\theta_t = \gamma_\mu B_1 t_{\rm w}$. If $t_{\rm w}$ is chosen such that $\theta_t = \pi/2$, the RF pulse simply turns $\mathbf{S}_\mu$ by 90°. We are then dealing with a so-called 90° pulse in NMR language (Slichter 1996) and $\mathbf{S}_\mu$ is perpendicular to $\mathbf{B}_{\rm ext}$.

Figure 4.1 presents an example. Note that this experiment has been performed at a pulsed muon source, triggering the RF source on the arrival of the muon beam burst. An offset is obviously possible. We recall that the frequency bandwidth of a pulsed

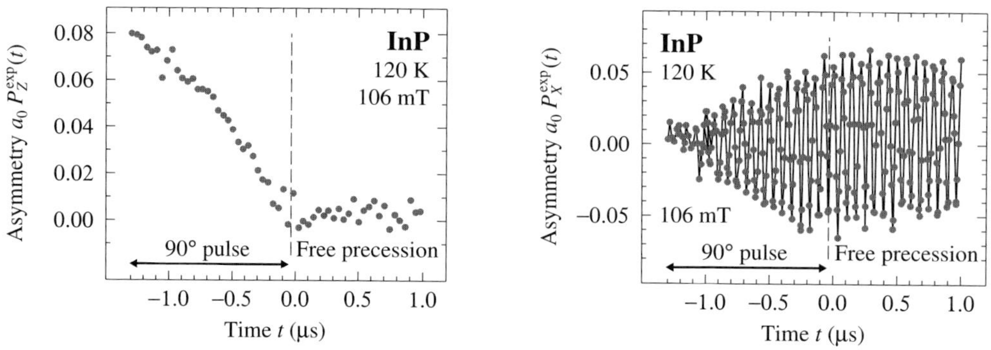

Fig. 4.1 The asymmetry measured for muons in InP at 120 K during and following a 90° RF pulse. The RF frequency is tuned at 14 MHz which corresponds to the Larmor frequency in a field of 106 mT. The left (right) panel displays the longitudinal (transverse) asymmetry. The dashed line indicates the end of the application of the 90° pulse. The origin of time is chosen a short while afterwards to allow for the ring-down of the tuned RF circuit. Adapted from Cottrell *et al.* (2000).

muon beam is not very large (6 MHz at ISIS at the time of writing) because of the width of the muon pulses. Turning $\mathbf{S}_\mu$ by 90°, and therefore getting it perpendicular to the Z axis, allows one to extend the bandwidth. At the time when the RF field is switched off, provided the muon spins have not been depolarised since they have been implanted, they all act coherently and a conventional transverse measurement starts. Of course, we recall that transverse-field measurements with a large bandwidth can be performed at a continuous source with no special trick.

The unique advantage of a pulsed muon source combined with a RF pulse sequence can be fully appreciated from the following simple example for the diamagnet $Ca(OH)_2$ (Cottrell *et al.* 2000). Here the muon signal damping is expected to arise from the proton nuclear magnetic moments. The transverse polarisation function is measured following a 90° pulse which turns $\mathbf{S}_\mu$ by 90°, as explained earlier. In addition, a long (about 20 μs) high power RF pulse is applied at the resonance frequency of the protons to "stir" the proton spins. The distribution of fields at the muon site is then motionally narrowed by the RF field.

In Fig. 4.2 are displayed the spectra recorded with the long-lasting RF field on and off. There is a drastic reduction of the damping when the decoupling field is applied. For a given applied static transverse field, each type of nuclei is characterised by a resonance frequency. Therefore the described RF sequence offers the possibility of distinguishing the different sources of muon depolarisation in a system of nuclei with different families of dipole moments.

We anticipate that other types of RF sequences will be used in the future, extending the possibility of the RF μSR techniques.

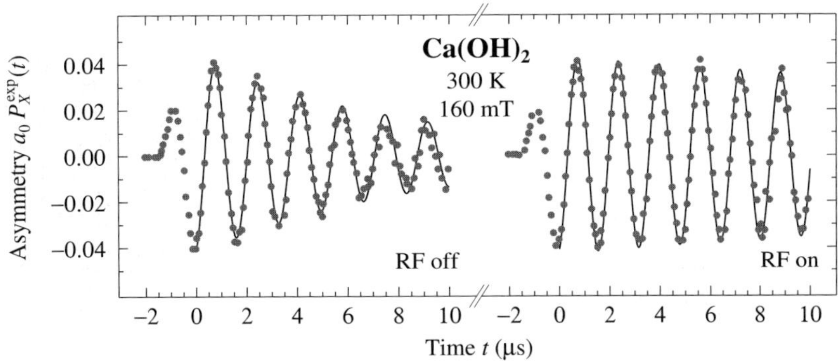

Fig. 4.2 Transverse-field spectra recorded at 300 K for $Ca(OH)_2$ in a transverse field of 160 mT with the decoupling RF field off (left) and on (right). This field of 4 mT induces a dynamics of the proton magnetic moments which drastically reduces the damping through motional narrowing. The data are presented in a rotating reference frame in order to reduce the number of displayed oscillations; see Section A.2 for details. Adapted from Cottrell *et al.* (2000).

4.1.3 Time-integrated radio-frequency μSR

Instead of measuring a polarisation function versus the time spent by the muons in the sample, the total number of positron counts can be recorded as a function of the RF frequency. This is a so-called time-integral technique. Its advantage is the large counting rate that can be reached, much higher than that for the usual time-differential technique which requires only one muon at a time in the sample, at least for a pseudo-continuous muon beam.

According to Section 2.3, in the simplest experimental case, the positron count is given by the expression

$$N_\pm(t) = N_0 \exp\left(-t/\tau_\mu\right)\left[1 \pm a_0 P_\alpha(t)\right]. \tag{4.10}$$

Let us compute $\bar{N}_\pm = \tau_\mu^{-1} \int_0^\infty N_\pm(t) \mathrm{d}t$. Using Eq. 4.7, we derive for the longitudinal geometry

$$\bar{N}_\pm = N_0 \left\{ 1 \pm a_0 \mp a_0 \sin^2\vartheta \frac{\left[(\omega_{\rm stat} - \omega_{\rm RF})^2 + \gamma_\mu^2 B_1^2\right]\tau_\mu^2}{1 + \left[(\omega_{\rm stat} - \omega_{\rm RF})^2 + \gamma_\mu^2 B_1^2\right]\tau_\mu^2} \right\}. \tag{4.11}$$

The quantities $\bar{N}_\pm$ are the sum of a constant and a dispersive signal. Following the discussion given in Section 2.3, a quantity of interest is the weighted difference between forward and backward counts. Thus we define

$$\mathcal{R}_{\rm int} = \frac{1}{a_0} \frac{\bar{N}_+ - \bar{N}_-}{\bar{N}_+ + \bar{N}_-}, \tag{4.12}$$

and find

$$\mathcal{R}_{\text{int}} = 1 - \sin^2\vartheta \frac{\left[(\omega_{\text{stat}} - \omega_{\text{RF}})^2 + \gamma_\mu^2 B_1^2\right]\tau_\mu^2}{1 + \left[(\omega_{\text{stat}} - \omega_{\text{RF}})^2 + \gamma_\mu^2 B_1^2\right]\tau_\mu^2}. \tag{4.13}$$

As expected, $\mathcal{R}_{\text{int}}$ displays a resonance at $\omega_{\text{RF}} = \omega_{\text{stat}}$, that is, it exhibits a minimum value equal to

$$\mathcal{R}_{\text{int}} = \frac{1}{1 + \gamma_\mu^2 B_1^2 \tau_\mu^2}. \tag{4.14}$$

Far outside the resonance, $\mathcal{R}_{\text{int}} \simeq 1$.

The relaxation of the muon depolarisation arising from the spin–lattice and spin–spin relaxation channels is not yet accounted for. Phenomenologically we write

$$P_Z(t) = \cos^2\vartheta \exp(-\lambda_Z t) + \sin^2\vartheta \exp(-\lambda_X t)\cos(\omega_{\text{eff}} t), \tag{4.15}$$

where λ_Z and λ_X are the spin–lattice and spin–spin relaxation rates, respectively. Instead of Eq. 4.13, we then get

$$\mathcal{R}_{\text{int}} = \frac{1}{\tau_\mu}\left\{\cos^2\vartheta\,\bar{\tau}_Z + \sin^2\vartheta \frac{\bar{\tau}_X}{1 + \left[(\omega_{\text{stat}} - \omega_{\text{RF}})^2 + \gamma_\mu^2 B_1^2\right]\bar{\tau}_X^2}\right\}, \tag{4.16}$$

with $\bar{\tau}_Z^{-1} = \lambda_Z + \tau_\mu^{-1}$ and $\bar{\tau}_X^{-1} = \lambda_X + \tau_\mu^{-1}$. The ratio $\mathcal{R}_{\text{int}}$ is again the sum of a constant and a dispersive term, the width of which is now influenced by the spin–spin relaxation rate.

An expression for $\mathcal{R}_{\text{int}}$ can also be derived for the transverse geometry. We leave it as a homework problem.

4.2 Avoided-level-crossing spectroscopy

The observation and treatment of an avoided-level-crossing phenomenon were first performed in optical resonance fluorescence (Eck *et al.* 1963). The honour of introducing the ALC spectroscopy to μSR belongs to Abragam (1984). It is a convenient method for characterising energy levels of a system interacting with the muon or muonium energy levels. Nowadays it is mostly an analytical tool for studying chemical species in interaction with their environment. However, it is has been used to study the Ho^{3+} spin dynamics in $LiY_{0.998}Ho_{0.002}F_4$ *via* the temperature and field dependences of the transverse-field relaxation rate (Graf *et al.* 2007).

Here we explain the basic principles of ALC spectroscopy using a simplified model. It shows clearly the importance of the entanglement of the quantum states, i.e. their linear combinational nature as explicitly expressed in Eq. 4.19. Its importance is also pointed out in Chapter 7. Further discussions of ALC spectroscopy can be found in Chapters 7 and 8 and two examples are mentioned in Chapter 12.

We first describe the energy level scheme of our simple model before computing its polarisation functions and resonance structures.

4.2.1 Hybridisation of energy levels

To track the origin of the ALC phenomena and determine how it influences the polarisation functions, we consider the energy level scheme of a simplified model which is not realised in nature but has the advantage of mathematical simplicity.

We put the muon at the origin of the coordinates and a spin $\frac{1}{2}$ nucleus on the Z axis in the positive direction at $r = 1.5\,\text{Å}$ from the muon. The Hamiltonian is taken as the sum of four terms:

$$\mathcal{H}_{\text{sys}} = \mathcal{H}_{\text{Z},\mu} + \mathcal{H}_{\text{Z,n}} + \mathcal{H}_{\text{dip}} + \mathcal{H}_{\text{split}}. \tag{4.17}$$

The first two terms on the right account for the Zeeman energy of the muon and nuclear magnetic moments and $\mathcal{H}_{\text{dip}}$ describes the dipole interaction between the muon and nuclear spins. The last term introduces a splitting of the two nuclear levels which could be induced by an electric field gradient. However, since we assume the nuclear spin to be $\frac{1}{2}$, an electric field gradient cannot be coupled to that spin. The splitting is therefore purely phenomenological. Explicitly, we get

$$\begin{aligned}\mathcal{H}_{\text{sys}} &= BS_\mu^Z + CI^Z + A\left[-2S_\mu^Z I^Z + \frac{1}{2}\left(S_\mu^+ I^- + S_\mu^- I^+\right)\right] + DI^Z \\ &= A\left[-2S_\mu^Z I^Z + \frac{1}{2}\left(S_\mu^+ I^- + S_\mu^- I^+\right)\right] + BS_\mu^Z + FI^Z.\end{aligned} \tag{4.18}$$

The spins $\mathbf{S}_\mu$ and $\mathbf{I}$ refer to the spin operators for the muon and nucleus, respectively. In addition $B = -\gamma_\mu \hbar B_{\text{ext}}$, $C = -\gamma_{\text{n}} \hbar B_{\text{ext}}$, $A = (\mu_o \gamma_\mu \gamma_{\text{n}} \hbar^2)/(4\pi r^3)$, D is a constant, and $F = C + D$. The orders of magnitude are as follows: $|B/C| \simeq 10$, $|D| \gg |A|$, and for usual field intensities, $|B/A| \gg 1$ and $|C/A| \gg 1$. We are dealing with a two-energy-level system in interaction with the muon spin through the weak dipolar interaction. The Hamiltonian $\mathcal{H}_{\text{sys}}$ can be analytically diagonalised. The energy levels are $E_1 = (-A + B + F)/2$, $E_2 = \left[A + \sqrt{A^2 + (B - F)^2}\right]/2$, $E_3 = \left[A - \sqrt{A^2 + (B - F)^2}\right]/2$, and $E_4 = (-A - B - F)/2$, respectively. In Fig. 4.3 we present the field dependence of these energy levels. We assume $\gamma_{\text{n}} = 100\,\text{Mrad}\,\text{s}^{-1}\,\text{T}^{-1}$ and take D as the value of B when $B_{\text{ext}} = 10\,\text{mT}$. We observe a crossing of all the energy levels, except levels 2 and 3, which repel each other. Their closest approach occurs for $B = F$, i.e. $B = C + D$ or $B_{\text{ext}} = B_{\text{cri}} = -D/\left[\hbar(\gamma_\mu - \gamma_{\text{n}})\right]$. For our numerical values, the critical field value is $B_{\text{cri}} = 11.33\,\text{mT}$. The splitting between the two levels at B_{cri} is equal to A, which is the magnitude of the dipolar coupling. The fact that the two levels never cross justifies the name "avoided-level-crossing spectroscopy".

It is interesting to consider the structure of the wavefunctions. Whereas the functions associated with levels 1 and 4 are pure, that is $|\psi_1\rangle = |+,+\rangle$ and $|\psi_4\rangle = |-,-\rangle$, the functions for levels 2 and 3 are quantum superpositions:

$$\begin{aligned}|\psi_2\rangle &= -\frac{A}{K_-}|+,-\rangle + \frac{B - F - \sqrt{A^2 + (B - F)^2}}{K_-}|-,+\rangle, \\ |\psi_3\rangle &= -\frac{A}{K_+}|+,-\rangle + \frac{B - F + \sqrt{A^2 + (B - F)^2}}{K_+}|-,+\rangle,\end{aligned} \tag{4.19}$$

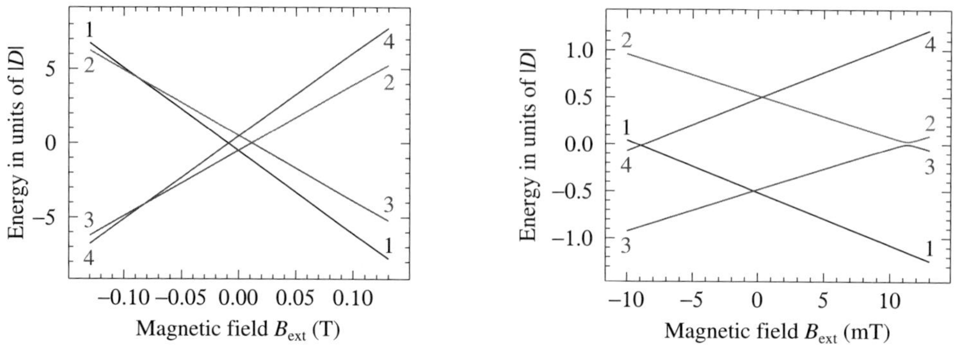

Fig. 4.3 Field dependence of the four energy levels of our simplified model. The right plot is the central part of the other one. While the levels labeled 2 and 3 never cross, any other pairs of levels do. The avoided-level-crossing phenomenon occurs at the critical field value $B_{\rm cri} = 11.33\,{\rm mT}$.

with $K_\pm = \sqrt{A^2 + \left(B - F \pm \sqrt{A^2 + (B-F)^2}\right)^2}$. The structure of Eq. 4.19 shows that the quantum superposition in the $|\psi_2\rangle$ and $|\psi_3\rangle$ states is intimately related to the finite value of A. Remarkably, at the critical field the absolute value of the amplitudes of the $|+,-\rangle$ and $|-,+\rangle$ states contributing to $|\psi_2\rangle$ and $|\psi_3\rangle$ are equal. The superposition of the $|+,-\rangle$ and $|-,+\rangle$ states arises because of the presence in $\mathcal{H}_{\rm dip}$ of the non-secular terms $S_\mu^+ I^-$ and $S_\mu^- I^+$. These terms induce transitions between states of pure Zeeman character with the condition (selection rule) that $\Delta M = 0$, where $M = m_\mu + m_{\rm n}$ and m_μ and $m_{\rm n}$ are the magnetic quantum numbers for $\mathbf{S}_\mu$ and $\mathbf{I}$, respectively.

4.2.2 Time-differential avoided-level-crossing μSR

We now compute the polarisation functions for our model. The thermal energy is assumed to be sufficiently large relative to the splitting of the involved energy levels that it is a good approximation to take the density operator proportional to the unit operator.

From Eq. 3.22 and the expressions of the four wavefunctions, we derive

$$P_Z(t) = \frac{1}{2}\left\{1 + \frac{(B-F)^2}{A^2 + (B-F)^2} + \frac{A^2}{A^2 + (B-F)^2}\cos\left[\frac{\sqrt{A^2+(B-F)^2}}{\hbar}t\right]\right\}. \tag{4.20}$$

The weight of the cosine term is equal to $2\left|\langle\psi_2\left|S_\mu^Z\right|\psi_3\rangle\right|^2$. At the critical field,

$$P_Z(t) = \frac{1}{2}\left[1 + \cos\left(\frac{At}{\hbar}\right)\right]. \tag{4.21}$$

In Fig. 4.4 we display $P_Z(t)$ for different intensities of the applied field. Clearly the field has a large effect on the shape of $P_Z(t)$ near $B_{\rm cri}$.

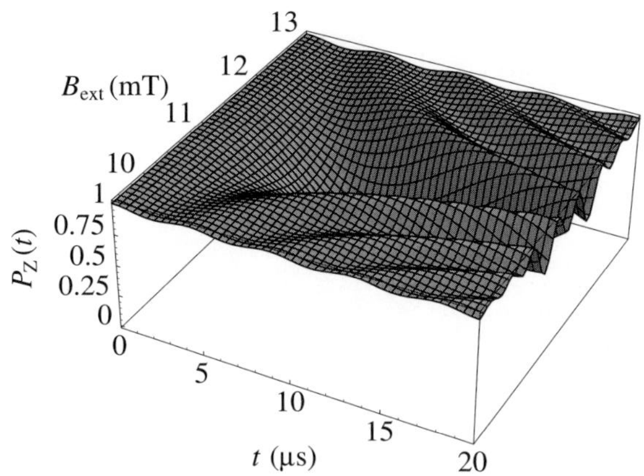

Fig. 4.4 The longitudinal polarisation function $P_Z(t)$ for different fields around the critical value $B_{\rm cri} = 11.33\,\text{mT}$.

The expression for $P_X(t)$ is more complicated:

$$2P_X(t) = a_-^2 \cos(b_- t) + a_+^2 \cos(b_+ t) + \left(\frac{A}{K_-}\right)^2 \cos(c_+ t) + \left(\frac{A}{K_+}\right)^2 \cos(c_- t)\,, \quad (4.22)$$

where we have introduced the notations

$$a_\pm = \frac{B - F \pm \sqrt{A^2 + (B-F)^2}}{K_\pm},$$
$$b_\pm = \frac{-2A + B + F \pm \sqrt{A^2 + (B-F)^2}}{2\hbar},$$
$$c_\pm = \frac{2A + B + F \pm \sqrt{A^2 + (B-F)^2}}{2\hbar}. \quad (4.23)$$

In Fig. 4.5 we illustrate the effect of the resonance on $P_X(t)$. For ease of representation we display $P_X(t)$ in a rotating reference frame; see Section A.2. For that purpose we have calculated $P_Y(t)$ which is given by

$$2P_Y(t) = a_-^2 \sin(b_- t) + a_+^2 \sin(b_+ t) + \left(\frac{A}{K_-}\right)^2 \sin(c_+ t) + \left(\frac{A}{K_+}\right)^2 \sin(c_- t)\,. \quad (4.24)$$

Again, as for $P_Z(t)$, the resonance effect is large near $B_{\rm cri}$.

4.2.3 Time-integrated avoided-level-crossing μSR

As we have done for the RF-μSR spectroscopy, we can compute the integrated signal and form the ratio $\mathcal{R}_{\rm int}$ defined by Eq. 4.12. We derive for the longitudinal geometry

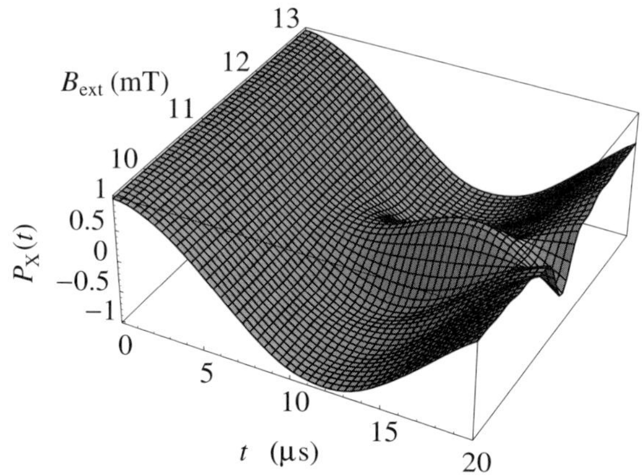

Fig. 4.5 The transverse polarisation function $P_X(t)$ for fields around $B_{\text{cri}} = 11.33\,\text{mT}$. It is plotted in a reference frame rotating with a frequency corresponding to B_{ext}. The resonance effect is seen at B_{cri}.

$$\mathcal{R}_{\text{int}} = 1 - \frac{1}{2}\left[1 - \frac{(B-F)^2}{A^2+(B-F)^2} - \frac{A^2}{A^2+(B-F)^2}\frac{1}{1+\dfrac{A^2+(B-F)^2}{\hbar^2}\tau_\mu^2}\right]. \tag{4.25}$$

For a particular material the relevant time scale is not τ_μ as in Eq. 4.25 but a time scale influenced by the dynamics of the muon–electron–nuclear species under study; see Section 4.1.3. It is convenient to define λ_{ALC}, which is just τ_μ^{-1} for the present model. Far outside the resonance, $\mathcal{R}_{\text{int}} \simeq 1$. At resonance,

$$\mathcal{R}_{\text{int}}(B_{\text{ext}} = B_{\text{cri}}) = 1 - \frac{1}{2}\frac{A^2}{A^2+\hbar^2\lambda_{\text{ALC}}^2}. \tag{4.26}$$

Although the weighting factor $\frac{1}{2}$ is a peculiarity of the model we are analysing, such a factor is always present for a more general formula as explained in Section 8.4.2. Scanning the field, a dip in $\mathcal{R}_{\text{int}}$ occurs in the vicinity of B_{cri}. In Fig. 4.6 we illustrate the resonance.

An expression for $\mathcal{R}_{\text{int}}$ can also be derived for the transverse geometry. We leave it to the interested reader. Nowadays the longitudinal time-integrated ALC spectroscopy is a standard tool for characterising the environment of muonium centres in matter; see Section 8.4.2.

4.3 Stroboscopic technique

The measurements of the so-called Knight shift provides a means for determining the local susceptibility of a system; see Section 5.3 for an introduction. The Knight shift in diamagnets is usually in the range of a few tens of parts per million. The stroboscopic method is well adapted for the measurement of such small shifts. Here we only give its principle. It is fully described by Klempt *et al.* (1982) and Schenck (1985).

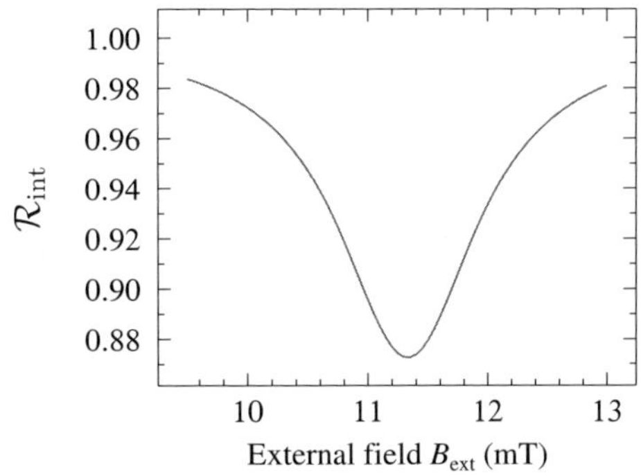

Fig. 4.6 The resonance for the longitudinal geometry arising from the ALC phenomena for our model. The weighted difference $\mathcal{R}_{\rm int}$ of forward and backward counts as defined in Eq. 4.12 and computed in Eq. 4.25 is shown. As expected, the resonance occurs at the critical field $B_{\rm cri} = 11.33\,\text{mT}$.

A prerequisite for this method is the availability of a pulsed beam of polarised muons. In practice, it has been applied at beamlines for which the separation between successive pulses is about 20 ns. At this level of separation, the beam can be considered as pseudo-continuous for usual μSR techniques; see Section 2.2.2. The short time structure derives from the cyclic proton acceleration induced by RF fields.

The method consists, in a transverse experimental geometry, of implanting muons from such a beam and letting them precess with a frequency equal to $\nu_\mu = \omega_\mu/(2\pi) = \gamma_\mu B_{\rm loc}/(2\pi)$, close to an integer multiple of the pulse repetition frequency $\nu_{\rm B} = \omega_{\rm B}/(2\pi)$. Therefore the muons of a given pulse stopping in the sample will have nearly the same phase as those stopped previously, which originate from the previous muon pulse. The decay positrons are observed in narrow time gates with fixed phases relative to the pulse structure of the muon beam. By sweeping $B_{\rm ext}$, the number of positrons detected in a positron telescope positioned in the plane of the muon precession, and registered in these time gates, will show a resonance when ν_μ corresponds to an integer multiple of $\nu_{\rm B}$.

The swiss muon source (SμS) at the Paul Scherrer Institute, formerly known as Schweizerisches Institut für Nuklearforschung (SIN), is the laboratory where the stroboscopic technique has most intensively been used. It happens that the maximum statistical accuracy is obtained there at the second harmonic. The number of positrons detected in the jth positron gate, N_j, can then be written

$$N_j(x) = N_0 \lambda_j \left[1 + A_{\rm eff} \frac{\sin(2\pi\lambda_j)}{2\pi\lambda_j \pi} \frac{\cos\psi_j - x \sin\psi_j}{1 + x^2} \right], \tag{4.27}$$

where N_0 and $A_{\rm eff}$ are constants and λ_j and ψ_j are the relative gate width and phase, respectively. The quantity x is defined by

$$x = \bar{\tau}_X \left[\left(\frac{\omega_\mu}{\omega_{\rm p}} \right) \omega_{\rm p} - 2\omega_{\rm B} \right] = \bar{\tau}_X \left[(1 + K_{\rm exp}) \frac{\gamma_\mu}{\gamma_{\rm p}} \omega_{\rm p} - 2\omega_{\rm B} \right]. \tag{4.28}$$

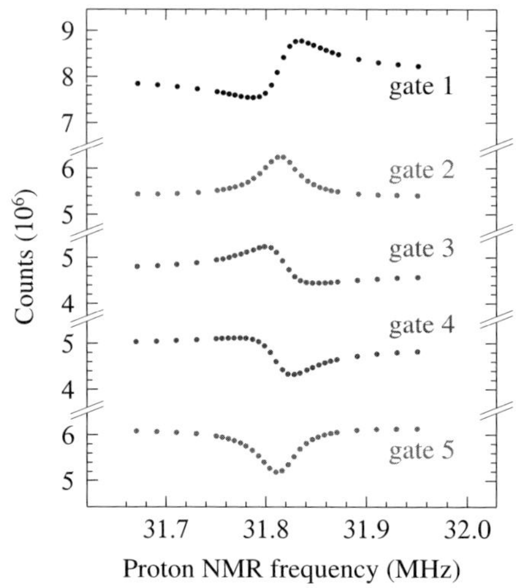

Fig. 4.7 Examples of stroboscopic signals from five sequentially ordered time gates. The signals are recorded sweeping the externally applied magnetic field which is monitored by a NMR probe. The field is given in proton NMR frequency unit, i.e. 31.8 MHz corresponds to a field of 0.725 T. Adapted from Camani *et al.* (1978).

The proton angular frequency denoted by $\omega_{\rm p}$ is measured with an NMR probe located near the sample. It serves to estimate $B_{\rm ext}$ since $\omega_{\rm p} = \gamma_{\rm p} B_{\rm ext}$ where $\gamma_{\rm p}$ is the proton gyromagnetic ratio. The envelope of the wiggles is assumed to be described by an exponential function characterised by a damping rate, λ_X. As in Section 4.1.3, $\bar{\tau}_X^{-1} = \tau_\mu^{-1} + \lambda_X$ where τ_μ is the muon lifetime. In Eq. 4.28 we have introduced $K_{\rm exp}$, which is a measure of the muon frequency shift as discussed in Section 5.3.

The positron number function $N_j(x)$ is the sum of a constant term and a function of x. This function is itself proportional to the weighted sum of a Lorentzian shape and a dispersion-like shape. Figure 4.7 presents some stroboscopic signals recorded as a function of the proton precession frequency, $\nu_{\rm p} = \omega_{\rm p}/(2\pi)$. The values of $K_{\rm exp}$ and λ_X can be extracted from such data. This method proves very powerful in measuring these parameters: the precision on $K_{\rm exp}$ can reach the part per million level, i.e. 10^{-6} in relative units.

This achievement is not only the result of the large number of events that can be accumulated relative to the usual time-differential technique, but also because it is not necessary to measure time intervals. In favour of this technique, $\omega_{\rm p}$ and $\omega_{\rm B}$ are accurately determined because their measurement relies on proton NMR and master quartz oscillator, respectively.

However, the stroboscopic technique has its own limitations: it is only applicable at selected fields and requires an adequate temporal structure of the muon beam. Stroboscopic measurements have only been carried out with high-energy muon beams and do not seem to be possible with surface muon beams (Schenck 1985).

4.4 Lattice steering

We recommend any person interested in the technique of lattice steering of muons and positrons to read the review paper of Carstanjen and Seeger (1984). A good review focusing on muons in semiconductors is provided by Patterson (1988).

The lattice-steering technique is not a μSR technique. It is briefly described here because it provides information complementary to μSR. In addition, it involves either the positive pion and muon or the positive muon and positron.

In the π^+/μ^+ lattice-steering technique, pions are implanted into a single crystal where they slow down. When thermalised the pions are expected to occupy interstitial sites of the crystal. They possibly undergo diffusive motion and can be trapped by impurities and defects like the muons. In semiconductors the pion may form a bound state with an electron to give the pionium state (Flick *et al.* 1986). We recall that a pion decays into a muon and a muon neutrino with a lifetime of 26 ns. When emitted along directions close to an open crystal direction, the muons are steered by the lattice atoms and can possibly emerge from the crystal surface. On the other hand, if emitted in random directions, they are blocked. Obviously, this method yields information on the trapping site of the pions and the sample crystalline quality. The μ^+/e^+ lattice-steering method works in the same way with the substitution of the muon for the pion and the positron for the muon.

For definitiveness, in the following we focus on the π^+/μ^+ technique. We note that only a small fraction ($\sim 10^{-5}$) of the muons undergoes channelling, giving rise to an angular distribution of muons which is recorded by a two-dimensional, position-sensitive, scintillation detector located far away from the sample (a few metres) to get a good angular resolution. This explains why a very high pion flux is needed. The channelled muons emanate from a depth of up to $\sim 50\,\mu$m.

In the left panel of Fig. 4.8 we illustrate a π^+/μ^+ lattice-steering profile for a high-quality tantalum crystal. It is characterised by its peak height and full-width at half-maximum. In the right panel of Fig. 4.8 we present this width versus temperature for a tantalum sample doped with 340 ppm of oxygen. Under the prevailing conditions the oxygen atoms should be atomistically dispersed. The enhancement of the width around 58 K is understood as follows. Below 30 K the pion diffusivity is too small for a detectable fraction of the pions to reach an oxygen-disturbed crystal region. As the temperature is raised, the probability for a pion to probe such a region increases. At the peak temperature, an appreciable number of pions are trapped around an oxygen atom during their lifetime. As the temperature is further increased, the thermal energy is getting larger than the free enthalpy of pion binding, so that an overwhelming majority of pions escape from the traps and diffuse in between them. Therefore they mainly probe the trap-free interstitial sites.

4.5 β-detected NMR technique

The present interest in the physical and chemical properties of thin films and heterostructures is mainly driven by their possible technological applications. Nevertheless, one does not expect these applications to develop without a basic understanding of the physical and chemical mechanisms involved.

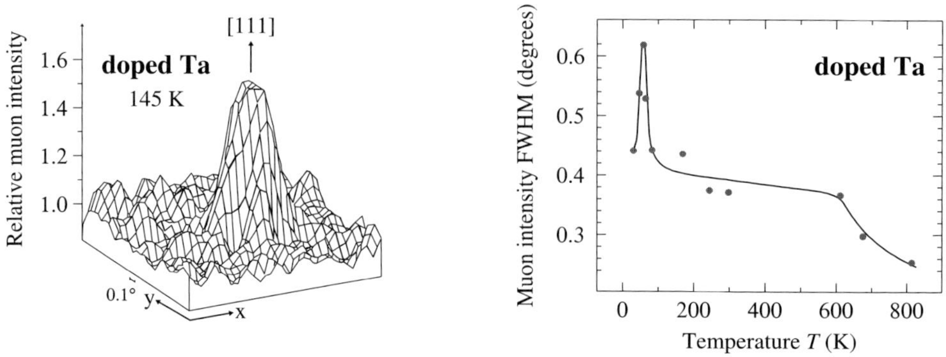

Fig. 4.8 (Left) Relative intensity distribution of decay muons as emitted along the [111] axis of a tantalum single crystal at a temperature of 145 K; the angular profile has been normalised by division with a spectrum obtained along a "random" direction. (Right) Full-width at half-maximum of the muon intensity peak around a [111] axis of a tantalum crystal doped with 340 ppm of oxygen as a function of the temperature of the sample. Both plots are adapted from Carstanjen and Seeger (1984).

Nuclear magnetic resonance is an efficient technique for the study of bulk condensed matter. However, because of the large number of nuclei required to perform a measurement, standard NMR cannot be used to probe the near-surface of systems. One powerful way to circumvent this limitation is to implant probes that are already highly spin-polarised into the material of interest. As explained in Section 2.2.2, the region near to the surface of a sample can be studied with a low-energy muon beam.

An additional experimental local probe method that is rapidly developing is the so-called β-detected NMR technique, which employs a highly spin-polarised radioactive ion beam. From the user point of view, a positive muon is substituted by a nucleus. Because of its low energy and relatively heavy mass, the ion does not diffuse a long way into the sample under study, and therefore only the near-surface is probed. A dedicated β-detected NMR instrument has been built with a ^{8}Li beam (Morris *et al.* 2004) at CMMS. As in the case of a low-energy muon beam, the energy of the implanted ^{8}Li nuclei can be controlled, allowing for depth-controlled β-detected NMR measurements. For examples of possibilities offered by this technique, we refer, e.g., to the work by Parolin *et al.* (2008) and the references therein.

The polarisation of the beam is achieved by in-flight optical pumping, resulting in a high beam polarisation — up to ~70%, Morris *et al.* (2004). Resonance measurements can be carried out with a continuous beam and a continuous wave RF magnetic field applied transverse to the beam polarisation. Spin–lattice relaxation measurements are usually performed with a pulsed beam (Parolin *et al.* 2008). The resonance and the relaxation of the probe are detected *via* the parity-violating β decay. In fact, apart from the production of the polarised beam, the principle of detection and the methods of analysis of β-detected NMR and μSR are quite similar.

Up to now, measurements have been routinely performed with ^{8}Li nuclei of spin $S = 2$, electric quadrupole moment of $+0.033$ barn, lifetime 1.21 s, and gyromagnetic ratio $6.3015\times10^6\,\mathrm{rad\,s^{-1}\,T^{-1}}$. Therefore, relative to μSR, the time range during which a measurement is performed is far larger. Using a pulse width on the order of 1 s does not pose any problem since the measurements can be easily extended up to at least 10 s. In addition, this enables the study of small relaxation rates, thanks partly to the high beam intensity resulting from the large pulse-width.

Since the spin of ^{8}Li nuclei is larger than $\frac{1}{2}$, quadrupolar splitting is a possibility, resulting in complicated spectra. However, β-detected NMR experiments seem possible with ^{11}Be. Unlike ^{8}Li, these $\frac{1}{2}$-spin nuclei behave as purely magnetic probes.

4.6 Experiments with negative muons

Here is the only section of this book dealing with negative muons. In fact, this type of work has so far been of a limited scope because the effective initial asymmetry is very small relative to that measured with positive muons. A typical example is given in Fig. 4.9: the initial asymmetry is about an order of magnitude smaller than expected for a positive muon spectrum.

A useful introduction to negative muon spin spectroscopy is given by Yamazaki (1984).

Nature provides positive and negative muons. The negative muon is a twin sister to the positive muon: it has the same mass and the same magnetic moment except for the sign. Obviously, its electric charge is also of opposite sign. However, its behaviour in matter is totally different. The fate of the negative muon ends up not only in the free decay

$$\mu^- \longrightarrow \mathrm{e}^- + \bar{\nu}_\mathrm{e} + \nu_\mu, \tag{4.29}$$

where $\bar{\nu}_e$ and ν_μ are the antineutrino and neutrino associated with the electron and muon, respectively, but also in muon capture by a nucleus. According to the Coulomb law, a negative muon is attracted by a positive nucleus: the nucleus and the muon form a bound state, a so-called muonic atom. This muonic atom is formed in an excited state and as the muon reaches the $1s$ state, its polarisation is only 15 to 20% of its initial value. This reduction of polarisation, which arises from the spin–orbit coupling in the muonic atom, sets a first limiting experimental factor: much more statistics than for the usual positive muon techniques is needed to get useful results. Secondly, the Bohr radius of the muonic atom is decreased by a factor $\simeq 200$ compared to the Bohr radius of a conventional nucleus–electron system. This reduction reflects the ratio between the muon and electron masses. For heavy nuclei the muon Bohr radius is even smaller than the nuclear radius. As a result the muon is usually captured by the nucleus. Then, the effective lifetime of a negative muon varies from $\tau_\mu^{\mathrm{eff}} \simeq 2.2\,\mu$s (i.e. about its value in free space) if implanted in hydrogen to only $\tau_\mu^{\mathrm{eff}} \simeq 70$ ns if implanted in a compound made of heavy nuclei. This restricted lifetime sets an upper time limit over which a negative muon can probe a compound. This is the second limitating experimental factor.

Another difference with usual μ^+SR measurements stems from the fact that the negative muon binds to a nucleus. The magnetic field distribution which is probed by μ^-SR measurements is then that of a subsitutional site rather than of an interstitial site as in μ^+SR.

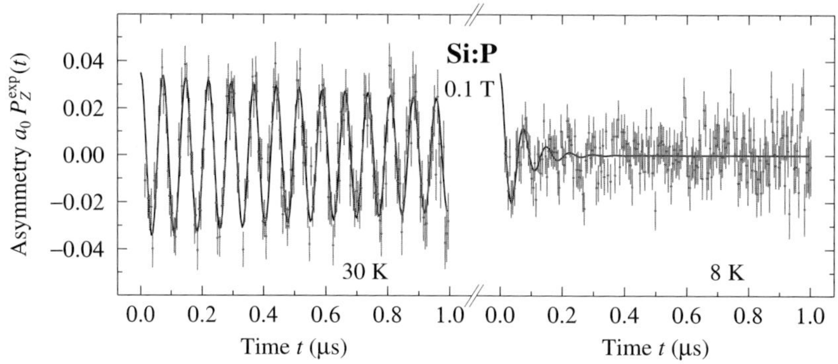

Fig. 4.9 Examples of two transverse-field μSR spectra recorded with negative muons in an applied field of 0.1 T at 30 and 8 K. The sample is phosphorus-doped silicon with a P concentration of 1.6×10^{13} cm^{-3}. The main difference between spectra recorded with positive and negative muons is the much reduced asymmetry for the latter case. Adapted from Mamedov *et al.* (1999).

Now, from the point of view of the system under study, the muonic atom can be regarded as an atom consisting of a pseudonucleus including Z protons and one μ^- with only Z − 1 electrons because the electronic density must respond to the $(Z-1)e$ nuclear electric charge. Up to now, the μ^-SR studies have focused on either the investigation of the magnetic field at the muonic atom as a probe of its electronic structure and of its interaction with the environment (Yamazaki 1984) or regarding the muonic atom as a doping element in, for example, a semiconductor such as silicon (Mamedov *et al.* 1999). Then the muonic atom is an analogue to the aluminium atom, i.e. a shallow acceptor centre in a silicon crystal lattice. With this point of view in mind, μ^-SR can be used as an alternative spectroscopic method for studying defects in semiconductors, complementary to traditional methods such as electron paramagnetic resonance. The methods developed for μ^+SR can be employed directly. For example, from Fig. 4.9 we note that the wiggles are damped at low temperature, meaning that, as the temperature is decreased, a slowing down of the magnetic fluctuations at the muonic (defect) site is observed.

References

Abragam A. (1984) *Comptes Rendus de l'Académie des Sciences* **299**, 95–100.

Brown J.A., Heffner R.H., Leon M., Dodds S.A., Vanderwater D.A., and Estle T.L. (1979) *Physical Review Letters* **43**, 1751–1754.

Camani M., Gygax F.N., Klempt E., Rüegg W., Schenck A., Schilling H., Schulze R., and Wolf H. (1978) *Physics Letters* **B 77**, 326–330.

Carstanjen H.D. and Seeger A. (1984) in *Muons and Pions in Materials Research*, edited by Chappert J. and Grynszpan R.I. (North-Holland, Amsterdam).

Coffin T., Garwin R.L., Penman S., Lederman L.M., and Sachs A.M. (1958) *Physical Review* **109**, 973–979.

Cottrell S.P., Cox S.F.J., Scott C.A., and Lord J.S. (2000) *Physica* **B 289-290**, 693–697.

Eck T.G., Foldy L.L., and Wieder H. (1963) *Physical Review Letters* **10**, 239–242.

Estle T.L. and Vanderwater D.A. (1983) *Physical Review* **B 27**, 3962–3979.

Flick G., Bradbury J.N., Cooke D.W., Heffner R.H., Leon M., Paciotti M.A., Schillaci M.E., Maier K., Rempp H., Reidy J.J., Boekema C., and Daniel H. (1986) *Physical Review Letters* **57**, 563–566.

Graf M.J., Lago J., Lascialfari A., Amato A., Baines C., Giblin S.R., Lord J.S., Tkachuk A.M., and Barbara B. (2007) *Physical Review Letters* **99**, 267203, 1–4.

Karlsson E.B. (1995) *Solid State Phenomena As Seen by Muons, Protons, and Excited Nuclei* (Clarendon Press, Oxford).

Klempt E., Schulze R., Wolf H., Camani M., Gygax F.N., Rüegg W., Schenck A., and Schilling H. (1982) *Physical Review* **D 25**, 652–676.

Kreitzman S.R., Williams D.Ll., Kaplan N., Kempton J.R., and Brewer J.H. (1988) *Physical Review Letters* **61**, 2890–2893.

Mamedov T.N., Chaplygin I.L., Duginov V.N., Gorelkin V.N., Herlach D., Major J., Stoykov A.V., Schefzik M., and Zimmermann U. (1999) *Journal of Physics: Condensed Matter* **11**, 2849–2860.

Morris G.D., MacFarlane W.A., Chow K.H., Salman Z., Arseneau D.J., Daviel S., Hatakeyama A., Kreitzman S.R., Levy C.D.P., Poutissou R., Heffner R.H., Elenewski J.E., Greene L.H., and Kiefl R.F. (2004) *Physical Review Letters* **93**, 157601, 1–4.

Nagamine K. (2003) *Introductory Muon Science* (Cambridge University Press, Cambridge).

Nagamine K. and Yamazaki T. (1981) mentioned by Schenck A. (1985) *Muon Spin Rotation Spectroscopy* (Adam Hilger, Bristol).

Parolin T.J., Salman Z., Chow K.H., Song Q., Valiani J., Saadaoui H., O'Halloran A., Hossain M.D., Keeler T.A., Kiefl R.F., Kreitzman S.R., Levy C.D.P., Miller R.I., Morris G.D., Pearson M.R., Smadella M., Wang D., Xu M., and MacFarlane W.A. (2008) *Physical Review* **B 77**, 214107, 1–8.

Patterson B.D. (1988) *Reviews of Modern Physics* **60**, 69–159.

Schenck A. (1985) *Muon Spin Rotation Spectroscopy* (Adam Hilger, Bristol).

Shiroka T., Bucci C., De Renzi R., Galli F., Guidi G., Eaton G.H., King P.J.C., and Scott C.A. (1999) *Physical Review Letters* **83**, 4405–4408.

Slichter C.P. (1996) *Principles of Magnetic Resonance*, third enlarged and updated edition (Springer, Berlin).

Yamazaki T. (1984) in *Muons and Pions in Materials Research*, edited by Chappert J. and Grynszpan R.I. (North-Holland, Amsterdam).

5
The magnetic field at the muon site

In this chapter we first discuss the interaction energies at the muon site, pointing out the importance of the local magnetic field at that site. This field partly arises from the dipole interaction between the magnetic moments of the muon and of the localised electrons. In the second section two procedures to carry out the lattice summation of this long-range field are exposed. In the next section the frequency shift observed in paramagnets and diamagnets is considered. Finally, in the last section, the characteristics of the field at the muon site in the ordered phase of a magnet are discussed. We assume that the muon does not form a bound state with an electron, i.e. there is no muonium formation. The physics of muonium centres is presented in Chapter 8.

5.1 The interaction energies at the muon site

The interaction energy between a muon and the compound in which it is implanted has both electric and magnetic origins. Although the electric forces are the strongest, it is the magnetic forces which determine the fate of the muon spin. Therefore we shall focus on the latter. However, before that, we discuss the effects of the electric forces when the muon is implanted into a metal using a case study.

5.1.1 Muon charge screening mechanism: the copper case

We first point out that, because of its positive electric charge, the muon is implanted in an interstitial rather than a substitutional site, at least if no vacancies are present. The muon which carries one unit of electric charge may displace atoms surrounding it and induce an electric field gradient on them. This must be included in the Hamiltonian describing the compound. In addition, we expect the electronic density around the muon to be perturbed by its electric charge. Since this perturbation has a direct influence on the magnetic field at the muon site, it will be dealt with in our discussion of the magnetic interactions at the muon site.

In order to grasp the physics involved when a muon interacts with its environment in a compound without forming a stable muonium state, we describe the simple case of a muon in metallic copper for which a detailed microscopic model is available (Teichler 1978); see also Section 1.2.2. A muon is described as a positive electric charge dressed by an electron screening cloud. Its interaction potential with a copper ion Cu^+ is the sum of two terms:

$$V(\mathbf{r}_0 - \mathbf{r}_i) = \frac{e^2 Z_{\rm eff}}{4\pi\epsilon_0 |\mathbf{r}_0 - \mathbf{r}_i|} + \int \delta_{\rm e}(\mathbf{r} - \mathbf{r}_0) v_{\rm ps}(\mathbf{r} - \mathbf{r}_i) {\rm d}^3 {\rm r}. \tag{5.1}$$

Here we have adopted the usual notation of this textbook which assumes the muon to be localised at site $\mathbf{r}_0$ and the atoms at $\mathbf{r}_i$ (see Section 5.2.4). In Eq. 5.1 $Z_{\rm eff}e$ is an effective charge. The first term of the equation accounts for the effective Coulomb interaction of the muon with the Cu^+ core at site $\mathbf{r}_i$ and the second models the interaction between the screening electron density $\delta_{\rm e}(\mathbf{r}-\mathbf{r}_0)$ around the muon and the same Cu^+ core *via* the local pseudopotential $v_{\rm ps}(\mathbf{r}-\mathbf{r}_i)$. The 4$s$ electrons, which do not belong to the core, are assumed to be distributed homogeneously. One has

$$\delta_{\rm e}(\mathbf{r}-\mathbf{r}_0) = n(\mathbf{r}-\mathbf{r}_0) - n_0(\mathbf{r}-\mathbf{r}_0), \tag{5.2}$$

where $n(\mathbf{r}-\mathbf{r}_0)$ is the electronic charge distribution around the position $\mathbf{r}_0$ with the muon present, and $n_0(\mathbf{r}-\mathbf{r}_0)$ is the corresponding distribution in the absence of the muon. The total potential energy $U(\mathbf{r}_0)$ of a muon in copper is the sum over the previous two-body interaction potentials, i.e. $U(\mathbf{r}_0) = \sum_{\mathbf{r}_i} V(\mathbf{r}_0 - \mathbf{r}_i)$. Because of its positive charge, one expects the muon to attract electrons. If one uses the linear-screening approximation the effective muon electric charge is computed to be about $0.7 \times e$. Teichler has evaluated $U(\mathbf{r}_0)$ with a nonlinear-screening jellium approach for $\delta_{\rm e}(\mathbf{r}-\mathbf{r}_0)$ and proper value for $Z_{\rm eff}$ and the function $v_{\rm ps}$. This work provides an explanation for the energetics of a muon in copper.

5.1.2 Interactions of a muon with a single electron

From now on in this section the muon is located at the origin of the coordinates and we consider a muon and a single electron. In the non-relativistic Pauli description of the electron, the Hamiltonian for the muon–electron system, in the case of a static muon, is (see for example Chapter 6 of Abragam's book (1961))

$$\mathcal{H}_{\mu-{\rm e}} = \frac{1}{2M_{\rm e}}\left(\mathbf{p}+e\mathbf{A}\right)^2 + g_{\rm e}\mu_{\rm B}\mathbf{S}_{\rm e}\cdot\mathbf{curl}\,\mathbf{A}, \tag{5.3}$$

where $M_{\rm e}$, $\mathbf{p}$, $g_{\rm e}$, and $\mathbf{S}_{\rm e}$ are the mass, linear momentum vector, spectroscopic splitting factor, and spin vector operator of the electron, respectively, and $\mu_{\rm B}$ is the Bohr magneton defined as $\mu_{\rm B} = e\hbar/(2M_{\rm e})$, where e is the unit charge (positive) and $\hbar$ the normalised Planck constant. The vector potential $\mathbf{A}$ at the electron due to the muon magnetic moment, $\mathfrak{m}_\mu$, has the well-known form

$$\mathbf{A} = \frac{\mu_0}{4\pi}\frac{\mathfrak{m}_\mu \times \mathbf{r}}{r^3} = \frac{\mu_0}{4\pi}\,\mathbf{curl}\left(\frac{\mathfrak{m}_\mu}{r}\right), \tag{5.4}$$

where μ_0 is the permeability of free space, $\mathbf{r}$ is the radius vector from the muon to the electron, and $\mathfrak{m}_\mu \times \mathbf{r}$ stands for the vector product of $\mathfrak{m}_\mu$ by $\mathbf{r}$. The vector $\mathbf{curl}\,\mathbf{V}$ has the components $(\mathbf{curl}\,\mathbf{V})^\alpha = \sum_{\beta\gamma}\epsilon^{\alpha\beta\gamma}\partial V^\gamma/\partial r^\beta$, where $\epsilon^{\alpha\beta\gamma}$ is an element of the antisymmetric Levi–Civita unit pseudotensor defined in Section C.1.

The development of the first term on the right-hand-side of Eq. 5.3 gives a sum of four terms. The kinetic energy of the electron (the term proportional to $\mathbf{p}^2$ in Eq. 5.3) will not be considered any longer since this energy does not couple to $\mathfrak{m}_\mu$. We also neglect the term proportional to $\mathbf{A}^2$ because it is quadratic in $\mathfrak{m}_\mu$. Retaining in Eq. 5.3

only the two terms which depend on both muon and electron variables and are linear in $\mathfrak{m}_\mu$, we get

$$\begin{aligned}\mathcal{H}_{\mu-\mathrm{e}} &= \frac{e}{2M_\mathrm{e}}(\mathbf{pA}+\mathbf{Ap}) + g_\mathrm{e}\mu_\mathrm{B}\mathbf{S}_\mathrm{e}\cdot\mathbf{curl\,A} \\ &= \frac{\mu_0}{4\pi}2\mu_\mathrm{B}\frac{\mathbf{l}_\mathrm{e}\cdot\mathfrak{m}_\mu}{r^3} + g_\mathrm{e}\mu_\mathrm{B}\mathbf{S}_\mathrm{e}\cdot\mathbf{curl\,A},\end{aligned} \tag{5.5}$$

where $\mathbf{l}_\mathrm{e}$ is the orbital momentum of the electron in units of $\hbar$: $\hbar\mathbf{l}_\mathrm{e} = \mathbf{r}\times\mathbf{p}$. We have used the relations $(\mathfrak{m}_\mu\times\mathbf{r})\cdot\mathbf{p} = \mathfrak{m}_\mu\cdot(\mathbf{r}\times\mathbf{p})$ and $\mathbf{p}\cdot(\mathfrak{m}_\mu\times\mathbf{r}) = \mathbf{r}\cdot(\mathbf{p}\times\mathfrak{m}_\mu)$. Note that $\mathbf{r}\times\mathbf{p} = -\mathbf{p}\times\mathbf{r}$. Whereas the first term on the right-hand-side of Eq. 5.5 only depends on the electron orbital operator (orbital current), the second term is only spin-dependent. With the help of Eqs. 5.4 and C.9, we derive

$$\mathcal{H}_{\mu-\mathrm{e}} = \frac{\mu_0}{4\pi}\mu_\mathrm{B}\left(\frac{2\mathbf{l}_\mathrm{e}\cdot\mathfrak{m}_\mu}{r^3} + g_\mathrm{e}\sum_{\alpha\beta}S^\alpha_\mathrm{e}\mathfrak{m}^\beta_\mu\frac{\partial^2}{\partial r^\alpha\partial r^\beta}\frac{1}{|\mathbf{r}|} - g_\mathrm{e}\mathbf{S}_\mathrm{e}\cdot\mathfrak{m}_\mu\sum_\gamma\frac{\partial^2}{\partial(r^\gamma)^2}\frac{1}{|\mathbf{r}|}\right). \tag{5.6}$$

It appears that the following equivalent form is useful:

$$\mathcal{H}_{\mu-\mathrm{e}} = \frac{\mu_0}{4\pi}\mu_\mathrm{B}\left(\frac{2\mathbf{l}_\mathrm{e}\cdot\mathfrak{m}_\mu}{r^3} + g_\mathrm{e}\sum_{\alpha\beta}S^\alpha_\mathrm{e}T^{\alpha\beta}(\mathbf{r})\mathfrak{m}^\beta_\mu - \frac{2}{3}g_\mathrm{e}\mathbf{S}_\mathrm{e}\cdot\mathfrak{m}_\mu\sum_\gamma\frac{\partial^2}{\partial(r^\gamma)^2}\frac{1}{|\mathbf{r}|}\right). \tag{5.7}$$

Here, we have introduced the tensor of second rank, $\boldsymbol{T}(\mathbf{r})$, such that

$$T^{\alpha\beta}(\mathbf{r}) = \left[\frac{\partial^2}{\partial r^\alpha\partial r^\beta} - \frac{1}{3}\delta^{\alpha\beta}\sum_\gamma\frac{\partial^2}{\partial(r^\gamma)^2}\right]\frac{1}{|\mathbf{r}|}. \tag{5.8}$$

5.1.3 Muon spin interactions in a solid

Using the previous results, we decompose the field at the muon site into its different contributions.

Up to now, we have considered the magnetic interaction of a single electron with the muon. However, when the muon is implanted in a solid, more than one electron must be considered. To describe this solid state effect the electronic wavefunction ψ_e must be taken into account. Therefore, we express $\mathcal{H}_{\mu-\mathrm{e}}$ in terms of ψ_e and denote the resulting Hamiltonian as $\mathcal{H}_\mu$. We shall distinguish two types of electrons, although this may not be always justified.

Let us first consider the electrons for which the probability to be at the muon site is negligible. Schematically, these electrons are orbiting around lattice sites. Hence, the first term of the right-hand-side of Eq. 5.6 can be neglected because, by hypothesis, there is no electron orbiting around the muon. The third term gives also a vanishingly small contribution since

$$\sum_\alpha \frac{\partial^2}{\partial (r^\alpha)^2} \frac{1}{|\mathbf{r}|} = -4\pi\delta(|\mathbf{r}|). \tag{5.9}$$

Then, we derive from Eq. 5.6

$$\mathcal{H}_\mu = \frac{\mu_0}{4\pi}\mu_{\rm B} g_{\rm e} \sum_j \sum_{\alpha\beta} S^\alpha_{{\rm e},j} \mathrm{m}^\beta_\mu \frac{\partial^2}{\partial r^\alpha_j \partial r^\beta_j} \frac{1}{|\mathbf{r}_j|}, \tag{5.10}$$

which gives the well-known formula

$$\mathcal{H}_\mu = \frac{\mu_0}{4\pi} \sum_j \left[\frac{\mathbf{m}^s_j \cdot \mathbf{m}_\mu}{r_j^3} - \frac{3\left(\mathbf{m}_\mu \cdot \mathbf{r}_j\right)\left(\mathbf{m}^s_j \cdot \mathbf{r}_j\right)}{r_j^5} \right], \tag{5.11}$$

where $\mathbf{m}^s_j = -g_{\rm e}\mu_{\rm B}\mathbf{S}_j$ (we write $\mathbf{m}^s_j$ instead of $\mathbf{m}^s_{{\rm e},j}$ for simplicity and since there is no risk of confusion) is the spin magnetic moment operator of electron j and $\mathbf{r}_j$ its vector radius. The Hamiltonian $\mathcal{H}_\mu$ is expressed as the sum of terms, each of them being the usual dipole interaction operator between the muon magnetic moment and an electron magnetic moment. The reader may have noticed that the spin–orbit interaction is not included in Eq. 5.3. This interaction is strong for $4f$ or $5f$ electronic shells (rare-earth and actinide ions, respectively) and we must consider the total angular momentum $\mathbf{J}_i$ instead of the spin and orbital angular momentum of the ions. Assuming the Russel–Saunders coupling scheme to hold (see Appendix H), $\mathbf{J}_i = \mathbf{L}_i + \mathbf{S}_i$, where the terms $\mathbf{L}_i$ and $\mathbf{S}_i$ are the total orbital and spin angular momenta of the unpaired electrons in the $4f$ or $5f$ shell of the atom at site i. For a transition element, the expectation value of the total orbital momentum may be small and then, to a good approximation, $\mathbf{J}_i = \mathbf{S}_i$.

To discuss the contribution to $\mathcal{H}_\mu$ of the electrons for which the density probability at the muon site is finite, it is practical to start from Eq. 5.7. The tensor components $T^{\alpha\beta}(\mathbf{r})$ (see Eq. 5.8) behave under a rotation of the coordinate system as spherical harmonics of rank 2. Hence, if $\psi_{\rm e}$ is expanded as a sum of spherical harmonics, $\psi_{\rm e} = \sum_{l,m} a_{l,m} Y_l^m$, the only non-vanishing contributions will come from terms such as $\langle Y_l^m | T^{\alpha\beta}(\mathbf{r}) | Y_{l'}^{m'} \rangle$ for which $l + l' \geq 2$ — see for example Messiah (1965). Let us assume that the electronic cloud around the muon is spherically symmetric. Then the contribution to $\mathcal{H}_\mu$ of the second term on the right-hand-side of Eq. 5.7 vanishes. Since there is no orbital magnetic moment for a spherical electronic cloud, the first term of Eq. 5.7 does not also contribute to $\mathcal{H}_\mu$. Thus, for a spherical electronic cloud, only the third term of this equation matters. We have

$$\mathcal{H}_\mu = -\frac{2\mu_0}{3}|\psi_{\rm s}(\mathbf{r}_0)|^2 \mathbf{m}^s_{\rm e} \cdot \mathbf{m}_\mu. \tag{5.12}$$

For this formula we have adopted the usual notation mentioned previously which defines the muon site as $\mathbf{r}_0$. The Hamiltonian $\mathcal{H}_\mu$ only depends on the electronic density at that site and is referred to as the Fermi contact interaction operator. It is a contact interaction in the sense that it is effective only if the electronic density is appreciable at the muon site. It should be noticed that the physical origin of the classical dipole interaction and the Fermi contact interaction is the same: the Pauli description of the electron.

According to Eq. 5.6 or 5.7, $\mathcal{H}_\mu$ is linear in $\mathfrak{m}_\mu$. Therefore the magnetic coupling between the muon magnetic moment and its environment can be described by an effective Zeeman interaction Hamiltonian, $\mathcal{H}_{Z,\mu}$. Thus, we write

$$\mathcal{H}_{Z,\mu} = -\mathfrak{m}_\mu \cdot \mathbf{B}_{\rm loc} = -\gamma_\mu \hbar \mathbf{S}_\mu \cdot \mathbf{B}_{\rm loc} = -\frac{1}{2}\gamma_\mu \hbar \boldsymbol{\sigma} \cdot \mathbf{B}_{\rm loc}, \qquad (5.13)$$

where $\mathbf{S}_\mu$ is the muon spin operator, $\boldsymbol{\sigma}$ the Pauli operator for the muon spin, and γ_μ the muon gyromagnetic ratio. We write $\mathbf{B}_{\rm loc}$ for the local magnetic field operator at the muon site to stress its local character. Obviously, in general terms, $\mathbf{B}_{\rm loc} = \langle \mathbf{B}_{\rm loc} \rangle + \delta \mathbf{B}_{\rm loc}$, where $\langle \mathbf{B}_{\rm loc} \rangle$ is its mean value (in the thermodynamical sense) and $\delta \mathbf{B}_{\rm loc}$ accounts for the fluctuations. In the magnetically ordered state of a compound, the mean value is usually different from zero: it is the spontaneous interstitial magnetic field. The field fluctuations drive the relaxation of the muon spin ensemble.

In summary, from the Pauli equation (see Eq. 5.3), the definition of $\mathbf{B}_{\rm loc}$ implied by the previous equation, and the definitions of the dipole and Fermi contact operators (see Eqs. 5.11 and 5.12 respectively), we have

$$\mathbf{B}_{\rm loc} = \mathbf{B}_{\rm ext} + \mathbf{B}_{\rm dip} + \mathbf{B}_{\rm Fermi}, \qquad (5.14)$$

where $\mathbf{B}_{\rm dip}$ and $\mathbf{B}_{\rm Fermi}$ are the dipole and Fermi magnetic field operators. They respectively arise from the localised and conduction electrons. The dipole term $\mathbf{B}_{\rm dip}$ is obtained from Eq. 5.11 and will be discussed at length in the next section. From Eq. 5.12, we get, for a spherical electronic cloud around the muon,

$$\mathbf{B}_{\rm Fermi} = -\frac{2\mu_0}{3}|\psi_{\rm s}(\mathbf{r}_0)|^2 \eta(\mathbf{r}_0) g_{\rm e} \mu_{\rm B}\, \mathbf{S}_{\rm e} = \frac{2\mu_0}{3}|\psi_{\rm s}(\mathbf{r}_0)|^2 \eta(\mathbf{r}_0)\, \mathfrak{m}_{\rm e}^{s}. \qquad (5.15)$$

We have introduced the multiplicative enhancement factor $\eta(\mathbf{r}_0)$ to account for the perturbation of the magnetic density at the muon site arising from the muon electric charge.

In s or p electron metals, the localised electronic contribution is considered to vanish and the Fermi contact term solely accounts for the magnetic interaction of the conduction electron spins with the muon spin except for the small contribution of the nuclear magnetic moments. However, diamagnetic effects may be important (see Section 5.3).

On the other hand, in compounds with d or f unpaired electrons, the magnetic density at each of the d and f ions is always large, even for itinerant magnets. In fact, the unpaired d electrons are usually depicted in a tight binding scheme in which the electronic density at the crystal sites exceeds the density in between the sites. Similarly, f electrons are mostly localised at the ion sites. This picture is supported by the measured magnetic form factors which show the magnetisation density to be usually mostly localised around the crystal sites. The localised magnetic density contributes to $\mathbf{B}_{\rm loc}$ in two ways: through the classical dipole interaction as discussed earlier (see

Eq. 5.11), and through the Ruderman–Kittel–Kasuya–Yosida (RKKY) interaction in metals or the transferred hyperfine field in insulators.

The RKKY field arises from the indirect interaction between the muon and the magnetic moments of the localised unpaired electrons mediated by the conduction electrons (Kittel 1963). This interaction is the result of a second order process: the conduction electron spins are coupled to the localised electron spins (and thus the total angular momenta) through the exchange interaction as written in Eq. 10.263, and to the muon spin *via* the Fermi contact interaction. In its simplest form, the corresponding field is written as

$$\mathbf{B}_{\mathrm{RKKY}} = \frac{\mu_0}{4\pi} \sum_{i=1}^{N'} \frac{(-g_i \mu_{\mathrm{B}}) H_{\mathbf{r}_i,\,\mathrm{RKKY}}}{v_c} \mathbf{J}_i, \tag{5.16}$$

where the sum is restricted to the N' nearest (and, possibly, next nearest) neighbour atoms to the muon. The reason for writing $\mathbf{B}_{\mathrm{RKKY}}$ with the coefficients $\mu_0/(4\pi)$ and $-g_i\mu_{\mathrm{B}}/v_c$ explicitly expressed will appear when discussing the dipole field in Section 5.2.1. The constant $H_{\mathbf{r}_i,\,\mathrm{RKKY}}$ depends on the electronic density of states at the Fermi level, the volume enclosed by the Fermi surface, the exchange interaction between $\mathbf{J}_i$ and the conduction electron spins, and the electronic density at the muon site. Therefore it contains valuable information. However, this constant also depends on $\eta(\mathbf{r}_0)$, which reflects the muon-induced change of the local electronic structure. So far it has never been estimated. However, this should be possible, at least in some cases, as the measured hyperfine field at the muon site is fairly well understood for Fe, Ni, and Co using a finite cluster method (Ellis and Lindgren 1984).

An additional source for $\mathbf{B}_{\mathrm{loc}}$ in covalent compounds stems from the possible overlap between the muon wavefunction and electronic wavefunctions. This leads to a so-called transferred hyperfine field which, to a first approximation, has the functional form of $\mathbf{B}_{\mathrm{RKKY}}$.

In summary, the magnetic field operator at the muon site for compounds with d or f unpaired electrons can be written as a sum of four terms:

$$\mathbf{B}_{\mathrm{loc}} = \mathbf{B}_{\mathrm{ext}} + \mathbf{B}_{\mathrm{dip}} + \mathbf{B}_{\mathrm{hyp}} + \mathbf{B}_{\mathrm{Fermi}}, \tag{5.17}$$

where

$$\mathbf{B}_{\mathrm{hyp}} = \frac{\mu_0}{4\pi} \sum_{i=1}^{N'} \frac{(-g_i \mu_{\mathrm{B}}) H_{\mathbf{r}_i}}{v_c} \mathbf{J}_i. \tag{5.18}$$

The hyperfine field $\mathbf{B}_{\mathrm{hyp}}$ accounts for the previously discussed two short-range magnetic interactions between the muon and the localised magnetic moments located nearby the muon site. We would expect the hyperfine constant $H_{\mathbf{r}_i}$ to be temperature independent. However, interpretation of nuclear magnetic resonance (NMR) data suggests the constant to be temperature dependent for some intermetallic compounds; see e.g. Clark *et al.* (1987) and Ohama *et al.* (1995). The same type of interpretation leads to a similar conclusion for the μSR case. The constant might even be orientation dependent. We refer to the work of Schenck *et al.* (2003) and references therein. We point out that the possible temperature and orientation dependences of the hyperfine constant is a unsettled subject and future development may reveal some surprises. According to our simple analysis, the Fermi contact term should not appear in Eq. 5.17: the hyperfine term should be sufficient. However, the interpretation of frequency shift data for the paramagnetic state of some compounds requires the Fermi term. This discussion is further pursued in Section 5.3.

In contrast to the situation in NMR, there is no interaction between the nuclear probes, because of the extreme dilution of the muons.

As pointed out at the beginning of this section, the muon, being an elementary particle of spin $\frac{1}{2}$, can only couple to a magnetic field. In quantum mechanics this field is represented by a vector operator, i.e. an axial tensor operator of rank one. Let us denote this operator as $\mathbf{V}^1$ with components $\{V^1_{-1}, V^1_0, V^1_1\}$. According to the coupling rule between spherical tensor operators (see, e.g. Messiah 1965),

$$V^K_Q = \sum \langle k_1\, k_2\, q_1\, q_2 | K\, Q \rangle \, W^{k_1}_{q_1}\, U^{k_2}_{q_2}, \tag{5.19}$$

with $K = 1$. This expression describes the contribution of the electronic magnetic density of an atom to $\mathbf{B}_{\rm loc}$. The measured field is the sum of such terms, one for each atom surrounding the muon. In this expression the Clebsch–Gordan coefficient $\langle k_1\, k_2\, q_1\, q_2 | K\, Q \rangle$ couples two spherical tensors, $\boldsymbol{W}^{k_1}$ and $\boldsymbol{U}^{k_2}$, of rank k_1 and k_2, to form a tensor of rank K and projection Q. This coefficient is zero unless the triangular condition $|k_1 - k_2| \leq K \leq k_1 + k_2$ is fulfilled and $Q = q_1 + q_2$. For example, if $\boldsymbol{U}^{k_2}$ represents a dipole moment, i.e. $k_2 = 1$, and therefore k_1 takes one of the values $\{0,1,2\}$. The case $k_1 = 0$ models the hyperfine interaction and $k_1 = 2$ accounts for the classical dipolar interaction. These results are entirely consistent with those already obtained in this chapter, where $\boldsymbol{U}^{k_2}$ with $k_2 = 1$ stands for the dipole moment of the localised unpaired electrons of an atom. However, Eq. 5.19 also predicts the coupling of $\boldsymbol{W}^{k_1}$ with $k_1 = 1$ to $\boldsymbol{U}^{k_2}$ with $k_2 = 1$ (therefore two vector operators) to contribute to $\mathbf{B}_{\rm loc}$. This coupling appears in Eq. 5.5 through the orbital angular momentum. Practically, this is relevant if the symmetry of the electronic cloud around the muon is lower than spherical as explained before. Magnetic multipoles of the magnetic density of rank higher than the dipole's one may contribute to the field at the muon site, i.e. k_2 can be larger than 1. A particularly favourable situation for the detection of such a field arises if the field of dipole origin is negligible. An example is discussed in Section 10.4.

The contribution of the multipole development of the electronic shell of an atom to $\mathbf{B}_{\rm loc}$ can be analysed based on Eq. 5.19 which only derives from symmetry arguments. However, an analytical formula is needed for explicit computation. It was derived a long time ago by Marshall (1963); see also Santini and Amoretti (2000) for formulae for the spin and orbital contributions of an octupole magnetic moment using the Stevens operators.

5.2 The long-range dipole field at the muon site

Here we first point out some basic properties of the dipole field. Then we introduce the important notions of the secular and non-secular terms in a Hamiltonian. Finally, we explain how to compute this long-range dipole field using two distinct methods.

5.2.1 Some of its basic properties

The dipole field can be written in different equivalent forms. For electronic magnetic moments, each is characterised by a magnetic moment $\mathbf{m}_i = -g_i\mu_\mathrm{B}\mathbf{J}_i$, where $\mathbf{J}_i$ is the total angular momentum and g_i the spectroscopic splitting factor of atom at site i (see Appendix H), we write

$$\mathbf{B}_\mathrm{dip} = \frac{\mu_0}{4\pi}\sum_{i=1}^{N}(-g_i\mu_\mathrm{B})\left[-\frac{\mathbf{J}_i}{r_i^3} + \frac{3\,(\mathbf{J}_i\cdot\mathbf{r}_i)\,\mathbf{r}_i}{r_i^5}\right]. \tag{5.20}$$

The vector radius $\mathbf{r}_i$ links the muon to atom i. The sum runs over the N magnetic ions in the sample under investigation. It is convenient to express the components of the dipole field in terms of a tensor as

$$B^\alpha_\mathrm{dip} = \frac{\mu_0}{4\pi}\sum_{i=1}^{N}\frac{-g_i\mu_\mathrm{B}}{v_c}\sum_\beta D^{\alpha\beta}_{\mathbf{r}_i}J^\beta_i, \tag{5.21}$$

where $\boldsymbol{D}_{\mathbf{r}_i}$ is the dipole field tensor with elements

$$D^{\alpha\beta}_{\mathbf{r}_i} = v_c\left(-\frac{\delta_{\alpha,\beta}}{r_i^3} + \frac{3r_i^\alpha r_i^\beta}{r_i^5}\right). \tag{5.22}$$

The volume v_c has been introduced to define a unitless dipole tensor. We do not need to specify this volume here; this will be done in Section 5.2.4. It is easily seen that $D^{\alpha\beta}_{\mathbf{r}_i} = D^{\beta\alpha}_{\mathbf{r}_i}$. In addition, since $\sum_\alpha (r_i^\alpha)^2 = r_i^2$, $\sum_\alpha D^{\alpha\alpha}_{\mathbf{r}_i} = 0$, i.e. $D_{\mathbf{r}_i}$ is traceless. Therefore there are up to five independent $D^{\alpha\beta}_{\mathbf{r}_i}$ tensor components. It is sometimes useful to express the components of $\boldsymbol{D}_{\mathbf{r}_i}$ in terms of the spherical coordinates r_i, θ_i, φ_i of vector $\mathbf{r}_i$, where θ_i and φ_i denote its polar and azimuthal angles. These components read

$$\frac{v_c}{r_i^3}\begin{pmatrix} 3\sin^2\theta_i\cos^2\varphi_i - 1 & 3\sin^2\theta_i\cos\varphi_i\sin\varphi_i & 3\cos\theta_i\sin\theta_i\cos\varphi_i \\ 3\sin^2\theta_i\cos\varphi_i\sin\varphi_i & 3\sin^2\theta_i\sin^2\varphi_i - 1 & 3\cos\theta_i\sin\theta_i\sin\varphi_i \\ 3\cos\theta_i\sin\theta_i\cos\varphi_i & 3\cos\theta_i\sin\theta_i\sin\varphi_i & 3\cos^2\theta - 1 \end{pmatrix}. \tag{5.23}$$

Until now, our discussion has focused on the dipole interaction between the muon and the electronic magnetic moments. While much weaker, the dipole interaction between the muon and nuclear magnetic moments can be of importance, even in magnets since the electronic magnetism can be motionally narrowed up to the point that the measured μSR signal is dominated by the nuclear field. The nuclear dipole interaction plays an important role in the analysis of the position and diffusion of diamagnetic muon states in diamagnetic compounds (see Chapter 9). Relative to the dipole field as written in Eq. 5.20, we then have to replace the total angular momenta by the nuclear spins. In addition, it is convenient to use gyromagnetic ratios instead of spectroscopic splitting factors. The relation between these parameters is given by the formula

$$\gamma_i\hbar = -g_i\mu_\mathrm{B}. \tag{5.24}$$

The nuclear dipole field is thus written as

$$\mathbf{B}_{\rm dip} = \frac{\mu_0}{4\pi} \sum_{i=1}^{N} \gamma_i \, \hbar \left[-\frac{\mathbf{I}_i}{r_i^3} + \frac{3\,(\mathbf{I}_i \cdot \mathbf{r}_i)\,\mathbf{r}_i}{r_i^5} \right], \tag{5.25}$$

where $\mathbf{I}_i$ is the spin of nucleus i. For simplicity and because of the similar mathematical structure, we do not distinguish in notation the dipole field from the electronic and nuclear moments.

The dipole field is complex because of its anisotropy and long-range character. Since, in contrast to the usual situation in NMR, it may dramatically influence the measured signal, it is studied in some detail in this chapter. We shall first describe some of its basic properties. They are in fact independent of the type of spins. It is only when considering the energy involved that the difference between the nuclear and electronic spins matters. For definitiveness, we shall deal with nuclear spins.

It is particularly interesting to express the Hamiltonian $\mathcal{H}_{\rm dip}$ associated to the dipole field in terms of the raising and lowering spin operators (see Sections C.3 and C.5). Following Abragam (1961), it is customary to write

$$\mathcal{H}_{\rm dip} = \sum_{i=1}^{N} \alpha_i \left(A_i + B_i + C_i + D_i + E_i + F_i \right), \tag{5.26}$$

where

$$\alpha_i = \frac{\mu_0}{4\pi} \frac{\gamma_\mu \gamma_i \, \hbar^2}{2 r_i^3}, \tag{5.27}$$

and

$$\begin{aligned}
A_i &= \sigma^Z I_i^Z \left(1 - 3\cos^2\theta_i\right), \\
B_i &= -\frac{1}{4}\left(\sigma^+ I_i^- + \sigma^- I_i^+\right)\left(1 - 3\cos^2\theta_i\right), \\
C_i &= -\frac{3}{2}\left(\sigma^+ I_i^Z + \sigma^Z I_i^+\right)\sin\theta_i \cos\theta_i \exp\left(-i\varphi_i\right), \\
D_i &= -\frac{3}{2}\left(\sigma^- I_i^Z + \sigma^Z I_i^-\right)\sin\theta_i \cos\theta_i \exp\left(i\varphi_i\right), \\
E_i &= -\frac{3}{4}\sigma^+ I_i^+ \sin^2\theta_i \exp\left(-2i\varphi_i\right), \\
F_i &= -\frac{3}{4}\sigma^- I_i^- \sin^2\theta_i \exp\left(2i\varphi_i\right).
\end{aligned} \tag{5.28}$$

The parameter α_i fixes the scale of $\mathcal{H}_{\rm dip}$. For a proton located at 2 Å from the muon, α_i is of order 10^{-10} eV.

The dipole Hamiltonian can be written as

$$\mathcal{H}_{\rm dip} = -\frac{1}{2}\gamma_\mu \hbar \boldsymbol{\sigma} \cdot \mathbf{B}_{\rm dip} = -\frac{1}{2}\gamma_\mu \hbar \left(\sigma^Z B_{\rm dip}^Z + \frac{1}{2}\sigma^+ B_{\rm dip}^- + \frac{1}{2}\sigma^- B_{\rm dip}^+ \right), \tag{5.29}$$

with

$$
\begin{aligned}
B^Z_{\rm dip} &= \frac{\mu_0}{4\pi}\sum_{i=1}^{N}\frac{\gamma_i\hbar}{r_i^3}\Big[I_i^Z\left(3\cos^2\theta_i - 1\right) + \frac{3}{2}I_i^+\sin\theta_i\cos\theta_i\exp\left(-i\varphi_i\right) \\
&\qquad + \frac{3}{2}I_i^-\sin\theta_i\cos\theta_i\exp\left(i\varphi_i\right)\Big], \\
B^-_{\rm dip} &= \frac{\mu_0}{4\pi}\sum_{i=1}^{N}\frac{\gamma_i\hbar}{r_i^3}\Big[-\frac{1}{2}I_i^-\left(3\cos^2\theta_i - 1\right) + 3I_i^Z\sin\theta_i\cos\theta_i\exp\left(-i\varphi_i\right) \\
&\qquad + \frac{3}{2}I_i^+\sin^2\theta_i\exp\left(-2i\varphi_i\right)\Big], \\
B^+_{\rm dip} &= \frac{\mu_0}{4\pi}\sum_{i=1}^{N}\frac{\gamma_i\hbar}{r_i^3}\Big[-\frac{1}{2}I_i^+\left(3\cos^2\theta_i - 1\right) + 3I_i^Z\sin\theta_i\cos\theta_i\exp\left(i\varphi_i\right) \\
&\qquad + \frac{3}{2}I_i^-\sin^2\theta_i\exp\left(2i\varphi_i\right)\Big].
\end{aligned}
\tag{5.30}
$$

5.2.2 Secular versus non-secular terms

This section introduces the very important notions of secular and non-secular terms of a Hamiltonian. As a model system, we shall discuss the particular case of a transverse-field measurement in a nuclear spin system. We assume the existence of a relatively strong transverse field $\mathbf{B}_{\rm ext}$, which is parallel to Z as usual. Its magnitude is assumed to be sufficient for the components perpendicular to Z of the total field acting on the muon spin to be neglected (see for instance Section 6.2.1). Therefore, for the dipole field we only need to consider $B^Z_{{\rm dip},i}$ for each nucleus. According to Eq. 5.30,

$$
B^Z_{{\rm dip},i} = \frac{\mu_0}{4\pi}\frac{\gamma_i\,\hbar}{r_i^3}\left[I_i^Z\left(3\cos^2\theta_i - 1\right) + 3I_i^X\sin\theta_i\cos\theta_i\right], \tag{5.31}
$$

if the X axis is chosen in the plane containing $\mathbf{r}_i$ and $\mathbf{B}_{\rm ext}$.

In general terms, we are interested in matrix elements of $B^Z_{{\rm dip},i}$. They include terms such as $\langle k|I_i^\alpha(t)|k'\rangle$ where $|k\rangle$ denotes an eigenvector of the Hamiltonian for spin $\mathbf{I}_i$. We shall write E_k for the associated eigenvalue. This Hamiltonian will include the usual Zeeman term and a perturbation. If we write $|k\rangle = \sum_m c_m\,|m\rangle$ and $|k'\rangle = \sum_m c'_m\,|m\rangle$, where the $|m\rangle$ kets are the eigenvectors of the Zeeman Hamiltonian for spin $\mathbf{I}_i$, we obtain

$$
\langle k|I_i^\alpha(t)|k'\rangle = \exp\left[\frac{i}{\hbar}\left(E_k - E_{k'}\right)t\right]\sum_{m,m'} c^*_m c'_{m'}\,\langle m|I_i^\alpha|m'\rangle\,. \tag{5.32}
$$

The quantity $\langle k|I_i^\alpha(t)|k'\rangle$ is therefore expected to oscillate. Let us assume the nucleus to be a proton and $B_{\rm ext} = 0.1\,$T. Then $\omega_{\rm p}t = \gamma_{\rm p}B_{\rm exp} = 27$ if $t = 1\,\mu$s. Therefore we expect any oscillation to average out to 0 in the microsecond time range if $B_{\rm ext}$ is at least a few 10 mT. This leads to the approximation $\langle k|B^Z_{{\rm dip},i}|k'\rangle = \langle k|B^Z_{{\rm dip},i}|k\rangle\delta_{k,k'}$, where the non-diagonal terms are neglected. This neglect is called the secular approximation. Therefore,

$$\langle k|B^Z_{\mathrm{dip},i}|k'\rangle = \frac{\mu_0}{4\pi}\frac{\gamma_i \hbar}{r_i^3}\delta_{k,k'}\left[\langle k|I_i^Z|k\rangle\left(3\cos^2\theta_i - 1\right) + 3\langle k|I_i^X|k\rangle \sin\theta_i\cos\theta_i\right]. \tag{5.33}$$

We recall that this expression is correct if the reference frame is adapted for each nucleus such that $\varphi_i = 0$. In the case where the spins are subjected only to the Zeeman interaction, the set of vectors $|k\rangle$ is simply the $|m\rangle$ set: this implies that the second term on the right-hand-side sum of Eq. 5.33 vanishes (see Section C.3) and only the matrix element of I_i^Z is of importance. On the other hand, if a perturbation such as an electric field gradient is acting on the nuclei (and if $I_i > \frac{1}{2}$), we have to keep the last term in Eq. 5.33.

5.2.3 The summation of the field in direct space

The dipole field is expressed as a sum of contributions from magnetic moments over the whole crystal. Although the field arising from a moment at a distance r decreases as r^{-3}, the number of moments at that distance increases with r^2. Therefore the series is expected to be conditionally convergent. At first sight, a reliable evaluation of the dipole field seems hopeless. However, as first pointed out by Lorentz, for the crystal site of interest (in our case the muon site) a continuous rather than a discrete sum is appropriate as soon as the distance between the muon and the magnetic moments is sufficiently large.

In this section we evaluate the electronic dipole field $\mathbf{B}_{\mathrm{dip}}$. The results are directly applicable for polarised paramagnets. They also serve to describe the magnetic field in ordered magnets. Section 5.4 will show that some care must nevertheless be exercised.

Here the field is a classical vector rather than a vector operator. We compute the average value of $\mathbf{B}_{\mathrm{dip}}$ and neglect the fluctuation $\delta\mathbf{B}_{\mathrm{dip}}$.

Following Lorentz — see any textbook on elementary magnetostatics or Uritskiĭ and Irkhin (2000) — we write $\mathbf{B}_{\mathrm{dip}}$ as the sum of three fields:

$$\mathbf{B}_{\mathrm{dip}} = \mathbf{B}'_{\mathrm{dip}} + \mathbf{B}_{\mathrm{Lor}} + \mathbf{B}_{\mathrm{dem}}. \tag{5.34}$$

The term $\mathbf{B}'_{\mathrm{dip}}$ is the dipole field obtained by restricting the sum in Eq. 5.20 to the moments inside a sphere, the Lorentz sphere, centred at the muon site and with a radius sufficiently large that the lattice sum convergence is reached. Typically a very good accuracy is obtained when a few thousand moments are considered. The symmetry at the muon site may be of great help for the evaluation of the discrete sum as discussed in Appendix D.7.

The remaining contribution to $\mathbf{B}_{\mathrm{dip}}$ arises from the dipoles located in the sample but outside the Lorentz sphere (see Fig. 5.1). Since these moments are located relatively far away from the muon, the corresponding field can be computed in the continuum approximation. This evaluation is conveniently performed introducing magnetic charges by analogy with the electric charges as the source of the electrostatic field. Two terms, the Lorentz field $\mathbf{B}_{\mathrm{Lor}}$ and the demagnetisation field $\mathbf{B}_{\mathrm{dem}}$ which

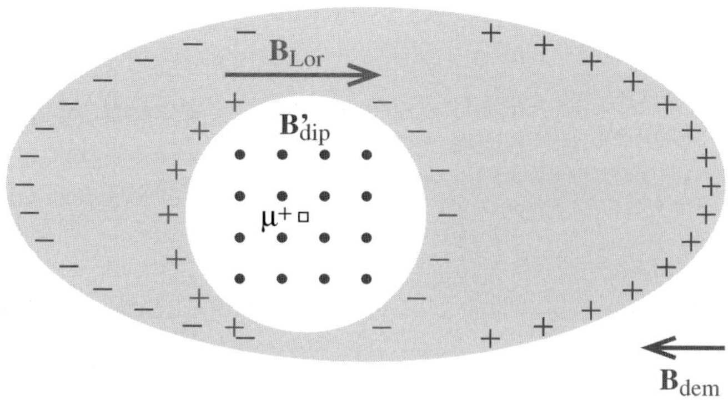

Fig. 5.1 Schematic representation of the three contributions to $\mathbf{B}_{\text{dip}}$ for a magnet at saturation: the field at the muon site (□) obtained from a discrete sum performed over the magnetic moments (•) within the Lorentz sphere and denoted as $\mathbf{B}'_{\text{dip}}$, and the Lorentz and demagnetisation fields, $\mathbf{B}_{\text{Lor}}$ and $\mathbf{B}_{\text{dem}}$ respectively. The direction and magnitude of $\mathbf{B}'_{\text{dip}}$ crucially depend on the muon site and on the arrangement of the magnetic moments. The two latter macroscopic fields can be understood as arising from the magnetic charges at the surface of the Lorentz sphere and of the sample, respectively. The drawing is done for an ellipsoidal sample and the external field applied along one of its principal axes.

result from the charges located at the surface of the Lorentz sphere and of the sample, respectively, are obtained. The former one reads

$$\mathbf{B}_{\text{Lor}} = \frac{\mu_0}{3}\mathbf{M}_{\text{Lor}}, \tag{5.35}$$

where $\mathbf{M}_{\text{Lor}}$ is the vector sum of the magnetic moments inside the Lorentz sphere divided by its volume.

In general cases a simple analytical expression for $\mathbf{B}_{\text{dem}}$ cannot be given. It is only for an ellipsoidal sample that the demagnetisation field becomes uniform and a demagnetisation tensor can be defined (Chikazumi 1964, Morrish 1965):

$$\mathbf{B}_{\text{dem}} = -\mu_0 \boldsymbol{N} \mathbf{M}_{\text{meas}}. \tag{5.36}$$

The magnetisation $\mathbf{M}_{\text{meas}}$ is the bulk magnetisation of the sample. In the case of a magnetically saturated sample it is equal to $\mathbf{M}_{\text{Lor}}$. However, $\mathbf{M}_{\text{meas}}$ might be much smaller than $\mathbf{M}_{\text{Lor}}$ if $\mathbf{B}_{\text{ext}}$ is small because, for example, of a large distribution in orientation of the domains magnetisation. The demagnetisation tensor $\boldsymbol{N}$ is diagonal in the principal axes of the ellipsoid which we denote by a, b, and c. Its trace is equal to one, i.e. $N^a + N^b + N^c = 1$ (for simplicity we write N^α instead of $N^{\alpha\alpha}$) and the elements N^α are positive. Therefore the components of $\mathbf{B}_{\text{dem}}$ and $\mathbf{M}_{\text{meas}}$ along the principal axes are of opposite sign. This justifies the name given to this field. For a sphere, since the three demagnetisation coefficients must be equal by symmetry, $N^a = N^b = N^c = \frac{1}{3}$.

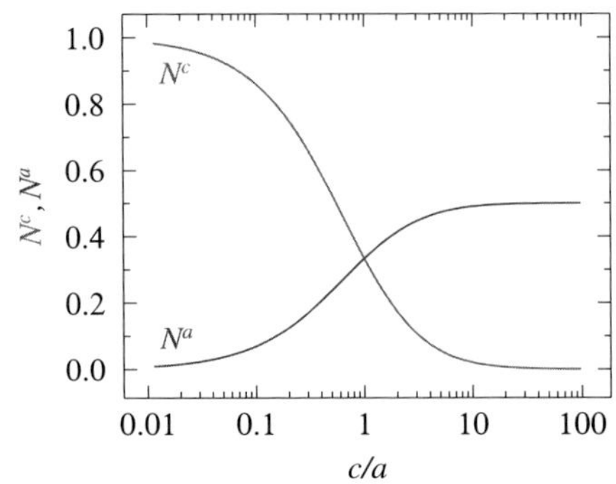

Fig. 5.2 The axial N^c and transverse $N^a = N^b$ demagnetisation coefficients for an ellipsoid of revolution as a function of the axial ratio $\gamma = c/a$. Asymptotically, $N^c \simeq [\ln(2\gamma) - 1]/\gamma^2$ for $\gamma \gg 1$ and $N^c \simeq 1 - 2^{1/2}\gamma$ for $\gamma \ll 1$.

Let us consider an ellipsoid of revolution, the axis of revolution being along the c axis, i.e. $a = b$. Denoting $\gamma = c/a$, we have (Chikazumi 1964, Zijlstra 1967)

$$N^c = \frac{1}{\gamma^2 - 1}\left[\frac{\gamma}{\sqrt{\gamma^2 - 1}}\operatorname{arccosh}\gamma - 1\right] \quad \text{for} \quad \gamma > 1,$$

$$N^c = \frac{1}{1 - \gamma^2}\left[1 - \frac{\gamma}{\sqrt{1 - \gamma^2}}\arccos\gamma\right] \quad \text{for} \quad \gamma < 1. \tag{5.37}$$

Obviously, $N^a = (1 - N^c)/2$. The coefficients N^c and N^a are plotted in Fig. 5.2 against γ. If $\gamma \simeq 1$ the sample shape is almost a sphere and as seen in this figure, the values of the coefficients are about equal and close to $\frac{1}{3}$ as expected. For a specimen whose length along the revolution axis is large relative to its transverse dimensions, N^c is very small. This means that if the magnetisation vector is parallel to the c axis, the demagnetisation field is negligible. Conversely, in a thin ellipsoid of axis of revolution c, N^c tends to 1, i.e. $N^a = N^b \simeq 0$. In that case, even if the magnetisation perpendicular to c is large, the component of the demagnetisation field in that direction is small.

Following Kittel (1996), for easy reference we list in Table 5.1 the value of the demagnetisation coefficients for simple sample shapes. It is a usual approximation to consider cylinders with a large length compared to their radius and very flat disks as prolate and oblate ellipsoids, respectively. Nevertheless, it must be realised that the field is approximately homogeneous in these volumes only far from the borders.

It is worthwhile to notice that if the sample is a sphere, and if $\mathbf{M}_{\rm Lor} = \mathbf{M}_{\rm meas}$, which is the case for instance for a paramagnet or a saturated magnet, $\mathbf{B}_{\rm dem} = -\mathbf{B}_{\rm Lor}$ and therefore $\mathbf{B}_{\rm dip} = \mathbf{B}'_{\rm dip}$.

When the sample is not ellipsoidal, the demagnetisation field is not homogeneous. This leads to a field distribution. In a transverse field μSR experiment, it manifests itself through an inhomogeneous broadening, i.e. a damping of the oscillations which reflects this field distribution. In a first approximation it is well approximated by a Gaussian function, the width of which is proportional to the bulk susceptibility and

Table 5.1 The demagnetisation coefficient for simple sample shapes. The given coefficient concerns the direction of the magnetisation.

Shape	Magnetisation direction	N^a, N^c
Sphere	Any	$\frac{1}{3}$
Thin slab	Normal	1
Thin slab	In plane	0
Long circular cylinder	Longitudinal	0
Long circular cylinder	Transverse	$\frac{1}{2}$

to $B_{\rm ext}$. For a given value of $B_{\rm ext}$, the damping rate contribution associated with the sample shape is therefore expected to be stronger as a sample is cooled down towards its magnetic critical temperature, as illustrated by the work of Hartmann *et al.* (1986). It is justified to multiply the aforementioned Gaussian function by the damping function intrinsic to the compound under study, provided the origins for both damping effects are independent. This is true for instance when the intrinsic damping is due to magnetic fluctuations.

5.2.4 The Fourier summation technique

In the previous section we have exposed a method for summing the dipole field, focusing on the Lorentz and demagnetisation fields. In this section we present a second method for the same task. It is much more complicated. However, it offers more opportunities, notably the possibility of computing the wave vector dependence of the dipole field. This dependence, as explained in Chapter 10, is usually required when, for example, a quantitative interpretation of the spin–lattice relaxation rate data recorded in magnetic materials is attempted.

One of the key properties of most solid state systems is the periodic array formed by the atoms. Our understanding of the physical properties of solids owes much to this periodicity which allows the use of the reciprocal lattice concept and the related Brillouin zone.

As a matter of fact, the collective excitations in solids (phonon, magnons, ...) and the conduction electrons are characterised by their energy and wave vector in the Brillouin zone. Therefore, it seems important to have the possibility of describing the μSR data recorded for magnetic compounds in terms of their excitations. This means that we must describe the data using the Brillouin zone concept; i.e. the lattice Fourier transform of physical quantities related to the measured parameters by μSR should be considered. As a first step, in this section we express the operator $\mathbf{B}_{\rm loc}$ in terms of its wave vector components. Some translational symmetry properties of periodic crystals and the lattice Fourier transform method are recalled in Appendix E.

One of the main conclusions reached in Section 5.1.3 for compounds with d or f unpaired electrons is that $\mathbf{B}_{\rm loc}$ is the sum of at least the long-range dipole and the short-range hyperfine fields. Although this latter field is easily treated in the direct

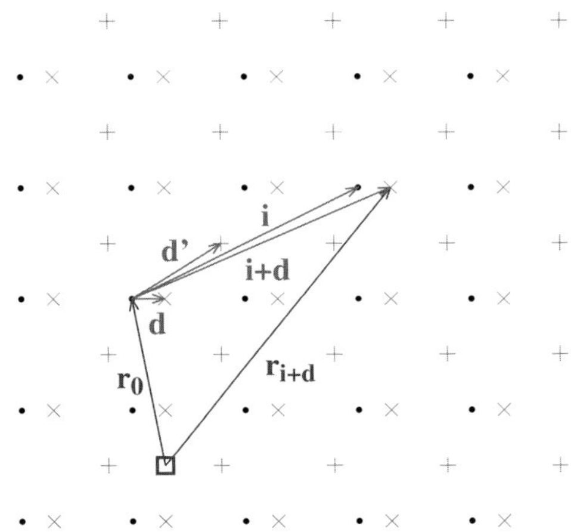

Fig. 5.3 Definition of vectors relative to the crystallographic (+, ×) and muon (□) sites. The points (•) specify the origin of each cell. The point at the extremity of $\mathbf{r}_0$ is the origin of the periodic crystal lattice. The drawing is done for a two-dimensional square lattice with two magnetic atoms per cell. The positions of these atoms in the unit cell are defined by vectors $\mathbf{d}$ and $\mathbf{d}'$. Two families of vectors are present in the problem: the vectors, as for example $\mathbf{i} + \mathbf{d}$, which define the atom sites in the crystal lattice, and the vectors linking the muon site to these atom sites, $\mathbf{r}_{i+d}$, for instance. Reproduced with kind permission from Dalmas de Réotier and Yaouanc (1997).

space, here we shall assimilate it to the dipole field and therefore compute the Fourier components of their sum.

The component $B^{\alpha}_{\rm loc}$ is conveniently expressed in terms of a tensor denoted as $\boldsymbol{G}_{\mathbf{r}_{i+d}}$:

$$B^{\alpha}_{\rm loc} = -\frac{\mu_0}{4\pi}\frac{g\mu_{\rm B}}{v_c}\sum_{\beta=X,Y,Z}\sum_i\sum_d G^{\alpha\beta}_{\mathbf{r}_{i+d}} J^{\beta}_{i+d}. \tag{5.38}$$

Accounting for the periodical arrangement of the magnetic atoms, we have replaced the sum over i as in Section 5.2.1 by a sum over the unit cells of the crystalline structure and a sum over the positions d of the magnetic atoms in the unit cell. Index i denotes now the unit cell of position $\mathbf{i}$ in the lattice. In the same spirit $\mathbf{d}$ is the vector characterising position d in the unit cell. The different relevant vectors are defined in Fig. 5.3. For simplicity, from now on we assume the localised electronic moments to be characterised by identical angular momenta and spectroscopic splitting factors. This assumption can be lifted at the expense of a more complicated notation. Volume v_c is the volume of the unit cell which contains n_d magnetic atoms; i.e. $v_c = n_d v_0$, where v_0 is the volume per magnetic ion. We denote by n_c the number of unit cells in the sample; this latter contains therefore $N = n_c n_d$ magnetic ions and its volume is $V = N v_0$.

The tensor components $G^{\alpha\beta}_{\mathbf{r}_{i+d}}$ defined in Eq. 5.38 are expressed as sums of two terms, the dipole and hyperfine contributions:

$$G^{\alpha\beta}_{\mathbf{r}_{i+d}} = D^{\alpha\beta}_{\mathbf{r}_{i+d}} + H^{\alpha\beta}_{\mathbf{r}_{i+d}}, \tag{5.39}$$

where

$$D^{\alpha\beta}_{\mathbf{r}_{i+d}} = v_c \left. \frac{\partial^2}{\partial r_\alpha \partial r_\beta} \frac{1}{|\mathbf{r}|} \right|_{\mathbf{r}=\mathbf{r}_{i+d}}; \tag{5.40}$$

see Eq. 5.22.

It is useful to work with the coupling tensor written in Fourier space. We set

$$G^{\alpha\beta}_{d,\mathbf{q}} = \sum_i G^{\alpha\beta}_{\mathbf{r}_{i+d}} \exp(i\mathbf{q}\cdot\mathbf{r}_{i+d}), \tag{5.41}$$

where $\mathbf{q}$ is a wave vector in the Brillouin zone. From Eq. E.6 we then derive

$$G^{\alpha\beta}_{\mathbf{r}_{i+d}} = \frac{1}{n_c} \sum_{\mathbf{q}} G^{\alpha\beta}_{d,\mathbf{q}} \exp(-i\mathbf{q}\cdot\mathbf{r}_{i+d}). \tag{5.42}$$

Note that $G^{\alpha\beta}_{d,\mathbf{q}}$ is not the space Fourier transform of $G^{\alpha\beta}_{\mathbf{r}_{i+d}}$ since the arguments in the exponential functions are proportional to $\mathbf{r}_{i+d}$ ($\mathbf{r}_{i+d} = \mathbf{r}_0 + \mathbf{i} + \mathbf{d}$) instead of $\mathbf{i}+\mathbf{d}$ and also because of the sign of $\mathbf{q}$; see e.g. Eq. E.9. The density of points in the Brillouin zone is extremely large. It is thus practical for the evaluation of sums over $\mathbf{q}$ or for the study of their analyticity to consider $\mathbf{q}$ as a continuous variable. Then, for instance, instead of writing $G^{\alpha\beta}_{d,\mathbf{q}}$, we shall use the notation $G^{\alpha\beta}_d(\mathbf{q})$.

For convenience we introduce the average $G^{\alpha\beta}_{\mathbf{q}} = (1/n_d)\sum_d G^{\alpha\beta}_{d,\mathbf{q}}$. It follows that $G^{\alpha\beta}_{\mathbf{q}} = D^{\alpha\beta}_{\mathbf{q}} + H^{\alpha\beta}_{\mathbf{q}}$ with obvious notations.

Because the hyperfine interaction is short-ranged, the $\mathbf{q}$-representation of the related tensor is easily obtained. Within the continuous notation for $\mathbf{q}$ and assuming the muon to sit halfway between two nearest neighbours,

$$H^{\alpha\beta}(\mathbf{q}) = \frac{1}{n_d}\sum_d H^{\alpha\beta}_d(\mathbf{q}) = 2H\delta^{\alpha\beta}\cos\left(\frac{\mathbf{q}\cdot\mathbf{d}_{\rm pair}}{2}\right). \tag{5.43}$$

Here, $\mathbf{d}_{\rm pair}$ is the vector which links the two atoms of the pair, and we have written $H^{\alpha\beta}_{\mathbf{r}_{i+d}} = n_d H\delta^{\alpha\beta}$. For simplicity, we have assumed that only the nearest neighbours to the muon are important for $\mathbf{B}_{\rm hyp}$. This approximation can be relaxed if necessary. In the same way more than two nearest neighbours to the muons can be considered. In the limit $\mathbf{q}\to 0$ we have

$$H^{\alpha\beta}(\mathbf{q}) = r_\mu H\delta^{\alpha\beta}, \tag{5.44}$$

where r_μ is the number of neighbour magnetic ions nearest to the muon localisation site. This equation includes a simple generalisation of Eq. 5.43: $r_\mu = 2$ in that equation.

The calculation of $D^{\alpha\beta}_d(\mathbf{q})$ is more involved, basically because the dipolar interaction is long range. We have

$$D^{\alpha\beta}_d(\mathbf{q}) = \sum_i D^{\alpha\beta}_{\mathbf{r}_{i+d}} \exp(i\mathbf{q}\cdot\mathbf{r}_{i+d})$$

$$= \exp\left(i\mathbf{q}\cdot\mathbf{r}_0\right)\, v_c \sum_i \left.\frac{\partial^2}{\partial r_\alpha \partial r_\beta}\frac{1}{|\mathbf{r}|}\right|_{\mathbf{r}=\mathbf{r}_{i+d}} \exp[i\mathbf{q}\cdot(\mathbf{i}+\mathbf{d})]. \tag{5.45}$$

We point out that

$$\left.\frac{\partial^2}{\partial r_\alpha \partial r_\beta}\frac{1}{|\mathbf{r}|}\right|_{\mathbf{r}=\mathbf{r}_{i+d}} = \left.\frac{\partial^2}{\partial r_\alpha \partial r_\beta}\frac{1}{|\mathbf{i}+\mathbf{d}-\mathbf{r}|}\right|_{\mathbf{r}=-\mathbf{r}_0}. \tag{5.46}$$

Therefore, as a first step towards the evaluation of Eq. 5.45, we study $\sum_{\mathbf{i}}(1/|\mathbf{i}+\mathbf{d}-\mathbf{r}|)\exp(i\mathbf{q}.\mathbf{i})$. Following Ewald — see, for example, Born and Huang (1954) — we note that

$$\frac{1}{|\mathbf{i}+\mathbf{d}-\mathbf{r}|} = \frac{2}{\sqrt{\pi}}\int_0^\infty \exp\left(-|\mathbf{i}+\mathbf{d}-\mathbf{r}|^2\rho^2\right) \mathrm{d}\rho\,, \tag{5.47}$$

and therefore

$$\begin{aligned}
&\sum_{\mathbf{i}} \frac{1}{|\mathbf{i}+\mathbf{d}-\mathbf{r}|}\exp(i\mathbf{q}\cdot\mathbf{i}) \\
&= \int_0^{\varrho}\left\{\frac{2}{\sqrt{\pi}}\sum_i \exp\left[-|\mathbf{i}+\mathbf{d}-\mathbf{r}|^2\rho^2 + i\mathbf{q}\cdot(\mathbf{i}+\mathbf{d}-\mathbf{r})\right]\right\}\exp[i\mathbf{q}\cdot(\mathbf{r}-\mathbf{d})]\,\mathrm{d}\rho \\
&\quad + \sum_{\mathbf{i}}\left\{\frac{2}{\sqrt{\pi}}\int_{\varrho}^{\infty} \exp\left(-|\mathbf{i}+\mathbf{d}-\mathbf{r}|^2\rho^2\right)\,\mathrm{d}\rho\right\}\exp(i\mathbf{q}\cdot\mathbf{i}).
\end{aligned} \tag{5.48}$$

We have used Eq. C.63 to obtain Eq. 5.47. Following again Ewald, we have divided the integrals into two regions: from 0 to ϱ and from ϱ to ∞. Now the expression in braces ({...}) in the first term is a function of $\mathbf{r}$ with the lattice periodicity. Hence, it can be expressed as a Fourier series over the $\mathbf{K}$ vectors of the reciprocal lattice. The Fourier weight factors can be written as (see Eq. E.5)

$$\frac{1}{v_c}\int_{v_c}\frac{2}{\sqrt{\pi}}\sum_i \exp\left[-|\mathbf{i}+\mathbf{d}-\mathbf{r}|^2\rho^2 + i\mathbf{q}\cdot(\mathbf{i}+\mathbf{d}-\mathbf{r})\right]\exp(-\mathbf{K}\cdot\mathbf{r})\mathrm{d}^3\mathbf{r}. \tag{5.49}$$

Noting that the lattice sum of the integral over the unit cell can be replaced by a single integral over the whole space for a specific lattice site, say $\mathbf{i}_0$, we derive using Eq. C.63

$$\begin{aligned}
\sum_{\mathbf{i}} \frac{1}{|\mathbf{i}+\mathbf{d}-\mathbf{r}|}\exp(i\mathbf{q}\cdot\mathbf{i}) &= \frac{4\pi}{v_c q^2}\exp\left(\frac{-q^2}{4\varrho^2}\right)\exp[i\mathbf{q}\cdot(\mathbf{r}-\mathbf{d})] \\
&\quad + \frac{\pi}{v_c\varrho^2}\sum_{\mathbf{K}\neq\mathbf{0}}\varphi_0\left[\frac{(\mathbf{K}+\mathbf{q})^2}{4\varrho^2}\right]\exp[i(\mathbf{K}+\mathbf{q})\cdot(\mathbf{r}-\mathbf{d})] \\
&\quad + \frac{\varrho}{\sqrt{\pi}}\sum_{\mathbf{i}}\varphi_{-1/2}\left(\varrho^2|\mathbf{i}+\mathbf{d}-\mathbf{r}|^2\right)\exp(i\mathbf{q}\cdot\mathbf{i}).
\end{aligned} \tag{5.50}$$

Here, we have used Misra's functions (Born and Huang 1954),

$$\varphi_m(x) = \int_1^\infty d\beta\, \beta^m \exp(-\beta x), \tag{5.51}$$

which have the useful transformation properties

$$\varphi_m(x) = \varphi_0(x) + (m/x)\varphi_{m-1}(x), \tag{5.52}$$

$$\frac{\partial}{\partial x}\varphi_m(x) = -\varphi_{m+1}(x). \tag{5.53}$$

The second and final step towards the derivation of $D_d^{\alpha\beta}(\mathbf{q})$ is made by taking the derivatives of Eq. 5.50 with respect to components of $\mathbf{r}$ (with the help of Eq. 5.53) and then setting $\mathbf{r} = -\mathbf{r}_0$. In addition Eq. 5.45 must be used. We then derive

$$\begin{aligned} D_d^{\alpha\beta}(\mathbf{q}) = & -4\pi \frac{q_\alpha q_\beta}{q^2} \exp\left(\frac{-q^2}{4\varrho^2}\right) \\ & - \frac{\pi}{\varrho^2} \sum_{\mathbf{K}\neq\mathbf{0}} (K_\alpha + q_\alpha)(K_\beta + q_\beta)\, \varphi_0\left(\frac{(\mathbf{q}+\mathbf{K})^2}{4\varrho^2}\right) \exp\left(-i\mathbf{K}\cdot\mathbf{r}_{0+d}\right) \\ & + \frac{2v_c\varrho^3}{\pi^{1/2}} \sum_i \left[2\varrho^2 r_{i+d,\alpha} r_{i+d,\beta} \varphi_{3/2}\left(\varrho^2 r_{i+d}^2\right) - \delta^{\alpha\beta}\varphi_{1/2}\left(\varrho^2 r_{i+d}^2\right)\right] \exp\left(i\mathbf{q}\cdot\mathbf{r}_{i+d}\right). \end{aligned} \tag{5.54}$$

This expression gives the same result for all values of the Ewald parameter ϱ, but for numerical applications a value of ϱ is chosen which ensures both series of the previous equation to converge rapidly.

When the interest is in the behaviour of $D_d^{\alpha\beta}(\mathbf{q})$ close the the zone centre, it is convenient to introduce $P_{\mathrm{L}}^{\alpha\beta}(\mathbf{q}) = q^\alpha q^\beta / q^2$, which is an element of the longitudinal projection operator (see Section C.1), to get

$$D_d^{\alpha\beta}(\mathbf{q}) = -4\pi\left[P_{\mathrm{L}}^{\alpha\beta}(\mathbf{q}) - C_d^{\alpha\beta}(\mathbf{q})\right], \tag{5.55}$$

with

$$\begin{aligned} C_d^{\alpha\beta}(\mathbf{q}) = & \frac{q_\alpha q_\beta}{q^2}\left[1 - \exp\left(\frac{-q^2}{4\varrho^2}\right)\right] \\ & - \frac{1}{4\varrho^2} \sum_{\mathbf{K}\neq\mathbf{0}} (K_\alpha + q_\alpha)(K_\beta + q_\beta)\, \varphi_0\left(\frac{(\mathbf{q}+\mathbf{K})^2}{4\varrho^2}\right) \exp\left(-i\mathbf{K}\cdot\mathbf{r}_{0+d}\right) \\ & + \frac{v_c\varrho^3}{2\pi^{3/2}} \sum_i \left[2\varrho^2 r_{i+d,\alpha} r_{i+d,\beta} \varphi_{3/2}\left(\varrho^2 r_{i+d}^2\right) - \delta^{\alpha\beta}\varphi_{1/2}\left(\varrho^2 r_{i+d}^2\right)\right] \exp\left(i\mathbf{q}\cdot\mathbf{r}_{i+d}\right). \end{aligned} \tag{5.56}$$

The limit of $D_d^{\alpha\beta}(\mathbf{q})$ at the centre of the Brillouin zone is given by

$$D_d^{\alpha\beta}(\mathbf{q}\to\mathbf{0}) = -4\pi\left[P_{\mathrm{L}}^{\alpha\beta}(\mathbf{q}) - C_d^{\alpha\beta}(\mathbf{q}=\mathbf{0})\right]. \tag{5.57}$$

The low $\mathbf{q}$ limit shows the benefit of introducing the longitudinal projector in Eq. 5.55. It allows the isolation of the non-analytical term of the dipole tensor in that limit:

whereas $\boldsymbol{P}_{\rm L}(\mathbf{q})$ is only piecewise continuous near $\mathbf{q} = \mathbf{0}$, $\boldsymbol{C}_d(\mathbf{q})$ is analytical for all values of $\mathbf{q}$.

From the definition of $D_d^{\alpha\alpha}(\mathbf{q} = \mathbf{0})$ (see Eq. 5.45) and from Eq. 5.22 or 5.23 it can be shown that the tensor $\boldsymbol{D}_d(\mathbf{q} = \mathbf{0})$ is traceless, i.e. $\sum_\alpha D_d^{\alpha\alpha}(\mathbf{q} = \mathbf{0}) = 0$. Consequently, the following relation holds:

$$\sum_\alpha C_d^{\alpha\alpha}(\mathbf{q} = \mathbf{0}) = 1. \tag{5.58}$$

Instead of performing the lattice sum for $D^{\alpha\beta}(\mathbf{q}) = (1/n_d) \sum_d D_d^{\alpha\beta}(\mathbf{q})$ we could have made a continuum approximation with the result:

$$D^{\alpha\beta}(\mathbf{q}) = -4\pi \left(\frac{q_\alpha q_\beta}{q^2} - \frac{1}{3}\delta^{\alpha\beta} \right). \tag{5.59}$$

This approximation is usually not justified because it predicts the $\boldsymbol{C}(\mathbf{q})$ tensor to be always scalar and independent of $\mathbf{q}$ (in the continuum approximation the index d is irrelevant).

Because $\boldsymbol{D}_d(\mathbf{q})$ given in Eq. 5.54 is analytical outside the $\mathbf{q} = \mathbf{0}$ limit, it can be developed in a Taylor series around any point of the Brillouin zone, except the zone centre.

For the rest of this section it is convenient to go back to the discrete description of the wave vector dependence. The field $\mathbf{B}_{\rm loc}$ can be expressed in terms of $\boldsymbol{G}_{d,\mathbf{q}}$ using Eqs. 5.38 and 5.41:

$$\mathbf{B}_{\rm loc} = \frac{\mu_0}{4\pi} \frac{1}{\sqrt{n_c}\, v_c} \sum_d \sum_{\mathbf{q}} \exp(-i\mathbf{q} \cdot \mathbf{r}_0)\, \boldsymbol{G}_{d,\mathbf{q}}\, \mathbf{m}_{d,\mathbf{q}}, \tag{5.60}$$

with

$$\mathbf{m}_{d,\mathbf{q}} = \frac{1}{\sqrt{n_c}} \sum_i \mathbf{m}_{i+d} \exp[-i\mathbf{q} \cdot (\mathbf{i} + \mathbf{d})]. \tag{5.61}$$

The field $\mathbf{B}_{\rm loc}$ is now written in terms of the Fourier components of the magnetic structure, $\mathbf{m}_{d,\mathbf{q}}$. In most cases, a magnetic structure is characterised by a single so-called propagation wave vector $\mathbf{k}$ and only terms $\mathbf{q} = \mathbf{k}$ and $-\mathbf{k}$ are present in the sum over $\mathbf{q}$. This method may seem complicated for a result which can be obtained by a simple lattice sum as shown in Section 5.2.3. However, it is more powerful: $\mathbf{B}_{\rm loc}$ can be computed for any magnetic structure specified by its wave vector components.

One may wonder how the expression of $\mathbf{B}_{\rm loc}$ derived from the Fourier method is related to the basic result given in Eq. 5.17. The hyperfine contribution does not pose any problem; see e.g. Eq. 5.43. This is also true for the Fermi field, see Eq. 5.15. The case of the dipolar field is more involved.

We first discuss a magnet for which all the moments have the same magnitude and are parallel to each other. Then, according to Eq. 5.61, the only non-vanishing $\mathbf{m}_{d,\mathbf{k}}$ term is $\mathbf{m}_{d,\mathbf{k}=\mathbf{0}}$. As already pointed out, $\boldsymbol{D}_d(\mathbf{q})$ splits into two parts. The first term on the right-hand-side of Eq. 5.55 yields the demagnetisation field. Therefore

$$\mathbf{B}'_{\rm dip} + \mathbf{B}_{\rm Lor} = \mu_0 \frac{1}{\sqrt{n_c}\, v_c} \sum_d \boldsymbol{C}_{d,\mathbf{k}=\mathbf{0}}\, \mathbf{m}_{d,\mathbf{k}=\mathbf{0}}. \tag{5.62}$$

In contrast to the real space lattice sum method, with the Fourier method the concept of the Lorentz sphere is not needed. Therefore, it is not essential to take the Lorentz volume as a sphere. It is just convenient for the evaluation of the Lorentz field.

For an antiferromagnet in zero field, the Lorentz field vanishes; see Section 5.4. The tensor $\boldsymbol{D}_{d,\mathbf{q}}$ does not present any singularities and therefore according to Eq. 5.54

$$\mathbf{B}_{\rm dip} = \frac{\mu_0}{4\pi} \frac{1}{\sqrt{n_c}\, v_c} \sum_d \exp(-i\mathbf{k}\cdot\mathbf{r}_0)\, \boldsymbol{D}_{d,\mathbf{k}}\, \mathbf{m}_{d,\mathbf{k}} + \text{c.c.} \tag{5.63}$$

For simplicity, we have assumed a single wave vector to be sufficient for the description of the magnet structure. The symbol "c.c." stands for the complex conjugate of the expression on its left.

5.3 Frequency shift in paramagnets or diamagnets

Let us consider a sample which is not magnetically ordered at the temperature of interest. It can be a magnetic compound in its paramagnetic state or a diamagnet. A magnetic field applied on the sample induces a magnetic polarisation, proportional to $B_{\rm ext}$ provided the field intensity is not too large. The response of the system is described by a susceptibility tensor. This discussion suggests that the field at the muon site, $\mathbf{B}_{\rm loc}$, is not equal to $\mathbf{B}_{\rm ext}$. As in NMR, a convenient method for quantifying the μSR magnetic response is to define the normalised scalar product

$$K_{\rm exp} = \frac{\mathbf{B}_{\rm ext}\cdot(\mathbf{B}_{\rm loc} - \mathbf{B}_{\rm ext})}{B_{\rm ext}^2}. \tag{5.64}$$

The parameter $K_{\rm exp}$ measures the muon frequency shift. The measurement is always performed in a field sufficiently large that $|\mathbf{B}_{\rm loc} - \mathbf{B}_{\rm ext}|$ is small relative to $B_{\rm ext}$. The shift $K_{\rm exp}$ characterises the magnetic response along the direction of $\mathbf{B}_{\rm ext}$, which is the Z axis by definition. It is, in fact, the local magnetic susceptibility at the muon site. It is usually in the few tens of parts per million range for a diamagnet and at most a few percent for a paramagnet.

As previously explained in this chapter, $\mathbf{B}_{\rm loc}$ is the sum of fields of different origins, including the two macroscopic fields, $\mathbf{B}_{\rm Lor}$ and $\mathbf{B}_{\rm dem}$. It is convenient to define the so-called muon Knight shift, which is independent of these two latter fields, by the relation

$$K_\mu = K_{\rm exp} - \frac{\mathbf{B}_{\rm ext}\cdot(\mathbf{B}_{\rm Lor} + \mathbf{B}_{\rm dem})}{B_{\rm ext}^2}. \tag{5.65}$$

The existence of a frequency shift at a nuclear probe is well known in NMR, following the pioneering work of Knight (1949). The expression of K_μ is further simplified using Eqs. 5.35 and 5.36. Since here $\mathbf{M}_{\rm Lor} = \mathbf{M}_{\rm meas}$, we have

$$K_\mu = K_{\rm exp} - \left(\frac{1}{3} - N^Z\right)\chi_0. \tag{5.66}$$

For simplicity, we have implicitly assumed the sample to be an ellipsoid, with one of its principal axis oriented along the laboratory Z axis. As before, N^Z is a demagnetisation coefficient, and χ_0 is the bulk susceptibility which characterises the compound. Then χ_0 is related to the measured susceptibility,

$$\chi_{\text{meas}} = \frac{\mu_0 M_{\text{meas}}}{B_{\text{ext}}}, \tag{5.67}$$

as

$$\chi_0 = \frac{\mu_0 M_{\text{meas}}}{B_{\text{ext}} - \mu_0 N^Z M_{\text{meas}}} = \frac{\chi_{\text{meas}}}{1 - N^Z \chi_{\text{meas}}}. \tag{5.68}$$

Note that

$$\chi_{\text{meas}} = \frac{\chi_0}{1 + N^Z \chi_0}. \tag{5.69}$$

For s and p metals, we derive from Section 5.1.3 the contribution to the shift as

$$K_\mu = K_{\text{Fermi}}, \tag{5.70}$$

where K_{Fermi} accounts for the shift resulting from the conduction electrons at the muon site:

$$K_{\text{Fermi}} = \frac{\mathbf{B}_{\text{ext}} \cdot \mathbf{B}_{\text{Fermi}}}{B_{\text{ext}}^2}. \tag{5.71}$$

Two additional shifts of completely different origin have been recognised (Schenck 1985). They stem from diamagnetic screening. First there is the contribution of Landau diamagnetism (orbital magnetism) from the conduction electrons at the Fermi surface. Typically it amounts to a few parts per million. The second and more important contribution arises from the charges piled up around the muon due to the screening of its electric charge.

For d and f electron compounds, the dipole and hyperfine field terms and possibly the Fermi contribution to K_μ are important relative to the small diamagnetic screening terms. Accordingly, to a good approximation, from Eq. 5.17

$$K_\mu = K'_{\text{dip}} + K_{\text{hyp}} + K_{\text{Fermi}}, \tag{5.72}$$

where K'_{dip} accounts for the shift from the dipole field arising from the magnetic moments inside the Lorentz sphere and K_{hyp} is the frequency shift induced by the hyperfine field:

$$K'_{\text{dip}} = \frac{\mathbf{B}_{\text{ext}} \cdot \mathbf{B}'_{\text{dip}}}{B_{\text{ext}}^2}, \qquad K_{\text{hyp}} = \frac{\mathbf{B}_{\text{ext}} \cdot \mathbf{B}_{\text{hyp}}}{B_{\text{ext}}^2}. \tag{5.73}$$

Let us first consider K'_{dip} and K_{hyp} and assume for definitiveness a f electron compound. The response of the f moments is described by their susceptibility tensor, $\boldsymbol{\chi}_f$, the elements of which are such that

$$m^\beta = \frac{v_0}{\mu_0} \chi_f^{\beta Z} B_{\text{ext}}. \tag{5.74}$$

Then, from Eqs. D.23 and 5.73, we derive

$$K'_{\text{dip}} = \frac{1}{4\pi} \sum_\beta \left(\sum_{i=1}^{N_{\text{L}}} D_{\mathbf{r}_i}^{Z\beta} \right) \chi_f^{\beta Z}. \tag{5.75}$$

Note that we have again considered here that there is only one magnetic atom per unit cell and therefore dropped the d index. It is straightforward to generalise. The shift $K_{\rm hyp}$ is obtained from $K'_{\rm dip}$ substituting $H^{Z\beta}_{\mathbf{r}_i}$ for $D^{Z\beta}_{\mathbf{r}_i}$ and restricting the lattice summation to the nearest and possibly next nearest neighbour atoms to the muon. A general expression for $K'_{\rm dip}$ in terms of the angles defining the orientation of $\mathbf{B}_{\rm ext}$ in the crystal frame is given in Section D.6. As explained in that section, the same type of expression is applicable for $K_{\rm hyp}$.

As a simple example, we assume the susceptibility to be uniaxial; i.e. the susceptibility tensor is diagonal with two of the three terms equal. We denote as c the principal axis characterised by a single value. This means that the f electron susceptibility measured for $\mathbf{B}_{\rm ext}$ parallel (perpendicular) to c is $\chi_f^{\|}$ ($\chi_f^{\perp}$). In addition, we assume the tensors describing the muon coupling to be also uniaxial in the same coordinate frame. From Eqs. D.31 and D.32, and using the notation defined in Section D.6, we derive

$$K'_{\rm dip} + K_{\rm hyp} = (\mathfrak{D}^{zz} + \mathfrak{H}^{zz})\, \chi_f^{\|}, \tag{5.76}$$

for $\mathbf{B}_{\rm ext}$ applied along the c axis and

$$K'_{\rm dip} + K_{\rm hyp} = \left(-\frac{\mathfrak{D}^{zz}}{2} + \mathfrak{H}^{\perp} \right) \chi_f^{\perp}, \tag{5.77}$$

if $\mathbf{B}_{\rm ext}$ is perpendicular to the c axis. If the hyperfine coupling tensor $\mathfrak{H}$ is scalar, as is often the case, $\mathfrak{D}^{zz}$ and $\mathfrak{H}^{zz}$ can be determined from measurements with $\mathbf{B}_{\rm ext}$ parallel and perpendicular to the c axis, respectively, provided of course the corresponding susceptibilities to be known. By comparing the value of $\mathfrak{D}^{zz}$ with the result of its calculation for different muon sites of axial symmetry, the site can be inferred.

It is a simple matter to derive from Eq. 5.15,

$$K_{\rm Fermi} = \frac{2}{3}\, v_0\, |\psi_{\rm s}(\mathbf{r}_0)|^2\, \eta(\mathbf{r}_0)\, \chi_{\rm cond}, \tag{5.78}$$

where $\chi_{\rm cond}$ is the susceptibility of the conduction electron assumed to be isotropic. In simple metals, this is the Pauli susceptibility, which is temperature independent.

This discussion and Eq. 5.72 imply that K_μ for a d or f metal is a weighted sum of the conduction electron susceptibility and d or f susceptibility, the former being approximately temperature independent while the latter depends strongly on the temperature. In general, $K_{\rm Fermi}$ is small relative to $K'_{\rm dip}$ and $K_{\rm hyp}$, reflecting the fact that $\chi_{\rm cond} \ll \chi_{d,f}$. Since $\chi_{d,f} \simeq \chi_{\rm bulk}$, the dependence of K_μ against $\chi_{\rm bulk}$ is expected to be linear. Such a drawing of K_μ versus $\chi_{\rm bulk}$ with the temperature as an implicit parameter is commonly called a Clogston–Jaccarino plot.

The spin–orbit interaction of the electrons has been neglected until now. Its strength increases with the atomic number of the atoms since it has a relativistic origin. A clear signature of such interaction would be, for example, the existence of an anisotropic Knight shift for a muon in a cubic site; see Carter *et al.* (1977) and references therein. An analysis of the Knight shift for the ferromagnetic superconductor UGe_2 shows that the predicted squared cosine angular dependence inferred from the results of Section D.6 is effectively observed. However, the mean value and amplitude of the Knight shift oscillation are not compatible with the results of the analysis of Clogston–Jaccarino plots (Sakarya *et al.* 2010). This breakdown of the conventional analysis is attributed to the importance of the spin–orbit interaction in uranium compounds.

It is generally taken for granted that the Knight shift is proportional to the susceptibility, with a constant of proportionality independent of the temperature. However, since this constant depends on the hyperfine constant, which is sometimes temperature and orientation dependent as already mentioned in Section 5.1.3, the reader should be aware that the Knight shift might not be proportional to the bulk susceptibility for some compounds. In fact, the deviation from a linear dependence of the Knight shift versus the susceptibility has been used as a probe of the strong electronic correlations in heavy-fermion systems, i.e. in strongly correlated metals (Ohishi *et al.* 2009).

Obviously, the above discussion has been based on the assumption that the muon is static and that the muon electric charge does not modify the electric field at the magnetic ions. Still, it should be noted that such influence has been assumed for the interpretation of Knight shift measurements performed for singlet ground state systems — see, e.g. Feyerherm *et al.* (1995).

Although we are not aware of any muon frequency shift measurements in magnetically ordered phases, their determination should be useful to characterise field-induced phases in magnets.

We now turn our attention to the local field in magnets.

5.4 Local field in magnets

At least for conventional magnets, magnetism stems from d and f electrons. Thus, according to the discussion in Sections 5.1.3 and 5.2, the local field at the muon site splits into five contributions:

$$\mathbf{B}_{\rm loc} = \mathbf{B}_{\rm ext} + \mathbf{B}'_{\rm dip} + \mathbf{B}_{\rm Lor} + \mathbf{B}_{\rm dem} + \mathbf{B}_{\rm hyp}. \tag{5.79}$$

Although, at first sight, the application of this formula is not expected to be more difficult than for a paramagnet or a saturated ferromagnet, we stress that this is not so (Smilga and Belousov 1994). The expression for the demagnetisation field given in Eq. 5.36 is still valid. However, the external field dependence of that field can be complicated if $B_{\rm ext}$ is smaller than the saturation field, because the magnetisation process is driven in part by the field response of the domains and domain walls. A drastic simplification applies if the susceptibility is large, that is, $N^Z\chi_0 \gg 1$. From Eq. 5.69 we obtain

$$\chi_{\rm meas} \simeq (N^Z)^{-1}. \tag{5.80}$$

Therefore, using Eqs. 5.36 and 5.67, $\mathbf{B}_{\rm dem} \simeq -\mathbf{B}_{\rm ext}$, that is, two out of the five terms in the expression of $\mathbf{B}_{\rm loc}$ (Eq. 5.79) cancel out if the susceptibility is large, i.e. for a high permeability magnet. This implies that $\mathbf{B}_{\rm loc}$ is independent of $\mathbf{B}_{\rm ext}$. The evaluation of $\mathbf{B}_{\rm loc}$ with a field applied to the sample is further discussed in Section 10.3 through a specific example.

We now focus our attention on the description of a zero-field measurement. In this case $\mathbf{B}_{\rm loc}$ is nothing else but $\mathbf{B}_0$, the spontaneous field at the muon site. Considering a ferromagnet which is not macroscopically polarised due to the presence of domains, Eq. 5.79 reduces to

$$\mathbf{B}_0 = \mathbf{B}'_{\rm dip} + \mathbf{B}_{\rm Lor} + \mathbf{B}_{\rm hyp}, \tag{5.81}$$

where the Lorentz field is proportional to the saturation magnetisation of a domain, $\mathbf{B}_{\rm Lor} = \mu_0\mathbf{M}_{\rm sat}/3$. This is so because the Lorentz sphere is embedded in a magnetic domain, the magnetisation of which is $\mathbf{M}_{\text{sat}}$. As explained in Section D.7, symmetry considerations are then quite useful for simplifying the numerical evaluation of $\mathbf{B}'_{\text{dip}}$. On the other hand, the expression of $\mathbf{B}_0$ for an antiferromagnet in zero field is quite simple:

$$\mathbf{B}_0 = \mathbf{B}'_{\text{dip}} + \mathbf{B}_{\text{hyp}}. \tag{5.82}$$

The Lorentz field is zero since the magnetisation vanishes.

References

Abragam A. (1961) *The Principles of Nuclear Magnetism* (Clarendon, Oxford).

Born M. and Huang K. (1954) *Dynamical Theory of Crystal Lattices* (Clarendon, Oxford).

Carter G.C., Bennett L.H., and Kahan D.J. (1977) *Metallic Shifts in NMR*, part I (Pergamon Press, Oxford).

Chikazumi S. (1964) *Physics of Magnetism* (Wiley, New York).

Clark W.G., Lan M.D., van Kalkeren G., Wong W.H., Tien Cheng, MacLaughlin D.E., Smith J.L., Fisk Z., and Ott H.R. (1987) *Journal of Magnetism and Magnetic Materials* **63 & 64**, 396–399.

Dalmas de Réotier P. and Yaouanc A. (1997) *Journal of Physics: Condensed Matter* **9**, R9113–R9166.

Ellis D.E. and Lindgren B. (1984) *Hyperfine Interactions* **17-19**, 279–286.

Feyerherm R., Amato A., Grayevski A., Gygax F.N., Kaplan N., and Schenck A. (1995) *Zeitschrift für Physik* **B 99**, 3–13

Hartmann O., Karlsson E., Wäppling R., Chappert J., Yaouanc A., Asch L., and Kalvius G.M. (1986) *Journal of Physics F: Metal Physics* **16**, 1593–1615.

Kittel C. (1963) *Quantum theory of solids* (Willey, New York).

Kittel C. (1996) *Introduction to Solid State Physics,* seventh edition (Wiley, New York).

Knight W.D. (1949) *Physical Review* **76**, 1259–1260.

Marshall W. (1963) in *Paramagnetic Resonance*, edited by Low W. (Academic Press, New York).

Messiah A. (1965) *Quantum Mechanics*, third reprint, vol. 2 (North-Holland, Amsterdam).

Morrish A.H. (1965) *The Physical Principles of Magnetism* (Wiley, New York).

Ohama T., Yasuoka H., Mandrus D., Fisk Z., and Smith J.L. (1995) *Journal of the Physical Society of Japan* **64**, 2628–2635.

Ohishi K., Heffner R.H., Ito T.U., Higemoto W., Morris G.D., Hur N., Bauer E.D., Sarrao J.L., Thompson J.D., MacLaughlin D.E., and Shu L. (2009) *Physical Review* **B 80**, 125104, 1–7.

Sakarya S., Gubbens P.C.M., Yaouanc A., Dalmas de Réotier P., Andreica D., Amato A., Zimmermann U., van Dijk N.H., Brück E., Huang Y., and Gortenmulder T. (2010) *Physical Review* **B 81**, 024429, 1-19.

Santini P. and Amoretti G. (2000) *Physical Review Letters* **85**, 2188–2191.

Schenck A. (1985) *Muon Spin Rotation Spectroscopy* (Adam Hilger, Bristol).

Schenck A., Gygax F.N., and Ōnuki Y. (2003) *Physical Review* **B 68**, 104422, 1–18.

Smilga V.P. and Belousov Yu.M. (1994) *The Muon Method in Science*, Proceedings of the Lebedev Physics Institute, vol. 219 (Nova Science, Commack, NY).
Teichler H. (1978) Physics Letters **A 67**, 313–315
Uritskiĭ M.Z. and Irkhin V.Yu. (2000) *Physics of the Solid State* **42**, 399–405.
Zijlstra H. (1967) *Experimental Methods in Magnetism*, part 2 (Interscience Publishers, New York).

6 Muon polarisation functions: field distribution approach

In this chapter and the following one we describe methods for computing the transverse- and longitudinal-field muon polarisation functions, $P_X(t)$ and $P_Z(t)$, respectively. Schematically, two families of methods are available. In one of them, which is the subject of the current chapter and which is somewhat phenomenological, the field at the muon site $\mathbf{B}_{\rm loc}$ is assumed to belong to a static field distribution and the time evolution of the muon polarisation is driven by this distribution. If relevant, i.e. if $\mathbf{B}_{\rm loc}$ evolves on the muon lifetime scale, the dynamics of the local field can also be accounted for. With proper hypotheses, this approach yields expressions which can be mathematically handled and, in addition, it provides a clear physical picture of the physics involved since only a few parameters relevant to the basic process are introduced. Using quantum mechanics or even classical mechanics which is sufficient in most cases, one can in a second step express these parameters in terms of microscopic properties of the compound under study. This family of methods describes the dephasing of the spin precession induced by the field distribution experienced by the ensemble of the muon spins. This dephasing leads to the so-called inhomogeneous broadening in nuclear magnetic resonance (NMR) — see for example Abragam (1961) and Slichter (1996). Microscopically, the distribution reflects the magnetic disorder in the compound under study.

The field distribution methods have been used for a long time in physics — see for example Chandrasekhar (1943). Their application to condensed matter has been nicely reviewed by Dattagupta (1987). A short introduction for μSR has been given by Nagamine (2003). Experimentally, with this procedure one measures the characteristics of the field distribution through the study of $P_X(t)$ and $P_Z(t)$. In the absence of $\mathbf{B}_{\rm loc}$ dynamics, a transverse-field experiment allows one to determine uniquely the distribution of the field component parallel to the external field.

In the second family of methods, which is the subject of the next chapter, the muon polarisation functions are computed using the Heisenberg (or Schrödinger) evolution equation. Here, $\mathbf{B}_{\rm loc}$ is a vector operator rather than a classical vector. This is *a priori* a better method since the spin-flip processes, which lead to the muon spin relaxation, are naturally taken into account. But the resulting formulae for the polarisation functions are usually extremely complicated and drastic approximations are needed to get useful results. For an insight into the significance and consequences of these approximations it is helpful to refer to the results obtained with the field distribution method.

A few numerical works have been presented in the literature. We will discuss them in this chapter or the following one, depending on the subject tackled by the works.

If a crystal contains magnetic moments, the muon spins will experience in general a field distribution, $D_{\rm v}(\mathbf{B}_{\rm loc})$, and not a unique field because the moments do not have in general the same orientation — see Eq. C.32 for a mathematical definition of $D_{\rm v}(\mathbf{B}_{\rm loc})$. This is the case for the nuclear magnetic moments which are disorderd and static on the muon time scale: their orientation can be viewed as frozen and random. These moments provide a source of field distribution. Electronic moments can be another source of distribution. Notice that these distributions exist even in a crystal without structural disorder or impurity. Obviously, if there is a structural disorder as in an amorphous or spin-glass material, or if randomly placed impurities are present, an additional source of distribution appears. Its physics is different and must be treated in a theoretical framework adapted to disordered magnets.

In the first section of this chapter, we shall recall some properties of field distributions. In the next two sections we successively describe $P_X(t)$ and $P_Z(t)$ which result from a static magnetic field distribution in a perfect crystal and from a time-dependent field distribution. In the fourth section we shall deal with the effect of structural disorder or impurities. The difficult problem of computing the polarisation functions when two independent sources of field are at play is discussed in the fifth section. Finally, in the last section, the shape of the distributions will be justified and their parameters expressed in terms of material parameters.

Before investigating the effects of a field distribution on the polarisation functions, we could have discussed, as is usually done, its microscopic origin. We have chosen to proceed the other way around because when performing a μSR experiment, one first needs to describe the measured spectra before attempting to relate the extracted parameters to the characteristics of the compound under study.

In this chapter we shall use well-known integrals and other formulae. They are listed in Section C.10.

6.1 Characteristics of field distributions

For the sake of completeness, we shall first recall some definitions and concepts relevant to field distributions. The field distribution is fully defined when we specify the set of possible values and their probability, $D_{\rm v}(\mathbf{B}_{\rm loc})$, over this set — see for example van Kampen (1981). It is usually possible, but not always, to express $D_{\rm v}(\mathbf{B}_{\rm loc})$ as the product of the distributions of the field components:

$$D_{\rm v}\left(\mathbf{B}_{\rm loc}\right) = D_{\rm c}\left(B^X_{\rm loc}\right) D_{\rm c}\left(B^Y_{\rm loc}\right) D_{\rm c}\left(B^Z_{\rm loc}\right). \tag{6.1}$$

The distributions $D_{\rm c}(B^\alpha_{\rm loc})$, formally defined in Eq. C.33, are normalised to unity. Among parameters which characterise them are their moments. Their definitions are recalled in Eq. C.37.

One of the function defined in statistics is the characteristic function, $G(t)$. Focusing for definitiveness on the component field distribution, the characteristic function of $B^\alpha_{\rm loc}$ is related to the Fourier transform of $D_{\rm c}(B^\alpha_{\rm loc})$, i.e.

$$G(t) = \int \exp\left(i\gamma_\mu B^\alpha_{\rm loc} t\right) D_{\rm c}\left(B^\alpha_{\rm loc}\right) {\rm d}B^\alpha_{\rm loc}. \tag{6.2}$$

Its series expansion is expressed in terms of the moments of $D_{\rm c}(B^{\alpha}_{\rm loc})$, denoted as $\overline{(B^{\alpha}_{\rm loc})^m}$:

$$G(t) = 1 + \sum_{m=1}^{\infty} \frac{(i\gamma_\mu t)^m}{m!} \overline{(B^{\alpha}_{\rm loc})^m}. \tag{6.3}$$

The characteristic function $G(t)$ serves to generate the cumulants κ_m which are defined through the relation

$$\ln G(t) = \sum_{m=1}^{\infty} \frac{(i\gamma_\mu t)^m}{m!} \kappa_m. \tag{6.4}$$

The cumulant κ_2 is equal to the variance and κ_3 is related to the skewness ϑ through the relation $\vartheta^3 = \kappa_3/\kappa_2^{3/2}$. The skewness parameter is defined in Eq. C.38. A distribution for which all cumulants beyond $m = 2$ are zero is, by definition, a Gaussian distribution. This describes the superposition of a large number of small effects. For example, the influence of the thermal fluctuations on the X-ray diffracted intensity is accounted for by the Debye–Waller factor which depends on the wave vector as a Gaussian function.

6.2 Static field distributions in a perfect crystal

As stated in Chapter 3, a μSR experiment consists of performing an ensemble average over a large number of muon decay events. Referring to the discussion of that chapter, we have

$$\begin{aligned} P_X(t) &= \int M^{XX}(t) D_{\rm v}(\mathbf{B}_{\rm loc})\, {\rm d}^3\mathbf{B}_{\rm loc} \\ &= \int \left\{ \left(\frac{B^X_{\rm loc}}{B_{\rm loc}}\right)^2 + \left[1 - \left(\frac{B^X_{\rm loc}}{B_{\rm loc}}\right)^2\right] \cos(\omega_\mu t) \right\} D_{\rm v}(\mathbf{B}_{\rm loc})\, {\rm d}^3\mathbf{B}_{\rm loc}, \end{aligned} \tag{6.5}$$

and

$$\begin{aligned} P_Z(t) &= \int M^{ZZ}(t) D_{\rm v}(\mathbf{B}_{\rm loc})\, {\rm d}^3\mathbf{B}_{\rm loc} \\ &= \int \left\{ \left(\frac{B^Z_{\rm loc}}{B_{\rm loc}}\right)^2 + \left[1 - \left(\frac{B^Z_{\rm loc}}{B_{\rm loc}}\right)^2\right] \cos(\omega_\mu t) \right\} D_{\rm v}(\mathbf{B}_{\rm loc})\, {\rm d}^3\mathbf{B}_{\rm loc}. \end{aligned} \tag{6.6}$$

We recall that $\omega_\mu = \gamma_\mu B_{\rm loc} = \gamma_\mu \left[(B^X_{\rm loc})^2 + (B^Y_{\rm loc})^2 + (B^Z_{\rm loc})^2\right]^{1/2}$. The integrations extend over all the possible values of $\mathbf{B}_{\rm loc}$.

It is a simple matter to check that if there is only one possible field value $\mathbf{B}_0$, i.e. $D_{\rm v}(\mathbf{B}_{\rm loc}) = \delta(\mathbf{B}_{\rm loc} - \mathbf{B}_0)$, and if $\mathbf{B}_0$ is oriented along the Z axis, $P_X(t) = \cos(\gamma_\mu B_0 t)$ and $P_Z(t) = 1$. Therefore we recover results already obtained in Chapter 3. On the other hand, if $\mathbf{B}_0$ is parallel to the X axis, we derive $P_X(t) = 1$ and $P_Z(t) =$

$\cos(\gamma_\mu B_0 t)$. Defining θ as the polar angle of $\mathbf{B}_0$ in the laboratory reference frame, we have from Eq. 6.6 the very important formula

$$P_Z(t) = \cos^2\theta + \sin^2\theta \cos(\omega_\mu t) . \tag{6.7}$$

In the simplest cases, this polarisation function can serve to model the μSR response of a magnet in its ordered phase, provided the sample is a single crystal. If instead the sample is polycrystalline the orientation of $\mathbf{B}_0$ will differ from one crystallite to another and accordingly the angle θ. For a random orientation of these crystallites, $D_\mathrm{v}(\mathbf{B}_\mathrm{loc})$ is equal to $\delta(B_\mathrm{loc} - B_0)/(4\pi B_0^2)$. The integration of Eq. 6.6 then leads to

$$P_Z(t) = \frac{1}{3} + \frac{2}{3}\cos(\omega_\mu t) , \tag{6.8}$$

which was already given in Eq. 1.5. In the case of a sample with a crystallite preferred orientation, $D_\mathrm{v}(\mathbf{B}_\mathrm{loc})$ is modified, resulting in a change in the relative weights of the oscillating and non-oscillating terms in Eq. 6.8.

We now consider the frequent case where an external field $\mathbf{B}_\mathrm{ext}$ is applied. The field distribution at the muon site is then denoted as $D_{\mathrm{v},\mathbf{B}_\mathrm{ext}}(\mathbf{B}_\mathrm{loc})$. We have introduced the subscript $\mathbf{B}_\mathrm{ext}$ in the label to stress the presence of the external field. In the absence of this latter we shall generally omit this subscript as we have done so far. Assuming $\mathbf{B}_\mathrm{ext}$ to have no effect on the field distribution apart from shifting it, and neglecting a possible Knight shift, the distribution of $\mathbf{B}_\mathrm{loc}$ is given by $D_{\mathrm{v},\mathbf{B}_\mathrm{ext}}(\mathbf{B}_\mathrm{loc}) = D_\mathrm{v}(\mathbf{B}_\mathrm{loc} - \mathbf{B}_\mathrm{ext})$.

In this chapter we shall only consider field distributions of relatively general characters. More specialised distributions will be considered in Chapters 10 and 11.

In general terms, in a system with a large concentration of uncorrelated magnetic moments the field distributions of the three components are expected to be Gaussian. This result is discussed in Section 6.6.2 where statistical techniques are used to derive it. As a starting point, we assume the three Gaussian distributions to have the same width; i.e. the vector distribution is isotropic. This hypothesis is valid if the symmetry at the muon site is cubic. In addition, it is a useful approximation for the description of measurements performed on non-perfect samples. The isotropic Gaussian model does not only provide an academic exercise; with a minimum of mathematics the type of information one can extract from a μSR experiment is determined.

We shall first discuss the polarisation functions for the case of an isotropic Gaussian field distribution at the muon site. We shall also suppose that there is possibly an additional magnetic field of fixed magnitude and orientation relative to the laboratory reference frame. For definitiveness, we shall suppose this additional field to be the applied external field, $\mathbf{B}_\mathrm{ext}$. But it could also be the internal spontaneous magnetic field in a magnetically ordered single-crystal compound.

In this chapter and the following, for a given physical model, we shall first analyse the transverse and then the longitudinal polarisation functions.

6.2.1 Isotropic distribution

If the three components of $\mathbf{B}_\mathrm{loc}$ are Gaussian distributed with the same variance Δ_G^2, we get

$$
\begin{aligned}
D_{\mathrm{v}}\left(\mathbf{B}_{\mathrm{loc}}\right)\mathrm{d}^3\mathbf{B}_{\mathrm{loc}} &= D_{\mathrm{c}}\left(B_{\mathrm{loc}}^X\right)D_{\mathrm{c}}\left(B_{\mathrm{loc}}^Y\right)D_{\mathrm{c}}\left(B_{\mathrm{loc}}^Z\right)\mathrm{d}B_{\mathrm{loc}}^X\,\mathrm{d}B_{\mathrm{loc}}^Y\,\mathrm{d}B_{\mathrm{loc}}^Z \\
&= \left(\frac{1}{\sqrt{2\pi}\Delta_{\mathrm{G}}}\right)^3 \exp\left[-\frac{\left(B_{\mathrm{loc}}^X\right)^2}{2\Delta_{\mathrm{G}}^2}\right]\exp\left[-\frac{\left(B_{\mathrm{loc}}^Y\right)^2}{2\Delta_{\mathrm{G}}^2}\right] \\
&\quad\times \exp\left[-\frac{\left(B_{\mathrm{loc}}^Z\right)^2}{2\Delta_{\mathrm{G}}^2}\right]\mathrm{d}B_{\mathrm{loc}}^X\,\mathrm{d}B_{\mathrm{loc}}^Y\,\mathrm{d}B_{\mathrm{loc}}^Z \\
&= \left(\frac{1}{\sqrt{2\pi}\Delta_{\mathrm{G}}}\right)^3 \exp\left(-\frac{B_{\mathrm{loc}}^2}{2\Delta_{\mathrm{G}}^2}\right)B_{\mathrm{loc}}^2\sin\theta\,\mathrm{d}\theta\,\mathrm{d}\varphi\,\mathrm{d}B_{\mathrm{loc}}, \qquad (6.9)
\end{aligned}
$$

where θ and φ are the polar and azimuthal angles of $\mathbf{B}_{\mathrm{loc}}$ in the laboratory reference frame. Since the argument of the exponential in the last line of Eq. 6.9 is independent of these two angles, the distribution is isotropic. Note that a three-dimensional distribution given as a product of the distributions for each component is isotropic only if the individual components are Gaussian distributed and with identical second moment.

Now, in the presence of an external field, the distribution $D_{\mathrm{v},\mathbf{B}_{\mathrm{ext}}}(\mathbf{B}_{\mathrm{loc}})$ must be considered. Recalling that in our convention $\mathbf{B}_{\mathrm{ext}}$ is oriented along the Z axis and sticking to a Gaussian isotropic field distribution we have

$$
\begin{aligned}
D_{\mathrm{v},\mathbf{B}_{\mathrm{ext}}}(\mathbf{B}_{\mathrm{loc}})\mathrm{d}^3\mathbf{B}_{\mathrm{loc}} &= D_{\mathrm{c}}\left(B_{\mathrm{loc}}^X\right)D_{\mathrm{c}}\left(B_{\mathrm{loc}}^Y\right)D_{\mathrm{c}}\left(B_{\mathrm{loc}}^Z - B_{\mathrm{ext}}\right)\mathrm{d}B_{\mathrm{loc}}^X\,\mathrm{d}B_{\mathrm{loc}}^Y\,\mathrm{d}B_{\mathrm{loc}}^Z \\
&= \left(\frac{1}{\sqrt{2\pi}\Delta_{\mathrm{G}}}\right)^3 \exp\left[-\frac{\left(B_{\mathrm{loc}}^X\right)^2}{2\Delta_{\mathrm{G}}^2}\right]\exp\left[-\frac{\left(B_{\mathrm{loc}}^Y\right)^2}{2\Delta_{\mathrm{G}}^2}\right] \\
&\quad\times \exp\left[-\frac{\left(B_{\mathrm{loc}}^Z - B_{\mathrm{ext}}\right)^2}{2\Delta_{\mathrm{G}}^2}\right]\mathrm{d}B_{\mathrm{loc}}^X\,\mathrm{d}B_{\mathrm{loc}}^Y\,\mathrm{d}B_{\mathrm{loc}}^Z. \qquad (6.10)
\end{aligned}
$$

Strictly speaking, the muon ensemble practically never probes a static field distribution. There is always a certain degree of field dynamics. Therefore the results obtained in this section are intermediate. To distinguish between the polarisation functions for static and dynamic field distributions, we denote the functions for the former case as $P_X^{\mathrm{stat}}(t)$ and $P_Z^{\mathrm{stat}}(t)$.

An analytical form for $P_X^{\mathrm{stat}}(t)$ can be found if the intensity of $\mathbf{B}_{\mathrm{ext}}$ is sufficiently large, i.e. if B_{ext} is much larger than the values of B_{loc} for which $D_{\mathrm{v}}(B_{\mathrm{loc}})$ has a sizeable weight. In this limit, which is verified in many cases of interest, the resultant magnetic field at the muon site is, to a good approximation, directed only along the Z axis. Then, Eq. 6.5 simplifies to

$$
\begin{aligned}
P_X^{\mathrm{stat}}(t) &\simeq \int_{-\infty}^{\infty} D_{\mathrm{c},\mathbf{B}_{\mathrm{ext}}}\left(B_{\mathrm{loc}}^Z\right)\cos(\gamma_\mu B_{\mathrm{loc}}^Z t)\mathrm{d}B_{\mathrm{loc}}^Z \\
&= \int_{-\infty}^{\infty} D_{\mathrm{c}}\left(B_{\mathrm{loc}}^Z - B_{\mathrm{ext}}\right)\cos(\gamma_\mu B_{\mathrm{loc}}^Z t)\mathrm{d}B_{\mathrm{loc}}^Z \\
&= \left[\int_{-\infty}^{\infty} D_{\mathrm{c}}(x)\cos\left(\gamma_\mu t x\right)\mathrm{d}x\right]\cos\left(\gamma_\mu B_{\mathrm{ext}} t\right)
\end{aligned}
$$

$$
\begin{aligned}
& - \left[\int_{-\infty}^{\infty} D_{\rm c}(x) \sin(\gamma_\mu t x)\, {\rm d}x \right] \sin(\gamma_\mu B_{\rm ext} t) \\
& = \left[\int_{-\infty}^{\infty} D_{\rm c}(x) \cos(\gamma_\mu t x)\, {\rm d}x \right] \cos(\gamma_\mu B_{\rm ext} t) , \qquad (6.11)
\end{aligned}
$$

where the change of variable $B^Z_{\rm loc} \rightarrow x + B_{\rm ext}$ has been performed. Equation 6.11 is a high-transverse-field approximation result and in fact it is quite a general expression: it depends neither on the isotropy of the field distribution nor on the precise form of the distribution of $B^Z_{\rm loc}$, which is therefore not necessarily a Gaussian. Only the last step in Eq. 6.11 requires $D_{\rm c}(x) = D_{\rm c}(-x)$. Hence, within these conditions the envelope of $P^{\rm stat}_X(t)$ is the cosine Fourier transform of the distribution of the field along $\mathbf{B}_{\rm ext}$. Referring to the introduction of this chapter, the envelope of $P^{\rm stat}_X(t)$ is the characteristic function of $B^Z_{\rm loc}$ if the field distribution is an even function. The integration in Eq. 6.11 is easily carried out for a Gaussian distribution, with the result

$$
P^{\rm stat}_X(t) = \exp\left(-\frac{\gamma_\mu^2 \Delta^2_{\rm G,TF} t^2}{2} \right) \cos(\gamma_\mu B_{\rm ext} t) . \qquad (6.12)
$$

An example for this function is plotted in Fig. 1.4. From now on, we denote the field component variance $\Delta^2_{\rm G,TF}$ instead of $\Delta^2_{\rm G}$ to distinguish it from the zero-field case. The reason of doing so will appear when we shall express the variances in terms of microscopic parameters in Section 6.6. The polarisation function oscillates with a precession frequency proportional to the external field. Its decay reflects a partial loss of the phase coherence of the muon spin precession. Since the envelope of the oscillation is a Gaussian function, the damping is quadratic in time at small times.

It is interesting to test the validity of the high-transverse-field approximation, i.e. how the result of Eq. 6.11 compares with the three-dimensional integration of Eq. 6.5. For definitiveness we take an isotropic Gaussian distribution and compare in Fig. 6.1 the results of the full integration with Eq. 6.12. As seen in the graphs, the high-field approximation is reasonable for $B_{\rm ext}$ larger than $\sim 5\Delta_{\rm G,TF}$.

Although in practice the transverse geometry is only used with $\mathbf{B}_{\rm ext} \neq 0$, for the sake of the discussion it is useful to consider $P^{\rm stat}_X(t)$ in zero field. Since the field distribution is isotropic, $P^{\rm stat}_X(t) = P^{\rm stat}_Z(t)$. Therefore, in zero field $P^{\rm stat}_X(t)$ is described by the Kubo–Toyabe formula given below in Eq. 6.17. In this respect, it is interesting to note the continuity of $P^{\rm stat}_X(t)$ towards the Kubo–Toyabe function when $B_{\rm ext}$ goes to 0. This can be appreciated from an inspection of Figs. 6.1 and 6.2.

Note that the expression of $P^{\rm stat}_X(t)$ written as a sum of two terms in the third and fourth lines of Eq. 6.11, each of them being the product of a decay term and a cosine or sine function, the frequency of which being characterised by $\gamma_\mu B_{\rm ext}$, is not a general feature. It arises from the high-field approximation: according to it, one can neglect the contribution of the field distribution in the plane perpendicular to $\mathbf{B}_{\rm ext}$.

It is important to note that our present discussion is relevant for analysing zero-field μSR spectra recorded deep in the magnetically ordered state of a single crystal when the direction of the internal field is perpendicular to the initial muon polarisation. In this case $B_{\rm ext}$ should be understood as the spontaneous magnetic field. Some care

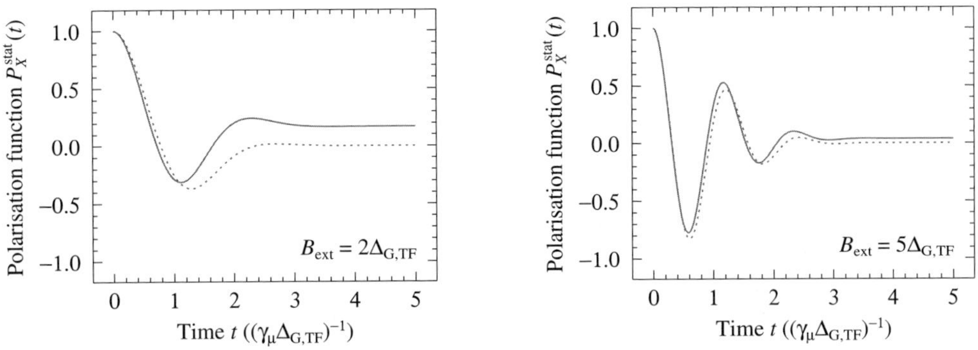

Fig. 6.1 Static transverse polarisation function for an isotropic Gaussian field distribution. The full curves are obtained from numerical integration of Eq. 6.5 with the distribution function given in Eq. 6.10, while the dashed line curves are results from the high-field approximation (Eq. 6.12). The left and right panels corresponds to different values of B_{ext}. When $B_{\mathrm{ext}} \gtrsim 10\Delta_{\mathrm{G,TF}}$ the difference between the two models is on the order or less than the line thickness. Reproduced with kind permission from Dalmas de Réotier and Yaouanc (1992).

must be exercised close to the critical temperature where $\Delta_{\mathrm{G,TF}}$ is not necessarily negligible relative to B_{ext}. Then the simple analytical result given by Eq. 6.12 cannot be trusted. The numerical solution of Eq. 6.5, with Eq. 6.10, should be used (see also Fig. 6.1). One must also be aware that, in fact, this solution only provides a description of the effect of the inhomogeneous broadening in the Gaussian field approximation. As explained in the next chapter and in Chapter 10, the spin–lattice relaxation process may also have to be taken into account. In addition, if the distribution originates from dilute defects, the field distribution may not be Gaussian. This is further discussed later on in this chapter.

We now consider $P_Z^{\mathrm{stat}}(t)$ in zero field. It is practical to write Eq. 6.6 in spherical coordinates:

$$P_Z^{\mathrm{stat}}(t) = \int \left[\cos^2\theta + \sin^2\theta \cos(\gamma_\mu B_{\mathrm{loc}} t)\right] D_{\mathrm{v}}\left(\mathbf{B}_{\mathrm{loc}}\right) \mathrm{d}^3\mathbf{B}_{\mathrm{loc}}. \tag{6.13}$$

We first assume $D_{\mathrm{v}}\left(\mathbf{B}_{\mathrm{loc}}\right) = D_{\mathrm{v}}\left(B_{\mathrm{loc}}, \theta, \varphi\right)$ to be independent of θ and φ, i.e. the distribution is isotropic. The two angular integrations in Eq. 6.13 can be performed. Since the spherical averages of $\cos^2\theta$ and $\sin^2\theta$ are respectively $\frac{1}{3}$ and $\frac{2}{3}$, we have

$$P_Z^{\mathrm{stat}}(t) = \frac{1}{3} + \frac{2}{3}\int_0^\infty \left[4\pi B_{\mathrm{loc}}^2 D_{\mathrm{v}}\left(B_{\mathrm{loc}}, \theta, \varphi\right)\right] \cos\left(\gamma_\mu B_{\mathrm{loc}} t\right) \mathrm{d}B_{\mathrm{loc}}, \tag{6.14}$$

which we write as

$$P_Z^{\mathrm{stat}}(t) = \frac{1}{3} + \frac{2}{3}\int_0^\infty D_{\mathrm{m}}\left(B_{\mathrm{loc}}\right) \cos\left(\gamma_\mu B_{\mathrm{loc}} t\right) \mathrm{d}B_{\mathrm{loc}}, \tag{6.15}$$

where $D_{\mathrm{m}}(B_{\mathrm{loc}})$ is the distribution of the modulus of $\mathbf{B}_{\mathrm{loc}}$ which is formally defined in Section C.7.1 and which is also called the field density distribution. This functional form is independent of the shape of $D_\mathrm{v}(\mathbf{B}_\mathrm{loc})$, but it is only valid if the distribution is isotropic. Remarkably, $P_Z^\mathrm{stat}(t)$ is the sum of two terms, one of them being proportional to the cosine Fourier transform of the density distribution.

We now specify the distribution and as before we choose an isotropic Gaussian distribution for the field components. The distribution term reads

$$D_{\mathrm{v},\mathbf{B}_\mathrm{ext}}(\mathbf{B}_\mathrm{loc})\,\mathrm{d}^3\mathbf{B}_\mathrm{loc} = \exp\left\{-\frac{1}{2\Delta^2_\mathrm{G,ZF}}\left[B^2_\mathrm{loc}\sin^2\theta + (B_\mathrm{loc}\cos\theta - B_\mathrm{ext})^2\right]\right\}$$
$$\times\left(\frac{1}{\sqrt{2\pi}\Delta_\mathrm{G,ZF}}\right)^3 B^2_\mathrm{loc}\sin\theta\,\mathrm{d}\theta\,\mathrm{d}\varphi\,\mathrm{d}B_\mathrm{loc}, \tag{6.16}$$

where we have introduced an external field B_ext for future use. We first consider the zero-field case, i.e. $B_\mathrm{ext} = 0$ in Eq. 6.16, and note that for a Gaussian distribution, the density distribution is the Maxwellian distribution which exhibits a maximum at $\sqrt{2}\Delta_\mathrm{G,ZF}$. It is widely used in statistical mechanics since it describes the velocity distribution of the particles of an ideal gas in the canonical ensemble. The integration over B_loc in Eq. 6.14 is easily performed:

$$P_Z^\mathrm{stat}(t) = \frac{1}{3} + \frac{2}{3}\left(1 - \gamma_\mu^2\Delta^2_\mathrm{G,ZF}t^2\right)\exp\left(-\frac{\gamma_\mu^2\Delta^2_\mathrm{G,ZF}t^2}{2}\right). \tag{6.17}$$

This is the so-called Kubo–Toyabe polarisation function; see Fig. 1.4 for a plot. In fact, Eq. 6.17 and the shape of the zero-field NMR spectrum, which is its Fourier transform, were first discussed by Kubo and Toyabe (1967). However, it is only with the advent of the μSR technique, and therefore the possibility of easily performing measurements in zero field, that the importance of this function was fully appreciated (Hayano *et al.* 1979). It exhibits a dip at $t_d = \sqrt{3}/(\gamma_\mu\Delta_\mathrm{G,ZF})$ for which $P_Z^\mathrm{stat}(t_d) = 0.03583$. If $\gamma_\mu\Delta_\mathrm{G,ZF}t$ is sufficiently large, $P_Z^\mathrm{stat}(t)$ tends towards $\frac{1}{3}$ while if $\gamma_\mu\Delta_\mathrm{G,ZF}t$ is small, $P_Z^\mathrm{stat}(t)$ is well approximated by the parabolic form $P_Z^\mathrm{stat}(t) = 1 - \gamma_\mu^2\Delta^2_\mathrm{G,ZF}t^2$. Relative to the transverse-field case, the initial depolarisation is stronger in zero field by a factor 2 since for the latter geometry the two components of $\mathbf{B}_\mathrm{loc}$ perpendicular to the Z axis contribute to the depolarisation. Qualitatively, it is worth noticing that the Kubo–Toyabe function displays an underdamped oscillation (possessing one local minimum) of period corresponding approximately to the field magnitude at which the Maxwellian distribution has its maximum.

The function $P_Z^\mathrm{stat}(t)$ in a longitudinal external field is the sum of two triple integrals (see Eq. 6.13), one of them being independent of time. Therefore it is interesting to compute the time derivative of $P_Z^\mathrm{stat}(t)$ since only one triple integral needs to be evaluated. Then, setting $x = \cos\theta$ and $y = [1/(\sqrt{2}\Delta_\mathrm{G,ZF})](B_\mathrm{loc} - xB_\mathrm{ext})$, and using the initial condition $P_Z^\mathrm{stat}(t=0) = 1$, we derive (Hayano *et al.* 1979)

$$P_Z^\mathrm{stat}(t) = 1 - \frac{2\Delta^2_\mathrm{G,ZF}}{B^2_\mathrm{ext}}\left[1 - \cos(\gamma_\mu B_\mathrm{ext}t)\exp\left(-\frac{\gamma_\mu^2\Delta^2_\mathrm{G,ZF}t^2}{2}\right)\right]$$

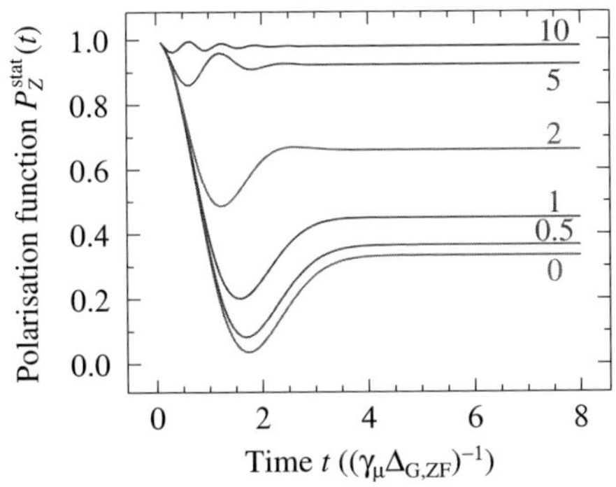

Fig. 6.2 External field dependence of the static longitudinal polarisation function $P_Z^{\rm stat}(t)$ for an isotropic Gaussian field distribution. The numbers next to the curves indicate the external field intensity in units of $\Delta_{\rm G,ZF}$, which is the standard deviation of the field component distribution. The zero-field curve corresponds to the so-called Kubo–Toyabe function (see Eq. 6.17). Reproduced with kind permission from Dalmas de Réotier and Yaouanc (1997).

$$+\frac{2\gamma_\mu \Delta^4_{\rm G,ZF}}{B^3_{\rm ext}} \int_0^t \sin(\gamma_\mu B_{\rm ext}\tau) \exp\left(-\frac{\gamma_\mu^2 \Delta^2_{\rm G,ZF}\tau^2}{2}\right) {\rm d}\tau. \tag{6.18}$$

This function is plotted in Fig. 6.2 for a few values of $B_{\rm ext}$. The two triple integrals in Eq. 6.18 can alternatively be calculated without resorting to the time derivative of $P_Z^{\rm stat}(t)$ but still using the x and y variables introduced above. We then obtain a second equivalent expression:

$$\begin{aligned} P_Z^{\rm stat}(t) = 1 &- \frac{2\Delta^2_{\rm G,ZF}}{B^2_{\rm ext}}\left[1 - \cos(\gamma_\mu B_{\rm ext} t)\exp\left(-\frac{\gamma_\mu^2\Delta^2_{\rm G,ZF} t^2}{2}\right)\right] \\ &+ \frac{2\Delta^3_{\rm G,ZF}}{B^3_{\rm ext}}\exp\left(-\frac{B^2_{\rm ext}}{2\Delta^2_{\rm G,ZF}}\right) \\ &\times \int_0^{B_{\rm ext}/\Delta_{\rm G,ZF}}\left[1 - \cos(\gamma_\mu \Delta_{\rm G,ZF} t u)\exp\left(-\frac{\gamma_\mu^2\Delta^2_{\rm G,ZF} t^2}{2}\right)\right]\exp\left(\frac{u^2}{2}\right){\rm d}u. \end{aligned} \tag{6.19}$$

The latter formula has the advantage that it provides an easy derivation of the asymptotic limit $P_Z^{\rm stat}(t \to \infty)$:

$$P_Z^{\rm stat}(t\to\infty) = 1 - \frac{2\Delta^2_{\rm G,ZF}}{B^2_{\rm ext}} + \frac{2\Delta^3_{\rm G,ZF}}{B^3_{\rm ext}}\exp\left(-\frac{B^2_{\rm ext}}{2\Delta^2_{\rm G,ZF}}\right)\int_0^{B_{\rm ext}/\Delta_{\rm G,ZF}} \exp\left(\frac{u^2}{2}\right){\rm d}u. \tag{6.20}$$

It is plotted in Fig. 6.3. As expected, the field repolarises the muon ensemble. An inspection of Figs. 6.2 and 6.3 shows that a field larger than $\sim 10\Delta_{\rm G,ZF}$ quenches

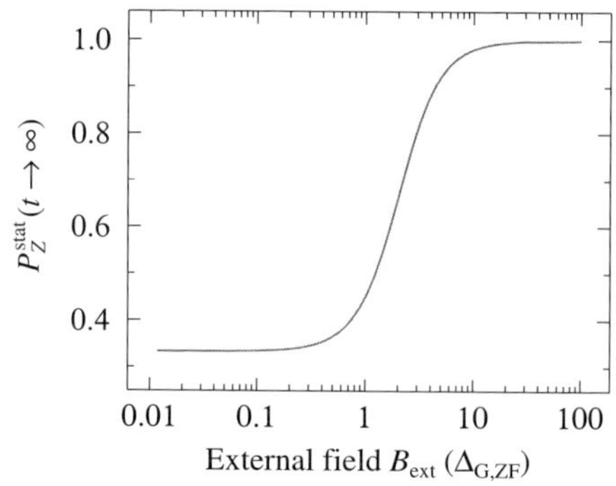

Fig. 6.3 Asymptotic limit $P_Z^{\rm stat}(t\to\infty)$ (Eq. 6.20) as a function of the external field intensity $B_{\rm ext}$ in units of $\Delta_{\rm G,ZF}$. The field distribution is assumed isotropic and Gaussian.

the depolarisation induced by the field distribution. At a sufficiently large field, the following asymptotic expression may be useful:

$$P_Z^{\rm stat}(t) \simeq P_Z^{\rm stat}(t\to\infty) + \left[1 - P_Z^{\rm stat}(t\to\infty)\right] \cos\left(\gamma_\mu B_{\rm ext} t\right) \exp\left(-\frac{\gamma_\mu^2 \Delta_{\rm G,ZF}^2 t^2}{2}\right). \tag{6.21}$$

As expected, the period of the oscillation observed at high field is only determined by the intensity of the external field. This formula provides an excellent approximation of Eq. 6.18 or 6.19 for $B_{\rm ext} \gtrsim 3\Delta_{\rm G,ZF}$ and can advantageously be used in numerical calculations.

6.2.2 Uniaxial distribution

For many systems of interest, the point symmetry at the muon site is lower than cubic. In such cases the muon field distribution is not expected to be isotropic. In this textbook we restrict ourselves to a simple extension: we discuss the polarisation functions for an uniaxial Gaussian distribution, the axis being parallel to Z, but the method of calculation can be extended to any distribution. Usually closed expressions cannot be derived.

As far as the transverse polarisation function is concerned, we showed in the previous pages that $P_X^{\rm stat}(t)$ is determined by the one-dimensional field distribution parallel to $\mathbf{B}_{\rm ext}$, provided that $B_{\rm ext}$ is sufficiently large. Therefore, for our uniaxial Gaussian distribution Eq. 6.12 applies where $\Delta_{\rm G,TF}$ is the standard deviation of the Z component of the field.

The calculation of $P_Z^{\rm stat}(t)$ is more involved. For simplicity we only consider the zero-field case. The reader can easily write $P_Z^{\rm stat}(t)$ for the case of a longitudinal field using the material of the present and previous sections. The field distribution in zero field is such that

$$
\begin{aligned}
D_{\rm v}\left(\mathbf{B}_{\rm loc}\right) {\rm d}^3\mathbf{B}_{\rm loc} &= \left(\frac{1}{\sqrt{2\pi}\Delta_{\rm G,ZF,\perp}}\right)^2 \left(\frac{1}{\sqrt{2\pi}\Delta_{\rm G,ZF,\parallel}}\right) \\
&\quad \times \exp\left[-\frac{\left(B^X_{\rm loc}\right)^2 + \left(B^Y_{\rm loc}\right)^2}{2\Delta^2_{\rm G,ZF,\perp}}\right] \exp\left[-\frac{\left(B^Z_{\rm loc}\right)^2}{2\Delta^2_{\rm G,ZF,\parallel}}\right] {\rm d}B^X_{\rm loc}\,{\rm d}B^Y_{\rm loc}\,{\rm d}B^Z_{\rm loc} \\
&= \left(\frac{1}{\sqrt{2\pi}\Delta_{\rm G,ZF,\perp}}\right)^2 \left(\frac{1}{\sqrt{2\pi}\Delta_{\rm G,ZF,\parallel}}\right) \qquad (6.22) \\
&\quad \times \exp\left(-\frac{B^2_{\rm loc}\sin^2\theta}{2\Delta^2_{\rm G,ZF,\perp}}\right) \exp\left(-\frac{B^2_{\rm loc}\cos^2\theta}{2\Delta^2_{\rm G,ZF,\parallel}}\right) B^2_{\rm loc}\sin\theta\,{\rm d}\theta\,{\rm d}\varphi\,{\rm d}B_{\rm loc}.
\end{aligned}
$$

We characterise the field distribution by the standard deviation $\Delta_{\rm G,ZF} \equiv \Delta_{\rm G,ZF,\perp}$ and the uniaxial anisotropy parameter $\alpha_{\rm ani} = \Delta_{\rm G,ZF,\parallel}/\Delta_{\rm G,ZF,\perp}$. We derive

$$
\begin{aligned}
P_Z^{\rm stat}(t) = \alpha^2_{\rm ani} \int_0^1 & \frac{1-x^2}{\left[\alpha^2_{\rm ani} + (1-\alpha^2_{\rm ani})x^2\right]^{3/2}} \left[1 - \frac{\alpha^2_{\rm ani}\gamma^2_\mu \Delta^2_{\rm G,ZF} t^2}{\alpha^2_{\rm ani} + (1-\alpha^2_{\rm ani})x^2}\right] \\
& \times \exp\left\{-\frac{\alpha^2_{\rm ani}\gamma^2_\mu \Delta^2_{\rm G,ZF} t^2}{2\left[\alpha^2_{\rm ani} + (1-\alpha^2_{\rm ani})x^2\right]}\right\} {\rm d}x + C_{\rm asym}(\alpha_{\rm ani}). \qquad (6.23)
\end{aligned}
$$

The function $C_{\rm asym}(\alpha_{\rm ani})$ is such that

$$
\begin{aligned}
C_{\rm asym}(\alpha_{\rm ani}) &= \frac{\alpha^2_{\rm ani}}{\alpha^2_{\rm ani}-1}\left(1 - \sqrt{\frac{1}{\alpha^2_{\rm ani}-1}} \arcsin\sqrt{\frac{\alpha^2_{\rm ani}-1}{\alpha^2_{\rm ani}}}\right) && \text{if } \alpha_{\rm ani} > 1, \\
&= \frac{\alpha^2_{\rm ani}}{1-\alpha^2_{\rm ani}}\left[-1 + \sqrt{\frac{1}{1-\alpha^2_{\rm ani}}} \ln\left(\frac{1+\sqrt{1-\alpha^2_{\rm ani}}}{\alpha_{\rm ani}}\right)\right] && \text{if } \alpha_{\rm ani} < 1, \\
&= \frac{1}{3} && \text{if } \alpha_{\rm ani} = 1.
\end{aligned}
\qquad (6.24)
$$

The asymptotic limit of $P_Z^{\rm stat}(t)$ at long times is given by $C_{\rm asym}(\alpha_{\rm ani})$. Some examples of $P_Z^{\rm stat}(t)$ for different values of the anisotropy parameter are presented in Fig. 6.4. At sufficiently long times, $P_Z^{\rm stat}(t)$ reaches a plateau with a value given by $C_{\rm asym}(\alpha_{\rm ani})$ (also plotted in Fig. 6.4) and at short times $P_Z^{\rm stat}(t) = 1 - \gamma^2_\mu \Delta^2_{\rm G,ZF} t^2 + \mathcal{O}(t^4)$. In between, $P_Z^{\rm stat}(t)$ always displays a dip at t_d. In Fig. 6.5 are shown t_d and $P_Z^{\rm stat}(t_d)$ as a function of $\alpha_{\rm ani}$. It is obvious that the shape of $P_Z^{\rm stat}(t)$ is strongly sensitive to the anisotropy. Interestingly, $P_Z^{\rm stat}(t)$ takes negative values around t_d if $\alpha_{\rm ani}$ is sufficiently small. The reader may consult the paper of Solt (1995) for more information.

6.2.3 Detecting a magnetic transition

The purpose of a large fraction of μSR experiments is to determine whether a compound exhibits a magnetic phase transition. In many systems of current interest, such

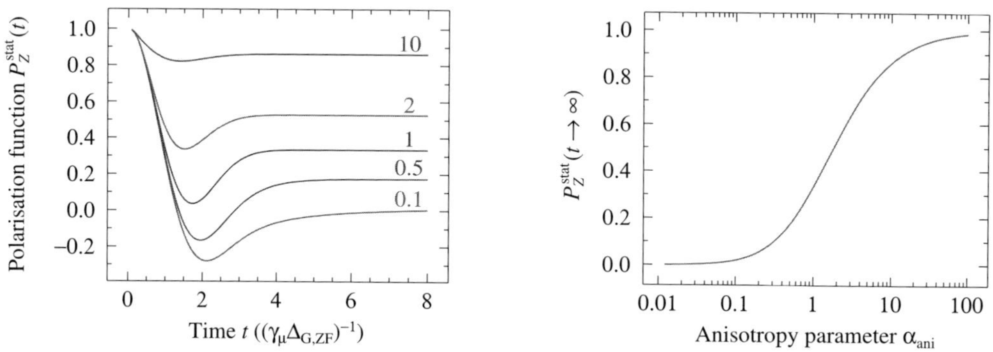

Fig. 6.4 (Left) Static zero-field polarisation function $P_Z^{stat}(t)$ for an uniaxial Gaussian field distribution (Eq. 6.23). The standard deviation of B_{loc}^X and B_{loc}^Y is $\Delta_{G,ZF}$, whereas that of B_{loc}^Z is $\Delta_{G,ZF,\parallel}$. The number next to each curve gives the value for the anisotropy parameter $\alpha_{ani} \equiv \Delta_{G,ZF,\parallel}/\Delta_{G,ZF}$. The Kubo–Toyabe function corresponds to $\alpha_{ani} = 1$. (Right) Long time value of $P_Z^{stat}(t)$, i.e. parameter $C_{asym}(\alpha_{ani})$ in Eq. 6.24, as a function of α_{ani}.

a signature is sometimes difficult to obtain by other experimental techniques, the usual reasons being that the magnetic moments are small or the magnetic order is relatively short range. In practice, the sample is polycrystalline for such a test measurement. Then, the model discussed in Section 6.2.1 is not appropriate since there, the spontaneous magnetic field at the muon site is assumed to have a fixed spatial orientation relative to the laboratory reference frame.

An analytical expression of $P_Z^{stat}(t)$ for a zero-field measurement on a polycrystalline sample below its magnetic phase transition temperature has been given by Koptev and Tarasov (1987). It has been rederived later on (Kornilov and Pomjakushin 1991). Since the crystallites are randomly oriented, the average field at the muon site has a random orientation with respect to $\{X, Y, Z\}$. This is described by splitting $\mathbf{B}_{loc}$ into two components, $\mathbf{B}_{loc} = \mathbf{B}_{loc,1} + \mathbf{B}_{loc,2}$, where the first field accounts for the

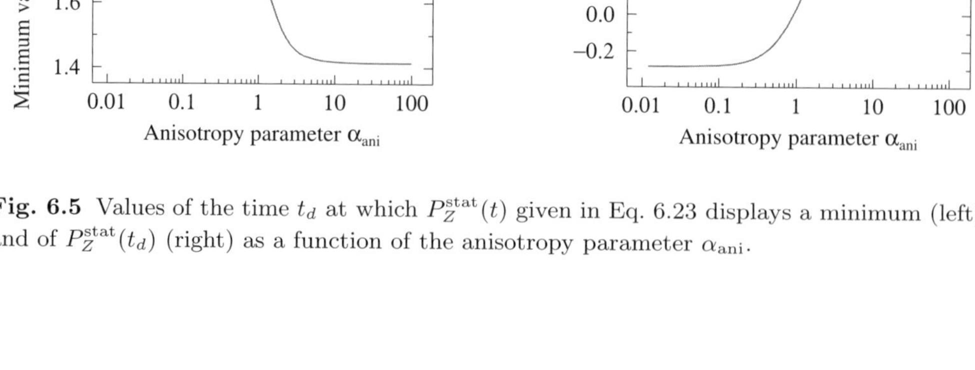

Fig. 6.5 Values of the time t_d at which $P_Z^{stat}(t)$ given in Eq. 6.23 displays a minimum (left) and of $P_Z^{stat}(t_d)$ (right) as a function of the anisotropy parameter α_{ani}.

spatial variations relative to the second field which is the spontaneous mean field at the muon site. The first field $\mathbf{B}_{\mathrm{loc},1}$ is assumed to follow an isotropic Gaussian distribution $D_{\mathrm{v},1}(\mathbf{B}_{\mathrm{loc},1})$ characterised by the component standard deviation $\Delta_{\mathrm{G,TF}}$. The second field $\mathbf{B}_{\mathrm{loc},2}$ has a uniform magnitude B_0 but is oriented randomly in space. Its distribution is expressed using a Dirac delta function (see Section C.7.2):

$$D_{\mathrm{v},2}(\mathbf{B}_{\mathrm{loc},2}) = \frac{1}{4\pi B_0^2}\delta(B_{\mathrm{loc},2} - B_0). \tag{6.25}$$

The prefactor ensures that $D_{\mathrm{v},2}(\mathbf{B}_{\mathrm{loc},2})$ is normalised to one as it should be. This distribution depends only on the modulus of the field since it is isotropic. Because the two distributions are statistically independent, the distribution probed by the muon spins is their convolution product. This means that

$$D_{\mathrm{v}}(\mathbf{B}_{\mathrm{loc}}) = \int D_{\mathrm{v},1}(\mathbf{B}_{\mathrm{loc},1})\, D_{\mathrm{v},2}(\mathbf{B}_{\mathrm{loc}} - \mathbf{B}_{\mathrm{loc},1})\, \mathrm{d}^3\mathbf{B}_{\mathrm{loc},1}. \tag{6.26}$$

Explicitly,

$$D_{\mathrm{v}}(\mathbf{B}_{\mathrm{loc}}) = \frac{1}{\left(\sqrt{2\pi}\Delta_{\mathrm{G,TF}}\right)^3}\frac{1}{2B_0^2}\int_0^\infty B_{\mathrm{loc},1}^2 \exp\left(-\frac{B_{\mathrm{loc},1}^2}{2\Delta_{\mathrm{G,TF}}^2}\right) J_{\mathbf{B}_{\mathrm{loc}},B_0}(\mathbf{B}_{\mathrm{loc},1})\, \mathrm{d}B_{\mathrm{loc},1}, \tag{6.27}$$

with

$$J_{\mathbf{B}_{\mathrm{loc}},B_0}(\mathbf{B}_{\mathrm{loc},1}) = \int_0^\pi \sin\theta\, \delta(\mid \mathbf{B}_{\mathrm{loc}} - \mathbf{B}_{\mathrm{loc},1} \mid - B_0)\, \mathrm{d}\theta. \tag{6.28}$$

Without loss of generality, we take $\mathbf{B}_{\mathrm{loc}}$ parallel to the local polar axis and thus

$$|\mathbf{B}_{\mathrm{loc}} - \mathbf{B}_{\mathrm{loc},1}| = \sqrt{B_{\mathrm{loc}}^2 + B_{\mathrm{loc},1}^2 - 2\cos\theta B_{\mathrm{loc}} B_{\mathrm{loc},1}}. \tag{6.29}$$

Carrying out the integration over θ in Eq. 6.28 with the help of Eq. C.43, we get

$$J_{\mathbf{B}_{\mathrm{loc}},B_0}(\mathbf{B}_{\mathrm{loc},1}) = \frac{B_0}{B_{\mathrm{loc}} B_{\mathrm{loc},1}}, \tag{6.30}$$

with the condition $(B_0 - B_{\mathrm{loc}})^2 \leq B_{\mathrm{loc},1}^2 \leq (B_0 + B_{\mathrm{loc}})^2$ and $J_{\mathbf{B}_{\mathrm{loc}},B_0}(\mathbf{B}_{\mathrm{loc},1}) = 0$ otherwise. With this result, the integration over $B_{\mathrm{loc},1}$ in Eq. 6.27 is easily performed with the result

$$D_{\mathrm{v}}(\mathbf{B}_{\mathrm{loc}}) = \frac{1}{(2\pi)^{3/2}\Delta_{\mathrm{G,TF}} B_0 B_{\mathrm{loc}}} \exp\left(-\frac{B_0^2 + B_{\mathrm{loc}}^2}{2\Delta_{\mathrm{G,TF}}^2}\right) \sinh\left(\frac{B_0 B_{\mathrm{loc}}}{\Delta_{\mathrm{G,TF}}^2}\right). \tag{6.31}$$

This distribution is isotropic since it depends only on the modulus of the local field. Introducing this distribution in Eq. 6.14, the Koptev–Tarasov polarisation function is found:

$$P_Z^{\mathrm{stat}}(t) = \frac{1}{3} + \frac{2}{3}\exp\left(-\frac{\gamma_\mu^2 \Delta_{\mathrm{G,TF}}^2 t^2}{2}\right)\left[\cos(\gamma_\mu B_0 t) - \frac{\gamma_\mu \Delta_{\mathrm{G,TF}}^2 t}{B_0}\sin(\gamma_\mu B_0 t)\right]. \tag{6.32}$$

We write $\Delta_{\mathrm{G,TF}}^2$ instead of $\Delta_{\mathrm{G,ZF}}^2$. This is justified since the second term of the right-hand-side of the equation describes a damped oscillation which accounts for the effect on $P_Z^{\rm stat}(t)$ of the components of the field transverse to the Z direction. The Koptev–Tarasov polarisation function must be used for data recorded on a polycrystalline sample with no preferential orientation of the crystallites. It does not include any spin–lattice relaxation mechanism.

A phenomenological extension for a sample with preferential orientation is easily written:

$$P_Z^{\rm stat}(t) = a_\parallel + a_\perp \exp\left(-\frac{\gamma_\mu^2 \Delta_{\rm G,TF}^2 t^2}{2}\right)\left[\cos(\gamma_\mu B_0 t) - \frac{\gamma_\mu \Delta_{\rm G,TF}^2 t}{B_0}\sin(\gamma_\mu B_0 t)\right], \tag{6.33}$$

with $a_\parallel + a_\perp = 1$ since $P_Z^{\rm stat}(t) = 1$ by definition. The effect of the preferential orientation is only approximately modelled because $D_{\rm v,2}(\mathbf{B}_{\rm loc,2})$ is still supposed to be isotropic. However, this expression enables a first step description of spectra recorded for such samples to be performed. A more elaborate treatment requires a model for the distribution of orientations.

For an insight into the physics of Eq. 6.32, we investigate two of its asymptotic limits. If B_0 vanishes, we recover the Kubo–Toyabe formula given in Eq. 6.17. In order to determine the asymptotic limit for large B_0 values, we first write Eq. 6.32 in an equivalent form:

$$P_Z^{\rm stat}(t) = \frac{1}{3} + \frac{2}{3}\sqrt{1 + \left(\frac{\gamma_\mu \Delta_{\rm G,TF}^2 t}{B_0}\right)^2}\exp\left(-\frac{\gamma_\mu^2 \Delta_{\rm G,TF}^2 t^2}{2}\right) \times \cos\left[\gamma_\mu B_0 t + \arctan\left(\frac{\gamma_\mu \Delta_{\rm G,TF}^2 t}{B_0}\right)\right]. \tag{6.34}$$

Expanding Eq. 6.34 in first order in $\gamma_\mu \Delta_{\rm G,TF}^2 t/B_0$ we get

$$P_Z^{\rm stat}(t) \simeq \frac{1}{3} + \frac{2}{3}\exp\left(-\frac{\gamma_\mu^2 \Delta_{\rm G,TF}^2 t^2}{2}\right)\cos\left[\left(\gamma_\mu B_0 + \frac{\gamma_\mu \Delta_{\rm G,TF}^2}{B_0}\right)t\right]. \tag{6.35}$$

This formula is the sum of two terms. The first one, which stands for one-third of the polarisation, is time independent. The origin of the second term is evident when the result of the high-field approximation is considered: it describes a damped oscillation with a slightly shifted angular frequency relative to the natural angular frequency $\gamma_\mu B_0$. Its relative weight is $\frac{2}{3}$. The oscillations arise from the coherent precession of the muon spins in the magnetic field arising from the ordered magnetic structure. The degree of coherence is measured by the damping rate $\gamma_\mu \Delta_{\rm G,TF}$. These characteristic damped oscillations appear right at the magnetic phase transition when cooling down the sample. If B_0 is really large compared to $\Delta_{\rm G,TF}$, Eq. 6.35 can be further simplified as

$$P_Z^{\rm stat}(t) \simeq \frac{1}{3} + \frac{2}{3}\exp\left(-\frac{\gamma_\mu^2 \Delta_{\rm G,TF}^2 t^2}{2}\right)\cos(\gamma_\mu B_0 t). \tag{6.36}$$

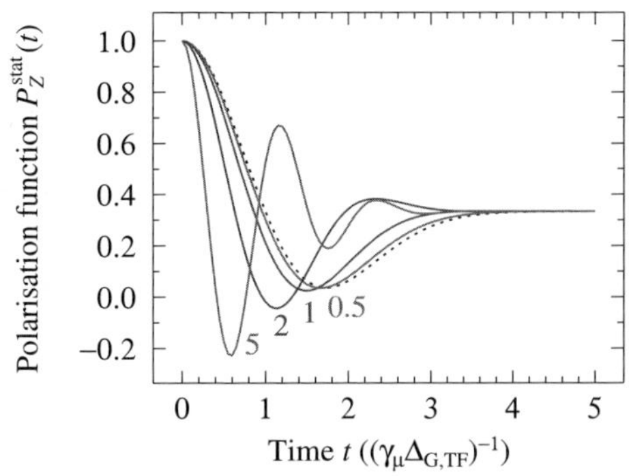

Fig. 6.6 Zero-field polarisation functions due to a static isotropic Gaussian field distribution to which is superposed a randomly oriented field of uniform modulus B_0. The dashed line is the Kubo–Toyabe function corresponding to Eq. 6.17 and the solid lines are the predictions of Eq. 6.32 for different values of the ratio $B_0/\Delta_{\mathrm{G,TF}}$ marked next to the curves. As usual. $\Delta_{\mathrm{G,TF}}$ is the standard deviation of the component field distributions. The comparison shows that a spontaneous field B_0 and therefore a magnetic order affects the shape of the spectra only if it is large enough: $B_0/\Delta_{\mathrm{G,TF}} \gtrsim 2$. For $B_0/\Delta_{\mathrm{G,TF}} \lesssim 1$ the spectra are very similar to the Kubo–Toyabe function, admittedly with a slightly rescaled $\Delta_{\mathrm{G,TF}}$ value.

This formula is understood intuitively: the terms $\frac{1}{3}$ and $\frac{2}{3}$ are the signature of the random orientation of a powdered sample and the $\exp\left(-\gamma_\mu^2 \Delta_{\mathrm{G,TF}}^2 t^2/2\right)$ factor accounts for the decay of the wiggles. Although this latter formula has been used intensively to analyse zero-field data recorded in a powder sample, our discussion shows that it is only valid in the very-high-field limit, i.e. when the spontaneous field is large. Analysing μSR spectra in the ordered state of a magnet near its ordering temperature with Eq. 6.36 may lead to a flawed interpretation because $\Delta_{\mathrm{G,TF}}$ is expected to be on the same order of magnitude as B_0. In fact, when looking for a magnetic phase transition, the μSR spectra should be first analysed with Eq. 6.32 and the asymptotic limits considered thereafter.

Note that if B_0 is small relative to $\Delta_{\mathrm{G,TF}}$, it is not possible to determine whether the observed polarisation arises from a disordered magnetic ground state characterised only by a field distribution or a spontaneous field for which a coherent damped oscillation is expected. In this limit Eqs. 6.17 and 6.32 lead to the same spectral shape. It is only when B_0 is sufficiently large relative to $\Delta_{\mathrm{G,TF}}$ that a presence of coherent oscillations can be safely inferred as seen in Fig. 6.6.

We have supposed that a Gaussian field distribution is superposed on a single spontaneous field. This latter hypothesis may not be valid. A simple example is a magnet with an incommensurate amplitude-modulated antiferromagnetic structure. This will be discussed in Section 10.1.1, which deals with the investigation of magnetic materials.

Sometimes the wiggles of a zero-field spectrum recorded in the ordered state of a magnet cannot be observed because of the limited resolution of the spectrometer (see Section 10.3.1). However, if the spin–lattice relaxation rate is not too large, it should

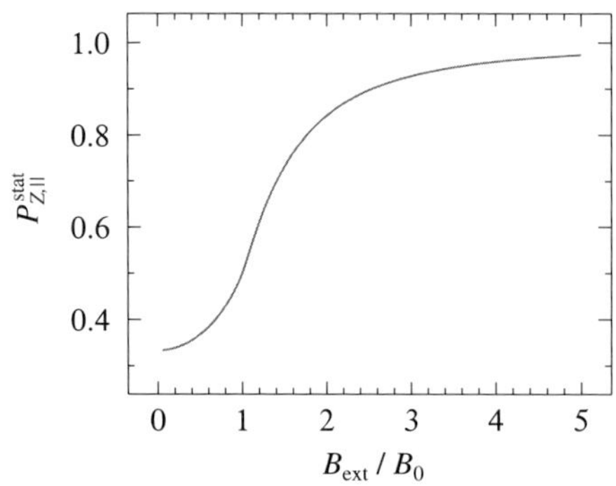

Fig. 6.7 Amplitude $P_{Z,\parallel}^{\text{stat}}$ of the spin–lattice relaxation channel as a function of B_{ext}/B_0, in the case of a polycrystalline sample in its magnetically ordered phase (see Eq. 6.38). The field $\mathbf{B}_0$ denotes the spontaneous field at the muon site.

be possible to observe the fraction of the signal arising from the spin–lattice relaxation channel, that is, the non oscillating component. In this case it might be possible to determine the magnitude of the spontaneous field through a study of the longitudinal field dependence of the signal.

Let us first consider measurements performed on a single crystal. In the same way as in the beginning of Section 6.2, $D_{\text{v},\mathbf{B}_{\text{ext}}}(\mathbf{B}_{\text{loc}})$ can be written as $\delta(\mathbf{B}_{\text{loc}} - \mathbf{B}_{\text{ext}} - \mathbf{B}_0)$, where $\mathbf{B}_0$ is the spontaneous field to which the muons are submitted. We compute $P_{Z,\parallel}^{\text{stat}}$, i.e. the non-oscillating term in Eq. 6.6:

$$P_{Z,\parallel}^{\text{stat}} \equiv \int \left(\frac{B_{\text{loc}}^Z}{B_{\text{loc}}}\right)^2 D_{\text{v},\mathbf{B}_{\text{ext}}}(\mathbf{B}_{\text{loc}})\, \mathrm{d}^3\mathbf{B}_{\text{loc}} = \frac{(\cos\theta + b_{\text{ext}})^2}{\sin^2\theta + (\cos\theta + b_{\text{ext}})^2}. \tag{6.37}$$

The reduced field b_{ext} is defined as B_{ext}/B_0 and θ denotes the polar angle of $\mathbf{B}_0$. As expected, $P_{Z,\parallel}^{\text{stat}} = 1$ for $\theta = 0$ no matter the value of b_{ext}. On the other hand, if $\theta = \pi/2$, $P_{Z,\parallel}^{\text{stat}} = 0$ for $b_{\text{ext}} = 0$ and $P_{Z,\parallel}^{\text{stat}}$ saturates to a value of 1 if b_{ext} is large enough, i.e. the external field repolarises the muon signal. The field magnitude B_0 and the angle θ are *a priori* unknown parameters. A measurement of the field dependence of $P_{Z,\parallel}^{\text{stat}}$ gives a way of determining them through the use of Eq. 6.37.

For a randomly oriented powder grain sample, the average over the solid angle must be performed. We obtain

$$P_{Z,\parallel}^{\text{stat}} = \frac{3}{4} - \frac{1}{4b_{\text{ext}}^2} + \frac{(b_{\text{ext}}^2 - 1)^2}{16b_{\text{ext}}^3} \ln \frac{(b_{\text{ext}} + 1)^2}{(b_{\text{ext}} - 1)^2}. \tag{6.38}$$

This function is plotted in Fig. 6.7. As expected, $P_{Z,\parallel}^{\text{stat}} = \frac{1}{3}$ at $b_{\text{ext}} = 0$ and it increases with the external field. This effect is often referred to as the repolarisation of the muon ensemble. The measurement of $P_{Z,\parallel}^{\text{stat}}$ as a function of b_{ext}, i.e. B_{ext}, thus provides a simple way of estimating the magnitude of $\mathbf{B}_0$.

We should note that the model of Eq. 6.6 which has been used to derive Eqs. 6.37 and 6.38 does not account for the spin–lattice relaxation. In practice this relaxation, often observed, may result in a decay of the non-oscillating term in Eq. 6.6 from its initial value $P_{Z,\parallel}^{\rm stat}$.

6.3 Time-dependent field distribution in a perfect crystal

6.3.1 Stochastic account of dynamics

In the previous section we implicitly supposed the muon to be at rest in the compound under study and the magnetic moments at the origin of the field distribution at the muon site to be static. In practice, these two hypotheses are rarely valid. In fact, the zero-field μSR technique is quite sensitive to a weak dynamical character of the magnetic moments or to a slight muon mobility. Therefore we should always *a priori* consider that the field distribution is dynamic.

Of special interest are time-dependent random events, that is, *stochastic* processes, whose properties do not change with the origin of time. They are called stationary. There is an important subclass of such stochastic processes: the Markovian processes. Their time evolution proceeds in jumps and the hopping probability does not depend on the state of the system before the jump. The Brownian motion is the oldest known Markovian process. An example of an *a priori* non-Markovian process is the motion of a positive muon between interstitial sites by a phonon-assisted tunnelling mechanism. As discussed in more details in Chapter 9, it occurs when the lattice vibrations equalise the energy of the two sites involved. At the instant when the target site is reached, the probability for the muon to tunnel back to its original site is much larger than it is to tunnel to other neighbouring sites; i.e. the process is not Markovian. However, up to now, no convincing signature of the breakdown of the Markovian hypothesis has been presented. Therefore, in this textbook we shall mainly use the Markovian approach and, for special cases, the type of Bloch equations first introduced in the μSR field by Kehr *et al.* (1978).

We shall use results which were found for spin resonance techniques and which have been proved quite successful. The theories dealing with the narrowing of their spectral lines by spin exchange or Browian motion have a long history. An early application was made by Anderson (1954) and the subject also owes much to Kubo (1954, 1969). This method has the advantage of simplicity and it can easily be adapted to any physical situation. As shown for μSR spectroscopy by Schenck (1985), the comparison between the results of different theoretical methods shows that they are in agreement in limiting cases. Because of the inherent approximations in these theoretical approaches, when a difference is noticed, it is not meaningful. Nowadays, an efficient and simple method for testing the validity of the Markovian approach is to compare its results with those obtained from the Monte Carlo technique. With this latter method, for a given model, the polarisation functions can be numerically computed without any approximation.

To proceed further, in addition to the Markovian hypothesis, one more piece of information on the dynamic process is needed. We shall suppose a Gaussian process. The significance of this approximation will be clarified in the course of the discussion.

Considering a stationary-Gaussian–Markovian process, the following useful property holds

$$\overline{B_{\rm loc}^{\alpha}(t_0)B_{\rm loc}^{\alpha}(t_0+t)} = \overline{(B_{\rm loc}^{\alpha})^2} \exp\left(-\nu_{\rm c}|t|\right), \tag{6.39}$$

where ν_c is the correlation frequency of B^{α}_{loc}. Here, the overbar stands for the weighted average of the term below it according to the distribution $D_c\left(B^{\alpha}_{\text{loc}}\right)$. The result written in Eq. 6.39 is known as the Doob theorem. As pointed out by Abragam — Chapter 10 in Abragam (1961) — the exponential decay which follows from the stationary-Gaussian–Markovian hypothesis is not satisfactory: indeed, the first derivative of the correlation function with respect to t does not vanish at $t = 0$. However, the result of Eq. 6.39 has the advantage of simplicity. Other choices for the time decay, which will be discussed in Chapter 10, show that the results obtained with the exponential approximation are quite useful.

Before proceeding further, we need to specify three points.

First, we note that a single crystallographic muon site may correspond to more than one type of magnetic site. The muon in metallic iron provides a good example (Seeger 1978): although the octahedral (and tetrahedral) interstitial sites are all crystallographically equivalent in a body-centred-cubic crystal structure, they are not magnetically equivalent. This is further discussed in Section D.7.

Second, we shall consider the one-state models, for which the muon spins probe only one type of magnetic environment. The two-state models, which are the simplest approximation to the multi-state environment, will be discussed in the section dealing with the polarisation functions for a partially disordered system (see Section 6.4). The reason for this choice is to avoid duplication because the two-state models are used for both perfect and disordered systems.

Third, for definitiveness, we shall use the language of diffusion. However, the reader should be aware that this does not mean that we restrict ourselves only to the description of the effects of muon diffusion on the polarisation functions. The results apply equally well for the magnetic moment dynamics.

We now consider the transverse polarisation function, assuming a Gaussian process to be at work. We take the external field to be strong enough that $P_X(t)$ is solely determined by $D_{c,\mathbf{B}_{\text{ext}}}(B^Z_{\text{loc}})$ (see Eq. 6.11). As the muon jumps from site to site, B^Z_{loc} becomes time dependent. The total precession angle for the muon spin between times 0 and t can be written as $\int_0^t \gamma_\mu B^Z_{\text{loc}}(t')\mathrm{d}t'$ and $P_X(t)$ is expressed as

$$P_X(t) = \mathcal{R}e\left\{\overline{\exp\left[i\int_0^t \gamma_\mu B^Z_{\text{loc}}(t')\mathrm{d}t'\right]}\right\}, \tag{6.40}$$

where $\mathcal{R}e\{\overline{A}\}$ ($\mathcal{I}m\{\overline{A}\}$) stands for the real (imaginary) part of the statistical average of A. If B^Z_{loc} is not distributed, we recover the expected result: $P^{\text{stat}}_X(t) = \cos\left(\gamma_\mu B^Z_{\text{ext}}t\right)$. In fact $B^Z_{\text{loc}}(t) = B^Z_{\text{loc}} + \delta B^Z_{\text{loc}}(t)$, where $\delta B^Z_{\text{loc}}(t)$ is the field fluctuation term. Here, assuming the Knight shift to be negligible, we set $B^Z_{\text{loc}} = B_{\text{ext}}$. Hence,

$$\begin{aligned} P_X(t) = {} & \mathcal{R}e\left\{\overline{\exp\left[i\int_0^t \gamma_\mu \delta B^Z_{\text{loc}}(t')\mathrm{d}t'\right]}\right\}\cos\left(\gamma_\mu B_{\text{ext}}t\right) \\ & - \mathcal{I}m\left\{\overline{\exp\left[i\int_0^t \gamma_\mu \delta B^Z_{\text{loc}}(t')\mathrm{d}t'\right]}\right\}\sin\left(\gamma_\mu B_{\text{ext}}t\right). \end{aligned} \tag{6.41}$$

Assuming $B^Z_{\text{loc}}(t)$ to follow a Gaussian process, the following relation holds:

$$\overline{\exp\left[i\int_0^t \gamma_\mu \delta B_{\rm loc}^Z(t')\mathrm{d}t'\right]} = \exp\left[-\int_0^t \mathrm{d}t' \int_0^t \gamma_\mu^2 \overline{\delta B_{\rm loc}^Z(t')\,\delta B_{\rm loc}^Z(t'')}\mathrm{d}t''\right]. \qquad (6.42)$$

As, in addition, the process is stationary, there is a further simplification:

$$\overline{\exp\left[i\int_0^t \gamma_\mu \delta B_{\rm loc}^Z(t')\mathrm{d}t'\right]} = \exp\left[-\int_0^t \mathrm{d}t' \int_0^t \gamma_\mu^2 \overline{\delta B_{\rm loc}^Z \delta B_{\rm loc}^Z(t'-t'')}\mathrm{d}t''\right]. \qquad (6.43)$$

Thus the second term on the right-hand-side of Eq. 6.41 vanishes. As the double integral of Eq. 6.43 can be reduced to a single integral (see Eq. C.91) and the Doob theorem is applicable (see Eq. 6.39), we finally obtain

$$\begin{aligned} P_X(t) &= \exp\left[-\gamma_\mu^2 \Delta_{\rm G,TF}^2 \int_0^t (t-\tau)\exp(-\nu_{\rm c}\tau)\mathrm{d}\tau\right] \cos(\gamma_\mu B_{\rm ext} t) \\ &= \exp\left\{-\frac{\gamma_\mu^2\Delta_{\rm G,TF}^2}{\nu_{\rm c}^2}\left[\exp(-\nu_{\rm c}t) - 1 + \nu_{\rm c}t\right]\right\}\cos(\gamma_\mu B_{\rm ext} t), \end{aligned} \qquad (6.44)$$

where we have set $\Delta_{\rm G,TF}^2 = \overline{(\delta B_{\rm loc}^Z)^2}$. Although the correlation time of the local field, $\tau_{\rm c}$ ($\tau_{\rm c} = 1/\nu_{\rm c}$), is not the mean residence time $\bar\tau$ of the muon in a site (Kehr *et al.* 1978), it is usually not very much different. This is further discussed in Sections 7.1.4.2 and 9.1.

The envelope for the polarisation function in Eq. 6.44, first derived by Anderson (1954), is often referred to as the Abragam formula in the μSR literature. Two asymptotic limits are of interest. If $\nu_{\rm c}t \ll 1$, we recover the result obtained for a static field distribution in the high-field approximation (Eq. 6.12). In the other extreme limit, $\nu_{\rm c}t \gg 1$, we get the motional narrowing expression

$$P_X(t) = \exp(-\lambda_X t)\cos(\gamma_\mu B_{\rm ext} t), \qquad \lambda_X = \gamma_\mu^2\Delta_{\rm G,TF}^2/\nu_{\rm c} = \gamma_\mu^2\Delta_{\rm G,TF}^2\tau_{\rm c}. \qquad (6.45)$$

Hence, when the muon dynamics is fast, the envelope of the oscillations is an exponential function with a damping rate λ_X proportional to the field correlation time. This means that the dynamics reduces the damping of the oscillations; i.e. the effective field distribution probed by the muon spins is narrowed by the dynamics. It is important to note that the time scale t appearing in the inequalities $\nu_{\rm c}t \ll 1$ and $\nu_{\rm c}t \gg 1$ is not the time window of an experiment ($t_w \simeq 15\,\mu$s) but is set by $(\gamma_\mu\Delta_{\rm G,TF})^{-1}$. Thus the motional narrowing limit is reached when $\nu_{\rm c}/(\gamma_\mu\Delta_{\rm G,TF}) \gg 1$.

Remarkably, the shape of the envelope strongly depends on the muon mobility. An estimate of the width of the effective field distribution probed by the spin of fast-moving muons can be obtained by recalling the basic property of a stochastic variable (Karlsson 1995). In our hypothesis of a strong enough external field, the distribution of $B_{\rm loc}^Z$ determines the polarisation function. The damping of $P_X(t)$ is related to the difference in the fields acting on the muon spins with respect to the average field. This difference is a stochastic variable and for simplicity we assume it can take only two different values $\pm\Delta B_1^Z$. The corresponding phase difference in the muon spin precession is $\Delta\varrho = \pm\gamma_\mu\Delta B_1^Z t$ after time t. Because of the random walk nature of the field difference, the mean square deviation angle after n steps is $\overline{(\Delta\varrho)^2} = n(\gamma_\mu \Delta B_1^Z \tau_c)^2$, where the mean time between two steps is τ_c. Denoting t_n the time elapsed after n steps, i.e. $t_n \equiv n\tau_c$, we have $\overline{(\Delta\varrho)^2} = \gamma_\mu^2 \Delta_{G,TF}^2 \tau_c t_n$ where we have set $\Delta_{G,TF} = \Delta B_1^Z$. Therefore this simple random walk model qualitatively illustrates the motional narrowing of the field distribution sensed by the muon spins. The faster the field fluctuations are, the smaller τ_c is and therefore the smaller $\overline{(\Delta\varrho)^2}$ is, which means that $P_X(t)$ is less damped.

It is possible to describe the effect of the muon mobility on $P_Z(t)$ as we have done for $P_X(t)$, assuming the field modulation to follow a Gaussian–Markovian process. This is the method originally chosen by Kubo and Toyabe. However, nowadays $P_\alpha(t)$, where α stands for X or Z, is usually described by the so-called strong-collision model, which is appealing since it is quite simple and can easily be generalised. An example of the general nature of the method will be given at the end of this section. Strong collision means that when the muon jumps from site i with field $\mathbf{B}_{\mathrm{loc},i}$ to site j with field $\mathbf{B}_{\mathrm{loc},j}$ it experiences a sudden change in $\mathbf{B}_{\mathrm{loc}}$, not a gradual change as assumed in the Gaussian–Markovian approach (Kubo 1954), which is sometimes referred to as the weak-collision model.

Two different equivalent formulations of the strong-collision model are possible: either a diffusion equation of the Fokker–Planck type is introduced — see Kubo and Toyabe (1967) — or the field modulation is seen as a random walk process. We will work with the second formulation. A nice advantage is that the description of the effect of the field modulation on a polarisation function is based on a simple physical picture. Its basic feature is that $\mathbf{B}_{\mathrm{loc}}$ keeps a certain value for a given time interval. At a later times it takes a new value not correlated to the first one. During the time spent in the sample, the muon spin senses ℓ different fields. Using this random walk approach (Kehr *et al.* 1978),

$$P_\alpha(t) = \sum_{\ell=0}^{+\infty} R_{\alpha,\ell}(t), \tag{6.46}$$

where $R_{\alpha,\ell}(t)$ is the average contribution of the muons which have performed ℓ jumps until time t. For example we have

$$R_{\alpha,0}(t) = \overline{M_i^{\alpha\alpha}(t)} \exp(-\nu_c t) = P_\alpha^{\mathrm{stat}}(t) \exp(-\nu_c t), \tag{6.47}$$

which is the product of two factors. In the first one, $M_i^{\alpha\alpha}(t)$ accounts for the evolution of the α component of a unit vector colinear to the muon spin submitted to the local field $\mathbf{B}_{\mathrm{loc},i}$ assumed to act during the time interval $[0, t]$. The expression for $M_i^{\alpha\alpha}(t)$ can be obtained from the material given in Section 3.1.1. Compared to the notation used in this latter section, here we have added the subscript i to emphasise that we consider the evolution of the muon spin submitted to a specific local field $\mathbf{B}_{\mathrm{loc},i}$. This field belongs to the distribution that the muon spins probe, and accordingly $\overline{M_i^{\alpha\alpha}(t)}$ denotes the average when $\mathbf{B}_{\mathrm{loc},i}$ runs over the distribution. Following Eqs. 6.5 and 6.6 this average is simply $P_\alpha^{\mathrm{stat}}(t)$. Note here that, in the case of the transverse polarisation function $P_X(t)$, we only consider the distribution of B_{loc}^Z. This means that only the envelope of the oscillations can be studied, since the actual polarisation function is the product of this envelope function by the precession term $\cos(\gamma_\mu B_{\rm ext} t)$. This latter term is indeed not modified by the dynamical character of $\mathbf{B}_{\rm loc}$. The second factor in Eq. 6.47 is the function $\exp(-\nu_{\rm c} t)$: it gives the probability for the muon spins to sense the same static field distribution until t; i.e. no field change has occurred until t. The fact that the static field distribution is the same for all the sites means that $P_\alpha^{\rm stat}(t)$ does not depend on the muon site. In the same way,

$$R_{\alpha,1}(t) = \overline{\int_0^t M_i^{\alpha\alpha}(t-t') \exp[-\nu_{\rm c}(t-t')] \nu_{\rm c} M_j^{\alpha\alpha}(t') \exp(-\nu_{\rm c} t') {\rm d}t'}. \tag{6.48}$$

The last factor to the right of the integrand represents the probability that the muon spin has precessed in $\mathbf{B}_{{\rm loc},j}$ until t', $\nu_{\rm c}$ the rate of jump from $\mathbf{B}_{{\rm loc},j}$ to $\mathbf{B}_{{\rm loc},i}$ and the fourth factor from the end, the probability of retaining this latter field from t' to t. Note that the ensemble average denoted by $\overline{\ldots}$ in Eq. 6.48 is here twofold since the distributions described by $\mathbf{B}_{{\rm loc},j}$ and $\mathbf{B}_{{\rm loc},i}$ must be considered. The assumption of strong collisions means that we can decouple the terms of $R_{\alpha,1}(t)$ as

$$\begin{aligned} R_{\alpha,1}(t) &= \int_0^t \overline{M_i^{\alpha\alpha}(t-t')} \exp[-\nu_{\rm c}(t-t')] \nu_{\rm c} \overline{M_j^{\alpha\alpha}(t')} \exp(-\nu_{\rm c} t') {\rm d}t' \\ &= \int_0^t P_\alpha^{\rm stat}(t-t') \exp[-\nu_{\rm c}(t-t')] \nu_{\rm c} P_\alpha^{\rm stat}(t') \exp(-\nu_{\rm c} t') {\rm d}t' \\ &= \nu_{\rm c} \int_0^t R_{\alpha,0}(t-t') R_{\alpha,0}(t') {\rm d}t'. \end{aligned} \tag{6.49}$$

With the same method we get the recursion relation:

$$R_{\alpha,\ell+1}(t) = \nu_{\rm c} \int_0^t R_{\alpha,0}(t-t') R_{\alpha,\ell}(t') {\rm d}t' = \nu_{\rm c} \int_0^t R_{\alpha,0}(t') R_{\alpha,\ell}(t-t') {\rm d}t'. \tag{6.50}$$

It is useful to consider the following identity which is derived using Eqs. 6.46 and 6.50:

$$\begin{aligned} \nu_{\rm c} \int_0^t P_\alpha(t-t') R_{\alpha,0}(t') {\rm d}t' &= \nu_{\rm c} \int_0^t \sum_{\ell=0}^{+\infty} R_{\alpha,\ell}(t-t') R_{\alpha,0}(t') {\rm d}t' \\ &= \sum_{\ell=0}^{+\infty} R_{\alpha,\ell+1}(t) = P_\alpha(t) - R_{\alpha,0}(t). \end{aligned} \tag{6.51}$$

A close look at this identity shows that we can write an integral equation for $P_\alpha(t)$ in terms of $P_\alpha^{\rm stat}(t)$:

$$P_\alpha(t) = P_\alpha^{\rm stat}(t) \exp(-\nu_{\rm c} t) + \nu_{\rm c} \int_0^t P_\alpha(t-t') P_\alpha^{\rm stat}(t') \exp(-\nu_{\rm c} t') {\rm d}t'. \tag{6.52}$$

If the field distribution is static ($\nu_{\rm c} = 0$), the second term on the right-hand side vanishes and we recover, as expected, the polarisation function for a static field distribution.

The derivation of $P_\alpha(t)$ does not require $P_\alpha^{\rm stat}(t)$ to be specified. In fact, that latter function can be computed using the field distribution method as in this chapter or directly by quantum mechanics as will be detailed in the next chapter. Therefore $P_\alpha(t)$ given by Eq. 6.52 may look quite general. However, it should be realised that a key physical hypothesis has been made. We have assumed a Markovian process for the field sequence probed by the muons as they diffuse; see beginning of the current section for a definition of a Markovian process. This can be a bad approximation if the probability for the muon to return to its original site is large or if adjacent sites share too many common neighbours. For example, let us consider the extreme case of a muon jumping between two sites with the same neighbouring spins. The actual shape of the polarisation function will still be the same as obtained for a static muon, even for a very fast mobility, in contrast to the prediction of the strong-collision model. This is so because the field probed by the muon is independent of its mobility. Since the strong-collision model neglects the revisiting of the local field which was probed just before the jump, its breakdown should be important in a one-dimensional ($d = 1$) lattice, and less pronounced as the dimensionality increases: the strong-collision model is exact for a $d = \infty$ lattice. The strong-collision model is, in fact, a kind of mean-field approximation. Using analytical and numerical methods, it has been shown for a Gaussian field distribution that the envelope of $P_X(t)$ is an exponential function at large times (Czech and Kehr 1986) for a $d = 1$ lattice. The damping rate is proportional to $\Delta_{\rm G,TF}^{\beta}$ with $\beta = \frac{4}{3}$ instead of $\beta = 2$ for $d = 3$ (see Eq. 6.45). For a $d = 2$ lattice $\beta \simeq 1.77$. Remarkably, there is no finite magnetic correlation time for $d = 1$ and $d = 2$. The muon zero-field polarisation function for a one-dimensional chain is further discussed by Czech (1989). The case of a muonium centre for which the muonium electron is performing a one-dimensional random walk is discussed in Section 8.4.3.

We recall that, while for definitiveness we have used the language of diffusion, the present results apply equally well to the description of the physics of static muons probing the spin dynamics of their environment.

6.3.2 Applications of the strong collision model

In this section we apply the previous results to different physical cases.

First we note that the polarisation functions obey a simple scaling law. The parameters $\Delta_{\rm G}$ (equal to $\Delta_{\rm G,TF}$ or $\Delta_{\rm G,ZF}$), $B_{\rm ext}$, and $\nu_{\rm c}$ appear only in products as $\gamma_\mu \Delta_{\rm G} t$, $\gamma_\mu B_{\rm ext} t$, and $\nu_{\rm c} t$, respectively. Let us multiply these three parameters by a real positive number $f_{\rm r}$. Then $P_\alpha(t)$ depends on $f_{\rm r}\gamma_\mu \Delta_{\rm G} t$, $f_{\rm r}\gamma_\mu B_{\rm ext} t$, and $f_{\rm r}\nu_{\rm c} t$ or $\gamma_\mu \Delta_{\rm G}(f_{\rm r} t)$, $\gamma_\mu B_{\rm ext}(f_{\rm r} t)$, and $\nu_{\rm c}(f_{\rm r} t)$. In other words, as first noticed by Uemura *et al.* (1994),

$$P_\alpha(f_{\rm r}\Delta_{\rm G}, f_{\rm r}B_{\rm ext}, f_{\rm r}\nu_{\rm c}, t) = P_\alpha(\Delta_{\rm G}, B_{\rm ext}, \nu_{\rm c}, f_{\rm r} t). \tag{6.53}$$

This scaling law tells us that if the muon ensemble senses a local field of significant magnitude for only a fraction $f_{\rm r}$ ($0 < f_{\rm r} < 1$) of the time, an observer performing a zero-field measurement will find a reduced width and correlation time; i.e. he will measure $f_{\rm r}\Delta_{\rm G}$ and $f_{\rm r}\nu_{\rm c}$ because his time scale is t and not $f_{\rm r} t$, assuming of course that the observer is not aware that $f_{\rm r} < 1$. Longitudinal-field measurements help to understand the system under study since if $f_{\rm r}$ is indeed smaller than 1, the effect of the field on $P_Z(t)$ will be as if $B_{\rm ext}$ was reduced by the factor $f_{\rm r}$.

The analysis of a zero-field spectrum may yield $\Delta_{\rm G,ZF}$ and $\nu_{\rm c}$ in the most favourable cases, or only the ratio $\Delta^2_{\rm G,ZF}/\nu_{\rm c}$ if the dynamics is fast. The investigation of the effect of an applied field on the polarisation function is always recommended. If the zero-field spectrum yields both $\Delta_{\rm G,ZF}$ and $\nu_{\rm c}$, the study of the field dependence will give the possibility of checking the consistency of the analysis, i.e. to determine whether $f_{\rm r} = 1$. In the motional narrowing limit, a sufficiently intense longitudinal field will quench the relaxation. The analysis of the quenching curves give access to both $\Delta_{\rm G,ZF}$ and $\nu_{\rm c}$; see Section 6.3.2.5.

6.3.2.1 Numerical evaluation of $P_\alpha^{\rm stat}(t)$. Even for the simplest possible $P_\alpha^{\rm stat}(t)$ functional form, an analytical expression for $P_\alpha(t)$ is usually not available and therefore it must be numerically evaluated. In the past, a method based on Laplace transform was introduced. It is numerically unstable. An efficient and simple method is the iterative method for which the expression of an integral is explicitly used. For the sake of simplicity, we express here an integral with the trapezium rule — see, for example, James *et al.* (1999) or Press *et al.* (2007). More efficient estimates of an integral are obviously possible. We subdivide the time interval t in n sub-intervals of the same length δt. The expression for $P_\alpha(\ell\delta t)$ with $2 \leq \ell \leq n$ reads

$$P_\alpha(\ell\delta t) = P_\alpha^{\rm stat}(\ell\delta t)\exp(-\nu_{\rm c}\ell\delta t) + \nu_{\rm c}\delta t\sum_{k=1}^{\ell-1} P_\alpha[(\ell-k)\delta t]P_\alpha^{\rm stat}(k\delta t)\exp(-\nu_{\rm c}k\delta t)$$
$$+ \frac{\nu_{\rm c}\delta t}{2}\left[P_\alpha(\ell\delta t) + \exp(-\nu_{\rm c}\ell\delta t)\,P_\alpha^{\rm stat}(\ell\delta t)\right]. \qquad (6.54)$$

To start the iteration procedure, $P_\alpha(\delta t)$ is needed:

$$P_\alpha(\delta t) = P_\alpha^{\rm stat}(\delta t)\exp(-\nu_{\rm c}\delta t) + \frac{\nu_{\rm c}\delta t}{2}\left[P_\alpha(\delta t) + \exp(-\nu_{\rm c}\delta t)\,P_\alpha^{\rm stat}(\delta t)\right]. \qquad (6.55)$$

The computation is performed as follows. First, a value of n is chosen. Then, $P_\alpha(\delta t)$ is computed from Eq. 6.55. Thereafter, the other $P_\alpha(\ell\delta t)$ terms with $\ell = 2, ..., n$ are deduced using Eq. 6.54. Since it is an iterative method, the first terms should be evaluated with good precision.

Figure 6.8 illustrates the behaviour of the transverse- and zero-field polarisation functions in the strong-collision model assuming an isotropic Gaussian distribution of the field components, i.e. $P_X^{\rm stat}(t)$ and $P_Z^{\rm stat}(t)$, respectively, given by Eqs. 6.12 and 6.17.

While the Laplace transform is not helpful for the numerical computation of $P_\alpha(t)$, it is of interest to analytically study its behaviour. A particularly simple expression for $P_\alpha(s)$ (see Eq. C.53 for the definition of the Laplace transform) is obtained from Eq. 6.52:

$$P_\alpha(s) = \frac{P_\alpha^{\rm stat}(s+\nu_{\rm c})}{1-\nu_{\rm c}P_\alpha^{\rm stat}(s+\nu_{\rm c})}. \qquad (6.56)$$

We shall use this equation in five physical cases.

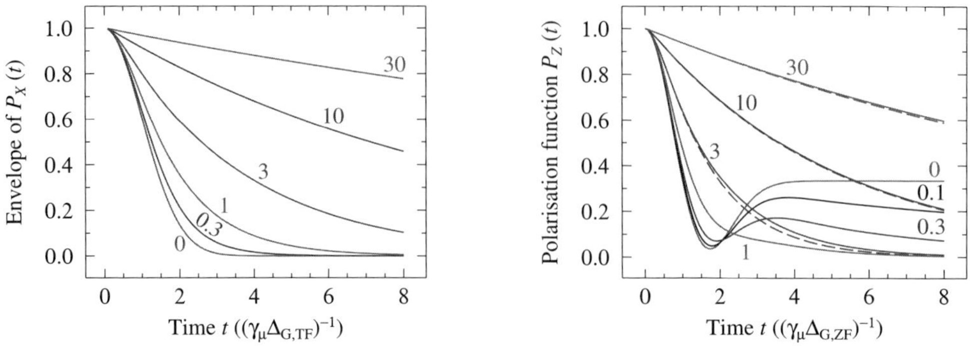

Fig. 6.8 (Left) Envelope of the high transverse-field polarisation function $P_X(t)$ for an isotropic Gaussian distribution computed with the strong-collision model. (Right) Zero–field longitudinal polarisation function $P_Z(t)$ calculated within the same model. The number written next to each curve of both panels corresponds to the value of $\nu_c/(\gamma_\mu\Delta_{G,TF})$ or $\nu_c/(\gamma_\mu\Delta_{G,ZF})$. The time scale is given in units of $(\gamma_\mu\Delta_{G,TF})^{-1}$ or $(\gamma_\mu\Delta_{G,ZF})^{-1}$. In the transverse-field case we implicitly assume $\Delta_{G,TF}$ to be much smaller than the mean field value at the muon site; see the approximation leading to Eq. 6.11. The actual polarisation function $P_X(t)$ is the product of the envelope and the oscillating term $\cos(\gamma_\mu B_{ext}t)$. On the right the dashed lines plotted for $\nu_c/(\gamma_\mu\Delta_{G,ZF}) \geq 3$ are the prediction of the analytical formula given in Eq. 6.62. Note that they differ very little from the full lines for $\nu_c/(\gamma_\mu\Delta_{G,ZF}) \geq 10$. Equation 6.62 is therefore a very reasonable approximation for $P_Z(t)$ in the $\nu_c/(\gamma_\mu\Delta_{G,ZF}) \gtrsim 3$ range. Its use saves the computation time needed to solve the integral equation in Eq. 6.52. Reproduced with kind permission from Dalmas de Réotier and Yaouanc (1997).

6.3.2.2 $P_X(t)$ **in the high-field limit.** We now analyse the high-field approximation of $P_X(t)$ within the strong-collision model. First we need to provide $P_X^{stat}(s)$. Again, we only consider the envelope of the polarisation function since the $\cos(\gamma_\mu B_{ext}t)$ oscillating term is not modified by the field dynamics. We find

$$P_X^{stat}(s+\nu_c) = \frac{1}{\gamma_\mu\Delta_{G,TF}}\sqrt{\frac{\pi}{2}}\left[1-\mathrm{erf}\left(\frac{s+\nu_c}{\sqrt{2}\gamma_\mu\Delta_{G,TF}}\right)\right]\exp\left[\left(\frac{s+\nu_c}{\sqrt{2}\gamma_\mu\Delta_{G,TF}}\right)^2\right], \tag{6.57}$$

where $\mathrm{erf}(x)$ is the error function (see Eq. C.105). To proceed further, we consider the limit $(s+\nu_c)/(\gamma_\mu\Delta_{G,TF}) \gg 1$. Since the argument of the error function is large in this limit, we use the $x \to \infty$ expansion of $\mathrm{erf}(x)$ given in Eq. C.106. Retaining the terms up to order x^{-2} we have

$$P_X^{stat}(s+\nu_c) \simeq \frac{1}{s+\nu_c}\left[1-\frac{\gamma_\mu^2\Delta_{G,TF}^2}{(s+\nu_c)^2}\right]. \tag{6.58}$$

Accounting for Eq. 6.56 and with $(s/\nu_c)[(s+\nu_c)/(\gamma_\mu\Delta_{G,TF})] \gg 1$, consistent with the assumption $(s+\nu_c)/(\gamma_\mu\Delta_{G,TF}) \gg 1$, we obtain

$$P_X(s) \simeq \frac{1}{s}\left[1 - \frac{\gamma_\mu^2 \Delta_{\rm G,TF}^2}{(s+\nu_{\rm c})^2}\left(1+\frac{\nu_{\rm c}}{s}\right)\right]. \tag{6.59}$$

We apply the method of partial fractions to this result and obtain

$$P_X(s) = \left(1 + \frac{\gamma_\mu^2 \Delta_{\rm G,TF}^2}{\nu_{\rm c}^2}\right)\frac{1}{s} - \frac{\gamma_\mu^2 \Delta_{\rm G,TF}^2}{\nu_{\rm c}}\frac{1}{s^2} - \frac{\gamma_\mu^2 \Delta_{\rm G,TF}^2}{\nu_{\rm c}^2}\frac{1}{s+\nu_{\rm c}}. \tag{6.60}$$

Performing the inverse Laplace transform of $P_X(s)$ gives

$$P_X(t) = 1 + \frac{\gamma_\mu^2 \Delta_{\rm G,TF}^2}{\nu_{\rm c}^2} - \frac{\gamma_\mu^2 \Delta_{\rm G,TF}^2}{\nu_{\rm c}}t - \frac{\gamma_\mu^2 \Delta_{\rm G,TF}^2}{\nu_{\rm c}^2}\exp(-\nu_{\rm c} t). \tag{6.61}$$

We recover the first order expansion of the envelope of Eq. 6.44. This result suggests that the so-called Abragam formula might be a reasonable approximation for the envelope derived from the strong-collision model. In fact, it has numerically been shown (Kehr *et al.* 1978) that the strong-collision model and the Abragam formula (weak-collision model) give very similar results even when $\nu_{\rm c}$ is small, i.e. when the muon is quasi-static. In practice, data analysis should be performed with the Abragam formula which is numerically quite convenient.

6.3.2.3 $P_Z(t)$ **in the fast fluctuation limit.** A study of the asymptotic expansion of $P_Z(t)$ in zero field for an isotropic Gaussian distribution can also be performed. First, we consider the fast dynamic limit, i.e. $(s+\nu_{\rm c})/(\gamma_\mu \Delta_{\rm G,TF}) \gg 1$. We express the Laplace transform $P_Z^{\rm stat}(s)$ of the static Kubo–Toyabe formula (Eq. 6.17) and using again the expansion of Eq. C.106, we find that $P_Z^{\rm stat}(s+\nu_{\rm c})$ is given by Eq. 6.58 with the substitution $\Delta_{\rm G,TF}^2 \rightarrow 2\Delta_{\rm G,ZF}^2$. Therefore we get in the motional narrowing limit, i.e. when $\nu_{\rm c}/(\gamma_\mu \Delta_{\rm G,ZF})$ is sufficiently large (Dalmas de Réotier and Yaouanc 1992),

$$P_Z(t) = \exp\left\{-2\frac{\gamma_\mu^2 \Delta_{\rm G,ZF}^2}{\nu_{\rm c}^2}\left[\exp(-\nu_{\rm c} t) - 1 + \nu_{\rm c} t\right]\right\}. \tag{6.62}$$

This is the formula for the envelope of the high-field $P_X(t)$ function (the Abragam formula) where $\Delta_{\rm G,TF}^2$ is replaced with $2\Delta_{\rm G,ZF}^2$. The factor 2 is related to the fact that the two components of $\mathbf{B}_{\rm loc}$ perpendicular to Z depolarise the muon spin in the longitudinal geometry in contrast to the high-transverse-field geometry for which only one component depolarises. The same argument explains the depolarisation rate near $t = 0$ in the static case; see the discussion following Eq. 6.17.

As we have obtained for $P_X(t)$, an expression for $P_Z(t)$ in the extreme motional narrowing limit can be derived:

$$P_Z(t) = \exp(-\lambda_Z t), \qquad \lambda_Z = 2\gamma_\mu^2 \Delta_{\rm G,ZF}^2/\nu_{\rm c} = 2\gamma_\mu^2 \Delta_{\rm G,ZF}^2 \tau_{\rm c}. \tag{6.63}$$

Hence, $P_Z(t)$ is an exponential function with a damping rate, λ_Z, proportional to the field correlation time.

6.3.2.4 $P_Z(t)$ in the quasi-static limit. We now study the Laplace transform $P_Z(s)$ in the limit $(s+\nu_c)/(\sqrt{\pi}\gamma_\mu \Delta_{G,ZF}) \ll 1$, i.e. in the quasi-static limit (ν_c small) and for small values of the Laplace parameter s. From the expansion $P_Z^{\rm stat}(s) = 1/(3s)$ of the Kubo–Toyabe function Laplace transform, the following result is found:

$$P_Z(s) \simeq \frac{1}{3}\frac{1}{s+\frac{2}{3}\nu_c}. \tag{6.64}$$

The inverse Laplace transform of this expression is simply $\frac{1}{3}\exp\left(-\frac{2}{3}\nu_c t\right)$. The exponential function describes the damping effect of the slow dynamics on the $\frac{1}{3}$ long-time tail. Note that the argument of the exponential is proportional to ν_c rather than $1/\nu_c$, as in the motional narrowing limit. The $\frac{2}{3}$ factor in front of ν_c reflects the fact that two out of the three components of $\mathbf{B}_{\rm loc}$ induce the relaxation; *i.e* the transverse components $B_{\rm loc}^X$ and $B_{\rm loc}^Y$ relax the $\frac{1}{3}$ tail. The second term of the Kubo–Toyabe function (Eq. 6.17) vanishes in the limit $\gamma_\mu \Delta_{G,ZF} t \gg 1$, and it is not influenced by a slow muon mobility. In view of this analysis, we write $P_Z(t)$ as the sum of the following two terms in the quasi-static limit:

$$P_Z(t) \simeq \frac{1}{3}\exp\left(-\frac{2}{3}\nu_c t\right) + \frac{2}{3}\left(1-\gamma_\mu^2\Delta_{G,ZF}^2 t^2\right)\exp\left(-\frac{\gamma_\mu^2\Delta_{G,ZF}^2 t^2}{2}\right). \tag{6.65}$$

A comparison of the curves plotted in Fig. 6.8 shows that the zero-field technique is far more sensitive than the transverse-field one to slow dynamics. This property has been used intensively to analyse either muon diffusion or magnetic fluctuations characterised by correlation times on the order of a microsecond or longer.

6.3.2.5 $P_Z(t)$ in a longitudinal field. A common method for measuring the characteristic fluctuation rate of a field distribution detected by the zero-field method is to analyse the field response of the system to a longitudinal field. The static function for an isotropic Gaussian distribution has been given in Eq. 6.18 and equivalent Eq. 6.19. The numerical evaluation of $P_Z(t)$ using the strong-collision model is time consuming when ν_c gets large. Hence, an analytical approximation is useful. A Laplace transform analysis of $P_Z(t)$ does not yield a tractable formula as found in the zero-field case. Fortunately, as explained Section 7.1.4.2, it is possible to derive an asymptotic formula from quantum mechanics. For completeness, we give it here:

$$P_Z(t) = \exp\left[-2\gamma_\mu^2\Delta_{G,ZF}^2\Gamma(t)\right], \tag{6.66}$$

with

$$\Gamma(t) = \frac{\left(\omega_\mu^2+\nu_c^2\right)\nu_c t + \left(\omega_\mu^2-\nu_c^2\right)\left[1-e^{-\nu_c t}\cos\left(\omega_\mu t\right)\right] - 2\nu_c\omega_\mu e^{-\nu_c t}\sin\left(\omega_\mu t\right)}{\left(\omega_\mu^2+\nu_c^2\right)^2}, \tag{6.67}$$

and $\omega_\mu = \gamma_\mu B_{\rm ext}$. It can numerically be shown that this formula gives a very good approximation to the result of the strong-collision model when the dynamics is relatively fast. Figure 6.9 shows examples.

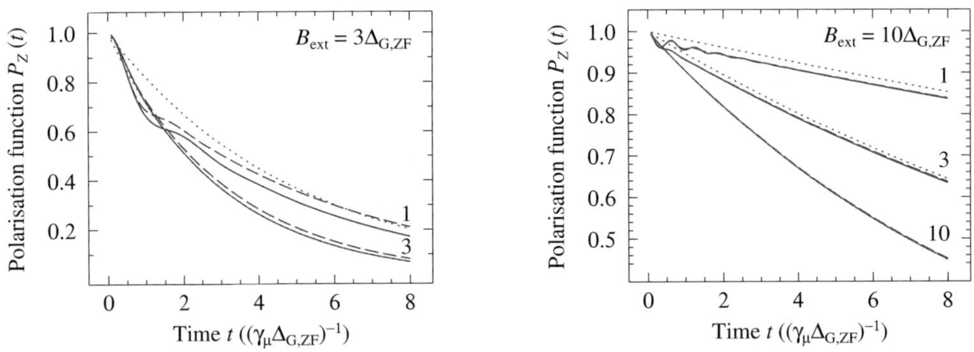

Fig. 6.9 Comparison of the longitudinal-field polarisation function $P_Z(t)$ for an isotropic Gaussian distribution computed either with the strong-collision model (full line) or with the approximation of Eq. 6.66 with Eq. 6.67 (dashed line). In addition, the exponential function with the relaxation rate calculated in the extreme motional narrowing limit (Eq. 6.68) is shown as a dotted line. The number next to each set of curves corresponds to the value of $\nu_c/(\gamma_\mu \Delta_{\text{G,ZF}})$. When the dynamics is fast enough, the strong-collision model result cannot be distinguished from the exponential. The left and right panels differ by the value of the external field. Note that $P_Z(t)$ decays faster as $\nu_c/(\gamma_\mu \Delta_{\text{G,ZF}})$ increases. This is not the usual behaviour. Yet it is observed in longitudinal fields but only if the field is strong enough ($\gamma_\mu B_{\text{ext}} \gtrsim \nu_c$), as indicated by Eq. 6.68.

In the extreme motional narrowing limit ($\nu_c t \gg 1$), $P_Z(t)$ is an exponential function with a damping rate λ_Z derived from Eq. 6.67:

$$\lambda_Z = \frac{2\gamma_\mu^2 \Delta_{\text{G,ZF}}^2 \nu_c}{\omega_\mu^2 + \nu_c^2}. \tag{6.68}$$

This equation is sometimes referred to as the Redfield formula. The validity of this limit can be judged from Fig. 6.9. According to Eq. 6.68, $\lambda_Z(\omega_\mu)$ — and consequently $\lambda_Z(B_{\text{ext}})$ — is therefore expected to follow a Lorentzian dependence. The half-width at half-maximum is a direct measure of ν_c since this width is equal to ν_c. Note that our analysis neglects any effect of B_{ext} on the system under study. In the zero-field case ($\omega_\mu = 0$) we recover the previous result given in Eq. 6.63. If $\omega_\mu \gg \nu_c$, $\lambda_Z(\omega_\mu)$ is predicted to scale to $1/\omega_\mu^2$. In fact, as discussed in Chapter 10, the scaling law followed by $\lambda_Z(\omega_\mu)$ is an efficient method for investigating the functional form of the field correlation function which does not always exponentially decay as taken in Eq. 6.39.

6.3.2.6 $P_Z(t)$ **in the case of a discrete stochastic field.** Until now we have discussed the time evolution of the polarisation function under the influence of a continuous field distribution. However, a discrete distribution can also have a profound influence. As an example, here we compute $P_Z(t)$ resulting from the flipping of an internal field, from parallel to antiparallel to the X axis and conversely. For convenience, here we denote this field as $\pm\mathbf{B}_{\text{fluc}}$. We use the framework of the strong-collision model as discussed earlier. An exposition is also given by Keren (2004). Since $P_Z^{\text{stat}}(t) = \cos(\gamma_\mu B_{\text{fluc}} t)$, from Eq. 6.56 we derive

$$P_Z(s) = \frac{s + \nu_c}{s^2 + \nu_c s + \gamma_\mu^2 B_{\rm fluc}^2}. \tag{6.69}$$

The evaluation of the inverse of $P_Z(s)$ does not pose any problem. In the case $\nu_c < 2\gamma_\mu B_{\rm fluc}$ we define $\omega_{\rm eff} = \sqrt{\gamma_\mu^2 B_{\rm fluc}^2 - \nu_c^2/4}$ and find

$$P_Z(t) = \frac{1}{\cos\varphi} \exp\left(-\frac{\nu_c t}{2}\right) \cos\left(\omega_{\rm eff} t - \varphi\right) \quad \text{with} \quad \tan\varphi = \frac{\nu_c}{2\omega_{\rm eff}}. \tag{6.70}$$

The function $P_Z(t)$ exhibits an exponentially damped oscillation. The spontaneous angular frequency is reduced from $\gamma_\mu B_{\rm fluc}$ to $\omega_{\rm eff}$ and remarkably, a negative phase shift is predicted. For $\nu_c = 2\gamma_\mu B_{\rm fluc}$,

$$P_Z(t) = \left(1 + \frac{\nu_c t}{2}\right) \exp\left(-\frac{\nu_c t}{2}\right). \tag{6.71}$$

Finally, if $\nu_c > 2\gamma_\mu B_{\rm fluc}$, setting $\nu_{\rm eff} = \sqrt{\nu_c^2/4 - \gamma_\mu^2 B_{\rm fluc}^2}$, we derive

$$P_Z(t) = \exp\left(-\frac{\nu_c t}{2}\right) \left[\cosh(\nu_{\rm eff} t) + \frac{\nu_c}{2\nu_{\rm eff}} \sinh(\nu_{\rm eff} t)\right]. \tag{6.72}$$

In the motional narrowing limit, that is, $\nu_c \gg \gamma_\mu B_{\rm fluc}$, $P_Z(t)$ is an exponential function as expected:

$$P_Z(t) = \exp\left(-\frac{\gamma_\mu^2 B_{\rm fluc}^2}{\nu_c} t\right). \tag{6.73}$$

6.4 Field distributions in a partially disordered crystal

Condensed-matter systems can be strongly disordered as in amorphous materials. Even a crystallised system may exhibit a certain degree of disorder arising, for example, from randomly placed impurities or defects. In these cases the distances or angles between the muon and the compound magnetic moments are distributed. Schematically, two situations can be foreseen. The disorder uniformly varies from site to site (homogeneous disorder, Ziman 1979) or is strongly site dependent. For the former case, it is described by performing a proper average over disorder of $P_\alpha(t)$ computed for a perfect crystal. In the latter case, a discrete multi-state model is more suitable: averaging would be meaningless because the polarisation functions for the different states are too different. In practice, for inhomogeneous disorder, we will restrict ourselves to the two-state models.

In the first step we discuss models for homogeneous disorder. For this purpose we use the so-called one-state models.

6.4.1 One-state models

It is possible to derive analytical formulae for the polarisation functions of disordered systems in two extreme limits: the diluted and the concentrated ones. We shall first present calculations for polarisation functions in the case of randomly placed diluted spin impurities in a lattice. Since in the motional narrowing limit $P_Z(t)$ is an exponential square-root function, it is tempting to generalise this form to the stretched exponential function. We shortly present a possible theoretical support for this function in the subsequent section. Then we introduce a phenomenological model to account for homogeneous disorder in the concentrated limit. Finally we compare the shape of different polarisation functions which have been derived.

6.4.1.1 Diluted magnetic impurities in a non-magnetic matrix. It is shown in Section 6.6.2 that the field distribution along the $\mathbf{B}_{\rm ext}$ direction has a Cauchy or Lorentzian-like profile for uncorrelated randomly diluted magnetic impurities if the range dependence of the muon–impurity interaction energy scales as r^{-3}. This is the case for the μSR technique since the dipolar interaction is the dominating interaction. We have therefore

$$D_{\rm c}\left(B^Z_{\rm loc}\right) = \frac{1}{\pi}\frac{\Delta_{\rm L,TF}}{\Delta^2_{\rm L,TF} + \left(B^Z_{\rm loc}\right)^2}, \tag{6.74}$$

where $\Delta_{\rm L,TF}$ is the half-width at half-maximum. This distribution displays a long tail on both sides of $B^Z_{\rm loc} = 0$, which is reflected in the fact that its second moment is undefined. This will have an important consequence when discussing the effect of dynamics. Following the method explained in Section 6.2.1, it is a simple matter to derive the expression for $P^{\rm stat}_X(t)$ in the high-field approximation:

$$P^{\rm stat}_X(t, \Delta_{\rm L,TF}) = \exp\left(-\gamma_\mu \Delta_{\rm L,TF} t\right) \cos\left(\gamma_\mu B_{\rm ext} t\right). \tag{6.75}$$

Therefore the envelope of the oscillation is an exponential rather than a Gaussian function as found in Section 6.2.1 for a Gaussian distribution of the field component along $\mathbf{B}_{\rm ext}$. Note that, to avoid confusion, we give the parameter $\Delta_{\rm L,TF}$ as an argument of the polarisation function.

The formula for $P^{\rm stat}_Z(t, \Delta_{\rm L,ZF})$ can be determined using Eq. 6.14 if the distribution $D_{\rm v}\left(B_{\rm loc}, \theta, \varphi\right)$ is assumed to be isotropic. Qualitatively, this is expected to be the case since the field at the muon site is the superposition of many impurity (or defect) fields, the distance between the muon and the impurity as well as the relative orientation of their spins being at random. The isotropic character of the distribution is theoretically justified in Section 6.6.2. Inverting Eq. 6.179 with the help of Eq. C.52, $D_{\rm v}\left(\mathbf{B}_{\rm loc}\right) = D_{\rm v}\left(B_{\rm loc}, \theta, \varphi\right)$ is shown to have a squared-Lorentzian shape in the dilute disordered limit:

$$D_{\rm v}\left(B_{\rm loc}\right) = \frac{1}{\pi^2}\frac{\Delta_{\rm L,ZF}}{\left(\Delta^2_{\rm L,ZF} + B^2_{\rm loc}\right)^2}. \tag{6.76}$$

Obviously, the distribution does not depend on θ and φ. This justifies the simplified notation. Its magnitude is divided by 2 compared to its maximum, for $B_{\rm loc} = (2^{1/2} - 1)^{1/2}\Delta_{\rm L,ZF}$. As it should be, the density is normalised as

$$\int_0^\infty \left[4\pi B^2_{\rm loc} D_{\rm v}\left(B_{\rm loc}\right)\right] {\rm d}B_{\rm loc} = 1. \tag{6.77}$$

We point out that this shape of $D_{\rm v}\left(B_{\rm loc}\right)$ is derived under the assumption that the impurity magnetic moments are uncorrelated. Inserting Eq. 6.76 into Eq. 6.14 the corresponding longitudinal polarisation function is quite simple:

$$P_Z^{\rm stat}(t, \Delta_{\rm L,ZF}) = \frac{1}{3} + \frac{2}{3}\left(1 - \gamma_\mu \Delta_{\rm L,ZF} t\right) \exp\left(-\gamma_\mu \Delta_{\rm L,ZF} t\right). \tag{6.78}$$

It is obtained using Eq. C.71. Monte Carlo simulations (Fiory 1981) support this functional form as well as the expression of $\Delta_{\rm L,ZF}$ given in terms of the material parameters in Section 6.6.2. We stress that the distribution of the modulus of $\mathbf{B}_{\rm loc}$ is derived to be squared-Lorentzian in zero field without any information on the shape of the distributions of the field components.

We now compare $P_Z^{\rm stat}(t)$ computed with the squared-Lorentzian distribution for the modulus of the local field and the Gaussian component field distribution. Their plots are shown in Fig. 6.15 (p. 138) together with a third polarisation function to be introduced later on. The two functions are clearly different: (i) the initial decay of $P_Z^{\rm stat}(t)$ is faster for the squared-Lorentzian case since $P_Z^{\rm stat}(t, \Delta_{\rm L,ZF}) = 1 - (4/3)\gamma_\mu \Delta_{\rm L,ZF} t + \mathcal{O}(t^2)$ instead of $P_Z^{\rm stat}(t, \Delta_{\rm G,ZF}) = 1 - \gamma_\mu^2 \Delta_{\rm G,ZF}^2 t^2 + \mathcal{O}(t^4)$ for the Kubo–Toyabe function (note that $P_Z^{\rm stat}(t, \Delta_{\rm L,ZF})$ defined in Eq. 6.78 has a non-vanishing slope at $t = 0$); (ii) the dip occurs at $t_d = 2/(\gamma_\mu \Delta_{\rm L,ZF})$ for which $P_Z^{\rm stat}(t_d, \Delta_{\rm L,ZF}) = 0.2431$ instead of $t_d = \sqrt{3}/(\gamma_\mu \Delta_{\rm G,ZF})$ with $P_Z^{\rm stat}(t_d, \Delta_{\rm G,ZF}) = 0.03583$ for the former and latter case, respectively. On the other hand, as expected since the distributions are isotropic, the two polarisation functions tend to $\frac{1}{3}$ at long time. If $\gamma_\mu \Delta_{\rm L,ZF} t \ll 1$, $P_Z^{\rm stat}(t, \Delta_{\rm L,ZF})$ can be written as an exponential function

$$P_Z^{\rm stat}(t, \Delta_{\rm L,ZF}) = \exp\left(-\frac{4\gamma_\mu \Delta_{\rm L,ZF} t}{3}\right). \tag{6.79}$$

For completeness, we give an analytical expression for $P_Z^{\rm stat}(t)$ in the case where the local field is the superposition of two contributions: the first one is isotropic and distributed according to Eq. 6.76 and the second field contribution has a fixed modulus but random orientation. This is an analogue of the physical case which was considered in Section 6.2.3, where the Gaussian distribution is replaced with a squared-Lorentzian one. Following a similar methodology we find

$$D_{\rm v}\left(B_{\rm loc}\right) = \frac{1}{4\pi^2} \frac{1}{B_{\rm loc} B_0 \Delta_{\rm L,ZF}} \left(\frac{1}{1 + D_-} - \frac{1}{1 + D_+}\right), \tag{6.80}$$

where we have defined

$$D_\pm = \frac{1}{\Delta_{\rm L,ZF}^2} \left(B_0 \pm B_{\rm loc}\right)^2. \tag{6.81}$$

Introducing this form of $D_{\rm v}(B_{\rm loc})$ in Eq. 6.14 and using Eq. C.73, we obtain

$$\begin{aligned} P_Z^{\rm stat}(t, \Delta_{\rm L,ZF}) &= \frac{1}{3} + \frac{2}{3} \exp\left(-\gamma_\mu \Delta_{\rm L,ZF} t\right) \left[\cos\left(\gamma_\mu B_0 t\right) - \frac{\Delta_{\rm L,ZF}}{B_0} \sin\left(\gamma_\mu B_0 t\right)\right] \\ &= \frac{1}{3} + \frac{2}{3} \sqrt{1 + \left(\frac{\Delta_{\rm L,ZF}}{B_0}\right)^2} \exp\left(-\gamma_\mu \Delta_{\rm L,ZF} t\right) \cos\left[\gamma_\mu B_0 t + \arctan\left(\frac{\Delta_{\rm L,ZF}}{B_0}\right)\right]. \end{aligned} \tag{6.82}$$

In contrast to the formula for the Gaussian distribution, the phase shift is constant. The reader will easily find the asymptotic limits.

We have previously described the effect of the field dynamics on $P_\alpha(t)$ using the strong-collision model. As first noticed a long time ago by Kubo (1954), there is no motional narrowing effect for the squared-Lorentzian distribution which is probed by zero-field measurements. This is illustrated by Fiory (1981) and Leon (1981) for $P_Z(t)$: while the recovery of $P_Z(t)$ to $\frac{1}{3}$ at large t is damped out as $\nu_{\rm c}$ increases, its initial slope is not affected by $\nu_{\rm c}$ and thus no motional narrowing occurs. We note that the initial steep decay corresponds to the full range of fields being accessible to the muon spin at each step, including the large field values of the distribution tail. This is unrealistic: an appreciable density of large field is physically improbable since each of its occurrence corresponds to a muon close to the magnetic moment (the dominant field is of dipolar origin and therefore scales as r^{-3} where r is the distance between the muon and the moment at the origin of the large field at the muon site).

If we compute $P_Z(t)$ with the strong-collision model using $P_Z^{\rm stat}(t, \Delta_{\rm L,ZF})$ as given by Eq. 6.78, we actually perform the average over the disorder before the complete description of the effect of the field dynamics on $P_Z(t)$ has been performed. This is not correct since the average over all the possible configurations should be the last mathematical operation (see for example Fisher and Hertz 1991). Focusing on $P_Z^{\rm stat}(t, \Delta_{\rm L,ZF})$, we must write

$$P_Z^{\rm stat}(t, \Delta_{\rm L,ZF}) = \int_0^\infty P_Z^{\rm stat}(t, \Delta_{\rm G,ZF})\rho_{\Delta_{\rm L,ZF}}(\Delta_{\rm G,ZF}){\rm d}\Delta_{\rm G,ZF}, \tag{6.83}$$

where $P_Z^{\rm stat}(t, \Delta_{\rm L,ZF})$ is the longitudinal polarisation function for the disordered system and $P_Z^{\rm stat}(t, \Delta_{\rm G,ZF})$ the function for a perfect crystal discussed above in Section 6.2. The weight function $\rho_{\Delta_{\rm L,ZF}}(\Delta_{\rm G,ZF})$ models the distribution of the interaction between the muon and the lattice magnetic moments, i.e. the distribution of $\Delta_{\rm G,ZF}$. Uemura (1980) has found the function

$$\rho_{\Delta_{\rm L,ZF}}(\Delta_{\rm G,ZF}) = \sqrt{\frac{2}{\pi}}\frac{\Delta_{\rm L,ZF}}{\Delta^2_{\rm G,ZF}}\exp\left(-\frac{\Delta^2_{\rm L,ZF}}{2\Delta^2_{\rm G,ZF}}\right) \tag{6.84}$$

to fulfill our purpose. Indeed, using Eq. C.63 it is first shown that $\rho_{\Delta_{\rm L,ZF}}(\Delta_{\rm G,ZF})$ is normalised to unity. Secondly, with the help of Eq. C.67, $P_Z^{\rm stat}(t, \Delta_{\rm L,ZF})$ computed from Eq. 6.83 with this weight function is equal to the polarisation function given in Eq. 6.78. The function $\rho_{\Delta_{\rm L,ZF}}(\Delta_{\rm G,ZF})$ exhibits a maximum at $\Delta_{\rm L,ZF}/\sqrt{2}$ with a value $\rho_{\Delta_{\rm L,ZF}}(\Delta_{\rm L,ZF}/\sqrt{2}) = 0.5871/\Delta_{\rm L,ZF}$. Rather than $\rho_{\Delta_{\rm L,ZF}}(\Delta_{\rm G,ZF})$, it is convenient for some applications to express this distribution with the dimensionless quantity $r \equiv \Delta_{\rm G,ZF}/\Delta_{\rm L,ZF}$. For future reference, we have

$$\rho_{\Delta_{\rm L,ZF}}(\Delta_{\rm G,ZF}){\rm d}\Delta_{\rm G,ZF} = \sqrt{\frac{2}{\pi}}\frac{1}{r^2}\exp\left(-\frac{1}{2r^2}\right){\rm d}r. \tag{6.85}$$

A natural extension of Eq. 6.83 for $P_Z(t, \Delta_{\rm L,ZF})$ is

$$P_Z(t, \Delta_{\rm L,ZF}) = \int_0^\infty P_Z(t, \Delta_{\rm G,ZF})\rho_{\Delta_{\rm L,ZF}}(\Delta_{\rm G,ZF}){\rm d}\Delta_{\rm G,ZF}. \tag{6.86}$$

Therefore we reach the following conclusion: if the longitudinal polarisation function is known for a crystal with no disorder, the function for the related disordered system is obtained from a weighted sum of the former function. This conclusion is supported by numerical works in the diluted limit (Leon 1981, Fiory 1981). The same method is assumed to apply for the transverse-field polarisation function. Implicit in Eq. 6.86 is the assumption that ν_c is not a distributed quantity. Although a common hypothesis, it is not of general validity. Note that the polarisation function given in Eq. 6.86 has been used to model spin-glass systems.

It is interesting to derive asymptotical analytical formulae for $P_\alpha(t, \Delta_L)$, both for getting a feeling of how disorder influences the functions and for numerical computational purposes (the numerical computation of $P_\alpha(t, \Delta_L)$ in the motional narrowing limit is relatively slow).

Let us first consider $P_X(t, \Delta_{L,TF})$. Recalling Eq. 6.44 and using the prescription of integrating over $\Delta_{G,TF}$ with the weight function $\rho_{\Delta_{L,TF}}$ (defined similarly to $\rho_{\Delta_{L,ZF}}$; see Eq. 6.84), we derive from Eq. C.67

$$P_X(t, \Delta_{L,TF}) = \exp\left\{-\sqrt{\frac{2\gamma_\mu^2\Delta_{L,TF}^2}{\nu_c^2}\left[\exp(-\nu_c t) - 1 + \nu_c t\right]}\right\}\cos(\gamma_\mu B_{ext} t). \quad (6.87)$$

In the extreme motional narrowing limit,

$$P_X(t, \Delta_{L,TF}) = \exp\left(-\sqrt{\frac{2\gamma_\mu^2\Delta_{L,TF}^2 t}{\nu_c}}\right)\cos(\gamma_\mu B_{ext} t). \quad (6.88)$$

As shown in Section 6.2, if the dynamics is sufficiently fast, $P_Z(t, \Delta_{G,ZF})$ is equal to the expression of the envelope of $P_X(t, \Delta_{G,TF})$ with the substitution $\Delta_{G,TF}^2 \longrightarrow 2\Delta_{G,ZF}^2$. Therefore

$$P_Z(t, \Delta_{L,ZF}) = \exp\left\{-\sqrt{\frac{4\gamma_\mu^2\Delta_{L,ZF}^2}{\nu_c^2}\left[\exp(-\nu_c t) - 1 + \nu_c t\right]}\right\} \quad (6.89)$$

in zero field if ν_c is sufficiently large relative to $\gamma_\mu\Delta_{L,ZF}$. In the extreme limit, $\nu_c t \gg 1$,

$$P_Z(t, \Delta_{L,ZF}) = \exp\left(-\sqrt{\frac{4\gamma_\mu^2\Delta_{L,ZF}^2 t}{\nu_c}}\right). \quad (6.90)$$

This functional form of relaxation was already observed in an NMR experiment back in 1968 (Tse and Hartmann 1968; see also Blumberg 1960 and McHenry *et al.* 1972). We stress that the square-root behaviour is an asymptotic limit valid at long times. Since the long-time tail of $P_Z(t, \Delta_{G,ZF})$ is independent of $\Delta_{G,ZF}$, referring to Eq. 6.65,

$$P_Z(t, \Delta_{L,ZF}) \simeq \frac{1}{3}\exp\left(-\frac{2}{3}\nu_c t\right) + \frac{2}{3}(1 - \gamma_\mu\Delta_{L,ZF} t)\exp(-\gamma_\mu\Delta_{L,ZF} t), \quad (6.91)$$

is valid in the quasi-static limit.

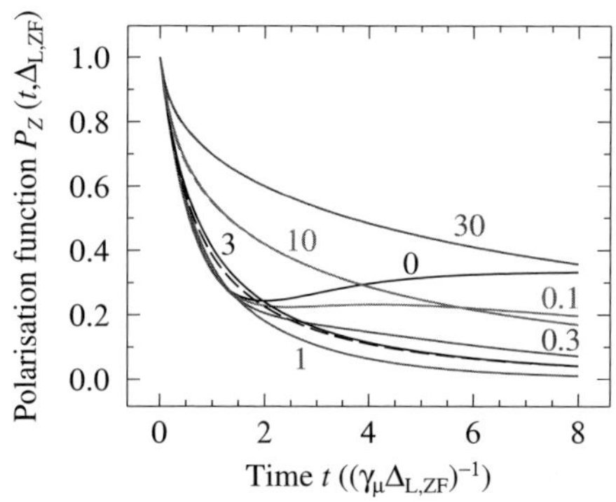

Fig. 6.10 The zero-field longitudinal polarisation function $P_Z(t, \Delta_{L,ZF})$ as a function of $\nu_c/(\gamma_\mu\Delta_{L,ZF})$ computed with the strong-collision model. The dashed lines shown for $\nu_c/(\gamma_\mu\Delta_{L,ZF}) \geq 3$ are the prediction of the analytical formula given in Eq. 6.89. Note that the dashed and full lines cannot be distinguished from each other for $\nu_c/(\gamma_\mu\Delta_{L,ZF}) \geq 10$. Therefore it is obvious that Eq. 6.89 is a reasonable approximation for $P_Z(t)$ if $\nu_c/(\gamma_\mu\Delta_{L,ZF}) \gtrsim 3$. The time scale is given in units of $(\gamma_\mu\Delta_{L,ZF})^{-1}$.

In Fig. 6.10 we present $P_Z(t, \Delta_{L,ZF})$ computed numerically as a function of the ratio $\nu_c/(\gamma_\mu\Delta_{L,ZF})$ in the zero-field case. The validity of the analytical asymptotic form given in Eq. 6.89 is tested.

The investigation of the field dependence of $P_Z(t, \Delta_{L,ZF})$ does not present any difficulty since the dependence for $P_Z(t, \Delta_{G,ZF})$ is known. A formula for the field dependence of $P_Z^{\rm stat}(t, \Delta_{L,ZF})$ can be derived (Uemura 1980 and Uemura *et al.* 1985):

$$\begin{aligned} P_Z^{\rm stat}(t, \Delta_{L,ZF}) = 1 &- \frac{\Delta_{L,ZF}}{B_{\rm ext}} j_1(\gamma_\mu B_{\rm ext} t) \exp(-\gamma_\mu \Delta_{L,ZF} t) \\ &- \frac{\Delta^2_{L,ZF}}{B^2_{\rm ext}} \left[j_0(\gamma_\mu B_{\rm ext} t) \exp(-\gamma_\mu \Delta_{L,ZF} t) - 1 \right] \\ &- \left(1 + \frac{\Delta^2_{L,ZF}}{B^2_{\rm ext}} \right) \gamma_\mu \Delta_{L,ZF} \int_0^t j_0(\gamma_\mu B_{\rm ext} \tau) \exp(-\gamma_\mu \Delta_{L,ZF} \tau) {\rm d}\tau, \end{aligned} \tag{6.92}$$

where $j_0(x)$ and $j_1(x)$ are spherical Bessel functions of the first kind; see Eq. C.96. Since the local magnetic field at the muon site usually has a dynamical character, the weighted average defined at Eq. 6.86 has generally to be performed. This is only possible numerically.

The expression of $P_Z(t, \Delta_{L,ZF})$ in the motional narrowing limit in the presence of a longitudinal field follows from Eq. 6.66:

$$P_Z(t, \Delta_{L,ZF}) = \exp\left(-\sqrt{4\gamma_\mu^2 \Delta^2_{L,ZF} \Gamma(t)} \right), \tag{6.93}$$

with $\Gamma(t)$ given in Eq. 6.67.

Strictly speaking, all the expressions of $P_\alpha(t)$ derived in the dilute impurity limit are not entirely satisfactory because the slope of $P_\alpha(t)$ near $t = 0$ is not zero even in

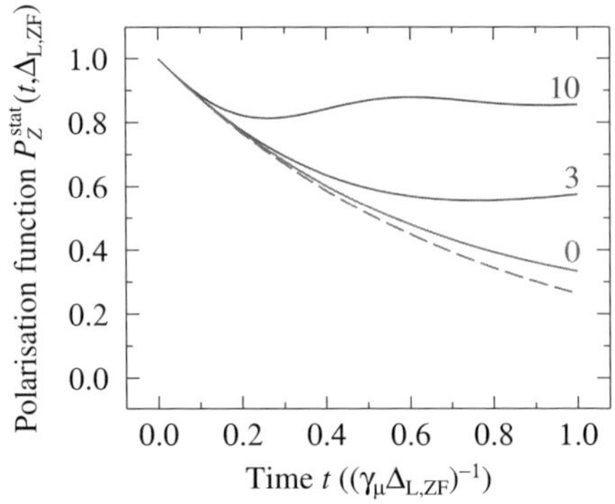

Fig. 6.11 Effect of an external field B_{ext} on the shape of the longitudinal polarisation function for a squared-Lorentzian distribution of the field modulus. We assume no dynamics and use Eq. 6.92. The numbers close to each full line are B_{ext} in units of $\Delta_{L,ZF}$. The dashed line is the exponential function $\exp(-4\gamma_\mu \Delta_{L,ZF} t/3)$ (Eq. 6.79).

the quasi-static regime. This means that they cannot be valid at short times. However, the time range for which they are not reliable is usually too small to be accessible to experiments. This is further discussed in Section 6.6.2.

Finally, we like to make an important remark. In general terms, the shape of the polarisation function at small times is usually expected to provide information on the type of field distribution probed by the muon spins, and the possible dynamical nature of the field. For the Gaussian distribution we have seen that at short times the function is quadratic in time if the distribution is static (see Section 6.2.1) and it is linear in the case of a strongly dynamical distribution (see Eq. 6.63). We already know that this criterion is justified if the strong-collision hypothesis applies; see Section 6.3.1 for a discussion. However, the polarisation function is linear in time at small times in the case of a squared-Lorentzian distribution (see Eq. 6.79). Therefore the question which immediately arises when an experimental polarisation function is initially t linear is the following: how to distinguish between the two types of distributions, or in other words, does the field distribution sensed by the muon spins stem from a concentrated magnetic moment system or from a diluted one? The question is especially relevant when only the short time part of the polarisation is accessible to the measurement, which is the case when $\Delta_{G,ZF}$ or $\Delta_{L,ZF}$ is small. A method for selecting the proper model is to investigate the field dependence of the polarisation function. If the exponential form of $P_Z(t)$ stems from the extreme motional narrowing of a Gaussian field distribution, $P_Z(t)$ should still be exponential in applied fields with the damping rate λ_Z following the Lorentzian law given in Eq. 6.68. However, if the diluted impurity model applies, the polarisation function measured in a field is clearly not exponential as illustrated in Fig. 6.11.

6.4.1.2 Distribution of relaxation processes. In the extreme motional narrowing limit, we have shown that $P_Z(t, \Delta_{L,ZF})$ is an exponential square-root function in zero field (Eq. 6.90) as well as in longitudinal fields (see Eq. 6.93 together with Eqs. 6.67 and 6.68). As a generalisation of this square-root relaxation, it is common to fit data with a stretched exponential function of formula

$$P_Z(t) = \exp\left[-(\lambda_Z t)^{\beta_{\rm se}}\right], \tag{6.94}$$

where $\beta_{\rm se}$ is a positive exponent. In fact such a relaxation of a physical quantity with $\beta_{\rm se} \leq 1$ is commonly observed in nature and is known as the Kohlrausch law. A possible interpretation assumes the existence of many independent relaxation channels for the system under study, each of which decays exponentially in time. We are thus in the motional narrowing limit in the case of μSR. Then the observed relaxation is taken as a sum of pure exponential decays, each one with a particular probability. Following Johnston (2006), we write

$$P_Z(t) = \int_0^\infty P(s, \beta_{\rm se}) \exp(-s\lambda_Z t)\, {\rm d}s, \tag{6.95}$$

where $P(s, \beta_{\rm se})$ is the distribution function of the relaxation rate which contains the physics of interest. The quantity $P(s, \beta_{\rm se})$ is the inverse Laplace transform of $P_Z(t)$, given by

$$P(s, \beta_{\rm se}) = \frac{1}{2\pi i} \int_{-i\infty}^{i\infty} \exp\left(-z^{\beta_{\rm se}}\right) \exp(sz)\, {\rm d}z. \tag{6.96}$$

In order to work with a real variable we introduce the change of variable $u = -iz$ and get

$$P(s, \beta_{\rm se}) = \frac{1}{\pi} \int_0^\infty \exp\left[-u^{\beta_{\rm se}} \cos(\pi\beta_{\rm se}/2)\right] \cos\left[su - u^{\beta_{\rm se}} \sin(\pi\beta_{\rm se}/2)\right] {\rm d}u. \tag{6.97}$$

It is possible to numerically evaluate $P(s, \beta_{\rm se})$ for arbitrary values of $\beta_{\rm se} < 1$; see Fig. 6.12. Obviously $P(s, 1) = \delta(s - 1)$. An analytical expression can be easily derived for $P(s, \frac{1}{2})$, with the change of variable $su = 2x^2$ and using Eq. C.75:

$$P\left(s, \frac{1}{2}\right) = \frac{1}{\sqrt{4\pi s^3}} \exp\left(-\frac{1}{4s}\right). \tag{6.98}$$

This distribution is notably different from the one introduced in Eq. 6.85. Both of them lead to a square-root $P_Z(t)$ function but the physical models underneath are different. In the first model it is the width of the field distribution at the muon sites which is assumed to be distributed following the distribution of the coupling of the muon spin with its neighbouring electronic spins. A single relevant correlation time is at play. In the present discussion the measured relaxation has been taken to arise from an ensemble of independent relaxation channels in the motional narrowing limit. It is also possible to derive Eq. 6.95 in the frame of an homogeneous picture in which the $P(s, \beta_{\rm se})$ distribution reflects the correlation between different degrees of freedom; see, for example, the paper of Palmer *et al.* (1984), and the review by Ehlers (2006) for a discussion. Here each muon spin probes a spectrum of relaxation channels because of the hierarchically constrained dynamics involved in the relaxation.

From a measured longitudinal polarisation function $P_Z^{\rm exp}(t)$ one might try to deduce numerically the distribution function $P(s, \beta_{\rm se})$. This obviously assumes the deviation

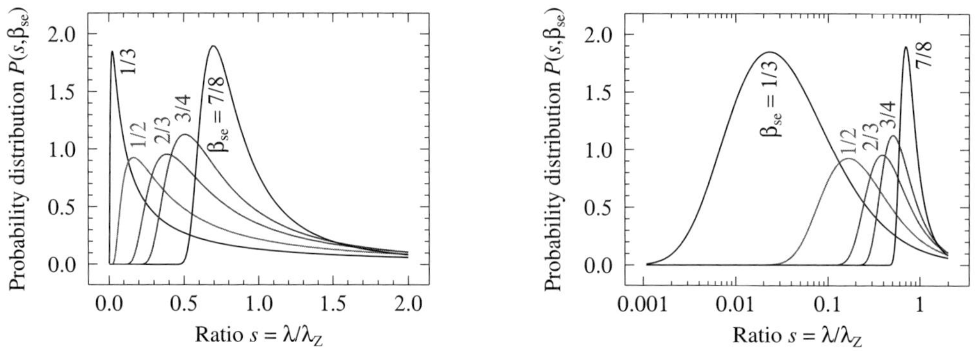

Fig. 6.12 Examples of distribution $P(s, \beta_{\rm se})$ of exponential functions $\exp(-\lambda t)$ leading to the stretched relaxation $\exp\left[-(\lambda_Z t)^{\beta_{\rm se}})\right]$. The parameter $s = \lambda/\lambda_Z$ is the normalised relaxation rate. The two panels differ by the nature of the horizontal scale: linear or logarithmic. When $\beta_{\rm se}$ is small, the distribution is shifted towards small s values and its asymmetry is more pronounced.

of $P_Z^{\rm exp}(t)$ from an exponential function to stem from a distribution of relaxation rates. However, we note that the numerical inversion of Eq. 6.95 is an ill-posed problem. Noise in $P_Z^{\rm exp}(t)$ results in a poorly defined $P(s, \beta_{\rm se})$ — see, for example, Press *et al.* (2000).

6.4.1.3 Phenomenological model for homogeneous disorder. Here the disorder is not due to diluted impurities, as in the previous section, but rather to a large concentration of impurities. We account for the disorder by replacing the single Gaussian distribution by a collection of distributions. We introduce a Gaussian weight function for the standard deviation, centred around $\Delta_0 \neq 0$:

$$\rho_{\Delta_0,\Delta_{\rm GbG}}(\Delta_{\rm G}) = \frac{1}{\sqrt{2\pi}}\frac{1}{\Delta_{\rm GbG}} \exp\left[-\frac{(\Delta_{\rm G} - \Delta_0)^2}{2\Delta_{\rm GbG}^2}\right]. \tag{6.99}$$

The variance of the distribution function of $\Delta_{\rm G}$ ($= \Delta_{\rm G,TF}$ or $\Delta_{\rm G,ZF}$) is denoted by $\Delta_{\rm GbG}^2$. The resulting polarisation function $P_\alpha(t, \Delta_0, \Delta_{\rm GbG})$ can be said to arise from a Gaussian-broadened Gaussian distribution (Noakes and Kalvius 1997). Although the integral over the weight function in Eq. 6.86 runs only over positive $\Delta_{\rm G}$ values, for the sake of mathematical simplicity, we extend the sum to minus infinity for the analytical analysis. This approximation, which avoids the use of error functions, is expected to be reasonable since the dominant contribution of $\rho_{\Delta_0,\Delta_{\rm GbG}}(\Delta_{\rm G})$ to $P_\alpha(t, \Delta_0, \Delta_{\rm GbG})$ arises from large positive $\Delta_{\rm G}$ values ($\rho_{\Delta_0,\Delta_{\rm GbG}}(\Delta_{\rm G})$ is centred at $\Delta_0 > 0$). In any case, the approximation can always be tested by comparing the analytical formulae with the results of the numerical evaluation of Eq. 6.86 with zero as the lower bound of the integral as it should be. Simple analytical integrations give

$$P_X^{\rm stat}(t, \Delta_0, \Delta_{\rm GbG}) = \frac{1}{\sqrt{1 + \gamma_\mu^2 \Delta_{\rm GbG}^2 t^2}} \exp\left[-\frac{\gamma_\mu^2 \Delta_0^2 t^2}{2\left(1 + \gamma_\mu^2 \Delta_{\rm GbG}^2 t^2\right)}\right] \cos(\gamma_\mu B_{\rm ext} t)$$

$$= \sqrt{\frac{1+R_{\rm b}^2}{1+R_{\rm b}^2+R_{\rm b}^2\gamma_\mu^2\Delta_{\rm eff}^2 t^2}} \exp\left[-\frac{\gamma_\mu^2\Delta_{\rm eff}^2 t^2}{2\left(1+R_{\rm b}^2+R_{\rm b}^2\gamma_\mu^2\Delta_{\rm eff}^2 t^2\right)}\right]\cos(\gamma_\mu B_{\rm ext} t), \tag{6.100}$$

and

$$\begin{aligned} P_Z^{\rm stat}(t,\Delta_0,\Delta_{\rm GbG}) &= \frac{1}{3}+\frac{2}{3}\frac{1}{\left(1+\gamma_\mu^2\Delta_{\rm GbG}^2 t^2\right)^{3/2}}\left(1-\frac{\gamma_\mu^2\Delta_0^2 t^2}{1+\gamma_\mu^2\Delta_{\rm GbG}^2 t^2}\right) \\ &\quad \times \exp\left[-\frac{\gamma_\mu^2\Delta_0^2 t^2}{2\left(1+\gamma_\mu^2\Delta_{\rm GbG}^2 t^2\right)}\right] \\ &= \frac{1}{3}+\frac{2}{3}\left(\frac{1+R_{\rm b}^2}{1+R_{\rm b}^2+R_{\rm b}^2\gamma_\mu^2\Delta_{\rm eff}^2 t^2}\right)^{3/2}\left(1-\frac{\gamma_\mu^2\Delta_{\rm eff}^2 t^2}{1+R_{\rm b}^2+R_{\rm b}^2\gamma_\mu^2\Delta_{\rm eff}^2 t^2}\right) \\ &\quad \times \exp\left[-\frac{\gamma_\mu^2\Delta_{\rm eff}^2 t^2}{2\left(1+R_{\rm b}^2+R_{\rm b}^2\gamma_\mu^2\Delta_{\rm eff}^2 t^2\right)}\right]. \end{aligned} \tag{6.101}$$

Here we have introduced two new parameters: $\Delta_{\rm eff}^2 = \Delta_0^2 + \Delta_{\rm GbG}^2$ and the ratio $R_{\rm b} = \Delta_{\rm GbG}/\Delta_0$ and from now on we shall use the notations $P_X^{\rm stat}(t, R_{\rm b}, \Delta_{\rm eff})$ and $P_Z^{\rm stat}(t, R_{\rm b}, \Delta_{\rm eff})$ for $P_X^{\rm stat}(t, \Delta_0, \Delta_{\rm GbG})$ and $P_Z^{\rm stat}(t, \Delta_0, \Delta_{\rm GbG})$, respectively. As expected, we recover the standard results (Eqs. 6.12 and 6.17) if $\Delta_{\rm GbG} = 0$. Since $P_X^{\rm stat}(t, R_{\rm b}, \Delta_{\rm eff}) = 1-\gamma_\mu^2\Delta_{\rm eff}^2 t^2/2+\mathcal{O}\left(t^4\right)$ and $P_Z^{\rm stat}(t, R_{\rm b}, \Delta_{\rm eff}) = 1-\gamma_\mu^2\Delta_{\rm eff}^2 t^2+\mathcal{O}\left(t^4\right)$, $\Delta_{\rm eff}^2$ governs the initial rate of damping and, as expected, the rate is twice as large for $P_Z^{\rm stat}(t, R_{\rm b}, \Delta_{\rm eff})$ as for $P_X^{\rm stat}(t, R_{\rm b}, \Delta_{\rm eff})$. In contrast to the polarisation functions computed for a Lorentzian distribution, the initial slope of $P_\alpha^{\rm stat}(t, R_{\rm b}, \Delta_{\rm eff})$ vanishes.

In Figs. 6.13 and 6.14 we display the two polarisation functions for different values of the ratio $R_{\rm b}$. Note that the time scale is given in terms of $\Delta_{\rm eff}^{-1}$ and not Δ_0^{-1}. The distribution of the variance of the field distribution has a strong influence on the shape of the functions, in particular on $P_Z^{\rm stat}(t, R_{\rm b}, \Delta_{\rm eff})$. There is no longer a minimum for $R_{\rm b} \geq 1$ as can be shown from the expression of the time derivative of $P_Z^{\rm stat}(t, R_{\rm b}, \Delta_{\rm eff})$. Increasing $R_{\rm b}$ above that value has little influence on $P_Z^{\rm stat}(t, R_{\rm b}, \Delta_{\rm eff})$.

We turn now to the dynamical case. As with the models previously considered in this chapter, the average over the different configurations must be performed in the last instance. Accordingly the dynamical polarisation function is

$$P_\alpha(t, R_{\rm b}, \Delta_{\rm eff}) = \int_{-\infty}^{\infty} P_\alpha(t, \Delta_{\rm G})\rho_{\Delta_0,\Delta_{\rm GbG}}(\Delta_{\rm G}){\rm d}\Delta_{\rm G}, \tag{6.102}$$

where $\rho_{\Delta_0,\Delta_{\rm GbG}}(\Delta_{\rm G})$ is defined in Eq. 6.99.

For the transverse polarisation function we take

$$P_X(t, \Delta_{\rm G}) = \exp[-\gamma_\mu^2\Delta_{\rm G}^2\Theta(t)]\cos(\gamma_\mu B_{\rm ext} t), \tag{6.103}$$

where we have introduced the auxiliary function

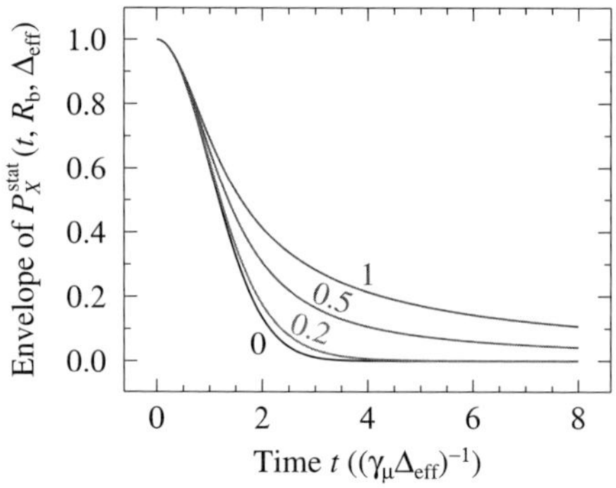

Fig. 6.13 The envelope of $P_X^{\rm stat}(t, R_{\rm b}, \Delta_{\rm eff})$ in the Gaussian-broadened Gaussian distribution model for different values of $R_{\rm b}$ indicated in the plot. This envelope is computed from an analytical approximation (Eq. 6.100). The case $R_{\rm b} = 0$ corresponds to the absence of broadening. The time scale is given in terms of $(\gamma_\mu \Delta_{\rm eff})^{-1}$.

$$\Theta(t) = \frac{\exp(-\nu_{\rm c} t) - 1 + \nu_{\rm c} t}{\nu_c^2}. \tag{6.104}$$

The envelope of $P_X(t, \Delta_{\rm G})$ is given by the so-called Abragam formula (Eq. 6.44). After integration we end up with

$$P_X(t, R_{\rm b}, \Delta_{\rm eff}) = \sqrt{\frac{1 + R_{\rm b}^2}{1 + R_{\rm b}^2 + 2R_{\rm b}^2\gamma_\mu^2\Delta_{\rm eff}^2\Theta(t)}} \exp\left[-\frac{\gamma_\mu^2\Delta_{\rm eff}^2\Theta(t)}{1 + R_{\rm b}^2 + 2R_{\rm b}^2\gamma_\mu^2\Delta_{\rm eff}^2\Theta(t)}\right]$$
$$\times \cos(\gamma_\mu B_{\rm ext} t). \tag{6.105}$$

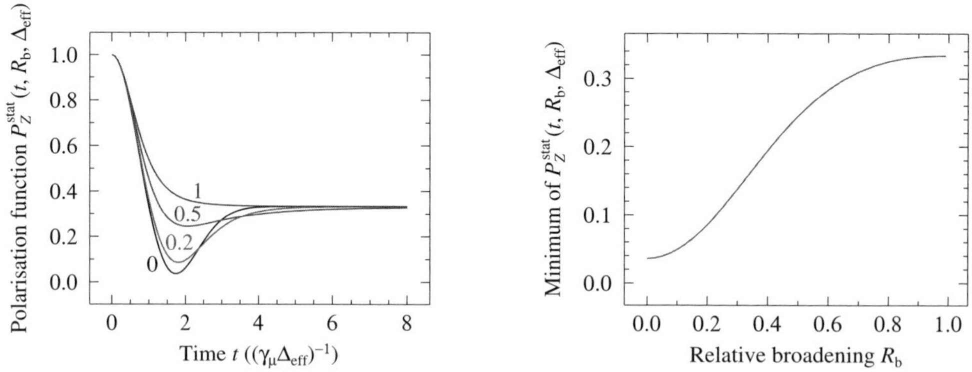

Fig. 6.14 (Left) The zero-field polarisation function $P_Z^{\rm stat}(t, R_{\rm b}, \Delta_{\rm eff})$ in the Gaussian-broadened Gaussian distribution model for different values of $R_{\rm b}$ indicated in the plot. The function is computed from an analytical approximation (Eq. 6.101). The plot labelled $R_{\rm b} = 0$ is the Kubo–Toyabe function. (Right) Value of the minimum of $P_Z^{\rm stat}(t, R_{\rm b}, \Delta_{\rm eff})$ as a function of $R_{\rm b}$.

For the zero-field polarisation function we take $P_Z(t, \Delta_{\rm G})$ as given by Eq. 6.62 and the integration yields

$$P_Z(t, R_{\rm b}, \Delta_{\rm eff}) = \sqrt{\frac{1 + R_{\rm b}^2}{1 + R_{\rm b}^2 + 4R_{\rm b}^2 \gamma_\mu^2 \Delta_{\rm eff}^2 \Theta(t)}} \exp\left[-\frac{2\gamma_\mu^2 \Delta_{\rm eff}^2 \Theta(t)}{1 + R_{\rm b}^2 + 4R_{\rm b}^2 \gamma_\mu^2 \Delta_{\rm eff}^2 \Theta(t)} \right]. \tag{6.106}$$

The longitudinal-field polarisation function follows immediately from the integration of Eq. 6.102 where $P_Z(t, \Delta_{\rm G})$ is given by Eq. 6.66. The result is similar to Eq. 6.106 upon the substitution $\Theta(t) \longrightarrow \Gamma(t)$ where $\Gamma(t)$ is given in Eq. 6.67.

The Gaussian-broadened Gaussian model is a purely phenomenological description of concentrated disordered systems. Using a Monte Carlo method, Noakes (1999) has shown that short-range correlations of the magnetic moments may induce the shallow static zero-field polarisation function which is found with the Gaussian-broadened Gaussian model. For an application of the model for describing experimental data, we refer to Section 10.4.

As explained at the end of Section 6.6.2 the distributions of the $\mathbf{B}_{\rm loc}$ components are found Gaussian when the number of dipole moments at their origin is large and their correlations negligible. In fact, the Gaussian distributions derive from the central limit theorem. However, when the correlations are appreciable there is no reason to expect the central limit theorem to apply (Bramwell *et al.* 2002). According to Section 6.1 a Gaussian distribution means that the only non-zero cumulant is κ_2. As a first step to account for the correlations, in addition to the variance, the term proportional to the cumulant parameter κ_3 in the expansion of the logarithm of the characteristic function of the distribution can be kept. A component distribution can then be obtained by a numerical inverse Fourier transform of the characteristic function. It is then a simple matter to numerically compute a polarisation function. Equations 6.130 and 6.131, to be shown below, can be used to evaluate $P_Z^{\rm stat}(t)$. Assuming such a procedure to be valuable for the description of experimental data, it would be interesting to justify on microscopic terms the extension of the expansion of the characteristic function beyond second order in t.

6.4.1.4 Comparison of the shape of $P_Z^{\rm stat}(t)$ for selected models. We have introduced in this chapter static zero-field polarisation functions arising from three types of field distributions: Gaussian, squared-Lorentzian, and Gaussian-broadened Gaussian distributions. In Fig. 6.15 we compare the shape of the related polarisation functions. We take $R_{\rm b} = 1$ for the Gaussian-broadened Gaussian distribution. The three polarisation functions are strongly different in two respects: the initial slope and the possible existence of a minimum around 1.7. On the other hand, the three functions tend to $\frac{1}{3}$ at large times, reflecting the isotropic character of $D_{\rm v}(\mathbf{B}_{\rm loc})$ in each case.

In order to understand the origin of the difference in shape for the three zero-field polarisation functions, it is interesting to compare the density distributions at the origin of the decay of the polarisation; we mean $D_{\rm m}(B_{\rm loc}) = 4\pi B_{\rm loc}^2 D_{\rm v}(B_{\rm loc})$. Whereas this density distribution can be computed readily using the distributions given in this chapter for the Gaussian and squared-Lorentzian models (Eqs. 6.16 and

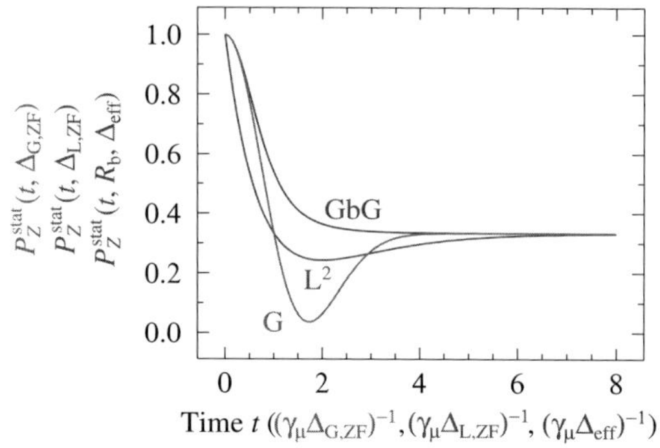

Fig. 6.15 The three static zero-field polarisation functions discussed in this chapter: $P_Z^{\mathrm{stat}}(t, \Delta_{\mathrm{G,ZF}})$, $P_Z^{\mathrm{stat}}(t, \Delta_{\mathrm{L,ZF}})$, and $P_Z^{\mathrm{stat}}(t, R_{\mathrm{b}}, \Delta_{\mathrm{eff}})$. The last function is plotted for $R_{\mathrm{b}} = 1$. The time scale is given in units of the relevant damping rate: either $(\gamma_\mu \Delta_{\mathrm{G,ZF}})^{-1}$, $(\gamma_\mu \Delta_{\mathrm{L,ZF}})^{-1}$, or $(\gamma_\mu \Delta_{\mathrm{eff}})^{-1}$.

6.76, respectively), we still need to determine the density distribution for the Gaussian-broadened Gaussian model. From Eq. 6.15 we obtain

$$3P_Z^{\mathrm{stat}}(t, R_{\mathrm{b}}, \Delta_{\mathrm{eff}}) - 1 = 2\int_0^\infty D_{\mathrm{m}}(B_{\mathrm{loc}}) \cos(\gamma_\mu B_{\mathrm{loc}} t)\, \mathrm{d}B_{\mathrm{loc}}. \tag{6.107}$$

This equation can be inverted with the help of the Fourier theorem to give

$$D_{\mathrm{m}}(B_{\mathrm{loc}}) = \frac{\gamma_\mu}{\pi}\int_0^\infty \left[3P_Z^{\mathrm{stat}}(t, R_{\mathrm{b}}, \Delta_{\mathrm{eff}}) - 1\right] \cos(\gamma_\mu B_{\mathrm{loc}} t)\, \mathrm{d}t. \tag{6.108}$$

Then, using Eq. 6.101 the density distribution of the Gaussian-broadened Gaussian model can be evaluated numerically.

As seen in Fig. 6.15, the oscillation damping of the Gaussian-broadened Gaussian polarisation is more pronounced than that of the Kubo–Toyabe function. According to Fig. 6.16, this corresponds to an increase of the weight at low fields. A remarkable difference between the three models is the fact that $D_{\mathrm{m}}(B_{\mathrm{loc}} = 0)$ is not zero for the Gaussian-broadened Gaussian polarisation function, whereas it vanishes for the other two. From an inspection of $P_Z^{\mathrm{stat}}(t, R_{\mathrm{b}}, \Delta_{\mathrm{eff}})$ (Fig. 6.14), say with $R_{\mathrm{b}} = 1$, it is indeed clear that the integral $\int_0^\infty [P_Z^{\mathrm{stat}}(t, R_{\mathrm{b}}, \Delta_{\mathrm{eff}}) - 1/3]\mathrm{d}t$ is strictly positive. Therefore $D_{\mathrm{m}}(B_{\mathrm{loc}} = 0)$, which is proportional to this integral (see Eq. 6.108), is also strictly positive.

6.4.2 Two-state models

Until now, we have discussed at length how to model $P_X(t)$ and $P_Z(t)$ when there is a continuity in the different magnetic environments that the muon spins sense; i.e. we have worked in the framework of the so-called one-state models. However, it may occur that the sites probed by the muon ensemble are magnetically so different that the one-state models break down. As an example, we refer to the study of the muon diffusion in a doped niobium sample in Section 1.2.1. Here we discuss two-state models

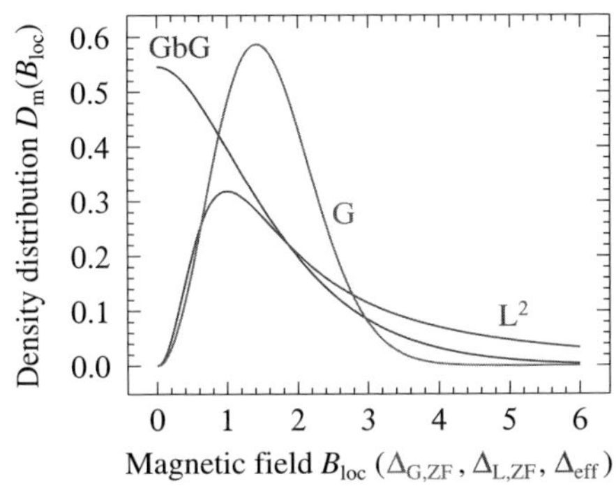

Fig. 6.16 Density distributions $D_{\rm m}(B_{\rm loc})$ for the three models discussed in this chapter. They correspond to $P_Z^{\rm stat}(t, \Delta_{\rm G,ZF})$, $P_Z^{\rm stat}(t, \Delta_{\rm L,ZF})$ and $P_Z^{\rm stat}(t, R_{\rm b}, \Delta_{\rm eff})$, respectively, which are shown in Fig. 6.15. For the last model we have taken $R_{\rm b} = 1$. The scales of the axes are given in terms of the relevant unit: either $\Delta_{\rm G,ZF}$, $\Delta_{\rm L,ZF}$ or $\Delta_{\rm eff}$ for the horizontal axis and their inverse for the vertical one.

for which one assumes the system to be best described by two clearly different states. We denote the two possible environments of the muon spins as states 1 and 2. They may differ by their mean field or magnetic field inhomogeneity or simply because the mobility of the muons is not the same in the two states. Although we shall sometimes continue to use the language of diffusion, we recall that the application of the results is not restricted to the discussion of the effect of the muon diffusion on $P_\alpha(t)$.

As a first step we consider a simple model. We assume that the muon has finite probabilities f_1 and f_2 to be implanted in states 1 and 2 respectively at the initial time. Then, as time evolves, it may move from state 1 to state 2 where it is trapped. We first analyse this model with the strong-collision algorithm. The polarisation function is then the sum of two terms of weight factors f_1 and f_2, with $f_1 + f_2 = 1$. The polarisation function relative to muons implanted initially in state 1 can be computed using the methodology of Section 6.3.1. More specifically, we use Eq. 6.46 where the sum over ℓ is restricted to the terms $R_{\alpha,0}(t)$ and $R_{\alpha,1}(t)$. Adapting Eqs. 6.47 and 6.49 to the current case, we arrive at

$$\tilde{P}_\alpha(t) = f_1 \left[\tilde{P}_\alpha^{(1)}(t) \exp(-\nu_1 t) + \nu_1 \int_0^t \tilde{P}_\alpha^{(2)}(t-t') \tilde{P}_\alpha^{(1)}(t') \exp(-\nu_1 t') {\rm d}t' \right] + f_2 \tilde{P}_\alpha^{(2)}(t). \tag{6.109}$$

The function $\tilde{P}_\alpha^{(i)}(t)$ stands for the α polarisation function in state i written in a complex notation satisfying $P_\alpha^{(i)}(t) = \mathcal{R}e\{\tilde{P}_\alpha^{(i)}(t)\}$. We introduce here this complex polarisation function because it greatly simplifies the equations as will be seen later in this section. The trapping rate to state 2 is denoted as ν_1; i.e. $\tau_1 = 1/\nu_1$ is the lifetime in state 1. The lifetime in state 2 is infinite according to our prescription, i.e. $\nu_2 = 0$. The second term in the square-bracket on the right-hand-side of Eq. 6.109 describes the transition from state 1 to state 2: it vanishes with the trapping rate ν_1.

Considering the transverse polarisation function, the integration in Eq. 6.109 can be carried out analytically in the extreme motional narrowing limit (see Section 6.3) where the envelope of $\tilde{P}_X^{(i)}(t)$ is an exponential function of damping rate $\lambda_X^{(i)}$. We obtain

$$P_X(t) = f_1 \left\{ \exp\left(-\breve{\lambda}_X^{(1)} t\right) \cos\left(\omega_\mu^{(1)} t\right) + \nu_1 \left[\left(\breve{\lambda}_X^{(1)} - \lambda_X^{(2)}\right)^2 + 4\left(\delta\omega_\mu\right)^2\right]^{-1/2} \right.$$
$$\left. \times \left[\exp\left(-\lambda_X^{(2)} t\right) \cos\left(\omega_\mu^{(2)} t + \theta_{\rm m}\right) - \exp\left(-\breve{\lambda}_X^{(1)} t\right) \cos\left(\omega_\mu^{(1)} t + \theta_{\rm m}\right)\right] \right\}$$
$$+ f_2 \exp\left(-\lambda_X^{(2)} t\right) \cos\left(\omega_\mu^{(2)} t\right), \qquad (6.110)$$

with the definitions

$$\breve{\lambda}_X^{(1)} = \lambda_X^{(1)} + \nu_1, \qquad \delta\lambda_X = \breve{\lambda}_X^{(1)} - \lambda_X^{(2)}, \qquad \delta\omega_\mu = \frac{\omega_\mu^{(1)} - \omega_\mu^{(2)}}{2}, \qquad (6.111)$$

and

$$\cos\theta_m = \frac{\delta\lambda_X}{\left[\left(\delta\lambda_X\right)^2 + 4\left(\delta\omega_\mu\right)^2\right]^{1/2}}, \qquad \sin\theta_m = \frac{2\delta\omega_\mu}{\left[\left(\delta\lambda_X\right)^2 + 4\left(\delta\omega_\mu\right)^2\right]^{1/2}}. \qquad (6.112)$$

The expression of $P_X(t)$ written in Eq. 6.110 was first given by Möslang *et al.* (1983) for the model $f_1 = 1$. Let us now study that case and assume $\lambda_X^{(1)}$ and $\lambda_X^{(2)}$ to be small relative to ν_1. We consider $P_X(t)$ for $t \geq 1/\nu_1$; i.e. the observation time is sufficiently long that the muons have left the site in which they were implanted. Within these approximations,

$$P_X(t) \simeq \left(1 + \tan^2\theta_m\right)^{-1/2} \exp\left(-\lambda_X^{(2)} t\right) \cos\left(\omega_\mu^{(2)} t + \theta_m\right), \qquad (6.113)$$

with

$$\tan\theta_m = \left(\omega_\mu^{(1)} - \omega_\mu^{(2)}\right)/\nu_1. \qquad (6.114)$$

The polarisation function $P_X(t)$ displays a shift of the wiggles and the asymmetry is reduced. If the angular frequencies of the two sites are very different, the absolute value of the shift tends to $\pm\pi/2$ and the asymmetry of the muon signal shrinks and may not be detectable. The sign of the angular shift depends on which of the two angular frequencies is higher. Note that a shift is also predicted for a discrete stochastic field; see Section 6.3.2.6.

Within the same approach, in the extreme motional narrowing limit,

$$P_Z(t) = f_1 \left\{ \exp\left(-\breve{\lambda}_Z^{(1)} t\right) + \frac{\nu_1}{\breve{\lambda}_Z^{(1)} - \lambda_Z^{(2)}} \left[\exp\left(-\lambda_Z^{(2)} t\right) - \exp\left(-\breve{\lambda}_Z^{(1)} t\right)\right] \right\}$$
$$+ f_2 \exp\left(-\lambda_Z^{(2)} t\right), \qquad (6.115)$$

with the definition $\breve{\lambda}_Z^{(1)} = \lambda_Z^{(1)} + \nu_1$.

It is interesting to consider the case for which the muon dynamics is so fast in state 1 that $P_Z^{(1)}(t) = 1$ and, in contrast, we assume the muon to be static in state 2; i.e. $P_Z^{(2)}(t)$ is given by the Kubo–Toyabe function (Eq. 6.17). With the hypothesis $f_1 = 1$, from Eq. 6.109

$$P_Z(t) = \frac{1}{3} + \frac{2}{3}\exp(-\nu_1 t) + \frac{2\nu_1}{3}\int_0^t \left[1 - \left(\gamma_\mu \Delta_{\rm G,ZF}^{(2)}\right)^2 (t-t')^2\right] \times \exp\left[-\frac{\left(\gamma_\mu \Delta_{\rm G,ZF}^{(2)}\right)^2 (t-t')^2}{2}\right] \exp(-\nu_1 t')\,{\rm d}t'. \tag{6.116}$$

The integration involves the error function (Eq. C.105); see Petzinger (1980). When $\nu_1/(\gamma_\mu \Delta_{\rm G,ZF}^{(2)}) \gg 1$, the Kubo–Toyabe formula is recovered, while for $\nu_1/(\gamma_\mu \Delta_{\rm G,ZF}^{(2)}) \ll 1$, $P_Z(t) \simeq \frac{1}{3} + \frac{2}{3}\exp(-\nu_1 t)$. In either limit the asymptotic value $\frac{1}{3}$ arises from the isotropic and static nature of the field at the muon site in state 2. Conversely, if the muon was diffusing away from site 2, the asymptotic value of $P_Z(t)$ would vanish. Therefore it is possible to distinguish between diffusion away from a trap from diffusion towards a trap.

At this stage we point out that for a zero-field measurement performed in a magnetically ordered compound, the expression of $P_X(t)$ given in Eqs. 6.110 through 6.112 is valid if the spontaneous field $\mathbf{B}_0$ is parallel to the Z axis. In that case $\omega_\mu^{(i)} = \gamma_\mu B_0^{(i)}$, where $B_0^{(i)}$ is the spontaneous magnetic field at the muon site in state i.

Following Kehr *et al.* (1978), we introduce a new notation which allows us to write Eq. 6.109 in a compact form. We denote by $\tilde{Q}_\alpha^{(ij)}(t)$ the polarisation of the muon ensemble which is in state i at time t, while it was in state j at $t = 0$. Note the initial conditions: $\tilde{Q}_\alpha^{(ij)}(t=0) = \delta_{i,j}$ (Kronecker symbol). For the model discussed until now, we have

$$\tilde{Q}_\alpha^{(11)}(t) = \tilde{P}_\alpha^{(1)}(t)\exp(-\nu_1 t), \qquad \tilde{Q}_\alpha^{(21)}(t) = \nu_1 \int_0^t \tilde{P}_\alpha^{(2)}(t-t')\tilde{Q}_\alpha^{(11)}(t'){\rm d}t',$$
$$\tilde{Q}_\alpha^{(22)}(t) = \tilde{P}_\alpha^{(2)}(t), \qquad \tilde{Q}_\alpha^{(12)}(t) = 0. \tag{6.117}$$

The total polarisation function is then written as

$$\tilde{P}_\alpha(t) = f_1\left[\tilde{Q}_\alpha^{(11)}(t) + \tilde{Q}_\alpha^{(21)}(t)\right] + f_2\tilde{Q}_\alpha^{(22)}(t). \tag{6.118}$$

In the extreme motional narrowing limit, $\tilde{P}_X^{(i)}(t) = \exp\left[-\left(i\omega_\mu^{(i)} + \lambda_X^{(i)}\right)t\right]$ and $\tilde{P}_Z^{(i)}(t) = \exp\left(-\lambda_Z^{(i)} t\right)$. Using this functional form for $\tilde{P}_X^{(i)}(t)$ and Eq. 6.117 we find the two $\tilde{Q}_X^{(i1)}(t)$ functions ($i = 1$ and 2) to obey the following set of two coupled linear differential equations:

$$\frac{{\rm d}\tilde{Q}_X^{(11)}(t)}{{\rm d}t} = -\left(i\omega_\mu^{(1)} + \lambda_X^{(1)} + \nu_1\right)\tilde{Q}_X^{(11)}(t),$$

$$\frac{\mathrm{d}\tilde{Q}_X^{(21)}(t)}{\mathrm{d}t} = -\left(i\omega_\mu^{(2)} + \lambda_X^{(2)}\right)\tilde{Q}_X^{(21)}(t) + \nu_1\tilde{Q}_X^{(11)}(t). \tag{6.119}$$

A similar set of equations can be written for $\tilde{Q}_Z^{(i1)}(t)$ ($i = 1$ and 2). Relative to Eq. 6.119, $\lambda_X^{(i)}$ must be replaced by $\lambda_Z^{(i)}$ and $\omega_\mu^{(i)} = 0$.

It is not difficult to generalise the expression of $\tilde{P}_\alpha(t)$ when $\nu_2 \neq 0$:

$$\tilde{P}_\alpha(t) = f_1\left[\tilde{Q}_\alpha^{(11)}(t) + \tilde{Q}_\alpha^{(21)}(t)\right] + f_2\left[\tilde{Q}_\alpha^{(22)}(t) + \tilde{Q}_\alpha^{(12)}(t)\right]. \tag{6.120}$$

The functions $\tilde{Q}_\alpha^{(11)}(t)$ and $\tilde{Q}_\alpha^{(21)}(t)$ obey the following set of coupled integral equations which are derived similarly to Eq. 6.52

$$\tilde{Q}_\alpha^{(11)}(t) = \tilde{P}_\alpha^{(1)}(t)\exp(-\nu_1 t) + \nu_2\int_0^t \exp\left[-\nu_1(t-t')\right]\tilde{P}_\alpha^{(1)}(t-t')\tilde{Q}_\alpha^{(21)}(t')\mathrm{d}t',$$

$$\tilde{Q}_\alpha^{(21)}(t) = \nu_1\int_0^t \exp\left[-\nu_2(t-t')\right]\tilde{P}_\alpha^{(2)}(t-t')\tilde{Q}_\alpha^{(11)}(t')\mathrm{d}t'. \tag{6.121}$$

An analogous set of equations is followed by $\tilde{Q}_\alpha^{(22)}(t)$ and $\tilde{Q}_\alpha^{(12)}(t)$: the subscripts 1 and 2 in Eq. 6.121 have simply to be interchanged. In the extreme motional narrowing limit we have

$$\frac{\mathrm{d}\tilde{Q}_X^{(11)}(t)}{\mathrm{d}t} = -\left(i\omega_\mu^{(1)} + \lambda_X^{(1)} + \nu_1\right)\tilde{Q}_X^{(11)}(t) + \nu_2\tilde{Q}_X^{(21)}(t),$$

$$\frac{\mathrm{d}\tilde{Q}_X^{(21)}(t)}{\mathrm{d}t} = -\left(i\omega_\mu^{(2)} + \lambda_X^{(2)} + \nu_2\right)\tilde{Q}_X^{(21)}(t) + \nu_1\tilde{Q}_X^{(11)}(t). \tag{6.122}$$

Following the same procedure, it is not difficult to write down the set of linear differential equations followed by $\tilde{Q}_Z^{(ij)}(t)$ in the fast diffusion limit.

While the treatment with the two coupled differential equations may seem useless since the integral equation method gives the same result, the description of the evolution of the polarisation with the differential equations is physically quite transparent. These equations are in fact similar to the Bloch equations. Let us consider the first line in Eq. 6.122. In an infinitesimal time interval, the polarisation $\tilde{Q}_X^{(11)}(t)$ decreases (i) by the decay arising from the effective field distribution at the muon site, and (ii) by transitions to state 2 governed by the rate ν_1, and increases by transitions from state 2 described by ν_2. We recall that this is equivalent to the strong-collision picture in the extreme motional narrowing limit. Interestingly, even if this limit is not fully reached, which means that $\tilde{P}_\alpha^{(i)}(t)$ is given by the so-called Abragam formula (see Eq. 6.44 and Eq. 6.62 for $\tilde{P}_X^{(i)}(t)$ and $\tilde{P}_Z^{(i)}(t)$, respectively), the system of differential equations is still a reasonable approximation to the complete system of integral equations (Kehr *et al.* 1978). It has the practical advantage that it is easily solved numerically using the Runge–Kutta–Merson algorithm — see for example, James *et al.* (1999) or Press *et al.* (2007).

The system of differential equations given in Eq. 6.122 and its extension for the longitudinal geometry can analytically be solved. If the muon precession frequency is the same for the two muon sites, the result derived for $P_Z(t)$ applies equally well for $P_X(t)$. Here, we shall restrict ourselves to the study of $P_Z(t)$. Let us consider that only state 1 is populated at initial time ($f_2 = 0$) and that we are still in the motional narrowing limit. We denote as $\mathbf{Q}_Z(t)$ the vector of components $\tilde{Q}_Z^{(11)}(t)$ and $\tilde{Q}_Z^{(21)}(t)$. With this vector notation, and when $\omega_\mu^{(i)}$ vanishes, Eq. 6.122 reads

$$\frac{\mathrm{d}\mathbf{Q}_Z(t)}{\mathrm{d}t} = -A_Z\,\mathbf{Q}_Z(t), \tag{6.123}$$

with

$$A_Z = \begin{pmatrix} \lambda_Z^{(1)} + \nu_1 & -\nu_2 \\ -\nu_1 & \lambda_Z^{(2)} + \nu_2 \end{pmatrix}. \tag{6.124}$$

The two eigenvalues of A_Z are

$$\ell_\pm = \frac{1}{2}\left[\lambda_Z^{(1)} + \nu_1 + \lambda_Z^{(2)} + \nu_2 \pm \sqrt{\left(\lambda_Z^{(1)} + \nu_1 - \lambda_Z^{(2)} - \nu_2\right)^2 + 4\nu_1\nu_2}\right]. \tag{6.125}$$

Because their product is positive, they are of the same sign. In addition, since their sum is positive, we conclude that they are both positive. With the initial conditions, $Q_Z^{(11)}(t=0) = 1$ and $Q_Z^{(21)}(t=0) = 0$, we derive from Eq. 6.120

$$P_Z(t) = A_+ \exp\left(-\ell_+ t\right) + A_- \exp\left(-\ell_- t\right), \tag{6.126}$$

with

$$A_\pm = \frac{\pm\lambda_Z^{(1)} \mp \nu_1 \mp \lambda_Z^{(2)} \mp \nu_2 + \sqrt{\left(\lambda_Z^{(1)} + \nu_1 - \lambda_Z^{(2)} - \nu_2\right)^2 + 4\nu_1\nu_2}}{2\sqrt{\left(\lambda_Z^{(1)} + \nu_1 - \lambda_Z^{(2)} - \nu_2\right)^2 + 4\nu_1\nu_2}}. \tag{6.127}$$

As expected, $P_Z(t)$ is the weighted sum of two exponential functions.

6.5 Polarisation functions for two field sources

In Section 6.2.3 (see also Section 6.4.1.1) we derived an expression for $P_Z^{\mathrm{stat}}(t)$ if the field at the muon site arises from two sources: an isotropic Gaussian distribution and an isotropic field with constant modulus. This model was built to describe the polarisation function measured for a powder sample in its magnetically ordered phase. In this section we extend our discussion to more general physical cases.

First, we recall that the distribution resulting from two independent field distributions is given by their convolution product (Eq. 6.26). According to Section 6.2.1, if B_{ext} is sufficiently large relative to the width of the resulting field distribution and if this distribution is even, the transverse-field polarisation function is the product of a cosine function, $\cos(\gamma_\mu B_{\mathrm{ext}} t)$, by an envelope function. That latter function is proportional to the cosine Fourier transform of the distribution of the component along $\mathbf{B}_{\mathrm{ext}}$; i.e. it is a one-dimensional Fourier transform. Since the Fourier transform of a convolution product is the product of the Fourier transforms, we deduce that the envelope function is the product of the envelopes related to the two field distributions. For example, if the two sources of fields are characterised by Gaussian-field distributions for the components, $P_X^{\rm stat}(t)$ is given by Eq. 6.12 with $\Delta^2_{\rm G,TF} = \left(\Delta^{(1)}_{\rm G,TF}\right)^2 + \left(\Delta^{(2)}_{\rm G,TF}\right)^2$, where $\left(\Delta^{(i)}_{\rm G,TF}\right)^2$ is the variance for the field distribution denoted by i. As mentioned in Section 6.4.1.1, the field component distribution is Lorentzian for diluted magnetic impurities. If the muons experience such a source plus an additional Gaussian component distribution,

$$P_X^{\rm stat}(t) = \exp\left[-\left(\frac{\gamma_\mu^2 \Delta^2_{\rm G,TF} t^2}{2} + \gamma_\mu \Delta_{\rm L,TF} t\right)\right] \cos\left(\gamma_\mu B_{\rm ext} t\right), \tag{6.128}$$

where $\Delta_{\rm L,TF}$ is the half-width at half-maximum of the Lorentzian distribution of the field component parallel to $\mathbf{B}_{\rm ext}$.

The derivation of a formula for $P_Z^{\rm stat}(t)$ when two sources of fields are at play is more difficult. It requires the determination of $D_{\rm v}(\mathbf{B}_{\rm loc})$. In a first step, we assume the two field sources to be Gaussian and isotropic. It is not difficult to show that $D_{\rm c}(B^\alpha_{\rm loc})$ is then a Gaussian distribution for each α component with its variance equal to the sum of the two variances. The resulting distribution $D_{\rm v}(\mathbf{B}_{\rm loc})$ follows immediately and also the polarisation function:

$$\begin{aligned} P_Z^{\rm stat}(t) = \frac{1}{3} + \frac{2}{3}&\left\{1 - \gamma_\mu^2 \left[\left(\Delta^{(1)}_{\rm G,ZF}\right)^2 + \left(\Delta^{(2)}_{\rm G,ZF}\right)^2\right] t^2\right\} \\ &\times \exp\left\{-\gamma_\mu^2 \frac{\left[\left(\Delta^{(1)}_{\rm G,ZF}\right)^2 + \left(\Delta^{(2)}_{\rm G,ZF}\right)^2\right]}{2} t^2\right\}. \end{aligned} \tag{6.129}$$

However, for more involved distributions, $P_Z^{\rm stat}(t)$ cannot be obtained this way since no analytical form for $D_{\rm v}(\mathbf{B}_{\rm loc})$ is usually available. We then resort to another method which turns out to be extremely powerful. According to Kubo (1981), the longitudinal polarisation function is given by the general formula

$$\begin{aligned} P_Z^{\rm stat}(t) = 1 + &\frac{2}{\gamma_\mu^2 B_{\rm ext}^2} \lim_{t\to 0}\left[\frac{1}{t}\frac{{\rm d}Q(t)}{{\rm d}t}\right] - 2t \frac{{\rm d}Q(t)}{{\rm d}t}\frac{\cos(\gamma_\mu B_{\rm ext} t)}{\gamma_\mu^2 B_{\rm ext}^2 t^2} \\ &+ 2\int_0^t \frac{\sin(\gamma_\mu B_{\rm ext}\tau)}{\gamma_\mu^3 B_{\rm ext}^3 \tau}\frac{\rm d}{{\rm d}\tau}\left[\frac{1}{\tau}\frac{{\rm d}Q(\tau)}{{\rm d}\tau}\right]{\rm d}\tau, \end{aligned} \tag{6.130}$$

where

$$Q(t) = \int_{-\infty}^{\infty} D_{\rm c}\left(B^Z_{\rm loc}\right) \cos(\gamma_\mu B^Z_{\rm loc} t) {\rm d}B^Z_{\rm loc}. \tag{6.131}$$

The function $Q(t)$ is proportional to the cosine Fourier transform of the field distribution along the Z axis. Note that the structures of Eqs. 6.18 and 6.130 are quite similar. This suggests a method for deriving the previous result. The zero-field limit of Eq. 6.130 can be obtained without any problem:

$$P_Z^{\rm stat}(t) = \frac{1}{3} + \frac{2}{3}\frac{\rm d}{{\rm d}t}\left[tQ(t)\right]. \tag{6.132}$$

Because of the factors $\frac{1}{3}$ and $\frac{2}{3}$, the field distribution leading to $P_Z^{\rm stat}(t)$ is obviously assumed to be isotropic.

Since $Q(t)$ is given as a one-dimensional Fourier transform, it can easily be computed when two sources of fields are acting on the muon spins. It is a problem similar to that discussed previously for $P_X^{\rm stat}(t)$. For example, for two Gaussian distributions at play, $Q(t) = \exp\left\{-\gamma_\mu^2\left[\left(\Delta_{\rm G,TF}^{(1)}\right)^2 + \left(\Delta_{\rm G,TF}^{(2)}\right)^2\right]t^2/2\right\}$, and we recover the result given in Eq. 6.129 if we replace $\Delta_{\rm G,TF}^{(i)}$ by $\Delta_{\rm G,ZF}^{(i)}$. This substitution is justified when considering the formal distribution characteristics, but it is not when computing these standard deviations from microscopic material parameters; see Section 6.6 below. More interesting, if the muon spins probe a Gaussian distribution in an environment of diluted magnetic impurities, $Q(t) = \exp\left(-\gamma_\mu^2\Delta_{\rm G,TF}^2 t^2/2 - \gamma_\mu\Delta_{\rm L,TF}t\right)$ and a simple evaluation gives

$$P_Z^{\rm stat}(t) = \frac{1}{3} + \frac{2}{3}\left(1 - \gamma_\mu^2\Delta_{\rm G,TF}^2 t^2 - \gamma_\mu\Delta_{\rm L,TF}t\right)\exp\left(-\frac{\gamma_\mu^2\Delta_{\rm G,TF}^2 t^2}{2} - \gamma_\mu\Delta_{\rm L,TF}t\right). \tag{6.133}$$

Again, substituting $\Delta_{\rm G,TF}$ by $\Delta_{\rm G,ZF}$ and $\Delta_{\rm L,TF}$ by $\Delta_{\rm L,ZF}$ we find the result that we would expect. Obviously more work is needed to justify Eq. 6.132. Additional examples of its applications are given by Larkin *et al.* (2000).

The effect of dynamics on the polarisation functions has not yet been investigated and therefore will not be discussed here.

When one of the field sources is not of physical interest (i.e. if it arises from the nuclear magnetic moments of a magnetic compound under study), it is convenient to apply a weak longitudinal field to quench its effect on $P_Z(t)$. However, the reader should be aware that this method might not be suitable in systems with a very weak electronic component if the characteristic correlation frequencies of the electronic and nuclear correlations are in the same range.

6.6 Static field distributions and material parameters

In the previous sections of this chapter we have studied the polarisation functions associated with different simple field distributions. In this section we shall investigate the validity of these distributions and express their parameters in terms of characteristics of the material under study. We shall neglect the interaction between the magnetic moments at the origin of the distribution; i.e. we shall assume the moments to be uncorrelated — the interaction effects will be treated in the next chapter and in Chapter 10. The second assumption that we make is that the fields acting on the muon spin are static; i.e. we shall only discuss $P_\alpha^{\rm stat}(t)$. To a good approximation, these two assumptions are fulfilled when nuclei are responsible for the field sensed by the muons. The formulae in this section will therefore be given in terms of nuclear parameters. A table of nuclear dipole magnetic and quadrupole electric moments is given in Appendix G for the naturally occuring isotopes.

The material given in this section can be generalised to dynamical fields, usually arising from electrons, with the help of Section 6.3. Again, the restriction to the generalisation is the absence of correlations among the moments. The results consequently apply to the paramagnetic state of a compound at a temperature far above any magnetic phase transition. Since the magnetism of an electronic shell is most usually characterised by its spectroscopic factor, the reader will use Eq. 5.24 to express the formulae written in this section in terms of the gyromagnetic ratio. Appendix H provides relevant spectroscopic factors.

We start our discussion by considering a perfect crystal. The case of a crystal with magnetic impurities is treated thereafter.

6.6.1 Perfect crystal

A good knowledge of the material contained in Chapters 1, 2, and 10 of Slichter (1996) and in Chapters 4 and 6 of Abragam's textbook (1960) is useful for understanding this section.

We shall mainly use quantum mechanics although, from time to time, we shall refer to classical mechanics since most of the results can be nicely obtained with it. We recall Eq. 3.22 (see Section 3.1.2):

$$P_X(t) = \frac{1}{2}\mathrm{Tr}\left\{\rho_{\mathrm{sys}}\sigma^X\sigma^X(t)\right\}, \quad P_Z(t) = \frac{1}{2}\mathrm{Tr}\left\{\rho_{\mathrm{sys}}\sigma^Z\sigma^Z(t)\right\}, \tag{6.134}$$

where ρ_{sys} is the density operator of the system. For definitiveness, we suppose that the field distribution at the muon site arises from nuclear magnetic moments located at site i with spin $\mathbf{I}_i$. Since the interaction energy between the nuclear spins is negligible relative to the thermal energy, the high temperature approximation is justified (see Slichter (1996)). This means that $\rho_{\mathrm{sys}} = \mathcal{I}/\Pi_{i=1}^{N_{\mathrm{ns}}}(2I_i+1)$, where $\mathcal{I}$ is the unit operator and N_{ns} the number of nuclear spins contributing to the field distribution. In the Heisenberg representation

$$\sigma^\alpha(t) = \exp\left(i\frac{\mathcal{H}_{\mathrm{tot}}}{\hbar}t\right)\sigma^\alpha\exp\left(-i\frac{\mathcal{H}_{\mathrm{tot}}}{\hbar}t\right). \tag{6.135}$$

The operator σ^α is the α component of the Pauli vector operator for the muon spin and $\mathcal{H}_{\mathrm{tot}}$ the Hamiltonian for the muon–system ensemble.

Our aim is now to compute the polarisation functions, or alternatively, when this is not possible, their short time expansion. In a first instance we shall neglect any electric field gradient acting on the nuclei, and then we shall account for it.

6.6.1.1 Basic results for the polarisation functions. We express $\mathcal{H}_{\mathrm{tot}}$ as the sum of three terms:

$$\mathcal{H}_{\mathrm{tot}} = \mathcal{H}_{\mathrm{Z},\mu} + \mathcal{H}_{\mathrm{Z,sys}} + \mathcal{H}_{\mathrm{dip}}. \tag{6.136}$$

The first two terms on the right-hand-side account for the Zeeman energies of the muon and the nuclear magnetic moments and $\mathcal{H}_{\rm dip}$ describes the muon–nuclear dipole interaction. The interaction between the muon and conduction electron (if any) moments is neglected. This approximation is further discussed in Section 10.2.

The muon and nuclear Zeeman Hamiltonians are given by

$$\mathcal{H}_{{\rm Z},\mu} = -\frac{\hbar\omega_\mu}{2}\,\sigma^Z \quad \text{and} \quad \mathcal{H}_{\rm Z,sys} = -\sum_{i=1}^{N_{\rm ns}} \hbar\omega_i\, I_i^Z. \tag{6.137}$$

The muon and nuclear angular frequencies, ω_μ and ω_i, respectively, may be different from those corresponding to the applied field because of demagnetisation effects and due to the Knight shift relative to the muon and the nuclei; see Chapter 5.

Let us first discuss $P_X^{\rm stat}(t)$ in the high-field approximation. From the discussion of Section 6.2.1, we know that only the dipole field component parallel to $\mathbf{B}_{\rm ext}$ contributes to the damping of the oscillations. As explained in Section 5.2.2, for a sufficiently high field we only need to consider the so-called secular part of the dipolar Hamiltonian denoted hereafter with the "ˇ" symbol. Altogether, this means that we can restrict our attention to the truncated dipole Hamiltonian

$$\check{\mathcal{H}}_{{\rm dip},\parallel} = \sum_{i=1}^{N_{\rm ns}} \alpha_i \left(1 - 3\cos^2\theta_i\right) \sigma^Z I_i^Z, \tag{6.138}$$

with

$$\alpha_i = \frac{\mu_0}{4\pi}\frac{\gamma_\mu \gamma_i\, \hbar^2}{2 r_i^3}. \tag{6.139}$$

If $\check{\mathcal{H}}_{{\rm dip},\parallel}$ is substituted for $\mathcal{H}_{\rm dip}$ in Eq. 6.136, $\mathcal{H}_{\rm tot}$ is expressed as the sum of three diagonal operators. Therefore it is possible to derive an analytical formula for $P_X^{\rm stat}(t)$. These three operators commute with each other. Hence, according to the operator properties listed in Section C.2, we obtain

$$\begin{aligned} P_X^{\rm stat}(t) = \frac{1}{2\prod_{i=1}^{N_{\rm ns}}(2I_i+1)} \,{\rm Tr} \Bigg\{ & \exp\left(i\frac{\omega_\mu t}{2}\sigma^Z\right)\sigma^X \exp\left(-i\frac{\omega_\mu t}{2}\sigma^Z\right) \\ & \times \exp\left(i\frac{\check{\mathcal{H}}_{{\rm dip},\parallel}}{\hbar}t\right)\sigma^X \exp\left(-i\frac{\check{\mathcal{H}}_{{\rm dip},\parallel}}{\hbar}t\right)\Bigg\}. \end{aligned} \tag{6.140}$$

It is practical to express $P_X^{\rm stat}(t)$ in terms of the raising and lowering Pauli operators. From Eq. C.29

$$\begin{aligned} P_X^{\rm stat}(t) = & \frac{1}{8\prod_{i=1}^{N_{\rm ns}}(2I_i+1)} \\ & \times \Bigg[\exp\left(i\omega_\mu t\right) {\rm Tr}\left\{\sigma^+ \exp\left(i\frac{\check{\mathcal{H}}_{{\rm dip},\parallel}}{\hbar}t\right)\sigma^- \exp\left(-i\frac{\check{\mathcal{H}}_{{\rm dip},\parallel}}{\hbar}t\right)\right\} \end{aligned} \tag{6.141}$$

$$+ \exp(-i\omega_\mu t)\,\mathrm{Tr}\left\{\sigma^- \exp\left(i\frac{\check{\mathcal{H}}_{\mathrm{dip},\parallel}}{\hbar}t\right)\sigma^+ \exp\left(-i\frac{\check{\mathcal{H}}_{\mathrm{dip},\parallel}}{\hbar}t\right)\right\}\Bigg].$$

As a first step, let us evaluate $P_X^{\mathrm{stat}}(t)$ for a single nucleus i interacting with the muon spin:

$$P_X^{\mathrm{stat}}(t) = \frac{1}{2I_i+1}\sum_{m_i=-I_i}^{I_i} \cos\left\{\left[\omega_\mu - 2\frac{\alpha_i}{\hbar}\left(1-3\cos^2\theta_i\right)m_i\right]t\right\} \tag{6.142}$$

$$-\left\{\frac{1}{2I_i+1}\sum_{m_i=-I_i}^{I_i} \cos\left[2\frac{\alpha_i}{\hbar}\left(1-3\cos^2\theta_i\right)m_i\,t\right]\right\}\cos(\omega_\mu t).$$

Since $\check{\mathcal{H}}_{\mathrm{dip},\parallel}$ is a sum of N_{ns} terms of the same functional form, using the previous result, it is not difficult to deduce $P_X^{\mathrm{stat}}(t)$ for N_{ns} nuclei:

$$P_X^{\mathrm{stat}}(t) = \left\{\prod_{i=1}^{N_{\mathrm{ns}}}\frac{1}{2I_i+1}\sum_{m_i=-I_i}^{I_i} \cos\left[2\frac{\alpha_i}{\hbar}\left(1-3\cos^2\theta_i\right)m_i\,t\right]\right\}\cos(\omega_\mu t). \tag{6.143}$$

In order to determine $\Delta_{\mathrm{G,TF}}$, we expand $\ln[P_X^{\mathrm{stat}}(t)]$ given by the field distribution and quantum theories for small t and identify the terms proportional to t^2. Using Eq. C.111, this gives

$$\Delta^2_{\mathrm{G,TF}} = \left(\frac{\mu_0}{4\pi}\right)^2 \sum_{i=1}^{N_{\mathrm{ns}}} \frac{\gamma_i^2\hbar^2}{r_i^6}\frac{I_i(I_i+1)}{3}\left(1-3\cos^2\theta_i\right)^2. \tag{6.144}$$

This result can be recast in the appealing simple form

$$\Delta^2_{\mathrm{G,TF}} = \sum_{i=1}^{N_{\mathrm{ns}}}\left\langle\left(B^Z_{\mathrm{dip},i}\right)^2\right\rangle. \tag{6.145}$$

Hence, as expected, the variance of the field distribution is equal to the sum of the variances arising from each of the nuclear moments surrounding the muon. This formula can be deduced from classical mechanics, although supplemented with the quantum mechanics result, $\left\langle\left(I_i^Z\right)^2\right\rangle = I_i(I_i+1)/3$. This quantity $\Delta_{\mathrm{G,TF}}$ depends on the distance between the muon site and the nuclei at the source of the field distribution. In addition, it depends on the angles between $\mathbf{B}_{\mathrm{ext}}$ and the radius vectors from the muon to the nuclei. This explains why the measurements of $\Delta_{\mathrm{G,TF}}$ as a function of the orientation of $\mathbf{B}_{\mathrm{ext}}$ relative to the crystal axes is an efficient method for determining the muon site in diamagnets. Note that the lattice sum converges very rapidly since its terms depend on the distance as $1/r_i^6$. In fact, the contribution of the nearest neighbour nuclei usually dominates the sum.

Our description of the polarisation functions so far assumes a single-crystal sample. Therefore, for analysing a measurement performed on a polycrystalline sample, we should use a polarisation function obtained from the average over the 4π-steradian solid angle of a function discussed in this chapter. Using a polarisation function for a single crystal with an effective variance obtained from the 4π-steradian solid angle average of the theoretical variance is, *a priori*, not a proper method. The variance is strongly anisotropic as noticed from Eq. 6.144 for the transverse geometry and this remark holds also for the zero-field case as we shall show below. Surprisingly, a numerical treatment suggests that the simple method of using the average of the variance can be a good first approximation in some cases (Noakes 1986). However, a more recent study, which admittedly was performed in the motional narrowing limit, presents an example of the breakdown of that approximation — Pinkpank *et al.* (2000); see also Section 10.3.4. Therefore, although it is wise to use the proper method, i.e. to average the polarisation function itself, it is useful to study the powder average of the variance. For the case under study, it is given by

$$\overline{\Delta^2_{\mathrm{G,TF}}} = \frac{4}{5}\left(\frac{\mu_0}{4\pi}\right)^2 \sum_{i=1}^{N_{\mathrm{ns}}} \frac{\gamma_i^2 \hbar^2}{r_i^6} \frac{I_i\,(I_i+1)}{3}. \tag{6.146}$$

Hence, even from a polycrystalline measurement, it is possible to get information on the muon site by comparing the measured and calculated $\overline{\Delta^2_{\mathrm{G,TF}}}$ values for different possible muon localisation sites.

An explicit analytical formula for $P_X^{\mathrm{stat}}(t)$ can be derived from Eq. 6.143 since the sum over m_i is easily performed:

$$P_X^{\mathrm{stat}}(t) = \left\{ \prod_{i=1}^{N_{\mathrm{ns}}} \frac{\sin\left[(2I_i+1)\,\alpha_i \hbar^{-1}\left(1-3\cos^2\theta_i\right)t\right]}{(2I_i+1)\sin\left[\alpha_i \hbar^{-1}\left(1-3\cos^2\theta_i\right)t\right]} \right\} \cos\left(\omega_\mu t\right). \tag{6.147}$$

This expression has first been noted by Gurevich *et al.* (1972). In Fig. 6.17 we compare the envelope of $P_X^{\mathrm{stat}}(t)$ computed with this analytical formula and with the Gaussian approximation. In the time region of interest the Gaussian function provides an excellent approximation. At longer times, this approximation breaks down because of the periodicity of the envelope of the analytical function in the physical situation considered. We recall that the field at the muon is supposed to be static.

To conclude our discussion of $P_X^{\mathrm{stat}}(t)$, we note that Eq. C.13 allows us to write an expression for $\Delta^2_{\mathrm{G,TF}}$ directly in terms of $\check{\mathcal{H}}_{\mathrm{dip},\parallel}$:

$$\Delta^2_{\mathrm{G,TF}} = -\frac{1}{2\gamma_\mu^2\hbar^2} \frac{\mathrm{Tr}\left\{[\check{\mathcal{H}}_{\mathrm{dip},\parallel},\, \sigma^X]^2\right\}}{2\prod_{i=1}^{N_{\mathrm{ns}}}(2I_i+1)}. \tag{6.148}$$

This is the famous result derived by Van Vleck for the width of the NMR linewidth of unlike nuclear spins (Van Vleck 1948).

In contrast to the case for $P_X^{\mathrm{stat}}(t)$, it seems there is no analytical expression for $P_Z^{\mathrm{stat}}(t)$. However, the Van Vleck method is still useful for calculating the second moment of the field distribution:

$$\Delta^2_{\mathrm{G,ZF}} = -\frac{1}{2\gamma_\mu^2\hbar^2} \frac{\mathrm{Tr}\left\{[\mathcal{H}_{\mathrm{dip}}, \sigma^Z]^2\right\}}{2\prod_{i=1}^{N_{\mathrm{ns}}}(2I_i+1)}. \tag{6.149}$$

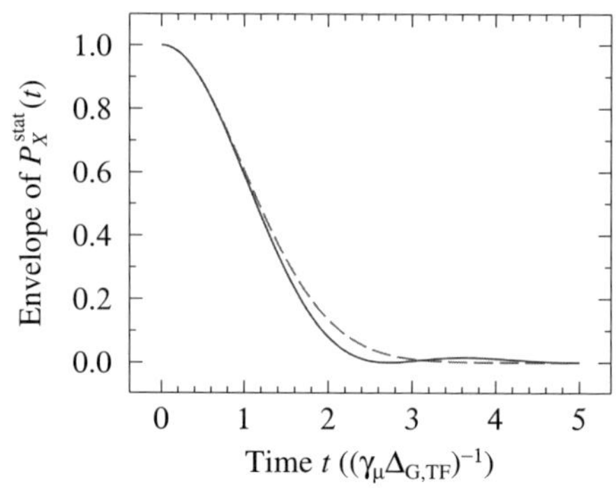

Fig. 6.17 Comparison of the envelope of the transverse polarisation function $P_X^{\rm stat}(t)$ in the high-field limit as computed from Eq. 6.147 (full line) with the Gaussian function $\exp(-\gamma_\mu^2\Delta^2_{\rm G,TF}/2)$ using $\Delta^2_{\rm G,TF}$ as given by Eq. 6.144 (dashed line). The muon is supposed to be located in an octahedral site of a face-centred-cubic compound; see left panel of Fig. 9.1 for a representation of this site. The field is applied along one of the cube edges and we assume $I=\frac{1}{2}$. The time scale is given in terms of $(\gamma_\mu\Delta_{\rm G,TF})^{-1}$. Note that for such physical conditions the envelope of $P_X^{\rm stat}(t)$ given by Eq. 6.147 is periodical with a period equal to $2\pi\sqrt{3}(\gamma_\mu\Delta_{\rm G,TF})^{-1}$.

Referring to the expression of $\mathcal{H}_{\rm dip}$ (see Eq. 5.29) we obtain

$$\Delta^2_{\rm G,ZF} = -\frac{1}{2\gamma_\mu^2\hbar^2}\frac{{\rm Tr}\left\{[\mathcal{H}_{\rm dip,\perp},\sigma^Z]^2\right\}}{2\prod_{i=1}^{N_{\rm ns}}(2I_i+1)}, \tag{6.150}$$

where $\mathcal{H}_{\rm dip,\perp} = -(\gamma_\mu\hbar/4)\left(\sigma^+B^-_{\rm dip}+\sigma^-B^+_{\rm dip}\right)$. We stress that Eq. 6.150 is written in the laboratory reference frame. The two dipole field components $B^\pm_{\rm dip}$ are listed in Eq. 5.30. This result is consistent with the fact that $P_Z^{\rm stat}(t)$ is damped by the two components of the dipole field perpendicular to Z; see Section 6.2.1.

In zero field the nuclei are not subject to a Zeeman interaction and therefore all the matrix elements of I_i are to be considered; i.e. a so-called secular approximation is not justified as explained in Section 5.2.2. Introducing the expression of $\mathcal{H}_{\rm dip,\perp}$ in Eq. 6.150, we derive

$$\Delta^2_{\rm G,ZF} = \frac{1}{2}\left\langle B^-_{\rm dip}B^+_{\rm dip}+B^+_{\rm dip}B^-_{\rm dip}\right\rangle = \left(\frac{\mu_0}{4\pi}\right)^2\sum_{i=1}^{N_{\rm ns}}\frac{\gamma_i^2\hbar^2}{r_i^6}\frac{I_i\left(I_i+1\right)}{3}\frac{\left(5-3\cos^2\theta_i\right)}{2}, \tag{6.151}$$

where we have used the fact that $\left\langle I_i^\alpha I_j^\beta\right\rangle$ vanishes unless $i=j$ for any set of Cartesian components α and β. In addition $\left\langle\left(I_i^Z\right)^2\right\rangle = \langle I_i^+I_i^-\rangle/2 = \langle I_i^-I_i^+\rangle/2 = I_i(I_i+1)/3$. The average over the 4π solid angle gives

$$\overline{\Delta^2_{\rm G,ZF}} = 2\left(\frac{\mu_0}{4\pi}\right)^2\sum_{i=1}^{N_{\rm ns}}\frac{\gamma_i^2\hbar^2}{r_i^6}\frac{I_i\left(I_i+1\right)}{3}. \tag{6.152}$$

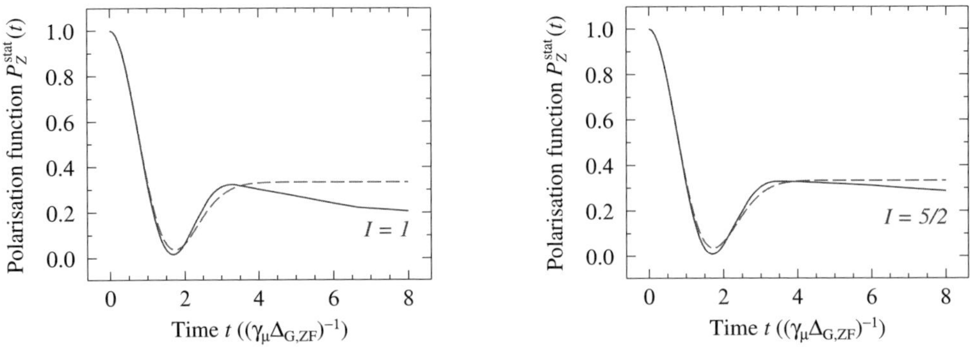

Fig. 6.18 Comparison of $P_Z^{\text{stat}}(t)$ obtained from the Kubo–Toyabe formula (dashed line) and from the numerical computation for a muon in an octahedral site of a face-centred-cubic compound with six nearest neighbour nuclei (full line). Relevant for this latter case, the Z axis defining the direction of the muon initial polarisation is oriented along a threefold axis (e.g. [111]). For the left and right panels, $I = 1$ and $I = \frac{5}{2}$, respectively. The time scale is given in terms of $1/(\gamma_\mu \Delta_{\text{G,ZF}})$. The Kubo–Toyabe model provides a very good approximation of the numerical results for $t \lesssim 4(\gamma_\mu \Delta_{\text{G,ZF}})^{-1}$. At longer times a deviation is found which is more important for small spins. Inspection of the right panel of Fig. 6.8 shows that the addition of weak dynamics to the Kubo–Toyabe model would better simulate the numerical curve. This dynamics is, however, purely phenomenological and not physical. Reproduced with kind permission from Dalmas de Réotier *et al.* (1992).

We note that $\overline{\Delta^2_{\text{G,ZF}}} = 2\left[(5/4)\overline{\Delta^2_{\text{G,TF}}}\right]$ for a randomly oriented powdered sample instead of the expected relation $\overline{\Delta^2_{\text{G,ZF}}} = 2\overline{\Delta^2_{\text{G,TF}}}$ if the simple geometrical argument given in Section 6.2.1 (two components of the dipole field contribute to $\Delta^2_{\text{G,ZF}}$ instead of one for $\Delta^2_{\text{G,TF}}$) was sufficient. The enhancement of the zero-field depolarisation relative to the transverse-field depolarisation is discussed further in Section 7.1.4.2. The $\Delta^2_{\text{G,ZF}}$ formula can be derived from classical mechanics although supplemented with the quantum result, $\left\langle \left(I_i^X\right)^2 \right\rangle = \left\langle \left(I_i^Y\right)^2 \right\rangle = I_i(I_i+1)/3$ (Hayano *et al.* 1979).

In Fig. 6.18 we compare $P_Z^{\text{stat}}(t)$ obtained from the Kubo–Toyabe formula with $\Delta^2_{\text{G,ZF}}$ given by Eq. 6.151 and from the numerical computation (see Section 7.2) for a muon in an octahedral site of a face-centred-cubic compound. Only the first six nearest neighbour nuclei are considered. The Kubo–Toyabe formula provides a reasonable model outside the $\frac{1}{3}$ plateau region.

6.6.1.2 Effect of the muon-induced electric field gradient. So far we have considered the Hamiltonian describing the muon–system ensemble to be given by Eq. 6.136. However, it is well known from NMR that a nucleus sitting in a site of point symmetry lower than cubic is influenced by the electric field gradient at this site for spins $I_i > \frac{1}{2}$. We account for this interaction by writing the muon–system ensemble Hamiltonian

$$\mathcal{H}_{\text{tot}} = \mathcal{H}_{\text{Z},\mu} + \mathcal{H}_{\text{Z,sys}} + \mathcal{H}_{\text{Q,sys}} + \mathcal{H}_{\text{dip}}, \tag{6.153}$$

where $\mathcal{H}_{\rm Q,sys} = \sum_i \mathcal{H}_{{\rm Q},i}$. Each term $\mathcal{H}_{{\rm Q},i}$ represents the interaction energy of the nuclear charge distribution of nucleus i in an electric potential V_i due to external sources. This interaction is nicely discussed by Slichter (1996). Expanding V_i in a Taylor series about the origin, it is possible to show that only the second order term is of importance: $V_i^{\alpha\beta} = \partial^2 V/\partial r_i^\alpha \partial r_i^\beta$, where r_i^α refers to a coordinate in the i nucleus reference frame. We associate here a distinct reference frame to each nucleus. It is always possible to find the principal axes of the potential: in this reference frame the components $V_i^{\alpha\beta}$ vanish unless $\alpha = \beta$. It is customary to choose the labels $\{x, y, x\}$ so that $|V_i^{zz}| \geq |V_i^{yy}| \geq |V_i^{xx}|$. Since V_i satisfies Laplace's equation, $V_i^{xx} + V_i^{yy} + V_i^{zz} = 0$, only two parameters are required to define the $V_i^{\alpha\beta}$ tensor: q_i and η_i, called, respectively, the electric field gradient and electric field gradient asymmetry parameter at site i. They follow the equations

$$eq_i = V_i^{zz}, \qquad \eta_i = \frac{V_i^{xx} - V_i^{yy}}{V_i^{zz}}, \tag{6.154}$$

where e is the proton electric charge. The asymmetry parameter fulfils the relation $0 \leq \eta_i \leq 1$. In this frame, $\mathcal{H}_{{\rm Q},i}$ is relatively simple:

$$\begin{aligned} \mathcal{H}_{{\rm Q},i} &= \frac{e^2 q_i Q_i}{4 I_i (2I_i - 1)} \left\{ \left[3 (I_i^z)^2 - (I_i)^2 \right] + \eta_i \left[(I_i^x)^2 - (I_i^y)^2 \right] \right\}, \\ &\equiv \hbar\omega_{\rm E} \left\{ \left[3 (I_i^z)^2 - (I_i)^2 \right] + \eta_i \left[(I_i^x)^2 - (I_i^y)^2 \right] \right\}, \end{aligned} \tag{6.155}$$

The quantity Q_i is the quadrupole moment of nucleus i and $\hbar\omega_{\rm E}$, which is proportional to Q_i, sets the energy scale of the quadrupole interaction energy. The subscript i on the energy has been dropped for simplicity.

For definitiveness, we shall assume the electric field gradient to only originate from the muon electric charge. Let us first consider a measurement performed at high field in the transverse geometry. Each nucleus surrounding the muon is then subject to the combined effect of the Zeeman and quadrupole interaction energies. Since the Z axis and the z axis associated with the reference frame of nucleus i are not parallel in general, the nuclear energy scheme is calculated either by transforming the Zeeman Hamiltonian to the $\{x, y, z\}$ coordinate system, or by writing the quadrupole Hamiltonian in the laboratory reference frame. Following Hartmann (1977), we take the second approach. The transformation can be performed with the help of rotation operators; see Section C.4. The diagonalisation of the sum of the two Hamiltonian operators yields the $(2I+1)$ eigenvalues and eigenvectors, E_k and $|k\rangle$, respectively (for simplicity we have again suppressed the subscript i since there is no risk of confusion). Using the result given in Eq. 6.145, we derive

$$\Delta^2_{\rm G,TF} = \sum_{i=1}^{N_{\rm ns}} \left\langle \left(B^Z_{{\rm dip},i}\right)^2 \right\rangle = \sum_{i=1}^{N_{\rm ns}} \left[\frac{1}{2I_i + 1} \sum_{k=-I_i}^{I_i} \left\langle k \left| \left(B^Z_{{\rm dip},i}\right)^2 \right| k \right\rangle \right], \tag{6.156}$$

with, as explained in the text leading to Eq. 5.33,

$$\left\langle k \left| \left(B^Z_{{\rm dip},i}\right)^2 \right| k \right\rangle = \left(\langle k \left| B^Z_{{\rm dip},i} \right| k \rangle \right)^2, \tag{6.157}$$

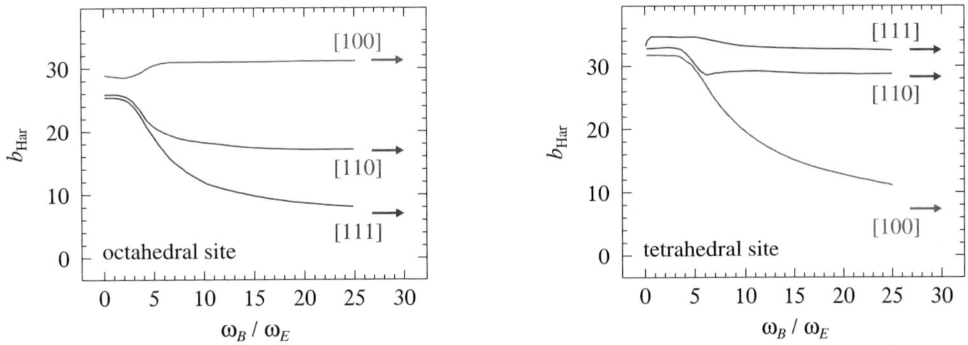

Fig. 6.19 Calculated reduced transverse damping rate b_{Har} for a face-centred-cubic structure as a function of the relative interaction strength $\omega_B/\omega_{\mathrm{E}}$ for different orientations of the external magnetic field. The arrows indicate the asymptotical limit of b_{Har} for large ω_B. By definition, $b_{\mathrm{Har}} \equiv \left(4\pi a^3 \Delta_{\mathrm{G,TF}}\right) / \left(\mu_0 \gamma_i \hbar\right)$, where a is the lattice parameter, $\omega_B = \gamma_i B_{\mathrm{ext}}$ and $\omega_{\mathrm{E}} = \left(e^2 q_i Q_i\right) / \left[4\hbar I_i (2I_i - 1)\right]$. The left and right panels present the results for the octahedral and tetrahedral interstitial muon sites, respectively. As examples, the reduced coordinates of such sites are respectively $(0, 0, \frac{1}{2})$ and $(\frac{1}{4}, \frac{1}{4}, \frac{1}{4})$. The computation assumes the nuclei spins to be $\frac{3}{2}$. Adapted from Hartmann (1977).

and

$$\left\langle k \left| B^Z_{\mathrm{dip},i} \right| k \right\rangle = \frac{\mu_0}{4\pi} \frac{\gamma_i \hbar}{r_i^3} \left[\left\langle k \left| I_i^Z \right| k \right\rangle \left(3\cos^2\theta_i - 1\right) + 3 \left\langle k \left| I_i^X \right| k \right\rangle \sin\theta_i \cos\theta_i \right]. \quad (6.158)$$

This result has first been derived by Hartmann (1977). The X axis is chosen to lie in the plane containing $\mathbf{r}_i$ and $\mathbf{B}_{\mathrm{ext}}$. In addition, B_{ext} needs to be sufficiently large that non-diagonal matrix elements such as $\left\langle k | B^Z_{\mathrm{dip},i} | k' \right\rangle$ average out to 0. This is the so-called secular approximation.

In the simple case of a face-centred-cubic crystal, the cubic symmetry at the lattice sites dictates the vanishing of any electric field gradient acting on the nuclei. Thus, the assumption we have made for our derivation that states that the only electric field gradient acting on the nuclei arises from the muon charge is justified. Hartmann has numerically computed the effect of that gradient using Eqs. 6.156 through 6.158 for octahedral or tetrahedral interstitial muon sites. In Fig. 6.19 we illustrate the dependence on the external field intensity of the reduced damping rate b_{Har}. More examples are given by Hartmann. It is obvious that the field has a strong effect on the width of a field distribution of nuclear origin when measured in transverse geometry. As expected, the damping rate reduces to the familiar Van Vleck expression (Eq. 6.144) at high field. The contribution to the damping rate from nucleus i vanishes if $1 - 3\cos^2\theta_i = 0$. For example, for a muon in an octahedral interstitial site of a face-centred-cubic crystal (see Fig. 9.1), the contribution of the first neighbour nuclei vanishes if the field is applied along a threefold axis. Hence, an expected dramatic decrease of the width is found for certain field directions.

We now consider the effect of the muon-induced electric field gradient on $\Delta_{\rm G,ZF}$. The calculation is conveniently carried out in the coordinate system for which the z axis is along the radius vector $\mathbf{r}_i$. In that system, only the first two terms of Eq. 5.28 contribute to $\mathcal{H}_{\rm dip}$. Expressing σ^Z in the $\{x, y, z\}$ reference frame for each nucleus, $\sigma^Z = \sigma^z \cos\theta - \sigma^x \sin\theta$, and using Eq. 6.150, we obtain

$$\Delta^2_{\rm G,ZF} = \frac{1}{2}\left(\frac{\mu_0}{4\pi}\right)^2 \sum_{i=1}^{N_{\rm ns}} \frac{\gamma_i^2\hbar^2}{r_i^6} \left\{\cos^2\theta_i \left[\left\langle (I_i^x)^2\right\rangle + \left\langle (I_i^y)^2\right\rangle\right] + \sin^2\theta_i \left[4\left\langle (I_i^z)^2\right\rangle + \left\langle (I_i^y)^2\right\rangle\right]\right\}. \quad (6.159)$$

First we note that taking $\left\langle (I_i^x)^2\right\rangle = \left\langle (I_i^y)^2\right\rangle = \left\langle (I_i^z)^2\right\rangle = I_i(I_i+1)/3$, we recover the result of Eq. 6.151, which was obtained in the absence of an electric gradient acting on the nuclei. Now we must compute $\left\langle (I_i^\alpha)^2\right\rangle$ in the presence of a quadrupole interaction. We assume that this interaction is sufficiently strong compared to the dipolar interaction between the muon and the nuclear spins so that the nature of the nuclear levels is solely determined by the quadrupole interaction. This situation differs from the large transverse-field case studied previously where the nuclear levels are split by a Zeeman Hamiltonian. The difference is that the quadrupole interaction does not fully lift the degeneracy of the energy levels. Recalling the discussion of Section 5.2.2, this implies that the terms $\langle k|I_i^\alpha(t)|k'\rangle$ do not necessarily average out to 0 if $|k\rangle$ and $|k'\rangle$ are different, but satisfy $E_k = E_{k'}$. In our assumption the electric field gradient acting on the nuclei arise from the muon charge. Hence it is axial and therefore $\eta_i = 0$. This means that the nuclear eigenvectors are the same as the Zeeman Hamiltonian eigenvectors. In addition, the eigenvalues associated to eigenvectors with the same absolute value of the magnetic quantum number m are the same. The computation leads to the simple result $\left\langle (I_i^z)^2\right\rangle = I_i(I_i+1)/3$. As far as $\left\langle (I_i^\alpha)^2\right\rangle$ with α equal to x and y are concerned, we must distinguish the integer and half-integer spin cases. For the former, all the matrix elements $\langle m|I_i^\alpha|m\rangle$ and $\langle m|I_i^\alpha|-m\rangle$ vanish. For half-integer spins some of them are not zero. Specifically, $\langle m|I_i^\alpha|-m\rangle\langle -m|I_i^\alpha|m\rangle = (I_i+1/2)^2/4$ for $|m\rangle = |\pm\frac{1}{2}\rangle$. This leads to $\left\langle (I_i^x)^2\right\rangle = \left\langle (I_i^y)^2\right\rangle = (2I_i+1)/8$.

Because of the difference in the analytical results for integer and half-integer spins, we introduce the contribution $\Delta^2_{{\rm G,ZF},i}$ of spin i to $\Delta^2_{\rm G,ZF}$:

$$\Delta^2_{\rm G,ZF} = \sum_i^{N_{\rm ns}} \Delta^2_{{\rm G,ZF},i}. \quad (6.160)$$

From Eq. 6.159 and the previous discussion, we obtain

$$\Delta^2_{{\rm G,ZF},i} = 2\left(\frac{\mu_0}{4\pi}\right)^2 \frac{\gamma_i^2\hbar^2}{r_i^6}\frac{I_i\,(I_i+1)}{3}\sin^2\theta_i, \quad (6.161)$$

for integer spins and

$$\Delta^2_{\mathrm{G,ZF},i} = \left(\frac{\mu_0}{4\pi}\right)^2 \frac{\gamma_i^2 \hbar^2}{r_i^6} \frac{I_i (I_i+1)}{3} \left[2 \sin^2 \theta_i + \frac{3(2I_i+1)}{16 I_i (I_i+1)} \left(2 - \sin^2 \theta_i\right)\right], \qquad (6.162)$$

for half-integer spins.

The powder averages of Eq. 6.161 and Eq. 6.162 are

$$\overline{\Delta^2_{\mathrm{G,ZF},i}} = \frac{4}{3} \left(\frac{\mu_0}{4\pi}\right)^2 \frac{\gamma_i^2 \hbar^2}{r_i^6} \frac{I_i (I_i+1)}{3}, \qquad (6.163)$$

and

$$\overline{\Delta^2_{\mathrm{G,ZF},i}} = \left(\frac{\mu_0}{4\pi}\right)^2 \frac{\gamma_i^2 \hbar^2}{r_i^6} \frac{I_i (I_i+1)}{3} \left[\frac{4}{3} + \frac{(2I_i+1)}{4 I_i (I_i+1)}\right], \qquad (6.164)$$

respectively. The formulae for the second moment in the zero-field case do not depend on the field gradient intensity and quadrupole moment. The reason is that they apply when the quadrupole energy is sufficiently large so that the secular approximation holds. Since the gradient is expected to decrease as fast as r_i^{-3}, the expressions of $\Delta^2_{\mathrm{G,ZF},i}$ in the large quadrupole energy limit may only be valid for the nearest neighbour nuclei to the muon. Equation 6.162 is a quantum mechanical formula: it cannot be derived from classical mechanics for which a half-integer spin is not defined (Hayano *et al.* 1979).

Let us consider the ratio $\mathcal{R}_{\mathrm{ZF}}$ of $\Delta_{\mathrm{G,ZF}}$ in the absence of quadrupole interaction over the same quantity computed for the case of a large quadrupole interaction. We assume $N_{\mathrm{ns}} = 1$ (and therefore $I_i = I$) and take the formulae for powder samples. For an integer spin, $\mathcal{R}_{\mathrm{ZF}} = \sqrt{3/2} = 1.2247$. For a half-integer spin

$$\mathcal{R}_{\mathrm{ZF}} = \sqrt{\frac{24 I (I+1)}{16 I (I+1) + 3(2I+1)}}. \qquad (6.165)$$

In Fig. 6.20 we present $\mathcal{R}_{\mathrm{ZF}}$ as a function of the spin value. The effect of the quadrupole interaction is not large. This is in contrast to $\Delta_{\mathrm{G,TF}}$ for which the effect of this interaction can be quite important as seen in Fig. 6.19.

When computing $\Delta_{\mathrm{G,TF}}$ or $\Delta_{\mathrm{G,ZF}}$ for a compound, one must sometimes cope with the presence of several isotopes for a given atom. In such a situation, the computation of the field variance is carried out as the sum of the contributions of the different muon neighbours in the same way as was done in Eq. 6.160 for the zero-field case in the large quadrupole interaction limit. With the notation used here, each term $\Delta^2_{\mathrm{G,ZF},i}$ is replaced by a weighted sum over the different isotopes present at site i with their own spins and gyromagnetic ratios. The weighting factor is of course the abundance of the corresponding isotope. The method described here obviously applies to all the variances derived in this section.

In Section 6.2.2 we studied $P_Z^{\mathrm{stat}}(t)$ for an anisotropic field distribution. With the $\Delta_{\mathrm{G,ZF}}$ formulae derived in the current section, it is a simple matter to compute the uniaxial anisotropy parameter α_{ani}. It is obtained from the ratio of $\Delta_{\mathrm{G,ZF}}$ computed with the Z axis parallel and perpendicular to the uniaxial axis of the crystal.

The reader interested by the symmetry properties of the various variances discussed in this section is referred to Appendix D.

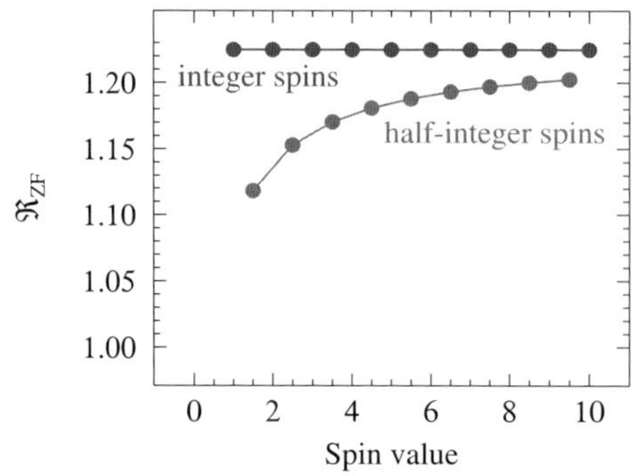

Fig. 6.20 Ratio $\mathcal{R}_{ZF}$ of $\Delta_{G,ZF}$ in the absence of quadrupole interaction over the same quantity computed for the case of a large quadrupole interaction. It is presented for integer and half-integer spins. The points for each species are linked by straight lines for the sake of clarity. Since quadrupolar interactions do not affect spins $\frac{1}{2}$, this spin value is not considered in the plot.

We cannot leave this section without mentioning the possible hyperfine enhancement of the nuclear magnetism. At the origin of the hyperfine-enhanced nuclear magnetism is a possible strong Van Vleck-like response for a non-Kramers ion with a non-magnetic $4f$ crystal-electric-field ground state. The hyperfine enhancement results from the coupling of the electronic and nuclear degrees of freedom. The admixture of $4f$ and nuclear wavefunctions leads to an effective nuclear moment with an enhancement factor proportional to the product of the atomic hyperfine coupling constant and the $4f$ susceptibility. This factor can be as large as 20. Up to now this effect has been identified in μSR data recorded on praseodymium-based compounds (see for example, MacLaughlin *et al.* 2000, and references therein).

6.6.2 Crystal with magnetic impurities

In this section we study the shape of the field distribution in a spatially homogeneous disordered system. We assume frozen magnetic impurities to be distributed at random in a non-magnetic matrix. Therefore they produce a static field distribution at the muon site. As usual in this chapter, we neglect the correlations between the moments. We stress that the model can be extended to other physical situations: for instance, the impurities can be defects in a well-ordered magnetic structure.

As always done in this chapter, we first study the distribution probed by the high-transverse-field technique and then consider the zero-field case.

Following Walstedt and Walker (1974), see also Chapter 4 of Abragam (1960), we write the distribution $D_{c,\mathbf{B}_{ext}}\left(B_{loc}^{Z}\right)$ investigated by the high-transverse-field technique in the form $D_{c,\mathbf{B}_{ext}}\left(B_{loc}^{Z}\right) = D_{c}\left(B_{loc}^{Z} - B_{ext}\right)$ with

$$D_{c}\left(B_{loc}^{Z}\right) = \int \cdots \int \delta\left(B_{loc}^{Z} - \sum_{i=1}^{N} B_{loc,i}^{z}\right) \tag{6.166}$$

$$\times \prod_{i=1}^{N} \left[(1 - c_{\rm imp})\, \delta\left(B^z_{{\rm loc},i}\right) + c_{\rm imp}\, w_i(B^z_{{\rm loc},i})\right] {\rm d}B^z_{{\rm loc},1} \cdots {\rm d}B^z_{{\rm loc},N}.$$

We recall that $B_{\rm ext}$ contributes a cosine term to $P_X(t)$ and does not have any effect on $D_{\rm c}\left(B^Z_{\rm loc}\right)$. In contrast to the unique quantisation axis hypothesis of Walstedt and Walker, we allow each impurity to have a distinct quantisation axis, z (for simplicity, we drop the subscript i for z). The unique quantisation axis hypothesis is good when the Zeeman energy of the magnetic moments in the applied field overcomes their anisotropy energy. Otherwise, each impurity should be considered with its own quantisation axis. We deal with $D_{\rm c}\left(B^Z_{\rm loc}\right)$ because this is the only distribution of interest for a high-transverse-field investigation as argued in Section 6.2.1. The summation and product are running over the N equivalent crystal positions which can be occupied by the impurity magnetic moments. The fractional impurity concentration, i.e. the occupation probability of an impurity at a given site, is denoted by $c_{\rm imp}$ $(0 \leq c_{\rm imp} \leq 1)$ and $w_i(B^z_{{\rm loc},i})$ is the distribution function of the field produced at the muon site by an impurity at site i. The distributions due to the impurities are assumed to be independent. This justifies the factorisation of the terms in the square bracket. This means that Eq. 6.166 cannot describe correlations between the moments. On the other hand, if needed, the factorisation does not prevent accounting for correlations between the moments and the crystal structure. The muon spins sense the sum of the fields: this justifies the integrations. In practice, the calculation is most conveniently carried out for the characteristic function of $D_{\rm c}\left(B^Z_{\rm loc}\right)$ (see Eq. 6.2):

$$G_{\rm TF}(t) = \int \exp\left(i\gamma_\mu B^Z_{\rm loc} t\right) D_{\rm c}\left(B^Z_{\rm loc}\right) {\rm d}B^Z_{\rm loc}, \qquad (6.167)$$

where the integral extends over all the $B^Z_{\rm loc}$ values contributing to $D_{\rm c}\left(B^Z_{\rm loc}\right)$. We have

$$\begin{aligned} G_{\rm TF}(t) &= \prod_{i=1}^{N} \left[1 - c_{\rm imp} + c_{\rm imp} \int \exp\left(i\gamma_\mu B^z_{{\rm loc},i} t\right) w_i(B^z_{{\rm loc},i}) {\rm d}B^z_{{\rm loc},i}\right] \qquad (6.168) \\ &= \exp\left\{\sum_{i=1}^{N} \ln\left[1 - c_{\rm imp} + c_{\rm imp} \int \exp\left(i\gamma_\mu B^z_{{\rm loc},i} t\right) w_i(B^z_{{\rm loc},i}) {\rm d}B^z_{{\rm loc},i}\right]\right\}. \end{aligned}$$

The characteristic function can thus be expressed as an infinite product. At low impurity concentration $(c_{\rm imp} \ll 1)$,

$$G_{\rm TF}(t) \simeq \exp\left\{-c_{\rm imp} \sum_{i=1}^{N} \left[1 - \int \exp\left(i\gamma_\mu B^z_{{\rm loc},i} t\right) w_i(B^z_{{\rm loc},i}) {\rm d}B^z_{{\rm loc},i}\right]\right\}. \qquad (6.169)$$

Since, in contrast to the usual NMR situation, the dipole interaction between the impurity and the muon magnetic moments is much more intense than the Ruderman–Kittel–Kasuya–Yosida (RKKY) interaction (see Chapter 5), it is usually sufficient to consider only the pure dipole interaction. If needed, the RKKY contribution can be included using the method of Walstedt and Walker. According to Eq. 5.31

$$B^z_{\mathrm{dip},i} = -\frac{\mu_0}{4\pi}\frac{g_i\mu_\mathrm{B}}{r_i^3}J_i^z\left(3\cos^2\theta_i - 1\right), \tag{6.170}$$

where J_i^z is the z component of the impurity (electronic) spin. To proceed further, we approximate the discrete sum over i by an integration with respect to $\mathbf{r}$:

$$\sum_{i=1}^{N} \longrightarrow \rho_\mathrm{vol}\int\int r^2\mathrm{d}\Omega\,\mathrm{d}r, \tag{6.171}$$

where ρ_vol^{-1} is the volume occupied per magnetic impurity and $\mathrm{d}\Omega = \sin\theta\,\mathrm{d}\varphi\,\mathrm{d}\theta$ the solid angle element. We implicitly assume that all the spins are identical. Equation 6.169 follows then from integrations over four variables performed in the following order: r, θ, φ, and B^z_loc. Using the identity

$$\int_0^\infty\left[1-\cos\left(\frac{\alpha}{r^3}\right)\right]r^2\,\mathrm{d}r = \frac{\pi|\alpha|}{6}, \tag{6.172}$$

derived using Eq. C.76, we obtain

$$G_\mathrm{TF}(t) = \exp\left(-\gamma_\mu\Delta_\mathrm{L,TF}|t|\right), \quad \Delta_\mathrm{L,TF} = K_\mathrm{L,TF}\frac{\mu_0}{4\pi}\rho_\mathrm{vol}c_\mathrm{imp}g\mu_\mathrm{B}\langle|m|\rangle, \tag{6.173}$$

where the m's are the eigenvalues of J^Z and with the constant $K_\mathrm{L,TF}$ defined as

$$K_\mathrm{L,TF} = \frac{\pi}{6}\frac{1}{4\pi}\int\int|\cos\theta'(3\cos^2\theta - 1)|\,\mathrm{d}\Omega'\mathrm{d}\Omega = \frac{4\pi^2}{9\sqrt{3}} \simeq 2.5325\,. \tag{6.174}$$

If the quantisation axis is unique, i.e. if B_ext is so large that the Zeeman energy overcomes the anisotropy energy of the impurity, the $\cos\theta'$ term which arises from the projection of the z axis into the Z axis is absent. The parameter $K_\mathrm{L,TF}$ is then twice as large. The product $\rho_\mathrm{vol}c_\mathrm{imp}$ which appears in Eq. 6.173 is the volume density of magnetic impurities. We have assumed that the moment magnitude and orientation are independent random variables. The imaginary part of the integral of Eq. 6.169 produces a combined demagnetisation and Lorentz correction which adds a term to the mean-field value at the muon site. We simply drop that term for the present discussion of the decay of the polarisation. The estimate of $K_\mathrm{L,TF}$ has first been given by Fiory (1981). The value of $\langle|m|\rangle$ depends of course on the density operator of the system. For instance, if the spins are frozen $g\mu_\mathrm{B}\langle|m|\rangle$ is equal to the magnetic moment of the impurity in its ground state. A detailed discussion of this point is given by Walstedt and Walker.

By inverse Fourier transform of Eq. 6.173 we derive

$$D_\mathrm{c}\left(B^Z_\mathrm{loc}\right) = \frac{1}{\pi}\frac{\Delta_\mathrm{L,TF}}{\Delta^2_\mathrm{L,TF} + \left(B^Z_\mathrm{loc}\right)^2}. \tag{6.175}$$

Numerical computations corroborate this analytical result for $c_\mathrm{imp} < 0.01$ (Walstedt and Walker 1974). The half-width at half-maximum introduced in Section 6.4.1 is now expressed in terms of material parameters of the compound.

We stress that the Lorentzian field distribution is a direct consequence of the r^{-3} range dependence of the dipole interaction: if the dependence were different we could not use Eq. 6.172, which is essential to get the exponential form of $G_{\mathrm{TF}}(t)$. In addition, to obtain $G_{\mathrm{TF}}(t)$ and therefore $D_{\mathrm{c}}\left(B_{\mathrm{loc}}^{Z}\right)$, the high-field approximation has been assumed to be valid.

One of the main approximations used for the derivation of Eq. 6.175 has been to replace the discrete sum by integrals over r and Ω (Mikaelyan and Smilga 1982). Actually this is legitimate if the argument of the cosine function of Eq. 6.172 is large, i.e.

$$\frac{\mu_0}{4\pi}\frac{\gamma_\mu g \mu_{\mathrm{B}} t}{r_0^3} \gtrsim 1, \tag{6.176}$$

where r_0 is a typical minimum value for the distance between the muon and an impurity, i.e. a few ångströms. Then the exact value of the lower bound of the integral in Eq. 6.172 is not important and the integration over the space including the origin is a good approximation. From this inequality, we can conclude that Eq. 6.173 is justified for $t \gtrsim t_m$ with

$$t_m = \frac{4\pi}{\mu_0}\frac{r_0^3}{\gamma_\mu g \mu_{\mathrm{B}}}. \tag{6.177}$$

For an estimate of t_m, we assume a typical moment of one Bohr magneton, i.e. $g = 2$ with a spin $J = \frac{1}{2}$. For $r_0 = 1\,\text{Å}$ we obtain $t_m = 0.6\,\mathrm{ns}$ and for $r_0 = 2\,\text{Å}$ we compute $t_m = 5\,\mathrm{ns}$. Therefore the result written in Eq. 6.78, which is directly derived from the Lorentzian field distribution given in Eq. 6.175, is usually valid in the entire time range accessible to μSR. If a small-time extension is needed, one must expand the polarisation function given in Eq. 6.11 and then take the configuration average as explained by Mikaelyan and Smilga. This yields, as expected, a parabolic time dependence for the envelope of the polarisation function at short times.

With the same methodology the distribution probed in a zero-field measurement can be determined. We recall that the vector distribution of $\mathbf{B}_{\mathrm{loc}}$ and not only the distribution of B_{loc}^{Z} must be considered:

$$\begin{aligned} D_{\mathrm{v}}\left(\mathbf{B}_{\mathrm{loc}}\right) = \int \cdots \int \delta\left(\mathbf{B}_{\mathrm{loc}} - \sum_{i=1}^{N} \mathbf{B}_{\mathrm{loc},i}\right) \\ \times \prod_{i=1}^{N}\left[(1 - c_{\mathrm{imp}})\,\delta\left(\mathbf{B}_{\mathrm{loc},i}\right) + c_{\mathrm{imp}}\, w_i(\mathbf{B}_{\mathrm{loc},i})\right] \mathrm{d}\mathbf{B}_{\mathrm{loc},1} \cdots \mathrm{d}\mathbf{B}_{\mathrm{loc},N}. \end{aligned} \tag{6.178}$$

The calculation is performed using the Fourier transform of the local field distribution which, in contrast to the previous calculation, depends on a vector $\mathbf{t}$ and not simply on a scalar t:

$$G_{\mathrm{ZF}}(\mathbf{t}) = \int \exp\left(i\gamma_\mu \mathbf{B}_{\mathrm{loc}} \cdot \mathbf{t}\right) D_{\mathrm{v}}\left(\mathbf{B}_{\mathrm{loc}}\right) \mathrm{d}^3\mathbf{B}_{\mathrm{loc}}. \tag{6.179}$$

The imaginary part of the integral vanishes when averaged over the directions of the quantisation axis. We find

$$G_{\mathrm{ZF}}(\mathbf{t}) = \exp\left(-\gamma_\mu \Delta_{\mathrm{L,ZF}} t\right), \quad \Delta_{\mathrm{L,ZF}} = K_{\mathrm{L,ZF}}\frac{\mu_0}{4\pi}\rho_{\mathrm{vol}} c_{\mathrm{imp}} g \mu_{\mathrm{B}} \langle |m| \rangle, \tag{6.180}$$

with the constant

$$K_{\mathrm{L,ZF}} = \frac{\pi}{6}\frac{1}{4\pi}\int\int |\cos\theta' - 3\cos\theta\,[\cos\theta\cos\theta' + \sin\theta\sin\theta'\cos(\phi-\phi')]\,|\mathrm{d}\Omega'\mathrm{d}\Omega$$
$$\simeq 4.5406. \qquad (6.181)$$

This numerical value of $K_{\mathrm{L,ZF}}$ has first been given by Fiory (1981); see also Gist and Dodds (1984). Since $G_{\mathrm{ZF}}(\mathbf{t})$ depends only on t, the field distribution computed from the inverse Fourier transform of $G_{\mathrm{ZF}}(\mathbf{t})$ is isotropic. It is found that $D_{\mathrm{v}}(B_{\mathrm{loc}})$ has a squared-Lorentzian profile; see Eq. 6.76. Numerical simulations support this analytical result (Walker and Walstedt 1980) and the ensuing polarisation function (McHenry *et al.* 1972, Fiory 1981) if the impurity concentration is small enough. We note that $\Delta_{\mathrm{L,TF}}/\Delta_{\mathrm{L,ZF}} = 0.55775$, whereas $\left(\overline{\Delta^2_{\mathrm{G,TF}}}/\overline{\Delta^2_{\mathrm{G,ZF}}}\right)^{1/2} = \sqrt{2/5} = 0.63246$ (see Eqs. 6.146 and 6.152) if we consider the polycrystalline average of $\Delta_{\mathrm{G,TF}}$ and $\Delta_{\mathrm{G,ZF}}$ and assume no electric field gradient acts on the field source. These two ratios are therefore quite similar.

Up to now, our concern has been the field distributions in the large dilution limit. In the following we set out to calculate them in the other extreme limit, i.e. $c_{\mathrm{imp}} \simeq 1$, for which analytical results can be obtained. The characteristic function for a transverse-field measurement in this limit is obtained from Eq. 6.168:

$$G_{\mathrm{TF}}(t) = \prod_{i=1}^{N}\left[\int \exp\left(i\gamma_\mu B^z_{\mathrm{loc},i}t\right) w_i(B^z_{\mathrm{loc},i})\mathrm{d}B^z_{\mathrm{loc},i}\right]. \qquad (6.182)$$

The field distribution no longer arises from the presence of magnetic impurities in the lattice. Its only origin is in the orientation or magnitude of the magnetic moments. The dominant contribution to $G_{\mathrm{TF}}(t)$ in the integral comes from values of t such that $|t| < 1/(\gamma_\mu B^z_{\mathrm{loc},i})$ because the factor $\exp\left(i\gamma_\mu B^z_{\mathrm{loc},i}t\right)$ oscillates rapidly at greater values of $|t|$ and therefore $G_{\mathrm{TF}}(t)$ vanishes in this limit. Accordingly, it is wise to study $G_{\mathrm{TF}}(t)$ at short t. This study is performed in two steps. First we write

$$\ln\left[G_{\mathrm{TF}}(t)\right] = \sum_{i=1}^{N}\ln\left[\int \exp\left(i\gamma_\mu B^z_{\mathrm{loc},i}t\right) w_i(B^z_{\mathrm{loc},i})\mathrm{d}B^z_{\mathrm{loc},i}\right] \qquad (6.183)$$
$$\simeq \sum_{i=1}^{N}\ln\left\{\int\left[1 + i\gamma_\mu B^z_{\mathrm{loc},i}t - \frac{\left(\gamma_\mu B^z_{\mathrm{loc},i}t\right)^2}{2}\right] w_i(B^z_{\mathrm{loc},i})\mathrm{d}B^z_{\mathrm{loc},i}\right\}.$$

The term proportional to $\gamma_\mu B^z_{\mathrm{loc},i}t$ does not influence the shape of the field profile. It contributes to the mean field at the muon site. Dropping this term, we finally obtain

$$G_{\mathrm{TF}}(t) \simeq \exp\left(-\frac{\gamma^2_\mu \check{\Delta}^2_{\mathrm{G,TF}}t^2}{2}\right), \qquad (6.184)$$

with

$$\check{\Delta}^2_{\mathrm{G,TF}} = \frac{1}{3}\left(\frac{\mu_0}{4\pi}\right)^2 \sum_{i=1}^{N} \frac{g^2\mu_{\mathrm{B}}^2}{r_i^6} \left\langle \left(J^Z\right)^2 \right\rangle \left(1 - 3\cos^2\theta_i\right)^2 . \tag{6.185}$$

We implicitly assume the magnitude of $\mathbf{J}$ to be independent of the site. If the $2J+1$ sublevels of $\mathbf{J}$ are equipopulated, $\left\langle \left(J^Z\right)^2 \right\rangle = J(J+1)/3$. From the inverse Fourier transform of $G_{\mathrm{TF}}(t)$ (see Eq. 6.167) we obtain the field component distribution

$$D_{\mathrm{c}}\left(B^Z_{\mathrm{loc}}\right) = \frac{\gamma_\mu}{2\pi}\int_{-\infty}^{\infty} \exp\left(-i\gamma_\mu B^Z_{\mathrm{loc}} t\right) G_{\mathrm{TF}}(t)\mathrm{d}t = \frac{1}{\sqrt{2\pi}\check{\Delta}_{\mathrm{G,TF}}} \exp\left[-\frac{\left(B^Z_{\mathrm{loc}}\right)^2}{2\check{\Delta}^2_{\mathrm{G,TF}}}\right]. \tag{6.186}$$

This is exactly the distribution needed to get the standard result defined by Eq. 6.12. The only difference is a factor $\frac{1}{3}$ in the field variance: compare Eqs. 6.144 and 6.185. This factor reflects our hypothesis of each impurity having its own quantisation axis. If it is common to all the impurities, the $\frac{1}{3}$ factor is no longer present and we recover the result obtained for nuclear spins, that is, $\check{\Delta}^2_{\mathrm{G,TF}} = \Delta^2_{\mathrm{G,TF}}$. The foregoing argument is essentially a rederivation of the central limit theorem (van Kampen 1981).

In the zero-field case, we need to focus on $G_{\mathrm{ZF}}(\mathbf{t})$. That quantity is expected to depend only on t^2 if the field distribution is isotropic. This obviously leads immediately to the Gaussian shape for $D_{\mathrm{v}}(B_{\mathrm{loc}})$. We leave the computation of $G_{\mathrm{ZF}}(\mathbf{t})$ to the interested reader.

The change of shape of the distributions when going from the diluted to the concentrated limits is not really surprising. It is well known that these shapes should be Gaussian in the concentrated limit. As pointed out by Abragam (1960), the ratio of the fourth moment over the square of the second moment favours a truncated Lorentzian shape in the diluted limit.

We have found formulae for the field distributions only in two limiting cases: c_{imp} tending to 0 and approaching 1. However, the distributions can always be determined numerically for any value of c_{imp} using Eqs. 6.166 and 6.178 for the transverse- and zero-field geometries, respectively. Some examples have been presented (Walstedt and Walker 1974, Walker and Walstedt 1980). The developments of Gist and Dodds (1984) are also of interest.

References

Abragam A. (1961) *The Principles of Nuclear Magnetism* (Clarendon, Oxford).

Anderson P.W. (1954) *Journal of the Physical Society of Japan* **9**, 316–339.

Blumberg W.E. (1960) *Physical Review* **119**, 79–84.

Bramwell S.T., Fennell T., Holdsworth P.C.W., and Portelli B. (2002) *Europhysics Letters* **57**, 310–314.

Chandrasekhar S. (1943) *Reviews of Modern Physics* **15**, 1–89.

Czech R. (1989) *Zeitschrift für Physik* **B 75**, 513–525.

Czech R. and Kehr K.W. (1986) *Physical Review* **B 34**, 261–277.

Dalmas de Réotier P. and Yaouanc A. (1992) *Journal of Physics: Condensed Matter* **4**, 4533–4556.

Dalmas de Réotier P. and Yaouanc A. (1997) *Journal of Physics: Condensed Matter* **9**, R9113–R9166.

Dalmas de Réotier P., Yaouanc A., and Meshkov S.V. (1992) *Physics Letters* **A 162**, 206–212.

Dattagupta S. (1987) *Relaxation Phenomena in Condensed Matter Physics* (Academic Press, Orlando).

Ehlers G. (2006) *Journal of Physics: Condensed Matter* **18**, R231–R244.

Fiory A.T. (1981) *Hyperfine Interactions* **8**, 777–780.

Fisher K.H. and Hertz J.A. (1991) *Spin Glasses* (Cambridge University Press, Cambridge).

Gist G.A. and Dodds S.A. (1984) *Physical Review* **B 30**, 2340–2344.

Gurevich I.I., Mel'eshko E.A., Muratova I.A., Nikol'skii B.A., Roganov V.S., Selivanov V.I., and Sokolov B.V. (1972) *Physics Letters* **A 40**, 143–144.

Hartmann O. (1977) *Physical Review Letters* **39**, 832–835.

Hayano R.S., Uemura Y.J., Imazato J., Nishida N., Yamazaki T., and Kubo R. (1979) *Physical Review* **B 20**, 850–859.

James G., Burley D., Dyke P., Searl J., Steele N., and Wright J. (1999) *Advanced Modern Engineering Mathematics*, second edition (Addison-Wesley, Harlow).

Johnston D.C. (2006) *Physical Review* **B 74**, 184430, 1–7.

Karlsson E.B. (1995) *Solid State Phenomena As Seen by Muons, Protons, and Excited Nuclei* (Clarendon Press, Oxford).

Kehr K.W., Honig G., and Richter D. (1978) *Zeitschrift für Physik* **B 32**, 49–58.

Keren A. (2004) *Journal of Physics: Condensed Matter* **16**, S4603–S4618.

Koptev V.P. and Tarasov N.A. (1987) preprint PNPI-1313, Leningrad.

Kornilov E.I. and Pomjakuskin V.Yu. (1991) *Physics Letters* **A 153**, 364–367.

Kubo R. (1954) *Journal of the Physical Society of Japan* **9**, 935–944.

Kubo R. (1969) in *Stochastic Processes in Chemical Physics*, 101–127, edited by Shuler K. (Wiley, New York).

Kubo R. (1981) *Hyperfine Interactions* **8**, 731–738.

Kubo R. and Toyabe T. (1967) in *Magnetic Resonance and Relaxation*, 810–823, edited by Blinc R. (North-Holland, Amsterdam).

Larkin M.I., Fudamoto Y., Gat I.M., Kinkhabwala A., Kojima K.M., Luke G.M., Merrin J., Nachumi B., Uemura Y.J., Azuma M., Saito T., and Takano M. (2000) *Physica* **B 289-290**, 153–156.

Leon M. (1981) *Hyperfine Interactions* **8**, 781–783.

MacLaughlin D.E., Heffner R.H., Nieuwenhuys G.J., Canfield P.C., Amato A., Baines C., Schenck A., Luke G.M., Fudamoto Y., and Uemura Y.J. (2000) *Physical Review* **B 61**, 555–563.

McHenry M.R., Sibernagel B.G., and Wernick J.H. (1972) *Physical Review* **B 5**, 2958–2972.

Mikaelyan A.L. and Smilga V.P. (1982) *Soviet Physics JETP* **53**, 1216–1220.

Möslang A., Graf H., Balzer G., Recknagel E., Weidinger A., Wichert Th., and Grynszpan R.I. (1983) *Physical Review* **B 27**, 2674–2681.

Nagamine K. (2003) *Introductory Muon Science* (Cambridge University Press, Cambridge).

Noakes D.R. (1986) *Hyperfine Interactions* **31**, 47–52.
Noakes D.R. (1999) *Journal of Physics: Condensed Matter* **11**, 1589–1600.
Noakes D.R. and Kalvius G.M. (1997) *Physical Review* **B 56**, 2352–2355.
Palmer R.G., Stein D.L., Abrahams E., and Anderson P.W. (1984) *Physical Review Letters* **53**, 958–961.
Petzinger K.G. (1980) *Physics Letters* **A 75**, 225–228.
Pinkpank M., Amato A., Andreica D., Gygax F.N., Ott H.R., and Schenck A. (2000) *Physica* **B 289-290**, 295–298.
Press W.H., Teukolsky S.A., Vetterling W.T., and Flannery B.P. (2000) *Numerical Recipes in C book set* (Cambridge University Press, Cambridge).
Press W.H., Teukolsky S.A., Vetterling W.T., and Flannery B.P. (2007) *Numerical Recipes 3rd Edition: The Art of Scientific Computing* (Cambridge University Press, Cambridge).
Schenck A. (1985) *Muon Spin Rotation Spectroscopy* (Adam Hilger, Bristol).
Seeger A. (1978) in *Hydrogen in Metals*, edited by Alefeld G. and Völkl J., Springer Topics in Current Physics, vol. 28 (Springer, Berlin).
Slichter C.P. (1996) *Principles of Magnetic Resonance*, third enlarged and updated edition (Springer, Berlin).
Solt G. (1995) *Hyperfine Interactions* **96**, 167–175.
Tse D. and Hartmann S.R. (1968) *Physical Review Letters* **21**, 511–514.
Uemura Y.J. (1980) *Solid State Communications* **36**, 369–372.
Uemura Y.J., Yamazaki T., Harshman D.R., Senba M., and Ansaldo E.J. (1985) *Physical Review* **B 31**, 546–563.
Uemura Y.J., Keren A., Kojima K., Le L.P., Luke G.M., Wu W.D., Ajiro Y., Asano T., Kuriyama Y., Mekata M., Kikuchi H., and Kakuri K. (1994) *Physical Review Letters* **73**, 3306–3309.
van Kampen N.G. (1981) *Stochastic Processes in Physics and Chemistry* (North-Holland, Amsterdam).
Van Vleck J.H. (1948) *Physical Review* **74**, 1168–1183.
Walker L.R. and Walstedt R.E. (1980) *Physical Review* **B 22**, 3816–3842.
Walstedt R.E. and Walker L.R. (1974) *Physical Review* **B 9**, 4857–4867.
Ziman J.M. (1979) *Models of Disorder* (Cambridge University Press, Cambridge).

7
Muon polarisation functions: quantum approach

In the previous chapter we introduced field distributions to compute the polarisation functions in a somewhat phenomenological approach. The parameters of these distributions were subsequently described in terms of microscopic parameters of the system under study. In the current chapter, the computation of the polarisation functions is performed entirely starting from the microscopic sources of the field at the muon site without the use of the field distribution concept. In addition, the functions are computed in the framework of quantum mechanics. As exceptions, at the end of the chapter, a few polarisation functions are calculated from a classical solution of the same model for the purpose of comparison.

We recall the expressions for the two polarisation functions of interest, $P_Z(t)$ and $P_X(t)$ (Section 3.1.2):

$$P_Z(t) = \frac{1}{2}\mathrm{Tr}\left\{\rho_{\mathrm{sys}}\,\sigma^Z\sigma^Z(t)\right\}, \quad P_X(t) = \frac{1}{2}\mathrm{Tr}\left\{\rho_{\mathrm{sys}}\,\sigma^X\sigma^X(t)\right\}, \tag{7.1}$$

with

$$\sigma^\alpha(t) = \exp\left(i\frac{\mathcal{H}_{\mathrm{tot}}\,t}{\hbar}\right)\sigma^\alpha \exp\left(-i\frac{\mathcal{H}_{\mathrm{tot}}\,t}{\hbar}\right), \tag{7.2}$$

where ρ_{sys} is the density operator for the system, σ^α a Pauli operator for the muon spin, and $\mathcal{H}_{\mathrm{tot}}$ the Hamiltonian of the muon–system ensemble. This Hamiltonian is assumed to be independent of time.

Because of its spin, the muon is basically a probe of the magnetic properties of condensed matter and, as quantum mechanics is the key for understanding magnetism (Van Vleck 1978), it is natural to compute the polarisation functions within the framework of quantum mechanics. The crucial entanglement of the quantum states — the fact that the wavefunction of a system with several degrees of freedom cannot be written as a product of wavefunctions for each degree of freedom — is taken into account. A simple example of its importance is given by Ghosh *et al.* (2003). However, since the rigorous computation of $P_Z(t)$ and $P_X(t)$ is usually not possible for systems of interest because the involved matrices are extremely large, approximations are needed. The results gathered in the previous chapter are quite useful as a guide, first to choose the most appropriate approximations and second, for a physical insight into the result of the computations.

This chapter contains two sections. The first section describes analytical methods for the computation of $P_\alpha(t)$. In order to get some feeling of the work involved, we first present some exact results. Then we focus on the Bloch equations as applied to μSR and on a rate equation method for $P_Z(t)$. This always leads to exponential polarisation functions. An expression for the decay rate of the longitudinal polarisation function is given in terms of field correlation functions. However, as explained in the previous chapter, $P_\alpha(t)$ is not always an exponential function. This justifies the introduction of an iteration method which provides results of more general validity.

The iteration method is not able to compute zero-field polarisation functions in the slow spin dynamics limit in the whole time range since it is basically a perturbation technique. The evaluation of the polarisation functions in this limit requires numerical methods. This is the subject of the second section of this chapter.

The practical knowledge of quantum mechanics needed to read this chapter is nicely summarised in the second chapter of the book of Smilga and Belousov (1994). Another extremely useful general reference is the textbook of Slichter (1996).

7.1 Analytical methods

7.1.1 Some exact results

The evaluation of $\sigma^\alpha(t)$ as given in Eq. 7.2 requires the knowledge of the eigenfunctions and eigenvalues. In general, their determination is a formidable task. For example, let us assume $N_{\rm ns}$ spins I in interaction with the muon spin. The dimension of the Hilbert space associated to $\mathcal{H}_{\rm tot}$ is $2 \times (2I+1)^{N_{\rm ns}}$. For a muon in an octahedral site interacting with only its six nearest neighbour one-half spins, the dimension is already $2^7 = 128$. A Hamiltonian of such a large dimension cannot analytically be diagonalised in general.

The number of physical cases for which $P_\alpha(t)$ can be analytically computed is quite small. An example is presented in Section 4.2. Here we consider a second example originally discussed by Vogel *et al.* (1986). It consists of a muon spin interacting with a single nucleus of arbitrary spin I. This latter nucleus is submitted to an axial electric field gradient whose principal axis is directed along the applied field, i.e. the Z axis. The vector linking the muon to the nucleus is also parallel to Z. The Hamiltonian of the muon–system ensemble is thus given by

$$\mathcal{H}_{\rm tot} = \hbar\left\{-\frac{\omega_\mu}{2}\sigma^Z - \omega_I I^Z + 3\omega_{\rm E}\left[\left(I^Z\right)^2 - \frac{I(I+1)}{3}\right] + \frac{\omega_{I,\rm d}}{2}\left[\boldsymbol{\sigma}\cdot\mathbf{I} - 3I^Z\sigma^Z\right]\right\}, \tag{7.3}$$

with

$$\omega_{I,\rm d} = \frac{\mu_0}{4\pi}\frac{\gamma_\mu\gamma_I\hbar}{r_I^3}, \tag{7.4}$$

$\omega_\mu = \gamma_\mu B_{\rm ext}$, and $\omega_I = \gamma_I B_{\rm ext}$. The constant $\omega_{\rm E}$ is defined in Eq. 6.155. Since we are considering a nuclear spin, the magnetic energy is negligible in comparison to the thermal energy; see Section 6.6.1. It follows that

$$P_Z(t) = \frac{1}{2}\frac{1}{2I+1}{\rm Tr}\left\{\sigma^Z\sigma^Z(t)\right\}. \tag{7.5}$$

With the help of the notations

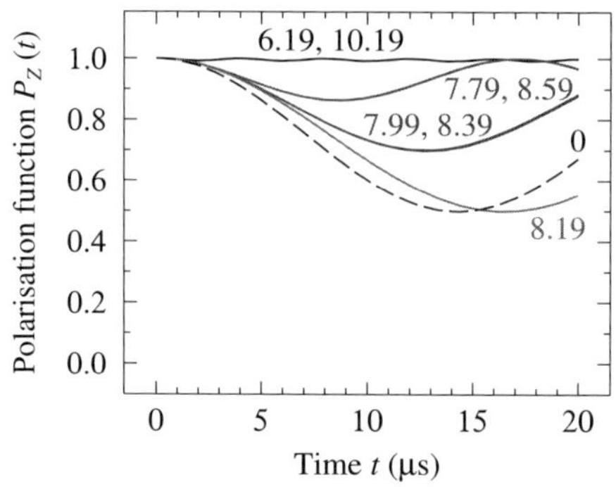

Fig. 7.1 Longitudinal polarisation function versus time for a muon interacting with a single nuclear spin $I = \frac{3}{2}$ plotted for various applied field strengths. The numbers in the plot indicate the field value in milliteslas. When two values are given separated by a comma, the corresponding curves can hardly be distinguished. The dashed line corresponds to the zero-field case. With the notation given in the main text, the model parameters are $\gamma_I = 72.5\,\mathrm{Mrad\,s^{-1}\,T^{-1}}$, $3\omega_\mathrm{E} = -3.3\,\mu\mathrm{s}^{-1}$, and $\omega_{I,\mathrm{d}} = 0.11\,\mu\mathrm{s}^{-1}$. They correspond to the case of metallic copper (Luke *et al.* 1991).

$$R_m = \frac{\omega_{I,\mathrm{d}}}{2}\sqrt{I(I+1) - m(m-1)}, \qquad S_m = (2m-1)\left(3\omega_\mathrm{E} + \frac{\omega_{I,\mathrm{d}}}{2}\right) - \omega_I + \omega_\mu, \tag{7.6}$$

we obtain

$$P_Z(t) = \frac{1}{2I+1}\sum_m \frac{S_m^2 + R_m^2 \cos\left(\sqrt{S_m^2 + R_m^2}\, t\right)}{S_m^2 + R_m^2}. \tag{7.7}$$

An example of $P_Z(t)$ for $I = \frac{3}{2}$ is plotted in Fig. 7.1. The avoided-level-crossing phenomenon is observed for B_ext such that $S_{3/2} = 0$; i.e. $B_\mathrm{ext} \simeq 8.19\,\mathrm{mT}$ for the parameters given in the figure caption. For this resonance field, the depolarisation is large. The resonance phenomenon is explained in Section 4.2.

The polarisation function $P_Z(t)$ in Eq. 7.7 is a weighted sum of cosine functions. Therefore $P_Z(t)$ may display oscillations, even at large times; see Fig. 7.1. This is quite distinct from the familiar Kubo–Toyabe form which derives from the field distribution approach; see Eq. 6.17. Large weakly damped oscillations have been measured for ionic fluoride crystals such as LiF (Brewer *et al.* 1986). In fact, the dimension of the Hilbert space needs to be large enough for the Kubo–Toyabe form to be a reasonable approximation as discussed in Section 7.2. This dimension is only $d_\mathrm{Hil} = 8$ for LiF (spin-$\frac{1}{2}$ muon and two spin-$\frac{1}{2}$ ^{19}F nuclei; see Appendix G).

7.1.2 Longitudinal and transverse relaxation rates

If the magnetisation of a system is driven out of thermodynamic equilibrium by a perturbation, as soon as the perturbation is switched off, the magnetisation relaxes back to its equilibrium value. The simplest description of the evolution of the magnetisation is given by a set of phenomenological equations, first proposed by Bloch (1946). While in the case of nuclear magnetic resonance (NMR), it is the nuclear magnetisation which relaxes, the muon polarisation is of interest for the μSR techniques. The relevant set of Bloch equations is then

$$\begin{aligned}
\frac{\mathrm{d}P_Z(t)}{\mathrm{d}t} &= -\lambda_Z P_Z(t), \\
\frac{\mathrm{d}P_X(t)}{\mathrm{d}t} &= -\lambda_X P_X(t) + \gamma_\mu \left[\mathbf{P}(t) \times \mathbf{B}_{\mathrm{loc}}\right]_X = -\lambda_X P_X(t) + \gamma_\mu B_{\mathrm{loc}} P_Y(t), \\
\frac{\mathrm{d}P_Y(t)}{\mathrm{d}t} &= -\lambda_X P_Y(t) + \gamma_\mu \left[\mathbf{P}(t) \times \mathbf{B}_{\mathrm{loc}}\right]_Y = -\lambda_X P_Y(t) - \gamma_\mu B_{\mathrm{loc}} P_X(t).
\end{aligned} \tag{7.8}$$

We have assumed the homogeneous field at the muon site $\mathbf{B}_{\mathrm{loc}}$ to be static and parallel to the Z axis and defined the polarisation vector $\mathbf{P}(t)$ of components $P_X(t)$, $P_Y(t)$, and $P_Z(t)$. Solving this set of equations we derive

$$\begin{aligned}
P_Z(t) &= P_Z(t=0)\exp(-\lambda_Z t), \\
P_X(t) &= P_X(t=0)\exp(-\lambda_X t)\cos(\gamma_\mu B_{\mathrm{loc}} t), \\
P_Y(t) &= -P_X(t=0)\exp(-\lambda_Y t)\sin(\gamma_\mu B_{\mathrm{loc}} t).
\end{aligned} \tag{7.9}$$

The value for $\mathbf{P}(t=0)$ depends on the experimental geometry (see Section 1.1). The parameters λ_Z and λ_X are, respectively, the longitudinal and transverse relaxation rates. They are denoted in NMR as $1/T_1$ and $1/T_2$, respectively. At first sight, the adopted form for Eq. 7.8 which leads to an exponential relaxation is rather arbitrary. It is already known to be valid in the extreme motional narrowing limit as explained in Section 6.3. This will be discussed at length in Section 7.1.4. The introduction of two relaxation rates follows from their different origin. The transverse relaxation rate arises from the spread in precession frequency produced by the field distribution as explained in Section 6.2. This distribution stems from the spins of the compound surrounding the muon. This justifies the name of spin–spin relaxation rate given to λ_X. On the other hand, λ_Z is of dynamical origin: an exchange of energy between the muon spin and the system is required for this type of relaxation to occur. This explains the name spin–lattice relaxation rate given to λ_Z, although the conduction electrons which can contribute to the relaxation are not attached to the crystal lattice.

Let us introduce a microscopic description for $P_Z(t)$. For this purpose, we shall closely follow the method introduced for NMR by Bloembergen *et al.* (1948). Its limitations are discussed at the end of Section 7.1.3. A detailed discussion of the different methods for NMR is given in the book of Wolf (1979). We note that the muon spin is a two-level quantum system and denote $N_{\mu,+}(t)$ ($N_{\mu,-}(t)$) the number of muon spins in state $|+\rangle$ ($|-\rangle$) at time t. Since each of these numbers is proportional to the magnetisation of the related energy level,

$$P_Z(t) = \frac{N_{\mu,+}(t) - N_{\mu,-}(t)}{N_{\mu,+}(t) + N_{\mu,-}(t)}. \tag{7.10}$$

Because $P_Z(t)$ depends on the relative population of the two Zeeman muon states, any change in their population will result in a change of $P_Z(t)$. Figure 7.2 illustrates

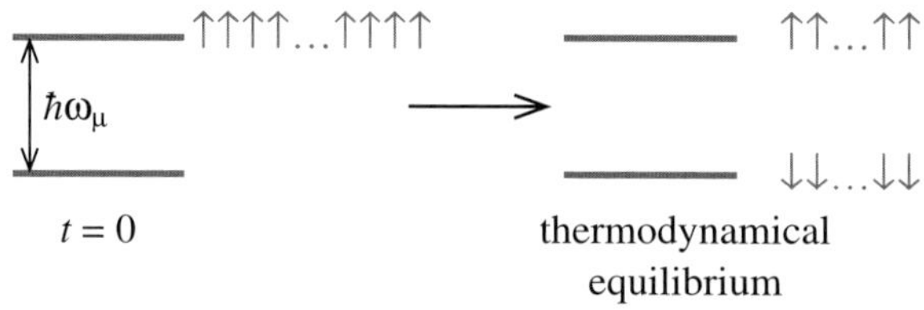

Fig. 7.2 Here we try to explain graphically the meaning of the decay of $P_Z(t)$ in the simplest possible physical case: the two muon energy levels are split by $\hbar\omega_\mu$. At initial time, i.e. $t = 0$, since the muons are fully polarised, they are all in the same quantum state. As time evolves, transitions between the two quantum states occur. If the relaxation is sufficiently fast, the muon polarisation is fully relaxed within the muon lifetime; that is, thermodynamical equilibrium is reached.

its decay. Let us consider a flipping from muon state $|+\rangle$ to $|-\rangle$. Because of energy conservation, the Zeeman energy absorbed for this flipping must be transferred from the system. Therefore the time evolution of the longitudinal polarisation depends on the availability of system levels which can exchange energy with the muon energy levels. Let us define $W_\uparrow$ to be the probability per unit time for the muon spin to flip from state $|+\rangle$ to state $|-\rangle$. The reverse probability is denoted by $W_\downarrow$. Then the change of population is accounted for by the following set of two rate equations:

$$\frac{\mathrm{d}N_{\mu,+}(t)}{\mathrm{d}t} = -W_\uparrow N_{\mu,+}(t) + W_\downarrow N_{\mu,-}(t),$$
$$\frac{\mathrm{d}N_{\mu,-}(t)}{\mathrm{d}t} = -W_\downarrow N_{\mu,-}(t) + W_\uparrow N_{\mu,+}(t). \tag{7.11}$$

As expected, $N_{\mu,+}(t) + N_{\mu,-}(t) = N_{\rm cst}$ is a constant. We could have labelled $W_\uparrow$ as $W(+ \to -)$. This would not be satisfactory since it would suggest that only the muon states are involved, which is not true. Introducing the notations $n(t) = N_{\mu,+}(t) - N_{\mu,-}(t)$ and $n_0 = N_{\rm cst}\left[(W_\downarrow - W_\uparrow)/(W_\uparrow + W_\downarrow)\right]$ (n_0 is a time independent constant), we derive

$$\frac{\mathrm{d}n(t)}{\mathrm{d}t} = -(W_\uparrow + W_\downarrow)\left[n(t) - n_0\right], \tag{7.12}$$

the solution of which is

$$n(t) = A\exp\left[-(W_\uparrow + W_\downarrow)\,t\right] + n_0, \tag{7.13}$$

where A is the constant of integration. Note that $P_Z(t) = n(t)/N_{\rm cst}$. At large times, the muon spin must be in thermal equilibrium with the system. Recalling that the Zeeman energy of a muon spin submitted to a field of 1 T corresponds to the thermal energy at 6.5 mK, the two muon level states are equipopulated under usual conditions. Therefore $n(t \to \infty) = n_0 = 0$. Because the muons are initially polarised, we have $N_{\mu,+}(t = 0) = 1$ and $N_{\mu,-}(t = 0) = 0$ and consequently $P_Z(t) = \exp(-\lambda_Z t)$ with

$$\lambda_Z = W_\uparrow + W_\downarrow. \tag{7.14}$$

The probabilities $W_\uparrow$ and $W_\downarrow$ can be computed with the Fermi golden rule, that is,

$$W_\uparrow = \frac{2\pi}{\hbar}\frac{1}{Z}\sum_{i,f} \exp(-\beta E_i)\left|\langle -, f|\mathcal{H}_{\text{pert}}|i, +\rangle\right|^2 \delta(E_i + E_+ - E_f - E_-), \qquad (7.15)$$

and

$$W_\downarrow = \frac{2\pi}{\hbar}\frac{1}{Z}\sum_{i,f} \exp(-\beta E_i)\left|\langle +, f|\mathcal{H}_{\text{pert}}|i, -\rangle\right|^2 \delta(E_i + E_- - E_f - E_+). \qquad (7.16)$$

The labels i and f stand for the initial and final states of the system under study and $\mathcal{H}_{\text{pert}}$ is the Hamiltonian which describes the perturbation inducing the transitions. As usual $Z = \sum_i \exp(-\beta E_i)$ is the partition function. The terms E_i and E_f are the energies of the system in states $|i\rangle$ and $|f\rangle$, respectively, while E_+ and E_- are the energies of the muon spin in states $|+\rangle$ and $|-\rangle$, respectively. The Dirac delta functions in Eqs. 7.15 and 7.16 ensure the energy conservation of the muon–system ensemble.

Since the transitions are induced by the fluctuations of the local field at the muon site, we simply have $\mathcal{H}_{\text{pert}} = -\frac{1}{2}\gamma_\mu \hbar \boldsymbol{\sigma}\cdot\delta\mathbf{B}_{\text{loc}}$. A derivation of this Zeeman Hamiltonian is given in Section 5.1.3. Here we consider only the fluctuating part. Using the integral representation of the Dirac distribution (Section C.7.2), the properties of the Pauli operators (Section C.5) and the solution of the Heisenberg equation, we obtain

$$W_\uparrow = \frac{\gamma_\mu^2}{4}\int_{-\infty}^{\infty} \exp\left(i\omega_\mu \tau\right)\left\langle \delta B^+_{\text{loc}}(\tau)\delta B^-_{\text{loc}}\right\rangle \mathrm{d}\tau, \qquad (7.17)$$

and

$$W_\downarrow = \frac{\gamma_\mu^2}{4}\int_{-\infty}^{\infty} \exp\left(-i\omega_\mu \tau\right)\left\langle \delta B^-_{\text{loc}}(\tau)\delta B^+_{\text{loc}}\right\rangle \mathrm{d}\tau. \qquad (7.18)$$

As usual we write $\langle A\rangle = \mathrm{Tr}\left\{\rho_{\text{sys}}A\right\}$, where the trace operation is performed over the Hilbert space of the system. Assuming the field correlations to be even functions of time, the probabiblities $W_\uparrow$ and $W_\downarrow$ are equal, consistent with the fact that $n_0 = 0$ in the absence of external field. In this case the sum of the two transition probabilities leads to the well-known NMR result (Moriya (1962) and, for example, Turov and Petrov (1972), or Slichter (1996)):

$$\lambda_Z = \pi\gamma_\mu^2 \int_{-\infty}^{\infty} \cos\left(\omega_\mu \tau\right)\left[\Phi^{XX}(\tau) + \Phi^{YY}(\tau)\right] \mathrm{d}\tau, \qquad (7.19)$$

where

$$\Phi^{\alpha\beta}(\tau) = \frac{1}{2\pi}\left\langle\left\{\delta B^\alpha_{\text{loc}}(\tau)\delta B^\beta_{\text{loc}}\right\}\right\rangle \qquad (7.20)$$

is, apart from a factor $1/(2\pi)$, the symmetrised correlation function of the magnetic field fluctuations at the muon site. The symbol $\langle\{AB\}\rangle$ indeed stands for the thermal average of the symmetrised correlation function of operators A and B: $2\langle\{AB\}\rangle = \langle AB\rangle + \langle BA\rangle$. The origin of the factor $1/(2\pi)$ follows from our definition of the

Fourier transform (see Eq. C.50) and the requirement of the normalisation to 1 of $\int_{-\infty}^{\infty} \Phi^{\alpha\beta}(\omega) \mathrm{d}\omega$.

Thus, the transitions between the muon energy levels are induced by the thermal fluctuations of the two field components perpendicular to Z. Note that λ_Z is computed from perturbation theory. This implies that the perturbation must be weak for the result to be valid. This is further discussed in the next sections.

7.1.3 Rate equation matrix method

Although the notation of the previous section is fully justified physically, it is interesting to introduce a new notation, which leads to a self-evident generalisation if more than two energy levels are at play as in the muonium case. We rewrite the set of equations given in Eq. 7.11 in a matrix form such as

$$\frac{\mathrm{d}\mathcal{N}(t)}{\mathrm{d}t} = -\mathcal{Y}\mathcal{N}(t), \tag{7.21}$$

where we have introduced the 2×2 matrix

$$\mathcal{Y} = \begin{pmatrix} Y_{12} & -Y_{21} \\ -Y_{12} & Y_{21} \end{pmatrix}, \tag{7.22}$$

and the column matrix

$$\mathcal{N}(t) = \begin{pmatrix} N_1(t) \\ N_2(t) \end{pmatrix} \quad \text{with} \quad \sum_{i=1}^{2} N_i(t) = 1. \tag{7.23}$$

The reader will easily identify the elements of $\mathcal{Y}$ and $\mathcal{N}(t)$ with physical quantities defined in the previous section. The transition probabilities, Y_{ij}, are calculated in the usual manner:

$$Y_{12} = \frac{J(\omega_\mu)}{Z} \sum_{i,j} \exp(-\beta E_i) \frac{\left|M_{ij}^{12}\right|^2}{\hbar^2},$$

$$Y_{21} = \frac{J(\omega_\mu)}{Z} \sum_{i,j} \exp(-\beta E_i) \frac{\left|M_{ij}^{21}\right|^2}{\hbar^2}, \tag{7.24}$$

with $|M_{ij}^{12}|^2 = |\langle -, i\,|\mathcal{H}_{\mathrm{pert}}|\, j, +\rangle|^2$ and $|M_{ij}^{21}|^2 = |\langle +, i\,|\mathcal{H}_{\mathrm{pert}}|\, j, -\rangle|^2$. The spectral density function $J(\omega)$ is introduced: this concept is well known in NMR (Bloembergen *et al.* 1948). It is proportional to the so-called spectral-weight function; see Section F.2.

We write the polarisation function,

$$P_Z(t) = 2\sum_{i=1}^{2} N_i(t) \left\langle \psi_i \left| S_\mu^Z \right| \psi_i \right\rangle, \tag{7.25}$$

where the vectors $|\psi_1\rangle$ and $|\psi_2\rangle$ are spanning the Hilbert space for the muon spin. Because of one null eigenvalue of the rate equation matrix, $P_Z(t)$ is a single exponential function, rather than the sum of two exponential functions. The non-zero eigenvalue is simply

$$\lambda_Z = Y_{12} + Y_{21}. \tag{7.26}$$

This immediately justifies the form $P_Z(t) = \exp(-\lambda_Z t)$. Let us assume $J(\omega)$ to be Lorentzian, that is,

$$J(\omega) = \frac{2\tau_c}{1 + \omega^2 \tau_c^2}, \tag{7.27}$$

where τ_c is the correlation time. As a function of τ_c, λ_Z starts to increase linearly with τ_c, reaching a maximum at $\tau_c = 1/\omega_\mu$ and then vanishes at large τ_c values. The maximum in λ_Z is analogous to the T_1 minima, which is well known in NMR.

A possible inadequacy of the microscopic rate equation approach to relaxation is the assumption that the population of the two muon eigenstates completely specifies the state of the muon spin system. Supporting this idea is the fact that a coherent oscillation cannot be described by the method. However, it is quite useful to get a picture of the spin–lattice relaxation mechanism as long as the splitting of the muon energy levels is appreciable so that the concept of population has a significance. The splitting is obviously well established if there is a large spontaneous field at the muon site (parallel to the Z axis by hypothesis). But even in a paramagnet in zero field, the concept is applicable if the Z component of the instantaneous field at the muon site is sufficiently strong for a reasonable time, so that the splitting is well defined. However, as shown in the next section, it is possible to define λ_Z in terms of field correlations even if the population concept cannot be applied. In the limit of very slow magnetic fluctuations a polarisation function of the Kubo–Toyabe type is always expected, as explained in the previous chapter. It is not described by the rate equation approach.

An additional fact, which shows that the validity of the present and previous sections for both $P_Z(t)$ and $P_X(t)$ is somewhat limited, is the observation that the polarisation functions are predicted to behave linearly in time at small times. This is unphysical as explained in Section 6.3. The origin of this issue arises from the implicit assumption that $\tau_c \ll t$; i.e. one should not trust the results obtained at too small times relative to the correlation time. The iteration method exposed in the next section provides a correct result in this limit.

7.1.4 Iteration method

7.1.4.1 Formal development. So far, using a qualitative argument (population of the two muon energy levels) an exponential form for $P_Z(t)$ has been derived from a system of two rate equations. The muon spin–lattice relaxation rate characterising $P_Z(t)$ has been computed from the Fermi golden rule. However, the exponential form is only expected in the extreme motional narrowing limit as explained in Section 6.3. Therefore the previous results are not entirely satisfactory. Their main advantage is their simplicity.

Here we study the polarisation functions using an iteration approach for the computation of the time evolution of the muon spin Pauli operator. This is entirely a quantum method for which the necessary approximations are well defined. We recover the results previously derived in the limit of extreme motional narrowing or long times. In addition, we also determine $P_\alpha(t)$ at times t that are comparable with, or less than, the magnetic correlation time τ_c. The first authors who properly emphasised this requirement are apparently Baryshevskiĭ and Kuten' (1976). A similar investigation was published soon after (McMullen and Zaremba 1978). The work of Hayano *et al.* (1979) contains most of the formulae needed to describe the polarisation functions. Here we shall closely follow the exposition of Dalmas de Réotier and Yaouanc (1992).

As noted from the expressions for the polarisation functions, see Eq. 7.1, we need to determine $\sigma^\alpha(t)$. Our approximate evaluation of this operator is based on the hypothesis that the Hamiltonian $\mathcal{H}_{\rm tot}$, which describes the muon–system ensemble, can be split into two parts: $\mathcal{H}_{\rm tot} = \mathcal{H}_0 + \mathcal{H}_1$. For the time being, we do not have to specify how the splitting is performed. We use the intermediate representation and define the following operators in this representation:

$$\tilde{\sigma}^\alpha(t) = \exp\left(-i\frac{\mathcal{H}_0 t}{\hbar}\right)\sigma^\alpha(t)\exp\left(i\frac{\mathcal{H}_0 t}{\hbar}\right), \tag{7.28}$$

and

$$\tilde{\mathcal{H}}_1(t) = \exp\left(-i\frac{\mathcal{H}_0 t}{\hbar}\right)\mathcal{H}_1\exp\left(i\frac{\mathcal{H}_0 t}{\hbar}\right). \tag{7.29}$$

Using Eq. 7.2, from the previous two equations we derive

$$\frac{\mathrm{d}\tilde{\sigma}^\alpha(t)}{\mathrm{d}t} = \frac{i}{\hbar}\left[\tilde{\mathcal{H}}_1(t), \tilde{\sigma}^\alpha(t)\right]. \tag{7.30}$$

The solution of this differential equation can be written as

$$\tilde{\sigma}^\alpha(t) = \tilde{\sigma}^\alpha(0) + \frac{i}{\hbar}\int_0^t \left[\tilde{\mathcal{H}}_1(t'), \tilde{\sigma}^\alpha(t')\right] \mathrm{d}t'. \tag{7.31}$$

We can substitute $\tilde{\sigma}^\alpha(t')$ in the previous equation by an expression similar to Eq. 7.31 and get

$$\begin{aligned}\tilde{\sigma}^\alpha(t) = \tilde{\sigma}^\alpha(0) &+ \frac{i}{\hbar}\int_0^t \left[\tilde{\mathcal{H}}_1(t'), \tilde{\sigma}^\alpha(0)\right] \mathrm{d}t' \\ &+ \left(\frac{i}{\hbar}\right)^2 \int_0^t \int_0^{t'} \left[\tilde{\mathcal{H}}_1(t'), \left[\tilde{\mathcal{H}}_1(t''), \tilde{\sigma}_\alpha(t'')\right]\right] dt''\mathrm{d}t'.\end{aligned} \tag{7.32}$$

The second and main hypothesis of the quantum computation (the first one consisted in neglecting the extremely small equilibrium polarisation of the muon states induced by the applied magnetic field, see Section 3.1.2) is to set $\tilde{\sigma}^\alpha(t'') = \tilde{\sigma}^\alpha(0)$ to terminate the iteration. The result is a kind of second order perturbation expression for $\tilde{\sigma}^\alpha(t)$ since the perturbing Hamiltonian, $\tilde{\mathcal{H}}_1(t)$, is acting twice. The range of validity of that approximation has been discussed in the past in connexion with the Redfield theory (see for example Slichter 1996). In the next section (see Section 7.1.4.2) we will establish that the aforementioned approximation is efficient for the description of the evolution of $P_\alpha(t)$ at short times in general, and for all t values in the motional narrowing limit.

We are now in a position to compute $\sigma^\alpha(t)$, within the approximation $\tilde{\sigma}^\alpha(t'') = \tilde{\sigma}^\alpha(0)$, using Eqs. 7.28, 7.29, and 7.32. It is then easy to get an expression for $P_Z(t)$ from Eq. 7.1:

$$P_Z(t) = P_Z^{(0)}(t) + P_Z^{(1)}(t) + P_Z^{(2)}(t), \tag{7.33}$$

with

$$\begin{aligned}
P_Z^{(0)}(t) &= \frac{1}{2}\mathrm{Tr}\left\{\rho_{\mathrm{sys}}\,\sigma^Z \exp\left(i\frac{\mathcal{H}_0 t}{\hbar}\right)\sigma^Z \exp\left(-i\frac{\mathcal{H}_0 t}{\hbar}\right)\right\},\\
P_Z^{(1)}(t) &= \frac{1}{2}\mathrm{Tr}\left\{\rho_{\mathrm{sys}}\,\sigma^Z \frac{i}{\hbar}\int_0^t \exp\left(i\frac{\mathcal{H}_0 t}{\hbar}\right)\right.\\
&\qquad \times\left.\left[\exp\left(-i\frac{\mathcal{H}_0 t'}{\hbar}\right)\mathcal{H}_1 \exp\left(i\frac{\mathcal{H}_0 t'}{\hbar}\right), \sigma^Z\right]\exp\left(-i\frac{\mathcal{H}_0 t}{\hbar}\right)\mathrm{d}t'\right\},\\
P_Z^{(2)}(t) &= \frac{1}{2}\mathrm{Tr}\left\{\rho_{\mathrm{sys}}\,\sigma^Z \left(\frac{i}{\hbar}\right)^2\int_0^t\int_0^{t'} \exp\left(i\frac{\mathcal{H}_0 t}{\hbar}\right)A\exp\left(-i\frac{\mathcal{H}_0 t}{\hbar}\right)\mathrm{d}t''\mathrm{d}t'\right\},
\end{aligned} \tag{7.34}$$

and

$$A \equiv \left[\exp\left(-i\frac{\mathcal{H}_0 t'}{\hbar}\right)\mathcal{H}_1\exp\left(i\frac{\mathcal{H}_0 t'}{\hbar}\right), \left[\exp\left(-i\frac{\mathcal{H}_0 t''}{\hbar}\right)\mathcal{H}_1\exp\left(i\frac{\mathcal{H}_0 t''}{\hbar}\right), \sigma^Z\right]\right]. \tag{7.35}$$

For the computation of the traces we need to specify $\mathcal{H}_0$ and $\mathcal{H}_1$. We set $\mathcal{H}_0 = \mathcal{H}_{\mathrm{sys}} + \bar{\mathcal{H}}_{\mathrm{Z},\mu}$, where $\mathcal{H}_{\mathrm{sys}}$ is the Hamiltonian for the system under study and $\bar{\mathcal{H}}_{\mathrm{Z},\mu}$ is the "mean field" muon Zeeman Hamiltonian:

$$\bar{\mathcal{H}}_{\mathrm{Z},\mu} = -\frac{1}{2}\hbar\omega_\mu\sigma^Z, \qquad \omega_\mu = \gamma_\mu\left\langle B^Z_{\mathrm{loc}}\right\rangle. \tag{7.36}$$

The wording "mean field" means that $\bar{\mathcal{H}}_{\mathrm{Z},\mu}$ is proportional to the thermal average $\left\langle B^Z_{\mathrm{loc}}\right\rangle$ rather than to the operator B^Z_{loc} (recall that $\mathbf{B}_{\mathrm{loc}} = \langle\mathbf{B}_{\mathrm{loc}}\rangle + \delta\mathbf{B}_{\mathrm{loc}}$, where $\delta\mathbf{B}_{\mathrm{loc}}$ is the fluctuating part of $\mathbf{B}_{\mathrm{loc}}$). The quantity $\left\langle B^Z_{\mathrm{loc}}\right\rangle$ includes an external field which is possibly applied. Note that $\mathbf{B}_{\mathrm{loc}}$ does not contain any muon spin operator. The Hamiltonian $\mathcal{H}_1$ is simply $\delta\mathcal{H}_{\mathrm{Z},\mu}$ and therefore

$$\mathcal{H}_1 = -\frac{1}{2}\gamma_\mu\hbar\boldsymbol{\sigma}\cdot\delta\mathbf{B}_{\mathrm{loc}}. \tag{7.37}$$

It is easily shown that $P_Z^{(0)}(t) = 1$ and $P_Z^{(1)}(t) = 0$. Using Eq. C.88 we get the following expression for $P_Z^{(2)}(t)$:

$$P_Z^{(2)}(t) = -\pi\gamma_\mu^2\int_0^t (t-\tau)\left[\exp\left(i\omega_\mu\tau\right)\Phi^{+-}(\tau) + \exp\left(-i\omega_\mu\tau\right)\Phi^{-+}(\tau)\right]\mathrm{d}\tau. \tag{7.38}$$

Note that we have

$$\delta B_{\rm loc}^{\alpha}(\tau) = \exp\left(i\frac{\mathcal{H}_{\rm sys}\tau}{\hbar}\right)\delta B_{\rm loc}^{\alpha}\exp\left(-i\frac{\mathcal{H}_{\rm sys}\tau}{\hbar}\right), \tag{7.39}$$

and $\Phi^{\alpha\beta}(\tau) = \Phi^{\beta\alpha}(-\tau)$. From the strong-collision model, we know that $P_Z(t)$ can be written in the following form in the motional narrowing limit (see Eq. 6.62):

$$P_Z(t) = \exp\left[-\psi_Z(t)\right]. \tag{7.40}$$

If we postulate that form, $\psi_Z(t)$ can be deduced from the second order iteration by identification of the expansion terms:

$$\psi_Z(t) = \pi\gamma_\mu^2\int_0^t (t-\tau)\left[\exp\left(i\omega_\mu\tau\right)\Phi^{+-}(\tau) + \exp\left(-i\omega_\mu\tau\right)\Phi^{-+}(\tau)\right]\mathrm{d}\tau, \tag{7.41}$$

or

$$\psi_Z(t) = 2\pi\gamma_\mu^2\int_0^t (t-\tau)\left\{\cos\left(\omega_\mu\tau\right)\left[\Phi^{XX}(\tau) + \Phi^{YY}(\tau)\right]\right.$$
$$\left. + \sin\left(\omega_\mu\tau\right)\left[\Phi^{XY}(\tau) - \Phi^{YX}(\tau)\right]\right\}\mathrm{d}\tau. \tag{7.42}$$

We recall the approximations that have been made. We have neglected on one hand the effect of the field on the equilibrium population of the two spin states of the muon, and on the other hand the time evolution of $\tilde{\sigma}^{\alpha}(t'')$; namely we have set $\tilde{\sigma}^{\alpha}(t'') = \tilde{\sigma}^{\alpha}(0)$. In addition we have assumed that the quantum calculation gives the first two terms of the expansion of the polarisation function.

This formula for $P_Z(t)$ reduces to an exponential relaxation function with the spin–lattice relaxation rate, λ_Z, given in Eq. 7.19 if the characteristic time of the correlation functions, $\tau_{\rm c}$, is much shorter than t and if the second term on the right-hand side of Eq. 7.42 can be neglected. The first condition means that the perturbation, which drives the relaxation, must only act during a relatively short time to get for $P_Z(t)$ an exponential function. The second term vanishes if $\Phi^{XY}(\tau) = \Phi^{YX}(\tau)$ or if the dynamics is sufficiently fast so that $\omega_\mu\tau_{\rm c} \ll 1$. Since $\Phi^{XY}(\tau) = \Phi^{YX}(-\tau)$, the condition $\Phi^{XY}(\tau) = \Phi^{YX}(\tau)$ requires $\Phi^{XY}(\tau)$ to be even in time. This is usually the case.

Using the same methodology, an expression for $P_X(t)$ can be derived. This is a tedious but straightforward computation (Dalmas de Réotier and Yaouanc 1992). As for $P_Z(t)$, $P_X(t)$ can be written with the $\{+,-,Z\}$ or $\{X,Y,Z\}$ coordinates. In the $\{+,-,Z\}$ coordinates we have

$$P_X(t) = \mathcal{R}e\left\{\exp\left[i\omega_\mu t - \Omega(t)\right]\right\}, \tag{7.43}$$

with

$$\Omega(t) = 2\pi\gamma_\mu^2\int_0^t (t-\tau)\Phi^{ZZ}(\tau)\mathrm{d}\tau + \pi\gamma_\mu^2\int_0^t (t-\tau)\Phi^{-+}(\tau)\exp\left(-i\omega_\mu\tau\right)\mathrm{d}\tau$$
$$+ i\frac{\pi\gamma_\mu^2}{2\omega_\mu}\int_0^t \left[\Phi^{++}(\tau) + \Phi^{--}(\tau)\right]\exp\left(-i\omega_\mu\tau\right)\mathrm{d}\tau. \tag{7.44}$$

Using the $\{X, Y, Z\}$ coordinates we write

$$P_X(t) = \exp\left[-\psi_X(t)\right] \cos\left[\omega_\mu t + \varphi(t)\right], \tag{7.45}$$

with

$$\begin{aligned}\psi_X(t) = \pi\gamma_\mu^2 \int_0^t (t-\tau)\left\{2\Phi^{ZZ}(\tau) + \cos(\omega_\mu\tau)\left[\Phi^{XX}(\tau) + \Phi^{YY}(\tau)\right]\right. \\ \left. + \sin(\omega_\mu\tau)\left[\Phi^{XY}(\tau) - \Phi^{YX}(\tau)\right]\right\} \mathrm{d}\tau \\ - \frac{\pi\gamma_\mu^2}{\omega_\mu}\int_0^t \sin(\omega_\mu\tau)\left[\Phi^{YY}(\tau) - \Phi^{XX}(\tau)\right] \mathrm{d}\tau,\end{aligned} \tag{7.46}$$

and

$$\begin{aligned}\varphi(t) = \pi\gamma_\mu^2 \int_0^t (t-\tau)\left\{\sin(\omega_\mu\tau)\left[\Phi^{XX}(\tau) + \Phi^{YY}(\tau)\right]\right. \\ \left. + \cos(\omega_\mu\tau)\left[\Phi^{YX}(\tau) - \Phi^{XY}(\tau)\right]\right\} \mathrm{d}\tau \\ + \frac{\pi\gamma_\mu^2}{\omega_\mu}\int_0^t \cos(\omega_\mu\tau)\left[\Phi^{YY}(\tau) - \Phi^{XX}(\tau)\right] \mathrm{d}\tau.\end{aligned} \tag{7.47}$$

The expression for $\varphi(t)$ given above is valid when $\varphi(t) \ll 1$. If the term proportional to $1/\omega_\mu$ in the equation for $\psi_X(t)$ is neglected, i.e. in the high-field approximation, and if the term in the second line of Eq. 7.46, which already appeared in the expression of $\psi_Z(t)$, can be neglected (see our previous discussion about $P_Z(t)$ concerning that term), we recover the NMR formula (Moriya 1962):

$$\psi_X(t) = \frac{1}{2}\psi_Z(t) + 2\pi\gamma_\mu^2 \int_0^t (t-\tau)\Phi^{ZZ}(\tau)\mathrm{d}\tau. \tag{7.48}$$

Therefore, as is well known in NMR, the decay of $P_X(t)$ is driven by the spin–lattice relaxation mechanism (the $\psi_Z(t)/2$ term), which involves the magnetic correlations of the transverse-field components, and the magnetic correlation of the field Z component.

In the motional narrowing limit and high-field approximations the phase term, $\varphi(t)$, is proportional to t and therefore leads to a frequency shift. We note that this shift is influenced by field correlations (the $\Phi^{\alpha\beta}(\tau)$ terms).

7.1.4.2 Basic applications. In this part we shall give a further insight into the results of Section 7.1.4.1 considering the off-diagonal terms of the field correlation tensor — e.g. $\Phi^{XY}(t)$ or $\Phi^{YX}(t)$ — to be negligible. In addition, we shall assume the thermal energy much larger than any interaction energy for the spins of the material at the origin of the time evolution of the polarisation functions. This approximation is relevant for the computation of the polarisation functions for diamagnetic metals for which the nuclear magnetic moments drive $P_\alpha(t)$. Moreover, the material of this section will give us the opportunity to deepen our understanding of the physics involved in the different terms of the polarisation functions. For definitiveness, we shall use the language of diffusion.

The expressions for $P_Z(t)$ and $P_X(t)$ given in Section 7.1.4.1 are general. They do not depend on the type of interactions included in the system Hamiltonian, $\mathcal{H}_{\rm sys}$, and in the coupling Hamiltonian $\mathcal{H}_1$, except that it has the Zeeman form. Here we assume that the fluctuations encapsulated in $\mathcal{H}_1$ arise from the nuclear spins of the lattice. Only the dipolar interaction between the muon and the nuclear spins is taken into account. The effect of the Fermi contact interaction, and therefore of the conduction electrons, on the spin dynamics will be discussed in Section 10.2. From Eqs. 5.29 and 5.30, we derive

$$\mathcal{H}_1 = -\frac{\gamma_\mu \hbar}{2}\left[\sigma^Z \delta B^Z_{\rm loc} + \frac{1}{2}\left(\sigma^+ \delta B^-_{\rm loc} + \sigma^- \delta B^+_{\rm loc}\right)\right], \tag{7.49}$$

where

$$\begin{aligned}
\delta B^Z_{\rm loc} &= \sum_{i=1}^{N}\left[\frac{\tilde{A}_i}{2}\exp(-i\varphi_i)I_i^+ + \frac{\tilde{A}_i}{2}\exp(i\varphi_i)I_i^- - 2\tilde{C}_i I_i^Z\right]\\
\delta B^+_{\rm loc} &= \sum_{i=1}^{N}\left[\tilde{C}_i I_i^+ + \tilde{B}_i \exp(2i\varphi_i)I_i^- + \tilde{A}_i \exp(i\varphi_i) I_i^Z\right]\\
\delta B^-_{\rm loc} &= \sum_{i=1}^{N}\left[\tilde{B}_i \exp(-2i\varphi_i)I_i^+ + \tilde{C}_i I_i^- + \tilde{A}_i \exp(-i\varphi_i) I_i^Z\right].
\end{aligned} \tag{7.50}$$

These equations are written taking into account that for a nuclear system $I_i^\alpha(\tau) = \delta I_i^\alpha(\tau)$. The geometrical factors are

$$\tilde{A}_i = 3\tilde{D}_i \cos\theta_i \sin\theta_i, \quad \tilde{B}_i = \frac{3}{2}\tilde{D}_i \sin^2\theta_i, \quad \tilde{C}_i = \tilde{D}_i \frac{1-3\cos^2\theta_i}{2}, \tag{7.51}$$

with

$$\tilde{D}_i = \frac{\mu_0}{4\pi}\frac{\gamma_i \hbar}{r_i^3}. \tag{7.52}$$

The quantities r_i, θ_i, and φ_i are the distance, polar and azimuthal angles of nucleus i relative to the muon and γ_i is the gyromagnetic ratio of nuclear spin i.

The muon diffusion, and therefore the spin dynamics, can be introduced by proper terms in the muon–system ensemble Hamiltonian. This quantum approach (McMullen and Zaremba 1978, Kondo 1986), which can describe the $P_\alpha(t)$ functions when the muon diffuses coherently (see Section 9.2), leads to a complicated formalism. However, a semi-classical approach seems sufficient in most cases. It consists in considering, like in NMR (Abragam 1961), that the muon coordinates are time dependent and in performing an ensemble average over these coordinates.

We have shown previously that the $P_\alpha(t)$ functions depend on magnetic field correlation functions and therefore, using Eq. 7.50, on terms such as

$$q^\alpha(\mathbf{r_i})\, q^\beta(\mathbf{r_j})\left\langle I_i^\alpha(\tau) I_j^\beta(0)\right\rangle, \tag{7.53}$$

where the q^α and q^β functions, which model the geometry around the muon, can be deduced by identification using Eq. 7.50. We shall neglect the cross-correlations and thus take $i = j$ in Eq. 7.53. As a result of muon diffusion, $\mathbf{r}_i$ is a stochastic variable, i.e. $\mathbf{r}_i = \mathbf{r}_i(\tau)$. Therefore we substitute the product function $q^\alpha(\mathbf{r}_i)\, q^\beta(\mathbf{r}_i)$ in Eq. 7.53 by its average over the stochastic diffusion process, $\overline{q_i^\alpha(\tau) q_i^\beta(0)}$. We have implicitly assumed the process to be stationary and therefore the origin of time is immaterial. To proceed further, we write $\overline{q_i^\alpha(\tau) q_i^\beta(0)} = q_i^\alpha(0) q_i^\beta(0) f_{\nu_c}(\tau)$ and assume the correlation function $f_{\nu_c}(\tau)$ to take the form

$$f_{\nu_c}(\tau) = \exp(-\nu_c|\tau|), \tag{7.54}$$

where ν_c is the muon correlation frequency. According to McMullen and Zaremba (1978) it is a good approximation to identify ν_c with the muon site transfer frequency, at least for a muon hopping or tunnelling in a simple cubic lattice; see Section 9.3. Basically with this simple form, the possibility for the muon to return to its original site is neglected. Therefore the validity of the approximation is questionable if the number of nearest neighbour interstitial sites is relatively small.

Since at usual experimental temperatures $\langle \beta \mathcal{H}_{\rm sys} \rangle \ll 1$, the density operator of the nuclear spin system is very simple:

$$\rho_{\rm sys} = \frac{\mathcal{I}}{\prod_{i=1}^{N} (2I_i + 1)}. \tag{7.55}$$

From the material gathered from the beginning of this section, the field correlation functions can be written in terms of a weighted sum of the symmetrised spin correlation functions, $\Lambda_{ii}^{\alpha\beta}(\tau)$:

$$\Lambda_{ii}^{\alpha\beta}(\tau) = \left\langle \left\{ I_i^\alpha(\tau) I_i^\beta \right\} \right\rangle. \tag{7.56}$$

As the interaction between the nuclear spins can be neglected,

$$I_i^\alpha(\tau) = \exp\left(i \frac{\mathcal{H}_i \tau}{\hbar}\right) I_i^\alpha \exp\left(-i \frac{\mathcal{H}_i \tau}{\hbar}\right), \tag{7.57}$$

where $\mathcal{H}_i$ is the Hamiltonian for nuclear spin i.

We first apply these results to the computation of the longitudinal polarisation function.

7.1.4.2.1 Longitudinal polarisation function. For the computation of $P_Z(t)$, we first assume the nuclear spins to only be subject to a Zeeman field. Therefore we have $\mathcal{H}_i = -\hbar\omega_{i,\rm Z} I_i^Z$ with $\omega_{i,\rm Z} = \gamma_i (1 + K_i) B_{\rm ext}$ where K_i is the Knight shift. With this simple Hamiltonian $I_i^\alpha(\tau)$ can be analytically computed using Eq. C.18. After some algebra we finally get

$$\psi_Z(t) = \sum_{i=1}^{N} \omega_{i,\rm d}^2 I_i (I_i + 1) \left[\frac{1}{6} \left(1 - 3\cos^2\theta_i\right)^2 g_{\nu_c}^{(c)}(\omega_\mu - \omega_{i,\rm Z}, t) \right. \tag{7.58}$$

$$\left. + 3 \sin^2\theta_i \cos^2\theta_i g_{\nu_c}^{(c)}(\omega_\mu, t) + \frac{3}{2} \sin^4\theta_i g_{\nu_c}^{(c)}(\omega_\mu + \omega_{i,\rm Z}, t) \right].$$

The dipole angular frequency $\omega_{i,\mathrm{d}} \equiv \omega_{I_i,\mathrm{d}}$ has been defined in Eq. 7.4. We have set

$$g_{\nu_\mathrm{c}}^{(\mathrm{c})}(\omega, t) = \int_0^t (t-\tau) f_{\nu_\mathrm{c}}(\tau) \cos(\omega\tau) \mathrm{d}\tau, \tag{7.59}$$

which integrates into

$$g_{\nu_\mathrm{c}}^{(\mathrm{c})}(\omega, t) = \frac{1}{(\nu_\mathrm{c}^2+\omega^2)^2} \left\{ \left[\nu_\mathrm{c}^2+\omega^2\right] \nu_\mathrm{c} t - \left[\nu_\mathrm{c}^2-\omega^2\right] \left[1-\exp\left(-\nu_\mathrm{c} t\right) \cos\left(\omega t\right)\right] \right.$$
$$\left. - 2\nu_\mathrm{c}\omega \exp\left(-\nu_\mathrm{c} t\right) \sin\left(\omega t\right) \right\}. \tag{7.60}$$

The static limit is

$$g_0^{(\mathrm{c})}(\omega, t) = \frac{1-\cos(\omega t)}{\omega^2}, \tag{7.61}$$

and

$$g_{\nu_\mathrm{c}}^{(\mathrm{c})}(\omega, t) = \frac{\nu_\mathrm{c} t}{\nu_\mathrm{c}^2+\omega^2} \tag{7.62}$$

is the motional narrowing limit. Equation 7.58 was originally written by Hayano *et al.* (1979). In zero field $\omega_{i,\mathrm{Z}} = \omega_\mu = 0$ and we recover the zero-field function given in Eq. 6.62 (the so-called Abragam function) with

$$\Delta_{\mathrm{G,ZF}}^2 = \frac{1}{2\gamma_\mu^2} \sum_{i=1}^{N} \omega_{i,\mathrm{d}}^2 \frac{I_i\left(I_i+1\right)}{3} \left(5-3\cos^2\theta_i\right). \tag{7.63}$$

This expression for $\Delta_{\mathrm{G,ZF}}^2$ was derived in Section 6.6 using a van Vleck type of method.

The nuclei around the muon with spins larger than one-half are sensitive to an electric field gradient, which is produced by the muon electric charge and the lattice electric charges for nuclei located at sites with symmetry lower than cubic (Hartmann 1977). To account for this electric field gradient, a new term must be added to $\mathcal{H}_i$ as explained in Section 6.6. In general this new term does not only depend on I_i^Z since the electric field gradient is not necessarily axial, and admitting it is, its principal axis does not necessarily coincide with the Z axis. Therefore the determination of $I_i^\alpha(\tau)$ (Eq. 7.57) is not so direct as before and a full analytical expression for $\psi_Z(t)$ cannot be written. Instead we start from Eq. 7.20 and use Eq. 7.50 to get

$$\Phi^{+-}(\tau) = \frac{f_{\nu_\mathrm{c}}(\tau)}{2\pi} \sum_{i=1}^{N} \left[\tilde{A}_i^2 \Lambda_{ii}^{ZZ}(\tau) + \tilde{C}_i^2 \Lambda_{ii}^{+-}(\tau) + \tilde{B}_i^2 \Lambda_{ii}^{-+}(\tau) \right.$$
$$+ \tilde{B}_i\tilde{C}_i \exp(-2i\varphi_i)\Lambda_{ii}^{++}(\tau) + \tilde{B}_i\tilde{C}_i \exp(2i\varphi_i)\Lambda_{ii}^{--}(\tau)$$
$$+ \tilde{A}_i\tilde{B}_i \exp(-i\varphi_i)\Lambda_{ii}^{Z+}(\tau) + \tilde{A}_i\tilde{C}_i \exp(i\varphi_i)\Lambda_{ii}^{Z-}(\tau)$$
$$\left. + \tilde{A}_i\tilde{C}_i \exp(-i\varphi_i)\Lambda_{ii}^{+Z}(\tau) + \tilde{A}_i\tilde{B}_i \exp(i\varphi_i)\Lambda_{ii}^{-Z}(\tau) \right]. \tag{7.64}$$

Obviously, $\Phi^{-+}(\tau)$ is the Hermitian conjugate of $\Phi^{+-}(\tau)$. The spin correlations $\Lambda_{ii}^{\alpha\beta}(\tau)$ are then numerically evaluated. Combining Eq. 7.41 and the previous equation, we note

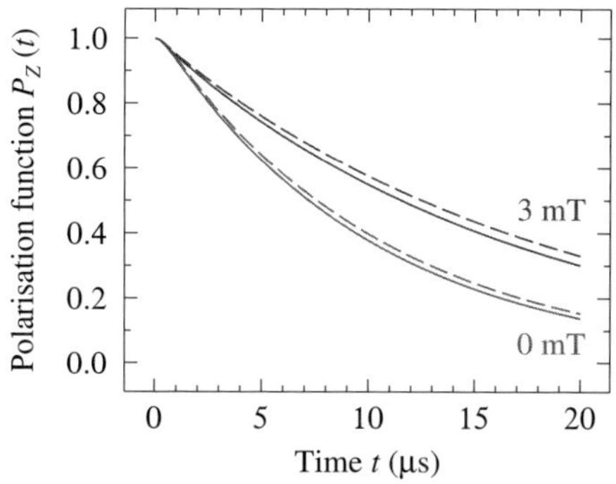

Fig. 7.3 Examples of the effect of the nuclear spin dynamics on the longitudinal polarisation function in the case of the muon in metallic copper. The examples are given for two values of the external magnetic field indicated in the plot. The full lines present the results obtained from either the second order iteration (i.e. Eqs. 7.40, 7.41, and 7.64) or the strong-collision model with the static polarisation functions computed numerically from quantum mechanics (see Section 7.2): the two sets of curves cannot be distinguished. The dashed lines display the results of the strong-collision model with the static functions given by the Kubo–Toyabe model. The parameters of the curves are given in the main text. Reproduced with kind permission from Dalmas de Réotier and Yaouanc (1992) and Yaouanc and Dalmas de Réotier (1995).

that the spin–lattice relaxation channel probes a weighted sum of parallel, perpendicular, and mixed magnetic fluctuations. This wording refers to the Z axis: parallel fluctuations only involve the Z component of $\mathbf{J}$, while perpendicular fluctuations only implicate its $+$ or $-$ components or alternatively X or Y.

In order to appreciate the effect of the nuclear spin dynamics arising from the electric field gradient on $P_Z(t)$, we present in Fig. 7.3 two examples (zero-field case and $B_{\rm ext} = 3\,{\rm mT}$) for this function computed from three different models. They are calculated either from the second order iteration method discussed here or from the strong-collision model (see Section 6.3). The static functions of the latter model are either taken from the Kubo–Toyabe theory or computed numerically using Celio's method (see Section 7.2). We have assumed the muon to diffuse between the octahedral sites of metallic copper; see Fig. 9.1 for a sketch of these sites in a face-centred-cubic crystallographic structure. We have chosen $\nu_{\rm c} = 3\,\mu{\rm s}^{-1}$, which should be roughly the value of the muon jump frequency at room temperature. We have taken the following values for the nuclear parameters of copper: the gyromagnetic ratio $\gamma_I = 72.5\,{\rm Mrad\,s^{-1}\,T^{-1}}$ and the quadrupole moment $Q_I = -0.205\,{\rm barn}$ (Luke *et al.* 1991). They roughly correspond to the averages over the two isotopes that can be performed from the data of Appendix G. With the magnitude of the axial component of the electric field gradient tensor $V^{zz} = 0.424\,{\rm V\,m^{-2}}$ and the distance between the muon and the copper neighbours $r_I = 1.81\,$Å, we compute $3\omega_{\rm E} = -3.3\,\mu{\rm s}^{-1}$ and $\omega_{I,{\rm d}} = 0.11\,\mu{\rm s}^{-1}$. We recall that $\hbar\omega_{\rm E}$ and $\hbar\omega_{I,{\rm d}}$ set the magnitude of quadrupolar electric and dipolar magnetic energies, respectively (see Eq. 7.3). Because of the cubic symmetry of the face-centred-cubic lattice sites, the quadrupolar interaction solely arises from the presence of the muon charge. The corresponding electric field gradient tensor is therefore axial. The crystal is oriented so that the Z axis defining the initial muon polarisation and the direction of the applied field is parallel to the cube edge [001]. Note that we consider only the six nearest neighbour copper nuclei to the muon because the exact numerical computation of the static $P_Z(t)$ functions can only be performed for this restricted number of nuclei. The result of this computation is reliable because, to a good approximation, the relaxation is only induced by the nearest neighbour copper nuclei. Regarding the functions plotted in Fig. 7.3, which are calculated from the model developed in Chapter 6, we compute $\Delta_{\rm G,ZF}$ from the material given in Section 6.6. Since the quadrupolar interaction energy is much larger than the dipolar one, we can safely use Eqs. 6.160 and 6.162 as the copper isotopes spins are $\frac{3}{2}$; i.e. they are half-integer. From the crystal orientation, the values of the polar angle for each of the six muon neighbours are 0, π, and four times $\pi/2$. Therefore $\gamma_\mu \Delta_{\rm G,ZF} = \omega_{I,\rm d}\sqrt{12} = 0.381\,\mu{\rm s}^{-1}$.

While $P_Z(t)$ computed either from the strong-collision model with the exact static functions or from the second order iteration method cannot be distinguished in the figure, the functions obtained from the strong-collision model with the static functions computed according to Kubo–Toyabe are somewhat different. The functions deduced from this latter method decrease more slowly.

Therefore, the comparison of $P_Z(t)$ deduced from the second order iteration or Celio's method and the Kubo–Toyabe method shows that the latter method does not describe the physics completely: in particular, it neglects the nuclear spin dynamics which is essential. Remarkably, a treatment of this dynamics using classical mechanics is sometimes possible; see Section 7.2. Interestingly, if the nuclear spin dynamics is ignored, i.e. if we set $\omega_{i,\rm Z} = 0$ in Eq. 7.58, an analytical result can be derived (Keren 1994):

$$\psi_Z(t) = 2\gamma_\mu^2 \Delta_{\rm G,ZF}^2 g_{\nu_{\rm c}}^{(\rm c)}(\omega_\mu, t)\,, \tag{7.65}$$

with $g_{\nu_{\rm c}}^{(\rm c)}(\omega_\mu, t)$ given by Eq. 7.60. However, this result should not be used if the value of $\omega_{i,\rm Z}$ is appreciable (Yaouanc and Dalmas de Réotier 1995). In the motional narrowing limit, $g_{\nu_{\rm c}}^{(\rm c)}(\omega_\mu, t)$ is proportional to t (Eq. 7.62). Therefore $P_Z(t)$ is an exponential function characterised by the relaxation rate

$$\lambda_Z = \frac{2\gamma_\mu^2 \Delta_{\rm G,ZF}^2}{\nu_{\rm c}} \frac{1}{1 + \omega_\mu^2/\nu_{\rm c}^2}. \tag{7.66}$$

In the zero-field case ($\omega_\mu = 0$) we recover the result derived from the field distribution approach; see Eq. 6.63. Equation 7.66 is sometimes referred to as the Redfield formula. A useful discussion of its predictions is found after Eq. 7.27. We recall that $\nu_{\rm c} = 1/\tau_{\rm c}$.

In this section we do not discuss in detail the relaxation induced by electronic magnetic moments because our current model ignores the correlations between these moments and because the thermal energy is assumed to be large relative to the magnetic energy of these moments. However, within these limitations, it is interesting to analyse the form of Eq. 7.58 for electronic moments. We shall take $\omega_{i,\rm Z} \gg \omega_\mu$. This is a valid assumption if a magnetic field is applied on the paramagnetic state of a magnetic material and in the magnetically ordered state of a material because the internal field

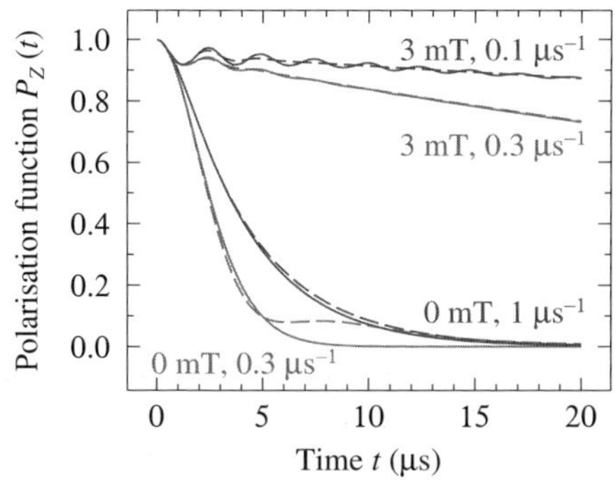

Fig. 7.4 Study of the validity of the second order iteration for the longitudinal polarisation function. The curves are drawn for the case of a muon in metallic copper. The full lines present the results of the second order iteration. The dashed lines show the prediction of the strong-collision model with the static functions fully computed from quantum mechanics. The values inserted next to each set of curves are the magnitude of the external field and of the field correlation frequency, respectively.

acting on an electronic magnetic moment is huge relative to the spontaneous field at the muon site. Within our assumption, $g^{(c)}_{\nu_c}(\omega_\mu - \omega_{i,Z}, t) \simeq g^{(c)}_{\nu_c}(\omega_\mu + \omega_{i,Z}, t)$ and these functions are negligible relative to $g^{(c)}_{\nu_c}(\omega_\mu, t)$ both in the quasi-static limit and in the motional narrowing limit. Note that the additional condition $\nu_c < \omega_\mu$ is needed for this latter approximation to be valid. Then $\psi_Z(t)$ reduces to

$$\psi_Z(t) = \left[3\sum_{i=1}^{N} \omega^2_{i,\mathrm{d}} I_i\,(I_i+1)\sin^2\theta_i\cos^2\theta_i\right] g^{(c)}_{\nu_c}(\omega_\mu, t)\,, \tag{7.67}$$

with $g^{(c)}_{\nu_c}(\omega_\mu, t)$ given in Eq. 7.60. Referring to Eq. 7.62, in the motional narrowing limit $P_Z(t)$ is an exponential function with a rate which follows a Lorentzian-field dependence. The measurement of such a dependence offers a possibility of determining both ν_c and the scale of the spin–lattice relaxation rate, i.e. the amplitude of the magnetic moments at the origin of the relaxation.

Up to now we have not established the smallest value of the field correlation frequency for which the iteration formula is reliable. In Fig. 7.4 we present $P_Z(t)$ computed for copper either from the strong-collision model with the static function computed numerically (Celio's method; see Section 7.2) or from the second order iteration. It is quite clear that even for $\nu_c = 1\,\mu\mathrm{s}^{-1}$ in zero field the iteration is a very reasonable approximation. This fact is interesting because it means that the iteration formula can be used to describe $P_Z(t)$ even for a relatively slowly diffusing muon. Thus, for most practical cases, the function can be obtained with a modest computing time. In the case of an external field the domain of validity of the second order iteration method extends even to lower values of ν_c. Actually this method correctly reproduces the average slope of the polarisation function induced by the muon jumps down to vanishing jump rates where it less successfully accounts for the oscillations.

The results described above might lead to the idea that the second order iteration gives reliable results only if an appreciable dynamics is present. This is not true. It always provides a good description of the initial part of $P_Z(t)$ in the whole range of ν_c. For example, when an electric field gradient or a magnetic field acts on the lattice nuclei, the iteration result has a less restricted time range validity than the usual parabola obtained from Van Vleck's method (see Section 6.6.1) because of the effect of the non-secular terms. This is clearly seen if one looks in Fig. 7.4 at the initial part of the numerically calculated $P_Z(t)$ function for the case of a quasi-static muon ($\nu_c = 0.3\,\mu s^{-1}$) in copper at zero field: the iteration formula gives a fair approximation up to approximately $5\,\mu s$.

7.1.4.2.2 Transverse polarisation function. The determination of $P_X(t)$ is not difficult but a bit long and tedious. Here we shall only give the result. A sketch of the derivation can be found elsewhere (Dalmas de Réotier and Yaouanc 1992). Assuming the nuclear spins to only be subject to Zeeman splitting, we derive from Eqs. 7.46 and 7.47

$$\psi_X(t) = \frac{1}{2}\psi_Z(t) + \sum_{i=1}^{N} \omega_{i,\mathrm{d}}^2 I_i (I_i + 1)\, v_i (\omega_\mu, \omega_{i,\mathrm{Z}}, t), \tag{7.68}$$

where

$$\begin{aligned} v_i (\omega_\mu, \omega_{i,\mathrm{Z}}, t) = {} & \frac{1}{3}\left(1 - 3\cos^2\theta_i\right)^2 g_{\nu_c}^{(c)}(0,t) + 3\sin^2\theta_i\cos^2\theta_i g_{\nu_c}^{(c)}(\omega_{i,\mathrm{Z}}, t) \\ & + \cos(2\varphi_i)\frac{3\sin^4\theta_i - 2\sin^2\theta_i}{2\omega_\mu} h_{\nu_c,\omega_\mu}^{(s)}(\omega_{i,\mathrm{Z}}, t) \\ & + \frac{3\cos(2\varphi_i)\sin^2\theta_i\cos^2\theta_i}{2\omega_\mu} k_{\nu_c}^{(s)}(\omega_\mu, t), \end{aligned} \tag{7.69}$$

and

$$\varphi(t) = \sum_{i=1}^{N} \omega_{i,\mathrm{d}}^2 I_i (I_i + 1)\, w_i (\omega_\mu, \omega_{i,\mathrm{Z}}, t), \tag{7.70}$$

where

$$\begin{aligned} 2w_i (\omega_\mu, \omega_{i,\mathrm{Z}}, t) = {} & \frac{1}{6}\left(1 - 3\cos^2\theta_i\right)^2 g_{\nu_c}^{(s)}(\omega_\mu - \omega_{i,\mathrm{Z}}, t) \\ & + 3\sin^2\theta_i\cos^2\theta_i g_{\nu_c}^{(s)}(\omega_\mu, t) + \frac{3}{2}\sin^4\theta_i g_{\nu_c}^{(s)}(\omega_\mu + \omega_{i,\mathrm{Z}}, t) \\ & - \cos(2\varphi_i)\frac{3\sin^4\theta_i - 2\sin^2\theta_i}{\omega_\mu} h_{\nu_c,\omega_\mu}^{(c)}(\omega_{i,\mathrm{Z}}, t) \\ & - \frac{3\cos(2\varphi_i)\sin^2\theta_i\ \cos^2\theta_i}{\omega_\mu} k_{\nu_c}^{(c)}(\omega_\mu, t). \end{aligned} \tag{7.71}$$

We have defined

$$g_{\nu_c}^{(s)}(\omega, t) = \int_0^t (t - \tau) f_{\nu_c}(\tau)\sin(\omega\tau)\mathrm{d}\tau,$$

$$h^{(\mathrm{s})}_{\nu_\mathrm{c},\omega_\mu}(\omega,t) = \int_0^t f_{\nu_\mathrm{c}}(\tau)\sin(\omega_\mu\tau)\cos(\omega\tau)\mathrm{d}\tau,$$

$$h^{(\mathrm{c})}_{\nu_\mathrm{c},\omega_\mu}(\omega,t) = \int_0^t f_{\nu_\mathrm{c}}(\tau)\cos(\omega_\mu\tau)\cos(\omega\tau)\mathrm{d}\tau,$$

$$k^{(\mathrm{s})}_{\nu_\mathrm{c}}(\omega,t) = \int_0^t f_{\nu_\mathrm{c}}(\tau)\sin(\omega\tau)\mathrm{d}\tau,$$

$$k^{(\mathrm{c})}_{\nu_\mathrm{c}}(\omega,t) = \int_0^t f_{\nu_\mathrm{c}}(\tau)\cos(\omega\tau)\mathrm{d}\tau. \tag{7.72}$$

We recover the expression given by Hayano *et al.* (1979) for $\psi_X(t)$ except for the terms proportional to $1/\omega_\mu$ which they neglected. In the high-field limit the $g^{(\mathrm{c})}_{\nu_\mathrm{c}}(\omega,t)$ and $g^{(\mathrm{s})}_{\nu_\mathrm{c}}(\omega,t)$ functions go to 0. Thus we find, as expected, the transverse-field Abragam formula (Eq. 6.44) with $\varphi(t) = 0$ and

$$\Delta^2_{\mathrm{G,TF}} = \frac{1}{\gamma_\mu^2}\sum_{i=1}^{N}\omega^2_{i,\mathrm{d}}\frac{I_i(I_i+1)}{3}\left(1-3\cos^2\theta_i\right)^2. \tag{7.73}$$

We are now going to investigate the effect of the functions defined in Eq. 7.72 as well as of $g^{(\mathrm{c})}_{\nu_\mathrm{c}}(\omega,t)$ (Eq. 7.59) on $P_X(t)$. Since in Eqs. 7.69 and 7.71 the ω argument of almost all these functions includes $\omega_{i,\mathrm{Z}}$, it is actually the nuclear spin precession which is accounted for by these functions. For definitiveness, we consider a system made of the muon and two copper nuclear magnetic moments with one located at 1.5 Å on the X axis and the other at 2 Å on the Y axis. Such a non-symmetric environment has been chosen to optimise the detection of the effect of the terms proportional to $1/\omega_\mu$. When the muon is in a highly symmetric environment, these terms do not contribute to $P_X(t)$ as it can be inferred from Eq. 7.46. A field of 2 mT is assumed to be applied along the Z axis. Since $B_\mathrm{ext} > 5\Delta_{\mathrm{G,TF}} = 1.37\,\mathrm{mT}$, we are in the high transverse-field limit; see Section 6.2.1 and Fig. 6.1. We set $\nu_\mathrm{c} = 3\,\mu\mathrm{s}^{-1}$ and assume that the nuclei do not feel any electric field gradient. In Fig. 7.5 we compare $P_X(t)$ computed with three different approximations. One curve (full line) has been calculated using the complete second order iteration formula. A computation of $P_X(t)$ has also been made with this formula but omitting the terms proportional to $1/\omega_\mu$. The results of these two computations cannot be distinguished. The small effect of these terms can actually be observed for smaller values of ν_c (not shown), and in fact, it is usually a good approximation to neglect them. Therefore we assume this approximation to hold from now on. The second curve (dashed line) is the result of the transverse-field Abragam formula with $\gamma_\mu\Delta_{\mathrm{G,TF}} = 0.234\,\mu\mathrm{s}^{-1}$. This value has been calculated using Eq. 7.73. The difference between the full and dashed lines shows that the nuclear spin dynamics has a strong influence on $P_X(t)$. Note that not only the damping is different in the two models, but also a small frequency shift is present.

As performed for $P_Z(t)$ we need to study the ν_c range of validity of the second order iteration result for $P_X(t)$. As a matter of fact, at least when the dimension of the Hilbert space for the nuclear spins is sufficiently large, this approximation is fair

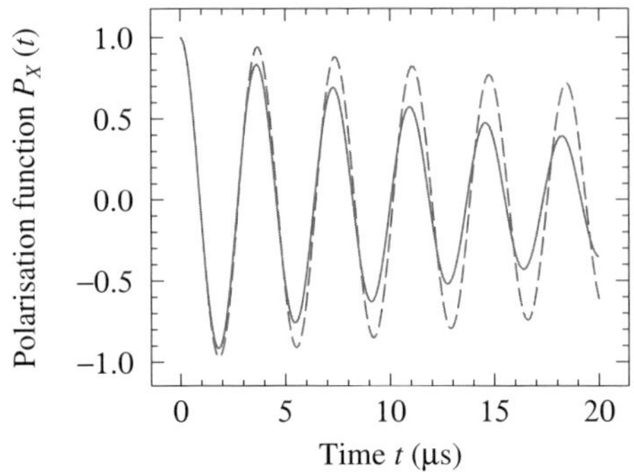

Fig. 7.5 Example of the effect of the nuclear spin dynamics on the transverse polarisation function for a small external magnetic field. The full line gives the result of the second order iteration. The dashed line shows the prediction of the so-called Abragam formula. The external applied magnetic field is $B_{\rm ext} = 2\,{\rm mT}$. The other parameters of the curves are given in the main text.

even for a static muon (Yaouanc and Dalmas de Réotier 1994). In Fig. 7.6 we illustrate this conclusion. The distribution model is obtained from Eq. 6.12, with $\gamma_\mu \Delta_{\rm G,TF} = \omega_{i,{\rm d}}\sqrt{15}/2 = 0.213\,\mu{\rm s}^{-1}$ according to the values of the polar angles of the copper spins and to Eq. 7.73. It improperly accounts for the evolution of the polarisation. The iteration solution is a very good approximation to the exact numerical result: a difference is only visible for times larger than about $7\,\mu$s. The fact that Eq. 6.12 is a bad approximation is understood looking at the structure of Eq. 7.69. The terms proportional to $\sin^2\theta_i \cos^2\theta_i$ do not contribute to $\psi_X(t)$ only when θ_i is 0, $\pi/2$ or π, and in our case we have four spins out of six with angle $\pi/4$ or $3\pi/4$. The other terms, except the one leading to the so-called Abragam formula (first term of the v_i function), are negligible if $B_{\rm ext}$ is large. Therefore the influence of the nuclear spin dynamics on $P_X(t)$ depends strongly on the geometrical conditions.

7.1.4.2.3 Relation between the longitudinal and transverse decay rates. According to Eq. 7.48, there is a relation between the longitudinal polarisation function and the envelope of the transverse polarisation function. Let us first investigate this relation in the static limit. Assuming that the polycrystalline average of the envelope of $P_X(t)$ is obtained by simply taking the angular average of the variance, from Eq. 7.73 we deduce for the envelope

$$\exp\left(-\frac{1}{2}\gamma_\mu^2 \overline{\Delta^2_{\rm G,TF}} t^2\right) \quad \text{with} \quad \gamma_\mu^2 \overline{\Delta^2_{\rm G,TF}} = \frac{4}{15}\sum_{i=1}^{N} \omega_{i,{\rm d}}^2 I_i\,(I_i+1)\,. \tag{7.74}$$

From Eq. 7.63, the polycrystalline average for the zero-field polarisation function can be written as

$$P_Z(t \to 0) = \exp\left(-5\frac{\gamma_\mu^2 \overline{\Delta^2_{\rm G,TF}}}{2} t^2\right). \tag{7.75}$$

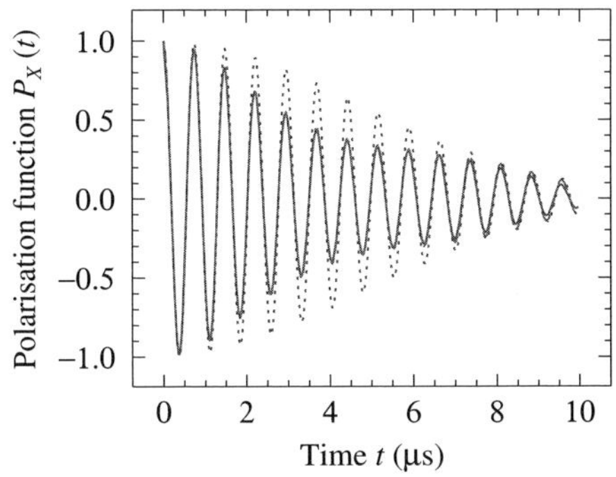

Fig. 7.6 Comparison between three methods for the computation of $P_X(t)$. The comparison is made for a static muon in an octahedral site of metallic copper. To avoid level crossing effects, the electric field gradient due to the muon is neglected. A magnetic field of 10 mT is applied along the [110] crystallographic direction. The full line gives the result of the numerical quantum mechanics method; see Section 7.2. The dashed line is the result of the iteration method described in this section. The difference between the two curves can only be observed at large time. The dotted line is the prediction of the distribution method. Reproduced with kind permission from Yaouanc and Dalmas de Réotier (1994).

The enhancement factor of $\sqrt{5}/2$ of the zero-field damping rate has already been noted in Section 6.6.1.1 (see discussion following Eq. 6.152).

The decay of both $P_X(t)$ and $P_Z(t)$ is expressed in terms of relaxation rates in the motional narrowing limit. Assuming the dynamics of the muon neighbouring spins to be negligible, i.e. $\omega_{i,Z} = 0$, and taking the powder average of the rates, we derive from Eqs. 7.66 and 7.68

$$\overline{\lambda_Z} = \frac{5\gamma_\mu^2 \overline{\Delta_{\mathrm{G,TF}}^2}}{\nu_{\mathrm{c}}} \frac{1}{1 + \omega_\mu^2/\nu_{\mathrm{c}}^2} \tag{7.76}$$

and

$$\overline{\lambda_X} = \frac{5}{2} \frac{\gamma_\mu^2 \overline{\Delta_{\mathrm{G,TF}}^2}}{\nu_{\mathrm{c}}} \left(1 + \frac{1}{1 + \omega_\mu^2/\nu_{\mathrm{c}}^2}\right), \tag{7.77}$$

respectively. When $\omega_\mu/\nu_{\mathrm{c}} \ll 1$,

$$\overline{\lambda_Z} = \frac{5\gamma_\mu^2 \overline{\Delta_{\mathrm{G,TF}}^2}}{\nu_{\mathrm{c}}} \quad \text{and} \quad \overline{\lambda_X} = \overline{\lambda_Z}. \tag{7.78}$$

We now examine the case where the dynamics of the muon neighbouring spins is faster than that of the muon. This is for instance the case of paramagnetic electronic moments. In the motional narrowing limit and in a transverse field we obtain from Eqs. 7.68 and 7.58

$$\overline{\lambda_X} = \frac{\gamma_\mu^2 \overline{\Delta_{\mathrm{G,TF}}^2}}{\nu_{\mathrm{c}}} \left(1 + \frac{3}{4} \frac{1}{1 + \omega_\mu^2/\nu_{\mathrm{c}}^2}\right), \tag{7.79}$$

since $\omega_{i,\mathrm{Z}} \gg \omega_\mu$. For the longitudinal-field case, performing the powder average of the spin–lattice relaxation rate given in Eq. 7.67 yields

$$\overline{\lambda_Z} = \frac{3}{2}\frac{\gamma_\mu^2\overline{\Delta^2_{\mathrm{G,TF}}}}{\nu_\mathrm{c}}\left(\frac{1}{1+\omega_\mu^2/\nu_\mathrm{c}^2}\right). \tag{7.80}$$

Therefore, in this limit

$$\overline{\lambda_X} = \frac{\overline{\lambda_Z}}{2} + \frac{\gamma_\mu^2\overline{\Delta^2_{\mathrm{G,TF}}}}{\nu_\mathrm{c}}. \tag{7.81}$$

For further discussions on λ_Z and λ_X we refer the reader to the paper of Hayano *et al.* (1979). Finally, note that there is a frequency shift in the motional narrowing limit since from Eqs. 7.70 and 7.71 we have

$$\varphi(t) = \frac{3}{4}\gamma_\mu^2\overline{\Delta^2_{\mathrm{G,TF}}}\frac{\omega_\mu}{\nu_\mathrm{c}^2+\omega_\mu^2}t, \tag{7.82}$$

assuming $\omega_{i,\mathrm{Z}} \gg \omega_\mu$. We have used the following formula which results from Eq. 7.72:

$$\begin{aligned} g^{(\mathrm{s})}_{\nu_\mathrm{c}}(\omega,t) = \frac{1}{(\nu_\mathrm{c}^2+\omega^2)^2}\Big\{&\left[\nu_\mathrm{c}^2+\omega^2\right]\omega t + \left[\nu_\mathrm{c}^2-\omega^2\right]\exp\left(-\nu_\mathrm{c}t\right)\sin\left(\omega t\right) \\ &+ 2\nu_\mathrm{c}\omega\left[\exp\left(-\nu_\mathrm{c}t\right)\cos\left(\omega t\right)-1\right]\Big\}. \end{aligned} \tag{7.83}$$

7.1.4.3 Fourier transforms. Here we first express the polarisation functions with the spectral density of magnetic field correlations using the time Fourier transformation. Next, we perform a space Fourier transformation of the field correlations. This is justified for crystalline compounds because of their translational symmetry. The interaction between the magnetic moments is then reflected in the wave vector dependence of the field correlations. This enables $P_\alpha(t)$ to be discussed, for example, in terms of the excitations of the system.

The space Fourier transform method is well known in NMR (Moriya 1962). However, a peculiarity of μSR relative to NMR arises because the dipole interaction is never negligible relative to the hyperfine interaction.

7.1.4.3.1 Time Fourier transform. For future use, let us consider the complex function $\zeta(t)$ such that

$$\zeta(t)t = \int_0^t (t-\tau)\exp\left(i\omega_\mu\tau\right)g(\tau)\mathrm{d}\tau, \tag{7.84}$$

where $g(t)$ is a real function. It is convenient to introduce its time Fourier transform denoted as $g(\omega)$. Within our definition (see Eq. C.50), we derive

$$\begin{aligned} \zeta(t)t &= \int_0^t (t-\tau)\left\{\int_{-\infty}^{\infty}\exp\left[-i\left(\omega-\omega_\mu\right)\tau\right]g(\omega)\frac{\mathrm{d}\omega}{2\pi}\right\}\mathrm{d}\tau \\ &= \int_0^t (t-\tau)\left[\int_{-\infty}^{\infty}\exp\left(-i\omega\tau\right)g(\omega+\omega_\mu)\frac{\mathrm{d}\omega}{2\pi}\right]\mathrm{d}\tau \end{aligned}$$

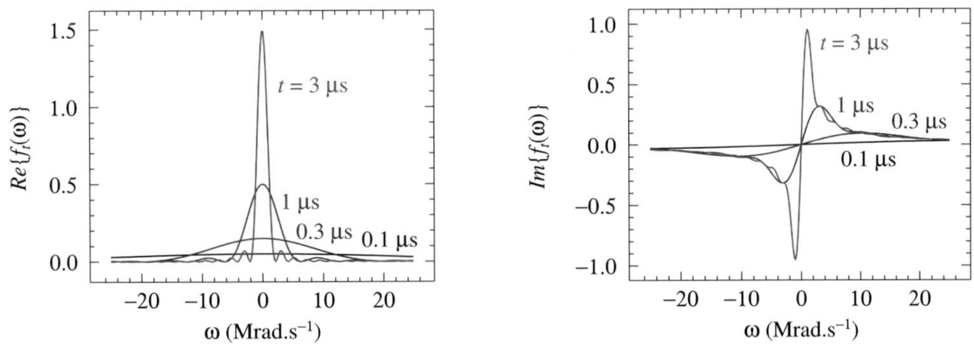

Fig. 7.7 The real and imaginary parts of the function $f_t(\omega)$ for different values of time t indicated in the plots. To fix the scale we note that a local field of 10 mT at the muon site corresponds to $\omega = \omega_\mu = 8.51\,\mathrm{Mrad\,s^{-1}}$.

$$= \int_{-\infty}^{\infty} g(\omega + \omega_\mu) \left[\int_0^t (t - \tau) \exp(-i\omega\tau)\, \mathrm{d}\tau \right] \frac{\mathrm{d}\omega}{2\pi}$$

$$= \int_{-\infty}^{\infty} g(\omega + \omega_\mu) \left[\frac{1 - \cos(\omega t)}{\omega^2} + i \frac{\sin(\omega t) - \omega t}{\omega^2} \right] \frac{\mathrm{d}\omega}{2\pi}. \tag{7.85}$$

Therefore, we can write (McMullen and Zaremba 1978)

$$\zeta(t) = \int_{-\infty}^{\infty} g(\omega + \omega_\mu) f_t(\omega) \frac{\mathrm{d}\omega}{2\pi}, \tag{7.86}$$

with the definition

$$f_t(\omega) = \frac{1 - \cos(\omega t)}{\omega^2 t} + i \frac{\sin(\omega t) - \omega t}{\omega^2 t}. \tag{7.87}$$

The real and imaginary parts of $f_t(\omega)$ are, respectively, even and odd functions of ω. They are plotted in Fig. 7.7.

An inspection of the formulae for $P_Z(t)$ (see Eqs. 7.40 and 7.42) shows that we can write

$$P_Z(t) = \exp\left[-\lambda_Z(t) t\right], \tag{7.88}$$

with

$$\lambda_Z(t) = \gamma_\mu^2 \int_{-\infty}^{\infty} \left[\Phi^{XX}(\omega + \omega_\mu) + \Phi^{YY}(\omega + \omega_\mu) \right] \mathcal{R}e\{f_t(\omega)\} \mathrm{d}\omega$$
$$+ \gamma_\mu^2 \int_{-\infty}^{\infty} \left[\Phi^{XY}(\omega + \omega_\mu) - \Phi^{YX}(\omega + \omega_\mu) \right] \mathcal{I}m\{f_t(\omega)\} \mathrm{d}\omega. \tag{7.89}$$

Obviously, we have introduced the time Fourier transform of $\Phi^{\alpha\beta}(t)$,

$$\Phi^{\alpha\beta}(\omega) = \int_{-\infty}^{\infty} \exp(i\omega t) \Phi^{\alpha\beta}(t) \mathrm{d}t, \tag{7.90}$$

following Eq. C.49.

With the same method, we obtain from Eqs. 7.45, 7.48, and 7.47

$$P_X(t) = \exp\left[-\lambda_X(t)t\right]\cos\left[\omega_\mu t + \rho(t)t\right], \tag{7.91}$$

with

$$\lambda_X(t) = \frac{\lambda_Z(t)}{2} + \gamma_\mu^2 \int_{-\infty}^{\infty} \Phi^{ZZ}(\omega) f_t(\omega)\mathrm{d}\omega, \tag{7.92}$$

and

$$\begin{aligned}\rho(t) = {} & \frac{\gamma_\mu^2}{2}\int_{-\infty}^{\infty}\left[\Phi^{XX}(\omega+\omega_\mu) + \Phi^{YY}(\omega+\omega_\mu)\right]\mathcal{Im}\{f_t(\omega)\}\mathrm{d}\omega \\ & + \frac{\gamma_\mu^2}{2}\int_{-\infty}^{\infty}\left[\Phi^{YX}(\omega+\omega_\mu) - \Phi^{XY}(\omega+\omega_\mu)\right]\mathcal{Re}\{f_t(\omega)\}\mathrm{d}\omega.\end{aligned} \tag{7.93}$$

We have neglected the terms proportional to $1/\omega_\mu$ since the discussion given after Eq. 7.73 suggests that they are negligible.

We now examine these results. First we note that the second line of Eqs. 7.89 and 7.93 can usually be neglected; this can be so either because $\Phi^{XY}(\tau)$ and $\Phi^{YX}(\tau)$ vanish for symmetry reason or because $\Phi^{XY}(t)$ is an even function, which implies $\Phi^{YX}(t) = \Phi^{XY}(t)$. Therefore

$$\lambda_Z(t) = \gamma_\mu^2 \int_{-\infty}^{\infty}\left[\Phi^{XX}(\omega) + \Phi^{YY}(\omega)\right]\mathcal{Re}\{f_t(\omega-\omega_\mu)\}\mathrm{d}\omega, \tag{7.94}$$

and

$$\rho(t) = \frac{\gamma_\mu^2}{2}\int_{-\infty}^{\infty}\left[\Phi^{XX}(\omega) + \Phi^{YY}(\omega)\right]\mathcal{Im}\{f_t(\omega-\omega_\mu)\}\mathrm{d}\omega. \tag{7.95}$$

If $\Phi^{ZZ}(t)$ is an even function of t,

$$\lambda_X(t) = \frac{\lambda_Z(t)}{2} + \gamma_\mu^2 \int_{-\infty}^{\infty} \Phi^{ZZ}(\omega)\mathcal{Re}\{f_t(\omega)\}\mathrm{d}\omega. \tag{7.96}$$

The functions $\mathcal{Re}\{f_t(\omega)\}$ and $\mathcal{Im}\{f_t(\omega)\}$ act as screening functions as they select the density of fluctuations and excitations described by the spectral functions $\Phi^{\alpha\alpha}(\omega)$ that contribute to $\lambda_Z(t)$, $\lambda_X(t)$ and $\rho(t)$. The term $\mathcal{Re}\{f_t(\omega)\}$ is zero except in a narrow window centred at $\omega = 0$ as shown in Fig. 7.7. This means that $\lambda_Z(t)$ probes fluctuations or excitations in a narrow range of energy: $\hbar\omega$ is on the order of 1 neV (for ω in the range of a few megaradians per second) around $\omega = \omega_\mu$. Only if $\left[\Phi^{XX}(\omega) + \Phi^{YY}(\omega)\right]$ presents an appreciable variation around $\omega = \omega_\mu$ will $\rho(t)$ be different from 0 since $\mathcal{Im}\{f_t(\omega)\}$ is an odd function of ω. As already recognised, $\lambda_X(t)$ is driven by fluctuations and excitations which are around $\omega = 0$ and $\omega = \omega_\mu$, respectively.

Denoting the characteristic time for the field fluctuations as τ_c and assuming $t \gg \tau_\mathrm{c}$, the ω width of $\Phi^{\alpha\beta}(\omega)$ is large compared to that of $\mathcal{Re}\{f_t(\omega-\omega_\mu)\}$. If in addition $\omega_\mu\tau_\mathrm{c} \ll 1$, $\lambda_Z(t)$ and $\lambda_X(t)$ are then t independent. From Eq. C.44 we find

$$\lambda_Z = \lim_{t\to\infty}\lambda_Z(t) = \pi\gamma_\mu^2\left[\Phi^{XX}(\omega_\mu) + \Phi^{YY}(\omega_\mu)\right], \tag{7.97}$$

and

$$\lambda_X = \lim_{t\to\infty} \lambda_X(t) = \frac{\lambda_Z}{2} + \pi\gamma_\mu^2 \Phi^{ZZ}(\omega = 0). \tag{7.98}$$

Denoting the spin–lattice relaxation rate λ_Z as $1/T_1$ and the spin–spin relaxation rate λ_X as $1/T_2$, we recover a well-known NMR result:

$$\frac{1}{T_2} = \frac{1}{2T_1} + \pi\gamma_\mu^2 \Phi^{ZZ}(\omega = 0). \tag{7.99}$$

If the field fluctuations are fast enough and isotropic, the two fluctuation rates are equal, that is

$$\lambda_Z = \lambda_X. \tag{7.100}$$

In the large t limit (relative to typical correlation times) the polarisation functions are therefore exponential functions and λ_Z and λ_X are well-defined relaxation rates since they are constant. In contrast for the small t limit we derive

$$\lambda_Z(t) = t\frac{\gamma_\mu^2}{2} \int_{-\infty}^{\infty} \left[\Phi^{XX}(\omega + \omega_\mu) + \Phi^{YY}(\omega + \omega_\mu)\right] \mathrm{d}\omega, \tag{7.101}$$

and

$$\lambda_X(t) = \frac{\lambda_Z(t)}{2} + t\frac{\gamma_\mu^2}{2} \int_{-\infty}^{\infty} \Phi^{ZZ}(\omega)\mathrm{d}\omega. \tag{7.102}$$

Therefore $\psi_Z(t)$ and $\psi_X(t)$ are proportional to t^2 at small t, as expected. In the small t limit, $\rho(t)$ is also proportional to t^2.

In fact, it is possible to provide a formula in the motional narrowing limit for $\lambda_Z(t)t$, which covers the range from small to large t if the spectral density of the magnetic field at the muon site is assumed to be a Lorentzian of half-width at half-maximum Γ, i.e.

$$\Phi^{\alpha\alpha}(\omega) = \left\langle (\delta B_{\rm loc}^{\alpha})^2 \right\rangle \frac{1}{\pi} \frac{\Gamma}{\Gamma^2 + \omega^2}. \tag{7.103}$$

In the time domain, $\Phi^{\alpha\alpha}(\omega)$ translates into

$$\Phi^{\alpha\alpha}(t) = \left\langle (\delta B_{\rm loc}^{\alpha})^2 \right\rangle \frac{1}{2\pi} \exp(-\Gamma|t|), \tag{7.104}$$

which we insert into Eq. 7.42 (of which we again neglect the second line) to obtain

$$\psi_Z(t) = \lambda_Z(t)t = \gamma_\mu^2 \left[\left\langle (\delta B_{\rm loc}^{X})^2 \right\rangle + \left\langle (\delta B_{\rm loc}^{Y})^2 \right\rangle\right] g_\Gamma^{(\mathrm{c})}(\omega_\mu, t). \tag{7.105}$$

This formula for $\psi_Z(t)$ can be identified with Eq. 7.65.

Other forms for the spectral density are considered in Section 10.2.1.

7.1.4.3.2 Space Fourier transform. The measured polarisation functions are expressed in terms of symmetrised correlation functions. These functions reflect the interactions between the magnetic moments. In a periodic crystal, the translational symmetry impose rules on these correlation functions, which are conveniently expressed through a space Fourier transform. This transformation and the related Brillouin zone concept are recalled in Appendix E.

Let us examine a symmetrised field correlation function $\Phi^{\alpha\beta}(\omega)$ defined as (see Eqs. 7.90 and 7.20)

$$\Phi^{\alpha\beta}(\omega) = \frac{1}{2\pi} \int_{-\infty}^{\infty} \exp(i\omega\tau) \left\langle \left\{ \delta B^{\alpha}_{\rm loc}(\tau) \delta B^{\beta}_{\rm loc} \right\} \right\rangle . \tag{7.106}$$

From Eq. 5.38

$$\Phi^{\alpha\beta}(\omega) = \frac{1}{2\pi} \left(\frac{\mu_0}{4\pi}\right)^2 \frac{g^2 \mu_{\rm B}^2}{v_c^2} \sum_{\gamma,\gamma'} \sum_{i,i'} \sum_{d,d'} G^{\alpha\gamma}_{\mathbf{r}_{i+d}} G^{\beta\gamma'}_{\mathbf{r}_{i'+d'}} \Lambda^{\gamma\gamma'}_{i+d,i'+d'}(\omega), \tag{7.107}$$

with

$$\Lambda^{\gamma\gamma'}_{i+d,i'+d'}(\omega) = \left\langle \left\{ \delta J^{\gamma}_{i+d}(\omega) \delta J^{\gamma'}_{i'+d'} \right\} \right\rangle . \tag{7.108}$$

As usual, μ_0 is the permeability of free space, g the spectroscopic factor, $\mu_{\rm B}$ the Bohr magneton, and v_c the volume of the unit cell. According to Eq. E.11

$$\delta J^{\alpha}_{i+d} = \frac{1}{\sqrt{n_c}} \sum_{\mathbf{q}} \exp[i\mathbf{q} \cdot (\mathbf{i} + \mathbf{d})] \delta J^{\alpha}_{d,\mathbf{q}} . \tag{7.109}$$

The sum is over the Brillouin zone and n_c is the number of unit cells. Then, taking into account that $\Lambda^{\gamma\gamma'}_{i+d,i'+d'}(\omega)$ depends only on $[(\mathbf{i}+\mathbf{d}) - (\mathbf{i}'+\mathbf{d}')]$ and that $[(\mathbf{i}+\mathbf{d}) - (\mathbf{i}'+\mathbf{d}')] = \mathbf{r}_{i+d} - \mathbf{r}_{i'+d'}$ (see, e.g. Fig. 5.3), we derive

$$\Lambda^{\gamma\gamma'}_{i+d,i'+d'}(\omega) = \frac{1}{n_c} \sum_{\mathbf{q}} \exp\left[i\mathbf{q}\left(\mathbf{r}_{i+d} - \mathbf{r}_{i'+d'}\right)\right] \Lambda^{\gamma\gamma'}_{dd',\mathbf{q}}(\omega), \tag{7.110}$$

with

$$\Lambda^{\gamma\gamma'}_{dd',\mathbf{q}}(\omega) = \left\langle \left\{ \delta J^{\gamma}_{d,\mathbf{q}}(\omega) \delta J^{\gamma'}_{d',-\mathbf{q}} \right\} \right\rangle . \tag{7.111}$$

Introducing Eq. 7.110 into Eq. 7.107 leads to

$$\Phi^{\alpha\beta}(\omega) = \frac{1}{2\pi} \left(\frac{\mu_0}{4\pi}\right)^2 \frac{g^2 \mu_{\rm B}^2}{n_c v_c^2} \sum_{\gamma,\gamma'} \sum_{d,d'} \sum_{\mathbf{q}} G^{\alpha\gamma}_{d,\mathbf{q}} G^{\beta\gamma'}_{d',-\mathbf{q}} \Lambda^{\gamma\gamma'}_{dd',\mathbf{q}}(\omega), \tag{7.112}$$

where, according to Eq. 5.41,

$$G^{\alpha\beta}_{d,\mathbf{q}} = \sum_{i} G^{\alpha\beta}_{\mathbf{r}_{i+d}} \exp(i\mathbf{q} \cdot \mathbf{r}_{i+d}). \tag{7.113}$$

It is sometimes convenient to consider $\mathbf{q}$ as a continuous variable (see Section E.2) and write

$$\Phi^{\alpha\beta}(\omega) = \frac{1}{2\pi} \left(\frac{\mu_0}{4\pi}\right)^2 \frac{g^2 \mu_{\rm B}^2}{v_c} \int \sum_{\gamma,\gamma'} \sum_{d,d'} G^{\alpha\gamma}_{d}(\mathbf{q}) G^{\beta\gamma'}_{d'}(-\mathbf{q}) \Lambda^{\gamma\gamma'}_{dd'}(\mathbf{q},\omega) \frac{{\rm d}^3\mathbf{q}}{(2\pi)^3} . \tag{7.114}$$

We are now in a position to write the longitudinal and transverse relaxation rates in terms of the wave vector-dependent spin correlation functions. We start from the expressions for the rates given in Eqs. 7.97 and 7.98 and use the $\Phi^{\alpha\beta}(\omega)$ formula derived earlier. This gives

$$\lambda_Z = \frac{\mathcal{D}}{2} \int \sum_{\beta,\gamma} \sum_{d,d'} \left[G_d^{X\beta}(\mathbf{q}) G_{d'}^{X\gamma}(-\mathbf{q}) + G_d^{Y\beta}(\mathbf{q}) G_{d'}^{Y\gamma}(-\mathbf{q}) \right] \Lambda_{dd'}^{\beta\gamma}(\mathbf{q}, \omega_\mu) \frac{\mathrm{d}^3\mathbf{q}}{(2\pi)^3}. \tag{7.115}$$

We have defined the constant $\mathcal{D} = (\mu_0/4\pi)^2 \gamma_\mu^2 (g\mu_B)^2 / v_c$. This equation was written by Yaouanc *et al.* (1993), however, with a different definition for the space and time Fourier transforms and therefore for $\mathcal{D}$. In the same way

$$\lambda_X = \frac{\lambda_Z}{2} + \frac{\mathcal{D}}{2} \int \sum_{\beta,\gamma} \sum_{d,d'} G_d^{Z\beta}(\mathbf{q}) G_{d'}^{Z\gamma}(-\mathbf{q}) \Lambda_{dd'}^{\beta\gamma}(\mathbf{q}, \omega = 0) \frac{\mathrm{d}^3\mathbf{q}}{(2\pi)^3}. \tag{7.116}$$

It is possible to write expressions for λ_Z and λ_X for which the correlation tensor is expressed in a frame adapted to the symmetry of the crystal; see Section D.3.

The expression for λ_α ($\alpha = X$ or Z) assumes the extreme motional narrowing limit to be valid. If only the motional narrowing limit is reached, $\lambda_\alpha(t)$ rather than λ_α should be considered.

The formulae given in Eqs. 7.115 and 7.116 are quite complicated. Fortunately, drastic simplifications are usually possible using simple arguments such as the symmetry of the muon site and of the system under study. These formulae, notably Eq. 7.115, are key results since they give the possibility to describe the μSR relaxation rates in terms of generalised susceptibilities or Green functions. These functions are commonly used by neutron practitioners. We refer to Chapter 10 and Appendix F for an extended discussion.

7.2 Numerical methods

One of the great advantages of the μSR techniques is that the zero-field technique can be applied to systems which have dynamics in the megahertz range. However, for a detailed analysis of the measured polarisation function, a mathematical model is needed. The physics is well understood and mathematically embedded in the Kubo–Toyabe formula; see Eq. 6.17. However, referring to Section 7.1.1, oscillations in $P_Z(t)$ not present in the Kubo–Toyabe formula could show up. To test the validity of the latter, an exact computation of $P_Z(t)$ is needed. It is the purpose of this section to provide a method for computing $P_Z(t)$. Obviously, the result can only be numerical. The method is based on Celio (1986). It works equally well for both $P_Z(t)$ and $P_X(t)$. For its exposition we consider $P_Z(t)$.

Our purpose is to compute $P_Z(t)$, as given in Eq. 7.1, with the evolution of the Pauli operator expressed by Eq. 7.2. For definitiveness our interest will be in nuclear magnetism, and therefore $\rho_{\rm sys}$ is proportional to the unit operator. In order to save space in the computer memory, we use the Schödinger rather than the Heisenberg representation. The time evolution of the system is indeed contained in eigenvectors of dimension $d_{\rm Hil}$ for the former representation instead of square matrices of dimension $d_{\rm Hil} \times d_{\rm Hil}$ for the latter one.

Denoting the basis vectors of the Hilbert space by $|u_k\rangle$, we write from Eq. 3.16,

$$P_Z(t) = \mathrm{Tr}\left\{\rho_{\mathrm{tot}}\sigma^Z(t)\right\} = \sum_{k=1}^{d_{\mathrm{Hil}}} \left\langle u_k \left| \rho_{\mathrm{tot}}\sigma^Z(t) \right| u_k \right\rangle = \sum_{k=1}^{d_{\mathrm{Hil}}/2} \frac{4}{d_{\mathrm{Hil}}} \left\langle u_k \left| S^Z(t) \right| u_k \right\rangle$$
$$= \sum_{k=1}^{d_{\mathrm{Hil}}/2} \frac{4}{d_{\mathrm{Hil}}} \left\langle u_k(t) \left| S^Z \right| u_k(t) \right\rangle, \qquad (7.117)$$

with the time-dependent eigenvector

$$|\, u_k(t)\rangle = \exp\left(-i\frac{\mathcal{H}_{\mathrm{tot}}\, t}{\hbar}\right) |\, u_k\rangle . \qquad (7.118)$$

In Eq. 7.117, ρ_{tot} is the density operator of the muon–system ensemble. As in Section 3.1.2 we factorise it into the direct product $\rho_\mu \rho_{\mathrm{sys}}$, the factors of which are the density operators for the muon and the system, respectively. According to Eq. 3.19, $\rho_\mu = (\mathcal{I} + \sigma^Z)/2$. Therefore, if the $|u_k\rangle$ vectors are ordered as

$$\{\, |{+1/2}, I_1, \ldots, I_{N_{\mathrm{ns}}}\rangle, |{+1/2}, I_1, \ldots, I_{N_{\mathrm{ns}}} - 1\rangle, \ldots, |{+1/2}, I_1, \ldots, -I_{N_{\mathrm{ns}}}\rangle; \ldots;$$
$$|{+1/2}, -I_1, \ldots, I_{N_{\mathrm{ns}}}\rangle, \ldots, |{+1/2}, -I_1, \ldots, -I_{N_{\mathrm{ns}}}\rangle\}, \qquad (7.119)$$

followed by a similar set of states in which the quantum state of muon spin $+\frac{1}{2}$, which appears in first position in a ket, is replaced with $-\frac{1}{2}$, the density matrix is diagonal with $\rho_{\mathrm{tot}}^{\alpha\alpha} = 2/d_{\mathrm{Hil}}$ if $\alpha \leq d_{\mathrm{Hil}}/2$ and 0 otherwise. Note that we have assumed N_{ns} identical nuclear spins I interacting with the muon spin. For each of them, there are $(2I+1)$ possible values for the nuclear magnetic quantum number, from I_i to $-I_i$ for spin I_i ($i = \{1, \ldots, N_{\mathrm{ns}}\}$). The generalisation to unlike spins is straightforward.

Referring to Eq. 7.118, the evaluation of $|u_k(t)\rangle$ requires the diagonalisation of a square matrix of dimension $d_{\mathrm{Hil}} \times d_{\mathrm{Hil}}$. For a static muon in aluminium, $d_{\mathrm{Hil}} = 2 \times (2 \times 5/2 + 1)^6 = 93312$ since the muon is believed to be implanted in an octahedral site and therefore it is surrounded by six nearest neighbour nuclei with spin $I = \frac{5}{2}$; see Fig. 9.1 for a sketch of the octahedral sites in face-centred-cubic crystallographic lattice. This is enormous. The computation can be performed with the help of two tricks.

First of all, we note that $\mathcal{H}_{\mathrm{tot}}$ can be written as the sum of N_{ns} Hamiltonians:

$$\mathcal{H}_{\mathrm{tot}} = \sum_{i=1}^{N_{\mathrm{ns}}} \mathcal{H}_i \quad \text{with} \quad \mathcal{H}_i = \frac{\mathcal{H}_{\mathrm{Z},\mu}}{N_{\mathrm{ns}}} + \mathcal{H}_{\mathrm{Z},i} + \mathcal{H}_{\mathrm{dip},i} + \mathcal{H}_{\mathrm{Q},i}. \qquad (7.120)$$

The expressions of the Hamiltonians in the sum can be deduced from Eqs. 5.13, 5.26, and 6.155. The size of the Hilbert space for each $\mathcal{H}_i$ is only $2 \times (2I_i + 1)$. For aluminium this is 12. Therefore, while the diagonalisation of $\mathcal{H}_{\mathrm{tot}}$ would be difficult, it is easy to diagonalise $\mathcal{H}_i$. This suggests the use of the Trotter formula (see Eq. C.109) for the computation of the exponential operator of Eq. 7.118:

$$\exp\left(-i\frac{\mathcal{H}_{\mathrm{tot}}\, t}{\hbar}\right) = \exp\left(-i\sum_{i=1}^{N_{\mathrm{ns}}} \mathcal{H}_i \frac{t}{\hbar}\right) = \lim_{\ell\to\infty}\left[\prod_{i=1}^{N_{\mathrm{ns}}} \exp\left(-i\frac{\mathcal{H}_i}{\ell}\frac{t}{\hbar}\right)\right]^{\ell}. \qquad (7.121)$$

The evaluation of $|u_k(t)\rangle$ requires the application of ℓ times the product of matrices $\prod_{i=1}^{N_{\rm ns}} \exp\left[-i\mathcal{H}_i t/(\ell\hbar)\right]$ on $|u_k\rangle$. This is feasible because the dimension of each of the matrices of the product is only $2 \times (2I_i + 1)$. We have neglected the dipole interaction between the different nuclear spins. It is indeed negligible in comparison with the muon–nucleus dipole interaction since the magnetic moment of the muon is much larger than that of any nucleus. However, according to Eq. 7.117, we still need to evaluate $d_{\rm Hil}/2$ wave vectors $|u_k(t)\rangle$ for each time. The number of wave vectors can be so large as to prevent the computation of $P_Z(t)$.

Now comes the second and last trick: the random-phase approximation. Following Bohm and Pines (1953) we define the initial state of the system with the linear combination

$$|\phi\rangle = \sum_{m=1}^{d_{\rm Hil}/2} \sqrt{\frac{4}{d_{\rm Hil}}} \exp\left(i\lambda_m\right)|u_m\rangle\,, \tag{7.122}$$

where λ_m is a random number of uniform probability in the interval $[0, 2\pi[$. The Schrödinger equation is linear, so that at later times t we have

$$|\phi(t)\rangle = \exp\left(-i\frac{\mathcal{H}_{\rm tot}\, t}{\hbar}\right)|\phi\rangle\,. \tag{7.123}$$

Remarkably,

$$\begin{aligned}\langle\phi(t)|\, S^Z\, |\phi(t)\rangle = & \sum_{m=1}^{d_{\rm Hil}/2} \frac{4}{d_{\rm Hil}} \langle u_m(t)|\, S^Z\, |u_m(t)\rangle \\ & + \sum_{m\neq n} \frac{4}{d_{\rm Hil}} \exp\left[i\left(\lambda_m - \lambda_n\right)\right]\langle u_n(t)|\, S^Z\, |u_m(t)\rangle\,. \end{aligned} \tag{7.124}$$

The first term on the right-hand side is just $P_Z(t)$. The second term is a random correction which tends to 0 as $d_{\rm Hil}$ increases (Bohm and Pines 1953). This suggests computing the average of a number of $\langle\phi(t)|\, S^Z\, |\phi(t)\rangle$ terms. It turns out to be 2 for the aforementioned case of aluminium. Therefore we need to evaluate a very small number of $|\phi(t)\rangle$ vectors (using the Trotter formula) instead of $d_{\rm Hil}/2$ different $|u_k(t)\rangle$ vectors. This is indeed a very efficient method.

Mathematically, the Trotter parameter ℓ should be extended to infinity. However, in practice a relatively small value of ℓ is needed to reach convergence. This value depends on the time t and the intensity of the external field. For $t = 1\,\mu$s, ℓ is about 2 to 4 and at $t = 16\,\mu$s it can reach a value of about 100 for the field magnitudes to be considered thereafter.

Altogether the evaluation of $P_Z(t)$ can be achieved in a couple of minutes with currently available computers when $d_{\rm Hil}$ is around 10^5. It may be worthwhile for analysing high-quality data. It is also possible to take into account muon hopping or tunnelling as explained in Section 6.3.

We now present some numerical results. Figure 6.18 shows the cases of a muon in an octahedral site of a face-centred-cubic lattice, i.e. with six nearest neighbours. For the nuclear spins which have been chosen, namely $I = 1$ and $I = \frac{5}{2}$, the dimensions of the

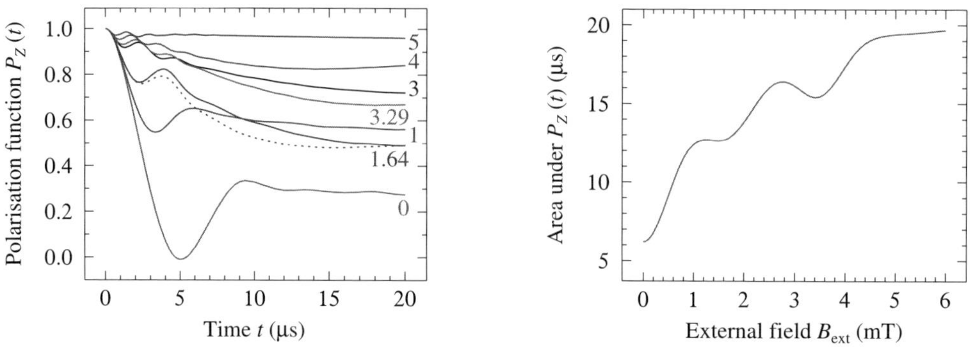

Fig. 7.8 (Left) Longitudinal polarisation as a function of the external field, the intensity of which is given next to each curve in units of milliteslas. The muon is assumed to be static in an octahedral site of a face-centred-cubic crystal. The six nearest neighbours which are considered have spins $\frac{5}{2}$ and the intensity of the dipolar interaction is $\omega_{I,{\rm d}} = 0.065\,\mu{\rm s}^{-1}$. The field is applied along the [111] crystal direction. The avoided-level-crossing (ALC) phenomenon happens to be around 1.64 and 3.29 mT. The quadrupole interaction is assumed positive ($3\omega_{\rm E} = 0.70\,\mu{\rm s}^{-1}$) except for the dashed line for which $3\omega_{\rm E} = -0.70\,\mu{\rm s}^{-1}$ and $B_{\rm ext} = 1.64\,{\rm mT}$. (Right) Area defined by the integral of $P_Z(t)$ between 0 and 20 μs versus $B_{\rm ext}$. The resonances are clearly visible. Reproduced with kind permission from Dalmas de Réotier *et al.* (1990).

Hilbert space are $d_{\rm Hil} = 1458$ and 93312, respectively. The latter case corresponds to aluminium in which the effect of the quadrupole interaction due to the muon charge has been neglected. The comparison with the result of the Kubo–Toyabe function shows that the deviations are basically observed around the minimum and at long times where $P_Z(t)$ does not reach the $\frac{1}{3}$ plateau. The Kubo–Toyabe function appears to be better matched when the dimension of the Hilbert space is large.

Other examples for $P_Z(t)$ as well as for $P_X(t)$ are given in Figs. 7.3, 7.4, and 7.6.

In Section 4.2 the avoided-level-crossing (ALC) phenomenon was introduced. It takes place if a resonant energy exchange occurs between the two muon energy levels and energy levels of the system under study. Going back to the model used to produce the right-hand plot of Fig. 6.18 which we have just discussed, the ALC phenomenon is expected for two values of the external field, if we include the effect of the muon-induced electric field gradient on the nearest nuclei to the muon. At these fields the muon Zeeman splitting equals the nuclear quadrupole splitting between either the $m_i = \pm\frac{1}{2}$ and $m_i = \pm\frac{3}{2}$ or $m_i = \pm\frac{3}{2}$ and $m_i = \pm\frac{5}{2}$ energy levels. From the diagonalisation of the electric quadrupolar Hamiltonian alone (Eq. 6.155) these splittings are $6\hbar\omega_{\rm E}$ and $12\hbar\omega_{\rm E}$, respectively. When $3\omega_{\rm E} = 0.70\,\mu{\rm s}^{-1}$ these energies match the muon Zeeman energy for $B_{\rm ext} = 1.64$ and 3.29 mT, respectively. Note that this evaluation of the resonance positions is only approximate since the perturbation due to the nuclear Zeeman interaction is neglected. In the left panel of Fig. 7.8 the two resonances are effectively visible as sudden increase of the relaxation. The right panel, which displays the integral $\int_0^{20\mu s} P_Z(t){\rm d}t$ as a function of $B_{\rm ext}$, indicates that their positions are close to these estimated field values. Note that the detection of these resonances in a time-integrated experiment (see Section 4.2.3) from the field dependence of $\mathcal{R}_{\rm int}$ (Eq. 4.12) is much more difficult than in time differential experiment. This is due to the relatively small value of the quadrupole interaction assumed here: the resonances occur in the field range where, out of the resonance, $P_Z(t)$ departs strongly from one. Finally, the left panel of Fig. 7.8 shows that the sign of the quadrupole interaction has a clear influence on the shape of $P_Z(t)$ as noticed by comparing the two curves drawn for $B_{\rm ext} = 1.64\,{\rm mT}$.

In Section 7.1.4.2, using the iteration method we have shown the importance of the nuclear spin dynamics on the shape of the polarisation function. However, in zero field the iteration method provides a good estimate for $P_Z(t)$ only if $\nu_{\rm c}$ is large enough. We shall now probe the effect of this nuclear spin dynamics in zero field and when $\nu_{\rm c} = 0$.

First we remark that the polarisation functions can be computed in the framework of classical mechanics at least when the quadrupole interaction acting on the muon neighbouring spins is neglected. The calculation was performed by Dalmas de Réotier *et al.* (1992) using a Monte Carlo technique. The orientation for each nuclear spin at initial time is randomly chosen. The value of the local field $\mathbf{B}_{\rm loc}$ acting on the muon is the vector sum of the nuclear dipole fields and the external field if any. The evolution of the muon spin is then evaluated from the solution of the Larmor equation (see Section 3.1.1). To account for the evolution of the nuclear spins in the field produced by the muon, the time range over which the polarisation function is calculated is divided into steps. At each step the new orientation for the nuclear spins is computed. This yields a new value of $\mathbf{B}_{\rm loc}$ which determines the muon spin motion during the subsequent time interval. Figure 7.9 shows an example of the results for this model. Note that $P_Z(t)$ depends on the spin value for the reason given at the end of next paragraph. The difference between the quantum and classical computations is only seen at relatively large time and in the case of spins $\frac{1}{2}$. For spins equal to 1 the agreement is excellent. This is also the case for larger spins (not shown).

With the algorithm that we have just described, it is an easy matter to prove the effect of the nuclear spin dynamics. For this purpose, we simply freeze the nuclear spins in their original orientations and therefore neglect their motion in the field produced by the muon. The result is presented in Fig. 7.10. It clearly indicates that the slope observed at large time in the "exact" model, which was presented earlier in the current section, is actually due to the dynamics of the neighbouring nuclear spins submitted to the muon dipole field. This conclusion is corroborated by the following remark. From the Larmor equation (Eq. 3.4) the precession rate of a spin is proportional to the gyromagnetic ratio. However, for a given magnetic moment (and therefore for a given $\Delta_{\rm G,ZF}$), the gyromagnetic ratio is inversely proportional to the spin; see Eq. G.2. Therefore one expects the effect of the nuclear spin dynamics to be more important for small spin values. A comparison between the plots of Fig. 6.18 confirms this prediction.

Note that within the framework of Chapter 6 it is possible to simulate a slope at large time in $P_Z(t)$ through the introduction of a finite muon magnetic correlation time, the physical origin being clearly different from the dynamics that we evoke now. Therefore, for a complete description of the physics, it is important to account for the effect of the muon dipole field on the nuclei.

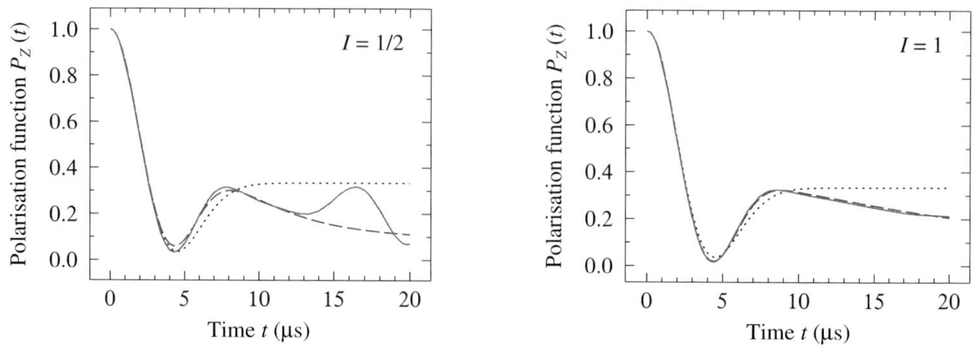

Fig. 7.9 Comparison of the zero-field polarisation functions obtained from the quantum model described in this section (full line) and its classical counterpart (dashed line). The six nearest neighbours to a muon in an octahedral site of a face-centred-cubic lattice are considered. The Z axis is taken parallel to the [111] direction and the effect of the quadupolar interaction is neglected. The left (right) panel shows the prediction for spins $\frac{1}{2}$ (1) with $\omega_{I,\mathrm{d}} = 0.222\,\mu\mathrm{s}^{-1}$ ($\omega_{I,\mathrm{d}} = 0.136\,\mu\mathrm{s}^{-1}$), which both lead to $\gamma_\mu \Delta_{\mathrm{G,ZF}} = 0.3845\,\mu\mathrm{s}^{-1}$ (Eq. 6.151). This latter value has been used for the Kubo–Toyabe function (dotted line), which is drawn for reference. In the case of the classical solution, the same parameters apply although with a renormalisation of the product of the gyromagnetic ratio with the spin value: $\gamma_i I_i \to \gamma_i \sqrt{I_i(I_i+1)}$. Reproduced with kind permission from Dalmas de Réotier *et al.* (1990, 1992).

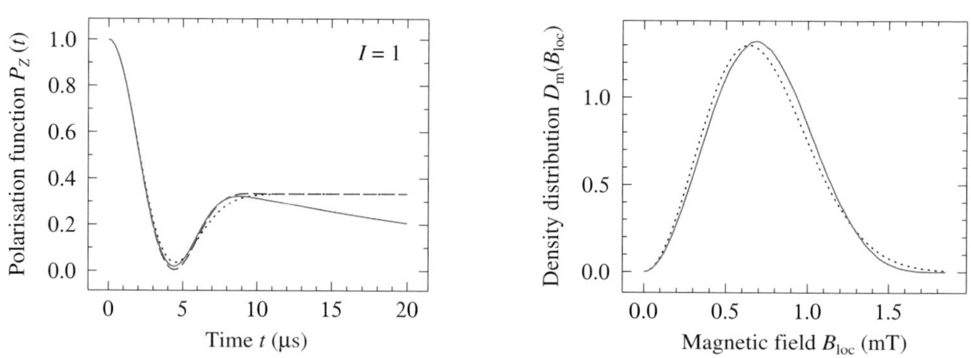

Fig. 7.10 (Left) Effect of the nuclear spin dynamics on the shape of the polarisation functions. While the full line presents the function obtained from classical mechanics with the nuclear spin dynamics included, this dynamics is neglected for the dashed line curve. The model parameters are the same as those for the right panel of Fig. 7.9. The dotted line shows again the Kubo–Toyabe function. (Right) Density distribution functions of the local field for the classical computation (full line) and the Kubo–Toyabe function (dotted line). The latter density distribution is the Maxwell–Boltzmann distribution. Reproduced with kind permission from Dalmas de Réotier *et al.* (1992).

The comparison of the curves for $P_Z(t)$ in Fig. 7.9 shows that deviations from the Kubo–Toyabe shape are present even at relatively early time, i.e. around the minimum of the function or slightly later. This can be tracked back to the non-Maxwellian shape of the distribution of the field modulus as shown in the right panel of Fig. 7.10. This means that the Gaussian form for the field component distribution is in fact only an approximation.

References

Abragam A. (1961) *The Principles of Nuclear Magnetism* (Clarendon, Oxford).

Baryshevskiĭ V.G. and Kuten' S.A. (1976) *Soviet Physics Solid State* **18**, 1677–1681.

Bloch F. (1946) *Physical Review* **70**, 460–474.

Bloembergen N., Purcell E.M., and Pound R.V. (1948) *Physical Review* **73**, 679–712.

Bohm D. and Pines D. (1953) *Physical Review* **92**, 609–625.

Brewer J.H., Kreitzman S.R., Noakes D.R., Ansaldo E.J., Harshman D.R., and Keitel R. (1986) *Physical Review* **B 33**, 7813–7816.

Celio M. (1986) *Physical Review Letters* **56**, 2720–2723.

Dalmas de Réotier P. and Yaouanc A. (1992) *Journal of Physics: Condensed Matter* **4**, 4533–4556.

Dalmas de Réotier P., Yaouanc A., and Boucher J.P. (1990) *Hyperfine Interactions* **65**, 1121–1128.

Dalmas de Réotier P., Yaouanc A., and Meshkov S.V. (1992) *Physics Letters* **A 162**, 206–212.

Ghosh S., Rosenbaum T.F., Aeppli G., and Coppersmith S.N. (2003) *Nature* **425**, 48–51.

Hartmann O. (1977) *Physical Review Letters* **39**, 832–835.

Hayano R.S., Uemura Y.J., Imazato J., Nishida N., Yamazaki T., and Kubo R. (1979) *Physical Review* **B 20**, 850–859.

Keren A. (1994) *Physical Review* **B 50**, 10039–10042.

Kondo J. (1986) *Hyperfine Interactions* **31**, 117–133.

Luke G.M., Brewer J.H., Kreizman S.R., Noakes D.R., Celio M., Kadono R., and Ansaldo E.J. (1991) *Physical Review* **B 43**, 3284–3297.

McMullen T. and Zaremba E. (1978) *Physical Review* **B 18**, 3026–3040.

Moriya T. (1962) *Progress in Theoretical Physics (Kyoto)* **28**, 371–400.

Slichter C.P. (1996) *Principles of Magnetic Resonance*, third enlarged and updated edition (Springer, Berlin).

Smilga V.P. and Belousov Yu.M. (1994) *The Muon Method in Science*, Proceedings of the Lebedev Physics Institute, vol. 219 (Nova Science, Commack, NY).

Turov E.A. and Petrov M.P. (1972) *Nuclear Magnetic Resonance in Ferro and Antiferromagnets* (Halsted Press, New York).

Van Vleck J.H. (1978) *Reviews of Modern Physics* **50**, 181–189.

Vogel S., Celio M., and Meier P.F. (1986) *Hyperfine Interactions* **31**, 35–40.

Wolf D. (1979) *Spin-Temperature and Nuclear-Spin Relaxation in Matter* (Clarendon Press, Oxford).

Yaouanc A. and Dalmas de Réotier P. (1994) *Hyperfine Interactions* **87**, 1147–1151.

Yaouanc A. and Dalmas de Réotier P. (1995) *Physical Review* **B 51**, 12011–12012.

Yaouanc A., Dalmas de Réotier P., and Frey E. (1993) *Physical Review* **B 47**, 796–809.

8
Muonium polarisation functions

When positive muons are implanted into insulators, semiconductors, or molecular systems, there is a high probability for them to form muonium atoms, i.e. muon–electron bound states that are analogous to hydrogen atoms. Since the μSR techniques only give access to the polarisation functions, it is important to understand the relation between the muonium physical and chemical properties and these functions. This is the purpose of this chapter. The type of information which can be extracted from the study of muonium centres in materials is the subject of Chapter 12.

There is a vast literature on the relation between muonium properties and the polarisation functions, $P_\alpha(t)$: the books of Walker (1983), Schenck (1985), Smilga and Belousov (1994), Karlsson (1995), and Nagamine (2003), as well as the review papers of Patterson (1988), Roduner (1988, 1999), and Cox (2003, 2009).

After a presentation of muonium in terms of atomic physics, we shall focus our attention on the isotropic muonium state, neglecting its interaction with the environment. In the third section we shall extend the discussion to the muonium state of axial symmetry. Finally, the effects of the muonium environment on the muonium state and therefore on $P_\alpha(t)$ will be accounted for.

8.1 Atomic properties of muonium

A muonium atom is a bound state of a positive muon with an electron. It is closely analogous to a hydrogen atom. In vacuum, the electron ground-state wavefunction is

$$\psi_{1s}(r) = \frac{1}{\sqrt{\pi}} \frac{1}{a_{\mathrm{Mu},0}^{3/2}} \exp\left(-\frac{r}{a_{\mathrm{Mu},0}}\right), \tag{8.1}$$

with the ground-state radius

$$a_{\mathrm{Mu},0} = 4\pi\varepsilon_0 \frac{\hbar^2}{e^2} \frac{1}{M_{\mathrm{e}}^*}, \tag{8.2}$$

where ε_0 is the permittivity of free space and M_{e}^* the reduced electron mass defined by the relation

$$\frac{1}{M_{\mathrm{e}}^*} = \frac{1}{M_{\mathrm{e}}} + \frac{1}{M_\mu}, \tag{8.3}$$

where M_{e} and M_μ are the electron and muon masses, respectively. In Eq. 8.1 the origin of the coordinates is taken at the muonium mass centre. The reduced electron mass in muonium is only about 0.4% smaller than in hydrogen. Similarly, the ground-state

Table 8.1 Some parameters for a muonium atom in vacuum. The units are indicated in parentheses. The reduced electron mass, ground-state radius, and hyperfine frequency follow from Eqs. 8.3, 8.2, and 8.6, respectively. The ground-state energy is given by $e^2/(8\pi\varepsilon_0 a_{\mathrm{Mu},0})$ according to the Bohr model. The muonium mass is the sum of the mass of its constituents diminished by the mass equivalent of the bond energy. In fact, with the precision we adopt in this book, the muonium and muon mass are equal. Data from the numerical values of fundamental constants tabulated at the begining of the book after the list of important symbols have also been used.

Quantity	Value
Mass (M_{e})	206.768
Reduced electron mass (M_{e})	0.995187
Ground-state radius $a_{\mathrm{Mu},0}$ (pm)	53.1736
Ground-state energy (eV)	−13.5403
Hyperfine frequency ν_0 (GHz)	4.4642

radius is 0.4% larger than the Bohr radius and the binding energy of the electron is smaller by the same amount. Hence, the basic parameters for muonium and hydrogen in vacuum are quite similar, except for the mass which is much less for muonium than for hydrogen — the mass of the muon is about $\frac{1}{9}$ of the proton mass. Some parameters of a muonium atom in vacuum are listed in Table 8.1.

8.2 Isotropic muonium state

We shall first describe the Hamiltonian for a muonium atom and continue with a discussion of the ensuing polarisation functions.

8.2.1 Hamiltonian

Similarly to atomic hydrogen, the electron spin $\mathbf{S}_{\mathrm{e}}$ and the muon spin $\mathbf{S}_\mu$ are coupled through the Fermi contact interaction (see Eq. 5.12):

$$\mathcal{H}_{\mathrm{iso}} = -\frac{2\mu_0}{3}|\psi_{1s}(r=0)|^2\hbar^2\gamma_{\mathrm{e}}\gamma_\mu\, \mathbf{S}_{\mathrm{e}}\cdot\mathbf{S}_\mu = \hbar\omega_{\mathrm{iso}}\, \mathbf{S}_{\mathrm{e}}\cdot\mathbf{S}_\mu. \tag{8.4}$$

The hyperfine coupling ω_{iso} is linked to the ground-state radius a_{Mu} of the muonium state:

$$\omega_{\mathrm{iso}} = -\frac{2\mu_0}{3}\frac{1}{\pi a_{\mathrm{Mu}}^3}\gamma_{\mathrm{e}}\hbar\gamma_\mu. \tag{8.5}$$

In vacuum, the hyperfine coupling is denoted as ω_0 and we have

$$\omega_0 = -\frac{2\mu_0}{3}\frac{1}{\pi a_{\mathrm{Mu},0}^3}\gamma_{\mathrm{e}}\hbar\gamma_\mu = 2\pi\nu_0. \tag{8.6}$$

As for any magnetic moment — see the discussion after Eq. 3.4 for the muon — the electron gyromagnetic ratio γ_{e} ($\gamma_{\mathrm{e}} = -1.76086\times10^{11}\,\mathrm{s}^{-1}\,\mathrm{T}^{-1}$; see the table of numerical values of fundamental constants given at the beginning of this book) is the ratio of the magnetic moment over the angular momentum of the electron. Note that γ_e is a negative number. The value of the so-called hyperfine frequency ν_0 is listed in Table 8.1. The Hamiltonian $\mathcal{H}_{\rm iso}$ is referred to as the muonium hyperfine Hamiltonian. We write $\mathcal{H}_{\rm iso}$ rather than $\mathcal{H}_\mu$, as in Eq. 5.12, to stress that we are dealing with isotropic muonium.

The isotropic muonium has been discovered in a number of semiconductors and dielectrics, and in organic compounds, mostly in its $1s$ electronic state. It is formed when the free electron density at the muon site in the material is small enough so that the screening of the Coulomb interaction is weak. In addition, the point symmetry at the muonium site must be high. Relative to the state in vacuum, the hyperfine frequency is usually found to be much smaller in materials. Since the frequency varies as the inverse of the ground-state radius to the third power, the muonium wavefunction is usually more extended in materials than in vacuum.

It is interesting to consider a muonium atom subject to a magnetic field, $\mathbf{B}_{\rm ext}$, applied as usual along the Z axis of the laboratory reference frame. The total Hamiltonian for this system, $\mathcal{H}_{\rm tot}$, is the sum of the hyperfine Hamiltonian and two Zeeman Hamiltonians, one for the electron spin and one for the muon spin:

$$\mathcal{H}_{\rm tot} = \hbar\omega_{\rm iso}\,\mathbf{S}_{\rm e}\cdot\mathbf{S}_\mu + g_{\rm e}\mu_{\rm B}B_{\rm ext}\,S_{\rm e}^Z - \hbar\gamma_\mu B_{\rm ext}S_\mu^Z. \tag{8.7}$$

For simplicity we shall assume the vacuum value for the electron spectroscopic factor, and for convenience in the following we shall use the electron gyromagnetic ratio γ_e which is related to g_e through the relation $g_e\mu_B = -\gamma_e\hbar$; see also Eq. 5.24.

To derive the eigenvalues and the eigenvectors of $\mathcal{H}_{\rm tot}$, we first introduce the states $\{|m_{\rm e}, m_\mu\rangle\}$, which are built up from the direct products of the muon and electron Zeeman states. There is no risk of confusion with this notation, which is a simplified one. Specifically, the basis states are

$$|\chi_1\rangle = \left|\frac{1}{2},\frac{1}{2}\right\rangle,\quad |\chi_2\rangle = \left|\frac{1}{2},-\frac{1}{2}\right\rangle,\quad |\chi_3\rangle = \left|-\frac{1}{2},\frac{1}{2}\right\rangle,\quad |\chi_4\rangle = \left|-\frac{1}{2},-\frac{1}{2}\right\rangle. \tag{8.8}$$

Unlike in Section 7.2, the muon state appears in the last instance. It is convenient to define a dimensionless external field parameter, namely $x_{\rm B} = B_{\rm ext}/\tilde{B}_{\rm iso}$, with the scaling field

$$\tilde{B}_{\rm iso} = \frac{\omega_{\rm iso}}{-\gamma_{\rm e}+\gamma_\mu}, \tag{8.9}$$

and the ratio

$$d_\gamma = \frac{-\gamma_{\rm e}-\gamma_\mu}{-\gamma_{\rm e}+\gamma_\mu} = 0.99037, \tag{8.10}$$

which is equal to 1 to a good approximation. For muonium in vacuum, $\tilde{B}_{\rm iso} = \tilde{B}_0 = 0.15853\,$T. The evaluation of $\mathcal{H}_{\rm tot}$ in the $\{|m_e, m_\mu\rangle\}$ basis yields the 4×4 matrix:

$$\frac{\hbar\omega_{\rm iso}}{4}\begin{pmatrix} 1+2d_\gamma\,x_{\rm B} & 0 & 0 & 0\\ 0 & -1+2x_{\rm B} & 2 & 0\\ 0 & 2 & -1-2x_{\rm B} & 0\\ 0 & 0 & 0 & 1-2d_\gamma\,x_{\rm B}\end{pmatrix}. \tag{8.11}$$

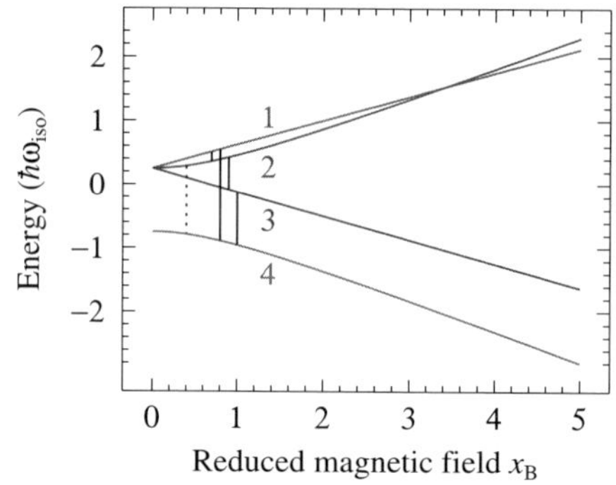

Fig. 8.1 Breit–Rabi diagram for an isotropic muonium state. The eigenvalues are plotted versus the dimensionless external field parameter $x_{\rm B} = B_{\rm ext}/\tilde{B}_{\rm iso}$. For reference, $\tilde{B}_{\rm iso} = \tilde{B}_0 = 0.15853\,{\rm T}$ in the case of muonium in vacuum. The energy scale is $\hbar\omega_{\rm iso}$. For graphical clarity, a fictitious value for the ratio d_γ is used to generate the plot: here $d_\gamma = \frac{3}{4}$. With the actual value of d_γ ($d_\gamma = 0.99037$), levels 1 and 2 cross for $x_{\rm B} = 103.38$. The vertical full (dotted) lines indicate the transitions observed in transverse- (zero- or longitudinal-) field experiments.

The four eigenvalues are given by

$$E_1 = \frac{\hbar\omega_{\rm iso}}{4}\left(1 + 2d_\gamma\, x_{\rm B}\right), \qquad E_3 = \frac{\hbar\omega_{\rm iso}}{4}\left(1 - 2d_\gamma\, x_{\rm B}\right),$$

$$E_2 = \frac{\hbar\omega_{\rm iso}}{4}\left(-1 + 2\sqrt{1 + x_{\rm B}^2}\right), \qquad E_4 = \frac{\hbar\omega_{\rm iso}}{4}\left(-1 - 2\sqrt{1 + x_{\rm B}^2}\right). \tag{8.12}$$

Note that $E_1(x_{\rm B})$ and $E_2(x_{\rm B})$ cross at $x_{\rm B} = 2d_\gamma/(1 - d_\gamma^2) = 103.38$, corresponding to $B_{\rm ext} = 16.389\,{\rm T}$. The associated eigenvectors have the form

$$|\psi_1\rangle = \left|\frac{1}{2}, \frac{1}{2}\right\rangle, \qquad |\psi_3\rangle = \left|-\frac{1}{2}, -\frac{1}{2}\right\rangle, \tag{8.13}$$

$$|\psi_2\rangle = \frac{1}{\sqrt{2}}\left(\sqrt{1 + \frac{x_{\rm B}}{\sqrt{1 + x_{\rm B}^2}}}\left|\frac{1}{2}, -\frac{1}{2}\right\rangle + \sqrt{1 - \frac{x_{\rm B}}{\sqrt{1 + x_{\rm B}^2}}}\left|-\frac{1}{2}, \frac{1}{2}\right\rangle\right),$$

$$|\psi_4\rangle = \frac{1}{\sqrt{2}}\left(\sqrt{1 - \frac{x_{\rm B}}{\sqrt{1 + x_{\rm B}^2}}}\left|\frac{1}{2}, -\frac{1}{2}\right\rangle - \sqrt{1 + \frac{x_{\rm B}}{\sqrt{1 + x_{\rm B}^2}}}\left|-\frac{1}{2}, \frac{1}{2}\right\rangle\right).$$

Note the coherent mixture nature of the $|\psi_2\rangle$ and $|\psi_4\rangle$ quantum states.

The four energy levels versus $x_{\rm B}$ are shown in Figs. 1.6 and 8.1 for different field scales. These are so-called Breit–Rabi diagrams. In Fig. 1.6 the diagram is plotted for free muonium, i.e. $\omega_{\rm iso} = \omega_0 = 2\pi\nu_0$.

In zero field, the singlet ground state is split from the excited triplet state ("triplet muonium") by an energy of $\hbar\omega_{\rm iso}$. This is understood by the following simple argument.

Since the muonium total spin, $\mathbf{S}_{\rm Mu} = \mathbf{S}_e + \mathbf{S}_\mu$, commutes with $\mathcal{H}_{\rm iso}$ defined in Eq. 8.4, $S_{\rm Mu}$ is a good quantum number. From the relation

$$\mathbf{S}_{\rm e} \cdot \mathbf{S}_\mu = \frac{1}{2}\left(\mathbf{S}_{\rm Mu}^2 - \mathbf{S}_{\rm e}^2 - \mathbf{S}_\mu^2\right), \tag{8.14}$$

we have

$$\begin{aligned} &\mathbf{S}_{\rm e} \cdot \mathbf{S}_\mu \left|S_{\rm e}, S_\mu, S_{\rm Mu}, M_{\rm Mu}\right\rangle \\ &\quad = \frac{1}{2}\left[S_{\rm Mu}\left(S_{\rm Mu}+1\right) - S_{\rm e}\left(S_{\rm e}+1\right) - S_\mu\left(S_\mu+1\right)\right]\left|S_{\rm e}, S_\mu, S_{\rm Mu}, M_{\rm Mu}\right\rangle, \end{aligned} \tag{8.15}$$

with $M_{\rm Mu} = m_{\rm e} + m_\mu$. This leads to a singlet state, $S_{\rm Mu} = 0$, with an energy of $-3\hbar\omega_{\rm iso}/4$ and a triplet state, $S_{\rm Mu} = 1$, with energy $\hbar\omega_{\rm iso}/4$. The eigenvectors have the form

$$\begin{aligned} \left|S_{\rm e}, S_\mu, S_{\rm Mu} = 1, M_{\rm Mu} = -1\right\rangle &= \left|-\frac{1}{2}, -\frac{1}{2}\right\rangle, \\ \left|S_{\rm e}, S_\mu, S_{\rm Mu} = 1, M_{\rm Mu} = 0\right\rangle &= \frac{1}{\sqrt{2}}\left(\left|\frac{1}{2}, -\frac{1}{2}\right\rangle + \left|-\frac{1}{2}, \frac{1}{2}\right\rangle\right), \\ \left|S_{\rm e}, S_\mu, S_{\rm Mu} = 1, M_{\rm Mu} = 1\right\rangle &= \left|\frac{1}{2}, \frac{1}{2}\right\rangle, \\ \left|S_{\rm e}, S_\mu, S_{\rm Mu} = 0, M_{\rm Mu} = 0\right\rangle &= \frac{1}{\sqrt{2}}\left(\left|\frac{1}{2}, -\frac{1}{2}\right\rangle - \left|-\frac{1}{2}, \frac{1}{2}\right\rangle\right). \end{aligned} \tag{8.16}$$

The total spin $\mathbf{S}_{\rm Mu}$ does not commute with the Hamiltonian if a field is applied. In high field, according to Eqs. 8.12 and 8.13, the energy levels vary linearly with the field and the $|\psi_i\rangle$ eigenvectors set is asymptotically equal to the $|\chi_i\rangle$ set, which is made of pure Zeeman states.

8.2.2 Polarisation functions

Since the eigenvalues and eigenvectors have been determined, the polarisation functions can be evaluated. Contrary to what has been done in Chapter 6 but in the same way as in Chapter 7, here we shall first compute $P_Z^{\rm stat}(t)$ because it is easier than $P_X^{\rm stat}(t)$. Because the hyperfine splitting is small ($\hbar\omega_{\rm iso}/k_{\rm B} = 0.2\,{\rm K}$), it is usually a good approximation to use the high-temperature limit for the density operator. Therefore, according to Eq. 3.29,

$$P_\alpha(t) = \frac{1}{4}{\rm Tr}\left\{\sigma_\mu^\alpha \sigma_\mu^\alpha(t)\right\}, \tag{8.17}$$

with $\sigma_\mu^\alpha(t)$ given in Eq. 3.23. Since no dynamical effect is taken into account, here we are concerned with $P_\alpha^{\rm stat}(t)$. Although tedious, the evaluation of the polarisation functions is not difficult. Denoting $\hbar\omega_{ij} = E_i - E_j$, we obtain

$$P_Z^{\rm stat}(t) = \frac{1}{2}\left\{1 + \frac{1}{1+x_{\rm B}^2}\left[x_{\rm B}^2 + \cos\left(\omega_{24}t\right)\right]\right\}, \tag{8.18}$$

with $\omega_{24} = \omega_{\rm iso}\sqrt{1+x_{\rm B}^2}$. The angular frequency ω_{24} can be rather large and therefore it may not be possible to detect the oscillation with a conventional spectrometer;

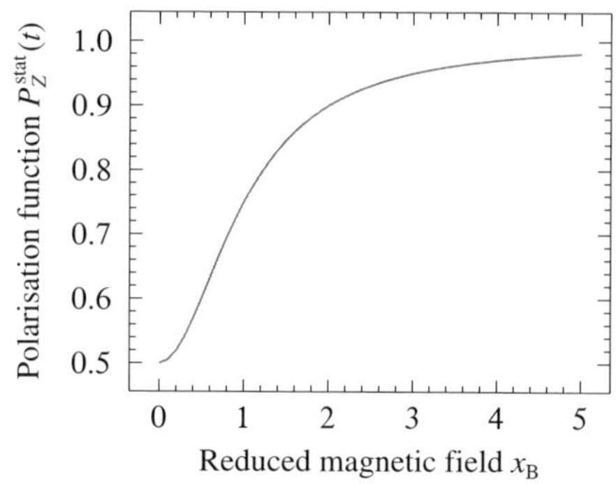

Fig. 8.2 Repolarisation curve for isotropic muonium as a function of the dimensionless external field parameter x_B.

see Section 2.2.2.1. In this case a time average of $P_Z^{stat}(t)$ must be performed. As a consequence $P_Z^{stat}(t)$ becomes time independent with a value increasing with B_{ext} up to saturation:

$$P_Z^{stat}(t) = \frac{1 + 2x_B^2}{2\,(1 + x_B^2)}. \tag{8.19}$$

We plot the repolarisation curve in Fig. 8.2. Noting that $P_Z^{stat}(t) = \frac{1}{2}$ in zero field, half of the repolarisation is achieved when $x_B = 1$, that is, when $B_{ext} = \tilde{B}_{iso}$. Therefore, the measurement of $P_Z^{stat}(t)$ versus B_{ext} provides a method for the estimation of ω_{iso}, although it is not very precise. The repolarisation arises at high field because the field purifies the eigenvectors such that they asymptotically become the $|\chi_i\rangle$ vectors. This is the so-called Paschen–Back limit.

The polarisation function $P_Z^{stat}(t)$ is remarkably simple in zero field:

$$P_Z^{stat}(t) = \frac{1}{2}[1 + \cos(\omega_{24}t)]. \tag{8.20}$$

This result is easily understood on physical grounds. If the initial state of the muonium state is given by $\left|-\frac{1}{2}, -\frac{1}{2}\right\rangle$ or $\left|\frac{1}{2}, \frac{1}{2}\right\rangle$, it remains in its spin state. According to Eq. 8.16, the initial states $\left|\frac{1}{2}, -\frac{1}{2}\right\rangle$ and $\left|-\frac{1}{2}, \frac{1}{2}\right\rangle$ are not eigenvectors. If formed in one of these two states, the muonium will execute singlet–triplet transitions with the angular frequency ω_{24}. The weight of these transitions is one-half because there is 50% chance of forming muonium in the $\left|\frac{1}{2}, -\frac{1}{2}\right\rangle$ or $\left|-\frac{1}{2}, \frac{1}{2}\right\rangle$ state. These transitions correspond to the selection rules

$$\Delta S_{Mu} = \pm 1 \quad \text{and} \quad \Delta M_{Mu} = 0. \tag{8.21}$$

We now derive $P_X^{stat}(t)$ which is measured in a transverse-field experiment. From Eq. 8.17 we find

$$P_X^{stat}(t) = \frac{1}{4}\left\{ \cos(\omega_{12}t) + \cos(\omega_{14}t) + \cos(\omega_{34}t) + \cos(\omega_{23}t) \right.$$

$$+ \frac{x_\mathrm{B}}{\sqrt{1+x_\mathrm{B}^2}} \left[\cos(\omega_{12}t) - \cos(\omega_{14}t) + \cos(\omega_{34}t) - \cos(\omega_{23}t)\right] \Bigg\} . \tag{8.22}$$

The four transition angular frequencies are

$$\omega_{12} = \frac{\omega_\mathrm{iso}}{2}\left(1 + d_\gamma x_\mathrm{B} - \sqrt{1+x_\mathrm{B}^2}\right), \qquad \omega_{14} = \frac{\omega_\mathrm{iso}}{2}\left(1 + d_\gamma x_\mathrm{B} + \sqrt{1+x_\mathrm{B}^2}\right),$$
$$\omega_{34} = \frac{\omega_\mathrm{iso}}{2}\left(1 - d_\gamma x_\mathrm{B} + \sqrt{1+x_\mathrm{B}^2}\right), \qquad \omega_{23} = \frac{\omega_\mathrm{iso}}{2}\left(-1 + d_\gamma x_\mathrm{B} + \sqrt{1+x_\mathrm{B}^2}\right). \tag{8.23}$$

An inspection of Fig. 8.1 shows that these angular frequencies are all positive with the exception of ω_{12} whose sign depends on the value of x_B: ω_{12} is positive at low field and negative at high field.

Experimentally, the determination of $P_X^\mathrm{stat}(t)$ does not provide the sign of ω_{12}. This is so, because, as noted in Eq. 8.22, $P_X^\mathrm{stat}(t)$ only depends on $\cos(\omega_{12}t)$. To establish this sign it is possible to examine the dependence of ω_{12} upon x_B in order to determine whether x_B is lower or higher than the critical field where ω_{12} vanishes.

It may not be easy to distinguish four frequencies in a spectrum. However, simplifications usually occur. If ω_iso is beyond detection capability, ω_{14} and ω_{34} cannot be observed and only one pair of frequencies is left:

$$P_X^\mathrm{stat}(t) \simeq \frac{1}{4}\left\{\cos(\omega_{12}t) + \cos(\omega_{23}t) + \frac{x_\mathrm{B}}{\sqrt{1+x_\mathrm{B}^2}}\left[\cos(\omega_{12}t) - \cos(\omega_{23}t)\right]\right\}. \tag{8.24}$$

In low field, that is for $x_\mathrm{B} \ll 1$, and still if ω_{14} and ω_{34} cannot be detected,

$$P_X^\mathrm{stat}(t) \simeq \frac{1}{4}\left[\cos(\omega_{12}t) + \cos(\omega_{23}t)\right]. \tag{8.25}$$

From Eq. 8.23,

$$\omega_{23} + \omega_{12} = d_\gamma x_\mathrm{B}\omega_\mathrm{iso} = (-\gamma_\mathrm{e} - \gamma_\mu)B_\mathrm{ext}, \tag{8.26}$$

and

$$\omega_{23} - \omega_{12} = \omega_\mathrm{iso}\left(-1 + \sqrt{1+x_\mathrm{B}^2}\right) \simeq \frac{(-\gamma_\mathrm{e} + \gamma_\mu)^2 B_\mathrm{ext}^2}{2\,\omega_\mathrm{iso}}. \tag{8.27}$$

The hyperfine frequency can be determined from the two measured muonium frequencies:

$$\nu_\mathrm{iso} = \frac{1}{2}\left[\frac{(\nu_{23} + \nu_{12} + 2\nu_\mu)^2 - (\nu_{23} - \nu_{12})^2}{\nu_{23} - \nu_{12}}\right]. \tag{8.28}$$

No approximation has been made to derive this formula.

In the extreme low-field limit,

$$P_X^\mathrm{stat}(t) \simeq \frac{1}{2}\cos(\omega_\mathrm{eff}t). \tag{8.29}$$

Hence, only one angular frequency (in fact two almost degenerate frequencies, ω_{12} and ω_{23}) is measured with $\omega_\mathrm{eff} \simeq (\gamma_\mathrm{e}/2)B_\mathrm{ext}$. This polarisation function does not provide any information on the muonium state since $\omega_{\rm eff}$ does not depend on a muonium parameter. However, the detection of an oscillation with a precession about two orders of magnitude faster than that expected for muon $((|\gamma_{\rm e}|/2)/\gamma_\mu = 103.38)$ together with a 50% missing fraction is a clear signature of muonium formation.

According to Eq. 8.22, only up to two frequencies are detected at high field, that is, for $x_{\rm B} \gg 1$:

$$P_X^{\rm stat}(t) = \frac{1}{2}\left[\cos\left(\omega_{12}t\right) + \cos\left(\omega_{34}t\right)\right]. \tag{8.30}$$

At the field for which ω_{12} vanishes, i.e. $x_{\rm B} = 2d_\gamma/(1-d_\gamma^2) = 103.38$, only one frequency contributes to $P_X^{\rm stat}(t)$, that is, $\omega_{34} = \omega_{\rm iso}$. Although this condition cannot be reached for muonium in vacuum because a field of more than 16 T would be required, this is possible for some semiconductors or organic compounds. For instance, taking $\nu_{\rm iso} =$ 300 MHz, we compute for the crossing field a value of 1.10 T.

In the high-field limit ω_{12} is negative as already mentionned. It is therefore convenient to consider $\omega_{21} = -\omega_{12}$. From Eq. 8.23, without approximation,

$$\omega_{34} + \omega_{21} = \omega_{\rm iso}\left(\sqrt{1 + x_{\rm B}^2} - d_\gamma x_{\rm B}\right) \quad \text{and} \quad \omega_{34} - \omega_{21} = \omega_{\rm iso}. \tag{8.31}$$

The former expression reduces to

$$\omega_{34} + \omega_{21} \simeq \omega_{\rm iso}(1 - d_\gamma)x_{\rm B} = 2\omega_\mu. \tag{8.32}$$

Combining these results we get

$$\omega_{21} \simeq \omega_\mu - \frac{\omega_{\rm iso}}{2} \quad \text{and} \quad \omega_{34} \simeq \omega_\mu + \frac{\omega_{\rm iso}}{2}. \tag{8.33}$$

Hence, the two angular frequencies are centred at ω_μ and separated by $\omega_{\rm iso}$.

The observed transitions in a transverse field obey the selection rules

$$\Delta S_{\rm Mu} = 0, \pm 1 \quad \text{and} \quad \Delta M_{\rm Mu} = \pm 1, \tag{8.34}$$

provided the field is small enough for Eq. 8.16 to be a good approximation for the eigenstates. At larger fields, this approximation breaks down: quantum numbers $S_{\rm Mu}$ and $M_{\rm Mu}$ can no longer be defined. Nevertheless, if the field is large enough the $\{|\psi_i\rangle\}$ eigenstates (Eq. 8.13) tends to the Zeeman states (Eq. 8.8). Then, the following selection rules apply,

$$\Delta m_{\rm e} = 0, \pm 1 \quad \text{and} \quad \Delta m_\mu = 0, \pm 1, \tag{8.35}$$

which reduce to

$$\Delta m_{\rm e} = 0 \quad \text{and} \quad \Delta m_\mu = \pm 1, \tag{8.36}$$

if the two high-energy transitions, namely between states 1 and 4, and 2 and 3, respectively, cannot be detected.

Interestingly, muonium spectroscopy can be viewed as an electron–muon double resonance spectroscopy. The transitions of angular frequencies ω_{12} and ω_{34} are analogous to those used in the well-known electron–nuclear double resonance (ENDOR) spectroscopy — see e.g. Atherton (1993). This remark may help the reader who is accustomed to this technique.

8.3 Anisotropic muonium state

In the seventies, weak-transverse-field measurements performed on silicon revealed, besides the pair of frequencies corresponding to the well-known muonium state, an additional frequency pair with a much reduced splitting; see Fig. 1.19. It has been shown that this latter pair arises from an anisotropic muonium state. The purpose of this section is to discuss the signature of this state on $P_\alpha(t)$. Its physical origin is dealt with in Sections 12.1 and 12.4. Following the same procedure as for isotropic muonium, we start with the description of the system Hamiltonian.

8.3.1 Hamiltonian

The total Hamiltonian describing an anisotropic muonium state in a field is written as

$$\mathcal{H}_{\rm tot} = \hbar \mathbf{S}_{\rm e} \boldsymbol{A} \mathbf{S}_\mu + g_{\rm e}\mu_{\rm B} B_{\rm ext}\, S_{\rm e}^Z - \hbar\gamma_\mu B_{\rm ext} S_\mu^Z, \tag{8.37}$$

where $\boldsymbol{A}$ is now a tensor. If the tensor is of rank zero, i.e. a scalar, we recover in zero field the Hamiltonian given in Eq. 8.4 with a single parameter $A = \omega_{\rm iso}$. Here we have neglected the anisotropy of the electronic spectroscopic splitting factor. This is a good approximation for germanium and silicon for which it is found to be only about 1 and 0.01%, respectively (Blazey *et al.* 1986). Likewise, the spectroscopic factor is also isotropic and close to 2 for many organic radicals. However, Lord *et al.* (2004) have reported strong deviations from the free electron value for a number of semiconductors of the II–VI family.

Fortunately, in most cases of physical interest (see Chapter 12), a model Hamiltonian with two parameters is sufficient for the description of the zero-field hyperfine Hamiltonian of an anisotropic muonium state, $\mathcal{H}_{\rm ani}$. Quartz at low temperature is a counterexample for which this minimum model does not apply (Brewer *et al.* 1981). Here we shall only consider the two-parameter model which can be written in two equivalent forms:

$$\mathcal{H}_{\rm ani} = \hbar\omega_\parallel S_{\rm e}^z S_\mu^z + \hbar\omega_\perp \left(S_{\rm e}^x S_\mu^x + S_{\rm e}^y S_\mu^y\right), \tag{8.38}$$

and

$$\mathcal{H}_{\rm ani} = \hbar\omega_{\rm dip}\left[(\mathbf{S}_{\rm e}\cdot\mathbf{n})(\mathbf{S}_\mu\cdot\mathbf{n}) - \frac{1}{3}\mathbf{S}_{\rm e}\cdot\mathbf{S}_\mu\right] + \hbar\omega_{\rm iso}\mathbf{S}_{\rm e}\cdot\mathbf{S}_\mu. \tag{8.39}$$

The unit vector $\mathbf{n}$ is taken parallel to the symmetry axis of the muonium state which is the z axis, since the reference frame $\{x, y, z\}$ is chosen according to the symmetry of the muonium state. The parameters of the two Hamiltonians are related:

$$\omega_{\rm iso} = \frac{\omega_\parallel + 2\omega_\perp}{3} \quad \text{and} \quad \omega_{\rm dip} = \omega_\parallel - \omega_\perp. \tag{8.40}$$

The second form (Eq. 8.39) is appealing, since the anisotropic part of the Hamiltonian is found to be a dipolar-like Hamiltonian that can be represented as a spherical tensor operator of rank two. Such a tensor transforms under rotation according to a well-known rule; see e.g. Edmonds (1957) or Messiah (1965). The second term on the right-hand side of Eq. 8.39 is a scalar tensor operator and so is invariant under rotation. Therefore this model assumes an axial symmetry for the muonium state. This reflects the symmetry at the muonium site. This is further discussed in Chapter 12.

We first take $\mathbf{n}$ to be parallel to the Z axis, i.e. to the applied field. This is also the axis of quantisation used in this monograph. The matrix representation of $\mathcal{H}_{\rm tot}$ in the $\{|m_e, m_\mu\rangle\}$ basis is easily obtained:

$$\mathcal{H}_{\rm tot} = \frac{\hbar\omega_{\rm iso}}{2}\begin{pmatrix} \frac{1}{2} + d_\gamma x_{\rm B} + p_{\rm ani} & 0 & 0 & 0 \\ 0 & -\frac{1}{2} + x_{\rm B} - p_{\rm ani} & 1 - p_{\rm ani} & 0 \\ 0 & 1 - p_{\rm ani} & -\frac{1}{2} - x_{\rm B} - p_{\rm ani} & 0 \\ 0 & 0 & 0 & \frac{1}{2} - d_\gamma x_{\rm B} + p_{\rm ani} \end{pmatrix}. \tag{8.41}$$

We have introduced the anisotropy ratio parameter $p_{\rm ani} = \omega_{\rm dip}/(3\omega_{\rm iso})$. The four eigenvalues are given by

$$E_1 = \frac{\hbar\omega_{\rm iso}}{2}\left(\frac{1}{2} + p_{\rm ani} + d_\gamma\, x_{\rm B}\right), \qquad E_3 = \frac{\hbar\omega_{\rm iso}}{2}\left(\frac{1}{2} + p_{\rm ani} - d_\gamma\, x_{\rm B}\right),$$

$$E_2 = \frac{\hbar\omega_{\rm iso}}{2}\left[-\frac{1}{2} - p_{\rm ani} + \sqrt{(1-p_{\rm ani})^2 + x_{\rm B}^2}\right],$$

$$E_4 = \frac{\hbar\omega_{\rm iso}}{2}\left[-\frac{1}{2} - p_{\rm ani} - \sqrt{(1-p_{\rm ani})^2 + x_{\rm B}^2}\right]. \tag{8.42}$$

In zero field, the triplet excitation state of isotropic muonium is split into a doublet and a singlet by the dipole anisotropy, the doublet state of energy E_1 $(= E_3)$ being at higher energy than the singlet state with energy E_2 if $p_{\rm ani} > 0$. On the other hand, this latter state is higher in energy if $p_{\rm ani} < 0$. Breit–Rabi diagrams for an anisotropic muonium state are shown in Fig. 8.3 for two values of $p_{\rm ani}$.

The associated eigenvectors are derived as

$$|\psi_1\rangle = \left|\frac{1}{2}, \frac{1}{2}\right\rangle, \qquad |\psi_3\rangle = \left|-\frac{1}{2}, -\frac{1}{2}\right\rangle,$$

$$|\psi_2\rangle = \frac{1}{\sqrt{D_{\rm aux}^2 + (1-p_{\rm ani})^2}}\left[(1-p_{\rm ani})\left|\frac{1}{2}, \frac{-1}{2}\right\rangle + D_{\rm aux}\left|-\frac{1}{2}, \frac{1}{2}\right\rangle\right],$$

$$|\psi_4\rangle = \frac{1}{\sqrt{D_{\rm aux}^2 + (1-p_{\rm ani})^2}}\left[D_{\rm aux}\left|\frac{1}{2}, -\frac{1}{2}\right\rangle - (1-p_{\rm ani})\left|-\frac{1}{2}, \frac{1}{2}\right\rangle\right]. \tag{8.43}$$

We use the auxiliary parameter $D_{\rm aux} = \sqrt{(1-p_{\rm ani})^2 + x_{\rm B}^2} - x_{\rm B}$. A singular case is $p_{\rm ani} = 1$ (i.e. $\omega_{\rm dip} = 3\omega_{\rm iso}$) where the eigenvectors are the $|\chi_i\rangle$ vectors. This is easily understood since only the components $S_{\rm e}^z$ and S_μ^z are then present in $\mathcal{H}_{\rm ani}$ (Eq. 8.38). Besides this singular case, the model we are analysing is characterised by eigenvectors with coherent mixture of the $|\chi_i\rangle$ states similar to that for the isotropic muonium model. Therefore the same type of transitions are probed by μSR measurements. This is a peculiarity of the model under consideration: the observed transitions depend on the orientation of the principal axes of the muonium tensor relative to the laboratory reference frame axes. This is further discussed in the next section.

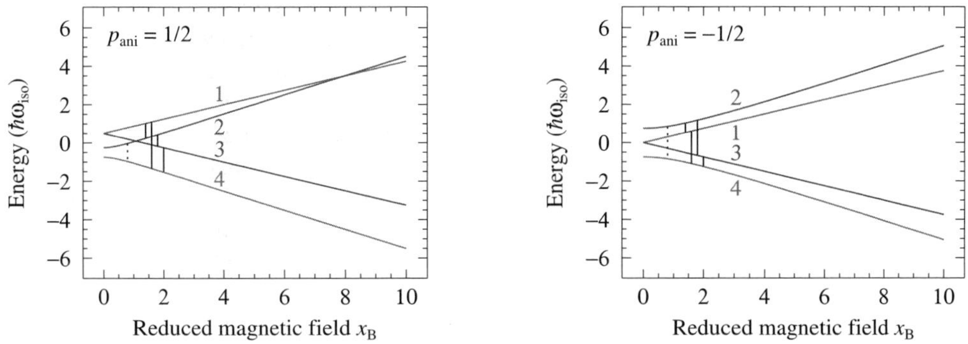

Fig. 8.3 Examples of Breit–Rabi diagrams for an axial muonium state. We assume an anisotropy parameter $p_{\rm ani} = \omega_{\rm dip}/(3\omega_{\rm iso})$ equal either to $\frac{1}{2}$ (left) or $-\frac{1}{2}$ (right). The direction of axial symmetry is taken along the applied field. The eigenvalues are plotted versus the dimensionless external field parameter $x_{\rm B} = B_{\rm ext}/\tilde{B}_{\rm iso}$. As in Fig. 8.1, for ease of presentation, we have used the non-physical value $d_\gamma = \frac{3}{4}$. The transition which can be detected by a longitudinal-field measurement is indicated by a dotted line and the four transitions contributing to the transverse-field polarisation are indicated by full lines.

8.3.2 Polarisation functions

We first provide derivations of the polarisation functions for muonium states in some simple cases, then we consider the high-field limit and complete with a summary and an outlook.

8.3.2.1 Simple physical cases. Let us first consider the polarisation functions when $\mathbf{n}$ is parallel to the Z axis.

Following the same methodology as in Section 8.2.2 we compute $P_Z^{\rm stat}(t)$:

$$P_Z^{\rm stat}(t) = \frac{1}{2}\left\{1 + \frac{1}{(1-p_{\rm ani})^2 + x_{\rm B}^2}\left[x_{\rm B}^2 + (1-p_{\rm ani})^2 \cos(\omega_{24} t)\right]\right\}, \tag{8.44}$$

with

$$\omega_{24} = \omega_{\rm iso}\sqrt{(1-p_{\rm ani})^2 + x_{\rm B}^2}. \tag{8.45}$$

As expected, Eq. 8.18 is recovered when $p_{\rm ani} = 0$. The case $p_{\rm ani} = 1$ (i.e. $\omega_{\rm dip} = 3\omega_{\rm iso}$) is again singular as no oscillation is present in $P_Z^{\rm stat}(t)$.

In Fig. 8.4 we plot the relative amplitude of the oscillating component for a few $p_{\rm ani}$ values. If $x_{\rm B}$ is large, i.e. at high field, the amplitude vanishes because the field quenches the time evolution of $P_Z^{\rm stat}(t)$. In zero field, 50% of the $P_Z^{\rm stat}(t)$ amplitude is oscillating, provided the anisotropy parameter is not equal to 1. In the case $p_{\rm ani} = 1$, $P_Z^{\rm stat}(t)$ is independent of time and equal to 1.

The repolarisation function for the model we are considering is simply obtained by dropping the cosine term of the previous equation. As for an isotropic muonium state, the polarisation is $\frac{1}{2}$ in zero field and monotonously increases with the field.

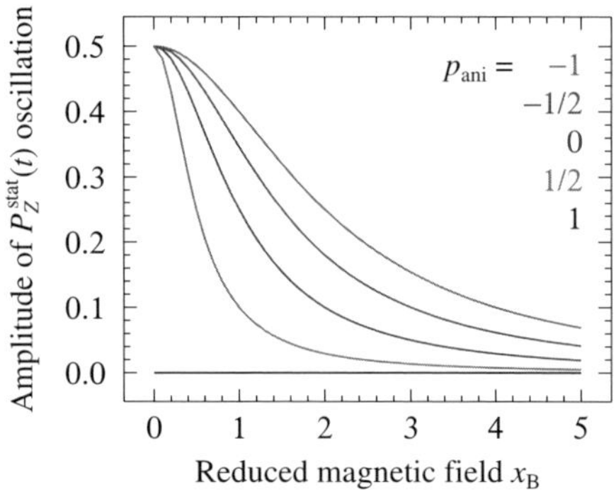

Fig. 8.4 Amplitude of the oscillating component of $P_Z^{\rm stat}(t)$ versus the dimensionless external field parameter $x_{\rm B} = B_{\rm ext}/\tilde{B}_{\rm iso}$ for a few values of the anisotropy parameter $p_{\rm ani} = \omega_{\rm dip}/(3\omega_{\rm iso})$, which are indicated in the same order as the lines. The muonium axis of symmetry and the external field are assumed to be parallel.

The evaluation of $P_X^{\rm stat}(t)$ gives

$$P_X^{\rm stat}(t) = \frac{1}{2}\left\{\frac{(1-p_{\rm ani})^2}{D_{\rm aux}^2 + (1-p_{\rm ani})^2}\left[\cos(\omega_{12}t) + \cos(\omega_{34}t)\right]\right.$$
$$\left. + \frac{D_{\rm aux}^2}{D_{\rm aux}^2 + (1-p_{\rm ani})^2}\left[\cos(\omega_{23}t) + \cos(\omega_{14}t)\right]\right\}, \tag{8.46}$$

with

$$\omega_{12} = \frac{\omega_{\rm iso}}{2}\left(1 + 2p_{\rm ani} + d_\gamma x_{\rm B} - \sqrt{(1-p_{\rm ani})^2 + x_{\rm B}^2}\right),$$
$$\omega_{14} = \frac{\omega_{\rm iso}}{2}\left(1 + 2p_{\rm ani} + d_\gamma x_{\rm B} + \sqrt{(1-p_{\rm ani})^2 + x_{\rm B}^2}\right),$$
$$\omega_{34} = \frac{\omega_{\rm iso}}{2}\left(1 + 2p_{\rm ani} - d_\gamma x_{\rm B} + \sqrt{(1-p_{\rm ani})^2 + x_{\rm B}^2}\right),$$
$$\omega_{23} = \frac{\omega_{\rm iso}}{2}\left(-1 - 2p_{\rm ani} + d_\gamma x_{\rm B} + \sqrt{(1-p_{\rm ani})^2 + x_{\rm B}^2}\right). \tag{8.47}$$

If $p_{\rm ani} = 1$, $P_X^{\rm stat}(t) = [\cos(\omega_{12}t) + \cos(\omega_{34}t)]/2$. If $p_{\rm ani}$ is not equal to 1, four transitions are detected as for an isotropic muonium state. A transition between the states labelled by 1 and 3 cannot be observed in zero and transverse fields because of the pure Zeeman character of the eigenvectors of these states.

As a homework for the interested reader we leave the task of deriving the low- and high-field limits of $P_X^{\rm stat}(t)$. It can be achieved following the framework we followed in Section 8.2.2 for the isotropic case. We note that the sign of the angular frequencies reported in Eq. 8.47 might change as a function of $x_{\rm B}$ and $p_{\rm ani}$. Figure 8.3 provides an illustration of this remark. Some care must therefore be exercised in the derivation of the limits. As a matter of fact, if

$$p_{\rm ani} < \frac{\sqrt{1-d_\gamma^2}-1}{2+\sqrt{1-d_\gamma^2}}, \tag{8.48}$$

i.e. $p_{\rm ani} \lesssim -0.4$, which is a common case, ω_{12} is negative for any value of $x_{\rm B}$.

We now study the general case where vector $\mathbf{n}$ is not parallel to the Z axis. We shall restrict ourselves to the determination of $P_Z^{\rm stat}(t)$ in zero field. We denote the polar and azimuthal angles of $\mathbf{n}$ in the $\{X,Y,Z\}$ frame as θ and φ, respectively, and the hyperfine Hamiltonian of the anisotropic muonium state as $\mathcal{H}_{\rm ani}(\varphi,\theta)$. Until now we considered the case $\theta = 0$. According to Eq. 8.17, $P_Z^{\rm stat}(t)$ is expressed as the trace of an operator:

$$P_Z^{\rm stat}(t) = \frac{1}{4}{\rm Tr}\left\{\sigma_\mu^Z \exp\left[i\frac{\mathcal{H}_{\rm ani}(\varphi,\theta)\,t}{\hbar}\right]\sigma_\mu^Z \exp\left[-i\frac{\mathcal{H}_{\rm ani}(\varphi,\theta)\,t}{\hbar}\right]\right\}. \tag{8.49}$$

Since we previously derived a matrix representation for $\mathcal{H}_{\rm ani} = \mathcal{H}_{\rm ani}(\theta = 0)$, it is interesting to express $P_Z^{\rm stat}(t)$ in terms of $\mathcal{H}_{\rm ani}$ rather than $\mathcal{H}_{\rm ani}(\varphi,\theta)$. Quite generally, this is performed with the help of the rotation operator, $R(\varphi,\theta,\psi = 0)$, defined in Appendix C (see also Fig. D.1): $R\mathcal{H}_{\rm ani}R^{-1} = \mathcal{H}_{\rm ani}(\varphi,\theta)$. Then, according to Eq. C.12,

$$R\exp\left(\pm i\frac{\mathcal{H}_{\rm ani}\,t}{\hbar}\right)R^{-1} = \exp\left[\pm i\frac{\mathcal{H}_{\rm ani}(\varphi,\theta)\,t}{\hbar}\right]. \tag{8.50}$$

Inserting this result into Eq. 8.49 and using Eq. C.10,

$$P_Z^{\rm stat}(t) = \frac{1}{4}{\rm Tr}\left\{R^{-1}\sigma_\mu^Z R\exp\left(i\frac{\mathcal{H}_{\rm ani}\,t}{\hbar}\right)R^{-1}\sigma_\mu^Z R\exp\left(-i\frac{\mathcal{H}_{\rm ani}\,t}{\hbar}\right)\right\}. \tag{8.51}$$

Because the trace of an operator is independent of the basis on which it is evaluated, we choose for convenience the basis formed by the eigenvectors $|\psi_i\rangle$ because $\mathcal{H}_{\rm ani}|\psi_i\rangle = E_i|\psi_i\rangle$:

$$\begin{aligned} P_Z^{\rm stat}(t) &= \frac{1}{4}\sum_i \langle\psi_i|\,R^{-1}\sigma_\mu^Z R\exp\left(i\frac{\mathcal{H}_{\rm ani}\,t}{\hbar}\right)R^{-1}\sigma_\mu^Z R\exp\left(-i\frac{\mathcal{H}_{\rm ani}\,t}{\hbar}\right)|\psi_i\rangle \\ &= \frac{1}{4}\sum_{i,j}\left|\langle\psi_i|\,R^{-1}\sigma_\mu^Z R\,|\psi_j\rangle\right|^2 \exp\left[i\frac{(E_j-E_i)\,t}{\hbar}\right]. \end{aligned} \tag{8.52}$$

The closure relation $\sum_j |\psi_j\rangle\langle\psi_j| = \mathcal{I}$ has been used. The rotation operator R is conveniently expressed as a direct product of rotation operators $R_{\rm e}$ and R_μ acting on the electron and muon spin states, respectively. Since $R_{\rm e}$ commutes with σ_μ^Z, only R_μ is to be considered, simplifying the evaluation of $P_Z^{\rm stat}(t)$. The matrix representation of R_μ in the space spanned by the muon states $\{|\frac{1}{2}\rangle, |-\frac{1}{2}\rangle\}$ is quite simple (see Eq. C.22):

$$R_\mu(\varphi,\theta,0) = \begin{pmatrix} \exp\left(-i\frac{\varphi}{2}\right)\cos\left(\frac{\theta}{2}\right) & -\exp\left(-i\frac{\varphi}{2}\right)\sin\left(\frac{\theta}{2}\right) \\ \exp\left(-i\frac{\varphi}{2}\right)\sin\left(\frac{\theta}{2}\right) & \exp\left(i\frac{\varphi}{2}\right)\cos\left(\frac{\theta}{2}\right) \end{pmatrix}. \tag{8.53}$$

Combining Eqs. 8.52 and 8.53, $P_Z^{\rm stat}(t)$ is then computed. Transitions between the following levels are involved: $1 \leftrightarrow 2$, $1 \leftrightarrow 4$, $2 \leftrightarrow 3$, $2 \leftrightarrow 4$, and $3 \leftrightarrow 4$. The corresponding differences in energy can be computed from Eqs. 8.45 and 8.47. Recalling that we consider the zero-field polarisation function, it is found that only three independent angular frequencies are involved irrespective of the sign of $(1 - p_{\rm ani})$. Altogether we arrive at

$$P_Z^{\rm stat}(t) = \frac{1}{2}\left\{\cos^2\theta\,[1+\cos(\Omega_1 t)] + \sin^2\theta\,[\cos(\Omega_2 t) + \cos(\Omega_3 t)]\right\}, \tag{8.54}$$

with

$$\Omega_1 = (1 - p_{\rm ani})\,\omega_{\rm iso}, \quad \Omega_2 = \frac{3p_{\rm ani}\omega_{\rm iso}}{2}, \quad \text{and} \quad \Omega_3 = (2 + p_{\rm ani})\,\frac{\omega_{\rm iso}}{2}. \tag{8.55}$$

As expected, $P_Z^{\rm stat}(t)$ does not depend on the azimuthal angle.

Of course, if $\mathbf{n}$ is along the Z axis, we are back to our previous model with a single precession frequency: we recover Eqs. 8.44 and 8.45 with $x_{\rm B} = 0$ and $|\Omega_1| = \omega_{24}$. When $\mathbf{n}$ is perpendicular to Z we have two precession frequencies, except if $p_{\rm ani} = 1$, in which case the two frequencies are degenerate. If $\mathbf{n}$ is neither parallel nor perpendicular to the Z axis, the time evolution of $P_Z^{\rm stat}(t)$ in zero field is characterised by three distinct frequencies.

The study of the angular dependence of $P_Z^{\rm stat}(t)$ provides a method for the determination of $\omega_{\rm iso}$ and $p_{\rm ani}$. Finally, note that

$$\Omega_2 = \frac{3p_{\rm ani}}{2}\omega_{\rm iso} = \frac{\omega_{\rm dip}}{2}. \tag{8.56}$$

Hence, zero-field experiments give a direct access to the absolute value of the dipole anisotropy of the hyperfine interaction.

In general the repolarisation curve for an anisotropic muonium state is quite different from that obtained for isotropic muonium. The shape of the curve depends on the type of anisotropy as well as on the orientation of the principal axes of the hyperfine tensor relative to the laboratory reference frame (Pratt 1997). It is quite common to observe a two-stage repolarisation. The polarisation vanishes in zero field in many cases. In addition to the anisotropy of the hyperfine interaction, the superhyperfine interaction (see Section 8.4 for a discussion of this interaction) may also contribute to the suppression of the low-field polarisation, as explained by Beck *et al.* (1975).

8.3.2.2 High-field limit. Many experiments are performed in the high-field limit, that is, when the electronic Zeeman energy is larger than the hyperfine energy. As we are going to explain now, a simple effective Hamiltonian valid in this limit can be written.

We start with the Hamiltonian given in Eq. 8.37. To proceed, we need to express the hyperfine part of the Hamiltonian written in the muonium frame in Eq. 8.38 into the laboratory reference frame $\{X, Y, Z\}$. For this purpose we shall use the $R^{-1} = R^{\dagger}$ rotation matrix where R is given in Section D.2. However, a couple of simplifications can be made beforehand.

On the one hand, to first order and in the high-field limit, the eigenvectors of the Hamiltonian are the $|\chi_i\rangle$ wavefunctions. Therefore, those terms involving $S_{\rm e}^X$ and $S_{\rm e}^Y$ possess zero diagonal elements, and may be omitted. On the other hand, because of the invariance of $\mathcal{H}_{\rm ani}$ in the plane perpendicular to z, and because we are interested in $P_Z^{\rm stat}(t)$, it is always possible to choose the axes Y and y parallel. The Euler angles φ and ψ defined in Fig. D.1 can therefore be set to 0. Hence, only the third angle, θ, i.e. the angle between the z and Z axes, must be considered. Then we have

$$\begin{aligned} S_{\rm e}^x &= \sin\theta\, S_{\rm e}^Z, & S_{\rm e}^y &= 0, & S_{\rm e}^z &= \cos\theta\, S_{\rm e}^Z, \\ S_\mu^x &= \cos\theta\, S_\mu^X + \sin\theta\, S_\mu^Z, & S_\mu^y &= S_\mu^Y, & S_\mu^z &= \cos\theta\, S_\mu^Z - \sin\theta\, S_\mu^X. \end{aligned} \tag{8.57}$$

Substituting these two sets of transformations in the expression of $\mathcal{H}_{\rm tot}$, we derive the effective Hamiltonian

$$\check{\mathcal{H}}_{\rm tot} = -\gamma_\mu \hbar \check{\mathbf{B}}_{\rm loc} \cdot \mathbf{S}_\mu + g_{\rm e}\mu_{\rm B} B_{\rm ext}\, S_{\rm e}^Z, \tag{8.58}$$

with the field operator $\check{\mathbf{B}}_{\rm loc}$ acting on the muon spin of components

$$\begin{aligned} \check{B}_{\rm loc}^X &= \frac{\omega_\parallel - \omega_\perp}{\gamma_\mu} \sin\theta\cos\theta\, S_{\rm e}^Z, \qquad \check{B}_{\rm loc}^Y = 0, \\ \check{B}_{\rm loc}^Z &= B_{\rm ext} - \frac{\omega_\parallel \cos^2\theta + \omega_\perp \sin^2\theta}{\gamma_\mu} S_{\rm e}^Z. \end{aligned} \tag{8.59}$$

The high-field approximation results in a drastic simplification: $\check{B}_{\rm loc}^X$ no longer depends on the transverse components of the electron spin operators; i.e. only the diagonal operator $S_{\rm e}^Z$ appears in $\check{B}_{\rm loc}^X$. This simplification is of great help for the study of $\check{\mathcal{H}}_{\rm tot}$. We leave it as an exercise to the interested reader.

It is worthwile to further examine $\check{\mathbf{B}}_{\rm loc}$. In general, this field is not parallel to the external field since $\check{B}_{\rm loc}^X$ is different from 0 if the muonium state is anisotropic and the anisotropy axis is not parallel or perpendicular to the Z axis. In fact, $\check{\mathbf{B}}_{\rm loc}$ strongly depends on the sign of $m_{\rm e}$. We note that $m_{\rm e}$ is positive, i.e. $m_{\rm e} = \frac{1}{2}$, with a probability of one-half. Let us consider the field value $B_{\rm ext} = (\omega_\parallel \cos^2\theta + \omega_\perp \sin^2\theta)/(2\gamma_\mu)$. At this critical field, there is a 50% of chance that the local effective field at the muonium site is fully transverse. This is a characteristic of anisotropic muonium states. The matrix representation of the Hamiltonian separates into two irreducible (independent) 2×2 submatrices in the $\{|\chi_i\rangle\}$ basis. Therefore it is always possible to derive analytical expressions for the polarisation functions. We again leave the exercise to the interested reader. The physics of anisotropic muonium states in a high field is further discussed at the end of this chapter when considering relaxation effects; see Sections 8.4.2 and 8.4.3.

8.3.2.3 Summary and outlook. We have derived expressions for the polarisation functions of an axial muonium state with its axial symmetry axis parallel to the Z axis of the laboratory reference frame. In addition, the zero-field response to a rotation of the axial symmetry axis as well as the high-field limit for any orientation of the symmetry axis has been studied. These three physical cases cover most experimental situations. It is also possible to derive analytical formulae for other experimental conditions. We refer the interested reader to the published literature, i.e. Patterson (1988), Smilga and Belousov (1994), and Senba (2001). In practice, the number of physical cases for which analytical expressions can be derived is limited. This is not much of an issue since the matrix representation of $\mathcal{H}_{\rm tot}$ leads to a square 4×4 matrix which can be numerically diagonalised.

8.4 Muonium centres in matter

Here, we discuss muonium states in matter where they are subject to interactions with their environment. We shall refer them to muonium centres.

Because of strong depolarisation, muonium centres in magnetic materials are usually not observed. Hence, they will not be considered in this section. With the exception of radiolysis products and charge carriers, the spins of the neighbouring atoms interacting with muonium will only be nuclear spins.

Except for the renormalisation of the strength of the hyperfine interaction and its anisotropy, the effect of the environment of the muonium centres on their properties has been neglected until now. The derived results cannot be entirely satisfactory. Muonium centres are submitted to interactions and, in addition, their dynamical character should be taken into account. The interactions of the most potential significance are of two origins. One is the so-called superhyperfine interaction. It arises from the Fermi contact interaction between the spin of the muonium electron and the spins of neighbouring nuclei if there is a finite probability density for the muonium electron to be at these sites. The second interaction is the classical dipole interaction of the muonium-electron spin with the neighbouring nuclear spins. Since $|\gamma_{\rm e}| = g_{\rm e}\mu_B/\hbar \gg \gamma_\mu$, the dipolar interaction between the muon spin and the neighbour nuclear spins can be safely neglected. A possible interaction between the quadrupole moment associated with the muonium electronic cloud and an electric field gradient has been considered (Baryshevskii *et al.* 1982). However, since that interaction has not been definitively recognised experimentally, it will not be considered further here. Finally, the finite chemical lifetime of muonium should be kept in mind: a muonium state may decay to a diamagnetic muon state or another muonium species before the muon decay.

The way the nuclear magnetic moments surrounding a muonium centre influence $P_\alpha(t)$ depends on their energetics. If the interaction of the nuclear moments with the muonium centre is sufficiently strong, they form with the muon and the muonium electron a well-defined multi-spin system, as found in some organic compounds. Then the polarisation functions may display additional transition frequencies, which characterise the muonium–nuclear coupling as explained in Section 8.4.1. Resonances between the muonium and the nuclear spins are possible at characteristic applied magnetic fields. The physics behind that mechanism has been physically explained in Section 4.2 and is detailed for muonium centres in Section 8.4.2. In the same section, muonium lifetime effects are considered. If the interaction of the nuclear moments with the muonium centre is weak, their directions are stochastically distributed, as in diamagnetic compounds. This induces depolarisation of $P_\alpha(t)$, which will be treated in Section 8.4.3. The muonium–phonon coupling which drives the muonium mobility will not be treated here. A discussion of that matter is deferred to Chapter 9.

8.4.1 Multi-spin systems

The strong interaction of a muonium centre with neighbouring nuclear spins can lead to complex μSR spectra as explained by Roduner (1999). Given a Hamiltonian for a multi-spin system, there is *a priori* no difficulty to numerically compute the μSR observables, that is, $P_\alpha^{\rm stat}(t)$. However, a good picture of what is occurring is already obtained by studying the simple model of an isotropic muonium interacting with a single nuclear spin $\frac{1}{2}$ such as a proton. As a first step, it is convenient to assume the electron–nuclear hyperfine interaction to be isotropic. The dipole interaction is usually weak relative to the hyperfine coupling. The additional spin doubles the number of states, and in zero field they form three groups, a quartet ($F = \frac{3}{2}$) and two doublet ($F = \frac{1}{2}$) states. Here F is the quantum number for the total spin angular momentum of the muonium–proton system. The nomenclature $|F, M_F\rangle$ is exact only in zero field but it is also often used in low fields. In high fields, we use the Zeeman states, $\{|m_{\rm e}, m_p, m_\mu\rangle\}$. This is in agreement with the results for the isotropic muonium model; see Eq. 8.13.

As usual, the allowed transitions between the eigenstates of the muonium–proton system must fulfill selection rules. These rules have been derived for the isotropic muonium model; see Eqs. 8.21 and 8.34 through 8.36 for $P_Z^{\rm stat}(t)$ in zero and low field and $P_X^{\rm stat}(t)$ in low and high field, respectively. Adapted to the muonium–proton system — see e.g. Roduner (1999) — they are

$$\Delta F = 0, \pm 1 \quad \text{and} \quad \Delta M_F = 0 \quad \text{in zero and low field} \tag{8.60}$$

for $P_Z^{\rm stat}(t)$ and

$$\begin{aligned} &\Delta F = 0, \pm 1 \quad \text{and} \quad \Delta M_F = \pm 1 \quad \text{in low field,} \\ &\Delta m_{\rm e} = 0 \quad \text{and} \quad \Delta m_\mu = \pm 1 \quad \text{in high field,} \end{aligned} \tag{8.61}$$

for $P_X^{\rm stat}(t)$. Similarly to the muonium case, certain transitions might not be observable if they are at too high energy. For $P_Z^{\rm stat}(t)$ we have added the condition $\Delta F = 0$ which is irrelevant for the isotropic muonium model since the three states of the triplet muonium are degenerate; hence no transition can be detected. As for isotropic muonium, there are no allowed transitions in high longitudinal fields.

With these rules, it is a simple matter to deduce that three transitions are allowed in zero field, and it is obvious that the highest frequency is the sum of the other two. Four transitions with non-negligible amplitude contribute at high field to the transverse polarisation function. However, these transitions are doubly degenerate so that only two frequencies can be detected. The sum of these two frequencies is exactly equal to $\nu_{\rm iso}$. Therefore, while a transverse-field measurement performed at high field is not sensitive to the coupling of the nuclear and muonium-electron magnetic moments, it provides a direct measurement of the muon–electron coupling constant. The number of allowed transitions at intermediate field intensity is large: 15 for our simple model. The muon polarisation is then a weighted sum of many oscillations, which means that the weight of each of them can be so reduced that it cannot experimentally be observed. A zero-field spectrum may become complex as well if too many nuclear moments are coupled to the muonium-electron magnetic moment.

The high-field limit can be reached with a moderate field intensity for most cases of interest. For isotropic muonium the field scale is set by the field ratio $x_{\rm B} = B_{\rm ext}/\tilde{B}_{\rm iso}$, where $\tilde{B}_{\rm iso}$ is proportional to $\nu_{\rm iso}$; see Eq. 8.9. Assuming $x_{\rm B} \gtrsim 10$ to be the high-field limit, a field larger than 1.6 T must be applied for muonium in vacuum. However, in organic compounds $\nu_{\rm iso}$ is at least an order of magnitude less than in vacuum and therefore the necessary field is much less than 1 T.

The spectroscopy of multi-spin systems needs optimum conditions to be successful. To spectrally resolve the transitions, the hyperfine coupling must be strong enough. In addition, for the amplitudes of the transitions to exceed the noise level, their number should not be too large. This explains why the measurements are usually restricted to high field or at least to a sufficiently strong field so as to avoid the distribution of the spectral weight over too many lines.

8.4.2 Resonance effects

As explained in Section 4.2 using a simplified model and in Sections 7.1.1 and 7.2, resonances can occur between the muon spin and its surrounding spins. A resonance may efficiently be detected by the time-integrated method running in longitudinal field. Scanning the field at relatively high fields, the μSR signal is field independent, except at some critical field, $B_{\rm cri}$, for which the muon spin flips and exchanges its energy with its environment. Physically, at $B_{\rm cri}$ two energy levels of the muon-surrounding system are strongly hybridised due to the presence of non-secular terms in the system Hamiltoninan. This hybridisation precludes the crossing of the two energy levels; i.e. an energy gap appears. This gap is directly proportional to the energy of the muon spin in the local field. The avoided-level-crossing technique (ALC) that we refer to here is mostly used in the study of muonium centres. It permits the analysis of the coupling between the muonium-electron spin and its surrounding nuclear spins. In practice, a single resonance is not detected, but rather an ensemble of them. They determine the energy levels associated with eigenstates of an appreciable amplitude. To detect a resonance a parameter must be scanned. Until now, it has been the applied magnetic field. However, it is conceivable that an applied pressure could trigger a resonance at a fixed applied field. A concise review on this subject has been written by Roduner (1999).

Relative to the multi-spin systems discussed in the previous section, the physics we are considering now is not very different, except that here we are interested in the resonance phenomena that occurs for specific conditions. We could simply write down the Hamiltonian of the system, compute numerically the integrated signal for the longitudinal geometry using Eq. 4.12, and look for resonances. Although the data are ultimately analysed based on the numerical computation, a simple analytical analysis of the Hamiltonian provides an estimate of the fields at which resonances can be expected as well as the corresponding energy gap.

We shall detail the derivation of the resonance condition for a simple system: a muon coupled to an electron by an isotropic hyperfine interaction, this electron interacting with a nearby nucleus through an isotropic nuclear hyperfine interaction, the so-called superhyperfine interaction. We are therefore restricting ourselves to a three-body problem: a muon, an electron, and a nucleus. This latter approximation is not really restrictive since the resonance condition selects the nucleus (or equivalent nuclei) which is probed. The resonances due to inequivalent nuclei do not usually overlap and therefore can separately be treated. Hence, according to Eq. 8.7,

$$\frac{\mathcal{H}_{\rm tot}}{\hbar} = \omega_{\rm iso}\,\mathbf{S}_{\rm e}\cdot\mathbf{S}_\mu + \frac{g_{\rm e}\mu_{\rm B}}{\hbar}B_{\rm ext}\,S^Z_{\rm e} - \gamma_\mu B_{\rm ext}S^Z_\mu + \omega_{\rm nhf}\,\mathbf{S}_{\rm e}\cdot\mathbf{I} - \gamma_n B_{\rm ext}I^Z, \qquad (8.62)$$

where we have added the nuclear hyperfine interaction term between the muonium electron and the nucleus, as well as the nuclear Zeeman term. In Eq. 8.62 we have neglected the interaction between the muon and nuclear spins. It is sometimes convenient to view the nuclear spin in interaction with the muonium electron as the source of an effective magnetic field acting on the muonium electron, with a field operator $\mathbf{B}_{\rm nhf} = \hbar\omega_{\rm nhf}\mathbf{I}/(g_{\rm e}\mu_{\rm B})$.

We are basically interested in the muon response at high field. Hence, according to our previous study of the isolated isotropic muonium state, see Section 8.2, the pure Zeeman states, i.e. $\{|m_{\rm e}, m_n, m_\mu\rangle\}$, are a good approximation to the actual states. At first order in perturbation theory, the energy of a state is given by $E_{m_{\rm e},m_n,m_\mu} = \langle m_\mu, m_n, m_{\rm e}|\mathcal{H}_{\rm tot}|m_{\rm e}, m_n, m_\mu\rangle$, that is,

$$\frac{E_{m_{\rm e},m_n,m_\mu}}{\hbar} = \frac{g_{\rm e}\mu_{\rm B}}{\hbar}B_{\rm ext}m_{\rm e} - B_{\rm ext}\left(\gamma_\mu m_\mu + \gamma_n m_n\right) + m_{\rm e}\left(\omega_{\rm iso}m_\mu + \omega_{\rm nhf}m_n\right). \qquad (8.63)$$

Because of a dominating electronic Zeeman energy, the energy levels form two well-separated groups according to the sign of $m_{\rm e}$. Therefore only the muon and nuclear spins can mutually flip. The observable resonance should follow the selection rule $\Delta M = \Delta(m_\mu + m_n) = 0$. This describes muon–nuclear spin flip-flop transitions in which the spin polarisation oscillates between the muon and the (magnetic) nucleus in resonance, mediated by the muonium electron. Scanning $B_{\rm ext}$, two energy levels computed to first order may cross at a so-called critical field, $B_{\rm cri}(\Delta M = 0)$. Hence, from the condition $E_{m_{\rm e},m_n^{(1)},m_\mu^{(1)}} = E_{m_{\rm e},m_n^{(2)},m_\mu^{(2)}}$ and the selection rule, we derive

$$B_{\rm cri}(\Delta M = 0) = \frac{1}{2}\frac{|\omega_{\rm iso} - \omega_{\rm nhf}|}{\gamma_\mu - \gamma_n}. \qquad (8.64)$$

In higher order the crossing is avoided by the non-secular terms of $\mathcal{H}_{\rm tot}$ as explained in Section 4.2, leading to an estimate for the gap angular frequency at resonance (Heming *et al.* 1986):

$$\omega_{\rm cri}(\Delta M = 0) = \pi\sqrt{I(I+1) - M(M-1)}\,\frac{\hbar\omega_{\rm iso}|\omega_{\rm nhf}|}{g_{\rm e}\mu_{\rm B}B_{\rm cri}(\Delta M = 0)}. \qquad (8.65)$$

There are $2I$ nearly degenerate ALC transitions with $\Delta M = 0$. If $\omega_{\rm cri}(\Delta M = 0)$ and $B_{\rm cri}(\Delta M = 0)$ can be estimated from ALC measurements and since $\omega_{\rm iso}$ is usually known from conventional longitudinal- or transverse-field measurements, the value of the angular frequency $\omega_{\rm nhf}$ can be deduced.

Other resonances are possible for the muon–electron–nucleus system if the hyperfine couplings are anisotropic as explained by Roduner (1999). As mentioned in Section 8.3, the anisotropy is usually dipole-like. Referring to the expression of the dipole Hamiltonian (Eqs. 5.26, 5.27, and 5.28), the terms C_i and D_i (E_i and F_i) correspond to $|\Delta M| = 1$ ($|\Delta M| = 2$).

Let us first focus on the $|\Delta M| = 1$ resonance. Since the muon spin must participate in it, only the C_i and D_i terms are at play. There is only one ALC transition of this sort, in agreement with the result of the model of Section 4.2. An approximate calculation gives

$$B_{\rm cri}(|\Delta M| = 1) \simeq \frac{\omega_{\rm iso}}{2\gamma_\mu} \left|1 + p_{\rm ani}\left(3\cos^2\theta - 1\right)\right|, \tag{8.66}$$

with

$$\omega_{\rm cri}(|\Delta M| = 1) = \frac{\omega_{\rm dip}}{2} \sin\theta\cos\theta. \tag{8.67}$$

We recall that θ is the angle between the Z axis and the axial symmetry axis of the muonium centre. Both expressions are orientation dependent. The resonance disappears if the anisotropy vanishes and for $\theta = n\pi/2$ where n is an integer. The detection of a $|\Delta M| = 1$ resonance is a clear signature of the anisotropy of the electron–muon hyperfine interaction. This resonance does not carry any direct information on the muonium environment, i.e. on the interaction of the muonium centre with the nuclear spins.

The $|\Delta M| = 2$ resonance involves the muon and the nucleus through a muon–nucleus mutual spin flip. Therefore the anisotropies of the hyperfine interaction for both the muon and the nucleus are of importance. The resonance turns out to be very narrow and weak and thus rarely observed in experiments. Approximately,

$$B_{\rm cri}(|\Delta M| = 2) = \frac{1}{2}\frac{\omega_{\rm iso} + \omega_{\rm nhf}}{\gamma_\mu + \gamma_n}. \tag{8.68}$$

The previous analytical results are valid at high field, i.e. if the Zeeman states are the eigenstates to a good approximation. There are no available formulae for the critical fields and corresponding gap frequencies for low-field resonances.

It may be necessary to include in $\mathcal{H}_{\rm tot}$ terms other than the hyperfine and superhyperfine interactions and the Zeeman Hamiltonians. For example, in semiconductors, it has been argued that the influence of the nuclear electric quadrupole interaction and the weak muon–nuclear dipole interaction cannot be neglected. An example of the latter influence is given in Section 12.1.

If the polarisation function is the sum of constant and cosine terms, it is a simple matter to show that the resonance is Lorentzian-shaped. The parameter $\mathcal{R}_{\rm int}$ (defined in Eq. 4.12) describing an ALC resonance is then written as

$$\mathcal{R}_{\rm int}(B_{\rm ext}) = 1 - \frac{w_{\rm cri}\omega_{\rm cri}^2}{\lambda_{\rm ALC}^2 + \omega_{\rm cri}^2 + \left\{2\pi\left(B_{\rm ext} - B_{\rm cri}\right)\left[\gamma_\mu + \left(|\Delta M| - 1\right)\gamma_n\right]\right\}^2}. \tag{8.69}$$

For $\Delta M = 0$ the weight factor $w_{\rm cri}$ is equal to $I/(2I+1)$ and for $|\Delta M| = 1$ it is $w_{\rm cri} = \frac{1}{2}$, in agreement with the result of a simplified model; see Section 4.2. The damping rate

$\lambda_{\rm ALC}$ describes any process that takes the muon–electron–nucleus system strictly out of resonance. The lower bound is provided by the muon lifetime: $\lambda_{\rm ALC} = \tau_\mu^{-1}$. If the muonium centre has a finite lifetime $\tau_{\rm R}$, after which it undergoes some process, e.g. a reaction into a diamagnetic muon state,

$$\lambda_{\rm ALC} = \frac{\tau_\mu + \tau_{\rm R}}{\tau_\mu \tau_{\rm R}}. \tag{8.70}$$

Regardless of the type of dynamics of the system under study, as the effective correlation time $\lambda_{\rm ALC}^{-1}$ decreases, the resonance gets broader and eventually vanishes, contributing to the baseline.

Accounting for the spin dynamics with only a single parameter such as $\lambda_{\rm ALC}$ may fail. Then one must have recourse to more sophisticated methods using, for example, the Monte Carlo (Tregenna-Piggott *et al.* 1996) or the density operator (Kreitzman and Roduner 1995) technique. The Monte Carlo approach is quite powerful since any type of dynamics can be implemented in the code. Note that Kreitzman and Roduner have rewritten the Hamiltonian pertaining to a high-field resonance as a Zeeman muon Hamiltonian with an effective field; see Section 8.3.2.2 for a discussion of the high-field limit. With that picture, a resonance is viewed as resulting from the existence of a field component perpendicular to the Z axis. Moreover, the dynamics can be easily introduced by considering the effective field as a stochastic variable, characterised by a correlation time. The effect of dynamics is further examined in the next section.

The Lorentzian expression of $\mathcal{R}_{\rm int}(B_{\rm ext})$ holds for single crystals in most favourable cases. For powders the signal must be integrated over all orientations. This often leads to characteristic non-Lorentzian line shapes.

8.4.3 Depolarisation and relaxation

As explained in Chapters 6 and 7, the time evolution of the spin ensemble of diamagnetic muons is influenced by the field distribution at the muon site and the flipping of the muon spin induced by its coupling to the lattice. The same mechanisms also drives the muon spin ensemble for a paramagnetic muon such as a muon bound to an electron. The difference between diamagnetic and paramagnetic muons is an increase in mathematical complexity, because a muon is a two-level system whereas an isolated muonium state is already a four-level system. Even worse, more than four levels are necessary for the description of a muonium centre coupled to its environment if the multi-spin system picture applies. In addition, experiments with muonium centres are more difficult than with muons because the spectral weight is distributed over numerous lines and the standard deviation of the field distribution is considerably larger. Despite these difficulties, valuable results based on the analysis of the depolarisation and relaxation of the polarisation functions have been obtained. Some examples are discussed in Chapters 9 and 12. Here we provide a framework for the data analysis. We shall first consider isotropic spin dynamics, then shortly discuss the case of anisotropic muonium centre and mention the possible transformation between different muonium states. Finally, we describe the case of a muon coupled to an electronic defect on a chain through an intermittent hyperfine interaction.

8.4.3.1 Isotropic muonium centre. We examine the depolarisation and relaxation arising from the isotropic spin Hamiltonian given in Eq. 8.62. If necessary, the discussion can be extended to anisotropic muonium centres (Patterson 1988). Qualitatively, a modulation of the nuclear interaction results in a depolarisation and a relaxation of the muonium-electron spin, which in turns leads to the depolarisation and relaxation of the muon spin *via* the muonium hyperfine interaction. It is the nuclear interaction which sets the time scale of the decay of $P_\alpha(t)$, since the muon–electron interaction is several orders of magnitude larger than the nuclear interaction.

The physics involved can be described by extending to muonium centres the rate equation matrix method introduced in Section 7.1.3 for the muon. This is not entirely satisfactory since it cannot take into account coherent oscillations. However, when the oscillating component is not of interest, the formalism is simple enough to provide useful insights into different relaxation mechanisms. For instance, it is easy to cover the case of intermittent super-hyperfine exchange interaction and reorientation of the muonium centre. We refer to the works of Cox and Sivia (1997) and Cox (2009) for more information.

The polarisation functions can also be computed using the physical idea of the strong-collision model introduced in Chapter 6 (Senba 1991). Here we use a third method based on the density operator. It is mathematically complex, but it has a certain general validity.

We attribute the time dependence of the polarisation function to the density operator and derive a simple enough expression for this operator such that the polarisation functions can be computed. According to Section 3.2

$$P^\mu_X(t) = \mathrm{Tr}\left\{\rho_{\mathrm{tot}}(t)\,\sigma^X_\mu\right\} \quad \text{and} \quad P^\mu_Z(t) = \mathrm{Tr}\left\{\rho_{\mathrm{tot}}(t)\,\sigma^Z_\mu\right\}, \tag{8.71}$$

with

$$i\hbar\frac{\mathrm{d}\rho_{\mathrm{tot}}(t)}{\mathrm{d}t} = \left[\mathcal{H}_{\mathrm{tot}}(t), \rho_{\mathrm{tot}}(t)\right]. \tag{8.72}$$

Note that we have here added the superscript μ to the muon polarisation functions: this is simply not to confuse them with the electron and mixed muon–electron polarisation functions which will also be involved in the following. Similarly, we have introduced the subscript μ to the Pauli operators to specify that these operators are related to the muon degrees of freedom. These superscripts and subscripts will only be used when necessary.

The starting point is first to neglect the interaction of the muonium centre with its environment. The Hamiltonian is therefore independent of time and is given by Eq. 8.7. The commutator of Eq. 8.72 can be evaluated using the density operator written in Eq. 3.34. This leads to a system of linear differential equations for the muon, electron, and mixed polarisation functions. Fortunately, this system can be separated into two irreducible (independent) subsystems, for both the transverse and longitudinal muon polarisation functions. Using Eq. C.23,

$$\frac{\mathrm{d}P^\mu_X}{\mathrm{d}t} = \frac{\omega_{\mathrm{iso}}}{2}\left(P^{\mu\mathrm{e}}_{ZY} - P^{\mu\mathrm{e}}_{YZ}\right) + \omega_\mu P^\mu_Y, \qquad \frac{\mathrm{d}P^\mu_Y}{\mathrm{d}t} = \frac{\omega_{\mathrm{iso}}}{2}\left(P^{\mu\mathrm{e}}_{XZ} - P^{\mu\mathrm{e}}_{ZX}\right) - \omega_\mu P^\mu_X,$$
$$\frac{\mathrm{d}P^\mathrm{e}_X}{\mathrm{d}t} = \frac{\omega_{\mathrm{iso}}}{2}\left(P^{\mu\mathrm{e}}_{YZ} - P^{\mu\mathrm{e}}_{ZY}\right) + \omega_\mathrm{e} P^\mathrm{e}_Y - 2\nu_{\mathrm{NY}} P^\mathrm{e}_X,$$

$$
\begin{aligned}
\frac{\mathrm{d}P^{\mathrm{e}}_{Y}}{\mathrm{d}t} &= \frac{\omega_{\mathrm{iso}}}{2}\left(P^{\mu\mathrm{e}}_{ZX} - P^{\mu\mathrm{e}}_{XZ}\right) - \omega_{\mathrm{e}} P^{\mathrm{e}}_{X} - 2\nu_{\mathrm{NY}} P^{\mathrm{e}}_{Y},\\
\frac{\mathrm{d}P^{\mu\mathrm{e}}_{XZ}}{\mathrm{d}t} &= \frac{\omega_{\mathrm{iso}}}{2}\left(-P^{\mu}_{Y} + P^{\mathrm{e}}_{Y}\right) + \omega_{\mu} P^{\mu\mathrm{e}}_{YZ} - 2\nu_{\mathrm{NY}} P^{\mu\mathrm{e}}_{XZ},\\
\frac{\mathrm{d}P^{\mu\mathrm{e}}_{YZ}}{\mathrm{d}t} &= \frac{\omega_{\mathrm{iso}}}{2}\left(P^{\mu}_{X} - P^{\mathrm{e}}_{X}\right) - \omega_{\mu} P^{\mu\mathrm{e}}_{XZ} - 2\nu_{\mathrm{NY}} P^{\mu\mathrm{e}}_{YZ},\\
\frac{\mathrm{d}P^{\mu\mathrm{e}}_{ZY}}{\mathrm{d}t} &= \frac{\omega_{\mathrm{iso}}}{2}\left(-P^{\mu}_{X} + P^{\mathrm{e}}_{X}\right) - \omega_{\mathrm{e}} P^{\mu\mathrm{e}}_{ZX} - 2\nu_{\mathrm{NY}} P^{\mu\mathrm{e}}_{ZY},\\
\frac{\mathrm{d}P^{\mu\mathrm{e}}_{ZX}}{\mathrm{d}t} &= \frac{\omega_{\mathrm{iso}}}{2}\left(P^{\mu}_{Y} - P^{\mathrm{e}}_{Y}\right) + \omega_{\mathrm{e}} P^{\mu\mathrm{e}}_{ZY} - 2\nu_{\mathrm{NY}} P^{\mu\mathrm{e}}_{ZX},
\end{aligned}
\tag{8.73}
$$

and

$$
\begin{aligned}
\frac{\mathrm{d}P^{\mu}_{Z}}{\mathrm{d}t} &= -\frac{\omega_{\mathrm{iso}}}{2}\left(P^{\mu\mathrm{e}}_{XY} - P^{\mu\mathrm{e}}_{YX}\right), \qquad \frac{\mathrm{d}P^{\mathrm{e}}_{Z}}{\mathrm{d}t} = \frac{\omega_{\mathrm{iso}}}{2}\left(P^{\mu\mathrm{e}}_{XY} - P^{\mu\mathrm{e}}_{YX}\right) - 2\nu_{\mathrm{NY}} P^{\mathrm{e}}_{Z},\\
\frac{\mathrm{d}P^{\mu\mathrm{e}}_{XX}}{\mathrm{d}t} &= \omega_{\mathrm{e}} P^{\mu\mathrm{e}}_{XY} + \omega_{\mu} P^{\mu\mathrm{e}}_{YX} - 2\nu_{\mathrm{NY}} P^{\mu\mathrm{e}}_{XX},\\
\frac{\mathrm{d}P^{\mu\mathrm{e}}_{YY}}{\mathrm{d}t} &= -\omega_{\mathrm{e}} P^{\mu\mathrm{e}}_{YX} - \omega_{\mu} P^{\mu\mathrm{e}}_{XY} - 2\nu_{\mathrm{NY}} P^{\mu\mathrm{e}}_{YY},\\
\frac{\mathrm{d}P^{\mu\mathrm{e}}_{XY}}{\mathrm{d}t} &= \frac{\omega_{\mathrm{iso}}}{2}\left(P^{\mu}_{Z} - P^{\mathrm{e}}_{Z}\right) - \omega_{\mathrm{e}} P^{\mu\mathrm{e}}_{XX} + \omega_{\mu} P^{\mu\mathrm{e}}_{YY} - 2\nu_{\mathrm{NY}} P^{\mu\mathrm{e}}_{XY},\\
\frac{\mathrm{d}P^{\mu\mathrm{e}}_{YX}}{\mathrm{d}t} &= -\frac{\omega_{\mathrm{iso}}}{2}\left(P^{\mu}_{Z} - P^{\mathrm{e}}_{Z}\right) + \omega_{\mathrm{e}} P^{\mu\mathrm{e}}_{YY} - \omega_{\mu} P^{\mu\mathrm{e}}_{XX} - 2\nu_{\mathrm{NY}} P^{\mu\mathrm{e}}_{YX},
\end{aligned}
\tag{8.74}
$$

where, as usual, we have used the notation $\omega_{\mu,\mathrm{e}} = \gamma_{\mu,\mathrm{e}} B_{\mathrm{ext}}$. These equations of the Wangsness–Bloch type (Wangsness and Bloch 1953) were written by Nosov and Yakovleva (1963). A damping term with decay rate $2\nu_{\mathrm{NY}}$ is included in the right-hand side of each of the equations describing the evolution of the pure electron and mixed muon–electron polarisation components. This term approximately accounts for the perturbation induced by the muonium environment on the muonium dynamics. In the absence of this term, the polarisation functions would not be damped (see e.g. Atherton 1993). Because the electronic relaxation is assumed to be fast, that is, $|\omega_{\mathrm{e}}|/\nu_{\mathrm{NY}} \ll 1$, and isotropic, its characterisation by a single parameter is justified as it is for the diamagnetic muon case.

To proceed further, it is convenient to write these two sets of equations in matrix form:

$$
\frac{\mathrm{d}}{\mathrm{d}t}\begin{pmatrix} P^{\mu}_{X} + iP^{\mu}_{Y} \\ P^{\mathrm{e}}_{X} + iP^{\mathrm{e}}_{Y} \\ P^{\mu\mathrm{e}}_{ZX} + iP^{\mu\mathrm{e}}_{ZY} \\ P^{\mu\mathrm{e}}_{XZ} + iP^{\mu\mathrm{e}}_{YZ} \end{pmatrix}
\tag{8.75}
$$

$$= \frac{\omega_{\rm iso}}{2} \begin{pmatrix} -a_- & 0 & -i & i \\ 0 & -\frac{4\nu_{\rm NY}}{\omega_{\rm iso}} + a_+ & i & -i \\ -i & i & -\frac{4\nu_{\rm NY}}{\omega_{\rm iso}} + a_+ & 0 \\ i & -i & 0 & -\frac{4\nu_{\rm NY}}{\omega_{\rm iso}} - a_- \end{pmatrix} \begin{pmatrix} P^\mu_X + iP^\mu_Y \\ P^{\rm e}_X + iP^{\rm e}_Y \\ P^{\mu {\rm e}}_{ZX} + iP^{\mu {\rm e}}_{ZY} \\ P^{\mu {\rm e}}_{XZ} + iP^{\mu {\rm e}}_{YZ} \end{pmatrix},$$

and

$$\frac{\rm d}{{\rm d}t} \begin{pmatrix} P^\mu_Z \\ P^{\rm e}_Z \\ P^{\mu {\rm e}}_{XX} + P^{\mu {\rm e}}_{YY} \\ P^{\mu {\rm e}}_{XY} - P^{\mu {\rm e}}_{YX} \end{pmatrix} = \frac{\omega_{\rm iso}}{2} \begin{pmatrix} 0 & 0 & 0 & -1 \\ 0 & -\frac{4\nu_{\rm NY}}{\omega_{\rm iso}} & 0 & 1 \\ 0 & 0 & -\frac{4\nu_{\rm NY}}{\omega_{\rm iso}} & -2x_{\rm B} \\ 2 & -2 & 2x_{\rm B} & -\frac{4\nu_{\rm NY}}{\omega_{\rm iso}} \end{pmatrix} \begin{pmatrix} P^\mu_Z \\ P^{\rm e}_Z \\ P^{\mu {\rm e}}_{XX} + P^{\mu {\rm e}}_{YY} \\ P^{\mu {\rm e}}_{XY} - P^{\mu {\rm e}}_{YX} \end{pmatrix}. \quad (8.76)$$

We have introduced the short-hand notation: $a_\pm = ix_{\rm B}(1 \pm d_\gamma)$. With these two matrix equations, the muon polarisation functions can be numerically computed taking into account that at $t = 0$ all the polarisation functions vanish except the relevant muon polarisation function, which is equal to 1. This gives four conditions for each set of equations. The roots of the characteristic equation of each of these systems can be analytically computed only in asymptotic cases (Smilga and Belousov 1994).

We first consider the fast electronic fluctuation limit, i.e. when $\nu_{\rm NY}/\omega_{\rm iso} \gg 1$. In the transverse-field case, from Eq. 8.75 we have ${\rm d}(P^{\mu {\rm e}}_{ZX} + iP^{\mu {\rm e}}_{ZY})/{\rm d}t \simeq -2\nu_{\rm NY}(P^{\mu {\rm e}}_{ZX} + iP^{\mu {\rm e}}_{ZY})$ leading to $P^{\mu {\rm e}}_{ZX}(t) = P^{\mu {\rm e}}_{ZY}(t) = 0$ since $P^{\mu {\rm e}}_{ZX}(t = 0) = P^{\mu {\rm e}}_{ZX}(t = 0) = 0$. Therefore, the first component of the vectorial Eq. 8.75 reads ${\rm d}(P^\mu_X + iP^\mu_Y)/{\rm d}t = -i\omega_\mu(P^\mu_X + iP^\mu_Y)$. Its solution, namely $P^\mu_X(t) = \cos(\omega_\mu t)$ and $P^\mu_Y(t) = -\sin(\omega_\mu t)$, was already found for the free muon; see Eqs. 3.8 and 3.9. Concerning the longitudinal-field case, using Eq. 8.76 and following an analogous procedure, we have $P^\mu_Z(t) = 1$. These results derived for the muon polarisation functions simply mean that frequent flipping of the electron spin results in the decoupling of the muon spin with the latter. With a higher order expansion in $\omega_{\rm iso}/\nu_{\rm NY}$, Nosov and Yakovleva (1963) found

$$P_X(t) = \exp(-\lambda_X t) \cos(\omega_\mu t) \quad \text{with} \quad \lambda_X = \frac{\omega^2_{\rm iso}}{4\nu_{\rm NY}}. \quad (8.77)$$

For a free muon, in the case of fast and isotropic fluctuations, it is known that $\lambda_Z = \lambda_X$; see e.g. Eq. 7.100. For a muonium centre, we therefore expect

$$P_Z(t) = \exp(-\lambda_Z t) \quad \text{with} \quad \lambda_Z = \frac{\omega^2_{\rm iso}}{4\nu_{\rm NY}}. \quad (8.78)$$

A simple formula can be derived for $P_X(t)$ if the electronic fluctuations are slow (Ivanter and Smilga 1968). Assuming the frequencies ω_{14} and ω_{34} are beyond detection capability,

$$P_X(t) = \frac{1}{2} \cos\left(\frac{\gamma_{\rm e} B_{\rm ext}}{2} t\right) \exp(-\nu_{\rm NY} t), \quad (8.79)$$

if the applied field is very small. It is rewarding to note that setting $\nu_{\rm NY} = 0$ in Eq. 8.79, the free muonium result given in Eq. 8.29 is recovered. In the same limit of slow electronic fluctuations, the expression for $P_Z(t)$ is relatively simple (Nosov and Yakovleva 1963). Averaging out the oscillating term it reads

$$P_Z(t) = \frac{1+2x_B^2}{2\,(1+x_B^2)} \exp\left(-\lambda_Z t\right) \qquad \text{with} \qquad \lambda_Z = \frac{\nu_{NY}}{1+x_B^2}. \tag{8.80}$$

Of course, the amplitude of the relaxing signal matches that found in the static case; see Eq. 8.18 or 8.19.

It is to be noted that the formula

$$\lambda_Z = \frac{\nu_{NY}}{\frac{4\nu_{NY}^2}{\omega_{iso}^2} + 1 + x_B^2}, \tag{8.81}$$

in the limiting cases of fast and slow dynamics, yields Eqs. 8.78 and 8.80, respectively.

8.4.3.2 Anisotropic muonium centre. In Section 8.3.2.2 we briefly discussed the particular case of an effective Hamiltonian resulting from the high-field approximation for a dipole-like muonium centre. We noted that at a specific critical value of B_{ext}, the local field at the muon site is fully perpendicular to the Z axis for the muonium states with a positive electron magnetic quantum number, m_e. This geometrical property has an immediate consequence if the spin-exchange rate is not too large such that λ_Z still depends on B_{ext}: to a good approximation $P_Z(t)$ is an exponential function with a relaxation rate exhibiting a pronounced maximum value at the critical field. This can be qualitatively understood referring to results of Chapter 7 which show that the spin–lattice relaxation is induced by the components of the field vector perpendicular to the Z axis. Since the modulus of this vector is maximum at the critical field, the relaxation channel $m_e = \frac{1}{2}$ dominates and therefore, to a good approximation, $P_Z(t)$ is a simple exponential function with a relaxation rate at a maximum value. Numerical computations support the conclusion of this qualitative discussion (Chow *et al.* 1994). An example is presented in Section 12.5.

8.4.3.3 Transformation between muonium states. The original muonium state may transform to a diamagnetic state or to another paramagnetic species before the muon decay. This transformation can even be cycling between the two states many times during the experimental window. The chemical reaction may be so rapid that the original muon state cannot be directly observed, and its existence can be inferred only from its effect on the measured signal amplitude and phase. The $P_\alpha(t)$ functions can be obtained with the two-state model developed in Section 6.4.2 for diamagnetic muons taking into account that now the polarisation functions form a vector, i.e. not just the muon polarisation function of interest must be taken into account. This vector is built from the functions $P_\alpha^\mu(t)$, $P_\alpha^e(t)$, and $P_{\alpha\beta}^{\mu e}(t)$, where $\{\alpha, \beta\} = \{X, Y, Z\}$. Since there is no special difficulty, we leave the task of deriving the relevant $P_\alpha(t)$ functions to the interested reader. The polarisation functions for muonium centres jumping between two states in a reversible or irreversible way are discussed by Patterson (1988) and Odermatt (1988).

8.4.3.4 Intermittent coupling to a defect diffusing on a chain. As already pointed out for the case of a diamagnetic muon in Section 6.3, the description of a random walk for a system with a restricted dimensionality such as a chain requires a special treatment since a correlation time cannot be defined. Risch and Kehr (1992) have presented the result of a stochastic theory for the relaxation of the muon spin coupled to the spin of an electronic defect by an intermittent hyperfine interaction. It is assumed that the interaction is "on" when the defect is nearby the muon (site 0) and "off" otherwise. The possibility of spin flips of the defect on sites other than site 0 with rate $\nu_{\rm flip}$ is included in the model. The rate of transfer of the defect to adjacent sites belonging to the chain is denoted by $\nu_{\parallel}$. The hyperfine interaction is taken to be isotropic. In contrast to the spin-flip mechanism previously considered in Section 8.4.3.1, here we focus on the influence of the one-dimensional defect motion on the measured relaxation. The nature of the defect will be clarified when discussing an application of the model to a polymer in Section 12.6.2.

The computation of $P_Z(t)$ is performed with the Liouville operator formalism and results in a rather complicated formula for the Laplace transform of $P_Z(t)$. Fortunately, for moderate values of $B_{\rm ext}$ and $\nu_{\rm flip}\, t_{\rm max} \gg 1$ where $t_{\rm max}$ is the experimental time scale, an approximate analytical expression is derived:

$$P_Z(t) = \exp\left(\lambda_{\rm chain} t\right) \operatorname{erfc}\left(\sqrt{\lambda_{\rm chain} t}\right), \tag{8.82}$$

where $\operatorname{erfc}(x)$ is the complementary error function defined in Eq. C.106 and $\lambda_{\rm chain}$ a relaxation rate. It is of interest to display the dependence of $P_Z(t)$ on time for small and large arguments:

$$\lim_{\lambda_{\rm chain} t \to 0} P_Z(t) = 1 - 2\sqrt{\frac{\lambda_{\rm chain} t}{\pi}}, \quad \text{and} \quad \lim_{\lambda_{\rm chain} t \to \infty} P_Z(t) = \frac{1}{\sqrt{\pi \lambda_{\rm chain} t}}. \tag{8.83}$$

Thus, for appreciable spin-flip and relaxation rates a power-law decay of $P_Z(t)$ is predicted for long times, rather than the exponential behaviour encountered in three-dimensional systems. The relaxation function of Risch and Kehr is plotted in Fig. 8.5. The relaxation rate $\lambda_{\rm chain}$ is given by

$$\lambda_{\rm chain} = \frac{\nu_{\rm flip}}{\left(1 + \nu_{\parallel}\sqrt{\dfrac{2 g_e \mu_{\rm B} B_{\rm ext} \nu_{\rm flip}}{\hbar \omega_{\rm iso}^4}}\right)^2}. \tag{8.84}$$

In the fast defect motion limit, i.e. $\nu_{\parallel} \gg \omega_{\rm iso}^2 / \sqrt{2 g_e \mu_{\rm B} B_{\rm ext} \nu_{\rm flip} / \hbar}$,

$$\lambda_{\rm chain} = \frac{\hbar \omega_{\rm iso}^4}{2 g_e \mu_{\rm B}\, B_{\rm ext}\, \nu_{\parallel}^2}. \tag{8.85}$$

Here, $\lambda_{\rm chain}$ is predicted to be inversely proportional to $B_{\rm ext}$. Since it is independent of $\nu_{\rm flip}$ and $\omega_{\rm iso}$ is usually easily estimated experimentally, the measurement of $\lambda_{\rm chain}$ gives access to $\nu_{\parallel}$. Note that the diffusion constant of the defect along the chain, $D_{\parallel}$, is directly proportional to $\nu_{\parallel}$: $D_{\parallel} = d_{\rm hop}^2 \nu_{\parallel}/2$, where $d_{\rm hop}$ is the distance between defect

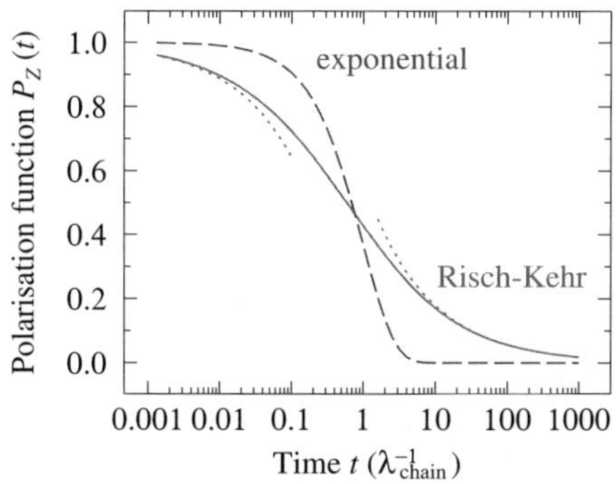

Fig. 8.5 The relaxation function $P_Z(t)$ for an intermittent isotropic muonium hyperfine interaction involving the diffusion of an electronic defect along a chain. The approximate solution given by Eq. 8.82 is plotted (full line). The muon is assumed to be static. The time scale is logarithmic and given in terms of $\lambda_{\text{chain}}^{-1}$, where λ_{chain} is a relaxation rate parameter (Eq. 8.84). The asymptotes of $P_Z(t)$ given by Eq. 8.83 for short and long times are shown as dotted lines. For reference the exponential function $\exp(-\lambda_{\text{chain}}t)$ is drawn with a dashed line.

potential sites: see Section 9.1.2. Here d_{hop} is the physical analog for the defect of the distance d_{nis} defined there for the muon or muonium centre. In addition, according to the Einstein–Smoluchowski relation, the mobility of the defect along the chain, $\mu_{\|}$, is directly proportional to $D_{\|}$: $\mu_{\|} = eD_{\|}/(k_{\text{B}}T)$. Therefore, from a determination of $\nu_{\|}$, a microscopic estimate of the defect mobility can be derived. This is further discussed in Section 12.6.2.

The relaxation function in zero field has also the form given in Eq. 8.82 (Risch and Kehr 1992). However, we note that apart from diffusion, other sources of relaxation can contribute to the zero-field muon spin relaxation. An obvious one stems from the nuclear spins of the specimen under investigation. Its effect can be quenched by a modest longitudinal field.

References

Atherton N.M. (1993) *Principles of Electron Spin Resonance* (Ellis Horwood and Prentice Hall, New York).

Baryshevskii V.G., Kuten S.A., and Rapoport V.I. (1982) Physics Letters **A 88**, 289–291.

Beck R., Meier P.F., and Schenck A. (1975) *Zeitschrift für Physik* **B 22**, 109–115.

Blazey K.W., Estle T.L., Holzschuh E., Meier P.F., Patterson B.D., and Richner M. (1986) *Physical Review* **B 33**, 1546–1552.

Brewer J.H., Spencer D.P., Fleming D.G., and Coope J.A.R. (1981) *Hyperfine Interactions* **8**, 405–408.

Chow K.H., Lichti R.L., Kiefl R.F., Dunsiger S., Estle T.L., Hitti B., Kadono R., MacFarlane W.A., Schneider J.W., Schumann D., and Shelley M. (1994) *Physical Review* **B 50**, 8918–8921.

Cox S.F.J. (2003) *Journal of Physics: Condensed Matter* **15**, R1727–R1780.
Cox S.F.J. (2009) *Reports on Progress in Physics* **72**, 116501 1–130.
Cox S.F.J. and Sivia D.S. (1997) *Applied Magnetic Resonance* **12**, 213–226.
Edmonds A.R. (1957) *Angular Momentum in Quantum Mechnics* (Princeton University Press, Princeton).
Heming M., Roduner E., Patterson B.D., Odermatt W., Schneider J., Baumeler H., Keller H., and Savić I.M. (1986) *Chemical Physics Letters* **128**, 100–106.
Ivanter I.G. and Smilga V.P. (1968) *Soviet Physics JETP* **27**, 301–306.
Karlsson E.B. (1995) *Solid State Phenomena As Seen by Muons, Protons, and Excited Nuclei* (Clarendon Press, Oxford).
Kreitzman S.R. and Roduner E. (1995) *Chemical Physics* **192**, 189–230.
Lord J.S., Cox S.F.J., Alberto H.V., Piroto Duarte J., and Vilão R.C. (2004) *Journal of Physics: Condensed Matter* **16**, S4707–S4720.
Messiah A. (1965) *Quantum Mechanics*, third reprint, vol. 2 (North-Holland, Amsterdam).
Nagamine K. (2003) *Introductory Muon Science* (Cambridge University Press, Cambridge).
Nosov V.G. and Yakovleva I.V. (1963) *Soviet Physics JETP* **16**, 1236–1246.
Odermatt W. (1988) *Helvetica Physica Acta* **61**, 1087–1129.
Patterson B.D. (1988) *Reviews of Modern Physics* **60**, 69–159.
Pratt F.L. (1997) *Philosophical Magazine Letters* **75**, 371–379.
Risch R. and Kehr K.W. (1992) *Physical Review* **B 46**, 5246–5257.
Roduner E. (1988) *The positive Muon as a Probe in Free Radical Chemistry*, Lecture Notes in Chemistry, vol. 49 (Springer, Berlin).
Roduner E. (1999) in *Muon Science: Muons in Physics, Chemistry, and Materials*, edited by Lee S.L., Kilcoyne S.H., and Cywinski R. (Institute of Physics, Bristol).
Schenck A. (1985) *Muon Spin Rotation Spectroscopy* (Adam Hilger, Bristol).
Senba M. (1991) *Journal of Physics B: Atomic, Molecular and Optical Physics* **24**, 3531–3549.
Senba M. (2001) *Journal of Physics B: Atomic, Molecular and Optical Physics* **34**, 4437–4454.
Smilga V.P. and Belousov Yu.M. (1994) *The Muon Method in Science*, Proceedings of the Lebedev Physics Institute, vol. 219 (Nova Science, Commack, NY).
Tregenna-Piggott P.L.W., Roduner E., and Santos S. (1996) *Chemical Physics* **203**, 317–337.
Walker D.C. (1983) *Muon and Muonium Chemistry* (Cambridge University Press, Cambridge).
Wangsness R.K and Bloch F. (1953) *Physical Review* **89**, 728–739.

Part III

Condensed matter applications

9
Muon and muonium centre diffusion

This chapter deals with the diffusion of the muons and muonium centres in condensed matter. The diffusion studies are performed with one of the two following aims: either to investigate the diffusion of the muon or muonium centre and the effects of defects and impurities on it as a primary purpose, or as an auxiliary work of, for example, a study of the magnetic properties of a compound. In the latter case, the purpose is to characterise a possible influence of the diffusion of the muon or muonium centre on the measured magnetic properties. Even if the diffusion itself is not the main interest of the reader, he or she should be aware that it may greatly influence the measured polarisation functions. Therefore all users of the μSR techniques are recommended to read this chapter.

We recall that the spirit of this book is neither to review experimental results nor to provide exhaustive bibliographies. Following this philosophy, we shall give a few typical results to illustrate the possibilities of the μSR techniques. The interested reader should consult the books of Karlsson (1995) and the review article of Storchak and Prokof'ev (1998) for more detailed information.

Quasielastic neutron scattering is a powerful method for studying the motion of atoms in condensed matter. It is described in detail in the monograph of Hempelmann (2000). When possible, it is useful to compare the results obtained by μSR and the quasielastic neutron technique.

We first describe the relation between the diffusion constant and the particle transfer rate and carry on with some basic properties of light interstitials in solids. In the third section we examine how the measured polarisation functions are influenced by the muon or muonium centre mobility. Finally we present some examples taken from the literature.

Note that, although different species of muonium states are observed (see Section 12.5), here they will be generically denoted as muonium centres. We shall also find it convenient to consider a diffusing muon or muonium centre simply as a diffusing particle. We shall distinguish between the two only when necessary.

9.1 Diffusion constant and site transfer rate

Because of its electric charge, a positive muon is expected to diffuse only between interstitial sites in a solid. Muons do not visit vacancies at sufficiently low temperature since the vacancy concentration at thermodynamical equilibrium, which is fixed by entropy conditions, is very small. They can explore vacancies only if their concentration is sufficiently high, as observed for iron between 930 and 1010 K (Fürderer *et al.* 1986) or for materials exposed to strong radiation doses. Good examples have been given for electron irradiated aluminium (Herlach 1981) and iron (Möslang *et al.* 1993, Weidinger 1984).

Muonium centres are also found to diffuse between interstitial sites.

Particle diffusion in a solid can be discussed at two levels. The diffusion process can be treated assuming the solid to be a continuum, that is, by ignoring its atomic structure. The problem then becomes one of obtaining and solving an appropriate differential equation. We shall first use this approach. It is not meant to detract from the importance of the study of the atomic diffusion mechanisms, but the macroscopic model leads to a description with the minimum of assumptions. In addition, macroscopic experimental techniques only provide macroscopic information, for example, on the diffusion constant. Thereafter we shall discuss the atomic processes involved in diffusion.

9.1.1 Macroscopic approach to diffusion

The macroscopic description of the diffusion of particles in solids dates back to Adolf Fick in 1855, and is the subject of many books, for example, Shewmon (1963), Philibert (1991), and Mehrer (2007). The purpose of this section is only to present a few basic ideas.

Let us consider the diffusion of a species of particles in a one-dimensional system. We denote the probability of finding a particle at time t in position x as $p(x,t)$. It obeys the second order partial differential equation

$$\frac{\partial p}{\partial t} = \frac{\partial}{\partial x}\left(D_{\rm diff}\frac{\partial p}{\partial x}\right). \tag{9.1}$$

It is assumed that there is no drifting effect, i.e. no mass flow. The number $D_{\rm diff}$ is the diffusion constant or diffusivity. Its dimension is the ratio of a squared length over a time. Equation 9.1 cannot be analytically solved if $D_{\rm diff}$ depends on p, and thus on x and t. However, for the cases of interest to us, $D_{\rm diff}$ is a constant. Then Eq. 9.1 can be written as

$$\frac{\partial p}{\partial t} = D_{\rm diff}\frac{\partial^2 p}{\partial x^2}. \tag{9.2}$$

This is the so-called Fick second law. To illustrate the involved physics, we consider the simple model of a one-dimensional system extending from $-\infty$ to $+\infty$ and particles deposited at initial time $t = 0$ in $x = 0$. For this model the initial condition is $p(x, t = 0) = \delta(x)$. The analytical solution of Eq. 9.2 is then

$$p(x,t) = \frac{1}{2\sqrt{\pi D_{\rm diff} t}}\exp\left(-\frac{x^2}{4D_{\rm diff} t}\right). \tag{9.3}$$

This is a Gaussian function of x which broadens as $\sqrt{4D_{\rm diff} t}$. Its value at $x = 0$ decays as the square root of time. The analysis of $p(x,t)$ gives access to $D_{\rm diff}$. Fick's second law in three dimensions can be obtained from Eq. 9.2 by substituting $\partial^2 p/\partial x^2$ with the Laplacian of p. Solutions for different initial conditions are given, e.g. by Philibert (1991).

Many μSR studies have focused on the investigation of the trapping of particles by impurities or defects. The purposes have been twofold: (i) to slow down the particle motion in order to detect its signature if the diffusion is so fast that the μSR polarisation function in the free state is not sensitive to the diffusion or (ii) to study the properties of the trapping sites themselves. Let us assume that the particles are implanted far away from a trap. As time evolves, the probability for a particle to be trapped increases. A relevant parameter for the description of the trapping process is the trapping rate, $\Gamma_{\rm trap}$. In the case of hopping or tunnelling of localised particles (incoherent process) the theory of diffusion-controlled trapping can be applied (Waite 1957), with the result:

$$\Gamma_{\rm trap} = 1/\tau_{\rm trap} = 4\pi R_{\rm trap} c_{\rm trap} D_{\rm diff}. \tag{9.4}$$

The trapping rate $\Gamma_{\rm trap}$ is therefore proportional to the product of the trapping radius, $R_{\rm trap}$, the concentration of trapping centres per volume unit, $c_{\rm trap}$, and $D_{\rm diff}$. A discussion of the trapping mechanisms and a comparison with experiments are given by Richter (1983).

9.1.2 Microscopic approach to diffusion

Particle diffusion results from individual jumps in a solid. Therefore we expect the diffusivity to be expressed in terms of physical quantities that describe these jumps.

Before dealing with that, it is useful to visualise the interstitial sites in the solid. Figure 9.1 shows the example of the octahedral interstices of a face-centred-cubic (fcc) crystal. Each particle can jump toward any one of the 12 neighbouring interstitial sites. The distance between the sites is $d = a/\sqrt{2}$, where a is the cube edge. To look more closely at the particle jump process, the right panel of Fig. 9.1 shows two octahedral sites in the (100) plane of a group of spheres packed into a fcc lattice. Before the transfer of the particle between the two interstitial sites, two host atoms labelled by the arrows should move apart enough to let it through. This means that an appreciable local deformation of the lattice has occurred before the particle transfer. This is one of the basic mechanisms for diffusion and not just a characteristic of octahedral sites in fcc crystals.

We now discuss the diffusion process from the microscopic point of view. For this purpose we first need to relate the probability function $p(x,t)$, which provides a macroscopic characterisation of the particle diffusion, to microscopic quantities. For a one-dimensional diffusion process (see, e.g. Mehrer (2007))

$$p(x, t+\tau) = \sum_X p(x-X, t) W(X, \tau). \tag{9.5}$$

This equation expresses the probability for the particle to be at time $t+\tau$ and position x in terms of a sum along the diffusion paths of the probability to find the particle at time t and position $x-X$ times the probability $W(X,\tau)$ for the particle to travel the distance X during the time interval τ. We now expand $p(x,t)$ with respect to variables t and x to get $p(x, t+\tau)$ and $p(x-X, t)$, respectively. We shall consider a time interval τ small enough, so that it is justified to keep only the term linear in τ in the time

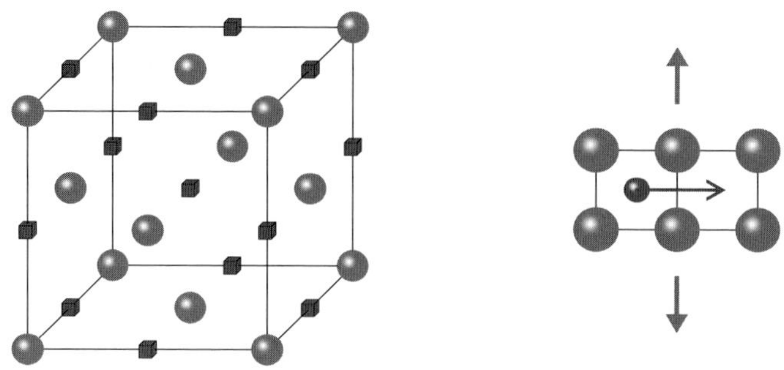

Fig. 9.1 The octahedral interstitial sites of a face-centred-cubic crystal (space group $Fm\bar{3}m$) with the large spheres symbolising the atoms. (Left) An overview is given of the sites in the crystal structure. The site positions are indicated by the small cubes. (Right) A plane parallel to a face of the cube. The horizontal arrow shows the path that the particle (small sphere) must follow to jump between octahedral sites. The pair of vertical arrows indicates the displacement of the atoms for a symmetric phonon mode.

expansion. With the same assumption of small τ, $W(X,\tau)$ has non-negligible values only for X close to 0. Inserting these expansions into Eq. 9.5 and keeping only the leading terms, we get

$$\frac{\partial p}{\partial t} = \frac{\langle X^2 \rangle}{2\tau} \frac{\partial^2 p}{\partial x^2}. \tag{9.6}$$

We have used the equality $\sum_X W(X,\tau) = 1$. In Eq. 9.6, $\langle X^2 \rangle$ is the second moment of $W(X,\tau)$; see Section C.7.1. The first moment has been set equal to 0 since we neglect any drifting effect. Referring to Eq. 9.2 we then derive

$$D_{\rm diff} = \frac{\langle X^2 \rangle}{2\tau}. \tag{9.7}$$

We are still using continuous variables. However, in the one-dimensional lattice of interstitial sites we are considering, the distance between sites can only take a few discrete values. Restricting ourselves to a simple chain of interstitial sites characterised by a single near interstitial distance $d_{\rm nis}$, we set $\langle X^2 \rangle = d^2_{\rm nis}$. Obviously the time scale is now the mean residence time $\bar{\tau}$ (or mean time of stay) of the particle in an interstitial site. Hence, we set $\tau = \bar{\tau}$. Therefore, for a chain of interstitial sites

$$D_{\rm diff} = \frac{d^2_{\rm nis}}{2\bar{\tau}}. \tag{9.8}$$

Until now we have discussed the diffusion of a particle along a chain of interstitial sites. However, in nature the diffusion process occurs generally in two- or three-dimensional lattices of interstitial sites. For example, the nearest neighbour interstitial sites may form a hexagon in a two-dimensional lattice. The parameter characterising the diffusion process is then $d_{\rm m} = 3$, rather than $d_{\rm m} = 1$ for the linear chain considered above. Obviously the value of $d_{\rm m}$ is not related to the space dimension. Assuming the jumps in the lattice of interstitial sites to be uncorrelated, we have

$$D_{\rm diff} = \frac{d_{\rm nis}^2}{2d_{\rm m}\bar{\tau}}. \tag{9.9}$$

The coefficient $2d_{\rm m}$ is in fact the number of nearest neighbour interstitial sites to a given site.

It is a simple matter to apply this formula for instance to face-centred- or body-centred-cubic crystal. As an example, for a particle diffusing between octahedral sites of the latter crystal structure, $D_{\rm diff} = a^2/(24\bar{\tau})$ since $d_{\rm m} = 3$ and $d_{\rm nis} = a/2$.

We stress that $1/\bar{\tau}$ is not the basic transfer rate between two given sites. This rate is given by

$$\Gamma_{\rm t} = \frac{1}{2d_{\rm m}\bar{\tau}}, \tag{9.10}$$

and therefore

$$D_{\rm diff} = \Gamma_{\rm t} d_{\rm nis}^2. \tag{9.11}$$

Depending on the circumstances, $\Gamma_{\rm t}$ is either the muon or muonium jump or tunnelling rate, Γ_μ or $\Gamma_{\rm Mu}$, respectively.

It is possible to express the diffusivity in terms of the velocity correlation function of the particle. For a one-dimensional system,

$$D_{\rm diff} = \int_0^{+\infty} \langle v_x(t) v_x(0)\rangle \, {\rm d}t. \tag{9.12}$$

The extension for a cubic symmetry is obviously

$$D_{\rm diff} = \frac{1}{3}\int_0^{+\infty} \langle \mathbf{v}(t)\cdot\mathbf{v}(0)\rangle \, {\rm d}t. \tag{9.13}$$

9.2 Diffusion of a light interstitial particle

In 1866 Thomas Graham discovered the ability of metallic palladium to absorb large amounts of hydrogen. Since then, metal–hydrogen systems have been the subject of many studies. One remarkable property is the large diffusion constant of hydrogen in several transition metals below room temperature (Fukai and Sugimoto 1985). It is many orders of magnitude larger than for other interstitial atoms. This experimental result has been a strong motivation to develop the theory of diffusion of light interstitials in solids (Kehr 1978). Since the seventies muons and muonium centres have been used to test it (Karlsson 1995). In fact, at the beginning of the eighties it became obvious from μSR results obtained on aluminium that the available theory was incomplete. This led Kondo (1984) and Yamada (1984) to a major modification of the theory of light interstitial diffusion in metals.

The problem of the basic particle jump frequency is essentially a many-body problem because the interstitial particle — the muon or the muonium centre but also proton, deuteron, and triton — is surrounded by other atoms with which it interacts; see for example the right panel of Fig. 9.1. In a classical picture, the particle diffusion occurs through a jump over an energy barrier determined by the saddle point located between the two interstices. From thermodynamic considerations,

$$\Gamma_{\rm t} = \bar{\Gamma}_{\rm t} \exp\left(-\frac{\Delta F}{k_{\rm B}T}\right), \tag{9.14}$$

where ΔF is the free energy needed to carry the particle from the initial interstice to the saddle point. Therefore $\Gamma_{\rm t}$ is predicted to follow an Arrhenius law. The prefactor $\bar{\Gamma}_{\rm t}$ is a frequency associated with the vibration of the particle in the potential well at the initial interstice. It depends on the particle mass and can be weakly (relative to the exponential term) temperature dependent (Vineyard 1957). Concerning the orders of magnitude, ΔF is expected to be on the order of the electronvolt and $\bar{\Gamma}_{\rm t}$ in the $10^{14}\,{\rm s}^{-1}$ range. In practice, while in many cases the measured transition rate can be fitted by an Arrhenius law at high temperature, the two parameters, $\bar{\Gamma}_{\rm t}$ and ΔF, do not have the expected values and $\bar{\Gamma}_{\rm t}$ does not depend on the particle mass as predicted (Vineyard 1957).

9.2.1 Particle tunnelling

In fact the transport of light interstitials in solids is a quantum mechanical problem. Basically, the particle tunnels through the energy barrier between the interstices in contrast to the classical picture in which the particle is assumed to hop over the barrier. Therefore the diffusion proceeds *via* a sub-barrier process. A first-principles calculation of the tunnelling matrix element J is difficult since it depends strongly on the details of the potential barrier. For the muon case, Kehr (1978) estimates J to be some tenths of a millielectronvolt for vanadium and niobium. For copper the computation gives $5.9\,\mu$eV (Richter 1983) while the analysis of experimental data yields $18.3\,\mu$eV. For aluminium, which crystallises in the face-centred-cubic structure as copper does, the experimental value is on the order of 1 meV (Kehr *et al.* 1982). Clearly J strongly depends on the material itself and not only on its crystal structure.

The tunnelling of an interstitial particle is not only determined by J since its interaction with the surrounding heavy atoms leads to its self-trapping. This is illustrated in the left panel of Fig. 9.2. The particle gains what is known as a self-trapping energy by locally distorting the lattice and forming a so-called small polaron. This concept was originally developed by Yamashita and Kurosawa (1958) and Holstein (1959a, 1959b) in order to describe the transfer of electrons in insulators. It was also introduced by Flynn and Stoneham (1970) in the field of light interstitial diffusion. A tunnelling transition to the final site is possible if the energy levels of the initial and final states happen to coincide; see middle panel of Fig. 9.2. This coincidence is induced by thermally excited phonons. Namely, as the heavy atoms of the material vibrate through a myriad of configurations, they occasionally establish positions for which the energy level associated with the light particle occupying the initial site momentarily coincides with the energy it would have if it occupied the final site. These transient degeneracies, termed coincidence events, are associated with a finite probability for the particle to swing between the sites. Then the self-trapping mechanism again comes into the game and stabilises the particle in the final interstitial site; see right panel of Fig. 9.2. This picture is valid at a temperature sufficiently high — typically a small fraction,

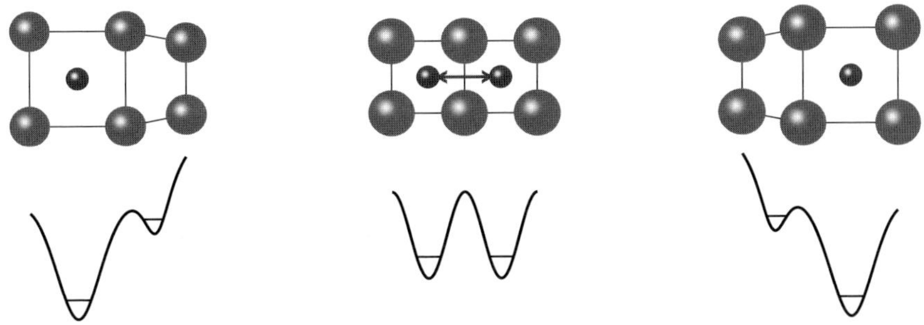

Fig. 9.2 Phonon-assisted diffusion process of a light interstitial particle. We schematically draw the potential energy at the initial and final sites and at the saddle point, that is, at initial time (left), at the coincidence (middle) and after the diffusion step (right).

about one-third, of the Debye temperature, Θ_{D} — that enough phonons are available to create the coincidence.

Because of the self-trapping effect, the tunnelling matrix element is reduced relative to its intrinsic value. The effective matrix element can be written as

$$J_{\mathrm{eff}}(T) = J \exp\left[-S(T)\right]. \tag{9.15}$$

At low temperature, the tunnelling matrix element is reduced because the whole small polaron (particle and distortion cloud) has to tunnel. The parameter $S(T)$ weakly depends on the temperature when the temperature is much less than Θ_{D}. For a simple model, $S(T=0) = 5E_a/(2\hbar\omega_{\mathrm{D}})$, where E_a is the energy needed to create the coincidence and $\hbar\omega_{\mathrm{D}}$ the Debye energy. The renormalisation of the tunnelling matrix element can be drastic since, for example, $S(T=0) = 6.3$ for copper.

As pointed out by Kondo (1984, 1986) and Yamada (1984), this picture is not complete for metals. At low temperature, it is not only the particle with its phonon cloud which must be transported between the two sites but the electron screening-cloud as well. The response of these electrons is very slow because electronic excitations of small energy are involved. This results in a small electronic overlap integral. This tunnelling rate reduction effect is particularly strong at low temperature for which the effective matrix element is given by

$$J_{\mathrm{eff}}(T) = J \exp\left[-S(T)\right] \left(\frac{\pi k_{\mathrm{B}} T}{\hbar\omega_{\mathrm{v}}}\right)^{K}. \tag{9.16}$$

The energy scale of the electronic effect is given by the vibrational energy of the particle in the potential well, $\hbar\omega_{\mathrm{v}}$ (Karlsson 1995). The conduction electrons within this energy interval cannot adiabatically follow the particle motion. The exponent K is a measure of the particle–electron interaction. In practice it is equal to $\sim$0.2 (Karlsson 1995). Considering the case of copper for which the theory gives $\hbar\omega_{\mathrm{v}} = 356\,\mathrm{meV}$ (Teichler 1978), the electronic overlap effect reduces the effective matrix element by an additional factor of ~4.2 at 1 K when $K = 0.2$. At high temperature the Kondo–Yamada electronic effect only leads to a temperature-independent renormalisation of the matrix element (Storchak and Prokof'ev 1998).

So far, while considering the tunnelling matrix element, we have not taken into account the actual lattice configuration around the particle. This is the so-called Condon approximation. However, as first stressed by Flynn and Stoneham (1970), this constitutes a severe approximation for some lattice geometries which can be favourable to phonon excitations which open up the path for the particle jump. For example, consider the transfer between octahedral sites in a face-centred-cubic structure pictured in the right panel of Fig. 9.1. The two arrows illustrate the symmetric displacement of the two heavy atoms which opens up the path between the two interstices. The effective tunnelling matrix element is drastically increased if this fluctuation happens to occur. Naturally, this requires the temperature to be relatively high. Explicit calculations of the diffusivity of light interstitial particles taking into account the symmetric displacements of the heavy atoms have been reported (Emin *et al.* 1979).

Two types of particle motion are possible. For the incoherent type considered up to now, the particle is at first at a interstice and then hops to a nearest neighbour interstice. In the coherent case, the particle is described as a Bloch wave; i.e. a phase coherence exists between the interstices. According to Anderson (1958), the latter picture is valid if the disorder in the lattice is not too strong. This means that the ratio $2d_{\rm m}J_{\rm eff}/\Delta E_{\rm dis}$ should be sufficiently large. In this ratio, $\Delta E_{\rm dis}$ is the width of the interstices energy distribution. This distribution arises from the defects and impurities contained in the sample which induce a long-range elastic distortion field (Eshelby 1956, Leibfried and Breuer 1978). In addition, Kohn–Friedel electron density oscillations may contribute to the disorder in a metal (Karlsson 1995). The Anderson condition can possibly be fulfilled far away from impurities and defects. For example, a numerical estimate is given by Richter (1983), who finds for aluminium doped with 100 parts per million of manganese that the width of the distribution is a few microelectronvolts at midpoint between the impurities.

In addition to the disorder, the coherence of the wavefunction can be destroyed by dynamical mechanisms (Kagan and Klinger 1974): the interaction of the particle with the phonons and electrons (in a metal) indeed leads to an effective broadening of the energy level of the particle, usually denoted by Ω. This can be viewed as a lifetime effect. We stress that these phononic and electronic effects should not be confused with the renormalisation of the tunnelling matrix element. In practice, coherent propagation seems to have been established at low temperature only for neutral muonium centres in insulators and maybe for muons in the superconducting phase of aluminium (Karlsson 1995). However, experiments performed on intermetallic compounds may have revealed muon states delocalised on ring-type arrangements of interstitial sites. This is further discussed in Section 9.4.

9.2.2 Transfer rate and diffusion constant

We have presented an overall picture of the mechanisms of transport of light interstitial particles in solids. In addition, quantitative information on the tunnelling matrix elements has been given. We now focus on the transfer rate and the diffusivity.

We first consider the incoherent motion for which the tunneling rate Γ_t is the basic physical quantity to deal with. At high temperature and in the Condon approximation, the rate is given by the asymptotic formula

$$\Gamma_t = \frac{J^2}{\hbar}\left(\frac{\pi}{4E_a k_B T}\right)^{1/2} \exp\left(-\frac{E_a}{k_B T}\right). \tag{9.17}$$

Here Γ_t is predicted to increase with the temperature provided that $k_B T \lesssim E_a$. This formula describes a phonon-assisted tunnelling process involving many phonons. It can be qualitatively understood as follows (Richter 1983). According to the Fermi golden rule, Γ_t is proportional to J^2 and also to the probability that a proper displacement of the heavy atoms occurs to create the coincidence. In the classical limit, this probability is proportional to $T^{-1/2}\exp(-E_a/k_B T)$. It is possible to extend the range of validity of Eq. 9.17 to lower temperatures by using the pre-asymptotic form. With this extension the multi-phonon mechanism seems to be valid down to about one-third of Θ_D. By fitting this model to experimental data, two parameters are extracted: the bare tunnelling matrix element and the energy necessary to create the coincidence, J and E_a, respectively. While it may be difficult to experimentally distinguish an Arrhenius law from the temperature behaviour predicted above because of the predominance of the exponential function, the value of the two parameters provides a means for deciding which diffusion mechanism is at play. This has already been discussed after Eq. 9.14. Concerning the tunnelling mechanism, refering to copper (see Section 1.2.1), the prefactor is several orders of magnitude smaller than the classically expected value, and the activation energy is about 0.1 eV.

When the coupling of the diffusing particle to the lattice is weak, and therefore the small polaron effect is not important, the breakdown of the Condon approximation might even appear at relatively low temperature. This has been observed for the diffusion of muonium centres. Only a detailed analysis can set apart polaron and barrier fluctuation mechanisms since Γ_t is predicted to increase with temperature as $\exp(-\alpha/T)$ where α is a constant, for both of them. Barrier fluctuations are expected to dominate at high temperature.

We now focus on diffusion at low temperature. For simplicity, we neglect any energy distribution of static origin at the interstitial site, which would arise from defects and impurities. However, the lifetime effect pointed out previously can never be discarded. A simple way to account for it is to assume the velocity correlation function introduced in Eq. 9.13 to damp away with a rate $\Omega/\hbar$. For a cubic diffusion symmetry, this leads to

$$D_{\rm diff} = \frac{1}{3}\langle v^2\rangle \frac{\hbar}{\Omega}, \tag{9.18}$$

where $\langle v^2\rangle$ is the variance of the particle velocity. Expressing $\langle v^2\rangle$ in terms of the characteristics of the particle energy band, we obtain

$$D_{\rm diff} = \frac{2d_{\rm m}d_{\rm nis}^2}{3}\frac{J_{\rm eff}^2}{\hbar\Omega}, \tag{9.19}$$

where the number $d_{\rm m}$ and the distance $d_{\rm nis}$ have been defined in Section 9.1.2. Admittedly, our derivation is crude. A detailed approach can justify this result; see for example Storchak and Prokof'ev (1998).

We are now going to discuss the implications of Eq. 9.19 for the interpretation of experimental results. Note that this equation predicts $\Gamma_t \propto J^2_{\rm eff}/\Omega$.

In a metal at low temperature, but outside the superconducting phase if any, the particle tunnelling rate can be written as (Kondo 1986)

$$\Gamma_t = \frac{J^2}{\hbar} \exp\left[-2S(0)\right] \frac{(K-1)!}{\sqrt{\pi}(K-1/2)!} \left(\frac{\pi k_B T}{\hbar \omega_v}\right)^{2K} \frac{1}{k_B T}, \tag{9.20}$$

where the ratio $(K-1)!/[\sqrt{\pi}(K-1/2)!]$ is on the order of unity. We recognise the factors entering the expression of $J_{\rm eff}$ given in Eq. 9.16. The additional factor proportional to T^{-1} accounts for the dynamical broadening of the energy levels. Equation 9.20 yields $\Gamma_t \propto T^{2K-1}$. Since $K \leq 1/2$, Γ_t is predicted to increase when the temperature decreases. This behaviour has been observed for muons in copper and aluminium; see, for example, the detailed report by Kehr *et al.* (1982) for the latter metal. Defects may lead to a flattening at low temperature. This point is further discussed by Karlsson (1995).

From the polaron and fluctuation barrier mechanisms we expect Γ_t to increase with temperature. The dynamical broadening effects lead to an increase of Γ_t as the system is cooled down. Therefore, between these two asymptotic thermal behaviours, Γ_t should display a minimum at an intermediate temperature. This is effectively observed; see e.g. Kehr *et al.* (1982).

Since low energy excitations in a superconductor are exponentially suppressed because of the superconducting energy gap at the Fermi surface (a gap without nodes is assumed), Ω is exponentially reduced below the superconducting temperature. This leads to an increase of Γ_t, and means that the interstitial particle is expected to diffuse faster in the superconducting phase than in the metallic phase. This is actually observed. An example is discussed in Section 1.2.1.

The dynamical broadening of phonon origin has experimentally been found only for neutral muonium centres in insulators. Since the phonon density of states accessible at low temperature is restricted, consequences on the transfer particle mechanism are expected. Because of energy conservation, two phonons are required if the interstices have the same energy. This gives Ω proportional to $T^{\nu_{\rm ph}}$ with $2 \leq \nu_{\rm ph} \leq 9$. The value of the exponent depends on the shape of the phonon density of states and the geometry of the path between the sites. The temperature range over which the two-phonon process is active is expected to be quite limited: below about one-tenth of Θ_D. The matrix element $J_{\rm eff}$, which is given in Eq. 9.15 for an insulator, is temperature independent at low temperature. Because Ω decreases with temperature, it is reasonable to expect the broadening due to the strain field, $\xi_{\rm strain}$, to dominate over the dynamical broadening at sufficiently low temperature. In fact, the simplest extension (see, for example, Storchak and Prokof'ev 1998) of Eq. 9.19 to account for the strain field is to substitute $1/\Omega$ by $\Omega/(\xi^2_{\rm strain} + \Omega^2)$. Then the site transfer rate becomes proportional to Ω in the limit $\xi_{\rm strain} \gg \Omega$.

In the low-temperature limit and if Ω is negligible, coherent propagation of the particle wave can be observed in the case of a strain field sufficiently weak that the mean free path is large enough relative to the distance between the interstitial sites. The transfer rate Γ_t is then essentially determined by $J_{\rm eff}$ and is therefore temperature independent. It is important to note that the condition for coherent propagation to be observed is so severe that such propagation is expected to be rare.

Finally, we point out that a one-phonon process can play a role at low temperature if the sites over which the muon or muonium centre diffuses are not equivalent. This non-equivalence can arise from the strain field due to defects but it can be intrinsic to the crystal structure as well. More details are given by Storchak and Prokof'ev (1998).

9.3 Diffusion and polarisation functions

The μSR techniques give access to polarisation functions. It is only through the analysis of these functions that information relevant to the localisation site of the muon and muonium centre as well as on the characteristics of their diffusion can be extracted.

The field distribution approach developed in Chapter 6 is usually sufficient to model the polarisation functions. For completeness, we note the detailed computations of the polarisation functions for muons diffusing in ferromagnets presented by Seeger and Monachesi (1982). Their results can be adapted for other cases of physical interest. At least for diffusion in a simple cubic lattice, it is a good approximation to take $\tau_c = \bar{\tau}$ (McMullen and Zaremba 1978), where $\tau_c = 1/\nu_c$ with ν_c the field correlation frequency (see Eq. 6.39), and $\bar{\tau}$ is the mean time of stay of the muon in an interstitial site. This latter time was introduced in Eq. 9.8. The reader should be aware that the equality $\tau_c = \bar{\tau}$ may not always be valid (Kehr *et al.* 1978).

9.4 Examples

The book of Karlsson (1995) and the review of Storchak and Prokof'ev (1998) cover most of the studies of the muon and muonium centre diffusion except for the work carried out on iron. This latter work can be studied from the papers of Yagi *et al.* (1984) and Fritzsche *et al.* (1990). Here we content ourselves with presenting selected examples which, we hope, give a feeling of the type of studies which can be performed.

For convenience the presentation is split into three parts: the first concerns metals, the second deals with semiconductors and dielectrics, and the third shows how useful information can be derived from μSR measurements on compounds of technological interest.

9.4.1 Metals

In a μSR experiment, when the muon polarisation starts to be monitored, i.e. usually around a few nanosecond after implantation, a muon implanted in a metal is in a diamagnetic state, i.e. it is not bound to an electron, and the muon localisation and diffusion properties can be studied.

As already noticed in Section 1.2.1, the first study of muon diffusion was performed for copper (Gurevich *et al.* 1972). It was properly interpreted as arising from a sub-barrier tunnelling rather than an over-barrier hopping. The copper nuclear spins are static on the μSR time scale as explained in Section 10.2.3.2. Thereafter many groups have investigated metallic copper. Schilling *et al.* (1981) derived from the orientation and field dependence of the root mean square of the field distribution at the muon site, $\Delta_{\mathrm{G,TF}}$, that the muon diffuses between octahedral interstitial sites. The electric field gradient induced by the muon electric charge at the copper nuclei was determined using the theory of Hartmann; see Section 6.6.1.2.

Schilling *et al.* (1981) have also recorded transverse-field spectra above 81 K, which have been analysed with the so-called Abragam formula. The temperature dependence of the magnetic correlation frequency, ν_c, has been determined. To a good approximation, it can be taken as the inverse of the mean time of stay of the muon in an octahedral site. As expected, ν_c is independent of the field intensity and of the direction of the applied field. The data have been interpreted with the model introduced by Teichler (1977); see also Yaouanc (1982). This model, which is based on the polaron theory, includes the asymptotic formula given in Eq. 9.17 and the pre-asymptotic regime. From the fits, the polaron activation energy and the tunnelling matrix element have been found to be 79.2 (3.1) meV and 18.3 (1.8) μeV, respectively. A 5% lattice dilation around the muon has been deduced from the value of $\Delta_{\rm G,TF}$ at low field. However, using longitudinal-field measurements Luke *et al.* (1991) have shown that the muon diffuses even in the temperature interval for which $\Delta_{\rm G,TF}$ was measured ($12 < T < 81$ K). This is in contrast to the assumption of a static muon used by Schilling and collaborators to derive a value for the muon-induced local deformation. Therefore, it is understandable that their estimate of the deformation was not correct. The upper limit for the local lattice dilation is now considered to be 2%. In terms of the strain energy involved, this is far more reasonable than the original 5%. In addition, these measurements determine the sign of the muon-induced electric field gradient, which is found to be positive.

The effect of impurities and defects, intrinsic or introduced on purpose, on the muon diffusion has been quite intensively studied. An example is given in Section 1.2.1. The spectra, usually recorded in transverse-field geometry, can be analysed with the system of two differential equations given in Eq. 6.122, at least if the two-state model is a reasonable approximation.

Perhaps the other metal which has been most studied by μSR techniques is aluminium. These studies have culminated with the experimental proof of the drastic effect of superconductivity on the muon mobility; see Section 1.2.1. This quite unexpected result is the definitive proof of the quantum nature and the role of the electrons in the muon diffusion in aluminium. In fact, tunnelling is not only restricted to aluminium and copper and not only to muons. Protons also exhibit tunnelling as discussed by Karlsson (1995).

For the sake of completeness and to illustrate the wide range of mean residence times that can be determined, we present in Fig. 9.3 the magnetic correlation frequency versus the inverse temperature for five face-centred-cubic metals.

The field orientation dependence of the Knight shift measured on a single crystal is a well-known means for the determination of the muon localisation site; see Fig. 1.14 and Section 5.3. In certain instances it also gives information on the muon diffusion. This is illustrated in Fig. 9.4 for the case of the hexagonal intermetallic ferromagnet $GdNi_5$ for which a ferromagnetic transition occurs at $T_{\rm C} = 30.85$ K. It is found that below $\sim$80 K a large fraction of the muons diffuses in an array of sites called either by 6k or 6m in Wyckoff notation, each type of them forming an hexagon. Since it is not known which of them the muon occupies, this site is referred to as the $6_{\rm ring}$

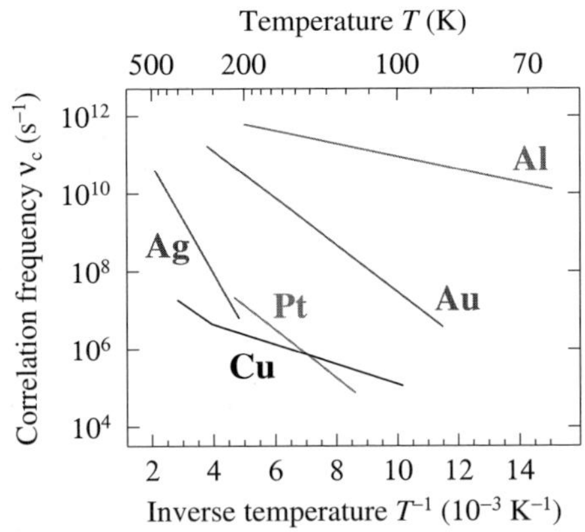

Fig. 9.3 Magnetic correlation frequency (approximately the inverse of the mean time of stay $\bar{\tau}$) in an interstitial site, for positive muons in several face-centred-cubic metals plotted in an Arrhenius diagram. Adapted from Karlsson (1995).

site. An illustration is given in Fig. 9.5. This does not seem to be a rare case since it has been recognised for other hexagonal and for cubic compounds as well. In fact it seems to come into play if the distance between each interstitial site of the ring state is sufficiently small, say about 1 Å.

In a series of compounds which crystallise within the same crystal structure, a diversity of muon sites can be found. At least for the RNi_5 series (R is a rare-earth element), this seems to be related to the relatively low density of states at the Fermi

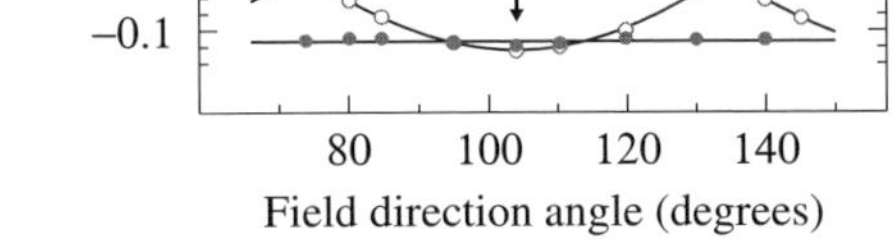

Fig. 9.4 Knight shift observed for different orientations of $\mathbf{B}_{\text{ext}}$ in the $(\mathbf{a}, \mathbf{a}^*)$ plane in the paramagnetic phase of $GdNi_5$ at 47 K. The angles where $\mathbf{B}_{\text{ext}} \parallel \mathbf{a}$ and $\mathbf{B}_{\text{ext}} \parallel \mathbf{a}^*$ are indicated in the figure. The solid lines through the open circles are fits using the model explained in Sections 5.3 and D.6. Besides the signature of the muon site already observed at 100 K (open circles), an additional muon signal (filled circles) is detected with almost no angular dependence (straight line). This latter site is the so-called 6_{ring} site. Reproduced with kind permission from Mulders *et al.* (2000).

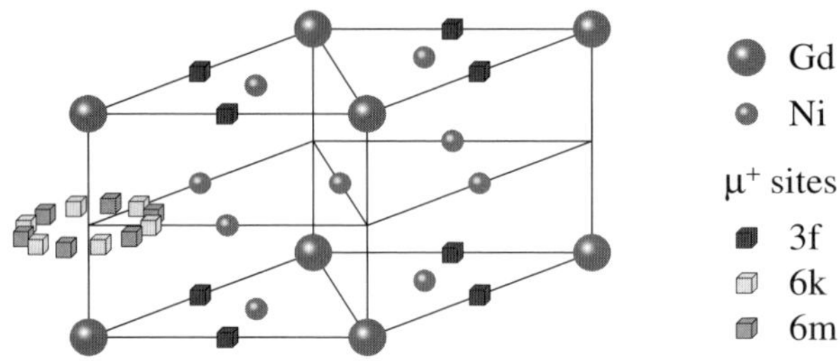

Fig. 9.5 Unit cell of $GdNi_5$ (hexagonal crystal structure, space group $P6/mmm$). The muon has a finite probability to localise at the 3f site in the whole temperature range and also below about 80 K in either the 6m (0.13, 0.26, $\frac{1}{2}$) or 6k (0.225, 0, $\frac{1}{2}$) sites. The muon occupies all positions denoted as 6k or 6m within the time range of the experiment. Adapted from Mulders *et al.* (2000).

level (Mulders *et al.* 2003).

9.4.2 Semiconductors and dielectrics

As examples of studies of diffusion of muonium centres we describe some results obtained for the semiconductor GaAs and for solid nitrogen.

As discussed in Chapter 12, muonium-charged centres are formed in heavily doped semiconductors. In GaAs doped with silicon a negatively charged muonium state is found. The left panel of Fig. 9.6 shows the measured standard deviation of the field distribution at the muonium site, $\Delta_{G,TF}$, as a function of the magnetic field. It is typical for a diamagnetic muon state. These data establish that muonium is located near a tetrahedral interstitial site with Ga nearest neighbours, that is, the T_{Ga} site. As in Chapter 12, we label this centre by Mu_T^-.

The Mu_T^- centre is static on the muon lifetime scale at room temperature and begins to diffuse above about 475 K with a hop rate described by an Arrhenius function as shown in the right panel of Fig. 9.6. The jump rate has been deduced from zero-field measurements. This technique is better suited than the transverse-field technique for measuring slow muon mobility as explained in Section 6.3. The comparison with the jump rate of Mu_T^0 in a high-resistivity GaAs sample shows the dramatic influence of the charge state. At room temperature, Mu_T^- moves more slowly by a factor of $\sim 10^{10}$. As expected, because of its electric charge, the interaction of Mu_T^- with the lattice is very strong, resulting in a large activation energy. Above 700 K, zero-field relaxation from charge-state fluctuations is observed. The analysis of these data implies that $Mu_T^- \longleftrightarrow Mu_T^0$ conversions occur *via* alternating releases and captures of an electron.

We now consider muonium diffusion in solid nitrogen (Storchak and Prokof'ev 1998). Triplet muonium precession corresponding to transitions between field split triplet states, see Section 8.2 for a discussion, has been observed in a transverse-field measurement under a field of 17 mT at 28 K. The hyperfine frequency 4.494 (5) GHz is slightly larger than the vacuum value $\nu_0 = 4.4642$ GHz. This suggests the muonium centre to be neutral and somewhat compressed by interactions with the lattice. This is

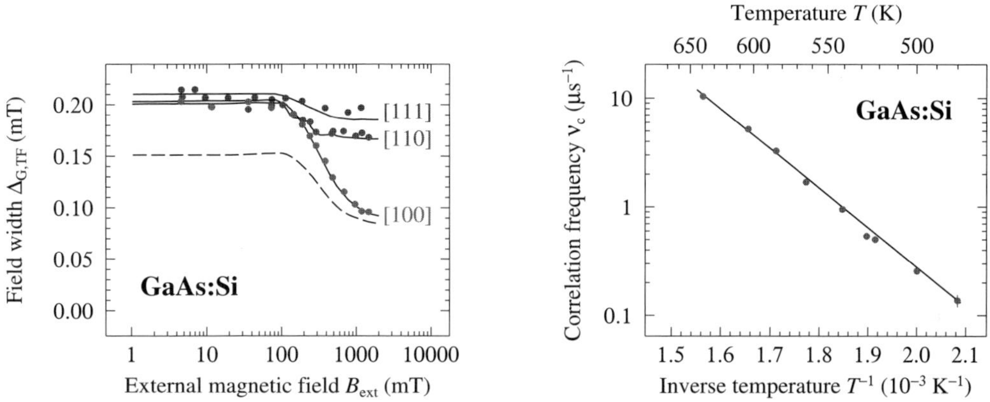

Fig. 9.6 Localisation and diffusion of a negatively charged muonium centre in a heavily doped semiconductor. (Left) The standard deviation of the field distribution as a function of the magnetic field. The data have been recorded at room temperature with the transverse–field technique. The Si concentration of the GaAs crystals used for the three field orientations are in the range $2.5–5\times10^{-18}$ cm^{-3}. The solid curves result from fits to the model presented in Section 6.6.1.2, assuming a T_{Ga} muon site (i.e. with four equivalent Ga neighbours) and allowing for symmetric lattice distortions. The dashed line is the best fit to the [100] direction data in the absence of distortion. (Right) The temperature dependence of the magnetic correlation frequency (approximately the inverse of the mean residence time of the muonium centre in an interstitial site) measured in a crystal of GaAs:Si with a net donor concentration of 4.5×10^{18} cm^{-3}. The solid line is the best fit assuming an Arrhenius law. It corresponds to an activation energy of 0.73 (1) eV and a prefactor of $5.6\,(5)\times10^{12}$ s^{-1}. These values indicate an over-barrier hopping mechanism is at play. Adapted from Chow *et al.* (1995, 1996).

in contrast to most results for which the hyperfine frequency is reduced compared to the free muonium value; see Fig. 12.1. Relaxation effects have been observed by transverse- and longitudinal-field measurements from above helium liquid temperature up to the melting temperature of nitrogen (63.1 K). They are entirely due to the modulation of the nuclear hyperfine field at the muonium site; i.e. the spin-exchange interaction with paramagnetic impurities (e.g. O_2) is negligible. The muonium site transfer rate $\Gamma_{\rm Mu}$ has been extracted using the conventional Redfield formula, Eq. 7.66, for the spin–lattice relaxation rate and setting $\Gamma_{\rm Mu} = \nu_c$. The expected equality of the spin–lattice and spin–spin ralaxation rates (see the discussion leading to Eq. 7.78) has been observed in the motional narrowing limit. The results are shown in Fig. 1.10. Note the drop of the rate from 18 K down to ~8 K. Fits give $\Gamma_{\rm Mu} \propto T^{\alpha}$ with $\alpha = 6.7\,(1)$ below 18 K and $\Gamma_{\rm Mu} \propto T^{-\alpha}$ with $\alpha = 7.3\,(2)$ in the temperature range $30 < T < 50$ K. The muonium coherent motion seems to be destroyed below 18 K. This happens, presumably, because of the orientational freezing of the nitrogen molecules, which induces static disorder. This disorder leads to an energy level distribution at the interstitial sites. If the width of this distribution is large relative to the dynamical level broadening Ω, the transfer rate is expected to be proportional to Ω; see Section 9.2.2. This is effectively observed below 18 K, with approximately $\Omega \propto T^7$, consistent with the two-phonon mechanism. Note that the disorder is relatively mild since the one-phonon process does not play any role. At first sight, the behaviour of Γ_{Mu} for $30 < T < 50\,\mathrm{K}$ is the signature of a two-phonon process in the limit of negligible disorder. However, this mechanism is only expected at a temperature much lower than the Debye temperature, which is 83 K for solid nitrogen. Storchak and Prokof'ev (1998) argue the strong temperature dependence of Γ_{Mu} near the minimum to be largely due to the renormalisation of the tunnelling amplitude.

9.4.3 Compounds of technological interest

Despite a long research history on materials for Li-ion batteries, one of their most important intrinsic physical properties, i.e. how Li^+ ions diffuse, has only been properly characterised recently. This has first been determined using the zero- and longitudinal-field μSR techniques for the cathode material $Li_x[Mn_{1.96}Li_{0.04}]O_4$ (Kaiser *et al.* 2000). The temperature at which the Li^+ ions start to hop was determined and their characteristic hopping frequency was measured. Technically, in order to extract reliable information from the measurements, it is recommended, for each temperature, to record a few spectra (typically three) as a function of the longitudinal-field intensity. The spectra are then simultaneously modeled using the Kubo–Toyabe function, with the dynamics described with the strong-collision model, i.e. the solution of Eq. 6.52 together with Eq. 6.17. Only the magnitude of the external field is changed between each spectrum; see Gubbens *et al.* (2006). In such a way, the variance of the field distribution at the muon and its correlation frequency can reliably be determined. In the case of $Li_x[Mn_{1.96}Li_{0.04}]O_4$ it was possible to distinguish the Li^+ and muon diffusions. More recently, a similar work has been published for Li_xCoO_2 for two values of x (Sugiyama *et al.* 2009). Using a generalisation of Eq. 9.11, these authors have been able to determine the Li^+ diffusion constant from their measured hopping rate. It favourably compares with the results of first-principles calculations. This clearly shows the definitive advantage of using μSR relative to other experimental techniques, such as quasielastic neutron scattering or NMR. For this latter technique the effect of the localised paramagnetic moments on the measured signal is so strong that it is not possible to determine the Li^+ transfer rate.

References

Anderson P.W. (1958) *Physical Review* **109**, 1492–1505.

Chow K.H., Hitti B., Kiefl R.F., Dunsiger S.R., Lichti R.L., and Estle T.L. (1996) *Physical Review Letters* **76**, 3790–3793.

Chow K.H., Kiefl R.F., MacFarlane W.A., Schneider J.W., Cooke D.W., Leon M., Paciotti M., Estle T.L., Hitti B., Lichti R.L., Cox S.F.J., Schwab C., Davis E.A. Morrobel-Sosa A., and Zavieh L. (1995) *Physical Review* **B 51**, R14762–R14765.

Emin D., Baskes M.I., and Wilson W.D. (1979) *Physical Review Letters* **42**, 791–794.

Eshelby H.J.D. (1956) in *Solid State Physics*, 79–144, edited by Seitz F. and Turnbill D., vol. 3 (Academic Press, New York).

Flynn C.P. and Stoneham A.M. (1970) *Physical Review* **B 1**, 3966–3978.

Fritzsche A., Hampele M., Herlach D., Maier K., Major J., Schimmele L., Seeger A., Staiger W., Templ W., and Baines C. (1990) *Hyperfine Interactions* **64**, 691–700.

Fukai Y. and Sugimoto H. (1985) *Advances in Physics* **34**, 263–326.

Fürderer K., Döring K.-P., Gladisch M., Haas N., Herlach D., Major J., Mundinger H.-J., Rosenkranz J., Schäfer W., Schimmele L., Schmolz M., Schwarz W., and Seeger A. (1986) *Hyperfine Interactions* **31**, 81–86.

Gubbens P.C.M., Wagemaker M., Sakarya S., Blaauw M., Yaouanc A., Dalmas de Réotier P., and Cottrell S.P. (2006) *Solid State Ionics* **177**, 145–147.

Gurevich I.I., Mel'eshko E.A., Muratova I.A., Nikol'skii B.A., Roganov V.S., Selivanov V.I., and Sokolov B.V. (1972) *Physics Letters* **A 40**, 143–144.

Hempelmann R. (2000) *Quasielastic Neutron Scattering and Solid State Diffusion* (Clarendon Press, Oxford).

Herlach D. (1981) in *Recent Developments in Condensed Matter Physics*, 93–110, edited by Devreese J.T. (Plenum Press, New York).

Holstein T. (1959a) *Annals of Physics* **8**, 325–342.

Holstein T. (1959b) *Annals of Physics* **8**, 343–389.

Kagan Yu. and Klinger M.J. (1974) *Journal of Physics C: Solid State Physics* **7**, 2791–2807.

Kaiser C.T., Verhoeven V.W., Gubbens P.C.M., Mulder F.M., de Schepper I., Yaouanc A., Dalmas de Réotier P., Cottrell S.P., Kelder E.M., and Schoonman J. (2000) *Physical Review* **B 62**, R9236–R9239.

Karlsson E.B. (1995) *Solid State Phenomena As Seen by Muons, Protons, and Excited Nuclei* (Clarendon Press, Oxford).

Kehr K.W. (1978) in *Hydrogen in Metals I*, 197–226, edited by Alefeld G. and Völkl J. Topics in Applied Physics, vol. 28 (Springer, Berlin).

Kehr K.W., Richter D., Welter J.M., Hartmann O., Karlsson E., Norlin L.O., Niinikoski T.O., and Yaouanc A. (1982) *Physical Review* **B 26**, 567–590.

Kondo J. (1984) *Physica* **B+C 126**, 377–384.

Kondo J. (1986) *Hyperfine Interactions* **31**, 117–133.

Leibfried G. and Breuer N. (1978) *Point Defects in Metals I. Introduction to the Theory*, Springer Tracks in Modern Physics, vol. 81 (Springer, Berlin).

Luke G.M., Brewer J.H., Kreitzman S.R., Noakes D.R., Celio M., Kadono R., and Ansaldo E.J. (1991) *Physical Review* **B 43**, 3284–3297.

McMullen T. and Zaremba E. (1978) *Physical Review* **B 18**, 3026–3040.

Mehrer H. (2007) *Diffusion in Solids: Fundamentals, Methods, Materials, Diffusion-Controlled Processes*, Springer Series in Solid-State Sciences, vol. 155 (Springer, Berlin).

Möslang A., Graf H., Balzer G., Recknagel E., Weidinger A., Wichert Th., and Grynszpan R.I. (1983) *Physical Review* **B 27**, 2674–2681.

Mulders A.M., Kaiser C.T., Harker S.J., Gubbens P.C.M., Amato A., Gygax F.N., Schenck A., Dalmas de Réotier P., Yaouanc A., Buschow K.H.J., and Menovsky A.A. (2003) *Physical Review* **B 67**, 014303, 1–10.

Mulders A.M., Kaiser C.T., Gubbens P.C.M., Amato A., Gygax F.N., Pinkpank M., Schenck A., Dalmas de Réotier P., Yaouanc A., Buschow K.H.J., Kayzel F., and Menovsky A. (2000) *Physica* **B 289-290**, 451–454.

Philibert J. (1991) *Atom Movements. Diffusion and Mass Transport in Solids*, translated from French by Rothman S.J., (Les Editions de Physique, Les Ulis).

Richter D. (1983) in *Neutron Scattering and Muon Spin Rotation. Transport Mechanisms of Light Interstitials in Metals*, 85–222, Springer Tracks in Modern Physics, vol. 101 (Springer, Berlin).

Schilling H., Camani M., Gygax F.N., Rüegg W., and Schenck A. (1981) *Hyperfine Interactions* **8**, 675–679.

Seeger A. and Monachesi P. (1982) *Philosophical Magazine* **B 46**, 283–312.

Shewmon P.G. (1963) *Diffusion in Solids* (McGraw-Hill, New York).

Storchak V.G. and Prokof'ev N.V. (1998) *Reviews of Modern Physics* **70**, 929–978.

Sugiyama J., Mukai K., Ikedo Y., Nozaki H., Månsson M., and Watanabe I. (2009) *Physical Review Letters* **103**, 147601, 1–4.

Teichler H. (1977) *Physics Letters* **A 64**, 78–80.

Teichler H. (1978) *Physics Letters* **A 67**, 313–315.

Vineyard G.H. (1957) *Journal of Physics and Chemistry of Solids* **3**, 121–127.

Waite T.R. (1957) *Physical Review* **107**, 463–470.

Weidinger A. (1984) *Hyperfine Interactions* **17-19**, 153–166.

Yagi E., Flik G., Fürderer K., Haas N., Herlach D., Major J., Seeger A., Jacobs W., Krause M., Krauth M., Mundinger H.J., and Orth H. (1984) *Physical Review* **30**, 441–443.

Yamada K. (1984) *Progress of Theoretical Physics* **72**, 195–201.

Yamashita J. and Kurosawa T. (1958) *Journal of Physics and Chemistry of Solids* **5**, 34–43.

Yaouanc A. (1982) *Physics Letters* **A 87**, 423–425.

10
Investigation of magnetic materials

The studies of magnetic materials are at the forefront of research thanks to the numerous uses of these materials in modern technology. A muon can be implanted in any material and its spin precesses in a magnetic field. These two simple facts are basically the reasons for the increasing use of the μSR techniques for probing magnetic materials. In this chapter we present the applications of μSR to their study. The investigation of the magnetic properties peculiar to superconductors is deferred to Chapter 11.

As already stressed in the previous chapter, our purpose is not to review the published μSR literature but to focus on the methodology, using published results as examples. Many other examples are discussed by Seeger (1978), Denison *et al.* (1979), Schenck (1985), Chappert and Yaouanc (1986), Smilga and Belousov (1994), Karlsson (1995), Schenck and Gygax (1995), Amato (1997), Dalmas de Réotier and Yaouanc (1997), Kalvius *et al.* (2001), and Nagamine (2003). In addition, chapters dedicated to magnetism and written by de Renzi (1999), Rainford (1999), Schenck (1999), and Uemura (1999) are gathered in the proceedings of a Summer School. The Institute of Physics (IOP) has published a large number of articles discussing results obtained for magnetic materials in a special issue of the *Journal of Physics: Condensed Matter* edited by Heffner and Nagamine in 2004.

Magnetism in condensed matter has been nicely described at the undergraduate level by Blundell (2001). A thorough exposition of the principles of magnetism and a discussion of the physical properties of magnetic materials can be found in the book edited by du Trémolet de la Lacheisserie *et al.* in 2002. The lectures of Barbara *et al.* (1988) are quite interesting since they focus on crystal symmetry for an insight into microscopic properties of magnetic materials. We also recommend the lecture notes of Fazekas (1999) for a modern presentation of the theoretical background on magnetism. Neutron scattering techniques are extremely valuable for the study of materials. An introduction is given by Squires (1997).

In this chapter, we shall first examine how static magnetic properties of materials influence the muon polarisation functions, referring to information discussed at length in Chapter 6. However, discussion of the polarisation function resulting from a spontaneous incommensurate amplitude-modulated magnetic structure, on the one hand, and the relation between the coherence length of a magnetic structure and the second moment of the field distribution at the muon site on the other hand, cannot be found elsewhere in this textbook. Hereafter we study the relationship between the polarisation functions and the magnetic correlation tensor of a system, and therefore focus on the dynamics. This leads us to survey relaxation and excitation processes at the origin of the decay of the muon polarisation. This is followed by the presentation of selected examples chosen to display how the μSR techniques enable the measurement of key physical properties of magnetic materials. We conclude this chapter by a comparison, through examples, of the information obtained by μSR and other microscopic experimental techniques.

Up to now, most of the systems studied by μSR have a three-dimensional magnetic character. Therefore we shall focus on them. Works on layered and chain magnetic systems have been reported more recently. We shall touch upon them at the end of Section 10.4. There is also an interest in high-spin molecules, which can be considered as magnetic systems in zero dimension. In this respect, we mention the work of Keren *et al.* (2007), in which it is argued that the muon spin relaxes through an indirect nuclear mechanism. An example of such a type of mechanism is given in Section 10.2.4.2, although the conduction electron spins rather than the nuclear spins are involved there.

10.1 Static properties

Before embarking on the discussion of the static properties which can be studied by μSR, we first discuss the characteristics of the spontaneous field distribution at the muon site.

In the ordered state of a magnetic material, the static field distribution at the muon site is primarily parametrised by two parameters: its mean value and its variance. In the paramagnetic state only the variance is relevant for a zero-field measurement since the mean value of the local field vanishes.

The mean-field value is measured by the frequency of the spontaneous muon spin precession: $\nu_\mu = \gamma_\mu B_0/(2\pi)$. The spontaneous field $\mathbf{B}_0$ is the sum of the dipole and hyperfine fields; see Section 5.4. Once a muon site is assumed, the former field can be computed as explained in Section 5.2. The temperature dependence of the dipole field and of the magnetic moments of a compound track each other, at least if there is only one type of magnetic moments and no spin reorientation. Therefore, if the hyperfine field is negligible, as for most insulators, the measurement of $\nu_\mu(T)$ gives access to the temperature dependence of the order parameter. The nice point is the possibility of performing these measurements in zero external field.

The hyperfine field cannot yet be neglected for a metal. We recall that it stems from the interaction between the muon spin and the spin of the unpaired electrons through the conduction electrons. Since this field is not strictly proportional to the magnetic moment of the unpaired electrons, $\nu_\mu(T)$ may not yield reliable information on the thermal behaviour of the order parameter. Examples are given in Section 10.3. According to Estreicher and Meier (1982) the hyperfine field crucially depends on the zero-point motion of the muon because of the Ruderman–Kittel–Kasuya–Yosida (RKKY)-type oscillating behaviour of the field. The dipole field, however, is not affected by the finite extent of the muon wavefunction, at least if the modulus of this function is spherically symmetric (Meier 1984). The effect of the lattice vibrations on $\mathbf{B}_0$ seems to be negligible.

As already mentionned, in the paramagnetic state the mean value of the field vanishes. If the field distribution at the muon site is Gaussian and isotropic, it is characterised by a single parameter, its variance. Likewise, only a single parameter is needed if the distribution is squared-Lorentzian. The Gaussian and squared-Lorentzian models apply to systems with concentrated and diluted magnetic moments, respectively. If the magnetic moments at the origin of the distribution are not correlated, simple analytical expressions for the parameters are available; see Section 6.6. On the other hand, we are not aware of any analytical formula available for correlated magnetic moments, except for the cases studied by Gist and Dodds (1984) for diluted magnetic moments. We expect the correlations to influence the shape of the distribution. This means that more than one parameter may be needed for its modelling; see the short discussion at the end of Section 6.4.1.3.

It is convenient to start by first examining the muon response to a long-range magnetic order. Then we consider the case where only short-range magnetic correlations are present. A discussion of the significance of long- versus short-range order from the μSR point of view follows in the next part. For simplicity, the effect of the nuclear magnetic moments on the polarisation functions will be neglected here. It is treated in Chapter 6.

10.1.1 Long-range correlations

Here we suppose we probe a magnetic compound in its ordered state characterised by a periodic magnetic structure; that is to say, magnetic Bragg reflections can be observed by neutron or X-ray magnetic scattering experiments.

10.1.1.1 Simple case: a unique spontaneous field with possibly a narrow distribution. We consider here the most common magnetic structures. More complex situations will be examined in Section 10.1.1.2. We shall assume that there is only one type of muon localisation site in the crystal structure. However, to such a single site may correspond more than one site from the magnetic point of view. For example, let us consider a body-centred-cubic lattice and a ferromagnetic collinear arrangement of the magnetic moments parallel to the cube edge. This is the case for metallic iron. In such a lattice, both the octahedral and the tetrahedral interstitial positions possess tetragonal symmetry. Inside a family of sites, a given site is characterised by the orientation of its tetragonal axis relative to the cube edges, and therefore to the magnetisation direction. This means that two muon precession frequencies can possibly be observed, although there is only one type of crystal sites. This point is further discussed by Seeger (1978).

Going back to the simple case, one type of crystal interstitial sites corresponds to a single type of magnetic sites. Then, for a measurement performed in zero field not too close to $T_{\rm c}$, the longitudinal polarisation function is given by (see Section 6.2)

$$P_Z^{\rm stat}(t) = \cos^2\theta + \sin^2\theta \exp\left(-\frac{\gamma_\mu^2 \Delta_{\rm G,TF}^2 t^2}{2}\right) \cos\left(\omega_\mu t\right), \tag{10.1}$$

where θ is the polar angle of $\mathbf{B}_0$. Here, the effect of the finite size of the correlation length of the magnetic structure on the field at the muon site is empirically described by the Gaussian envelope. In the case of an almost perfect structure, the component of the field distribution at the muon site along the direction of the spontaneous field is Lorentzian rather than Gaussian (see Section 6.4.1) and therefore

$$P_Z^{\rm stat}(t) = \cos^2\theta + \sin^2\theta \exp\left(-\gamma_\mu \Delta_{\rm L,TF} t\right) \cos\left(\omega_\mu t\right). \tag{10.2}$$

The factorisation of the second term of the right-hand side for both Eqs. 10.1 and 10.2 is valid if the mean field at the muon site is sufficiently large relative to the width of the field distribution; see Section 6.2.1.

Since the polarisation function obtained for example in Eqs. 10.1 or 10.2 is derived from a static distribution model it cannot account for the magnetic moment dynamics. To account for the spin–lattice relaxation (see Section 7.1.2), it is usually a good approximation to multiply the $\cos^2\theta$ term by an exponential function characterised by the spin–lattice relaxation rate, λ_Z. This rate is discussed at length in Section 10.2. The conditions for the validity of the exponential relaxation function approximation are spelt out in Section 7.1. As explained in Section 7.1.4.3, spin dynamics has also an influence on the envelope of the wiggles. The interplay between this dynamical relaxation and the static damping (supposed to account for the defects of the magnetic structure) in the formation of the wiggle envelope has never been investigated in detail. We note that the expression for the spin–spin relaxation rate λ_X given in Section 7.1.4.3 contains both dynamical and static terms. Phenomenologically, the oscillating term of $P_Z(t)$ is usually written as the product of the geometrical factor, $\sin^2\theta$, by a cosine function and a damping function which, in general terms, is taken as either an exponential or a Gaussian function. However, the reader should realise that this may not always be justified.

The magnitude, orientation in the crystal lattice, and temperature dependence of $\mathbf{B}_0$ carry significant information. While performing a single-crystal experiment, if the muon site symmetry is appropriate and if the magnetic domains are aligned along a single direction, it is usually possible to select the experimental geometry such that $\mathbf{B}_0$ is either parallel or perpendicular to the Z axis, i.e. $\theta = 0$ or $\pi/2$ in Eqs. 10.1 and 10.2. This allows one to study either the spin–lattice relaxation channel or the precession channel, which is characterised by an oscillation and its decay. Then the data analysis is considerably simplified and the amplitude of the signal of interest is optimised. The wiggle decay depends on the field distribution as well as on the spin dynamics, but the spin–lattice relaxation channel is simpler to interpret since only spin dynamics contributes to it. It is only when this spin dynamics is properly characterised that the decay of the wiggles can safely be interpreted.

When neutron or X-ray scattering measurements indicate the presence of Bragg magnetic reflections, common wisdom tells us that these reflections are the signature of a more or less long-range magnetic moment order. In zero-field μSR experiments wiggles should appear. They stem from the Larmor precession of the muon spins about the spontaneous magnetic field at the muon site. However, they may not be observed if the field vanishes because of the high magnetic symmetry at the muon site or if the spin dynamics is sufficiently fast. An example of the former situation is given in Section 10.4. The latter mechanism has already been discussed in Section 6.3.2 where the local field was assumed to be a stochastic variable taking the two values $\mathbf{B}_{\rm fluc}$ and $-\mathbf{B}_{\rm fluc}$, with $\mathbf{B}_{\rm fluc}$ aligned along the X axis. The resulting $P_Z(t)$ function was found oscillating only if the correlation time $\tau_{\rm c}$ satisfies the relation $2\gamma_\mu B_{\rm fluc}\tau_{\rm c} > 1$ (see Eq. 6.70). More intuitively, it can be noticed that the spontaneous field at the muon must last for a minimum time to enable the muon precession to be detected. A crude examination of the previous expression leads to the conclusion that the precession angle performed by the muon spin in the time interval of τ_c must be at least 0.5 radian. Numerically for a fluctuating spontaneous field of 0.1 T at the muon site, τ_c should be longer than $\sim$6 ns.

In a conventional magnet the magnetic moments are frozen on the time scale required to detect wiggles; i.e. their characteristic magnetic correlation time is far longer than a few nanoseconds. However, the reader must be aware that there are exceptions to this rule. For example, the geometrically frustrated magnetic compound $Tb_2Sn_2O_7$ undergoes a magnetic transition at $T_c \simeq 0.9$ K with magnetic susceptibility, specific heat, and neutron scattering signatures. However, neither wiggles nor asymmetry loss is detected for $T \leq T_c$; instead, an exponential decay of the zero-field μSR signal is observed. Dalmas de Réotier *et al.* (2006) interpreted this result in terms of the presence of a fast fluctuation mode in the dynamical response of $Tb_2Sn_2O_7$ with a characteristic time of 0.1 ns, a value derived from Eq. 6.73.

In general, two types of magnet are observed. Either the periodicity of the magnetic modulation can be expressed as a rational fraction of the periodicity of the underlying crystal lattice or as an irrational number. The former case corresponds to the so-called commensurate structure and the second one gives rise to an incommensurate magnetic structure. The object of the next part is the computation of $P_Z(t)$ in these systems. We shall restrict ourselves to amplitude-modulated magnetic structures.

10.1.1.2 Incommensurate magnetic structure. A well-known example for an incommensurate magnet is found for metallic chromium, whose magnetic structure below 123 K is described in terms of a longitudinal spin-density wave characterised by an incommensurate wave vector and magnetic moments parallel to the [100] axis of the body-centred-cubic structure; see for example Fawcett (1988). The physics of incommensurate phases has been reviewed by Izyumov and Syromyatnikov (1990) using Ginzburg–Landau functionals. The review of Blinc (1981) on magnetic resonance line shapes in structurally incommensurate systems is extremely valuable.

Here we shall consider the magnetic structure to be described by a single propagation vector $\mathbf{k}$. Possible generalisations will be mentioned at the end of this section. Instead of a single or a finite number of discrete values for $\mathbf{B}_0$ in the case of commensurate magnetic structures, we expect a continuous set of local fields for the incommensurate ones. In order to compute $P_Z^{\rm stat}(t)$, it is then natural to resort to the concept of field distributions which we introduced in Section 6.1.

As a first step, we assume $\mathbf{B}_0$ to be proportional to the magnetic moments and to be directed along the X axis. For definitiveness, we describe the modulation as a spin-density wave. We write

$$B_0^X(\mathbf{r}_0') = \cos\left[\alpha\left(\mathbf{r}_0'\right)\right] B_{\rm max}. \tag{10.3}$$

The amplitude $B_{\rm max}$ gives the scale of the field. In an experiment the several million implanted muons enter different equivalent sites which are related by the lattice translational symmetry. Since according to Eq. 10.3, the spontaneous field differ from site to site, rather than the single vector $\mathbf{r}_0$ the ensemble of vectors $\{\mathbf{r}_0'\}$, where $\mathbf{r}_0' = \mathbf{r}_0 + \mathbf{i}$, must be considered. We recall that $\mathbf{i}$ is a vector belonging to the direct lattice and $\mathbf{r}_0$ defines a muon site (see Fig. 5.3). As we focus on single-$\mathbf{k}$ sinusoidal modulations, the plane wave approximation holds, i.e. $\alpha(\mathbf{r}'_0) = \mathbf{k}\cdot\mathbf{r}'_0+\varphi$. Since the phase φ is unimportant for our purpose we drop it in the following. For a commensurate magnetic structure $\mathbf{k} = (t/s)\mathbf{K}$, where t and s are integers with $t < s$ and $\mathbf{K}$ is a reciprocal-lattice vector. Since $\mathbf{K} \cdot \mathbf{i} = 2\pi u$ where u is an integer (see Section E.2),

$$\exp\left[i\mathbf{k}\cdot\mathbf{r}'_0\right] = \exp\left(i\mathbf{k}\cdot\mathbf{r}_0\right)\exp\left(i2\pi\frac{tu}{s}\right). \qquad (10.4)$$

Because

$$\exp\left(i2\pi\frac{t}{s}u\right) = \exp\left[i2\pi\frac{t}{s}\left(u+s\right)\right], \qquad (10.5)$$

$\exp[i2\pi(tu/s)]$ as a function of u is periodic with the period s. Therefore this phase factor takes a maximum of s discrete different values, although u can take an enormous number of values. A maximum of s contributions for the muon polarisation function is then expected for a given interstitial site. However, for an incommensurate magnetic structure $\mathbf{k} = v_{\rm ir}\mathbf{K}$ where $v_{\rm ir}$ is an irrational number, i.e. $\mathbf{k}$ cannot be written as the ratio of two integer numbers times $\mathbf{K}$, and so a uniform and continuous distribution of phases is generated. In conclusion, because of the incommensurate nature of the modulation, $\alpha(\mathbf{r}'_0)$ is a continuous variable with values between 0 and 2π and therefore, according to Section C.7.1, the field distribution $D_{\rm c}(B_0^X)$ is given by

$$D_{\rm c}(B_0^X) = \frac{\displaystyle\int_0^{2\pi}\delta\left(\cos\alpha\, B_{\rm max} - B_0^X\right){\rm d}\alpha}{\displaystyle\int_0^{2\pi}{\rm d}\alpha}. \qquad (10.6)$$

With the help of Eq. C.43,

$$D_{\rm c}(B_0^X) = \frac{1}{\pi}\frac{1}{\sqrt{\left(B_{\rm max}\right)^2 - \left(B_0^X\right)^2}}, \qquad (10.7)$$

for $-B_{\rm max} < B_0^X < B_{\rm max}$ and zero otherwise. The distribution $D_{\rm c}(B_0^X)$, which was first derived by Overhauser (1960) and is displayed in Fig. 10.1, is normalised to 1, as it should be. In addition, $D_{\rm c}(B_0^X = 0) = 1/\left(\pi B_{\rm max}\right)$. It is therefore a symmetric continuous distribution with an appreciable weight at $B_0^X = 0$, that is, there is a finite probability for a vanishing local magnetic field at the muon site. This double-peak structure is also predicted under certain conditions for a vortex lattice moving as a whole past a fixed muon; see Section 11.1.1.4 for further discussion. In contrast to the cases considered in Chapter 6, the distribution is not the signature of a random distribution but arises from the incommensurability of the magnetic structure.

The polarisation function is obtained from Eq. 6.6 with the use of Eq. 6.1. Following our assumption that $\mathbf{B}_0$ is directed along the X axis we have set $D_c(B_0^Y) = \delta(B_0^Y)$ and $D_c(B_0^Z) = \delta(B_0^Z)$. Using Eqs. C.81 and C.102 we arrive at

$$P_Z^{\rm stat}(t) = \int_{-B_{\rm max}}^{B_{\rm max}} D_{\rm c}(B_0^X)\cos\left(\gamma_\mu B_0^X t\right){\rm d}B_0^X = J_0\left(\gamma_\mu B_{\rm max} t\right), \qquad (10.8)$$

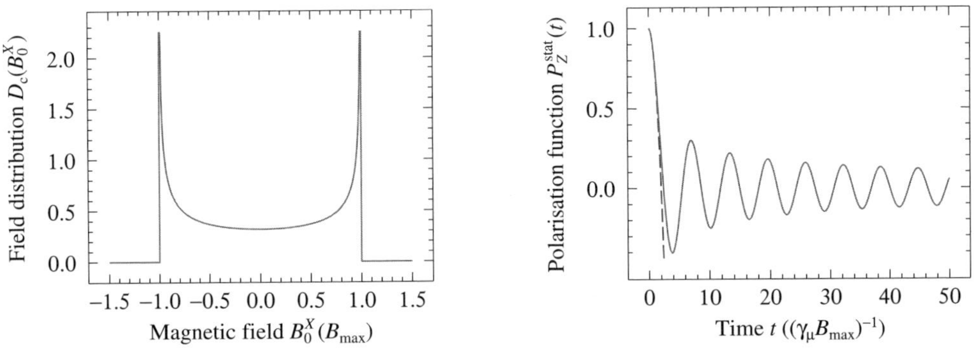

Fig. 10.1 (Left) Field distribution for a cosine-modulated magnetic field at the muon site. A one-dimensional model and a plane wave modulation are assumed for this field. Note the van Hove singularities at $\pm B_{\max}$. (Right) The related polarisation function. The fast decay of the polarisation at small time is remarkable. The time scale is given in terms of the inverse of $\gamma_\mu B_{\max}$, where $B_{\max}$ is the maximum value of the modulus of the field at the muon site. The dashed line is the small time approximation given by Eq. 10.10. The large time expansion of Eq. 10.9 (not shown) already matches the actual polarisation function within the line thickness for $\gamma_\mu B_{\max} t \gtrsim 1$.

where J_0 is the zeroth-order Bessel function of the first kind. We plot $J_0\left(\gamma_\mu B_{\max} t\right)$ in Fig. 10.1. When t is large relative to $1/(\gamma_\mu B_{\max})$, from Eq. C.95

$$J_0\left(\gamma_\mu B_{\max} t\right) \simeq \sqrt{\frac{2}{\pi \gamma_\mu B_{\max} t}} \cos\left(\gamma_\mu B_{\max} t - \frac{\pi}{4}\right). \tag{10.9}$$

Thus, a negative phase shift appears when the field at the muon site is modulated as given by Eq. 10.3. In the large t limit the wiggles are damped, although the damping is relatively weak. A numerical study shows that the previous form is an excellent approximation to $P_Z^{\rm stat}(t)$ even at fairly small time; see caption of Fig. 10.1.

Using Eq. C.93, we now give the short time expansion of Eq. 10.8:

$$J_0\left(\gamma_\mu B_{\max} t\right) \simeq 1 - \frac{\left(\gamma_\mu B_{\max}\right)^2}{4} t^2. \tag{10.10}$$

In this limit, the damping of the wiggles is strong as it is controlled by $B_{\max}$ and not the root mean square of the distribution as usual.

Instead of working directly with $\mathbf{B}_0$, we should deal with the magnetic moments themselves and derive the distribution of $\mathbf{B}_0$. We represent each of the moments $\mathbf{m}_i$ by a $\mathbf{k}$-Fourier component

$$\mathbf{m}_i = \mathbf{m}_0 \cos\left(\mathbf{k} \cdot \mathbf{i}\right). \tag{10.11}$$

We do not again account for a possible phase parameter since it does not directly affect our present discussion. In Eq. 10.11, we assume the structure to be collinear with a modulation of the magnitude of the magnetic moments. Other types of modulations, for instance in the orientation of the magnetic moments, are also found in magnetic materials. We leave to the interested reader the task of investigating the resulting polarisation functions in such cases.

Using Eqs. 5.20 and 5.18, $\mathbf{B}_0(\mathbf{r}'_0)$ is written as

$$\mathbf{B}_0(\mathbf{r}'_0) = \cos(\mathbf{k}\cdot\mathbf{r}'_0)\,\mathbf{B}_{0,\mathrm{c}} + \sin(\mathbf{k}\cdot\mathbf{r}'_0)\,\mathbf{B}_{0,\mathrm{s}}, \tag{10.12}$$

where

$$\mathbf{B}_{0,\mathrm{c}} = \frac{\mu_0}{4\pi}\left\{\sum_{i=1}^{N}\left[-\frac{\mathbf{m}_0}{r_i^3} + \frac{3\,(\mathbf{m}_0\cdot\mathbf{r}_i)\,\mathbf{r}_i}{r_i^5}\right]\cos(\mathbf{k}\cdot\mathbf{r}_i) + \mathbf{m}_0\frac{H}{v_0}\sum_{i'=1}^{N'}\cos(\mathbf{k}\cdot\mathbf{r}_{i'})\right\}, \tag{10.13}$$

and

$$\mathbf{B}_{0,\mathrm{s}} = \frac{\mu_0}{4\pi}\left\{\sum_{i=1}^{N}\left[-\frac{\mathbf{m}_0}{r_i^3} + \frac{3\,(\mathbf{m}_0\cdot\mathbf{r}_i)\,\mathbf{r}_i}{r_i^5}\right]\sin(\mathbf{k}\cdot\mathbf{r}_i) + \mathbf{m}_0\frac{H}{v_0}\sum_{i'=1}^{N'}\sin(\mathbf{k}\cdot\mathbf{r}_{i'})\right\}. \tag{10.14}$$

The hyperfine interaction, with coupling constant H is restricted to the N' nearest neighbours to the muon. The spontaneous field $\mathbf{B}_0$ describes an ellipse in the $(\mathbf{B}_{0,\mathrm{c}}, \mathbf{B}_{0,\mathrm{s}})$ plane (Schenck *et al.* 2001). As already explained, the argument of the cosine and sine functions is a continuous variable and therefore we can substitute Eq. 10.12 by

$$\mathbf{B}_0(\alpha) = \cos\alpha\,\mathbf{B}_{0,\mathrm{c}} + \sin\alpha\,\mathbf{B}_{0,\mathrm{s}}, \tag{10.15}$$

where α is a continuous variable as in Eq. 10.3 when the set of vectors $\{\mathbf{r}'_0\}$ is spanned. The lattice sums $\mathbf{B}_{0,\mathrm{c}}$ and $\mathbf{B}_{0,\mathrm{s}}$ are the same for all equivalent muon positions. When these positions are centres of symmetry, $\mathbf{B}_{0,\mathrm{s}} = \mathbf{0}$, and therefore $\mathbf{B}_0$ is parallel to $\mathbf{B}_{0,\mathrm{c}}$, i.e. $\mathbf{B}_0$ has a fixed direction in space. Then, from Eqs. 6.6 and 10.8

$$P_Z^{\mathrm{stat}}(t) = \cos^2\theta + \sin^2\theta J_0\left(\gamma_\mu B_{\mathrm{max}} t\right), \tag{10.16}$$

where θ is the polar angle of $\mathbf{B}_{0,\mathrm{c}}$ and $B_{\mathrm{max}} = B_{0,\mathrm{c}}$. Note that a μSR spectrum provides the amplitude of the modulation *via* the periodicity of the oscillation.

We now study the case for which the muon positions are not centres of symmetry.

Let us first consider $\mathbf{B}_{0,\mathrm{c}}$ and $\mathbf{B}_{0,\mathrm{s}}$ to be perpendicular to each other and in a first instance we admit their modulus differs. Since we shall assume below $\mathbf{B}_{0,\mathrm{c}}$ and $\mathbf{B}_{0,\mathrm{s}}$ to be both perpendicular to the Z axis, it is practical to consider the distribution of the modulus of $\mathbf{B}_0$, namely $D_{\mathrm{m}}(B_0)$; see Eq. C.34 for its definition. We find

$$D_{\mathrm{m}}(B_0) = \frac{2}{\pi}\frac{B_0}{\sqrt{(B_0)^2 - (B_{\mathrm{min}})^2}\sqrt{(B_{\mathrm{max}})^2 - (B_0)^2}}, \tag{10.17}$$

for $B_{\mathrm{min}} < B_0 < B_{\mathrm{max}}$, and zero otherwise. We have set $B_{\mathrm{min}} = \min\{B_{0,\mathrm{c}}, B_{0,\mathrm{s}}\}$ and $B_{\mathrm{max}} = \max\{B_{0,\mathrm{c}}, B_{0,\mathrm{s}}\}$. Since $\mathbf{B}_0$ depends on the single parameter α, see Eq. 10.15,

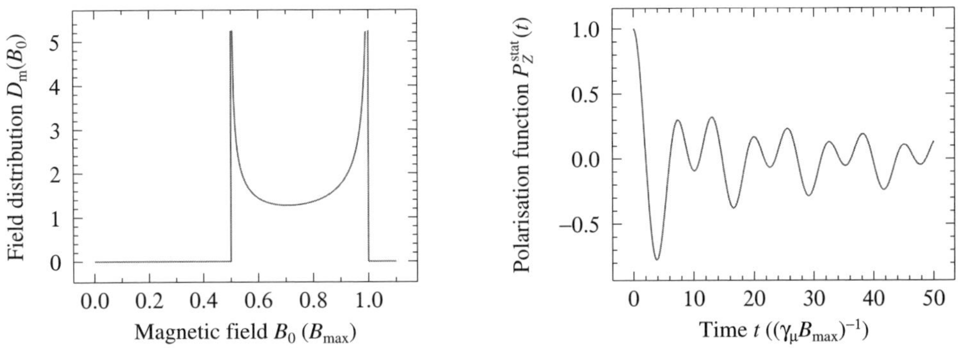

Fig. 10.2 (Left) Field distribution whose analytical expression is given in Eq. 10.17 with $B_{max} = 2B_{min}$. The van Hove singularities (divergence) are at $B_0 = B_{min}$ and $B_0 = B_{max}$. (Right) The associated polarisation function in the case where the local field lies in the plane perpendicular to the Z axis.

$D_m(B_0)$ is a one-dimensional distribution. It exhibits two peaks which can be experimentally misinterpreted as the signature of two different muon sites. The associated $P_Z^{stat}(t)$ function can be numerically computed. If $\mathbf{B}_{0,c}$ and $\mathbf{B}_{0,s}$ are both perpendicular to the Z axis, according to Eq. 6.6,

$$P_Z^{stat}(t) = \int_{B_{min}}^{B_{max}} D_m(B_0)\cos(\gamma_\mu B_0 t) dB_0. \tag{10.18}$$

An example of $D_m(B_0)$ and the associated polarisation function are presented in Fig. 10.2. We have so far discarded the case for which $B_{0,c} = B_{0,s}$. If so, $\mathbf{B}_0$ is located on a circle and $D_m(B_0) = \delta(B_0 - B_{0,c})$. The corresponding polarisation function is $P_Z^{stat}(t) = \cos(\gamma_\mu B_{0,c} t)$ if $\mathbf{B}_{0,c}$ and $\mathbf{B}_{0,s}$ are still both perpendicular to the Z axis.

The interested reader will derive $P_Z^{stat}(t)$ in the case where these vectors are not both perpendicular to the Z axis. He/She will start from Eq. 6.6 together with Eq. 10.17.

Let us now turn to the general case where the fields $\mathbf{B}_{0,c}$ and $\mathbf{B}_{0,s}$ are not perpendicular to each other. The path of $\mathbf{B}_0$ is still located on an ellipse of which we shall find semimajor and semiminor axes denoted by $\mathbf{B}_{max}$ and $\mathbf{B}_{min}$, respectively. Since these vectors belong to the ellipse they satisfy Eq. 10.15 and their corresponding angles α_{max} and α_{min} maximise and minimise B_0, respectively, i.e. obey the equation $d[B_0(\alpha)]^2/d\alpha = 0$. If $B_{0,c} = B_{0,s}$ the solutions of this latter equation are $\alpha = \pi/4 + k\pi/2$ where k is an integer; otherwise they satisfy

$$\tan(2\alpha) = 2\frac{\mathbf{B}_{0,c} \cdot \mathbf{B}_{0,s}}{B_{0,c}^2 - B_{0,s}^2}; \tag{10.19}$$

i.e. the solutions are $\alpha = (\arctan C + k\pi)/2$ where C is the right-hand-side term of Eq. 10.19. In any case, α_{max} and α_{min} can be chosen such that $\alpha_{max} = \alpha_{min} + \pi/2$.

Now setting $\alpha = \alpha_{max} - \beta$ in Eq. 10.15, $\mathbf{B}_0$ reads

$$\mathbf{B}_0 = \cos\beta\,\mathbf{B}_{\max} + \sin\beta\,\mathbf{B}_{\min}, \tag{10.20}$$

since, with our notations,

$$\begin{aligned}\mathbf{B}_{\max} &= \mathbf{B}_0(\alpha_{\max}) = \cos\alpha_{\max}\mathbf{B}_{0,\mathrm{c}} + \sin\alpha_{\max}\mathbf{B}_{0,\mathrm{s}},\\ \mathbf{B}_{\min} &= \mathbf{B}_0(\alpha_{\min}) = \cos\alpha_{\min}\mathbf{B}_{0,\mathrm{c}} + \sin\alpha_{\min}\mathbf{B}_{0,\mathrm{s}}.\end{aligned} \tag{10.21}$$

Because β depends linearly on α in the interval $[0, 2\pi[$, it takes uniformly distributed values if α does so. In addition $\mathbf{B}_{\max}$ and $\mathbf{B}_{\min}$, being semimajor and semiminor axes of the ellipse, those are perpendicular to each other. These conditions are required to derive Eq. 10.17, which therefore applies even if $\mathbf{B}_{0,\mathrm{c}}$ and $\mathbf{B}_{0,\mathrm{s}}$ are not perpendicular.

In addition to the distribution arising from the incommensurability of the magnetic structure, the muon spin may feel a field distribution due, for example, to the finite magnetic correlation length. Restricting ourselves to a muon located at a centre of symmetry and a local field perpendicular to the Z axis, the polarisation function $P_Z^{\mathrm{stat}}(t)$ is simply the cosine Fourier transform of the field distribution, $D_{\mathrm{c}}(B_0^X)$. This latter can be expressed as the convolution product of the two field distributions, that is,

$$D_{\mathrm{c}}\left(B_0^X\right) = \int D_{\mathrm{c},1}\left(B_{0,1}^X\right) D_{\mathrm{c},2}\left(B_0^X - B_{0,1}^X\right) \mathrm{d}B_{0,1}^X, \tag{10.22}$$

provided no correlations exist between the two distributions. The distribution $D_{\mathrm{c},1}(B_0^X)$ is given by Eq. 10.7 and $D_{\mathrm{c},2}(B_0^X)$ can be a Gaussian or a Lorentzian function as explained in Section 6.6. Rather than attempting to compute $D_{\mathrm{c}}(B_0^X)$, it is simpler to remember that the Fourier transform of a convolution product is the product of the Fourier transforms. Hence

$$\begin{aligned}P_Z^{\mathrm{stat}}(t) &= \exp\left(-\frac{\gamma_\mu^2\Delta_{\mathrm{G,ZF}}^2 t^2}{2}\right) \mathrm{J}_0\left(\gamma_\mu B_{\max} t\right),\\ P_Z^{\mathrm{stat}}(t) &= \exp\left(-\gamma_\mu\Delta_{\mathrm{L,ZF}} t\right) \mathrm{J}_0\left(\gamma_\mu B_{\max} t\right),\end{aligned} \tag{10.23}$$

in the concentrated and diluted limit, respectively. This justifies the usual practice of fitting data with one of these two functions. It may be a rough approximation and the factorisation of a Bessel modulation with an envelope function may not be valid. Improving over the present result may require numerical computations.

We have shown that the field distribution due to an incommensurate single-**k** collinear magnetic structure (with the additional hypothesis that the muon site is a centre of symmetry) leads to a phase shift of the wiggles of $-\pi/4$ at large time. It is interesting to generalise the distribution written in Eq. 10.7 such that the phase shift is controlled by a continuous parameter. A possibility is to consider the distribution:

$$D_{\mathrm{c}}\left(B_0^X\right) = \frac{\nu!}{\pi^{1/2}\left(\nu - 1/2\right)!\, B_{\max}\left[1 - \left(B_0^X/B_{\max}\right)^2\right]^{1/2-\nu}}, \tag{10.24}$$

with $\nu > -\frac{1}{2}$. The factorial function is defined through Eqs. C.108 and C.107. The reader can verify that the distribution is properly normalised using Eqs. C.80 and C.102. Examples of this distribution for different values of ν are displayed in Fig. 10.3.

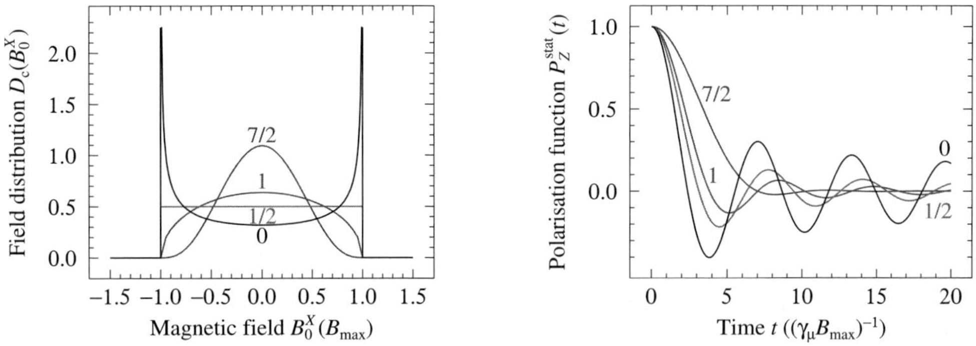

Fig. 10.3 Examples of one-dimensional field distributions given in Eq. 10.24 and the associated polarisation functions (Eq. 10.25). The latter functions have been plotted for a field polar angle $\theta = \pi/2$. The index ν is indicated for each curve. Note that the case $\nu = 0$ corresponds to the model in Fig. 10.1.

From this form of distribution, we derive

$$P_Z^{\rm stat}(t) = \cos^2\theta + \sin^2\theta\, \nu! \left(\frac{2}{\gamma_\mu B_{\rm max} t}\right)^\nu J_\nu\left(\gamma_\mu B_{\rm max} t\right), \tag{10.25}$$

where J_ν is a Bessel function of the first kind. This is obtained using Eq. C.81. According to Eq. C.93, near $t = 0$

$$\nu! \left(\frac{2}{\gamma_\mu B_{\rm max} t}\right)^\nu J_\nu\left(\gamma_\mu B_{\rm max} t\right) \simeq 1 - \frac{\nu!}{(\nu+1)!}\frac{\left(\gamma_\mu B_{\rm max}\right)^2}{4} t^2. \tag{10.26}$$

At long times, from Eq. C.95

$$\nu! \left(\frac{2}{\gamma_\mu B_{\rm max} t}\right)^\nu J_\nu\left(\gamma_\mu B_{\rm max} t\right) \simeq \frac{\nu!}{\sqrt{\pi}} \left(\frac{2}{\gamma_\mu B_{\rm max} t}\right)^{\nu+1/2} \cos\left[\gamma_\mu B_{\rm max} t - \frac{\pi}{2}\left(\nu + \frac{1}{2}\right)\right]. \tag{10.27}$$

Therefore the phase shift $-\pi/4$ is a peculiarity of the distribution with $\nu = 0$: the model distribution chosen in Eq. 10.24 can yield any phase shift. A 2π phase shift is then obtained for $\nu = \frac{7}{2}$, but note the severe damping of the oscillations. Although the function given in Eq. 10.25 does not derive from a known distribution, it may be practical for parametrising experimental spectra. It should be noticed that muon trapping can lead to a phase shift as well; see Section 6.4.2. In that case, the sign of the shift depends on which of the frequencies (associated with the trapping or the free state) is the larger.

We have shown that a single-**k** incommensurate amplitude-modulated magnetic structure leads to a J_0-like oscillation. Conversely, the observation of such oscillations does not guarantee that the probed magnetic structure is incommensurate. A counterexample is presented in Fig. 10 of the paper of Chiba *et al.* (1992), where a relatively complex yet commensurate structure yields a field distribution reminiscent of that plotted in Fig. 10.2.

Complex incommensurate structures have been treated in the hyperfine interaction limit; see Blinc (1981) and Koumoulis *et al.* (2010). Accounting for the dipole interaction using the material presented above, should not be a problem. Going beyond the first harmonic approximation in the development of the modulation is possible, at least theoretically. However, we note that the observation of these harmonics by diffraction techniques is quite uncommon.

McMillan (1976) — see also e.g. Chaikin and Lubensky (1995) — pointed out the possibility of inhomogeneous structures composed of commensurate regions separated by regions with large incommensurability called discommensurations. Convincing evidences for spin discommensurations, or more simply spin slips, have been gathered by diffraction techniques — for example, Gibbs *et al.* (1985). The ability of nuclear magnetic resonance (NMR) to distinguish a plane-wave-incommensurate charge density wave from one possessing a discommensurate structure has been demonstrated; see for example Ross *et al.* (1986). Technically, the phase of the modulation is no longer linear in wave vector but is strongly nonlinear. This results in asymmetric field distributions with smeared van Hove singularities. Hence, a discommensuration is an additional source of damping of the oscillations expected in the magnetically ordered state of a material.

As shown in Section 6.2.3 for a compound with a simple magnetic structure, the powder-averaged response may not necessarily result from a simple angular average of the formula derived for a single crystal. This result applies as well for systems with an incommensurate magnetic structure. Instead of the average, the convolution of distributions should be used. We leave the task to the interested reader.

So far the dynamical nature of the spin system has not been accounted for in our derivation of the polarisation function in incommensurate magnets; see for example Eqs. 10.16 or 10.25. This can phenomenologically be carried out in the same way as earlier in this section in the case of commensurate magnetic structures.

10.1.2 Short-range correlations

Much work has been devoted since the 1990s to the study of magnetic compounds for which a long-range magnetic order does not appear even in the millikelvin temperature range and only short-range correlations are detected. This absence of magnetic order can be due, for example, to the frustration of the magnetic interactions. A common characteristic of this type of magnetism is the unusual fluctuation rates involved. The zero-field μSR technique has played a crucial role in revealing them.

Depending on the value of the fluctuation rates, it is convenient to think about the decay of a muon polarisation in one of the two following ways. If the fluctuations are slow and therefore lead to a decay of $P_Z(t)$ of the Kubo–Toyabe type (see Chapter 6), it is appropriate to view that decay as arising from a field distribution at the muon site. On the other hand, if the magnetic dynamics is fast, $P_Z(t)$ is usually well approximated by an exponential function characterised by the relaxation rate λ_Z. The key point in that case is the possibility of interpreting the data in the framework of the linear response theory (see Appendix F), a framework constantly used in neutron scattering work.

Taking the point of view of the field distribution, the most important polarisation function in the static case is given by the Kubo–Toyabe formula, i.e.

$$P_Z^{\rm stat}(t) = \frac{1}{3} + \frac{2}{3}\left(1 - \gamma_\mu^2 \Delta_{\rm G,ZF}^2 t^2\right) \exp\left(-\frac{\gamma_\mu^2 \Delta_{\rm G,ZF}^2 t^2}{2}\right). \tag{10.28}$$

For the sake of completeness, we note that the dynamics is usually accounted for within the framework of the strong collision model. This is explained in Sections 6.3 and 10.2. In the quasi-static limit, the decay of the one-third tail is characterised by the parameter $2\nu_{\rm c}/3$ where $\nu_{\rm c} = 1/\tau_{\rm c}$; see Eq. 6.65. As well as λ_Z, this $\nu_{\rm c}$ parameter can be derived from linear response theory. We leave this derivation as an exercise to the interested reader.

In fact, it is common in the literature to find experimental spectra analysed by one of the following two simple formulae:

$$P_Z^{\rm stat}(t) = \frac{1}{3} + \frac{2}{3}\left[1 - (\gamma_\mu \Delta_{\rm G,ZF} t)^{\beta_{\rm se}}\right] \exp\left[-\frac{(\gamma_\mu \Delta_{\rm G,ZF} t)^{\beta_{\rm se}}}{\beta}\right], \tag{10.29}$$

$$P_Z(t) = \exp\left[-(\lambda_Z t)^{\beta_{\rm se}}\right]. \tag{10.30}$$

The first equation is a natural generalisation of the Kubo–Toyabe formula. The expression written in Eq. 10.30 is referred to as the stretched-exponential relaxation function when $0 < \beta_{\rm se} \leq 1$. This form of relaxation function is usually taken as the signature of a large number of relaxation channels or of an inhomogeneous spin system in the motional narrowing limit. We refer to Section 6.4.1.2 for more information. Interestingly, in analogy with particle diffusion for which the average path of the diffusing particle is proportional to $\sqrt{t}$, the evolution of a random field is also expected to give a stretched relaxation with $\beta_{\rm se} = \frac{1}{2}$ — see, e.g. Miyashita and Saito (2001). Still, we stress that the use of these two formulae with $\beta_{\rm se}$ as a free parameter is usually difficult to justify on physical grounds. Sometimes it is argued that a good fit of data with one of these two functions means that the magnetism is disordered. This may not be justified if there is no other convincing reason.

10.1.3 Long- versus short-range correlations

Common wisdom tells us that the observation of an oscillating μSR polarisation function for a compound, as it is cooled down in zero field, is a signature of the occurrence of a magnetic phase transition. This means that at least one symmetry element of the space group characterising the system at high temperature is broken at low temperature. On the other hand, if only a marked increase of the relaxation is observed at the transition and if the ordered moments are not extremely small, the compound under study is believed to display an inhomogeneous magnetic transition. In fact, the validity of this insight was never seriously questioned until the detection of a coherent μSR precession for a magnetic compound with a correlation length as small as 2.5 (3) nm

— Yaouanc *et al.* (2008); see also Akiyama *et al.* (2010) — led to the investigation of the significance of the observation of coherent oscillations. This is the purpose of the present section.

In the real world, a magnetic structure is never perfect. There are at least two reasons: the position of the magnetic moments in the lattice may slightly deviate from the space group prescription and their orientation and modulus may be distributed relative to their theoretical values. Here, we compute the variance of the field distribution arising from these imperfections. Physically, point defects, dislocations, and impurities can be at the origin of the disturbances.

The field variance arising from the nuclear moments describes an additional source of depolarisation which must be taken into account (see Section 6.5). But we shall not do it here. In addition, dynamical effects will be neglected; see Section 10.2.2 for more information on that matter. A study of the variance resulting from the distribution of the vortex-core positions around a mean value in a flux-line lattice has been published (Dalmas de Réotier *et al.* 2004); see also Section 11.1.1.3. We shall partly rely on that work.

Our physical intuition about the influence of imperfections on a lattice relies heavily on the way it is gauged. It is usually based on the study of the profile of Bragg reflections, in our case magnetic Bragg reflections recorded in a neutron or X-ray magnetic scattering experiment. These profiles are most commonly modelled by a convolution product of two functions: a Gaussian and a Lorentzian. The former accounts for the finite resolution of the diffractometer used for data recording and the latter describes the sample under study. Its relevant parameter is its half-width at half-maximum denoted by $\kappa_{\rm m}$. The magnetic structure correlation length $\xi_{\rm m}$ is the inverse of $\kappa_{\rm m}$: $\xi_{\rm m} = 1/\kappa_{\rm m}$. Therefore, a single parameter is introduced for the description of the translational symmetry breaking.

If the field at the muon is sufficiently large, only the Z component has to be considered; see Section 6.2.1. It is built up from the contribution of all the magnetic moments of the compound. To start with, we take the lattice to be perfect. Using Eq. 5.38, we rewrite $B^Z_{\rm loc}(\mathbf{r}_0)$ as

$$B^Z_{\rm loc}(\mathbf{r}_0) = \frac{\mu_0}{4\pi}\frac{1}{v_c}\sum_\beta \int_V G^{Z\beta}(\mathbf{r}_0+\mathbf{r})\varrho^\beta(\mathbf{r})\mathrm{d}^3\mathbf{r}, \tag{10.31}$$

where the integration is performed over the sample volume V and $\varrho^\beta(\mathbf{r})$ is the β component of the magnetic density:

$$\varrho^\beta(\mathbf{r}) = \sum_{\mathbf{i}} m_i^\beta \delta(\mathbf{r}-\mathbf{i}). \tag{10.32}$$

We recall that $\mathbf{m}_i = -g\mu_{\rm B}\mathbf{J}_i$. As usual $\mathbf{i}$ is the vector position for ion i in the lattice and $\mathbf{r}_0$ defines the muon location (see Fig. 5.3). Without serious loss of generality for our study of the influence of lattice imperfections on $B^Z_{\rm loc}(\mathbf{r}_0)$, we have assumed the magnetic ions to form a Bravais lattice and consequently we have dropped the d index in Eq. 5.38. For definitiveness the lattice is taken as three-dimensional. A two-dimensional lattice is probably better suited for magnetic layers. The present formalism can be adapted to this case in a straightforward manner.

Considering $G^{Z\beta}(\mathbf{r}_0+\mathbf{r})$ with discrete variables, i.e. $G^{Z\beta}_{\mathbf{r}_0+\mathbf{r}}$, we note from Eq. 5.42 that it can be expressed in terms of $G^{Z\beta}_{\mathbf{q}}$. Writing this previous equation with continuous variables with the help of the usual substitution $\sum_{\mathbf{q}} \longrightarrow V\int_{v_c^*} \mathrm{d}^3\mathbf{q}/(2\pi)^3$, where the integral is performed over the Brillouin zone, we arrive at

$$B^Z_{\mathrm{loc}}(\mathbf{r}_0) = \frac{\mu_0}{4\pi}\sum_{\beta}\int_{v_c^*}\int_V G^{Z\beta}(\mathbf{q})\exp\left[-i\mathbf{q}\cdot(\mathbf{r}+\mathbf{r}_0)\right]\varrho^\beta(\mathbf{r})\mathrm{d}^3\mathbf{r}\frac{\mathrm{d}^3\mathbf{q}}{(2\pi)^3}. \tag{10.33}$$

The muon–system coupling tensor $\boldsymbol{G}(\mathbf{q})$ is described in full details in Section 5.2.4. Using Eq. C.51, B^Z_{loc} can be expressed in terms of the Fourier transform $\varrho^\beta(\mathbf{q})$:

$$B^Z_{\mathrm{loc}}(\mathbf{r}_0) = \frac{\mu_0}{4\pi}\sum_{\beta}\int_{v_c^*} G^{Z\beta}(\mathbf{q})\varrho^\beta(\mathbf{q})\exp(-i\mathbf{q}\cdot\mathbf{r}_0)\frac{\mathrm{d}^3\mathbf{q}}{(2\pi)^3}. \tag{10.34}$$

For the sake of simplicity, we shall assume from now on that we are dealing with a ferromagnet, i.e. $\mathbf{m}_i = \mathbf{m}$. Since, for the time being, the magnet is free of defects, $\varrho^\beta(\mathbf{r})$ is a periodic function. It can therefore be expanded as a Fourier series (see Section E.2):

$$\varrho^\beta(\mathbf{r}) = \frac{m^\beta}{v_c}\sum_{\mathbf{K}}\exp(i\mathbf{K}\cdot\mathbf{r}). \tag{10.35}$$

The sum is performed over the reciprocal lattice vectors.

We now introduce disorder. As explained in the introduction of this section, it has two sources: the position of the magnetic ions and the orientation and modulus of their moments. For the sake of simplification we consider only the first source of disorder. Hence, the ions do not sit at the ideal positions $\{\mathbf{i}\}$. Instead they are located at the effective positions $\{\mathbf{i}+\tilde{\mathbf{u}}(\mathbf{i})\}$ which deviate from the equilibrium prescription by the displacement vectors $\{\tilde{\mathbf{u}}(\mathbf{i})\}$. Then the magnetic density is written as

$$\varrho^\beta(\mathbf{r}) = m^\beta\sum_{\mathbf{i}}\delta\left[\mathbf{r}-\mathbf{i}-\tilde{\mathbf{u}}(\mathbf{i})\right]. \tag{10.36}$$

Although this is not a periodic density function, it can be expanded in terms of Fourier components if $\tilde{\mathbf{u}}(\mathbf{i})$ is smooth (see for example Appendices A of Giamarchi and Le Doussal (1995) and Nattermann and Scheidl (2000)):

$$\varrho^\beta(\mathbf{r}) \simeq \frac{m^\beta}{v_c}\sum_{\mathbf{K}}\exp\left\{i\mathbf{K}\cdot[\mathbf{r}-\mathbf{u}(\mathbf{r})]\right\}, \tag{10.37}$$

where the displacement $\mathbf{u}(\mathbf{r})$ is defined in terms of the actual ion position $\mathbf{r}$. Note that in the absence of displacement ($\mathbf{u}(\mathbf{r}) = 0$), Eq. 10.35 is naturally recovered. This decomposition is strictly valid in the absence of dislocations. Since, according to Eq. C.51, $\varrho^\beta(\mathbf{q}) = \int_V \varrho^\beta(\mathbf{r})\exp(-i\mathbf{q}\cdot\mathbf{r})\mathrm{d}^3\mathbf{r}$ we derive from Eq. 10.34

$$B^Z_{\mathrm{loc}}(\mathbf{r}_0) = \frac{\mu_0}{4\pi}\frac{1}{v_c}\sum_{\mathbf{K}}\sum_{\beta}\int_{v_c^*} m^\beta G^{Z\beta}(\mathbf{q})\exp(-i\mathbf{q}\cdot\mathbf{r}_0) \tag{10.38}$$

$$\times \left\{ \int_V \exp\left[-i\left(\mathbf{q}-\mathbf{K}\right)\cdot\mathbf{r}\right]\exp\left[-i\mathbf{K}\cdot\mathbf{u}(\mathbf{r})\right]\mathrm{d}^3\mathbf{r} \right\} \frac{\mathrm{d}^3\mathbf{q}}{(2\pi)^3}.$$

To proceed further, we assume the point symmetry at the muon site to be sufficiently high and the crystal to be oriented such that the coupling tensor $\boldsymbol{G}(\mathbf{q})$ is diagonal in the reference frame $\{X, Y, Z\}$, i.e. $G^{Z\beta}(\mathbf{q}) = \delta_{Z,\beta}G^{ZZ}(\mathbf{q})$. This hypothesis allows us to simplify the notation without jeopardising the physics we are investigating. We recall that a measurement only gives access to the thermal average of physical quantities. An ensemble average must be performed to account for the disorder. It must be carried out after the thermal average. We shall assume the thermal average to be already performed and denote, as usual, the ensemble average of A as $\overline{A}$.

We first focus on the average field at the muon site which is measured through the angular frequency ω_μ. From the previous equation,

$$B_0^Z \equiv \overline{B_{\mathrm{loc}}^Z(\mathbf{r}_0)} = \frac{\mu_0}{4\pi}\frac{m^Z}{v_c}\sum_{\mathbf{K}}\int_{v_c^*} G^{ZZ}(\mathbf{q})\exp(-i\mathbf{q}\cdot\mathbf{r}_0) \tag{10.39}$$

$$\times \left\{ \int_V \exp\left[-i\left(\mathbf{q}-\mathbf{K}\right)\cdot\mathbf{r}\right]\overline{\exp\left[-i\mathbf{K}\cdot\mathbf{u}(\mathbf{r})\right]}\mathrm{d}^3\mathbf{r} \right\} \frac{\mathrm{d}^3\mathbf{q}}{(2\pi)^3}.$$

Guided by the derivation of the Debye–Waller factor — see e.g. Kittel (1996) — we obtain

$$\overline{\exp\left[-i\mathbf{K}\cdot\mathbf{u}(\mathbf{r})\right]} = \exp\left(-K^2\overline{u^2}/6\right). \tag{10.40}$$

With this result, we derive

$$B_0^Z = \frac{\mu_0}{4\pi}\frac{m^Z}{v_c}\sum_{\mathbf{K}}\exp\left(-K^2\overline{u^2}/6\right)\int_{v_c^*} G^{ZZ}(\mathbf{q})\exp(-i\mathbf{q}\cdot\mathbf{r}_0)\delta(\mathbf{q}-\mathbf{K})\mathrm{d}^3\mathbf{q}$$

$$= \frac{\mu_0}{4\pi}\frac{m^Z}{v_c}G^{ZZ}(\mathbf{q}=\mathbf{0}), \tag{10.41}$$

because the volume of integration is the Brillouin zone which contains a single reciprocal lattice vector, namely $\mathbf{K} = \mathbf{0}$; see Section E.2. We find, as expected, that disorder does not affect B_0^Z.

On the other hand, it has a strong influence on the variance of the field. This reflects the limited value of the magnetic coherence length. The variance is given by the formula

$$\Delta_{Z,\mathrm{m}}^2 = \overline{\left|B_{\mathrm{loc}}^Z(\mathbf{r}_0)\right|^2} - \left(B_0^Z\right)^2. \tag{10.42}$$

To start with, we consider the first term on the right-hand side of this equation. The following quantity appears in it:

$$\overline{\exp\left[-i\mathbf{K}'\cdot\mathbf{u}\left(\mathbf{r}'\right)\right]\exp\left[i\mathbf{K}''\cdot\mathbf{u}\left(\mathbf{r}''\right)\right]}. \tag{10.43}$$

Its form suggests to perform the change of variable $\mathbf{r}'' = \mathbf{r}+\mathbf{r}'$ in the integral over the $\mathbf{r}''$ variable. The expression in Eq. 10.43 transforms into

$$\overline{\exp\left[-i\mathbf{K}'\cdot\mathbf{u}\left(\mathbf{r}'\right)\right]\exp\left[i\mathbf{K}''\cdot\mathbf{u}\left(\mathbf{r}'+\mathbf{r}\right)\right]}, \tag{10.44}$$

that, as a correlation function, should only depend on the vector distance between the two considered crystal sites, i.e. should only depend on $\mathbf{r}$ and not on $\mathbf{r}'$. Since this correlation function is negligible unless $\mathbf{K}' = \mathbf{K}''$, we define

$$\mathcal{C}_{\mathbf{K}}(\mathbf{r}) = \overline{\exp\left\{i\mathbf{K}\cdot\left[\mathbf{u}\left(\mathbf{r}\right) - \mathbf{u}\left(\mathbf{0}\right)\right]\right\}}. \tag{10.45}$$

The first term on the right-hand side of Eq. 10.42 is then written as

$$\left(\frac{\mu_0}{4\pi}\frac{m^Z}{v_c}\right)^2 \int_{v_c^*} \left|G^{ZZ}(\mathbf{q})\right|^2 \sum_{\mathbf{K}} \int_V \exp\left[i\left(\mathbf{q}-\mathbf{K}\right)\cdot\mathbf{r}\right] \mathcal{C}_{\mathbf{K}}(\mathbf{r}) \mathrm{d}^3\mathbf{r} \frac{\mathrm{d}^3\mathbf{q}}{(2\pi)^3}. \tag{10.46}$$

Now, remarking that $\mathcal{C}_{\mathbf{K}=\mathbf{0}}(\mathbf{r}) = 1$ (see Eq. 10.45) and using Eq. C.48, the term $\mathbf{K} = \mathbf{0}$ in the sum of Eq. 10.46 is nothing else than $\left(B_0^Z\right)^2$. Therefore we get

$$\Delta^2_{Z,\mathrm{m}} = \left(\frac{\mu_0}{4\pi}\frac{m^Z}{v_c}\right)^2 \int_{v_c^*} \left|G^{ZZ}(\mathbf{q})\right|^2 \sum_{\mathbf{K}\neq\mathbf{0}} \int_V \exp\left[i\left(\mathbf{q}-\mathbf{K}\right)\cdot\mathbf{r}\right] \mathcal{C}_{\mathbf{K}}(\mathbf{r}) \mathrm{d}^3\mathbf{r} \frac{\mathrm{d}^3\mathbf{q}}{(2\pi)^3}. \tag{10.47}$$

An equivalent expression for $\Delta^2_{Z,\mathrm{m}}$ can be obtained. From Eq. 5.41, we note that

$$\left|G^{ZZ}(\mathbf{q})\right|^2 = \sum_{i,i'} G^{ZZ}_{\mathbf{r}_i} G^{ZZ}_{\mathbf{r}_{i'}} \exp[i\mathbf{q}\cdot(\mathbf{i}-\mathbf{i}')]. \tag{10.48}$$

Thus, integrating successively over $\mathbf{q}$ and $\mathbf{r}$ in Eq. 10.47, we end up with

$$\Delta^2_{Z,\mathrm{m}} = \left(\frac{\mu_0}{4\pi}\frac{m^Z}{v_c}\right)^2 \sum_{i,i'} G^{ZZ}_{\mathbf{r}_i} G^{ZZ}_{\mathbf{r}_{i'}} \sum_{\mathbf{K}\neq\mathbf{0}} \mathcal{C}_{\mathbf{K}}(\mathbf{i}-\mathbf{i}'), \tag{10.49}$$

since $\exp(i\mathbf{K}\cdot\mathbf{i}) = 1$.

It is interesting to note that an expression formally similar to Eq. 10.47 is found for the second moment of the field distribution arising from a non-perfectly ordered vortex lattice; see Eq. 11.65. The muon–system coupling tensor is replaced by a term related to the Fourier transform of the field of a single vortex and the volume of integration over $\mathbf{q}$ is changed from the Brillouin zone to the whole reciprocal space.

It is straightforward to examine the case where the magnetic structure is free of defects. This means that the displacement vectors $\mathbf{u}(\mathbf{r})$ uniformly vanish and accordingly $\mathcal{C}_{\mathbf{K}}(\mathbf{r}) = 1$. The integral over $\mathbf{r}$ in Eq. 10.47 is equal to $(2\pi)^3 \sum_{\mathbf{K}\neq\mathbf{0}} \delta(\mathbf{q}-\mathbf{K})$ and the integral over $\mathbf{q}$ cancels since the integration volume is the Brillouin zone which contains a single reciprocal lattice vector: $\mathbf{K} = \mathbf{0}$. Therefore $\Delta^2_{Z,\mathrm{m}}$ is zero in accord with the expectation that no field broadening arises from a perfect structure.

A formula for $\mathcal{C}_{\mathbf{K}}(\mathbf{r})$ in the general case is now needed. Giamarchi and Le Doussal (1995, 1998) derived an expression for the correlation, but their interest was mainly on flux-line lattices. Here we simply note that $\mathcal{C}_{\mathbf{K}}(\mathbf{r})$ is expected to be a Gaussian function of K, and therefore might be negligible for large K. In addition, in the limit of small $\mathbf{r}$ it behaves as an exponential function for a three-dimensional system. In order to get a flavour of the damping arising from the magnetic moment disorder we

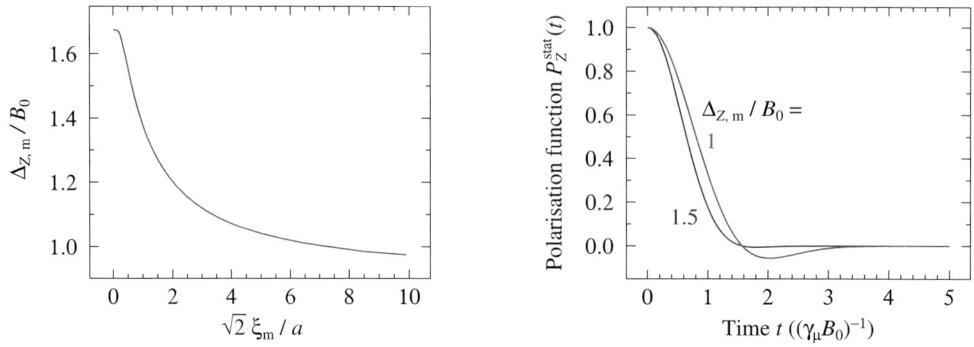

Fig. 10.4 (Left) Ratio of the standard deviation of the field at the muon site given by our simplified model, over the intensity of the mean spontaneous field versus ξ_{m}, the magnetic correlation length of the ferromagnetic order. The muon is assumed to sit in the octahedral site of a face-centred-cubic lattice of parameter a. The length ξ_{m} is shown in units of $a/\sqrt{2}$ which is the distance between nearest neighbour crystal sites. Note that for the chosen muon site the only contribution to the spontaneous field is the Lorentz field since $\mathbf{B}'_{\mathrm{dip}} = \mathbf{0}$ (see Section 5.2.3 for the definition of $\mathbf{B}'_{\mathrm{dip}}$). (Right) Polarisation functions $P_Z^{\mathrm{stat}}(t) = \exp(-\gamma_\mu^2\Delta_{Z,\mathrm{m}}^2 t^2/2)\cos(\gamma_\mu B_0 t)$ for two values of $\Delta_{Z,\mathrm{m}}/B_0$ indicated in the plot. From the two panels one can conclude that a local minimum is visible in the polarisation function when ξ_{m} is larger than about ten nearest neighbour distances.

assume the correlation function to take an exponential form and set $\sum_{\mathbf{K}\neq\mathbf{0}} \mathcal{C}_{\mathbf{K}}(\mathbf{r}) = \exp(-r/\xi_{\mathrm{m}})$. This is obviously a rough approximation, but sufficient for our purpose, which is to determine at which level of disorder the μSR oscillations reflecting the magnetic structure of a compound are overdamped. This is conveniently assessed using the ratio $\Delta_{Z,\mathrm{m}}/B_0$. Figure 10.4 shows the prediction of the model for the case of a muon in the octahedral site of a face-centred-cubic lattice; see left panel of Fig. 9.1 for a sketch of this site. It is computed from Eq. 10.49. A relatively short correlation length of about ten interactomic distances in a ordered magnet is enough to give rise to a detectable spontaneous precession of the muon polarisation. The oscillations are nevertheless quite heavily damped as expected for such a short-range magnetic order.

The whole study of the influence on the μSR oscillations of the effect of disorder has here been carried out in the case of a ferromagnetic collinear structure, where disorder arises from deviations of the positions of the magnetic atoms with respect to the prescription of the crystal structure. Although, qualitatively the conclusions of the study are not expected to be strongly dependent on the exact structure, for a quantitative work the proper magnetic structure must be taken into account. In addition, the muon localisation site is needed. We are not going to develop this subject further. However, to help the interested reader, we note that the specification of the magnetic structure is first given in Eq. 10.32. For a magnetic structure which can be seen as interpenetrating sublattices, $\delta(\mathbf{r}-\mathbf{i})$ in Eq. 10.32 must be substituted by $\exp(i\mathbf{k}\cdot\mathbf{r})\delta(\mathbf{r}-\mathbf{i})$ where $\mathbf{k}$ is the propagation vector of the magnetic structure. As an example, for a one-dimensional structure of lattice parameter a, $k = \pi/a$ describes a chain of alternating up and down moments.

10.2 Dynamical properties

In this section we are mainly concerned with the interpretation of experiments performed with the longitudinal-field technique, since this technique is particularly suited for the investigation of spin dynamics. First we consider the shape of the longitudinal-field polarisation function. Next we examine the spin–lattice and spin–spin relaxation rates, the latter being measured by the transverse-field technique. Finally a review of fluctuation and excitation processes is given.

We define a fluctuation as an excitation centred at zero energy. However, the distinction between fluctuation and excitation is somewhat arbitrary. For example, the relaxation of the muon spin by conduction electrons in a metal is taken as arising from a typical fluctuation process, whereas the relaxation of an electronic two crystal-field level system by the same electron density is thought to originate from an excitation process since the energy splitting is usually appreciable relative to the dynamical width (lifetime effect).

10.2.1 The longitudinal-field polarisation function

Starting from the longitudinal-field polarisation function in the case of a static field distribution, i.e. $P_Z^{\rm stat}(t)$, $P_Z(t)$ can be evaluated within the framework of the strong collision model for the field dynamics. This gives the integral equation derived in Eq. 6.52 that we reproduce here:

$$P_Z(t) = P_Z^{\rm stat}(t)\exp(-\nu_{\rm c}t) + \nu_{\rm c}\int_0^t P_Z(t-t')P_Z^{\rm stat}(t')\exp(-\nu_{\rm c}t'){\rm d}t', \qquad (10.50)$$

where $\nu_{\rm c}$ is the local field correlation frequency. This formula is a kind of mean-field approximation i.e. it breaks down when the spatial dimension of the system under study is restricted, as for example in the case of a chain. As explained in Section 6.3, this integral equation can be solved for $P_Z(t)$. Using a quantum mechanical approach, if the spin dynamics is sufficiently fast $P_Z(t)$ can be expressed in terms of field correlation functions; see Chapter 7. Because of the translational symmetry of crystals, it is shrewd to introduce a space Fourier transform as detailed in Appendix E. In addition, we find it convenient to work directly with the spin correlation functions expressed in the crystal reference frame; see Appendix D. Combining these results, we derive

$$P_Z(t) = \exp[-\Psi_Z(t)], \qquad (10.51)$$

with

$$\Psi_Z(t) = \lambda_Z(t)\,t = \mathcal{D}\sum_{\beta,\alpha} L^{\beta\alpha}(\theta,\varphi)\int_{v_c^*}\sum_{\gamma,\gamma'}\sum_{d,d'} G_d^{\alpha\gamma}(\mathbf{q})G_{d'}^{\gamma'\beta}(-\mathbf{q}) \qquad (10.52)$$

$$\times\int_0^t (t-\tau)\Lambda_{dd'}^{\gamma\gamma'}(\mathbf{q},\tau)\cos(\omega_\mu\tau){\rm d}\tau\frac{{\rm d}^3\mathbf{q}}{(2\pi)^3},$$

with the constant $\mathcal{D} = (\mu_0/4\pi)^2\,\gamma_\mu^2\,(g\mu_B)^2\,/v_c$. The elements of $L(\theta,\varphi)$ are defined in Eq. D.8. The field cross-correlations $\Phi^{XY}(\tau)$ and $\Phi^{YX}(\tau)$ which appeared e.g. in

Eq. 7.42 have been neglected for simplicity. This is justified in most cases of interest. For instance, in zero-field these functions are usually even and therefore their difference vanishes. More generally, these $\Phi^{XY}(\tau)$ and $\Phi^{YX}(\tau)$ terms can be expressed as a function of $\Phi^{\gamma\gamma'}(\tau)$ in the crystal frame. The symmetry is often such that $\Phi^{\gamma\gamma'}(\tau) = 0$ when $\gamma \neq \gamma'$. Then these cross-correlations do not influence the relaxation. The function $\Psi_Z(t)$ basically depends on three quantities: (i) the matrix $L(\theta, \varphi)$ which accounts for the orientation of the reference frame attached to the sample relative to the laboratory reference frame, (ii) the coupling between the muon spin and the spins of the compound modelled by $\boldsymbol{G}$, and (iii) the correlation tensor $\boldsymbol{\Lambda}$ we are interested in.

Although the previous expression may need to be used in its full extent for interpreting experimental results, the following simplified formula is sufficient to grasp the physics involved:

$$\Psi_Z(t) = \lambda_Z(t)\, t = \mathcal{D} \int_{v_c^*} \sum_{\gamma,\gamma'} \mathcal{A}^{\gamma\gamma'}(\mathbf{q}) \int_0^t (t-\tau)\Lambda^{\gamma\gamma'}(\mathbf{q},\tau)\cos(\omega_\mu \tau)\mathrm{d}\tau \frac{\mathrm{d}^3\mathbf{q}}{(2\pi)^3}, \quad (10.53)$$

where we have defined for convenience

$$\mathcal{A}^{\gamma\gamma'}(\mathbf{q}) = G^{X\gamma}(\mathbf{q})G^{\gamma'X}(-\mathbf{q}) + G^{Y\gamma}(\mathbf{q})G^{\gamma'Y}(-\mathbf{q}). \quad (10.54)$$

Here, the unit cell is assumed to contain only one magnetic atom, i.e. the subscripts d and d' are irrelevant, and the laboratory and crystal reference frames are taken to be identical.

A key result is derived from Eq. 10.53 if the correlations functions $\Lambda^{\gamma\gamma'}(\mathbf{q},\tau)$ decay sufficiently fast relative to the experimental time scale. The time τ can then be neglected relative to t in the factor $(t-\tau)$ and the upper bound of the integral can be extended to infinity. As expected in this motional narrowing limit, $\lambda_Z(t) = \lambda_Z$ and $P_Z(t)$ is an exponential function.

Different methods for tackling the functional time dependence of $\Psi_Z(t)$ are possible. As a first method, we introduce the time Fourier transform of the spin-correlation functions. Obviously, this can be performed only if the analytical properties of these functions are such that Fourier transforms can be defined. Referring to Section 7.1.4.3 and Eq. F.15,

$$\lambda_Z(t) = \frac{\mu_0}{4\pi}\frac{\gamma_\mu^2}{2} k_\mathrm{B} T \sum_{\gamma,\gamma'} \int_{v_c^*} \mathcal{A}^{\gamma\gamma'}(\mathbf{q})\chi^{\gamma\gamma'}(\mathbf{q}) \int_{-\infty}^{\infty} F^{\gamma\gamma'}(\mathbf{q},\omega+\omega_\mu)\frac{1-\cos(\omega t)}{\omega^2 t}\frac{\mathrm{d}\omega}{2\pi}\frac{\mathrm{d}^3\mathbf{q}}{(2\pi)^3}. \quad (10.55)$$

This result is basically derived using the fluctuation–dissipation theorem which provides a relation between $\boldsymbol{\Lambda}(\mathbf{q},\omega)$ and $\boldsymbol{\chi}(\mathbf{q},\omega)$. The susceptibility $\chi^{\gamma\gamma'}(\mathbf{q})$ is wave vector-dependent and $F^{\gamma\gamma'}(\mathbf{q},\omega)$ is a spectral-weight function. Because the second factor in the ω-integral tends to $\pi\delta(\omega)$ as $t \to \infty$, see Eq. C.44, $\lambda_Z(t)$ becomes a constant in this limit as already argued. The significance of the limit $t \to \infty$ is to be gauged relative to the extension of $F^{\gamma\gamma'}(\mathbf{q},\omega)$ in ω. We remind the reader that the function $[1-\cos(\omega t)]/(\omega^2 t)$ is plotted in Fig. 7.7. If $F^{\gamma\gamma'}(\mathbf{q},\omega+\omega_\mu)$ notably varies over the ω

range of Fig. 7.7, the $t \to \infty$ limit is not experimentally reached, and therefore $\Psi_Z(t)$ is not found linear in t. This point is clarified below through two model computations. We shall assume $\boldsymbol{F}(\mathbf{q},\omega)$ to be scalar for simplicity and introduce the intermediate function

$$\tilde{\Psi}_Z(\mathbf{q},t) = \int_{-\infty}^{\infty} F(\mathbf{q},\omega+\omega_\mu)\frac{1-\cos(\omega t)}{\omega^2}\frac{\mathrm{d}\omega}{2\pi}, \tag{10.56}$$

while $\Psi_Z(t) = \lambda_Z(t)t$ is proportional to a sum of integrals over $\mathbf{q}$ of weighted $\tilde{\Psi}_Z(\mathbf{q},t)$ functions.

The simplest function one can imagine for the spectral-weight function is the Lorentzian. This corresponds to an exponential correlation function. Using the notation of Eq. F.16, we need to evaluate

$$\tilde{\Psi}_Z(\mathbf{q},t) = \frac{1}{\pi}\int_{-\infty}^{\infty} \frac{\Gamma(\mathbf{q})}{\Gamma^2(\mathbf{q})+(\omega+\omega_\mu)^2}\frac{1-\cos(\omega t)}{\omega^2}\frac{\mathrm{d}\omega}{2\pi}. \tag{10.57}$$

Using Eq. C.84, we derive

$$\begin{aligned}2\pi\tilde{\Psi}_Z(\mathbf{q},t) = {} & \frac{\Gamma(\mathbf{q})}{\Gamma^2(\mathbf{q})+\omega_\mu^2}t + \frac{\{1-\cos(\omega_\mu t)\exp[-\Gamma(\mathbf{q})t]\}\left[\omega_\mu^2-\Gamma^2(\mathbf{q})\right]}{\left[\Gamma^2(\mathbf{q})+\omega_\mu^2\right]^2}\\ & - \frac{2\Gamma(\mathbf{q})\omega_\mu\sin(\omega_\mu t)\exp[-\Gamma(\mathbf{q})t]}{\left[\Gamma^2(\mathbf{q})+\omega_\mu^2\right]^2}.\end{aligned} \tag{10.58}$$

As expected, identifying $2\pi\tilde{\Psi}_Z(\mathbf{q},t)$ with $g^{(c)}_{\nu_c}(\omega_\mu,t)$, we recover Eq. 7.60 setting $\Gamma(\mathbf{q}) = \nu_c$. The advantage here is that the interaction between the moments is taken into account through the $\mathbf{q}$ dependence of $\Gamma(\mathbf{q})$. The limits of $\tilde{\Psi}_Z(\mathbf{q},t)$ are easily derived. If we set $\omega_\mu = 0$, i.e. the zero-field condition, we get

$$\tilde{\Psi}_Z(\mathbf{q},t) = \frac{\exp[-\Gamma(\mathbf{q})t] - 1 + \Gamma(\mathbf{q})t}{2\pi\Gamma^2(\mathbf{q})}. \tag{10.59}$$

Recalling Eq. 10.55, a $\mathbf{q}$-integrated Abragam function is derived. A numerical evaluation of this formula for specific models of $\mathcal{A}(\mathbf{q})$, $\chi(\mathbf{q})$, and $\Gamma(\mathbf{q})$ has been presented (Steer 2006). A key feature is the impossibility to describe $\Psi_Z(t)$ as an exponential-power function with an exponent larger than 2 or smaller than 1. The polarisation function $P_Z(t)$ is exponential when $\Gamma(\mathbf{q})t \gg 1$ and Gaussian in the other extreme limit, i.e. $\Gamma(\mathbf{q})t \ll 1$.

Under an applied field and when $\Gamma(\mathbf{q})t$ is large enough, $P_Z(t)$ becomes exponential and we find the expected Lorentzian-field dependence:

$$\tilde{\Psi}_Z(\mathbf{q},t) \simeq \frac{1}{2\pi}\frac{\Gamma(\mathbf{q})}{\Gamma^2(\mathbf{q})+\omega_\mu^2}t. \tag{10.60}$$

Conversely, when $\Gamma(\mathbf{q})t$ is small, $\tilde{\Psi}_Z(\mathbf{q},t)$ displays wiggles at the angular frequency ω_μ:

$$\tilde{\Psi}_Z(\mathbf{q},t) \simeq \frac{[1-\cos(\omega_\mu t)]\left[\omega_\mu^2-\Gamma^2(\mathbf{q})\right] - 2\Gamma(\mathbf{q})\omega_\mu\sin(\omega_\mu t)}{2\pi\left[\Gamma^2(\mathbf{q})+\omega_\mu^2\right]^2}. \tag{10.61}$$

The Lorentzian function for the spectral-weight function introduced in Eq. 10.57 only depends on one parameter and is symmetric about the origin in energy. Hence, only quasielastic processes are accounted for in this limit in zero field. The next simplest spectral-weight function models the damped harmonic oscillator. Restricting to the zero-field case and still assuming a scalar spectral-weight, with the notations of Eq. F.19,

$$F(\mathbf{q},\omega) = \frac{2}{\pi}\frac{\Gamma(\mathbf{q})\Omega^2(\mathbf{q})}{\left[\omega^2 - \Omega^2(\mathbf{q})\right]^2 + 4\omega^2\Gamma^2(\mathbf{q})}. \tag{10.62}$$

Applying the method of partial fractions, we find that $F(\mathbf{q},\omega)$ depends drastically on the value of the ratio $\Gamma(\mathbf{q})/\Omega(\mathbf{q})$. When $\Gamma(\mathbf{q})/\Omega(\mathbf{q}) < 1$, defining $\varpi = \sqrt{\Omega^2(\mathbf{q}) - \Gamma^2(\mathbf{q})}$, we have

$$F(\mathbf{q},\omega) = \frac{1}{2\pi}\frac{\Gamma(\mathbf{q})}{\varpi}\left[\frac{\omega + 2\varpi}{\omega^2 + 2\varpi\omega + \Omega^2(\mathbf{q})} + \frac{-\omega + 2\varpi}{\omega^2 - 2\varpi\omega + \Omega^2(\mathbf{q})}\right]. \tag{10.63}$$

The inverse Fourier transform $F(\mathbf{q},t)$ is interesting. Using Eqs. C.73 and C.74,

$$F(\mathbf{q},t) = \frac{1}{2\pi}\left[\cos(\varpi t) + \frac{\Gamma(\mathbf{q})}{\varpi}\sin(\varpi t)\right]\exp\left[-\Gamma(\mathbf{q})t\right]. \tag{10.64}$$

They are therefore well-defined excitations at $\pm\varpi$. This is confirmed by looking at the limit of $F(\mathbf{q},\omega)$ when $\Gamma(\mathbf{q})/\Omega(\mathbf{q}) \ll 1$:

$$F(\mathbf{q},\omega) = \frac{1}{2\pi}\frac{\Gamma(\mathbf{q})}{\Omega(\mathbf{q})}\left\{\frac{\omega + 2\Omega(\mathbf{q})}{\left[\omega + \Omega(\mathbf{q})\right]^2} + \frac{-\omega + 2\Omega^2(\mathbf{q})}{\left[\omega - \Omega(\mathbf{q})\right]^2}\right\}. \tag{10.65}$$

This corresponds to two well-defined excitations at $\omega = \pm\Omega(\mathbf{q})$. In the other limit, i.e. $\Gamma(\mathbf{q})/\Omega(\mathbf{q}) > 1$,

$$F(\mathbf{q},\omega) = \frac{2}{\pi}\frac{\Gamma(\mathbf{q})\Omega^2(\mathbf{q})}{\omega_-^2(\mathbf{q}) - \omega_+^2(\mathbf{q})}\left[\frac{1}{\omega^2 + \omega_+^2(\mathbf{q})} - \frac{1}{\omega^2 + \omega_-^2(\mathbf{q})}\right], \tag{10.66}$$

with

$$\begin{aligned}\omega_-^2(\mathbf{q}) &= 2\Gamma^2(\mathbf{q}) - \Omega^2(\mathbf{q}) - 2\Gamma(\mathbf{q})\sqrt{\Gamma^2(\mathbf{q}) - \Omega^2(\mathbf{q})},\\ \omega_+^2(\mathbf{q}) &= \Omega^4(\mathbf{q})/\omega_-^2(\mathbf{q}).\end{aligned} \tag{10.67}$$

Here, two quasielastic excitations are at play. The related $F(\mathbf{q},t)$ is easily computed with the help of Eq. C.69:

$$F(\mathbf{q},t) = \frac{1}{\pi}\frac{\Gamma(\mathbf{q})\Omega^2(\mathbf{q})}{\omega_-^2(\mathbf{q}) - \omega_+^2(\mathbf{q})}\left\{\frac{\exp\left[-\omega_+(\mathbf{q})t\right]}{\omega_+(\mathbf{q})} - \frac{\exp\left[-\omega_-(\mathbf{q})t\right]}{\omega_-(\mathbf{q})}\right\}. \tag{10.68}$$

Only one quasielastic excitation is left at the boundary $\Gamma(\mathbf{q}) = \Omega(\mathbf{q})$ since

$$F(\mathbf{q},\omega) = \frac{2}{\pi}\frac{\Omega^3(\mathbf{q})}{\left[\omega^2 + \Omega^2(\mathbf{q})\right]^2}. \tag{10.69}$$

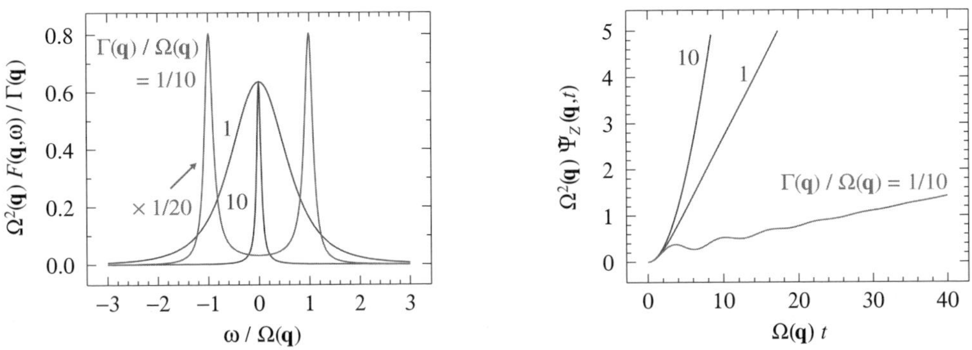

Fig. 10.5 Normalised damped-harmonic-oscillator spectral-weight function (left-hand side panel) and related normalised zero-field $\tilde{\Psi}_Z(\mathbf{q},t)$ function (right-hand side panel) for three values of the ratio $\Gamma(\mathbf{q})/\Omega(\mathbf{q})$ indicated in the plots. For the sake of clarity the spectral-weight function at $\Gamma(\mathbf{q})/\Omega(\mathbf{q}) = 1/10$ has been divided by a factor 20. Oscillations at short time for $\tilde{\Psi}_Z(\mathbf{q},t)$ are clearly visible when $\Gamma(\mathbf{q})/\Omega(\mathbf{q}) < 1$.

In this case, with the help of Eq. C.70 we compute

$$F(\mathbf{q},t) = \frac{1}{2\pi}\left[1+\Omega(\mathbf{q})t\right]\exp\left[-\Omega(\mathbf{q})t\right]. \tag{10.70}$$

For the evaluation of $\tilde{\Psi}_Z(\mathbf{q},t)$ it is convenient to directly deal with $F(\mathbf{q},\tau)$ and therefore to work with

$$\tilde{\Psi}_Z(\mathbf{q},t) = \int_0^t (t-\tau)F(\mathbf{q},\tau)\cos(\omega_\mu\tau)\mathrm{d}\tau, \tag{10.71}$$

setting $\omega_\mu = 0$ because we are only interested in the zero-field polarisation function.

We now discuss $\tilde{\Psi}_Z(\mathbf{q},t)$ for the three cases. The expression when $\Gamma(\mathbf{q})/\Omega(\mathbf{q}) > 1$ is simple:

$$\tilde{\Psi}_Z(\mathbf{q},t) = \frac{1}{\pi}\frac{\Gamma(\mathbf{q})\Omega^2(\mathbf{q})}{\omega_-^2(\mathbf{q})-\omega_+^2(\mathbf{q})}\left\{\frac{\exp\left[-\omega_+(\mathbf{q})t\right]-1+\omega_+(\mathbf{q})t}{\omega_+^3(\mathbf{q})}\right.$$
$$\left.-\frac{\exp\left[-\omega_-(\mathbf{q})t\right]-1+\omega_-(\mathbf{q})t}{\omega_-^3(\mathbf{q})}\right\}. \tag{10.72}$$

When $\Gamma(\mathbf{q})/\Omega(\mathbf{q}) = 1$, we have

$$\tilde{\Psi}_Z(\mathbf{q},t) = \frac{t}{2\pi\Omega(\mathbf{q})}\left\{2+\exp\left[-\Omega(\mathbf{q})t\right]\right\} + \frac{3}{2\pi\Omega^2(\mathbf{q})}\left\{\exp\left[-\Omega(\mathbf{q})t\right]-1\right\}. \tag{10.73}$$

In the case where $\Gamma(\mathbf{q})/\Omega(\mathbf{q}) < 1$ the expression for $\tilde{\Psi}_Z(\mathbf{q},t)$, although easy to derive, is quite long. Rather than reproducing it here we prefer to illustrate its behaviour in Fig. 10.5, together with the other two cases. At short time, oscillations appear at an angular frequency ϖ only for $\Gamma(\mathbf{q})/\Omega(\mathbf{q})] < 1$. On the other hand, in the limit

$t \to \infty$, $\tilde{\Psi}_Z(\mathbf{q}, t)$ is always linear in time. The integral over the Brillouin zone needed to compute $\lambda_Z(t)$ will not change this behaviour. The wiggles predicted when the damping rate $\Gamma(\mathbf{q})$ is small reflect the built-in inelastic excitations in the zero-field spectral-weight function of the damped oscillator.

The Lorentzian spectral-weight function is a consequence of the second Fick law. It models the spin-diffusion mechanism which is one among several possible mechanisms to account for quasielastic scattering. We are going to investigate an additional spectral-weight function for this scattering. Rather than working in ω-space as before, here we shall write $F(\mathbf{q}, \tau)$ directly and therefore we will have to use Eq. 10.71. We assume $F(\mathbf{q}, \tau)$ to depend on a single variable and to have the simple form: $F(\mathbf{q}, \tau) = f\left(q^{\theta_{\rm e}}\tau\right)$, where $\theta_{\rm e}$ is a positive exponent. To proceed further we need to specify f. We select a power law:

$$f\left(q^{\theta_{\rm e}}\tau\right) = \frac{\tau_e^x}{\left(q^{\theta_{\rm e}}\tau/\mathcal{E} - \tau_e\right)^x} \simeq \frac{\mathcal{E}^x \tau_e^x}{q^{x\theta_{\rm e}}\tau^x}, \tag{10.74}$$

with $x > 0$ and $\mathcal{E} > 0$. Implicit in our model is the condition $\tau \gg \tau_e$ which justifies the proposed approximation. Under zero-field condition, a simple computation gives

$$\tilde{\Psi}_Z(\mathbf{q}, t) = \frac{\mathcal{E}^x \tau_e^x q^{-x\theta_{\rm e}}}{(2-x)(1-x)} t^{2-x}, \tag{10.75}$$

with $0 < x < 1$. Because of the factorisation of the wave vector and time variables, the time scale of the spin dynamics has no influence on the shape of the relaxation function i.e. there is no motional narrowing limit. This is rooted in the time scale invariance of the power-law decay chosen for $F(\mathbf{q}, \tau)$.

It is interesting to investigate the possible quenching of the relaxation by a field. For this purpose we need to work with the Laplace transform $\Psi_Z(s)$ where s is the Laplace conjugate variable to t. The Laplace transform definition is given in Eq. C.53. Starting from Eq. 10.53,

$$\Psi_Z(s) = \frac{\mu_0}{4\pi}\frac{\gamma_\mu^2}{2} k_{\rm B}T \int \sum_{\beta,\gamma} \mathcal{A}^{\beta\gamma}(\mathbf{q})\chi^{\beta\gamma}(\mathbf{q})\tilde{\Psi}_Z^{\beta\gamma}(\mathbf{q}, s)\frac{{\rm d}^3\mathbf{q}}{(2\pi)^3}. \tag{10.76}$$

As for the Fourier transform, we shall neglect the anisotropy of the spin dynamics. From Eq. C.92, the intermediate function can be written as

$$\tilde{\Psi}_Z(\mathbf{q}, s) = \frac{1}{s^2}\int_0^\infty \exp(-st)F(\mathbf{q}, t)\cos(\omega_\mu t){\rm d}t. \tag{10.77}$$

Using Eq. C.78, we derive for the power-law decay introduced earlier

$$\tilde{\Psi}_Z(\mathbf{q}, s) = \frac{1}{s^2}\frac{\mathcal{E}^x \tau_e^x q^{-x\theta_{\rm e}}(-x)!}{(\omega_\mu^2 + s^2)^{(1-x)/2}}\cos\left[(1-x)\arctan(\omega_\mu/s)\right]. \tag{10.78}$$

The analytical evaluation of the image of $\tilde{\Psi}_Z(\mathbf{q}, s)$ requires the derivation of a simple enough form for this Laplace transform. Assuming $x \ll 1$,

$$\tilde{\Psi}_Z(\mathbf{q}, s) = \frac{\mathcal{E}^x \tau_e^x q^{-x\theta_e}}{s^2\left(\omega_\mu^2 + s^2\right)} \left\{ s \cos\left[x \arctan(\omega_\mu/s)\right] + \omega_\mu \sin\left[x \arctan(\omega_\mu/s)\right]\right\}. \quad (10.79)$$

To proceed further analytically, a new approximation is needed. It concerns the ratio ω_μ/s. When $\omega_\mu/s \gg 1$, i.e. in the limit $t \to \infty$, the time domain function can be analytically derived and we get

$$\tilde{\Psi}_Z(\mathbf{q}, t) = \mathcal{E}^x \tau_e^x q^{-x\theta_e} \left\{ \frac{1 - \cos(\omega_\mu t)}{\omega_\mu^2} + \frac{\pi x}{2\omega_\mu}\left[t - \frac{\sin(\omega_\mu t)}{\omega_\mu}\right]\right\}. \quad (10.80)$$

We have used the justified approximations, $\cos(\pi x/2) \simeq 1$ and $\sin(\pi x/2) \simeq \pi x/2$. If the applied field is large enough, it is a good approximation to write

$$\tilde{\Psi}_Z(\mathbf{q}, t) = \mathcal{E}^x \tau_e^x q^{-x\theta_e} \frac{\pi x}{2\omega_\mu} t, \quad (10.81)$$

Then $P_Z(t)$ is found to be exponential with the spin–lattice relaxation rate scaling as the inverse of the field, but this is true only if t is sufficiently large. This $\lambda_Z \propto 1/\omega_\mu$ scaling was first noticed by Keren *et al.* (2004). When $\omega_\mu/s \ll 1$, the $x \ll 1$ limit of Eq. 10.75 is reached.

The power-law spectral-weight function we have considered is the limiting case of the more general cutoff-power law. This is further discussed qualitatively at the beginning of the next section. Any quantitative prediction for $\Psi_Z(t)$ using this more general $F(\mathbf{q}, t)$ function requires a numerical computation.

An attempt to proceed further with the Laplace transform method has been made (Steer 2006). This led Steer to predict a stretched exponential-like $P_Z(t)$ with an exponent smaller than 1 for a homogeneous magnetic system. However, $\Psi_Z(t)$ was found to be proportional to a function which seems to be ill-defined.

In summary, we have shown in this section that the polarisation function $P_Z(t) = \exp[-\lambda_Z(t)t]$ is an exponential in the vast majority of the cases; i.e. $\lambda_Z(t)$ is time independent and equal to λ_Z, which is the spin–lattice relaxation rate.

10.2.2 Relaxation rates

Following the previous part, the most general formula for the relaxation rate is found to be

$$\lambda_Z = \frac{\mathcal{D}}{2} \sum_{\beta,\alpha} L^{\beta\alpha}(\theta, \varphi) \int_{v_c^*} \sum_{\gamma,\gamma'} \sum_{d,d'} G_d^{\alpha\gamma}(\mathbf{q}) G_{d'}^{\gamma'\beta}(-\mathbf{q}) \Lambda_{dd'}^{\gamma\gamma'}(\mathbf{q}, \omega_\mu) \frac{\mathrm{d}^3\mathbf{q}}{(2\pi)^3}. \quad (10.82)$$

Although this expression has sometimes been used to interpret experimental results — see for example Dalmas de Réotier and Yaouanc (1994) — we shall limit ourselves to the simplified formula

$$\lambda_Z = \frac{\mathcal{D}}{2} \int_{v_c^*} \sum_{\alpha,\beta} \mathcal{A}^{\alpha\beta}(\mathbf{q})\, \Lambda^{\alpha\beta}(\mathbf{q}, \omega_\mu) \frac{\mathrm{d}^3\mathbf{q}}{(2\pi)^3}, \quad (10.83)$$

as we have done for the study of the functional form of $P_Z(t)$. The definition of the correlation tensor $\mathbf{\Lambda}$ is found in Eq. F.8; see also Section 7.1.4.3. Interpreting λ_Z as the muon spin–lattice relaxation rate, Eq. 10.83 is a well-known NMR expression (see e.g. Moriya 1962).

An important result can be derived if the pair correlations are neglected, i.e. if the wave vector dependence of the correlation tensor is negligible. Assuming in addition this correlation tensor to be scalar for simplicity, we deduce from the previous section

$$\lambda_Z \propto \Lambda(\mathbf{q}=\mathbf{0}, \omega_\mu = 0)\frac{\nu_c^2}{\nu_c^2 + \omega_\mu^2}, \tag{10.84}$$

for a conventional exponential decay of the magnetic correlations. This decay leads to a Lorentzian-field dependence for λ_Z as predicted by Redfield (1957). If pair correlations cannot be neglected, a sum over the magnetic modes must be performed and therefore the previous result may not apply; see Section 10.2.3.2 for two examples. Assuming Eq. 10.84 to be valid, the following scaling law is obeyed in the high-field limit, i.e. $\omega_\mu = \gamma_\mu B_{\rm ext} \gg \nu_c$:

$$P_Z(t, B_{\rm ext}) = P_Z(t/B_{\rm ext}^\gamma), \tag{10.85}$$

with $\gamma = 2$. The exponential assumption for the time dependence of the spin correlation is reasonable for a three-dimensional system if there is a single relevant relaxation channel. For more complex systems, e.g. spin-glasses, a power law, a stretched-exponential, and a cutoff-power law, such as the Ogielski law (1985),

$$\Lambda_{ii}^{\alpha\alpha}(t) = c_{\rm cpl} t^{-\alpha_{\rm cpl}} \exp[-(\lambda t)^{\beta_{\rm cpl}}], \tag{10.86}$$

where $\alpha_{\rm cpl}$ and $\beta_{\rm cpl}$ are positive exponents, have been proposed. We recall that the symmetrised spin correlation function $\Lambda_{ii}^{\alpha\alpha}(t)$ has been defined in Eq. 7.56. All these laws asymptotically lead to scaling as in Eq. 10.85, but with the exponent γ differing from 2 (Keren *et al.* 1996, Keren 2004).

Although the study of spin dynamics is best carried out with the longitudinal-field technique, information relevant to spin dynamics can also be derived from the analysis of the decay of the wiggles (arising from a spontaneous field or induced by an applied magnetic field). According to Eq. D.9, the exponential decay rate of the wiggles in the extreme motional narrowing limit is given by the formula

$$\lambda_X = \frac{\lambda_Z}{2} + \frac{\mathcal{D}}{2}\int_{v_c^*}\sum_{\alpha,\beta} G^{Z\alpha}(\mathbf{q})G^{\beta Z}(-\mathbf{q})\Lambda^{\alpha\beta}(\mathbf{q}, \omega = 0)\frac{{\rm d}^3\mathbf{q}}{(2\pi)^3}, \tag{10.87}$$

where we have used the same simplifications as those needed to derive Eq. 10.83 from Eq. 10.82. Interestingly, while λ_Z probes the spin correlation function at $\omega = \omega_\mu$ (see Eq. 10.83), the second term on the right-hand side of the previous equation tests the spin correlations at $\omega = 0$.

We recall that a discussion of the wiggles envelope can be found in Section 7.1.4.3.

The above expressions for the two decay rates, λ_Z and λ_X, are rather complicated. Their structure shows that they depend on the correlation tensor $\mathbf{\Lambda}$ that we aim to investigate and a coupling tensor $\boldsymbol{G}$. The presence of the latter tensor may be rather disturbing but its importance is emphasised in the next two parts (Sections 10.2.3 and 10.2.4).

10.2.3 Relaxation by fluctuations

In this section we compute λ_Z for some fluctuation mechanisms which are all electronic in nature. Instead of working with correlation functions, it is usually more practical to use susceptibility functions. Assuming the correlation tensor to be a scalar, from Eqs. 10.83 and F.15

$$\lambda_Z = \frac{\mu_0 \gamma_\mu^2}{16\pi} k_B T \int_{v_c^*} \sum_\alpha \mathcal{A}^{\alpha\alpha}(\mathbf{q}) \chi(\mathbf{q}) F(\mathbf{q}, \omega_\mu) \frac{\mathrm{d}^3\mathbf{q}}{(2\pi)^3}. \tag{10.88}$$

The factors $\chi(\mathbf{q})$ and $F(\mathbf{q}, \omega)$ are, respectively, the susceptibility and spectral-weight functions at wave vector $\mathbf{q}$. With the usual hypothesis of a Lorentzian spectral-weight function,

$$\lambda_Z = \frac{\mu_0 \gamma_\mu^2}{16\pi^2} k_B T \int_{v_c^*} \sum_\alpha \mathcal{A}^{\alpha\alpha}(\mathbf{q}) \chi(q) \frac{\Gamma(q)}{\Gamma^2(q) + \omega_\mu^2} \frac{\mathrm{d}^3\mathbf{q}}{(2\pi)^3}, \tag{10.89}$$

where $\Gamma(q)$ is the decay rate of the fluctuation of wave vector $\mathbf{q}$. Following the common practice of the neutron scattering community, in the rest of this chapter we shall call $\Gamma(q)$ the linewidth. Note that in the last equation the static susceptibility and linewidth are taken to only depend on the modulus of the wave vector for simplicity. The anisotropy of the correlation tensor can be taken into account. An example is discussed in Section 10.2.3.3.

When the static susceptibility displays a sharp maximum at the zone centre, as for a ferromagnetic compound at a temperature not too far away from its Curie point, it is sufficient to evaluate the limiting form $\sum_\alpha \mathcal{A}^{\alpha\alpha}(\mathbf{q} \to \mathbf{0})$. We recall that this quantity depends on tensor elements such as $G^{\alpha\beta}(\mathbf{q})$. At first sight, we may simply think to set $\mathbf{q} = \mathbf{0}$. This is not possible because $\boldsymbol{G}$ is the sum of the dipolar and hyperfine tensors, the elements of the former tensor being nonanalytical near $\mathbf{q} = \mathbf{0}$ as explained in Section 5.2.4. Its limit as $\mathbf{q} \to \mathbf{0}$ indeed depends on the path used to reach the origin of reciprocal space. From the same Section 5.2.4,

$$G^{\alpha\beta}(\mathbf{q} \to \mathbf{0}) = -4\pi \left[P_L^{\alpha\beta}(\mathbf{q}) - C^{\alpha\beta}(\mathbf{q} = \mathbf{0}) - \frac{r_\mu H}{4\pi} \delta^{\alpha\beta} \right], \tag{10.90}$$

where $P_L^{\alpha\beta}(\mathbf{q}) = q^\alpha q^\beta / q^2$ is an element of the longitudinal projection operator, $C^{\alpha\beta}(\mathbf{q})$ is analytical for all value of $\mathbf{q}$, r_μ is the number of nearest neighbour magnetic ions to the muon, and H the hyperfine constant. Instead of using this general expression for the coupling tensor, for simplicity we assume here $\boldsymbol{C}(\mathbf{q} = \mathbf{0})$ to be scalar, which implies $C^{\alpha\beta}(\mathbf{q} = \mathbf{0}) = \delta^{\alpha\beta}/3$ since the trace of $\boldsymbol{C}(\mathbf{q} = \mathbf{0})$ is zero (Eq. 5.58). This particular form holds for example for a muon in a tetrahedral or octahedral site of a face-centred-cubic crystal; see Section D.7. This limitation can be lifted at the expense of a more involved algebra. Setting

$$p = \frac{1}{3} + \frac{r_\mu H}{4\pi}, \tag{10.91}$$

we compute

$$\sum_{\alpha} \mathcal{A}^{\alpha\alpha}(\mathbf{q} \to \mathbf{0}) = 16\pi^2 \left\{ 2p^2 + (1-2p)\left[1 - \frac{(q^Z)^2}{q^2}\right]\right\}. \tag{10.92}$$

With this result and Eq. 10.89, we derive

$$\lambda_Z = \frac{\mu_0 \gamma_\mu^2}{\pi^2}\left[\left(p - \frac{1}{3}\right)^2 + \frac{2}{9}\right] k_{\rm B}T \int_0^{q_{\rm BZ}} \chi_0(q) \frac{\Gamma_0(q)}{\Gamma_0^2(q) + \omega_\mu^2} q^2 {\rm d}q, \tag{10.93}$$

where $q_{\rm BZ}$ is the radius of the Brillouin zone taken to be a sphere. Referring to the data listed in Table 10.3 (p. 330) for elemental ferromagnetic metals, we note that $|p - 1/3| = r_\mu |H|/(4\pi)$ is clearly less than unity. If the tensor $\boldsymbol{C}(\mathbf{q} = \mathbf{0})$ is not a scalar, the expression in the bracket of Eq. 10.93 will be changed. However, its numerical value will not be drastically different. We have written $\chi_0(q)$ and $\Gamma_0(q)$, respectively, for the susceptibility and linewidth of the isotropic ferromagnetic fluctuation of wave vector $\mathbf{q}$. The reason for this notation will become clear in the discussion below which deals with antiferromagnets.

The susceptibility of an antiferromagnet in a temperature regime not too far away from its Néel point exhibits a strong peak at a wave vector $\mathbf{k}$, which is the propagation wave vector defining the magnetic structure. The dipolar and hyperfine tensors are both analytical near $\mathbf{k}$ and therefore can be developed in a Taylor series around that vector. The relaxation rate λ_Z is expressed as a weighted integral of correlations over the Brillouin zone. Since the integrand peaks at $\mathbf{k}$, we shift the origin of the reciprocal space to $\mathbf{k}$. We assume $\mathbf{k}$ to be not too close from the zone centre since it may not be a good idea to perform this shift if $\mathbf{k}$ is small. We derive from Eq. 10.89

$$\lambda_Z = \frac{\mu_0 \gamma_\mu^2}{16\pi^2} k_{\rm B}T \int_{v_c^*} \sum_\alpha \mathcal{A}^{\alpha\alpha}(\mathbf{k}+\mathbf{q}) \chi_{\mathbf{k}}(q) \frac{\Gamma_{\mathbf{k}}(q)}{\Gamma_{\mathbf{k}}^2(q) + \omega_\mu^2} \frac{{\rm d}^3\mathbf{q}}{(2\pi)^3}. \tag{10.94}$$

We use the notations $\chi_{\mathbf{k}}(\mathbf{q}) = \chi(\mathbf{k}+\mathbf{q})$ and $\Gamma_{\mathbf{k}}(\mathbf{q}) = \Gamma(\mathbf{k}+\mathbf{q})$ and assume again a cubic symmetry. A fair approximation is usually to only keep the constant terms of the Taylor series of $\mathcal{A}^{\alpha\alpha}(\mathbf{k}+\mathbf{q})$, that is,

$$\begin{aligned}\lambda_Z &\simeq \frac{\mu_0 \gamma_\mu^2}{16\pi^2} k_{\rm B}T \int_{v_c^*} \sum_\alpha \mathcal{A}^{\alpha\alpha}(\mathbf{k}) \chi_{\mathbf{k}}(q) \frac{\Gamma_{\mathbf{k}}(q)}{\Gamma_{\mathbf{k}}^2(q) + \omega_\mu^2} \frac{{\rm d}^3\mathbf{q}}{(2\pi)^3}, \\ &= \frac{\mu_0 \gamma_\mu^2}{32\pi^4} \sum_\alpha \mathcal{A}^{\alpha\alpha}(\mathbf{k})\, k_{\rm B}T \int_0^{q_{\rm BZ}} \chi_{\mathbf{k}}(q) \frac{\Gamma_{\mathbf{k}}(q)}{\Gamma_{\mathbf{k}}^2(q) + \omega_\mu^2} q^2 {\rm d}q. \end{aligned} \tag{10.95}$$

With the conventional NMR notation, $\lambda_Z = 1/T_1$, we recover a well-known formula. For simplicity, we shall mostly use it. However, some care should be exercised. For example, the symmetry at the muon site may be such that $\mathcal{A}^{\alpha\alpha}(\mathbf{k}) = 0$. In that case, $\mathcal{A}^{\alpha\alpha}(\mathbf{k})$ acts as a screening factor, suppressing the contribution to λ_Z of the fluctuations at $\mathbf{k}$; i.e. $\lambda_Z = 0$ if only the constant term of the Taylor expansion is kept. A famous example is given by some high-T_c superconductors for which the screening effect is important for NMR at oxygen sites (Mila and Rice 1989). As a matter of fact,

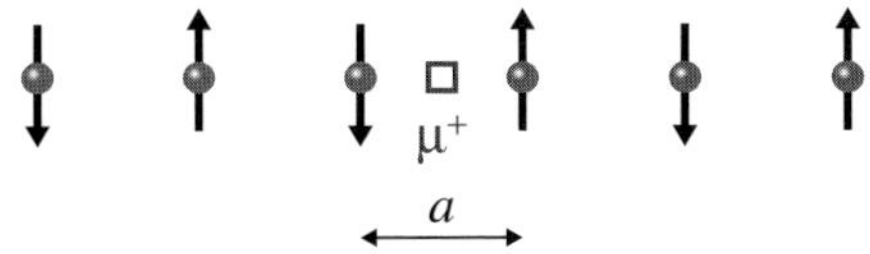

Fig. 10.6 A linear magnetic chain of lattice spacing a with magnetic moments localised on the lattice points and antiferromagnetically coupled. The muon is assumed to sit in between two lattice points.

fluctuations at a point different from $\mathbf{k}$ may drive the behaviour of λ_Z when $\mathcal{A}^{\alpha\alpha}(\mathbf{k}) = 0$.

The screening mechanism can be understood from a simple model. Let us consider a linear magnetic chain with the muon localised at equal distance from the lattice points. The magnetic moments carried by the atoms are coupled antiferromagnetically. This is illustrated in Fig. 10.6 where $k = \pi/a$. For convenience we neglect the dipole coupling and consider the hyperfine interaction between the muon spin and its nearest neighbour atomic spins only. Assuming the chain to be directed along the Z axis and using Eq. 5.43, we have

$$H(\mathbf{q}) = 2H \cos\left(\frac{aq^Z}{2}\right), \tag{10.96}$$

and therefore

$$\begin{aligned} H_{\mathbf{k}}(\mathbf{q}) &= 2H\left[\cos\left(\frac{ak^Z}{2}\right)\cos\left(\frac{aq^Z}{2}\right) - \sin\left(\frac{ak^Z}{2}\right)\sin\left(\frac{aq^Z}{2}\right)\right] \\ &= -2H\sin\left(\frac{aq^Z}{2}\right), \end{aligned} \tag{10.97}$$

with the notation $H_{\mathbf{k}}(\mathbf{q}) = H(\mathbf{k}+\mathbf{q})$. This means that $H_{\mathbf{k}}(0) = 0$ and therefore $\mathcal{A}^{\alpha\alpha}(\mathbf{k}) = 0$; that is, the antiferromagnetic correlations do not contribute to the relaxation. Let us consider explicitly λ_Z. From Eq. 10.83,

$$\begin{aligned} \lambda_Z &\propto \int H^2(\mathbf{q})\Lambda^{\alpha\alpha}(q,\omega_\mu)\,\mathrm{d}^3\mathbf{q} \\ &= 2H^2\int \Lambda^{\alpha\alpha}(q,\omega_\mu)\,\mathrm{d}^3\mathbf{q} + 2H^2\int \cos\left(aq^Z\right)\Lambda^{\alpha\alpha}(q,\omega_\mu)\,\mathrm{d}^3\mathbf{q}, \end{aligned} \tag{10.98}$$

When the fluctuations at $q^Z = k^Z = \pi/a$ have an overwhelming weight, the absolute value of the second term on the second line of the previous equation is about equal to that of the first term, but their signs are opposite, and thus the contribution of the fluctuations around $\mathbf{k}$ to λ_Z is vanishingly small.

The antiferromagnetic fluctuations are therefore filtered out at the muon site due to its position in the middle of neighbouring magnetic atoms. In that case, the ferromagnetic fluctuations, i.e. with $q = 0$, may determine λ_Z. This simple model computation clearly indicates that a bad approximation for the screening term may lead to an unreliable conclusion on the type of spin dynamics probed by μSR. A second conclusion is that the spin dynamics studied by different local nuclear techniques may not be comparable. For example, a nuclear probe embedded in a magnetic ion, as is sometimes the case for a NMR or Mössbauer nucleus, is mainly sensitive to self-correlations. On the other hand, the μSR technique may give information on self- and cross-correlations. A quantitative analysis requires the determination of the muon site.

Incidentally, we note here that, besides the screening effect that we have just described, a magnetic phase transition might also be overlooked for another reason. Let us consider two-dimensional systems for which the magnetic interactions are not of the Ising type. Since such systems do not order at finite temperature (Mermin and Wagner 1966), a dimensional crossover from $d = 2$ to $d = 3$ is required for a transition to occur. When this crossover takes place in a so narrow temperature region that it is experimentally unaccessible, no critical behaviour is observed and the transition is unnoticed from the dynamical response (Budnick *et al.* 1988, Ohira *et al.* 2004).

In the rest of this chapter Eq. 10.95 is assumed to be valid for antiferromagnets. We are now going to compute λ_Z for specific simple models. It is convenient to split this discussion according to whether we deal with localised, weakly itinerant, or critical fluctuations. We point out that this discussion will only consider a few possible relaxation mechanisms. In the future it may seem quite incomplete whence already scheduled improvements of μSR beams and spectrometers are bound to open possibilities for studying so far unknown mechanisms.

10.2.3.1 Ionic spin fluctuations. We label spin fluctuations arising from localised electron spins as ionic, in contrast to fluctuations due to conduction electrons which are labelled as itinerant. We shall study in this section the simplest possible physical case: paramagnetic fluctuations in a cubic crystal.

We first discuss the wave vector-dependent susceptibility for a Heisenberg paramagnet within the simplest approach, i.e. the mean-field approximation. For a ferromagnetic material it has the form (see, for example, Lovesey (1986))

$$\chi_0(\mathbf{q}) = \frac{\mu_0}{v_0} \frac{g^2 \mu_B^2 J(J+1)}{3 k_B T_C} \frac{1}{\tau + \left[1 - \dfrac{\mathcal{J}_0(\mathbf{q})}{\mathcal{J}_0}\right]}, \tag{10.99}$$

where τ is the reduced temperature,

$$\tau = |T - T_c| / T_c, \tag{10.100}$$

the critical temperature T_c being equal to the Curie temperature T_C. The absolute value in the definition of τ has been included for future use. We recall that v_0 is the volume per magnetic ion. The Fourier transform $\mathcal{J}(\mathbf{q})$ of the exchange interaction has been introduced:

$$\mathcal{J}(\mathbf{q}) = \mathcal{J} \sum_{\mathbf{i}}{}' \exp(-i\mathbf{q} \cdot \mathbf{i}). \tag{10.101}$$

Here, for simplicity, we consider a single exchange constant $\mathcal{J}$ and therefore the sum is limited to the first nearest neighbours of a given spin as indicated by the prime symbol.

The interaction Hamiltonian between the spins is assumed to be of the Heisenberg form,

$$\mathcal{H}_{\rm sys} = -\mathcal{J} \sum_{\langle i,i' \rangle} \mathbf{J}_i \cdot \mathbf{J}_{i'}, \tag{10.102}$$

where $\mathbf{J}_i$ and $\mathbf{J}_{i'}$ are total angular momenta at the nearest neighbour sites i and i', respectively. This means that the interaction energy between sites i and i' is $2\mathcal{J}$. In order to have a notation consistent with that of the antiferromagnets — see e.g. Eq. 10.104 below — we have written $\mathcal{J}_0(\mathbf{q})$ instead of $\mathcal{J}(\mathbf{q})$ in Eq. 10.99. In addition we have set $\mathcal{J}_0 \equiv \mathcal{J}(\mathbf{q} = \mathbf{0})$. If $\tau \gg 1$, i.e. if the temperature is much higher than the Curie temperature, $\chi_0(\mathbf{q})$ becomes wave vector independent and is given by the well-known Curie-Weiss law

$$\chi_0 = \frac{\mu_0}{v_0} \frac{g^2 \mu_{\rm B}^2 J(J+1)}{3k_{\rm B}\,(T - T_{\rm C})}. \tag{10.103}$$

Note that the high-temperature limit of this law is the Curie law which follows from the fluctuation–dissipation theorem; see Appendix F.

We now consider an antiferromagnetic compound. In the simplest but common case, its magnetic structure is described by a single propagation wave vector $\mathbf{k}$. The material is viewed as a two-lattice system such that the phase factor $\exp[i(\mathbf{i}+\mathbf{d})\cdot\mathbf{k}]$ has the value $+1$ on one sub-lattice and -1 on the other. For example, let us assume the compound to be a simple cubic compound of lattice parameter a and $\mathbf{k} = (\pi/a, 0, 0)$. Two $\mathbf{d}$ vectors are required: $\mathbf{d}_1 = (0,0,0)$ and $\mathbf{d}_2 = (a,0,0)$. We compute $\exp(i\mathbf{d}_1\cdot\mathbf{k}) = 1$ and $\exp(i\mathbf{d}_2\cdot\mathbf{k}) = -1$. It is easily seen that these two phase factors and only them are found, no matter which cell is considered. Within the mean-field approximation, the wave vector-dependent static susceptibility $\chi_{\mathbf{k}}(\mathbf{q})$ is of the same form as in Eq. 10.99, except that $\mathcal{J}_0(\mathbf{q}) = \mathcal{J}(\mathbf{q})$ is replaced by $\mathcal{J}_{\rm k}(\mathbf{q}) = \mathcal{J}(\mathbf{k}+\mathbf{q})$ since the vector $\mathbf{q}$ is defined relative to $\mathbf{k}$. In addition $T_{\rm C}$ is substituted by the Néel temperature, $T_{\rm N}$, and $\mathcal{J}_0$ by $\mathcal{J}_{\mathbf{k}} = \mathcal{J}(\mathbf{k})$. Explicitly,

$$\chi_{\mathbf{k}}(\mathbf{q}) = \frac{\mu_0}{v_0} \frac{g^2 \mu_{\rm B}^2 J(J+1)}{3k_{\rm B}T_{\rm N}} \frac{1}{\tau + \left[1 - \dfrac{\mathcal{J}_{\rm k}(\mathbf{q})}{\mathcal{J}_{\mathbf{k}}}\right]}, \tag{10.104}$$

where τ is defined by Eq.10.100 with $T_{\rm c} = T_{\rm N}$. As for a ferromagnet, $\chi_{\mathbf{k}}(\mathbf{q})$ becomes $\mathbf{q}$ independent at high temperature. We shall denote that susceptibility by $\chi_{\mathbf{k}}$. The subscript 0 in the susceptibility and $\mathcal{J}_0(\mathbf{q})$ for a ferromagnet acknowledges the fact that $\mathbf{k} = \mathbf{0}$ in that case.

There is at present no single unified theory for the dynamics of Heisenberg magnets that adequately covers the entire temperature range. Setting aside the treatment of the effect of the critical dynamics on λ_Z, we shall discuss the influence of spin dynamics on λ_Z in two different ways. We shall first use methods introduced in the 1950s and then consider the predictions given by the random-phase approximation.

10.2.3.1.1 Basic approach to ionic spin fluctuations. We first consider the $T \to \infty$ limit. From Eq. 10.88 λ_Z is found to be temperature independent because the susceptibility follows the Curie–Weiss law and the spectral-weight function is independent of the temperature; see, for example, Lovesey (1986). For a quantitative evaluation of λ_Z it is not convenient to start from Eqs. 10.83 and 10.54 since the cross-correlations vanish and therefore the space Fourier transform introduced in Section 7.1.4.3 is useless. Instead we start from the expression of λ_Z written in direct space, i.e. according to Eq. 7.97,

$$\lambda_Z = \pi\gamma_\mu^2 \left[\Phi^{XX}(\omega_\mu) + \Phi^{YY}(\omega_\mu)\right], \tag{10.105}$$

where $\Phi^{\alpha\alpha}(\omega)$ is given by Eq. 7.107. For simplicity, we will assume the μSR spectra to be recorded in zero field, i.e. $\omega_\mu = 0$. Since the Gaussian random model is suitable for the spin dynamics in the high-temperature limit (Anderson and Weiss 1953) and

$$\lim_{T\to\infty} \left\langle \delta J_i^\alpha \delta J_j^\beta \right\rangle = \delta_{\alpha,\beta}\delta_{i,j}\frac{1}{3}J(J+1), \tag{10.106}$$

if the spin dynamics is isotropic, we derive with the help of Eq. C.63,

$$\begin{aligned}\Lambda^{\gamma\gamma'}_{i+d,i'+d'}(\omega=0) &= \delta_{i,i'}\delta_{d,d'}\delta_{\gamma,\gamma'}\frac{J(J+1)}{3}\int_{-\infty}^{\infty}\exp\left(-\frac{\omega_{\rm ex}^2 t^2}{2}\right)\mathrm{d}t \\ &= \delta_{ii'}\delta_{dd'}\delta_{\gamma\gamma'}\frac{J(J+1)}{3}\frac{\sqrt{2\pi}}{\omega_{\rm ex}}.\end{aligned} \tag{10.107}$$

Here we have introduced $\omega_{\rm ex}$ which is the exchange angular frequency. Collecting the different terms,

$$\lambda_Z = \left(\frac{\mu_0}{4\pi}\right)^2 \gamma_\mu^2 g^2 \mu_{\rm B}^2 \frac{J(J+1)}{3}\frac{\sqrt{2\pi}}{\omega_{\rm ex}}\frac{1}{2v_c^2}\sum_{i,d}\sum_{\gamma=X,Y,Z}\left[\left(G^{X\gamma}_{\mathbf{r}_{i+d}}\right)^2 + \left(G^{Y\gamma}_{\mathbf{r}_{i+d}}\right)^2\right]. \tag{10.108}$$

This expression can be reduced if the hyperfine interaction is negligible. Using Eq. 5.22, we get

$$\lambda_Z = \left(\frac{\mu_o}{4\pi}\right)^2 \gamma_\mu^2 g^2 \mu_{\rm B}^2 \frac{J(J+1)}{3}\frac{\sqrt{2\pi}}{\omega_{\rm ex}}\sum_{i,d}\frac{1}{r_{i+d}^6}\left(\frac{5-3\cos^2\theta_{i+d}}{2}\right), \tag{10.109}$$

where θ_{i+d} is the polar angle of the radius vector $\mathbf{r}_{i+d}$. Note that the sums over i and d correspond in fact to a single sum over all the magnetic ions of the system under study. Defining

$$\tau_{\rm ex} = \sqrt{\frac{\pi}{2}}\frac{1}{\omega_{\rm ex}}, \tag{10.110}$$

the expression of λ_Z has an appealing form:

$$\lambda_Z = 2\left[\left(\frac{\mu_o}{4\pi}\right)^2 \gamma_\mu^2 g^2 \mu_{\rm B}^2 \frac{J(J+1)}{3}\sum_{i,d}\frac{1}{r_{i+d}^6}\left(\frac{5-3\cos^2\theta_{i+d}}{2}\right)\right]\tau_{\rm ex}. \tag{10.111}$$

The term in the square brackets is in fact identical to $\gamma_\mu^2\Delta^2_{\rm G,ZF}$ where $\Delta^2_{\rm G,ZF}$ is defined in Eq. 6.151. Hence, we have recovered the standard motional narrowing result, valid when the correlations between the spins are neglected.

To compute ω_{ex}, we use the Heisenberg Hamiltonian written in Eq. 10.102 and, following Moriya — Moriya (1956), see also Hartmann *et al.* (1986) — we expand $J^\alpha(t)$ up to second order in t with the result

$$\omega_{\text{ex}} = \sqrt{\frac{8\mathcal{J}^2 z_{\text{nn}} J(J+1)}{3\hbar^2}}, \tag{10.112}$$

where z_{nn} is the number of nearest magnetic ions surrounding each magnetic ion. In the mean-field approximation (see e.g. Kittel (1996)),

$$\mathcal{J} = \eta_{\text{l}} \frac{3\, k_{\text{B}}\, \theta_{\text{CW}}}{2 z_{\text{nn}} J(J+1)}. \tag{10.113}$$

The absolute value of the Curie–Weiss temperature θ_{CW} is usually approximately equal to the ordering temperature, except for frustrated compounds — see for example Ramirez (2001). The multiplicative factor η_{l} depends on the actual lattice and typical values range from 1.5 to 2. Note that, since the sign of the exchange parameter has no influence on the expression of λ_Z, Eq. 10.108 is equally valid for ferromagnets and antiferromagnets.

As a matter of fact, the predicted temperature independence of λ_Z at high temperature is not always observed, even at very high temperature. Other relaxation processes involving phonons or conduction electrons of the specimen may also be at play. This is further discussed in Section 10.2.4.2. In addition, short-range correlations may influence λ_Z as explained below.

As a compound is cooled down from a temperature much higher than its magnetic ordering temperature, short-range pair correlations are expected to appear. They were first experimentally detected by NMR as an increase (decrease) of the spin–lattice and spin–spin relaxation rates for a ferromagnetic (antiferromagnetic) compound as it is cooled down and the pair correlations become increasingly important (Silbernagel *et al.* 1968). This is easily understood qualitatively as ferromagnetic (antiferromagnetic) correlations correspond to the enhancement (reduction) of the local field at the nuclear probe. As noted earlier, see Eq. 10.106, there is no correlation between distinct spins in the $T \to \infty$ limit. Such correlations are found by expanding $\left\langle \delta J_i^\alpha \delta J_j^\beta \right\rangle$ to first order in $1/(k_{\text{B}}T)$:

$$\left\langle \delta J_i^\alpha \delta J_j^\beta \right\rangle = \delta_{\alpha,\beta} \frac{2\mathcal{J}}{k_{\text{B}}T} \left[\frac{1}{3} J(J+1) \right]^2 . \tag{10.114}$$

As usual in this chapter, a cubic compound is assumed. The key feature of high-temperature pair correlations is already evident from this formula: it depends on the sign of the exchange constant. Recalling our definition of the interaction Hamiltonian given in Eq. 10.102, the pair correlation is positive (negative) for a ferromagnetic (antiferromagnetic) exchange constant. This sign dependence immediately explains the drastically different effect of the two types of correlations on λ_Z. For a quantitative evaluation of λ_Z, two approaches are possible: either we stick to the real space formalism, as we have done for the description of the correlations in the high-temperature limit, or we use the more powerful $\mathbf{q}$-space representation. The first method is followed in the literature (Silbernagel *et al.* 1968 and Hartmann *et al.* 1986). Here, we choose the second method.

For definitiveness, we shall consider ferromagnets. Referring to Eq. 10.88, we first need an expression for the susceptibility $\chi_0(\mathbf{q})$. We start its evaluation in zeroth order to obtain $\chi_0^{(0)}(\mathbf{q})$, which corresponds to the self-correlation function of Eq. 10.106. This means that it is a very high-temperature approximation. Using the energy-integrated fluctuation–dissipation theorem which gives a prescription for getting a static susceptibility from an instantaneous correlation function (Eq. F.20) and the definition of the lattice Fourier transform given in Section E.2. we derive $\chi_0^{(0)}(\mathbf{q}) = \chi_\mathbf{0}$, which is found to be $\mathbf{q}$ independent, as expected for self-correlations at high temperatures. Proceeding in the same way, using Eq. 10.114 we get the first order correction

$$\chi_0^{(1)}(\mathbf{q}) = \frac{\mu_0}{v_0} g^2 \mu_B^2 \left[\frac{1}{3} J(J+1)\right]^2 \frac{2\mathcal{J}_0(\mathbf{q})}{(k_B T)^2} = \frac{v_0}{\mu_0} \frac{2}{g^2 \mu_B^2} \chi_0^2 \mathcal{J}_0(\mathbf{q}). \tag{10.115}$$

We note that (see, e.g. Lovesey (1986))

$$\frac{T_C}{T} = \frac{v_0}{\mu_0} \frac{2}{g^2 \mu_B^2} \chi_\mathbf{0} \mathcal{J}_\mathbf{0}. \tag{10.116}$$

In addition, we know that $|\mathcal{J}_0(\mathbf{q})|$ reaches its maximum value at $\mathbf{q} = \mathbf{0}$, for which we get $\chi_0^{(1)}(\mathbf{k}) = (T_C/T)\chi_0$. The first order term $\chi_0^{(1)}(\mathbf{q})$ is obviously $\mathbf{q}$ dependent because it accounts for pair correlations. At the level we are working, $\chi_0(\mathbf{q}) = \chi_0^{(0)}(\mathbf{q}) + \chi_0^{(1)}(\mathbf{q})$. It is easily proven that this is in fact the sum of the first two terms of the expansion of the mean-field result given in Eq. 10.99 and therefore, in practice, the mean-field result for $\chi_0(\mathbf{q})$ can be used. According to Eq. 10.88, an expression for the spectral-weight function is required to compute λ_Z. Its functional form depends on $\mathbf{q}$; see Lovesey (1986) for a full discussion. Here, for simplicity we assume a Gaussian form. This means that, in zero field, the case of most experimental interest, $F_0(\mathbf{q}, 0) = 1/\sqrt{2\pi\langle\tilde{\omega}_{ex}^2(\mathbf{q})\rangle}$, where $\langle\tilde{\omega}_{ex}^2(\mathbf{q})\rangle$ is the second moment of $F_0(\mathbf{q}, \omega)$. For a Heisenberg magnet, using Eq. F.14 in the limit of very high temperature,

$$\langle\tilde{\omega}_{ex}^2(\mathbf{q})\rangle = \frac{8\mathcal{J}^2 J(J+1)}{3\hbar^2} \sum_{\mathbf{d}_{pair}} \left[1 - \cos\left(\mathbf{q} \cdot \mathbf{d}_{pair}\right)\right], \tag{10.117}$$

where $\mathbf{d}_{pair}$ is a vector joining nearest neighbour magnetic atoms. The second moment $\langle\tilde{\omega}_{ex}^2(\mathbf{q})\rangle$ is temperature independent in the approximation we are using. If the $\mathbf{q}$ dependence of the moment is neglected, i.e. if we forget the cosine terms, we recover the formula for ω_{ex}^2; see Eq. 10.112. Since the expressions for the susceptibility and spectral-weight functions are now available, λ_Z can be computed from Eq. 10.88. This can only be achieved numerically.

We recall that Eq. 10.115 is written for a ferromagnet. As a homework the reader will derive the equivalent formula for an antiferromagnet.

Qualitatively the real space formalism indicates in first order in $1/(k_B T)$ that

$$\lambda_Z(T) = \lambda_Z^{(\infty)} \left(1 + \alpha \frac{\theta_{CW}}{T}\right), \tag{10.118}$$

where $\lambda_Z^{(\infty)}$ is the spin–lattice relaxation rate computed from Eq. 10.108 and α is a constant of order unity (Hartmann *et al.* 1986). The Curie–Weiss temperature $\theta_{\rm CW}$ is positive for ferromagnets and negative for antiferromagnets. It is easy to understand that the weight of the second term on the right-hand side depends drastically on the spin configuration around the muon and may even vanish. Using a more involved formalism, we shall give below (Eq. 10.144) the derivation of a formula that leads to an expression able also to account for the physics described by Eq. 10.118. Moreover, that formalism can be used in cases for which the simple method of the present section cannot be applied.

We now turn our attention to temperatures not far away from $T_{\rm C}$, but not in the critical regime, for which the susceptibility presents a well-developed maximum at the zone centre for ferromagnets. One of the points of this discussion is to unravel in a simple way an additional effect of the type of interaction, ferromagnetic versus antiferromagnet, on the spin dynamics. In the temperature range of interest here a proper approximation for the susceptibility is given by the Ornstein–Zernike form which is derived from Eq. 10.99 in the small wave vector limit

$$\chi_0(\mathbf{q}) = \frac{\chi_0}{1+(q/\kappa_{\rm m})^2}, \tag{10.119}$$

with $\kappa_{\rm m}^2 = \tau/\gamma$. The correlation length $\xi_{\rm m} = 1/\kappa_{\rm m}$ is defined in terms of the stiffness constant, γ, of the dispersion of the exchange energy:

$$1 - \frac{\mathcal{J}_0(\mathbf{q})}{\mathcal{J}_0} = \gamma\, q^2 + \mathcal{O}\left(q^4\right). \tag{10.120}$$

The definition of $\kappa_{\rm m}$ implies that

$$\kappa_{\rm m} = \kappa_{\rm m,0}\sqrt{\frac{|T-T_{\rm C}|}{T_{\rm C}}}, \tag{10.121}$$

where $\xi_{\rm m,0} = 1/\kappa_{\rm m,0}$ is on the order of the interatomic distance.

Following Moriya (1962), the linewidth is expanded up to fourth order in wave vector

$$\hbar\Gamma_0(q) = D_{\rm sd}\, q^2 + \Theta\, q^4, \tag{10.122}$$

where $D_{\rm sd}$ is proportional to τ, that is,

$$D_{\rm sd} = D_0 \frac{T-T_{\rm C}}{T_{\rm C}}, \tag{10.123}$$

and Θ is temperature independent. It is remarkable that $\Gamma_0(q)$ vanishes with q. This is a consequence of the conservation of the total magnetisation for a Heisenberg ferromagnet — see e.g. Mezei (1984). The lowest order term corresponds to spin diffusion: at time t the correlation extends over a distance $r = \sqrt{tD_{\rm sd}/\hbar}$ if the term proportional to Θ is negligible. From Eqs. 10.93, 10.103, 10.119, and 10.122

$$\lambda_Z = \frac{\mu_0^2}{3\pi^2}\frac{\gamma_\mu^2 g^2 \mu_{\rm B}^2 J(J+1)}{v_0}\left[\left(p-\frac{1}{3}\right)^2 + \frac{2}{9}\right]\frac{T}{T-T_{\rm C}}$$

$$\times \int_0^{q_{\rm BZ}} \frac{\hbar}{1+\left(\frac{q}{\kappa_{\rm m}}\right)^2} \frac{D_{\rm sd}\, q^2 + \Theta\, q^4}{[D_{\rm sd}\, q^2 + \Theta\, q^4]^2 + \hbar^2\omega_\mu^2} q^2 {\rm d}q. \qquad (10.124)$$

With the typical field intensity usually available, the muon-Zeeman energy is much smaller than the energy width corresponding to the decay of fluctuations and thus the field dependence of λ_Z is negligible. The evaluation of λ_Z for a zero-field measurement yields

$$\lambda_Z = \frac{\mu_0^2}{3\pi^2} \frac{\hbar\gamma_\mu^2 g^2 \mu_{\rm B}^2 J(J+1)}{v_0 D_0} \left[\left(p-\frac{1}{3}\right)^2 + \frac{2}{9}\right] \frac{T T_{\rm C}}{(T-T_{\rm C})^2} I(q_{\rm BZ}, \kappa_{\rm m}, \varrho), \qquad (10.125)$$

with the definition

$$I(q_{\rm BZ}, \kappa_{\rm m}, \varrho) = \int_0^{q_{\rm BZ}} \frac{1}{1+\left(\frac{q}{\kappa_{\rm m}}\right)^2} \frac{1}{1+\left(\frac{q}{\varrho}\right)^2} {\rm d}q. \qquad (10.126)$$

Although we have not written it explicitly, $I(q_{\rm BZ}, \kappa_{\rm m}, \varrho)$ depends on the temperature. In Eq. 10.126 we have introduced the notation $\varrho^2 = D_{\rm sd}/\Theta$. Whereas $1/\kappa_{\rm m}$ is the conventional correlation length characterising the wave vector-dependent static susceptibility, $1/\varrho$ is an additional correlation length which only concerns the dynamics. Since in the critical regime only one length scale exists for a Heisenberg magnet (see Section 10.2.3.3), it is reasonable to assume this restriction to extend outside the critical region and therefore set $\varrho = \kappa_{\rm m}$. The computation of $I(q_{\rm BZ}, \kappa_{\rm m}, \kappa_{\rm m})$ can analytically be performed with the result

$$I(q_{\rm BZ}, \kappa_{\rm m}, \kappa_{\rm m}) = \frac{\kappa_{\rm m}}{2} \left[\frac{q_{\rm BZ}}{\kappa_{\rm m}} \frac{1}{1+\left(\frac{q_{\rm BZ}}{\kappa_{\rm m}}\right)^2} + \arctan\left(\frac{q_{\rm BZ}}{\kappa_{\rm m}}\right)\right]. \qquad (10.127)$$

In the limit $\kappa_{\rm m} \ll q_{\rm BZ}$ we get the simple formula

$$I(q_{\rm BZ}, \kappa_{\rm m}, \kappa_{\rm m}) = \frac{\pi}{4}\kappa_{\rm m}. \qquad (10.128)$$

Using Eq. 10.121, we recover the result originally given by Moriya (1962): $\lambda_Z \propto [T_{\rm C}/(T-T_{\rm C})]^{3/2}$.

Note that Eq. 10.125 leads to $\lambda_Z \propto 1/T$ if $T \gg T_{\rm C}$ since in this limit $I(q_{\rm BZ}, \kappa_{\rm m}, \varrho)$ is temperature independent. This is consistent with the presence of short-range ferromagnetic correlations, as explained earlier. However, in this limit Moriya's result (Eq. 10.125) does not apply because it is based on the assumption of a sharply peaked wave vector-dependent susceptibility. In fact, as explained at the beginning of this section, λ_Z is expected to be temperature independent in the high-temperature limit for the Heisenberg Hamiltonian or any ionic (localised) spin Hamiltonian.

We now focus on antiferromagnetic compounds. At temperatures not too far above $T_{\rm N}$ the Ornstein–Zernike small $\mathbf{q}$ development of $\chi_{\mathbf{k}}(\mathbf{q})$ is suitable. It is obtained by substituting the subscript 0 of $\chi_0(\mathbf{q})$ and χ_0 with $\mathbf{k}$ in Eq. 10.119. However, the linewidth is basically different for ferromagnetic and antiferromagnetic fluctuations. Following Moriya (1962), up to second order in wave vector,

$$\hbar\Gamma_{\mathbf{k}}(q) = \Xi + \Upsilon\, q^2, \tag{10.129}$$

where Ξ is proportional to τ, that is,

$$\Xi = \Xi_0 \frac{T - T_{\rm N}}{T_{\rm N}}, \tag{10.130}$$

and Υ is temperature independent. Thus $\Gamma_{\mathbf{k}}(q)$ is finite at $q = 0$, reflecting that the staggered magnetisation is not a constant of motion, in contrast with the uniform magnetisation of a Heisenberg ferromagnet. In zero field, from Eq. 10.95

$$\lambda_Z = \frac{\mu_0^2}{96\pi^4} \frac{\hbar\gamma_\mu^2 g^2 \mu_{\rm B}^2 J(J+1)}{v_0\, \Xi_0} \sum_\alpha \mathcal{A}^{\alpha\alpha}(\mathbf{k}) \frac{T T_{\rm N}}{(T - T_{\rm N})^2} J(q_{\rm BZ}, \kappa_{\rm m}, \varrho), \tag{10.131}$$

where $\varrho^2 = \Xi/\Upsilon$ and

$$J(q_{\rm BZ}, \kappa_{\rm m}, \varrho) = \int_0^{q_{\rm BZ}} \frac{1}{1 + \left(\dfrac{q}{\kappa_{\rm m}}\right)^2} \frac{1}{1 + \left(\dfrac{q}{\varrho}\right)^2} q^2 {\rm d}q. \tag{10.132}$$

Like $I(q_{\rm BZ}, \kappa_{\rm m}, \varrho)$, $J(q_{\rm BZ}, \kappa_{\rm m}, \varrho)$ also depends on the temperature. Again, taking $\varrho = \kappa_{\rm m}$,

$$J(q_{\rm BZ}, \kappa_{\rm m}, \kappa_{\rm m}) = \frac{\kappa_{\rm m}^3}{2} \left[-\frac{q_{\rm BZ}}{\kappa_{\rm m}} \frac{1}{1 + \left(\dfrac{q_{\rm BZ}}{\kappa_{\rm m}}\right)^2} + \arctan\left(\frac{q_{\rm BZ}}{\kappa_{\rm m}}\right) \right]. \tag{10.133}$$

We get a simple formula in the $\kappa_{\rm m} \ll q_{\rm BZ}$ limit:

$$J(q_{\rm BZ}, \kappa_{\rm m}, \kappa_{\rm m}) = \frac{\pi}{4} \kappa_{\rm m}^3. \tag{10.134}$$

This means that $\lambda_Z \propto [T_{\rm N}/(T - T_{\rm N})]^{1/2}$, as given originally by Moriya (1962).

We stress that the results derived in this section cannot be applied in the critical regime. This is the subject of Section 10.2.3.3. However, we point out that, independent of the temperature regime of the fluctuations and their characteristics (ionic or itinerant), the linewidth of ferromagnetic fluctuations vanishes with their wave vector, whereas it saturates to a finite value for antiferromagnetic fluctuations in the same limit. Since μSR measures extremely slow fluctuations relative to neutron techniques, it is particularly adapted to probe ferromagnetic fluctuations. This type of fluctuations may exist in a magnetic material even if antiferromagnetic fluctuations have an overwhelming weight.

10.2.3.1.2 A more elaborate approach to ionic spin fluctuations: the random phase approximation. As announced earlier, we now focus on the predictions given by the random-phase approximation (RPA). A good reference to this approximation is the book of Jensen and Mackintosh (1991). This model directly considers the dynamical susceptibility and therefore one does not have to separately study the static susceptibility and the linewidth. This may suggest to the reader that the exposition of the results given above is of no interest. However, using the methods of the previous discussion we have unveiled in a simple way basic facts, such as the influence of the sign of the interaction on the temperature dependence of the relaxation rate. The RPA, which confirms the results derived above, is far more complicated to handle. Nonetheless, one of its advantages is the possibility of tackling relatively complex Hamiltonians. For example, a crystal-field Hamiltonian can be taken into account if needed.

We shall continue to assume the spin-correlation tensor to be a scalar, with its element denoted as $\Lambda(\mathbf{q},\omega)$. The fluctuation–dissipation theorem provides a relation between the elements of the correlation and susceptibility tensors; see Appendix F. Denoting the imaginary part of the susceptibility $\chi(\mathbf{q},\omega)$ as $\mathcal{I}m\{\chi(\mathbf{q},\omega)\}$,

$$\Lambda(\mathbf{q},\omega) = \frac{\hbar v_c}{\mu_0 g^2 \mu_{\mathrm{B}}^2} \coth\left(\frac{\beta\hbar\omega}{2}\right) \mathcal{I}m\{\chi(\mathbf{q},\omega)\}. \tag{10.135}$$

Therefore we need to compute $\mathcal{I}m\{\chi(\mathbf{q},\omega)\}$. According to the random-phase approximation,

$$\boldsymbol{\chi}(\mathbf{q},\omega) = \left\{\boldsymbol{\mathcal{I}} - \frac{2v_c\mathcal{J}(\mathbf{q})}{\mu_0 g^2 \mu_{\mathrm{B}}^2}\boldsymbol{\chi}(\omega)\right\}^{-1} \boldsymbol{\chi}(\omega). \tag{10.136}$$

The presence of the factor 2 reflects our definition of the Heisenberg Hamiltonian — see Eq. 10.102 — and $\mathcal{J}(\mathbf{q})$ is defined in Eq. 10.101. For the sake of generality Eq. 10.136 has been written in terms of tensors; i.e. $\boldsymbol{\chi}(\mathbf{q},\omega)$ is the susceptibility tensor for a lattice of magnetic ions, $\boldsymbol{\chi}(\omega)$ is the susceptibility tensor whence the intersite interactions are switched off, and $\{A\}^{-1}$ stands for the inverse of tensor A. For future use, we define the correlation tensor associated with $\boldsymbol{\chi}(\omega)$ as $\boldsymbol{\Lambda}(\omega)$. Section 10.2.4.2 provides an example of the derivation of an expression for this latter tensor when the spin relaxation is driven by crystal electric-field transitions. The exchange parameter $\mathcal{J}(\mathbf{q})$, defined in Eq. 10.101, is here a scalar, but a formula more general than Eq. 10.136 can be written down where the coupling between the spins is described by a tensor. In this way one can account for instance for the dipolar interaction between magnetic moments.

Focusing still on the scalar case and using again the fluctuation–dissipation theorem,

$$\mathcal{I}m\{\chi(\omega)\} = \frac{\mu_0 g^2 \mu_{\mathrm{B}}^2}{\hbar v_c} \tanh\left(\frac{\beta\hbar\omega}{2}\right) \Lambda(\omega). \tag{10.137}$$

The real part of $\chi(\omega)$ is readily expressed as a function of $\mathcal{I}m\{\chi(\omega)\}$ thanks to the Kramers–Kronig theorem (Eq. F.3). Thus, once $\Lambda(\omega)$ is known, $\chi(\omega)$ can be computed. Then $\mathcal{I}m\{\chi(\mathbf{q},\omega)\}$ can be determined and $\Lambda(\mathbf{q},\omega)$ is finally obtained:

$$\Lambda(\mathbf{q},\omega) = \boldsymbol{\Lambda}(\omega) \left\{\left[1 + \frac{2\mathcal{J}(\mathbf{q})}{\pi\hbar} \int_{-\infty}^{\infty} \frac{\Lambda(u)}{\omega - u} \tanh\left(\frac{\beta\hbar u}{2}\right) du\right]^2\right.$$

$$+ \left[\frac{2\mathcal{J}(\mathbf{q})}{\hbar} \tanh\left(\frac{\beta\hbar\omega}{2}\right) \Lambda(\omega) \right]^2 \Bigg\}^{-1} . \tag{10.138}$$

Therefore, if $\Lambda(\omega)$ and $\mathcal{J}(\mathbf{q})$ are known, it is possible to numerically compute $\Lambda(\mathbf{q}, \omega)$ and thus λ_Z for a given muon site.

We can proceed further analytically for the elastic case, that is, when $\omega = \omega_\mu = 0$. For definitiveness, we assume $\Lambda(\omega)$ to be a Lorentzian, i.e.

$$\Lambda(\omega) = \frac{2\langle(\delta J_z)^2\rangle\Gamma^{zz}}{\omega^2 + (\Gamma^{zz})^2}. \tag{10.139}$$

We derive

$$\Lambda(\mathbf{q}, \omega = \omega_\mu = 0) = \frac{\Lambda(\omega = \omega_\mu = 0)}{\left[1 - \frac{4\langle(\delta J_z)^2\rangle\mathcal{J}(\mathbf{q})}{\pi\hbar\Gamma^{zz}} \int_{-\infty}^{\infty} \frac{1}{x(x^2+1)} \tanh\left(\frac{\beta\hbar\Gamma^{zz}}{2}x\right) \mathrm{d}x\right]^2}. \tag{10.140}$$

For an insight into this expression we consider the consequences of a magnetic phase transition. Such a transition is characterised, among other properties, by its critical temperature T_c and a propagation wave vector $\mathbf{k}$ which sets the repetition length of the magnetic structure for $T < T_\mathrm{c}$. We recall that $T_\mathrm{c} = T_\mathrm{C}$ for a ferromagnet and $T_\mathrm{c} = T_\mathrm{N}$ for an antiferromagnet. At T_c the generalised susceptibility diverges for $\mathbf{q} = \mathbf{k}$, and so does $\Lambda(\mathbf{k}, \omega = 0)$. From Eq. 10.140, it diverges for

$$\frac{\hbar}{\mathcal{J}(\mathbf{k})} = \frac{4\langle(\delta J_z)^2\rangle}{\pi\Gamma^{zz}} \int_{-\infty}^{\infty} \frac{1}{x(x^2+1)} \tanh\left(\frac{\hbar\Gamma^{zz}}{2k_\mathrm{B}T_\mathrm{c}}x\right) \mathrm{d}x. \tag{10.141}$$

Let us study the function $I(\alpha) = \int_{-\infty}^{\infty} \frac{1}{x(x^2+1)} \tanh(\alpha x)\mathrm{d}x$. We numerically evaluate $I(1) = 2.08$. This is clearly smaller than $\int_{-\infty}^{\infty} \frac{1}{x^2+1}\mathrm{d}x = \pi$. However, the smaller α, the better the approximation of linearising the function $\tanh(\alpha x)$. In our case $\alpha = \hbar\Gamma^{zz}/(2k_\mathrm{B}T_\mathrm{c})$. Physically, we expect $\hbar\Gamma^{zz} \simeq k_\mathrm{B}T_\mathrm{c}$ for a conventional phase transition. Assuming the linearisation to be valid, we get

$$\mathcal{J}(\mathbf{k}) = \frac{k_\mathrm{B}T_\mathrm{c}}{2\langle(\delta J_z)^2\rangle}. \tag{10.142}$$

This is consistent with the mean-field result given in Eq. 10.113.

We know that λ_Z in zero field is given as an integral over the Brillouin zone of weighted $\Lambda(\mathbf{q}, \omega_\mu = 0)$. Approaching the critical point, the integral is expected to be dominated by $\Lambda(\mathbf{k}, \omega_\mu = 0)$ since that correlation diverges at T_c. How does it diverge? According to our RPA result in Eq. 10.140, upon linearisation of the hyperbolic tangent function, we derive

$$\Lambda(\mathbf{k}, \omega = \omega_\mu = 0) = \Lambda(\omega = \omega_\mu = 0) \left(\frac{T}{T - T_\mathrm{c}}\right)^2. \tag{10.143}$$

While Eq. 10.143 is obviously not valid in the critical regime since the RPA is a kind of mean-field approximation, it may provide a good description in the paramagnetic regime far from T_c.

Now we study $\Lambda(\mathbf{q}, \omega = \omega_\mu = 0)$ in the limit $T \gg T_c$. We find

$$\Lambda(\mathbf{q}, \omega = \omega_\mu = 0) = \Lambda(\omega = \omega_\mu = 0)\left(1 + \frac{\mathcal{J}(\mathbf{q})}{\mathcal{J}(\mathbf{k})}\frac{T_c}{T}\right). \tag{10.144}$$

Because $\mathcal{J}(\mathbf{k}) > 0$ (see Eq. 10.142), the sign of the thermal correction is given by the sign of $\mathcal{J}(\mathbf{q})$. Positive and negative $\mathcal{J}(\mathbf{q})$ values correspond to ferromagnetic and antiferromagnetic interactions at wave vector $\mathbf{q}$, respectively. This means that a ferromagnetic (antiferromagnetic) interaction increases (decreases) the correlation. This is consistent with our previous result obtained by a simpler method; see Eq. 10.114 and the discussion below it.

10.2.3.2 Itinerant spin fluctuations. In this section we first discuss the relaxation of the muon spin in a metal induced by an electron density at its site, and later on we analyse the relaxation arising from fluctuations in weak itinerant magnets.

It is well known that the conduction-electron density in a metal is able to relax a nuclear spin embedded in it. The measured nuclear spin–lattice relaxation rate is proportional to the temperature. This is taken nowadays as a signature of a Fermi liquid ground state. Since a muon spin is only a special nuclear spin, λ_Z is also expected to be proportional to the temperature for a metal with a Fermi liquid ground state.

According to Section 5.1.3, in the simplest physical assumption of a muon spin interacting with an electronic-spin density of spherical symmetry, the interaction spin Hamiltonian is given as

$$\mathcal{H}_\mu = -\frac{2\mu_0}{3}|\psi_{1s}(\mathbf{r}_0)|^2\eta(\mathbf{r}_0)\gamma_e\gamma_\mu\hbar^2\,\mathbf{S}_e\cdot\mathbf{S}_\mu. \tag{10.145}$$

The enhancement of the electron density at the muon site due to its electric charge is accounted for by the multiplicative constant $\eta(\mathbf{r}_0)$. When a magnetic field is applied, the magnetic field intensity at the muon site, measured by the transverse-field method, is not equal to the applied field. A field shift is detected, the so-called muon Knight shift K_μ, as explained in Section 5.3. A relationship between λ_Z and K_μ was derived a long time ago by Korringa (1950):

$$\lambda_Z = \frac{4\pi k_B T K_\mu^2}{\hbar}\left(\frac{\gamma_\mu}{\gamma_e}\right)^2. \tag{10.146}$$

The proportionality between the spin–lattice relaxation rate and the temperature is well established by NMR experiments. However, the measured coefficient of proportionality is usually different from the theoretical expectation. For a metal with an appreciable mean free path, the measured enhancement is thought to reflect the electronic correlations neglected in the simple computation. As an order of magnitude, the relaxation time of copper nuclei in metallic copper is measured to be $\sim$3 ms at room temperature (Redfield 1955) and of course longer at low temperature. This is a long time relative to the μSR time scale and this justifies of treating the copper nuclear spin as static when analysing μSR spectra recorded for metallic copper.

The muon–spin relaxation induced by a conduction electron density is usually vanishingly small. For example, assuming the smallest measurable λ_Z value to be

$1\,\mathrm{ms}^{-1}$, at 10 K the Korringa relation gives $K_\mu = 1.6 \times 10^{-3}$. Since this is more than one order of magnitude larger than the measured Knight shift for diamagnetic metals — see Schenck (1985) for a review — the muon-spin relaxation induced by the conduction electron spin density at the muon site is usually too small to be detected.

However, Knight shifts of the computed order of magnitude have been observed for some diamagnetic muon states. In this case an appreciable value for λ_Z is measured but the proportionality between λ_Z and the temperature seems to break down. The key point is that the conduction-electron density is insufficient to effectively screen the muon charge. This leads to a considerable enhancement of both Knight shift and relaxation rate. We refer to Section 12.5 for a further discussion.

A family of ferromagnetic and antiferromagnetic compounds, the so-called weak itinerant magnets, is characterised by relatively low ordering temperatures and small magnetic moments. Their magnetic properties arise from their conduction electrons and their long-range order is lost due to the vanishing amplitude of the magnetic moments at the ordering point. This behaviour is in marked contrast to localised magnets for which long-range order is destroyed by thermal excitations, i.e. by spin waves. Originally, these metals comprised some compounds based only on $3d$ elements. However, intermetallics made of f elements have been found to exhibit some similar magnetic properties. One should still note that this family of metals is somewhat restricted since, for example, metallic chromium does not belong to it (Fawcett 1988). The description of these itinerant magnets rests on the recognition of the importance of low-energy spin fluctuations, using either a self-consistent renormalisation theory (SCR, Moriya (1985)) or a phenomenological Landau–Ginzburg expansion for the spin fluctuations (Lonzarich and Taillefer 1985). Short reviews are given by Rainford (1999), Mohn and Khmelevskyi (2001), and Mishra (2001).

Since the energy of the magnetic fluctuations is small relative to the thermal energy for weak itinerant magnets, we can use Eq. F.25 which is a relationship between the variance of the fluctuating magnetic moments, $\overline{\langle(\delta m)^2\rangle}$, and $\chi_{\mathbf{k}}(q)$:

$$\frac{\overline{\langle(\delta m)^2\rangle}}{v_0^2} = \frac{3k_{\mathrm{B}}T}{2\pi^2\mu_0}\int_0^{q_u} \chi_{\mathbf{k}}(q)q^2\mathrm{d}q, \tag{10.147}$$

assuming again a system of cubic symmetry. The upper cutoff q_u sets the upper bound on the wave vector of the correlations to be considered. Since only long-wavelength fluctuations are of importance, the Ornstein–Zernike approximation for the susceptibility (Eq. 10.119) is adequate over the whole temperature range. The integration in Eq. 10.147 can be analytically performed with the result

$$\overline{\langle(\delta m)^2\rangle} = v_0^2\frac{3k_{\mathrm{B}}T\kappa_{\mathrm{m}}^3\chi_{\mathbf{k}}}{2\pi^2\mu_0}\left[\frac{q_u}{\kappa_{\mathrm{m}}} - \arctan\left(\frac{q_u}{\kappa_{\mathrm{m}}}\right)\right]. \tag{10.148}$$

Although of drastically different origin, the uniform susceptibility of both localised electron paramagnets and of weak itinerant paramagnets follows the Curie–Weiss law. Then, referring to the definition of κ_{m} (see Eq. 10.121), $\kappa_{\mathrm{m}}^2\chi_{\mathbf{k}}$ is deduced to be temperature independent. We shall argue that q_u is only weakly temperature dependent. Therefore $\overline{\langle(\delta m)^2\rangle}$ is almost linear in temperature insofar as $q_u \gg \kappa_{\mathrm{m}}$. This is a characteristic of weak itinerant paramagnets. In addition, their basic correlation length,

$\xi_{\mathrm{m},0} = 1/\kappa_{\mathrm{m},0}$, is particularly long: at least one order of magnitude longer than for localised paramagnets for transition metals.

As for compounds with localised electrons, the linewidth of paramagnetic fluctuations from itinerant electrons depends on their nature. For ferromagnetic fluctuations above T_{C},

$$\hbar\Gamma_0(q) = \mathcal{F} q \chi_0^{-1}(q) = \frac{\mathcal{F}}{\chi_0} q \left[1 + \left(\frac{q}{\kappa_{\mathrm{m}}}\right)^2\right], \tag{10.149}$$

where $\mathcal{F}$ is a constant which sets the scale of the linewidth. This linear decay at small q, known as the Landau damping, assumes a ballistic motion of the quasi-particles at the Fermi level. It may break down in the presence of strong disorder. The spin–orbit interaction is assumed to be negligible. Otherwise the total magnetisation would not be conserved and the expression for the linewidth given in Eq. 10.149 would not be justified.

For an antiferromagnet, we have

$$\hbar\Gamma_{\mathbf{k}}(q) = \mathcal{A}(\kappa_{\mathrm{m}}^2 + q^2). \tag{10.150}$$

As for ionic compounds, because the staggered magnetisation is not a conserved quantity, the linewidth associated with antiferromagnetic fluctuations does not vanish at small wave vectors. In fact, the dependence of $\Gamma_{\mathbf{k}}$ on q is the same for fluctuations from localised and itinerant electrons. The scaling constants $\mathcal{F}$ and $\mathcal{A}$ can be computed from the band structure near the Fermi surface.

The upper bound for q_u is of course q_{BZ}. However, q_u is usually smaller and is estimated from the condition (see Murata 1975):

$$\hbar\Gamma_{\mathbf{k}}(q_u) = k_{\mathrm{B}} T. \tag{10.151}$$

This means that, for our description of the spin dynamics, we keep the fluctuations modes which live longer than the time scale provided by the characteristic thermal excitation energy. The cutoff q_u does not naturally appear in the theory but is phenomenologically added to it. Anyway, we recall that we are using a Ginzburg–Landau expansion which is basically a phenomenological approach. Close to the ordering point, κ_{m} becomes negligible relative to q_u, and we get $q_u = (k_{\mathrm{B}} T \kappa_{\mathrm{m}}^2 \chi_0/\mathcal{F})^{1/3} \propto T^{1/3}$ and $q_u = (k_{\mathrm{B}} T/\mathcal{A})^{1/2} \propto T^{1/2}$ for ferromagnetic and antiferromagnetic fluctuations, respectively.

With the material collected above, we can evaluate λ_Z for weak itinerant paramagnets. Because the susceptibility is assumed to be peaked at $\mathbf{k}$ in the whole temperature range, λ_Z can be computed from the approximation given by Eqs. 10.93 and 10.95 for ferromagnets and antiferromagnets, respectively, with an upper bound for the q integration changed from q_{BZ} to q_u. Note that the prefactor in Eq. 10.93 has been computed assuming a high symmetry muon site.

We first deal with the zero-field case (Moriya and Ueda 1974). For ferromagnetic fluctuations

$$\lambda_Z = \frac{\mu_0 \gamma_\mu^2}{\pi^2} \left[\left(p - \frac{1}{3} \right)^2 + \frac{2}{9} \right] \frac{k_B T \hbar \kappa_m^2 \chi_0^2}{2\mathcal{F}} \frac{\left(\frac{q_u}{\kappa_m} \right)^2}{1 + \left(\frac{q_u}{\kappa_m} \right)^2}. \tag{10.152}$$

It follows that $\lambda_Z \propto T/(T - T_C)$ in the limit $\kappa_m \ll q_u$ if the uniform susceptibility follows a Curie–Weiss law. Within this approximation, λ_Z is predicted to become temperature independent in the very high-temperature limit. This level-off behaviour has already been encountered for compounds with localised fluctuations. For antiferromagnetic fluctuations

$$\lambda_Z = \frac{\mu_0 \gamma_\mu^2}{32\pi^4} \sum_\alpha \mathcal{A}^{\alpha\alpha}(\mathbf{k}) \frac{k_B T \hbar \kappa_m \chi_{\mathbf{k}}}{2\mathcal{A}} \left[\arctan\left(\frac{q_u}{\kappa_m} \right) - \frac{\frac{q_u}{\kappa_m}}{1 + \left(\frac{q_u}{\kappa_m} \right)^2} \right]. \tag{10.153}$$

When the staggered susceptibility shows a Curie–Weiss behaviour and $\kappa_m \ll q_u$, we get $\lambda_Z \propto T/(T - T_N)^{1/2}$ and therefore $\lambda_Z \propto T^{1/2}$ in the high-temperature limit.

A characteristic of weak itinerant paramagnets is the relatively low energy of their fluctuations, i.e. their relatively small fluctuation rates. Hence, it may occur that $\Gamma_{\mathbf{k}}(q) \leq \omega_\mu = \gamma_\mu B_{\rm ext}$ for some fluctuation modes. Then λ_Z becomes field dependent. The spectral-weight function for a single fluctuation mode is a Lorentzian function of $B_{\rm ext}$ and λ_Z is a weighted sum of spectral-weight functions. Therefore we do not expect λ_Z versus $B_{\rm ext}$ to display a Lorentzian behaviour if it is built up from contributions of a large number of fluctuation modes. For the computation of the field dependence of λ_Z, we introduce the reduced field variables $b_{\rm FM} = \hbar\gamma_\mu B_{\rm ext}\chi_0/(\mathcal{F}\kappa_m)$ and $b_{\rm AF} = \hbar\gamma_\mu B_{\rm ext}/(\mathcal{A}\kappa_m^2)$ and write

$$\lambda_Z = \frac{\mu_0 \gamma_\mu^2}{\pi^2} \left[\left(p - \frac{1}{3} \right)^2 + \frac{2}{9} \right] \frac{k_B T \hbar \kappa_m^2 \chi_0^2}{\mathcal{F}} I^{\rm FM}_{b_{\rm FM}} \left(\frac{q_u}{\kappa_m} \right), \tag{10.154}$$

for ferromagnetic fluctuations and

$$\lambda_Z = \frac{\mu_0 \gamma_\mu^2}{32\pi^4} \sum_\alpha \mathcal{A}^{\alpha\alpha}(\mathbf{k}) \frac{k_B T \hbar \kappa_m \chi_{\mathbf{k}}}{\mathcal{A}} I^{\rm AF}_{b_{\rm AF}} \left(\frac{q_u}{\kappa_m} \right) \tag{10.155}$$

for antiferromagnatic fluctuations. The analytical expression for $I_b^{\rm FM}(x)$ is not simple, whereas $I_b^{\rm AF}(x)$ can be written as $I_b^{\rm AF}(x) = I_b^+(x) + I_b^-(x)$ where

$$I_b^\pm(x) = \frac{\mp 1}{4\sqrt{2(p-1)}} \ln \left[1 + \left(\sqrt{\frac{2}{p+1}} x \pm \sqrt{\frac{p-1}{p+1}} \right)^2 \right] + \frac{1}{2\sqrt{2(p+1)}} \arctan \left(\sqrt{\frac{2}{p+1}} x \pm \sqrt{\frac{p-1}{p+1}} \right), \tag{10.156}$$

and $p = \sqrt{1 + b^2}$. In Fig. 10.7 we plot the field dependence of λ_Z for the two types of fluctuations in reduced variables. Marked deviations from the Lorentzian behaviour

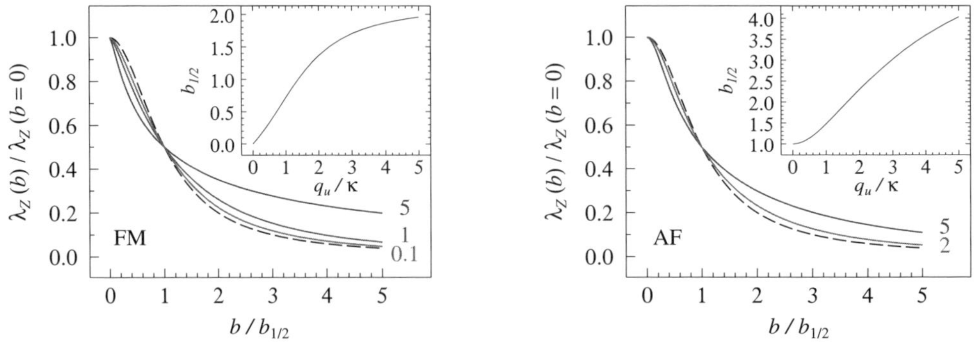

Fig. 10.7 Field dependence of λ_Z for weak itinerant ferromagnets and antiferromagnets in the left and right panels, respectively. For simplicity, we denote the variables $b_{\rm FM}$ and $b_{\rm AF}$ introduced in the text by the generic symbol b. In order to show the deviation from the standard Lorentzian-field dependence, the relaxation rate is normalised to its zero-field value and the field scale is normalised to the half-width at half-maximum $b_{1/2}$. The shown functions correspond therefore to $I_b^{\rm FM}(q_u/\kappa_{\rm m})/I_{b=0}^{\rm FM}(q_u/\kappa_{\rm m})$ and $I_b^{\rm AF}(q_u/\kappa_{\rm m})/I_{b=0}^{\rm AF}(q_u/\kappa_{\rm m})$, respectively. They are plotted as solid lines for the different values of $q_u/\kappa_{\rm m}$ indicated next to each curve. For reference the dashed lines represent Lorentzian functions. The insets show the variation of $b_{1/2}$ versus $q_u/\kappa_{\rm m}$.

are expected for $q_u/\kappa_{\rm m} \geq 2$. As $q_u/\kappa_{\rm m}$ increases, the field dependence of λ_Z gets stronger at low field. A tendency towards saturation appears at high field. Thus, the investigation of the field dependence of λ_Z may provide an estimate of q_u. Implicit is the assumption that the spin dynamics is not influenced by $B_{\rm ext}$.

In contrast to compounds with conventional localised magnetic moments (such as most of the rare-earth compounds), magnetic fluctuations are known to exist in the ordered state of weak itinerant metals, and so may contribute to the μSR spin–lattice relaxation. This relaxation channel will be discussed here. Obviously, excitations can also relax the muon spin. This subject is treated in Section 10.2.4. The perpendicular (to the easy axis taken as the Z axis) response is much larger than the parallel one, resembling the result for the Heisenberg model where the tilting of the localised spins away from the Z axis, which induces a perpendicular spin component, is the driving mechanism for spin dynamics. Writing $\Lambda^{\alpha\beta}(q,\omega_\mu) = \Lambda^{\perp}(q,\omega_\mu)(\delta^{X\alpha}\delta^{X\beta} + \delta^{Y\alpha}\delta^{Y\beta})$ for a cubic compound, and therefore neglecting $\Lambda^{ZZ}(q,\omega_\mu)$, we get from Eq. 10.83

$$\lambda_Z = \frac{\mathcal{D}}{2}\int_{v_c^*} \left[\mathcal{A}^{XX}(\mathbf{q}) + \mathcal{A}^{YY}(\mathbf{q}\right]\, \Lambda^{\perp}(q,\omega_\mu)\frac{{\rm d}^3\mathbf{q}}{(2\pi)^3}. \tag{10.157}$$

Proceeding as for the paramagnetic regime,

$$\lambda_Z = \frac{\mu_0\gamma_\mu^2}{\pi^2}\left[\left(p-\frac{1}{3}\right)^2 + \frac{7}{45}\right] k_{\rm B}T\int_{q_\ell}^{q_u} \chi_0^{\perp}(q)\frac{\Gamma_0^{\perp}(q)}{\left[\Gamma_0^{\perp}(q)\right]^2+\omega_\mu^2}q^2{\rm d}q, \tag{10.158}$$

for a ferromagnet and

$$\lambda_Z = \frac{\mu_0 \gamma_\mu^2}{32\pi^4} \left[\mathcal{A}^{XX}(\mathbf{k}) + \mathcal{A}^{YY}(\mathbf{k})\right] k_{\mathrm{B}} T \int_{q_\ell}^{q_u} \chi_{\mathbf{k}}^{\perp}(q) \frac{\Gamma_{\mathbf{k}}^{\perp}(q)}{\left[\Gamma_{\mathbf{k}}^{\perp}(q)\right]^2 + \omega_\mu^2} q^2 \mathrm{d}q, \qquad (10.159)$$

for an antiferromagnet. We have introduced the lower cutoff q_ℓ. When explicitly evaluating λ_Z, it will be obvious that this parameter is required.

The Hamiltonian of the system we consider is rotationally invariant. Because the rotational symmetry is spontaneously broken in the ordered magnetic state, Nambu–Goldstone modes appear — see e.g. Chaikin and Lubensky (1995). This is reflected by the divergence of $\chi_{\mathbf{k}}^{\perp}(q)$ near $q = 0$:

$$\chi_{\mathbf{k}}^{\perp}(q) = \frac{\chi_{\mathbf{k}}}{\left(\dfrac{q}{\kappa_{\mathrm{m}}}\right)^2} = \frac{\kappa_{\mathrm{m}}^2 \chi_{\mathbf{k}}}{q^2}. \qquad (10.160)$$

Note that $\kappa_{\mathrm{m}}^2 \chi_{\mathbf{k}}$ is temperature independent in the limit $T \ll T_{\mathrm{c}}$ since both κ_{m} and $\chi_{\mathbf{k}}$ become constant in this limit. According to Eq. 10.149, for a ferromagnet

$$\hbar \Gamma_0^{\perp}(q) = \mathcal{F} q \left[\chi_0^{\perp}(q)\right]^{-1} = \frac{\mathcal{F}}{\kappa_{\mathrm{m}}^2 \chi_0} q^3, \qquad (10.161)$$

and therefore, neglecting the muon-Zeeman energy,

$$\lambda_Z \simeq \frac{\mu_0 \gamma_\mu^2}{2\pi^2} \left[\left(p - \frac{1}{3}\right)^2 + \frac{7}{45}\right] \frac{\hbar \left(\kappa_{\mathrm{m}}^2 \chi_0\right)^2}{\mathcal{F}} \frac{k_{\mathrm{B}} T}{q_\ell^2}. \qquad (10.162)$$

We have assumed $q_\ell \ll q_u$. Had we taken $q_\ell = 0$, λ_Z would have been infinite. This is clearly unphysical and justifies a finite value for q_ℓ. How do we choose this lower cutoff? A given energy band near the Fermi energy may split under the action of the molecular field into two bands. A simple choice for q_ℓ is $q_\ell = k_{\mathrm{F}\uparrow} - k_{\mathrm{F}\downarrow}$, where $k_{\mathrm{F}\uparrow}$ ($k_{\mathrm{F}\downarrow}$) is the wave vector of the majority (minority) spin electrons near the Fermi level (Lonzarich and Taillefer 1985). Assuming this wave vector difference to be proportional to the magnetisation, M_0, we deduce that $\lambda_Z \propto T/M_0^2$. This is the thermal behaviour predicted by Moriya and Ueda (1974). Hence, at first sight, we do not expect λ_Z to be proportional to the temperature at relatively high temperature because the thermal dependence of M_0 cannot be neglected. However, an experiment on the long-range incommensurate helical weak magnet MnSi shows that $\lambda_Z \propto T$, suggesting q_ℓ to be a constant in a relatively large temperature range for $T < T_{\mathrm{C}}$ (Yaouanc *et al.* 2005).

Neglecting again the muon-Zeeman energy, assuming $q_\ell \ll q_u$ and dropping the term κ_{m}^2 which is negligible compared to q^2 in the linewidth expression (Eq. 10.150), we derive for antiferromagnetic fluctuations

$$\lambda_Z \simeq \frac{\mu_0 \gamma_\mu^2}{32\pi^4} \left[\mathcal{A}^{XX}(\mathbf{k}) + \mathcal{A}^{YY}(\mathbf{k})\right] \frac{\hbar \kappa_{\mathrm{m}}^2 \chi_{\mathbf{k}}}{\mathcal{A}} \frac{k_{\mathrm{B}} T}{q_\ell}. \qquad (10.163)$$

Therefore $\lambda_Z \propto T/M_{\mathbf{k}}$, where $M_{\mathbf{k}}$ is the staggered magnetisation if we assume $q_\ell \propto M_{\mathbf{k}}$ (Moriya and Ueda 1974).

10.2.3.3 Critical spin fluctuations. When a system is at or close to a critical point, anomalies occur in a wide variety of static and dynamical properties, most commonly discussed within the theory of critical phenomena. This theory concerns continuous phase transitions and therefore excludes first order phase transitions characterised, for example, by magnetic hysteresis and latent heat at the ordering point. We shall restrict our discussion to simple critical points, with no effort to deal with multi-critical points. A well theoretically documented review of critical phenomena is given by Pelissetto and Vicari (2002).

A central concept of the theory for critical phenomena is the universality class. Since the study of critical phenomena concerns the behaviour of a system whose correlation length $\xi_{\rm m}$ is very large compared to interatomic spacings, the many details of the microscopic Hamiltonian are unimportant. It follows that systems near a critical point can be divided into broad groups known as universality classes, such that all members of a given class have identical critical properties. It is experimentally found that static and dynamic physical properties usually follow simple power laws near critical points. A critical exponent is associated to each law. A universality class is characterised by a set of critical exponents — see e.g. Collins (1989).

When studying a magnetic material with μSR techniques, one is particularly interested in the temperature dependence of the spontaneous field, $\mathbf{B}_0$, at the muon site in the vicinity of the ordering point, and in the spin–lattice relaxation rate, λ_Z. Whereas $\mathbf{B}_0$ contains information concerning the static critical properties of the system under study, λ_Z probes its dynamics.

It is convenient to first discuss the universality classes for static physical properties. Originally, the static critical exponents were thought to depend on three, and only three, properties of the system: its spatial dimensionality, the dimensionality of its order parameter (i.e. one for an Ising Hamiltonian), and whether the interactions are short or long range. However, more recently the possible chiral nature of a non-collinear antiferromagnet has been recognised as a relevant characteristic for the definition of its universality class — see the review of Kawamura (1998). Here, for simplicity we shall only consider collinear magnets. As an example of a universality class, we cite the one comprising magnetic systems described by the Heisenberg model in three dimensions. Since the local field tracks the magnetisation, at least for insulators as explained at the beginning of Section 10.1 and in Section 10.3.2, the critical exponents for the magnetisation and the field are the same, that is,

$$B_0(\tau) \propto \tau^{\beta}, \tag{10.164}$$

where the reduced temperature τ has been defined in Eq. 10.100. The critical temperature $T_{\rm c}$ is either the Curie or Néel temperature. There is no need here to discuss the critical behaviour of B_0 any further. We refer to books and reviews, e.g. Collins (1989) and Hohenemser *et al.* (1989). For the purpose of completeness, we list in Table 10.1 the theoretical values of the static exponents.

It is important to note that there are linear relationships between the exponents of the static scaling laws: the six static critical exponents may be expressed in terms of just two independent exponents. Therefore, the knowledge of only two static exponents is theoretically sufficient to determine the static universality class of a compound.

Table 10.1 Theoretical values of static exponents for chemically ordered materials, assuming collinear magnetic structures and short-range interactions. The parameter d is the dimensionality of the system and n the dimensionality of the order parameter, i.e. the number of its components. There is no exponent listed for ($d = 2$, $n = 2$) and ($d = 2$, $n = 3$) since these systems do not undergo phase transitions with long-range order at finite temperature. In the same way no magnetic order is expected for chains, i.e. $d = 1$. The critical exponents β, γ, α, and ν are defined through the relations $\lim_{\tau\to 0} m_{\mathbf{k}} \propto \tau^{\beta}$, $\lim_{\tau\to 0} \chi_0 \propto \tau^{-\gamma}$, $\lim_{\tau\to 0} C_H \propto \tau^{-\alpha}$, and $\lim_{\tau\to 0} \xi_{\rm m} \propto \tau^{-\nu}$. The physical quantities $m_{\mathbf{k}}$ and C_H are the staggered magnetic moment and the constant-field specific heat, respectively. The critical exponents δ and η are defined as $\lim_{H\to 0} M \propto B_{\rm ext}^{1/\delta}$ and $\lim_{q\to 0} \chi^{\alpha\beta}(\mathbf{q}) \propto q^{-2+\eta}$ where the magnetisation M and the susceptibility tensor $\chi^{\alpha\beta}(\mathbf{q})$ are measured at $T_{\rm c}$. The parameter η is often quoted as the Fisher coefficient. The scaling theory predicts the equality of the critical exponents above and below the critical point. Of course this does not apply to β and δ. The data are taken from Hohenemser *et al.* (1989).

Univers. class d	n	β	γ	δ	α	ν	η
2	1	0.125	1.75	15.0	0.0	1.0	0.25
3	1	0.325 (2)	1.241 (2)	4.82 (2)	0.110 (5)	0.630 (2)	0.031 (4)
3	2	0.346 (2)	1.316 (3)	4.80 (2)	−0.007 (6)	0.669 (2)	0.033 (4)
3	3	0.365 (3)	1.386 (4)	4.80 (2)	−0.115 (9)	0.705 (3)	0.033 (4)
Mean field		$\frac{1}{2}$	1	3	0	$\frac{1}{2}$	0

There is one exponent relevant to dynamics, usually denoted as z. Assuming for simplicity that we are dealing with an isotropic magnetic system and that the spectral-weight function is Lorentzian (see Appendix F for a discussion of this function), dynamical scaling requires

$$\hbar\Gamma_{\mathbf{k}}(q) = \Omega q^z \hat{\Gamma}(x). \tag{10.165}$$

This equation defines z, while Ω sets the scale of the linewidth, $\hat{\Gamma}$ is a scaling function and $x = 1/(q\xi_{\rm m})$ where $\xi_{\rm m}$ is the correlation length. The scaling function $\hat{\Gamma}$ is normalised to 1 at criticality, i.e. $\hat{\Gamma}(x = 0) = 1$. From quite general scaling arguments, at a continuous phase transition a time correlation function is proportional to a power-law function of time. The Lorentzian assumption which corresponds to an exponential decay of the time correlation function does not follow the required form. However, it happens that this practical form is consistent with most available neutron diffraction data. On the other hand, works on spin glasses have unveiled the breakdown of this approximation; see Pappas *et al.* (2003) for a summary. μSR experimental data have played an important role; see Section 10.3.4 for an example.

The value of z is a characteristic of a universality class if classes defined for static properties are expanded to specify conservation laws — see e.g. Halperin and Hohenberg (1977). For example, this means that two classes are needed for the description of the physics of the Heisenberg model in three dimensions. The ferromagnets, for which the order parameter is a conserved quantity (it is the uniform magnetisation) form

Table 10.2 Theoretical values of exponent z for some spin models. We assume collinear magnetic structures. The parameter d is the dimensionality of the system and n is the dimensionality of the order parameter, i.e. the number of its components. The Wilson and Fisher parameter ϵ is defined as $\epsilon = 4 - d$. The data are from Hohenemser *et al.* (1989).

Model spin system	Static univers. class	Scaling law for z	Value of z for $d = 3$
Ferromagnet with phonons or dipole interaction	(d, n)	$2 + c\eta$ with $c = 0.72(1 - 1.69\epsilon)$	1.984 (4)
Ising antiferromagnet	$(d, 1)$	$2 + \alpha/\nu$	2.175 (7)
Ising ferromagnet	$(d, 1)$	$2 + \alpha/\nu$	2.175 (7)
Heisenberg antiferromagnet	$(d, 3)$	$d/2$	$\frac{3}{2}$
Heisenberg ferromagnet	$(d, 3)$	$(d + 2 - \eta)/2$	2.484 (4)

one class. In the other class are found all the collinear antiferromagnets for which the order parameter (the staggered magnetisation) is not conserved. Using the language of Halperin and Hohenberg, for the first class there is one conserved field, i.e. the uniform magnetisation. The second class is characterised by a non-conserved field, i.e. the staggered magnetisation. We list the theoretical values of z in Table 10.2 for some spin models.

As the critical temperature is approached, some fluctuation modes slow down, and the response time tends to infinity at the critical point. Because of the slowing down process, λ_Z is usually expected to diverge at the ordering point. The analysis of its behaviour in the vicinity of T_c provides information on the universality class of the system under study. However, as explained in Section 10.2.3, it may occur that the critical fluctuations are screened out, at least in a large part. In the following we shall neglect the screening mechanism. If necessary, it can be accounted for. Relative to scattering experiments, such as quasielastic neutron scattering, which theoretically allows one to determine the spatial and temporal development of the critical dynamics — see, e.g. Collins (1989) — local probe techniques such as μSR are only able to yield limited information. However, the experimental precision can be extremely good. In addition, they give access to regions in the parameter space, for example at extremely low energy, which cannot be probed by scattering techniques. A fine introduction to the study of critical phenomena by local probe techniques is given by Hohenemser *et al.* (1989).

The study of the behaviour of physical quantities in the vicinity of a critical point represents a challenge to experimentalists, since it implies that one must measure sufficiently close to $|\tau| = 0$ to ensure that true limiting behaviour is sampled. How close is close enough varies from system to system, and is best established by empirical tests of asymptotic behaviour. In practice the temperature regime with τ smaller than 0.1 must be probed. To get reliable information one needs to reach $\tau = 10^{-3}$. This is possible at a state-of-the-art muon spectrometer because a high-quality spectrum can be recorded in less than one hour and the sample does not need to be large: a sample of ~50 mg is common, at least at a pseudo-continuous muon source. Under these experimental conditions the thermal gradient is minimised and the temperature fluctuation during the measurement can be less than 0.01 K. One experimental limitation is the determination of T_c. Sometimes this can be achieved using the μSR data themselves. However, it is very helpful to perform an additional measurement while the μSR spectra are being recorded. For example, an a.c. susceptibility measurement enables the determination of the Curie temperature for a ferromagnet. Another limitation concerns the determination of B_0 close to T_c. Since the damping of the oscillation increases as the sample is heated up to T_c, it might get so strong that a complete oscillation cannot be even observed very close to T_c. This difficulty does not exist for λ_Z.

Before examining the paramagnetic critical behaviour expected for λ_Z we remark that up to now in this chapter, our discussion of the fluctuations has been split according to whether they arise from localised or itinerant electrons. This distinction is probably irrelevant in the critical regime, i.e. when ξ_m gets very large. However, a definitive experimental proof of this statement is not yet available.

We first limit ourselves to isotropic spin dynamics and therefore start from the expressions given in Eqs. 10.93 and 10.95 for ferro- and antiferromagnets, respectively, where we set $\omega_\mu = 0$. The scaling law for the static susceptibility reads

$$\chi_{\mathbf{k}}(q) = \Upsilon q^{-2+\eta} \hat{\chi}(x), \tag{10.166}$$

where $\hat{\chi}(x)$ is a scaling function of $x = 1/(q\xi_m)$. We recall that the ferromagnet case corresponds to a propagation vector $\mathbf{k} = \mathbf{0}$. Combining with the dynamical scaling relation (Eq. 10.165) we derive

$$\lambda_Z \propto \int_0^{q_{BZ}} q^{-2+\eta-z} \frac{\hat{\chi}[1/(q\xi_m)]}{\hat{\Gamma}[1/(q\xi_m)]} q^{d-1} dq, \tag{10.167}$$

where d is the dimensionality of the system. This result is valid both for ferro- and antiferromagnets. We have neglected the screening effect; i.e. in the case of antiferromagnets, we have discarded the wave vector dependence of $\mathcal{A}^{\alpha\alpha}(\mathbf{k}+\mathbf{q})$ on the grounds that $\chi_{\mathbf{k}}(q)/\Gamma_{\mathbf{k}}(q)$ is negligible if q is not close to 0. With the change of variable $y = 1/(q\xi_{m_0}\tau^{-\nu})$ in the integral of Eq. 10.167, it is easily shown that

$$\lambda_Z \propto \tau^{\varpi}, \tag{10.168}$$

with

$$\varpi = -\nu(z + 2 - \eta - d). \tag{10.169}$$

Since η can be safely neglected in the expression of ϖ (see Table 10.1) we end up with

$$\varpi = -\nu(z - 1), \tag{10.170}$$

for three-dimensional magnets. We recall that in this case $\nu \simeq 0.7$ (Table 10.1) and z is close to $\frac{5}{2}$ and $\frac{3}{2}$ for Heisenberg ferro- and antiferromagnets (Table 10.2), respectively.

We now focus on the behaviour of λ_Z under an applied field in the paramagnetic critical temperature regime. We use the λ_Z expression given in Eq. 10.88. As for the result derived just above, we shall assume the coupling tensor $\mathcal{A}(\mathbf{q})$ to have a smooth variation around the magnetic structure propagation vector so that its variation can be neglected. For simplicity, we still suppose the spin dynamics to be isotropic. We can write

$$\lambda_Z \propto \int_0^{q_{\rm BZ}} \chi(q) F(q, \omega_\mu) q^{d-1} {\rm d}q, \tag{10.171}$$

where we have implicitly assumed that $\chi(q)$ is field independent even at large field. Scaling as applied to critical dynamics permits the substitution of $F(q, \omega_\mu)$ with $[1/(\Omega q^z)]\hat{F}(x, y_{\rm b})$. The two scaling variables are $x = 1/(q\xi_{\rm m})$ as before and $y_{\rm b} = \Omega q^z/(\hbar\omega_\mu)$, which is a new scaling variable. This choice for $y_{\rm b}$ tells us that what matters is not ω_μ but the ratio of the angular frequency of the spin dynamics to ω_μ. The reader has probably noticed that the scale of the frequency is in fact given by that of the linewidth; see Eq. 10.165. At criticality $x = 0$ and $y_{\rm b} = 0$. Hence, the result we are going to derive is asymptotically valid in the high-field limit. With the above form for the spectral function, we get

$$\lambda_Z \propto \int_0^{q_{\rm BZ}} q^{-2+\eta-z} \hat{F}\left(\frac{1}{q\xi_{\rm m}}, \frac{\Omega q^z}{\hbar\omega_\mu}\right) q^{d-1} {\rm d}q. \tag{10.172}$$

After the change of variable $y_{\rm b} = \Omega q^z/(\hbar\omega_\mu)$ we arrive at

$$\lambda_Z \propto \omega_\mu^{-(z+2-d-\eta)/z}. \tag{10.173}$$

Again, as for Eq. 10.168, we stress that this new result only applies in the critical regime.

Equation 10.168 explains the observed critical dynamics in antiferromagnets; an example is given in Section 10.3.4. However, the theory breaks down for ferromagnets as simple as iron and nickel. The dipole interaction is at the origin of an anisotropy which has not been accounted for in our model which assumed the magnetic energy to be isotropic. We shall therefore describe the critical spin dynamics of dipolar Heisenberg ferromagnets, focusing first on the paramagnetic regime. In these systems a crossover is observed from the Heisenberg to the dipole regimes: it has a dramatic effect on the temperature dependence of λ_Z. In addition, since a quantitative theory based on mode-coupling theory is available — see the review of Frey and Schwabl (1994) — the crossover between the Heisenberg and dipole regimes is quantitatively described. In such a theory each four-spin correlation is factorised into all possible products of two-spin correlations. This type of approximation is very successful in many areas of condensed-matter theory — see e.g. Balucani *et al.* (2003). In the present case, it leads to coupled integral equations for the linewidths.

We follow the work of Yaouanc *et al.* (1993) with a slightly different definition of some variables. For simplicity we shall assume a material of cubic crystal structure. Note that the analysis has been extended to metallic gadolinium which crystallises in a double hexagonal compact structure (Dalmas de Réotier and Yaouanc 1994, Henneberger *et al.* 1999).

The Hamiltonian of the system is the sum of the Heisenberg and dipole contributions. Although the dipole interaction between the magnetic ions is weak relative to the Heisenberg interaction, it gains importance near T_C due to its long-range nature. This is so because only the long wavelengths matter in the critical regime. Mathematically, it means the system Hamiltonian needs to be expanded in small wave vectors and we only keep the terms quadratic in wave vectors, that is,

$$\mathcal{H}_{\rm sys} \simeq v_0 \int_{v_c^*} \sum_{\alpha,\beta} \Big\{ \left[-\mathcal{J}_0 + \mathcal{J}_1 q^2 a^2 \right] P_{\rm T_1}^{\alpha\beta}(\mathbf{q}) \qquad (10.174)$$

$$+ \left[-\mathcal{J}_0 + \mathcal{J}_1 q^2 a^2 + \mathcal{J}_1 q_{\rm D}^2 a^2 \right] P_{\rm L}^{\alpha\beta}(\mathbf{q}) \Big\} J^\alpha(\mathbf{q}) J^\beta(-\mathbf{q}) \frac{{\rm d}^3\mathbf{q}}{(2\pi)^3}.$$

In this expression a is the lattice parameter (the cube edge) and $\mathcal{J}_0$ and $\mathcal{J}_1$ are exchange parameters: $\mathcal{J}_0$ is specifically defined after Eq. 10.102 and $\mathcal{J}_1$ can be deduced from the Fourier transform of the Heisenberg interaction. The operators $P_{\rm T_1}^{\alpha\beta}(\mathbf{q})$ and $P_{\rm L}^{\alpha\beta}(\mathbf{q})$ are the transverse and longitudinal projectors defined in Section C.3. The ratio of the dipole to exchange interactions is characterised by the dimensionless parameter

$$(q_{\rm D} a)^2 = \frac{\mu_0}{v_0} \frac{(g\mu_{\rm B})^2}{2\mathcal{J}_1}, \qquad (10.175)$$

where $q_{\rm D}$ is the dipolar wave vector. The structure of Eq. 10.174 is remarkable. If the strength of the dipole interaction is neglected, i.e. $q_{\rm D} = 0$, the terms in the curly bracket sum up to $(-\mathcal{J}_0 + \mathcal{J}_1 \mathbf{q}^2 a^2)\delta^{\alpha\beta}$ since $P_{\rm T_1}^{\alpha\beta}(\mathbf{q}) + P_{\rm L}^{\alpha\beta}(\mathbf{q}) = \delta^{\alpha\beta}$. We recover the usual long-wavelength limit of the Heisenberg Hamiltonian with its single length scale $\xi_{\rm m}$ such that $\xi_{\rm m}^2 \propto \mathcal{J}_1$. The dipole interaction introduces a second length scale proportional to $1/q_{\rm D}$ and the Hamiltonian is no longer expressed in terms of the modulus of $\mathbf{q}$. Since the Hamiltonian is a weighted sum of $P_{\rm T_1}^{\alpha\beta}(\mathbf{q})$ and $P_{\rm L}^{\alpha\beta}(\mathbf{q})$ for each fluctuation mode, for symmetry reasons we can write the correlation tensor for a given wave vector as a weighted sum of these two operators:

$$\Lambda^{\alpha\beta}(\mathbf{q}, \omega = 0) = \Lambda^{\rm T_1}(\mathbf{q}, \omega = 0) P_{\rm T_1}^{\alpha\beta}(\mathbf{q}) + \Lambda^{\rm L}(\mathbf{q}, \omega = 0) P_{\rm L}^{\alpha\beta}(\mathbf{q}). \qquad (10.176)$$

The $\Lambda^{\rm T_1}(\mathbf{q}, \omega = 0)$ and $\Lambda^{\rm L}(\mathbf{q}, \omega = 0)$ functions are the transverse and the longitudinal correlation functions, respectively. From Eq. F.18, it follows that

$$\Lambda^{\alpha\beta}(\mathbf{q}, \omega = 0) = \frac{2v_c}{\mu_0 g^2 \mu_{\rm B}^2} k_{\rm B} T \left[\frac{\chi^{\rm T_1}(q)}{\Gamma^{\rm T_1}(q)} P_{\rm T_1}^{\alpha\beta}(\mathbf{q}) + \frac{\chi^{\rm L}(q)}{\Gamma^{\rm L}(q)} P_{\rm L}^{\alpha\beta}(\mathbf{q}) \right], \qquad (10.177)$$

where $\chi^{\rm T_1}(q)$ and $\chi^{\rm L}(q)$ are the static transverse and longitudinal wave vector-dependent susceptibilities, respectively, and $\Gamma^{\rm T_1}(q)$ and $\Gamma^{\rm L}(q)$ the linewidths of the related fluctuations. Assuming again, for simplicity, a muon in a tetrahedral or octahedral site of a face-centred-cubic crystal (the coupling tensor $\boldsymbol{G}(\mathbf{q} \to \mathbf{0})$ is particularly simple in these cases; see beginning of Section 10.2.3), we derive from Eqs. 10.83 and 10.54 and the previous result

$$\lambda_Z = \frac{\mu_0}{3\pi^2} \gamma_\mu^2 k_{\rm B} T \int_0^{q_{\rm BZ}} \left[2p^2 \frac{\chi^{\rm T_1}(q)}{\Gamma^{\rm T_1}(q)} + (1-p)^2 \frac{\chi^{\rm L}(q)}{\Gamma^{\rm L}(q)} \right] q^2 {\rm d}q, \qquad (10.178)$$

where p is defined in Eq. 10.91. To arrive at this result, we made use of the projector properties given in Eq. C.4. When the hyperfine interaction is large compared with the dipole interaction, i.e. $|p| \gg 1$, the previous result reduces to

$$\lambda_Z = \frac{\mu_0}{48\pi^4}\gamma_\mu^2 (r_\mu H)^2 k_{\rm B}T \int_0^{q_{\rm BZ}} \left[2\frac{\chi^{\rm T_1}(q)}{\Gamma^{\rm T_1}(q)} + \frac{\chi^{\rm L}(q)}{\Gamma^{\rm L}(q)}\right] q^2 {\rm d}q. \qquad (10.179)$$

This corresponds to the experimental situation for NMR, Mössbauer spectroscopy, and perturbed angular correlation techniques; that is, these techniques probe twice as much transverse fluctuations than longitudinal fluctuations (Frey and Schwabl 1994). On the other hand, in the case of μSR, a key insight into the experimental results is to remember that the dipole interaction between the muon spin and the spins of the compound is strong relative to the hyperfine interaction. In fact, in all the materials where it has been determined, the parameter p is found to be small. This means that zero-field μSR mainly probes the longitudinal fluctuations. As a matter of fact, the transverse fluctuations do not contribute to λ_Z if $r_\mu H/(4\pi) = -\frac{1}{3}$ (Yushankhai 1989) since, according to Eq. 10.178, from the expression for p given in Eq. 10.91 we get $p = 0$. This singles out μSR among the local probe techniques. The parameter p can be computed for simple ferromagnets using the data of Table 10.3 (p. 330).

To proceed further, we recall that the static susceptibilities and linewidths in the critical temperature regime must obey scaling laws. As already mentioned, two scaling lengths are required for a cubic dipolar Heisenberg ferromagnet. The two scaling variables are

$$x = \frac{1}{q\xi_{\rm m}} \quad \text{and} \quad y = \frac{q_{\rm D}}{q}. \qquad (10.180)$$

As a consequence one gets an extension of the usual static scaling law, that is,

$$\chi^{\rm L,T_1}(q) = \Upsilon q^{-2}\hat{\chi}^{\rm L,T_1}(x,y), \qquad (10.181)$$

where we have neglected the critical exponent η and where $\hat{\chi}^{\rm L,T_1}(x,y)$ are susceptibility scaling functions normalised to unity at $x = y = 0$. The nonuniversal constant Υ is not given by scaling theory. From mode-coupling theory one derives $\Upsilon = q_{\rm D}^2$.

We write down the two scaling relations for the linewidths using two dynamical scaling laws:

$$\hbar\Gamma^{\rm L,T_1}(q) = \Omega q^z \hat{\Gamma}^{\rm L,T_1}(x,y), \qquad (10.182)$$

with exponent $z = \frac{5}{2}$ (we approximate $z = 2.484$ (4) to $\frac{5}{2}$; see Table 10.2). The nonuniversal frequency scale can be computed again from mode-coupling theory. Assuming a Lorentzian spectral-weight function,

$$\Omega = \frac{P}{\pi^2\sqrt{2}}\sqrt{\frac{\mu_0}{4\pi}\frac{g\mu_{\rm B}}{q_{\rm D}}}\sqrt{k_{\rm B}T_{\rm C}}, \qquad (10.183)$$

where $P \simeq 5.1326$. The crossover of the critical dynamical exponent is contained in the scaling linewidth functions $\hat{\Gamma}^{\rm L,T_1}(x,y)$ also normalised to unity at $x = y = 0$. When q is sufficiently small $\Gamma^{\rm L,T_1}(q) \propto q^{z_{\rm eff}}$, where $z_{\rm eff}$ is an effective critical exponent. If the spin dynamics is driven by the Heisenberg interaction, i.e. when $q \gg q_{\rm D}$, $z_{\rm eff} = \frac{5}{2}$

for the longitudinal as well as the transverse linewidths. In contrast, in the dipolar regime, i.e. $q \ll q_{\rm D}$, $z_{\rm eff}$ is equal to either 0 or 2 depending on whether one refers to the longitudinal or transverse linewidth.

For a quantitative prediction of the thermal behaviour of λ_Z we need to completely specify the susceptibility and linewidth functions and not only their asymptotes given by the scaling theory. We use an Ornstein–Zernike form for the susceptibilities:

$$\chi^{\rm T_1}(q) = \frac{q_{\rm D}^2}{q^2 + \xi_{\rm m}^{-2}} \quad \text{and} \quad \chi^{\rm L}(q) = \frac{q_{\rm D}^2}{q^2 + q_{\rm D}^2 + \xi_{\rm m}^{-2}}. \tag{10.184}$$

The correlation length $\xi_{\rm m}$ $(= \kappa_{\rm m}^{-1})$ follows a power law,

$$\xi_{\rm m} = \xi_{\rm m,0} \left(\frac{T - T_{\rm C}}{T_{\rm C}} \right)^{-\nu}. \tag{10.185}$$

We again neglect Fisher's critical exponent η for the susceptibility. It is indeed very small in the case of a three-dimensional Heisenberg magnet; see Table 10.2. The static critical exponents for the dipolar and Heisenberg regions are similar. At the Brillouin zone centre Eq. 10.184 shows that while the transverse susceptibility diverges at $T_{\rm C}$, the longitudinal susceptibility flattens out. This is the key result which is at the origin of the thermal behaviour of λ_Z near $T_{\rm C}$.

The influence of the dipole interaction on the linewidths has been studied quantitatively on the basis of the mode-coupling theory. A short review is given by Yaouanc *et al.* (1993). Numerical values for the scaling functions $\hat{\Gamma}^{\rm L,T_1}(x,y)$ can be found in the review written by Frey and Schwabl (1994).

Starting from Eq. 10.178 and collecting all the results listed after this equation, we finally derive

$$\lambda_Z = \mathcal{W} \left[2p^2 I^{\rm T_1}(\varphi) + (1-p)^2 I^{\rm L}(\varphi) \right], \tag{10.186}$$

where the angle φ is a measure for the temperature through $\xi_{\rm m}$:

$$\varphi = \arctan\left(\frac{y}{x}\right) = \arctan\left(q_{\rm D}\xi_{\rm m}\right); \tag{10.187}$$

see Eq. 10.180. The nonuniversal constant $\mathcal{W}$ sets the scale of the relaxation:

$$\mathcal{W} = \frac{4\pi\sqrt{2}}{3P} \sqrt{\frac{\mu_0}{4\pi}} \frac{\gamma_\mu^2 \hbar q_{\rm D}^{3/2}}{g\mu_{\rm B}} \sqrt{k_{\rm B}T_{\rm C}}. \tag{10.188}$$

The two functions $I^{\rm L,T_1}(\varphi)$ defined as

$$I^{\rm L,T_1}(\varphi) = \left[1 + \tan^{-2}(\varphi)\right]^{-3/4} \int_{r_{\rm BZ}}^{\infty} \frac{\hat{\chi}^{\rm L,T_1}(r,\varphi)}{\hat{\Gamma}^{\rm L,T_1}(r,\varphi)} r^{1/2} {\rm d}r, \tag{10.189}$$

where $r = (x^2 + y^2)^{1/2}$ and $r_{\rm BZ} = (q_{\rm D}^2 + \xi_{\rm m}^{-2})^{1/2}/q_{\rm BZ}$, are displayed in Fig. 10.8. While $I^{\rm T_1}(\varphi)$ exhibits a strong temperature dependence in the whole temperature range of the critical region, $I^{\rm L}(\varphi)$ is practically temperature independent for $q_{\rm D}\xi_{\rm m} > 1$. A numerical investigation of the region in reciprocal space contributing to the integrals

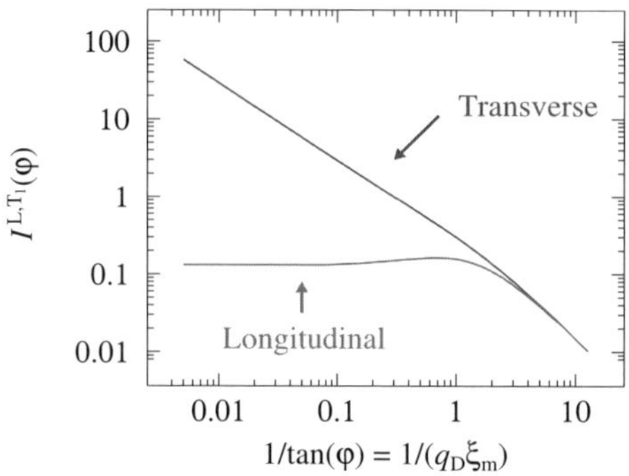

Fig. 10.8 The two functions $I^{L,T_1}(\varphi)$ versus $1/\tan(\varphi) = 1/(q_D\xi_m)$. The functions $I^L(\varphi)$ and $I^{T_1}(\varphi)$ account, respectively, for the contributions of the longitudinal and transverse fluctuations (relative to the wave vector $\mathbf{q}$) to the muon spin–lattice relaxation. The horizontal scale is actually a reduced temperature since $1/(q_D\xi_m) = (T - T_C)^\nu/(q_D\xi_{m,0}T_C^\nu)$. The slope of the two functions for $q_D\xi_m \ll 1$ is $\frac{3}{2}$. The slope of $I^{T_1}(\varphi)$ in the low-temperature regime ($q_D\xi_m \gg 1$) is 1. Reproduced with kind permission from Dalmas de Réotier *et al.* (2004); the original data are from Frey and Schwabl (1994).

defining $I^{L,T1}(\varphi)$ shows that λ_Z is mostly sensitive to fluctuation modes near the zone centre (Dalmas de Réotier *et al.* 1994). For example, in the case of metallic nickel at a temperature such that $T - T_C = 0.056\,\mathrm{K}$, 90% of λ_Z arises from fluctuation modes with wave vectors smaller than $0.02\,\text{Å}^{-1}$.

If the dynamics is driven by the Heisenberg interaction alone, i.e. when $q \gg q_D$, the distinction between the longitudinal and transverse fluctuations is meaningless. This is the temperature region where $I^L(\varphi)$ and $I^{T_1}(\varphi)$ merge into a single curve in Fig. 10.8. It is rewarding to note that we then obtain for λ_Z a power law of τ with the same critical exponent as that inferred from Eq. 10.170.

Equation 10.186 has been derived under the assumption that the tensor $\boldsymbol{C}(\mathbf{q} = \mathbf{0})$ is scalar. As a matter of fact, this hypothesis can be removed without difficulty, resulting in a relaxation rate of a similar mathematical structure, i.e.

$$\lambda_Z = \mathcal{W}\left[a_{T_1} I^{T_1}(\varphi) + a_L I^L(\varphi)\right]. \tag{10.190}$$

The weighting factors a_{T_1} and a_L only depend on the characteristics of the dipole and hyperfine fields at the muon site. Numerically a_{T_1} is found to be much smaller than a_L, resulting in a spin dynamics in the vicinity of T_C mainly driven by the longitudinal fluctuation modes (Dalmas de Réotier and Yaouanc 1997).

Up to now we have been concerned with the paramagnetic critical fluctuations. However, according to the dynamical scaling hypothesis, the critical spin dynamics in the paramagnetic and ferromagnetic regimes should be similar (Halperin and Hohenberg 1967). For a given fluctuation mode characterised by its wave vector $\mathbf{q}$, the key parameter is the product $q\xi_m$; i.e. the paramagnetic or ferromagnetic nature of the fluctuations is not relevant. Because the scale $\xi_{m,0}$ of the correlation length is smaller in the ferromagnetic state than in the paramagnetic state, the reduced temperature at which critical dynamics is observed is different for the two phases.

For the sake of completeness, we sketch now some properties expected for weak itinerant ferromagnets at criticality and for which μSR can give information. In the following we rely on the work on dipolar Heisenberg ferromagnets exposed just above. The reader is referred to Yaouanc *et al.* (2002) for more information. According to Eq. 10.147, $\sqrt{\overline{\langle(\delta m)^2\rangle}} \propto q_{\rm D} q_u^{1/2}$ where $\overline{\langle(\delta m)^2\rangle}$ is the variance of the fluctuating magnetic moments. Use has been made of the Ornstein–Zernike type of susceptibility; see Eq. 10.184. Since $\overline{\langle(\delta m)^2\rangle}$ is small in weak ferromagnets we infer that $q_{\rm D}$ is also small because q_u is not expected to be particularly small. This implies that the longitudinal and transverse susceptibilities are nearly equal. Since close enough to $T_{\rm C}$ the dynamics is controlled by modes with wave vectors of order $q_{\rm D}$, it follows from Eqs. 10.182 and 10.183, that $\nu_{\rm c} = \Gamma^{\rm L,T_1}(q_{\rm D}) \propto q_{\rm D}^{5/2} q_u^{1/2}$ and from Eq. 10.188 that $\mathcal{W} \propto q_{\rm D}^{1/2} q_u^{-1/2}$. If $\nu_{\rm c}$ and $\mathcal{W}$ are experimentally determined one can deduce $q_{\rm D}$ and q_u. Once $q_{\rm D}$ is known, the exchange energy can be estimated using Eq. 10.175. The fact that $q_{\rm D}$ is small implies that $\nu_{\rm c}$ is also small; i.e. the critical spin dynamics for these magnets is slow relative to usual ferromagnets.

10.2.4 Relaxation by excitations

The most common excitations in ordered magnetic materials are spin waves, the quanta of which are magnons. They are mostly observed by neutron scattering. The spin-wave modes are derived from the precession of magnetic moments in the molecular magnetic field. The transverse component of the moments evolves in a coherent fashion during such a precession. Additional magnetic excitations associated with transitions among local energy levels may exist in the paramagnetic and ordered state. These are the crystal-field excitations. They are quite common in f-electron magnetic materials. In this section, we shall first study the effect of the magnons on the relaxation functions. A similar study for the crystal-field excitations will follow.

Experimentally, the number of μSR studies interpreted in terms of magnons or crystal-field excitations is still limited. Therefore, we shall keep our discussion at a relatively elementary level.

As a good general reference on magnons and crystal-field excitations, we mention the textbook of Jensen and Mackintosh (1991). The theoretical discussion by Turov and Petrov (1972) of the NMR relaxation is quite useful.

10.2.4.1 Magnon-induced relaxation. An elementary introduction to spin-wave excitations is given by e.g. Kittel (1996) or Ashcroft and Mermin (2002). There are a large number of books and articles devoted to their description — e.g. Mattis (1965), Keffer (1966), Lovesey (1986), Izyumov and Chernoplekov (1994).

For simplicity, we shall always assume a crystal structure with a centre of inversion, see for example Kittel (1963), and restrict ourselves to collinear magnetic structures. We shall first consider the case of ferromagnets. This exposition will be followed by a discussion for antiferromagnets.

10.2.4.1.1 Ferromagnets. We start by analysing the relaxation induced by magnons in the simplest possible case: a ferromagnetic insulator with isotropic exchange interactions. Following Eq. 10.102, the Hamiltonian for the system is simply

$$\mathcal{H}_{\text{sys}} = -\mathcal{J} \sum_{\langle i,i' \rangle} \mathbf{J}_i \cdot \mathbf{J}_{i'} + g\mu_{\text{B}} B_{\text{ani}} \sum_i J_i^Z, \tag{10.191}$$

since the magnetic moment associated to $\mathbf{J}_i$ is $\mathbf{m}_i = -g\mu_{\text{B}}\mathbf{J}_i$. Here the first sum runs over pairs of nearest neighbours magnetic moments, $\mathcal{J}$ is positive, and the magnetic anisotropy of the compound is accounted for by the anisotropy field $\mathbf{B}_{\text{ani}}$ assumed to be parallel to the Z axis. We shall find that this term is physically important even though its intensity is usually much smaller than the exchange term. Other expressions for the anisotropy energy are possible. We have chosen the simplest form. The ground state of the system described by Eq. 10.191 is obtained when all the spins (in fact total angular momenta) are parallel. To determine its excitation spectrum, it is convenient to introduce the creation and annihilation boson operators, $a_i^\dagger$ and a_i, as first proposed by Holstein and Primakoff (1940). We are here only interested in the excitation states at low energy; i.e. we consider a system in a state with few excitations from the ground state in which the expectation value $\langle J_i^Z \rangle$ is $-J$ for all the spins. Hence, we shall consider the linear approximation to the Holstein–Primakoff transformation. The reader is referred to Mazenko (2003) for a simple introduction. Note, however, that we do not use the most frequent form of this transformation in order to account for the positive sign in front of the anisotropy term in the Hamiltonian (Eq. 10.191). The transformation reads

$$J_i^+ = \sqrt{2J}\, a_i^\dagger, \qquad J_i^- = \sqrt{2J}\, a_i, \qquad \text{and} \qquad J_i^Z = a_i^\dagger a_i - J \tag{10.192}$$

with the commutation relation

$$\left[a_i, a_j^\dagger\right] = \delta_{i,j}, \tag{10.193}$$

as expected for bosons. With these relations the spin operators J_i^X, J_i^Y, and J_i^Z satisfy the usual commutation rules (Eq. C.14). For future reference, the deviation of J_i^Z from its ground state expectation value is

$$\delta J_i^Z = J_i^Z + J = a_i^\dagger a_i. \tag{10.194}$$

Because of the translational invariance of the system under study, we introduce the lattice Fourier transform of the boson operators:

$$a_i = \frac{1}{\sqrt{n_c}} \sum_{\mathbf{q}} a_{\mathbf{q}} \exp(i\mathbf{q} \cdot \mathbf{i}) \quad \text{and} \quad a_i^\dagger = \frac{1}{\sqrt{n_c}} \sum_{\mathbf{q}} a_{\mathbf{q}}^\dagger \exp(-i\mathbf{q} \cdot \mathbf{i}), \tag{10.195}$$

where n_c is the number of unit cells which, in the present case, is equal to the number of magnetic atoms N in the sample. The reason for chosing the notation n_c will be clear when considering the antiferromagnets. The $\mathbf{q}$ sums run over the magnetic Brillouin zone which is here the crystallographic Brillouin zone. Operator $a_{\mathbf{q}}^\dagger$ creates a magnon of wave vector $\mathbf{q}$ and $a_{\mathbf{q}}$ annihilates the same magnon. As usual, we shall only keep

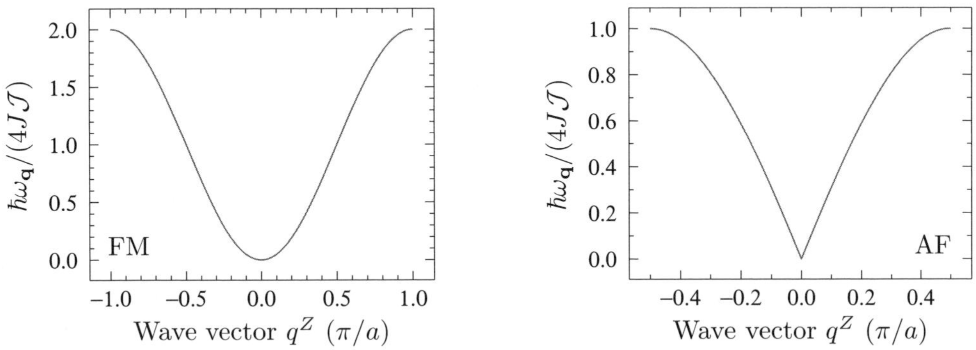

Fig. 10.9 Dispersion relations in the whole magnetic Brillouin zone for ferromagnetic (FM) and antiferromagnetic (AF) chains, respectively. The chains are oriented along the Z axis, the distance between atoms is a, and we only consider nearest neighbour interactions. The energy $\hbar\omega_{\mathbf{q}}$ is minimum at the zone centre and reaches its maximum value at the zone boundaries for both ferromagnetic and antiferromagnetic chains. For simplicity, the dispersion relations are drawn neglecting the energy gap.

the bilinear term of $\mathcal{H}_{\text{sys}}$. Then, written in terms of these latter operators, $\mathcal{H}_{\text{sys}}$ reads, apart from constant terms,

$$\mathcal{H}_{\text{sys}} = \sum_{\mathbf{q}} \hbar\omega_{\mathbf{q}} a^{\dagger}_{\mathbf{q}} a_{\mathbf{q}}. \tag{10.196}$$

The double sum of Eq. 10.191 is therefore reduced to a single one and $\mathcal{H}_{\text{sys}}$ is diagonal. The dispersion relation for the magnons is easily derived to be

$$\hbar\omega_{\mathbf{q}} = g\mu_{\text{B}}B_{\text{ani}} + 2J\left[\mathcal{J}_0 - \mathcal{J}(\mathbf{q})\right]. \tag{10.197}$$

The Fourier transform of the exchange interaction $\mathcal{J}(\mathbf{q})$ is defined in Eq. 10.101 and we recall the notation $\mathcal{J}_0 = \mathcal{J}(\mathbf{q} = \mathbf{0})$. For small $\mathbf{q}$ and a cubic lattice of lattice parameter a,

$$\hbar\omega_{\mathbf{q}} = g\mu_{\text{B}}B_{\text{ani}} + D_{\text{FM}}q^2, \tag{10.198}$$

with

$$D_{\text{FM}} = 2\mathcal{J}Ja^2. \tag{10.199}$$

For a ferromagnetic linear chain along the Z axis, the dispersion relation is given by

$$\hbar\omega_{\mathbf{q}} = g\mu_{\text{B}}B_{\text{ani}} + 4J\mathcal{J}\left[1 - \cos\left(q^Z a\right)\right]. \tag{10.200}$$

This dispersion relation is displayed in Fig. 10.9. The magnon excitation energy is minimum at the zone centre with a value of $g\mu_{\text{B}}B_{\text{ani}}$ and maximum at the boundaries of the Brillouin zone where it amounts to $g\mu_{\text{B}}B_{\text{ani}} + 8J\mathcal{J}$. The energy gap of the magnon dispersion is $g\mu_{\text{B}}B_{\text{ani}}$. It is interesting to compare the energy of an ion in the field B_{ani} with the energy of the muon in a typical spontaneous field B_0. A typical value for B_0 is 0.1 T (see Table 10.3 in p. 330), whereas the value of B_{ani} strongly depends on the material under study according to its anisotropy. A lower bound for

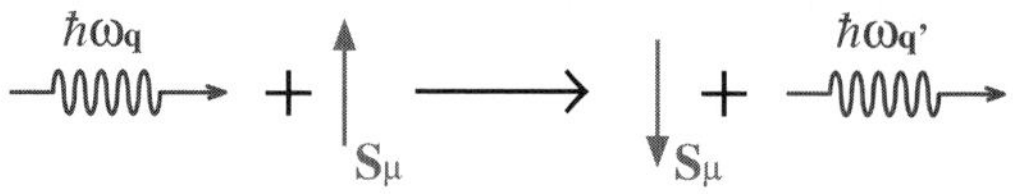

Fig. 10.10 Schematic view of a Raman scattering process in the case of μSR. Two magnons are involved: while one of energy $\hbar\omega_{\mathbf{q}}$ is absorbed and the second of energy $\hbar\omega_{\mathbf{q}'}$ is emitted, the muon spin flips.

it is, however, determined from the dipolar anisotropy and is also on the order of 0.1 T. The corresponding energies are then, respectively, $\hbar\gamma_\mu B_0 = 56\,\text{neV}$ and $\mu_\text{B} B_\text{ani} = 5.8\,\mu\text{eV}$. This means that the muon energy splitting in the magnetically ordered state is much smaller than any magnon energy gap.

Having summarised the properties of the magnons for our simple model of a ferromagnet, we are now in a position to investigate how they influence the muon spin. According to Eq. 7.37, the coupling of the muon spin to the system is described by the Zeeman Hamiltonian $\mathcal{H}_1$:

$$\mathcal{H}_1 = -\frac{1}{2}\gamma_\mu\hbar\left[\sigma^Z\delta B^Z_\text{loc} + \frac{1}{2}\left(\sigma^+\delta B^-_\text{loc} + \sigma^-\delta B^+_\text{loc}\right)\right]. \tag{10.201}$$

We stick here to the notation B^α_loc rather than B^α_0 since this latter notation refers to an averaged spontaneous field. Relaxation of the muon spin means that it flips under the action of $\delta\mathbf{B}_\text{loc}$, such as to tend to equalise the population of its up-and-down energy levels; see Fig. 7.2. Only the last two terms of $\mathcal{H}_1$ can induce such a flip. The components δB^+_loc and δB^-_loc are linear combinations of δJ^-_i, δJ^+_i, and δJ^Z_i:

$$\begin{aligned}\delta B^+_\text{loc} &= \frac{\mu_0}{4\pi}\sum_{i=1}^{N}\frac{(-g_i\mu_\text{B})}{r_i^3}\left[-\frac{1}{2}\delta J^+_i\left(3\cos^2\theta_i - 1\right) + \frac{3}{2}\delta J^-_i\sin^2\theta_i\exp\left(2i\varphi_i\right)\right.\\ &\qquad\left. + 3\delta J^Z_i\sin\theta_i\cos\theta_i\exp\left(i\varphi_i\right)\right] + \frac{\mu_0}{4\pi}\sum_{i=1}^{N'}\frac{(-g_i\mu_B)H_{\mathbf{r}_i}}{v_0}\delta J^+_i,\\ \delta B^-_\text{loc} &= \left(\delta B^+_\text{loc}\right)^\dagger.\end{aligned} \tag{10.202}$$

The first sum arises from the dipole interaction (see Eq. 5.30), and the second from the hyperfine interaction (see Eq. 5.18). Because of the requirement of energy conservation during a flipping process, the components of $\mathbf{J}_i$ perpendicular to Z cannot contribute to the muon spin relaxation. We recall indeed that the minimum magnon energy is much larger than the muon energy splitting in the local field. In other words, only spin fluctuations parallel to the quantisation axis can induce a muon-spin relaxation. We therefore only retain the term proportional to δJ^Z_i in δB^+_loc (and in δB^-_loc).

From Eq. 10.194 we infer that δJ^Z_i involves two excitations. A relaxation process occurring through two simultaneous excitations, one absorption and one emission, is usually called a Raman process: see Fig. 10.10 for a schematic view of this process. It probes the correlation function $\Lambda^{ZZ}(\mathbf{q}, 0)$. It is remarkable that an isotropic hyperfine interaction for which $\delta B^+_{\rm loc}$ and $\delta B^-_{\rm loc}$ only depend on δJ^+_i and δJ^-_i cannot relax the muon spin. This was noticed a long time ago for NMR (Mitchell 1957). Since the contribution of the dipole field to $\mathbf{B}_{\rm loc}$ is always important for μSR, we expect the Raman process to be particularly relevant for μSR as pointed out by Yaouanc and Dalmas de Réotier (1991). Because the magnon energy is minimum at the zone centre as can be seen from Fig. 10.9, the long-wavelength magnons, i.e. the magnons with small wave vectors, drive the relaxation of the muon spin at low temperature. Hence, it is sufficient to consider the small wave vector limit of the coupling tensor between the muon spin and the magnons; see Eq. 10.90. This discussion and the formulae given for λ_Z in Eqs. 10.83 and 10.54 lead to the following expression for the spin–lattice relaxation rate sometimes referred to as $1/T_1$:

$$\lambda_Z = 4\mathcal{D}\left\{\frac{2}{15} + \left[C^{XZ}(\mathbf{q}=\mathbf{0})\right]^2 + \left[C^{YZ}(\mathbf{q}=\mathbf{0})\right]^2\right\}\int_0^{q_{\rm BZ}} \Lambda^{ZZ}(q,\omega=0)q^2 {\rm d}q. \tag{10.203}$$

We have assumed $\Lambda^{ZZ}(\mathbf{q},\omega=0)$ to only depend on q. As usual, the Brillouin zone is taken to be a sphere. Its radius is equal to $q_{\rm BZ} = \left(6\pi^2\right)^{1/2}/a$ for a simple cubic lattice of parameter a. A formula for the spin–spin relaxation induced by a Raman process can also be written down. The corresponding relaxation rate λ_X is sometimes denoted by $1/T_2$. We leave the derivation of its expression to the interested reader. For the evaluation of the correlation function it is convenient to return to the discrete representation of the Fourier transform. From Eq. 7.111

$$\Lambda^{ZZ}_{\mathbf{q}}(\omega=0) = \frac{1}{2}\left[\langle \delta J^Z_{\mathbf{q}}(\omega=0)\delta J^Z_{-\mathbf{q}}\rangle + \langle \delta J^Z_{-\mathbf{q}}\delta J^Z_{\mathbf{q}}(\omega=0)\rangle\right]. \tag{10.204}$$

An expression for $\delta J^Z_{\mathbf{q}}$ is required. From Eq. E.9 and using the linear Holstein–Primakoff approximation (Eq. 10.194),

$$\delta J^Z_{\mathbf{q}} = \frac{1}{\sqrt{n_c}}\sum_{\mathbf{q}_1,\mathbf{q}_2} \delta_{\mathbf{q}+\mathbf{q}_1-\mathbf{q}_2,\mathbf{0}}\, a^\dagger_{\mathbf{q}_1} a_{\mathbf{q}_2}. \tag{10.205}$$

Taking into account that

$$a_{\mathbf{q}}(t) = \exp(-i\omega_{\mathbf{q}}t)a_{\mathbf{q}} \quad \text{and} \quad a^\dagger_{\mathbf{q}}(t) = \exp(i\omega_{\mathbf{q}}t)a^\dagger_{\mathbf{q}}, \tag{10.206}$$

we obtain

$$\delta J^Z_{\mathbf{q}}(\omega=0) = \frac{2\pi}{\sqrt{n_c}}\sum_{\mathbf{q}_1,\mathbf{q}_2} \delta_{\mathbf{q}+\mathbf{q}_1-\mathbf{q}_2,\mathbf{0}}\, \delta\left(\omega_{\mathbf{q}_1}-\omega_{\mathbf{q}_2}\right) a^\dagger_{\mathbf{q}_1} a_{\mathbf{q}_2}. \tag{10.207}$$

According to Eq. 10.204, $\Lambda^{ZZ}_{\mathbf{q}}(\omega=0)$ is the sum of two different correlations. Using the last two results, each of them is found to be expressed in terms of the thermal average of a product of four magnon operators, explicitly $\langle a^\dagger_{\mathbf{q}_1} a_{\mathbf{q}_2} a^\dagger_{\mathbf{q}'_1} a_{\mathbf{q}'_2}\rangle$.

To proceed further, we introduce the mode-coupling approximation which consists of breaking down $\langle a^\dagger_{\mathbf{q}_1} a_{\mathbf{q}_2} a^\dagger_{\mathbf{q}'_1} a_{\mathbf{q}'_2}\rangle$ into a sum of products of two two-point correlation functions; see for example Lovesey (1986). We get

$$\left\langle a^\dagger_{\mathbf{q}_1} a_{\mathbf{q}_2} a^\dagger_{\mathbf{q}'_1} a_{\mathbf{q}'_2}\right\rangle \simeq \delta_{\mathbf{q}_1,\mathbf{q}'_2}\delta_{\mathbf{q}_2,\mathbf{q}'_1}\left\langle a^\dagger_{\mathbf{q}_1} a_{\mathbf{q}_1}\right\rangle\left\langle a_{\mathbf{q}_2} a^\dagger_{\mathbf{q}_2}\right\rangle + \delta_{\mathbf{q}_1,\mathbf{q}_2}\delta_{\mathbf{q}'_1,\mathbf{q}'_2}\left\langle a^\dagger_{\mathbf{q}_1} a_{\mathbf{q}_1}\right\rangle\left\langle a^\dagger_{\mathbf{q}'_1} a_{\mathbf{q}'_1}\right\rangle . \tag{10.208}$$

It is easily seen that the second term on the right-hand side does not contribute to $\int \Lambda^{ZZ}_{\mathbf{q}}(q,0)q^2 \mathrm{d}q$ because it gives a term proportional to the Kronecker symbol $\delta_{\mathbf{q},\mathbf{0}}$. We recall that

$$\langle a^\dagger_{\mathbf{q}} a_{\mathbf{q}}\rangle = n_{\mathrm{P}}\left(\frac{\hbar\omega_{\mathbf{q}}}{k_{\mathrm{B}}T}\right), \tag{10.209}$$

where $n_{\mathrm{P}}\left(\hbar\omega_{\mathbf{q}}/k_{\mathrm{B}}T\right)$ is the Planck population factor; see Eq C.31. With this material at hand, the two correlation functions on the right-hand side of Eq. 10.204 can be evaluated. We finally derive

$$\int_0^{q_{\mathrm{BZ}}} \Lambda^{ZZ}(q,\omega=0)q^2\mathrm{d}q = \tag{10.210}$$

$$\frac{1}{2}\frac{v_c}{(2\pi)^3}\int_{v_c^*} n_{\mathrm{P}}\left(\frac{\hbar\omega(q)}{k_{\mathrm{B}}T}\right)\left[n_{\mathrm{P}}\left(\frac{\hbar\omega(q)}{k_{\mathrm{B}}T}\right)+1\right]\left[\int_{v_c^*}\delta\left[\omega(q)-\omega(q_1)\right]\mathrm{d}^3\mathbf{q}_1\right]\mathrm{d}^3\mathbf{q}.$$

We explicitly display the temperature dependence of the Planck function to indicate the origin of the temperature dependence of the spin–lattice relaxation. For simplicity, we recall that we assume all quantities to only depend on the modulus of wave vectors. To compute λ_Z, one can specify the dispersion relation and directly evaluate the two previous integrals over wave vectors. This is done in the literature; see for example Dalmas de Réotier and Yaouanc (1995). As already noted by Moriya (1956), a more fruitful method is first to recognise that $\int \Lambda^{ZZ}(q,\omega=0)q^2\mathrm{d}q$ can be written down in terms of the density of magnetic states $g_m(E)$ defined by the relation

$$g_m(E) = \int_{v_c^*}\delta\left[E-E(q_1)\right]\frac{\mathrm{d}^3\mathbf{q}_1}{(2\pi)^3}. \tag{10.211}$$

Since $E(q) = \hbar\omega(q)$, we derive

$$\int_0^{q_{\mathrm{BZ}}} \Lambda^{ZZ}(q,\omega=0)q^2\mathrm{d}q = \frac{\hbar v_c}{2}\int_{v_c^*} n_{\mathrm{P}}\left(\frac{\hbar\omega(q)}{k_{\mathrm{B}}T}\right)\left[n_{\mathrm{P}}\left(\frac{\hbar\omega(q)}{k_{\mathrm{B}}T}\right)+1\right] g_m[E(q)]\mathrm{d}^3\mathbf{q}. \tag{10.212}$$

Because, as shown in Section C.9, the integration over the Brillouin zone can be transformed to an integration over energy, we can write

$$\int_0^{q_{\mathrm{BZ}}} \Lambda^{ZZ}(q,\omega=0)q^2\mathrm{d}q = \frac{(2\pi)^3\hbar v_c}{2}\int n_{\mathrm{P}}\left(\frac{E}{k_{\mathrm{B}}T}\right)\left[n_{\mathrm{P}}\left(\frac{E}{k_{\mathrm{B}}T}\right)+1\right] g_m^2(E)\mathrm{d}E. \tag{10.213}$$

Hence,

$$\lambda_Z = 2(2\pi)^3 \hbar \mathcal{D} v_c \left\{ \frac{2}{15} + \left[C^{XZ}(\mathbf{q}=\mathbf{0})\right]^2 + \left[C^{YZ}(\mathbf{q}=\mathbf{0})\right]^2 \right\} \times \int n_\mathrm{P}\left(\frac{E}{k_\mathrm{B}T}\right)\left[n_\mathrm{P}\left(\frac{E}{k_\mathrm{B}T}\right)+1\right] g_m^2(E)\mathrm{d}E. \tag{10.214}$$

This result is written down in a physically appealing form: the two Planck functions stand for the creation and annihilation of the two excitations and a density of states is associated with each of them.

To proceed further, we evaluate $g_m(E)$ for our model Hamiltonian. We first note the relation

$$g_m(E)\mathrm{d}E = \frac{\mathrm{d}^3\mathbf{q}}{(2\pi)^3} = \frac{q^2\mathrm{d}q}{2\pi^2}. \tag{10.215}$$

Since the small wave vector limit of the dispersion relation is sufficient for our purpose, from Eq. 10.198 we deduce that $\mathrm{d}E = 2D_\mathrm{FM} q\mathrm{d}q$. Hence,

$$g_m(E) = \frac{1}{4\pi^2 D_\mathrm{FM}^{3/2}} \sqrt{E - g\mu_\mathrm{B} B_\mathrm{ani}}, \tag{10.216}$$

for $E > g\mu_\mathrm{B}B_\mathrm{ani}$ and $g_m(E) = 0$ otherwise. As expected, there is no excitation available in the spin gap and a continuum of excitations exists above it. Obviously there is an upper bound on the energy an excitation can acquire. We denote this bound as E_max. The integration over the energy in Eq. 10.214 can be performed with this simple form for the density of states. However, the result is physically relevant only for temperatures such that $g\mu_\mathrm{B}B_\mathrm{ani} \ll k_\mathrm{B}T \ll E_\mathrm{max}$. The following approximation holds within these bounds:

$$\lambda_Z = \frac{\hbar k_\mathrm{B}^2 \mathcal{D} v_c}{\pi} \left\{ \frac{2}{15} + \left[C^{XZ}(\mathbf{q}=\mathbf{0})\right]^2 + \left[C^{YZ}(\mathbf{q}=\mathbf{0})\right]^2 \right\} \frac{T^2}{D_\mathrm{FM}^3} \ln\left(\frac{k_\mathrm{B}T}{g\mu_\mathrm{B}B_\mathrm{ani}}\right). \tag{10.217}$$

The rate λ_Z is found to be approximately quadratic in temperature as already discovered by Mitchell in 1957 for NMR. In many cases of physical interest the tensor $\boldsymbol{C}(\mathbf{q}=\mathbf{0})$ is diagonal. Then λ_Z does not depend on parameters characterising the muon site since

$$\lambda_Z = \mathcal{N}_\mathrm{FM} \frac{g^2 T^2}{D_\mathrm{FM}^3} \ln\left(\frac{k_\mathrm{B}T}{g\mu_\mathrm{B}B_\mathrm{ani}}\right), \tag{10.218}$$

where $\mathcal{N}_\mathrm{FM} \equiv (2/15\pi)(\mu_0/4\pi)^2\gamma_\mu^2\mu_\mathrm{B}^2\hbar k_\mathrm{B}^2 = 129.39\,(\mathrm{meV})^3\,\text{Å}^6\,\mathrm{s}^{-1}\,\mathrm{K}^{-2}$. We have chosen to express $\mathcal{N}_\mathrm{FM}$ in these units because D_FM is often given in millielectronvolts per square ångström.

We briefly consider the case of itinerant ferromagnets. As for insulator ferromagnets, spin waves are expected; see for example Lovesey (1986) and Kübler (2000). In addition, at least for some itinerant ferromagnets such as the weak itinerant ferromagnets, fluctuations (i.e. excitations at zero energy) are also present as explained in Section 10.2.3.2. These fluctuations compete with the magnons in driving the relaxation of the muon spin.

10.2.4.1.2 Antiferromagnets. We now extend our discussion of magnon-induced relaxation to two-sublattice antiferromagnets. As a natural extension of the Hamiltonian for a ferromagnet given in Eq. 10.191, we write the Hamiltonian for an antiferromagnet in the form

$$\mathcal{H}_{\rm sys} = \sum_{i,i'}\sum_{d,d'} \mathcal{J}_{i+d,i'+d'}\mathbf{J}_{i+d}\cdot\mathbf{J}_{i'+d'} + g\mu_{\rm B}\sum_{i}\sum_{d}\mathbf{B}_{{\rm ani},d}\cdot\mathbf{J}_{i+d}, \tag{10.219}$$

where $\mathcal{J}_{i+d,i'+d'}$ is the exchange interaction and $\mathbf{B}_{\rm ani}$ an effective magnetic field which accounts for the magnetic anisotropy. We shall assume it to be parallel to the Z axis. The indices i and i' label the unit cells and d and d' the magnetic atoms of each cell. In the following we only consider the nearest neighbour interaction between the spins. The value of $\mathcal{J}_{i+d,i'+d'}$ vanishes unless spins at locations $\mathbf{i}+\mathbf{d}$ and $\mathbf{i}'+\mathbf{d}'$ are nearest neighbours. In this latter instance we simply denote $\mathcal{J}_{i+d,i'+d'}$ with the positive number $\mathcal{J}$. We focus our attention on a compound consisting of two compensating sublattices undergoing exchange interactions, as first studied by Néel. Such a compound can be viewed as built from unit cells containing two magnetic atoms. We label the corresponding atoms with d_1 and d_2. Since the magnetic moments are opposite on the two sublattices, the two anisotropy fields acting on each sublattice are taken opposite with the common magnitude $B_{\rm ani}$. Let us now consider the exchange part of $\mathcal{H}_{\rm sys}$ which is given by

$$\mathcal{J}\sum_{\langle i,i'\rangle}\mathbf{J}_{i+d_1}\cdot\mathbf{J}_{i'+d_2} + \mathcal{J}\sum_{\langle i,i'\rangle}\mathbf{J}_{i+d_2}\cdot\mathbf{J}_{i'+d_1}, \tag{10.220}$$

where the notation $\langle i,i'\rangle$ indicates that the double sum is restricted. For instance, in the first term only the (i,i') pairs for which $\mathbf{i}+\mathbf{d}_1$ and $\mathbf{i}'+\mathbf{d}_2$ are nearest neighbours are retained. The two terms in Eq. 10.220 yield the same contribution to $\mathcal{H}_{\rm sys}$. Therefore

$$\mathcal{H}_{\rm sys} = 2\mathcal{J}\sum_{\langle i,i'\rangle}\mathbf{J}_{i+d_1}\cdot\mathbf{J}_{i'+d_2} + g\mu_{\rm B}\sum_{i}B_{\rm ani}J^Z_{i+d_1} - g\mu_{\rm B}\sum_{i}B_{\rm ani}J^Z_{i+d_2}. \tag{10.221}$$

Our first purpose is to approximately diagonalise $\mathcal{H}_{\rm sys}$. As a first step, we introduce the linear Holstein–Primakoff transformation. As a matter of fact, two such transformations are needed, i.e. one for each sublattice denoted as sublattices 1 and 2, respectively:

$$\begin{aligned} J^+_{i+d_1} &= \sqrt{2J}\,a^\dagger_{i+d_1}, & J^-_{i+d_1} &= \sqrt{2J}\,a_{i+d_1}, & J^Z_{i+d_1} &= a^\dagger_{i+d_1}a_{i+d_1} - J, \\ J^+_{i+d_2} &= \sqrt{2J}\,b_{i+d_2}, & J^-_{i+d_2} &= \sqrt{2J}\,b^\dagger_{i+d_2}, & J^Z_{i+d_2} &= J - b^\dagger_{i+d_2}b_{i+d_2}. \end{aligned} \tag{10.222}$$

The boson operators obey the commutation relations

$$\left[a_{i+d_1}, a^\dagger_{j+d_1}\right] = \delta_{i,j} \quad \text{and} \quad \left[b_{i+d_2}, b^\dagger_{j+d_2}\right] = \delta_{i,j}. \tag{10.223}$$

The signs of the fluctuations $\delta J^Z_{i+d_1} = J^Z_{i+d_1} + J$ and $\delta J^Z_{i+d_2} = J^Z_{i+d_2} - J$ are opposite as are the spins. Applying the linear Holstein–Primakoff transformations to $\mathcal{H}_{\rm sys}$ and keeping only the terms bilinear in creation and annihilation operators gives, apart from constant terms,

$$\mathcal{H}_{\rm sys} = 2\mathcal{J}J \sum_{\langle i,i'\rangle} \left(a^\dagger_{i+d_1} b^\dagger_{i'+d_2} + a_{i+d_1} b_{i'+d_2} \right)$$
$$+ \left(2\mathcal{J} z_{\rm nn} J + g\mu_{\rm B} B_{\rm ani}\right) \sum_i \left(a^\dagger_{i+d_1} a_{i+d_1} + b^\dagger_{i+d_2} b_{i+d_2} \right), \qquad (10.224)$$

where $z_{\rm nn}$ is the number of nearest neighbour magnetic atoms to a magnetic atom. Since we are dealing with two sublattices, and therefore the translational symmetry plays a key role, it is wise to work in Fourier space. We introduce the transformations

$$a_{i+d_1} = \frac{1}{\sqrt{n_c}} \sum_{\mathbf{q}} a_{\mathbf{q}} \exp\left[i\mathbf{q}\cdot(\mathbf{i}+\mathbf{d}_1)\right], \quad b_{i+d_2} = \frac{1}{\sqrt{n_c}} \sum_{\mathbf{q}} b_{\mathbf{q}} \exp\left[-i\mathbf{q}\cdot(\mathbf{i}+\mathbf{d}_2)\right], \qquad (10.225)$$

where n_c is the number of unit cells, i.e. $n_c = N/2$, where N is the number of magnetic atoms in the sample. We assume N to be even. Note that the magnetic Brillouin zone is now half as large as the crystallographic Brillouin zone. Now in Fourier space the Hamiltonian has a simple expression:

$$\mathcal{H}_{\rm sys} = g\mu_{\rm B} B_{\rm ex} \sum_{\mathbf{q}} \gamma_{\mathbf{q}} \left(a^\dagger_{\mathbf{q}} b^\dagger_{\mathbf{q}} + a_{\mathbf{q}} b_{\mathbf{q}} \right) + g\mu_{\rm B} \left(B_{\rm ex} + B_{\rm ani}\right) \sum_{\mathbf{q}} \left(a^\dagger_{\mathbf{q}} a_{\mathbf{q}} + b^\dagger_{\mathbf{q}} b_{\mathbf{q}} \right), \qquad (10.226)$$

where we have introduced the usual definitions

$$B_{\rm ex} = \frac{2\mathcal{J} z_{\rm nn} J}{g\mu_{\rm B}} \quad \text{and} \quad \gamma_{\mathbf{q}} = \frac{1}{z_{\rm nn}} \sum_{\mathbf{r}} \exp\left(i\mathbf{q}\cdot\mathbf{d}_{\rm pair}\right). \qquad (10.227)$$

The field $B_{\rm ex}$ is the so-called exchange field. The set of $\mathbf{d}_{\rm pair}$ vectors corresponds to the distance vectors joining a given site to its $z_{\rm nn}$ nearest neighbours. Vector $\mathbf{d}_2 - \mathbf{d}_1$ is one of these vectors. To derive Eq. 10.226 we use the relation $\gamma_{\mathbf{q}} = \gamma_{-\mathbf{q}}$ since, by hypothesis, the crystal has an inversion centre. Note that $-1 \le \gamma_{\mathbf{q}} \le 1$ and $\gamma_{\mathbf{q}=\mathbf{0}} = 1$.

Whereas the second term on the right-hand side of Eq. 10.226 is written in a diagonalised form, the first term is not. To achieve the diagonalisation of $\mathcal{H}_{\rm sys}$, we perform a linear transformation defined by the Bogoliubov transformation

$$a_{\mathbf{q}} = u_{\mathbf{q}}\alpha_{\mathbf{q}} + v_{\mathbf{q}}\beta^\dagger_{\mathbf{q}}, \qquad b_{\mathbf{q}} = u_{\mathbf{q}}\beta_{\mathbf{q}} + v_{\mathbf{q}}\alpha^\dagger_{\mathbf{q}},$$
$$a^\dagger_{\mathbf{q}} = u_{\mathbf{q}}\alpha^\dagger_{\mathbf{q}} + v_{\mathbf{q}}\beta_{\mathbf{q}}, \qquad b^\dagger_{\mathbf{q}} = u_{\mathbf{q}}\beta^\dagger_{\mathbf{q}} + v_{\mathbf{q}}\alpha_{\mathbf{q}}, \qquad (10.228)$$

where $u_{\mathbf{q}}$ and $v_{\mathbf{q}}$ are taken to be real functions. We choose the operators $\alpha_{\mathbf{q}}$, $\alpha^\dagger_{\mathbf{q}}$, $\beta_{\mathbf{q}}$, and $\beta^\dagger_{\mathbf{q}}$ to obey the Bose commutation relations

$$\left[\alpha_{\mathbf{q}}, \alpha^\dagger_{\mathbf{q}'}\right] = \delta_{\mathbf{q},\mathbf{q}'} \quad \text{and} \quad \left[\beta_{\mathbf{q}}, \beta^\dagger_{\mathbf{q}'}\right] = \delta_{\mathbf{q},\mathbf{q}'}, \qquad (10.229)$$

whereas each of $\alpha_{\mathbf{q}_1}$ and $\alpha^\dagger_{\mathbf{q}_2}$ commutes with each of $\beta_{\mathbf{q}_3}$ and $\beta^\dagger_{\mathbf{q}_4}$. This implies the relation

$$u_{\mathbf{q}}^2 - v_{\mathbf{q}}^2 = 1. \tag{10.230}$$

This suggests the existence of a function $x_{\mathbf{q}}$ such that

$$u_{\mathbf{q}} = \cosh x_{\mathbf{q}} \qquad \text{and} \qquad v_{\mathbf{q}} = \sinh x_{\mathbf{q}}. \tag{10.231}$$

After introducing the transformations given by Eq. 10.228 into Eq. 10.226, the Hamiltonian is diagonal if the term proportional to $\alpha_{\mathbf{q}}^\dagger \beta_{\mathbf{q}}^\dagger + \alpha_{\mathbf{q}}\beta_{\mathbf{q}}$ vanishes, therefore if

$$\tanh(2x_{\mathbf{q}}) = -\gamma_{\mathbf{q}} \frac{B_{\text{ex}}}{B_{\text{ex}} + B_{\text{ani}}}. \tag{10.232}$$

With this material at hand, we derive the well-known result

$$\mathcal{H}_{\text{sys}} = \sum_{\mathbf{q}} \hbar\omega_{\mathbf{q}} \left(\alpha_{\mathbf{q}}^+ \alpha_{\mathbf{q}} + \beta_{\mathbf{q}}^+ \beta_{\mathbf{q}} + 1\right), \quad \text{with} \quad \hbar\omega_{\mathbf{q}} \equiv g\mu_{\text{B}} \sqrt{\left(B_{\text{ex}} + B_{\text{ani}}\right)^2 - B_{\text{ex}}^2 \gamma_{\mathbf{q}}^2}. \tag{10.233}$$

For each $\mathbf{q}$ there are two degenerate modes, one associated with the operator $\alpha_{\mathbf{q}}$ and the other with the operator $\beta_{\mathbf{q}}$. Hereafter we shall neglect the constant term in $\mathcal{H}_{\text{sys}}$ which arises from the zero-point energy. It reflects the fact that the Néel state is only an approximation to the ground state, although quite good: the magnitude of the expectation value $\langle J_{i+d}^Z \rangle$ is slightly reduced with respect to J. Since we can write

$$\hbar^2 \omega_{\mathbf{q}}^2 = g^2 \mu_{\text{B}}^2 B_{\text{ex}}^2 \left[\left(1 + \frac{B_{\text{ani}}}{B_{\text{ex}}}\right)^2 - \gamma_{\mathbf{q}}^2 \right], \tag{10.234}$$

and recalling $\gamma_{\mathbf{q}=\mathbf{0}} = 1$, the anisotropy field B_{ani} creates a gap in the spin-wave spectrum at $\mathbf{q} = \mathbf{0}$. Neglecting B_{ani},

$$\hbar\omega_{\mathbf{q}} = g\mu_{\text{B}} B_{\text{ex}} \sqrt{1 - \gamma_{\mathbf{q}}^2}. \tag{10.235}$$

At small $\mathbf{q}$ and for a cubic lattice of lattice parameter a, this dispersion relation reduces to

$$\hbar\omega_{\mathbf{q}} \simeq g\mu_{\text{B}} B_{\text{ex}} \sqrt{2} \sqrt{1 - \gamma_{\mathbf{q}}} \simeq D_{\text{AF}}\, q, \tag{10.236}$$

with

$$D_{\text{AF}} = 4\sqrt{3}\mathcal{J} J a. \tag{10.237}$$

Hence, at small wave vectors the dispersion relation is linear in wave vector for an antiferromagnet, whereas it is quadratic for a ferromagnet. The energy gap at the zone centre can be appreciable since

$$\Delta_{\text{gap}} = g\mu_{\text{B}} B_{\text{ani}} \sqrt{1 + 2B_{\text{ex}}/B_{\text{ani}}}. \tag{10.238}$$

The following approximation holds for a cubic compound at small wave vectors:

$$\hbar^2 \omega_{\mathbf{q}}^2 = D_{\text{AF}}^2 q^2 + \Delta_{\text{gap}}^2. \tag{10.239}$$

For an antiferromagnetic chain with no anisotropy the dispersion relation is given by

$$\hbar\omega_q = 4\mathcal{J} J \left|\sin\left(q^Z a\right)\right|. \tag{10.240}$$

We plot this dispersion relation in Fig. 10.9.

Therefore, as for a ferromagnet, the energy of a magnon for an antiferromagnet has its minimum value at the Brillouin zone centre at which it is equal to the gap. That gap is expected to be appreciable. Recalling our discussion for a ferromagnet, we deduce that the muon-spin relaxation observed in the magnetically ordered state at low temperatures is induced by a Raman process involving two magnons of long wavelength. This Raman process was noticed a long time ago for the spin–lattice relaxation process studied by NMR (Moriya 1956, van Kranendonk and Bloom 1956, Mitchell 1957). We recall that the Raman process to be effective requires the Hamiltonian coupling the spin of the nuclear probe to the spins of the system to be anisotropic. It should therefore be more common in μSR than it is in NMR. This is because the hyperfine interaction is usually isotropic, whereas the dipole interaction which contributes significantly to the coupling Hamiltonian in the case of μSR is strongly anisotropic. In the following we shall only study the spin–lattice relaxation rate. The interested reader will easily derive the expression for the spin–spin relaxation due to a Raman process in a Néel antiferromagnet. Because only the parallel correlations, i.e. the correlations along the Z axis, are of importance, we derive from Eq. 7.115

$$\lambda_Z = \frac{\mathcal{D}}{2}\frac{1}{V}\sum_{\mathbf{q}}\sum_{d,d'}\left(G^{XZ}_{d,\mathbf{q}}G^{XZ}_{d',-\mathbf{q}} + G^{YZ}_{d,\mathbf{q}}G^{YZ}_{d',-\mathbf{q}}\right)\Lambda^{ZZ}_{dd',\mathbf{q}}(\omega = \omega_\mu = 0). \tag{10.241}$$

We stress that we take $\omega_\mu = 0$, a well-justified approximation. The sums over d and d' are easily performed explicitly since there are only two sublattices. Introducing the self- and cross-correlations,

$$\begin{aligned}\Lambda^{ZZ}_{\mathrm{self},\mathbf{q}}(\omega = 0) &= \Lambda^{ZZ}_{d_1d_1,\mathbf{q}}(\omega = 0) = \Lambda^{ZZ}_{d_2d_2,\mathbf{q}}(\omega = 0),\\ \Lambda^{ZZ}_{\mathrm{cross},\mathbf{q}}(\omega = 0) &= \Lambda^{ZZ}_{d_1d_2,\mathbf{q}}(\omega = 0) = \Lambda^{ZZ}_{d_2d_1,\mathbf{q}}(\omega = 0),\end{aligned} \tag{10.242}$$

we obtain for the sum over wave vectors in Eq. 10.241

$$\begin{aligned}&\sum_{\mathbf{q}}\left(G^{XZ}_{d_1,\mathbf{q}}G^{XZ}_{d_1,-\mathbf{q}} + G^{YZ}_{d_1\mathbf{q}}G^{YZ}_{d_1,-\mathbf{q}} + G^{XZ}_{d_2,\mathbf{q}}G^{XZ}_{d_2,-\mathbf{q}} + G^{YZ}_{d_2\mathbf{q}}G^{YZ}_{d_2,-\mathbf{q}}\right)\Lambda^{ZZ}_{\mathrm{self},\mathbf{q}}(\omega = 0)\\ &+\sum_{\mathbf{q}}\left(G^{XZ}_{d_1,\mathbf{q}}G^{XZ}_{d_2,-\mathbf{q}} + G^{YZ}_{d_1\mathbf{q}}G^{YZ}_{d_2,-\mathbf{q}} + G^{XZ}_{d_2,\mathbf{q}}G^{XZ}_{d_1,-\mathbf{q}} + G^{YZ}_{d_2\mathbf{q}}G^{YZ}_{d_1,-\mathbf{q}}\right)\Lambda^{ZZ}_{\mathrm{cross},\mathbf{q}}(\omega = 0)\end{aligned} \tag{10.243}$$

We need to evaluate these two correlation functions. We easily get

$$\begin{aligned}\delta J^Z_{d_1,\mathbf{q}} &= \frac{1}{\sqrt{n_c}}\sum_i a^\dagger_{i+d_1}a_{i+d_1}\exp\left[-i\mathbf{q}\cdot(\mathbf{i}+\mathbf{d}_1)\right]\\ &= \frac{1}{\sqrt{n_c}}\sum_{\mathbf{q}_2}\Big(u_{\mathbf{q}_2-\mathbf{q}}u_{\mathbf{q}_2}\alpha^\dagger_{\mathbf{q}_2-\mathbf{q}}\alpha_{\mathbf{q}_2} + u_{\mathbf{q}_2-\mathbf{q}}v_{\mathbf{q}_2}\alpha^\dagger_{\mathbf{q}_2-\mathbf{q}}\beta^\dagger_{\mathbf{q}_2}\\ &\qquad + v_{\mathbf{q}_2-\mathbf{q}}u_{\mathbf{q}_2}\beta_{\mathbf{q}_2-\mathbf{q}}\alpha_{\mathbf{q}_2} + v_{\mathbf{q}_2-\mathbf{q}}v_{\mathbf{q}_2}\beta_{\mathbf{q}_2-\mathbf{q}}\beta^\dagger_{\mathbf{q}_2}\Big).\end{aligned} \tag{10.244}$$

We have made use of Eqs. E.7 and E.14 applied to $a_{\mathbf{q}}$ and $a^\dagger_{\mathbf{q}}$.

Since $\alpha_{\mathbf{q}}(t) = \exp(-i\omega_{\mathbf{q}}t)\alpha_{\mathbf{q}}$ and $\beta_{\mathbf{q}}(t) = \exp(-i\omega_{\mathbf{q}}t)\beta_{\mathbf{q}}$, we derive

$$\begin{aligned}
&\delta J^Z_{d_1,\mathbf{q}}(\omega = 0) \qquad (10.245)\\
&= \frac{2\pi}{\sqrt{n_c}} \sum_{\mathbf{q}_2} \Big[\delta\left(\omega_{\mathbf{q}_2-\mathbf{q}} - \omega_{\mathbf{q}_2}\right) u_{\mathbf{q}_2-\mathbf{q}} u_{\mathbf{q}_2} \alpha^\dagger_{\mathbf{q}_2-\mathbf{q}} \alpha_{\mathbf{q}_2} + \delta\left(\omega_{\mathbf{q}_2-\mathbf{q}} + \omega_{\mathbf{q}_2}\right) u_{\mathbf{q}_2-\mathbf{q}} v_{\mathbf{q}_2} \alpha^\dagger_{\mathbf{q}_2-\mathbf{q}} \beta^\dagger_{\mathbf{q}_2}\\
&\qquad + \delta\left(\omega_{\mathbf{q}_2-\mathbf{q}} + \omega_{\mathbf{q}_2}\right) v_{\mathbf{q}_2-\mathbf{q}} u_{\mathbf{q}_2} \beta_{\mathbf{q}_2-\mathbf{q}} \alpha_{\mathbf{q}_2} + \delta\left(\omega_{\mathbf{q}_2-\mathbf{q}} - \omega_{\mathbf{q}_2}\right) v_{\mathbf{q}_2-\mathbf{q}} v_{\mathbf{q}_2} \beta_{\mathbf{q}_2-\mathbf{q}} \beta^\dagger_{\mathbf{q}_2} \Big].
\end{aligned}$$

The two terms proportional to $\delta\left(\omega_{\mathbf{q}_2+\mathbf{q}} + \omega_{\mathbf{q}_2}\right)$ do not contribute to the relaxation, because they cannot conserve energy since $\omega_{\mathbf{q}_2+\mathbf{q}}$ and $\omega_{\mathbf{q}_2}$ are both positive quantities. After some algebra, using the same decoupling procedure as for the ferromagnets and the fact that the relaxation is driven by the long-wavelength magnons, we finally get for λ_Z a formula which is an extension of the result given in Eq. 10.214 for a ferromagnet:

$$\lambda_Z = 2(2\pi)^3 \hbar \mathcal{D} v_c \int n_{\mathrm{P}}\left(\frac{E}{k_{\mathrm{B}}T}\right)\left[n_{\mathrm{P}}\left(\frac{E}{k_{\mathrm{B}}T}\right) + 1\right] g_m(E)\,\mathcal{S}\,\mathrm{d}E, \qquad (10.246)$$

where $\mathcal{S}$ is a weighted sum of two functions. Explicitly,

$$\begin{aligned}
\mathcal{S} = h_{m,1}(E) &\left\{ \frac{4}{15} + \left[C^{XZ}_{d_1}(\mathbf{q}=\mathbf{0})\right]^2 + \left[C^{YZ}_{d_1}(\mathbf{q}=\mathbf{0})\right]^2 + \left[C^{XZ}_{d_2}(\mathbf{q}=\mathbf{0})\right]^2 \right.\\
&\left. + \left[C^{YZ}_{d_2}(\mathbf{q}=\mathbf{0})\right]^2 \right\} \qquad (10.247)\\
&- 4h_{m,2}(E)\left[\frac{2}{15} + C^{XZ}_{d_1}(\mathbf{q}=\mathbf{0})C^{XZ}_{d_2}(\mathbf{q}=\mathbf{0}) + C^{YZ}_{d_1}(\mathbf{q}=\mathbf{0})C^{YZ}_{d_2}(\mathbf{q}=\mathbf{0})\right].
\end{aligned}$$

These two functions are in fact two weighted density-of-states functions:

$$\begin{aligned}
h_{m,1}(E) = \int_{v_c^*} &\left(\cosh^2\{x[q(E)]\}\cosh^2[x(q')] + \sinh^2\{x[q(E)]\}\sinh^2[x(q')]\right)\\
&\times \delta\left[E - \hbar\omega(q')\right]\frac{\mathrm{d}^3\mathbf{q}'}{(2\pi)^3}, \qquad (10.248)
\end{aligned}$$

and

$$\begin{aligned}
h_{m,2}(E) = \int_{v_c^*} &\cosh\{x[q(E)]\}\cosh[x(q')]\sinh\{x[q(E)]\}\sinh[x(q')]\\
&\times \delta\left[E - \hbar\omega(q')\right]\frac{\mathrm{d}^3\mathbf{q}'}{(2\pi)^3}. \qquad (10.249)
\end{aligned}$$

The hypothesis that all quantities only depend on the modulus of wave vectors is implicit. We recall that in Eq. 10.246 only the low-energy part of the different density-of-states functions is of importance. This is so because the $n_{\mathrm{P}}[E/(k_{\mathrm{B}}T)]$ function acts as a cutoff factor. In turn, only the excitation modes close to the Brillouin zone centre matter.

To proceed further we investigate the function $x(q)$. From the definition given in Eq. 10.232 and the dispersion relation (Eq. 10.234) we find that

$$\tanh^2\left[2x(q)\right] = 1 - \frac{\hbar^2\omega^2(q)}{g^2\mu_{\mathrm{B}}^2\left(B_{\mathrm{ex}} + B_{\mathrm{ani}}\right)^2}. \tag{10.250}$$

It is useful at this point to estimate some parameters. We assume the anisotropy field to be small compared to the exchange field. Taking $B_{\mathrm{ex}} = 100\,\mathrm{T}$, $B_{\mathrm{ani}} = B_{\mathrm{exc}}/100$, and $g = 2$, we compute $\Delta_{\mathrm{gap}} = 1.64\,\mathrm{meV}$ from Eq. 10.238. This is a reasonable small value. Assuming a cubic compound of cube edge 3 Å with $z_{\mathrm{nn}} = 6$, we evaluate $D_{\mathrm{AF}} = 20\,\mathrm{meV}\,\text{Å}$ using Eqs. 10.227 and 10.237. We thus conclude that, for small wave vectors, the second term on the right-hand side of Eq. 10.250 is negligible relative to the first term, at least for the parameters we consider. This means that $\tanh^2[2x(q)]$ is in fact independent of q to a good approximation. We conclude therefore that

$$h_{m,1}(E) \simeq \frac{1}{2}\frac{2\left(B_{\mathrm{exc}} + B_{\mathrm{ani}}\right)^2 - B_{\mathrm{exc}}^2}{\left(B_{\mathrm{exc}} + B_{\mathrm{ani}}\right)^2 - B_{\mathrm{exc}}^2}\, g_m(E) \tag{10.251}$$

and

$$h_{m,2}(E) \simeq \frac{1}{4}\frac{B_{\mathrm{exc}}^2}{\left(B_{\mathrm{exc}} + B_{\mathrm{ani}}\right)^2 - B_{\mathrm{exc}}^2}\, g_m(E). \tag{10.252}$$

Hence, we arrive at the simple result that the expressions for λ_Z in terms of the density of states are the same for the ferromagnetic and antiferromagnetic cases, except for a multiplicative constant.

It is sometimes a fair approximation to neglect the non-diagonal components of $\boldsymbol{C}(\mathbf{q} = \mathbf{0})$. Within this approximation, like in the case of a ferromagnet, λ_Z is found independent of the parameters characterising the muon site. We derive

$$\lambda_Z = \frac{8}{15}(2\pi)^3\hbar\mathcal{D}v_c \int n_{\mathrm{P}}\left(\frac{E}{k_{\mathrm{B}}T}\right)\left[n_{\mathrm{P}}\left(\frac{E}{k_{\mathrm{B}}T}\right) + 1\right] g_m^2(E)\mathrm{d}E. \tag{10.253}$$

Within a factor of 2, we recover the formula obtained for a ferromagnet (Eq. 10.214). Another difference is, of course, the expression of the density of states. In the present case, using the antiferromagnet dispersion relation, we find that

$$g_m(E) = \frac{1}{2\pi^2}\frac{E}{D_{\mathrm{AF}}^3}\sqrt{E^2 - \Delta_{\mathrm{gap}}^2} \tag{10.254}$$

for $E > \Delta_{\mathrm{gap}}$ and $g_m(E) = 0$ otherwise; i.e. there is obviously no magnon with energy smaller than Δ_{gap}. To derive the formula for λ_Z using the approximation given in Eq. 10.253, we need to evaluate the integral

$$I(T) = \int_{\Delta_{\mathrm{gap}}}^{E_{\max}} \frac{E^2\left(E^2 - \Delta_{\mathrm{gap}}^2\right)\exp\left(\dfrac{E}{k_{\mathrm{B}}T}\right)}{\left[\exp\left(\dfrac{E}{k_{\mathrm{B}}T}\right) - 1\right]^2}\mathrm{d}E. \tag{10.255}$$

We consider the high-temperature limit relative to $\Delta_{\rm gap}/k_{\rm B}$, i.e. $\Delta_{\rm gap} \ll k_{\rm B}T \ll E_{\rm max}$. We get

$$I(T) = k_{\rm B}^3 T^3 \left(k_{\rm B}^2 T^2 I_4 - \Delta_{\rm gap}^2 I_2\right), \qquad (10.256)$$

where we have introduced the integral

$$I_{2m} = \int_0^\infty \frac{x^{2m}\exp(x)}{[\exp(x)-1]^2}{\rm d}x. \qquad (10.257)$$

This integral is proportional to an integral representation of the Bernoulli numbers. From Eq. C.103 we compute $I_2 = \pi^2/3$ and $I_4 = 4\pi^4/15$. Hence, it is justified to drop the second term on the right-hand side of Eq. 10.256 in the high-temperature limit we are considering. We finally derive

$$\lambda_Z = \mathcal{N}_{\rm AF}\frac{g^2T^5}{D_{\rm AF}^6}, \qquad (10.258)$$

where $\mathcal{N}_{\rm AF} \equiv (64\pi^3/225)(\mu_0/4\pi)^2\gamma_\mu^2\mu_{\rm B}^2\hbar k_{\rm B}^5 = 17.206\,({\rm meV})^6\,{\rm \AA}^6\,{\rm s}^{-1}\,{\rm K}^{-5}$. As the specimen is cooled down, λ_Z drops much faster in the present case, $\lambda_Z \propto T^5$, than for the ferromagnetic model studied previously, $\lambda_Z \propto T^2 \ln T$. In general terms, in the limit we are considering, $\lambda_Z(T)$ always follows a power law, the value of the exponent being strongly dependent on the functional form of the dispersion relation. For most compounds of interest $\Delta_{\rm gap}$ is in fact larger than the thermal energy at the temperature at which the measurements are carried out. Then the spin–lattice relaxation induced by the magnons vanishes following approximately an Arrhenius behaviour with a characteristic energy equal to $\Delta_{\rm gap}$. In fact, for a detailed analysis of experimental data the function $I(T)$ needs to be numerically evaluated.

As for metallic ferromagnets, magnon excitations are also observed for itinerant antiferromagnets. In addition, fluctuations (excitations at zero energy) might be present. They are able to induce a relaxation of the muon spin as explained in Section 10.2.3.2.

Here we have limited ourselves to the two-magnon relaxation process. This process, which is based on the linear spin-wave approximation, explains the available μSR data for conventional magnetic compounds at temperatures up to about two-thirds of the magnetic ordering temperature. This agreement for such a large temperature range is expected (Mattis 1965). It should be stressed that other processes, for instance involving three-magnon (Beeman and Pincus 1968) or magnon–phonon interactions — see e.g. Sinha and Kumar (1980) — are also possible. Neutron scattering studies usually focus on the one-magnon rather than the two-magnon process. Yet this latter mechanism is sometimes observed: we refer to the work of Huberman *et al.* (2005) and references therein.

10.2.4.2 Relaxation through crystal-field transitions. The energy levels of an ion embedded in a matrix may be split by the crystal electric field (CEF) arising from the electric charges of the neighbouring ions and also the conduction electrons in the case of a metal. The splitting can be quite large relative to the thermal energy, as usually found for transition element ions. In this case it cannot have any influence on the thermal dependence of the relaxation of the muon spin. However, when the splitting is comparable with the thermal energy, as is usually the case for f and sometimes d electron ions, it is important to take it into account. Obviously, although the CEF concerns a single ion, its importance is not restricted to impurities in a matrix; i.e. it should also be described when dealing with a homogeneous compound. A review of the physics of magnetic excitations in crystal-field split systems has been given by Fulde and Loewenhaupt (1986). The focus was on $4f$ systems, as most of the data have been gathered for such systems, usually by inelastic neutron scattering. The importance of the CEF for the magnetism of the rare-earth metals was recognised a long time ago. It is nicely documented in the book of Jensen and Mackintosh (1991).

In systems with CEF split energy levels, the relaxation of the muon spin may be driven by magnetic excitations involving well-defined CEF energy levels. The description of these levels is performed with a CEF Hamiltonian expressed in terms of polynomials of the total angular momentum of the magnetic ion — for definitiveness, we focus our discussion on f ions — i.e. by Stevens operators; see Appendix H. Fortunately, if the symmetry at the ion site is high enough, the independent free parameters in the CEF Hamiltonian are restricted to only a few. In favourable instances they can be determined from the analysis of specific heat and bulk magnetic measurements, and inelastic neutron scattering as well as light absorption data.

In addition to the position of the CEF energy levels, we also consider the linewidth of the crystal-field excitations. As we shall discover below, the physics driving the linewidth is also involved in understanding the relaxation of the muon spin. A theory for the CEF linewidth in metallic rare-earth systems was given a long time ago (Becker *et al.* 1977). The description of the single-ion dynamics was based on the memory function technique. More recently, the same problem was reconsidered by Richards (1984) using the more conventional iterative method of Redfield, the derived system of equations being solved by Laplace transform.

Here, we examine the contribution of CEF transitions to the spin dynamics. The single-ion physics is dealt with in this section. We heavily rely on the work of Yaouanc (2009). The interaction between the ions can be accounted for as explained at the end of Section 10.2.3.1.

We shall split the Hamiltonian of the system under study into two parts: we write $\mathcal{H}_{\rm sys} = \mathcal{H}_0 + \mathcal{H}_1$, where $\mathcal{H}_1$ describes the perturbation of the system relative to $\mathcal{H}_0$. We take $\mathcal{H}_0$ as the sum of two commutating operators, $\mathcal{H}_0 = \mathcal{H}_{\rm CEF} + \mathcal{H}_{\rm L}$, where the first term models the crystal-field energy levels of the f ions, see Appendix H, and the second the lattice with which the f ions exchange energy. The lattice is either the conduction electrons or phonon bath. We shall assume $[\mathcal{H}_{\rm L}, J^\alpha(t)] = 0$. We restrict ourselves to the case for which no magnetic field is applied. Physically the relaxation of a spin correlation function of a single ion is driven by the crystal-field transitions induced by the coupling of the crystal-field with the other degrees of freedom of the system. Neglecting this coupling as a first step, i.e. setting $\mathcal{H}_1 = 0$, we can determine which spin correlations are compatible with the point symmetry at the f ion site as shown in Section D.8. If a correlation vanishes when the coupling is neglected, it will be negligible when $\mathcal{H}_1$ is taken into account.

Since the local character of crystal-field transitions is strongly marked, as a first approximation it is convenient to neglect the wave vector dependence of the physical parameters related to them. We therefore discard the $\mathbf{q}$ dependence of the correlation functions.

Let us introduce the auxiliary function

$$\tilde{\Omega}^{\alpha\beta}(\omega) = \int_{-\infty}^{\infty} \left\langle J^\alpha(t) J^\beta \right\rangle \exp(i\omega t) \mathrm{d}t. \tag{10.259}$$

The condition of detailed balance (Eq. F.10) gives the relation

$$\left\langle J^\alpha(\omega) J^\beta \right\rangle = \exp\left(\beta\hbar\omega\right) \left\langle J^\beta J^\alpha(\omega) \right\rangle. \tag{10.260}$$

This implies

$$\Lambda^{+-}(\omega) = [1 + \exp\left(-\beta\hbar\omega\right)]\tilde{\Omega}^{+-}(\omega)/2,$$
$$\Lambda^{zz}(\omega) = [1 + \exp\left(-\beta\hbar\omega\right)]\tilde{\Omega}^{zz}(\omega)/2. \tag{10.261}$$

We recall that $\Lambda^{+-}(\omega)$ and $\Lambda^{zz}(\omega)$ are symmetrised correlation functions. Hence, once $\tilde{\Omega}^{zz}(\omega)$ and $\tilde{\Omega}^{+-}(\omega)$ are computed, $\Lambda^{+-}(\omega)$ and $\Lambda^{zz}(\omega)$ can be evaluated. These functions are sufficient for the description of λ_Z if the point group symmetry at the position of spin J is such that z is at least a threefold symmetry axis; see Section D.8.

Note that the inelastic neutron scattering intensity is directly proportional to $\tilde{\Omega}^{\alpha\beta}(\omega)$. The so-called scattering function $S^{\alpha\beta}(\omega)$ is within a constant of proportionality equal to $\tilde{\Omega}^{\alpha\beta}(\omega)$.

Before proceeding to the derivation of $\tilde{\Omega}^{\alpha\beta}(\omega)$ we find it convenient to detail the two perturbation Hamiltonians $\mathcal{H}_1$ that we shall consider, namely the coupling with the conduction electrons and the lattice vibrations.

Lattice correlation functions will be needed. They will be generically denoted as $f(t)$ and specifically $f_{\mathrm{el}}(t)$ and $f_{\mathrm{ph}}(t)$ for the two physical cases we will consider. Related to $f(t)$ are the two one-sided time Fourier transforms

$$g(\omega) = \int_0^{\infty} f(t) \exp(i\omega t) \mathrm{d}t, \qquad h(\omega) = \int_0^{\infty} f(-t) \exp(i\omega t) \mathrm{d}t. \tag{10.262}$$

We first focus on the conduction electron relaxation mechanism. The exchange interaction between an f ion and a conduction electron at the ion site is modelled by the Hamiltonian

$$\mathcal{H}_1 = \mathcal{H}_{\mathrm{el}} = -2I_{\mathrm{ex}}(g-1)\mathbf{J} \cdot \mathbf{S}_{\mathrm{e}}, \tag{10.263}$$

where I_{ex} is an exchange integral and $\mathbf{S}_{\mathrm{e}}$ the conduction electron spin operator. Here, the presence of the factor $(g-1)$ means that we have made the hypothesis that I_{ex} is independent of the f ion for a given series of compounds. We could have included the factor $(g-1)$ in the definition of the exchange integral. Obviously, in this case the exchange would have depended on the ion. By writing Eq. 10.263, we have implicitly assumed the coupling between the total angular momentum and the conduction electron spin to be isotropic. Therefore, it is sufficient to compute the spin correlation for only one component of $\mathbf{S}_{\rm e}$, say x, defined as $f_{\rm el}(t) = \langle S_{\rm e}^x(t) S_{\rm e}^x \rangle_{\rm el}$. The quantum average over the electron bath is carried out. We shall use the fermionic representation of $S_{\rm e}^x$:

$$S_{\rm e}^x = \frac{1}{2} \sum_{\mathbf{q},\mathbf{q}'} \left(c^+_{\mathbf{q}',\downarrow} c_{\mathbf{q},\uparrow} + c^+_{\mathbf{q}',\uparrow} c_{\mathbf{q},\downarrow} \right), \tag{10.264}$$

where $c^+_{\mathbf{q},\sigma}$ and $c_{\mathbf{q},\sigma}$ denote the creation and annihilation electron operators of wave vector $\mathbf{q}$ and spin σ. For simplicity, we assume free conduction electrons. Without loss of generality, the ion is taken to sit at the origin of the coordinates. Using the Hartree–Fock (mode-coupling) approximation to decouple the four fermion operator expectation values appearing in $f_{\rm el}(t)$ and neglecting the spin dependence of the energy of the conduction electron, we get

$$f_{\rm el}(t) = \frac{1}{2} \sum_{\mathbf{q},\mathbf{q}'} n_{\rm FD}(\beta \hbar \omega_{\mathbf{q}'}) \left[1 - n_{\rm FD}(\beta \hbar \omega_{\mathbf{q}})\right] \exp[i(\omega_{\mathbf{q}'} - \omega_{\mathbf{q}})t], \tag{10.265}$$

where $n_{\rm FD}(\beta \hbar \omega_{\mathbf{q}})$ is the Fermi–Dirac occupation factor (Eq. C.30). It is a simple matter to derive

$$g_{\rm el}(\omega) = \frac{\pi}{2} \sum_{\mathbf{q},\mathbf{q}'} n_{\rm FD}(\beta \hbar \omega_{\mathbf{q}'}) \left[1 - n_{\rm FD}(\beta \hbar \omega_{\mathbf{q}})\right] \delta\left(\omega_{\mathbf{q}'} - \omega_{\mathbf{q}} + \omega\right). \tag{10.266}$$

We have neglected the imaginary part of $g_{\rm el}(\omega)$ since it leads to a negligible energy shift. Because the product function $n_{\rm FD}(\beta E) \left[1 - n_{\rm FD}(\beta E)\right]$ has an appreciable weight only near the Fermi energy — see, for example, Slichter (1996) — we can write

$$g_{\rm el}(\omega) = \frac{\pi}{2} \hbar [N(E_{\rm F})]^2 \int_{-\infty}^{\infty} n_{\rm FD}(\beta E) \left\{1 - n_{\rm FD}\left[\beta\left(E + \hbar\omega\right)\right]\right\} {\rm d}E, \tag{10.267}$$

where $N(E_{\rm F})$ is the electron density at the Fermi level per spin. Performing the integration over the energy with the help of Eq. C.79, we finally obtain

$$g_{\rm el}(\omega) = \frac{\pi}{2} \hbar [N(E_{\rm F})]^2 \frac{\hbar\omega}{1 - \exp\left(-\beta\hbar\omega\right)}. \tag{10.268}$$

We turn our attention to the phonon mechanism. From numerous electron spin resonance studies, the phonon-induced relaxation between electronic energy levels is known to be driven by the modulation of the crystal field. We expand this field in powers of the strain and limit the expansion to the linear term denoted as $\mathcal{H}_{\rm sp}$. It accounts for the spin–phonon interactions, i.e. the magneto-elastic coupling, and represents an additional electric potential generated by the lattice vibrations. To determine its form, we should study in details the vibration-induced distortion of the environment of the ion of interest. Here we follow a much simpler path. We take an average strain ϵ and ignore any directional properties. This results in

$$\mathcal{H}_1 = \mathcal{H}_{\rm sp} = V\epsilon. \tag{10.269}$$

To second order in $\mathbf{J}$, or more practically using the Stevens operators O_2^n — these operators are listed in different references, e.g. Orbach (1961) and Abragam and Bleaney (1970) — we have

$$V = g_0 O_2^0 + g_1 O_2^1 + g_{-1} O_2^{-1} + g_2 O_2^2 + g_{-2} O_2^{-2}, \tag{10.270}$$

where the five g_n are unknown real parameters. Only the operators in this sum which break the point symmetry at the rare-earth site are effective in inducing crystal-field transitions. Practically, to limit the number of parameters, one may have to select one or two out of the sum in a somewhat arbitrary fashion. As far as ϵ is concerned, it is convenient to express it in terms of its Fourier components. Since the f ion can be taken at the origin of the coordinates,

$$\epsilon = i \sum_{\mathbf{q}} \sqrt{\frac{\hbar q^2}{2 n_c M_c \omega_{\mathbf{q}}}} \left(a_{\mathbf{q}} - a_{\mathbf{q}}^+ \right), \tag{10.271}$$

where $n_c M_c$ is the mass of the crystal and $a_{\mathbf{q}}^+$ and $a_{\mathbf{q}}$ are the creation and annihilation phonon operators. Note that, here, only a single branch of phonons is accounted for. The correlation function $f_{\rm ph}(t) = \langle \epsilon(t)\epsilon \rangle_{\rm ph}$, where the quantum average is over the phonon bath, can be written as

$$f_{\rm ph}(t) = \frac{1}{n_c} \sum_{\mathbf{q}} \frac{\hbar q^2}{2 M_c \omega_{\mathbf{q}}} \left\{ [n_{\rm P}(\beta\hbar\omega_{\mathbf{q}}) + 1] \exp(-i\omega_{\mathbf{q}} t) + n_{\rm P}(\beta\hbar\omega_{\mathbf{q}}) \exp(i\omega_{\mathbf{q}} t) \right\}. \tag{10.272}$$

As for magnons (see Section 10.2.4.1), $n_{\rm P}(\beta\hbar\omega_{\mathbf{q}})$ denotes the Planck population factor (Eq. C.31). Depending on the sign of ω only the former or latter term in the sum of Eq. 10.272 contributes to $g_{\rm ph}(\omega)$. This is so because $\omega_{\mathbf{q}}$ is positive and the energy conservation prescription must be fulfilled. For both signs of ω the computation leads to the single formula

$$g_{\rm ph}(\omega) = \frac{3}{4\pi} \frac{1}{\varrho v_s^5 \hbar^2} \frac{\hbar^3 \omega^3}{1 - \exp(-\beta\hbar\omega)}. \tag{10.273}$$

The imaginary part of $g_{\rm ph}(\omega)$, which leads to a slight energy shift, has again been neglected. The factor 3 has been added to account for the three phonon branches. We denote the crystal density by ϱ and assume a linear energy dispersion for the phonon field: $\omega_{\mathbf{q}} = v_s q$ where v_s is the sound velocity.

Finally, since the relation $h(\omega) = g(-\omega)$ holds in both the phonon and electron cases — see Eq. 10.262 for the definition of $h(\omega)$ — $h(\omega)$ is easily computed. In addition, the relation

$$g(-\omega) = \exp(-\beta\hbar\omega) g(\omega) \tag{10.274}$$

is obeyed for the functions $g_{\rm ph}(\omega)$ and $g_{\rm el}(\omega)$.

We are now going to give formulae for $\tilde{\Omega}^{zz}(\omega)$ and $\tilde{\Omega}^{+-}(\omega)$ derived from the iterative method of Richards (1984).

We first focus on the determination of the matrix elements of $J^\alpha(t)$. We work with the interaction representation and therefore introduce two new operators (see e.g. Section 7.1.4.1):

$$\tilde{J}^\alpha(t) = \exp(-i\mathcal{H}_0 t/\hbar) J^\alpha(t) \exp(i\mathcal{H}_0 t/\hbar),$$
$$\tilde{\mathcal{H}}_1(t) = \exp(-i\mathcal{H}_0 t/\hbar) \mathcal{H}_1 \exp(i\mathcal{H}_0 t/\hbar). \tag{10.275}$$

Note that $\tilde{J}^\alpha(t=0) = J^\alpha(t=0)$. Using the Heisenberg evolution equation for $J^\alpha(t)$, we obtain

$$\frac{\mathrm{d}J^\alpha(t)}{\mathrm{d}t} - \frac{i}{\hbar}[\mathcal{H}_0, J^\alpha(t)] = \exp(i\mathcal{H}_0 t/\hbar)\frac{\mathrm{d}\tilde{J}^\alpha(t)}{\mathrm{d}t}\exp(-i\mathcal{H}_0 t/\hbar), \tag{10.276}$$

with

$$\frac{\mathrm{d}\tilde{J}^\alpha(t)}{\mathrm{d}t} = \frac{i}{\hbar}[\tilde{\mathcal{H}}_1(t), \tilde{J}^\alpha(t)]. \tag{10.277}$$

The solution of the previous equation obviously reads

$$\tilde{J}^\alpha(t) = J^\alpha + \frac{i}{\hbar}\int_0^t [\tilde{\mathcal{H}}_1(t'), \tilde{J}^\alpha(t')]\mathrm{d}t'. \tag{10.278}$$

We now project out Eq. 10.276 into the crystal-field-state manifold, the eigenstates of which are labelled $|n\rangle$. The quantity $\hbar\omega_n$ is the eigenvalue of state $|n\rangle$, i.e. $\mathcal{H}_{\mathrm{CEF}}|n\rangle = \hbar\omega_n|n\rangle$. We also set $\omega_{mn} \equiv \omega_m - \omega_n$. In order to evaluate the right-hand side of Eq. 10.276, we use Eq. 10.277 together with Eq. 10.278. We drop J^α in this latter equation since this term would not conserve energy during a crystal-field transition process. The perturbation Hamiltonian is only taken to second order. Expanding the two commutators in Eqs. 10.277 and 10.278, averaging over the lattice quantum states and taking the Laplace transform of the matrix elements of each term in Eq. 10.276 using the definition $Z^\alpha_{mn}(s) = \int_0^\infty \exp(-st)\langle m|J^\alpha(t)|n\rangle \mathrm{d}t$, we arrive at

$$\begin{aligned}(s - i\omega_{mn})Z^\alpha_{mn}(s) - \langle m\,|J^\alpha|\,n\rangle = &-\sum_{p,q} G_{mp,pq}\, h\,(is + \omega_{pn})\, Z^\alpha_{qn}(s)\\ &-\sum_{p,q} G_{pq,qn}\, g\,(is + \omega_{mq})\, Z^\alpha_{mp}(s)\\ &+\sum_{p,q} G_{mp,qn}\left[h\,(is+\omega_{pn}) + g\,(is+\omega_{mq})\right] Z^\alpha_{pq}(s),\end{aligned} \tag{10.279}$$

where, depending on the case, $G_{mp,pq}$ is equal either to $G^{\mathrm{ph}}_{mn,pq} = \langle m\,|V|\,n\rangle\,\langle p\,|V|\,q\rangle\,/\hbar^2$ or $G^{\mathrm{el}}_{mn,pq} = \left[4I^2_{\mathrm{ex}}(g-1)^2/\hbar^2\right]\sum_\gamma \langle m\,|J^\gamma|\,n\rangle\,\langle p\,|J^\gamma|\,q\rangle$. Here, the index γ is one of the Cartesian coordinates, *i.e.* x, y, or z. Of course the set of functions $\{g(\omega),\, h(\omega)\}$ is either $\{g_{\mathrm{ph}}(\omega),\, h_{\mathrm{ph}}(\omega)\}$ or $\{g_{\mathrm{el}}(\omega),\, h_{\mathrm{el}}(\omega)\}$ depending on the case of interest. Setting $s = i\omega$, Eq. 10.279 reads

$$\begin{aligned}i\,(\omega - \omega_{mn})\, Z^\alpha_{mn}(i\omega) &+ \sum_{p,q} G_{mp,pq}\, h\,(-\omega + \omega_{pn})\, Z^\alpha_{qn}(i\omega)\\ &+ \sum_{p,q} G_{pq,qn}\, g\,(-\omega + \omega_{mq})\, Z^\alpha_{mp}(i\omega)\\ &- \sum_{p,q} G_{mp,qn}\left[h\,(-\omega+\omega_{pn}) + g\,(-\omega+\omega_{mq})\right] Z^\alpha_{pq}(i\omega)\\ &= \langle m\,|J^\alpha|\,n\rangle - \langle J^\alpha\rangle\delta_{m,n}.\end{aligned} \tag{10.280}$$

The inclusion of the last term — namely $\langle J^\alpha \rangle \delta_{m,n}$ — guarantees that $\langle J^\alpha(t) \rangle$ relaxes towards its thermal equilibrium value. With Eq. 10.280 we have a set of $(2J+1)^2$ coupled linear equations satisfied by the $Z^\alpha_{mn}(i\omega)$ functions.

In fact, as already stated, if the point symmetry at the f ion site is sufficiently high, the only correlation functions needed for the evaluation of λ_Z are $\Lambda^{+-}(\omega)$ and $\Lambda^{zz}(\omega)$, or alternatively $\tilde{\Omega}^{+-}(\omega)$ and $\tilde{\Omega}^{zz}(\omega)$ if we use Eq. 10.261. From the definitions of these latter functions (Eq. 10.259), it is a simple matter to get the relations

$$\tilde{\Omega}^{+-}(\omega) = \frac{1}{Z_{\rm CEF}} \sum_{m,n} \exp(-\beta E_m)\, Z^{+}_{mn}(-i\omega)\, \langle n\,|J^-|\,m\rangle + \text{c.c.},$$

$$\tilde{\Omega}^{zz}(\omega) = \frac{1}{Z_{\rm CEF}} \sum_{m,n} \exp(-\beta E_m)\, Z^{z}_{mn}(-i\omega)\, \langle n\,|J^z|\,m\rangle + \text{c.c.} \qquad (10.281)$$

The symbol c.c. stands for the complex conjugate of the expression on its left. The partition function $Z_{\rm CEF} = \sum_m \exp(-\beta E_m)$ for the crystal-field level scheme of an f ion has been introduced. The sum is over the $(2J+1)$ states.

As already noted, the evolution of the matrix elements $Z^\alpha_{mn}(i\omega)$ is described by a system of linear equations in energy space; see Eq. 10.280. Viewed in time space, we are dealing with a system of linear differential equations. Physically, the electronic or phonon perturbation induces transitions between crystal-field energy levels, and their effect on the magnetic correlations is modelled by a master equation. Focusing on the phonon-induced transitions, we note that only real phonons are involved; i.e. their energy is equal to the crystal-field splitting. However, it is well known that virtual phonons may also induce crystal-field transitions. This is the Raman phonon scattering process. It cannot be described within the framework we have just set up. For completeness we specify that it leads to a linewidth $\Gamma_{\rm ph} \propto T^{-n}$ with a large value for the exponent n, typically $n = 5$ or $n = 7$ (Orbach 1961). This mechanism can only appear at low temperature.

The linear system of equations in Eq. 10.280 seems formidable. It can be written as a matrix equation which can be numerically solved. However, to grasp its physical content, here we shall study its prediction for two simple CEF level schemes. We shall assume, for simplicity, λ_Z to be proportional to $\Lambda^{zz}(\omega = 0) = \tilde{\Omega}^{zz}(\omega = 0)$. Hence, we focus on $\tilde{\Omega}^{zz}(\omega)$ with ω small. This is consistent with the fact that we are interested in the analysis of zero-field measurements.

We take our first example from the paper of Becker *et al.* (1977): a Ce^{3+} ion sitting in a metal at a site of cubic symmetry. Only the coupling to the electron bath is considered. The six crystal-field levels split into a doublet Γ_7 and a quartet Γ_8. We assume the energy of the latter to be sufficiently high that it is a good approximation to consider only the Γ_7 doublet. We label its two states as $|\psi_{\Gamma_7}\rangle$ and $|\bar{\psi}_{\Gamma_7}\rangle$. They are written as $|\psi_{\Gamma_7}\rangle = \left(\frac{1}{6}\right)^{1/2}|\frac{5}{2}\rangle - \left(\frac{5}{6}\right)^{1/2}|-\frac{3}{2}\rangle$ and $|\bar{\psi}_{\Gamma_7}\rangle = \left(\frac{1}{6}\right)^{1/2}|-\frac{5}{2}\rangle - (\frac{5}{6})^{1/2}|\frac{3}{2}\rangle$. Since $\langle \psi_{\Gamma_7}|J^z|\bar{\psi}_{\Gamma_7}\rangle = \langle \bar{\psi}_{\Gamma_7}|J^z|\psi_{\Gamma_7}\rangle = 0$ and $\langle \bar{\psi}_{\Gamma_7}|J^z|\bar{\psi}_{\Gamma_7}\rangle = -\langle \psi_{\Gamma_7}|J^z|\psi_{\Gamma_7}\rangle$, we only need to study Eq. 10.280 for $m = n = 1$. Here, we have introduced the short-hand notations $|1\rangle = |\psi_{\Gamma_7}\rangle$ and $|2\rangle = |\bar{\psi}_{\Gamma_7}\rangle$. A further simplification is now made, consistent with our interest in the function $\tilde{\Omega}^{zz}(\omega)$ in the small ω limit: we neglect the dependence of the g and h functions on ω in Eq. 10.280. According to Eq. 10.281, this means that $\tilde{\Omega}^{zz}(\omega)$

is a Lorentzian function of ω. For simplicity, this low-energy approximation will be performed for the two examples treated here. Obviously, it can easily be overcome. Since $h_{\rm el}(\omega = 0) = g_{\rm el}(\omega = 0)$, we derive

$$i\,\omega Z^z_{11}(i\omega) + 4g_{\rm el}(\omega = 0)G^{\rm el}_{12,21}Z^z_{11}(i\omega) = \langle 1|J^z|2\rangle. \qquad (10.282)$$

The relaxation arises from the only $G^{\rm el}_{mn,pq}$ term that satisfies the secular condition, i.e. $\omega_{mn} = \omega_{pq}$; see Section 5.2.2. Using Eq. 10.268 and the definition of $G^{\rm el}_{mn,pq}$, we derive

$$\tilde{\Omega}^{zz}(\omega) = \frac{g^2_{\Gamma_7}}{2}\frac{\Gamma^{zz}_{\rm el}}{\omega^2 + \left(\Gamma^{zz}_{\rm el}\right)^2}, \qquad (10.283)$$

with the Korringa linewidth relation (Korringa 1950)

$$\Gamma^{zz}_{\rm el} = 4\pi g^2_{\Gamma_7}\left[I_{\rm ex}(g-1)N(E_{\rm F})\right]^2\frac{k_{\rm B}T}{\hbar}. \qquad (10.284)$$

The spectroscopic factor of $|\Gamma_7\rangle$ is $g_{\Gamma_7} = 2|\langle 1|J^z|2\rangle| = \frac{5}{3}$. From a straightforward computation,

$$\tilde{\Omega}^{zz}(\omega = 0) = \frac{1}{8\pi}\frac{\hbar}{\left[I_{\rm ex}(g-1)N(E_{\rm F})\right]^2 k_{\rm B}T}. \qquad (10.285)$$

Therefore this model predicts $\lambda_Z \propto 1/T$. This is in sharp contrast to the Korringa result given in Eq. 10.146, for which $\lambda_Z \propto T$. The basic reason for this difference stems from the assumption that the muon spin is directly coupled to the conduction electron spins in the Korringa case, whereas the former result reflects the fact that the muon spin is directly coupled to the ion spin which interacts with the conduction electron spins. Hence, the muon spin does not probe directly the conduction electron spins. The same type of indirect mechanism, although involving the nuclear spins rather than the conduction electron spins, seems to be at play in high-spin molecules (Keren *et al.* 2007).

In Eq. 10.285, $\tilde{\Omega}^{zz}(\omega = 0)$ is inversely proportional to the temperature and to $[I_{\rm ex}N(E_{\rm F})]^2$. A spin–lattice relaxation rate varying as the inverse temperature was first reported for the intermetallic compound $ErAl_2$ by Hartmann *et al.* (1986). The relaxation rate λ_Z is not only inversely proportional to the temperature, but also on the density of states at the Fermi level. This latter characteristics has been employed to detect the reduction of the density of states due to the formation of a charge-density wave in an intermetallic compound (Galli *et al.* 2002). According to inelastic neutron scattering results — see for example Fulde and Loewenhaupt (1986) — $|I_{\rm ex}N(E_{\rm F})|$ can be as large as 0.1. Note that our model neglects any Kondo effect.

Interestingly, the derivation of Eq. 10.283 shows the linewidth to be proportional to $g_{\rm el}(\omega = 0)$. Hence, had we considered the relaxation to arise from the coupling to the phonon bath, referring to Eq. 10.273, we would have found $\Gamma^{zz}_{\rm ph} = 0$. Therefore, for the present two CEF energy levels, in contrast to the electronic case, the coupling to the strain field does not broaden the transition line. When more than two levels are involved, a strain field broadening may occur as explained below.

Motivated by the experimental identification of an Orbach-like (Orbach 1961) muon relaxation (Dalmas de Réotier *et al.* 2003), we now investigate the relaxation arising

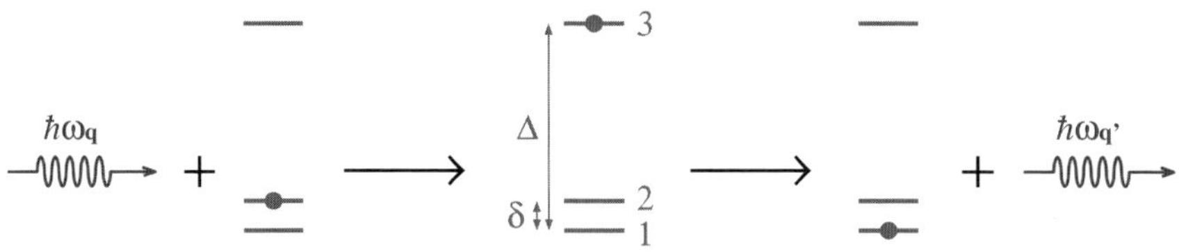

Fig. 10.11 A schematic illustration of the Orbach relaxation process. The system is made of two low-energy CEF states denoted 2 and 1, together with a third state (3). State 2 (3) lies at energy δ (Δ) above state 1 which is the ground state. The absorption of a phonon of energy $\hbar\omega_{\mathbf{q}} = \Delta - \delta$ induces a CEF transition from state 2 to state 3 by a direct one-phonon process. Thereafter, a phonon of energy $\hbar\omega_{\mathbf{q}'} = \Delta$ is emitted by a similar process. The net result is a transition between states 2 and 1. The rate of phonon absorption and emission is controlled by the phonon density of states. Here, only real phonons are involved. This is in contrast to the Raman scattering process which involves the simultaneous absorption and emission of two virtual phonons. This latter process is only effective at extremely low temperature.

from a simple crystal-field scheme of three levels at energies $\langle\psi_2|\mathcal{H}_{\rm CEF}|\psi_2\rangle = \delta$ and $\langle\psi_3|\mathcal{H}_{\rm CEF}|\psi_3\rangle = \Delta$ with $\Delta \gg \delta$. The ground state is the $|\psi_1\rangle$ state with energy equal to 0. We focus on the phonon relaxation mechanism. Since our interest is on the low-energy physics, we study Eq. 10.280 with $m = 1$ and $n = 2$. Here, we have again introduced the short-hand notations $|i\rangle = |\psi_i\rangle$. Assuming δ negligible,

$$\left(i\,\omega + \Gamma^{zz}_{\rm ph}\right) Z^z_{12}(i\omega) = \langle 1|J^z|2\rangle, \tag{10.286}$$

where

$$\Gamma^{zz}_{\rm ph} = h_{\rm ph}\,(\Delta/\hbar)\left(G^{\rm ph}_{13,31} + G^{\rm ph}_{23,32}\right). \tag{10.287}$$

Again, the contributions to Eq. 10.287 only come from the two terms of $G^{\rm ph}_{mn,pq}$ that satisfy the secular condition, i.e. $\omega_{mn} = \omega_{pq}$. The other terms oscillate with an angular frequency $\omega_\Delta = \Delta/\hbar \simeq 10^{13}\,{\rm rad\,s^{-1}}$ for $\Delta/k_{\rm B} = 100\,{\rm K}$, and therefore average out to 0 on the time scale of the measurements. This simple model predicts a quasielastic scattering. Assuming further $\beta\Delta \gg 1$ gives $\Gamma^{zz}_{\rm ph} \propto \exp(-\beta\Delta)$ and therefore an Arrhenius behaviour is found for $\lambda_Z(T)$. The Orbach-like relaxation we are discussing is illustrated in Fig. 10.11. The scattering occurs only if phonons of proper energy are available. For a compound with a Debye phonon spectrum, this means that the excitation energy Δ cannot be larger than $k_{\rm B}\Theta_{\rm D}$ where $\Theta_{\rm D}$ is the Debye temperature. To complete our analysis of the three-level system, we note that $\Gamma^{zz}_{\rm ph}$ is found to be proportional to T when $\beta\Delta \ll 1$. Therefore λ_Z is proportional to $1/T$ when the thermal energy exceeds the crystal-field energy.

We have just explained that the magneto-elastic coupling predicts an activated behaviour for λ_Z at relatively high temperature for a simple three-level energy scheme. Yet the same coupling cannot explain the observed plateau at intermediate temperature for the garnet compound $Yb_3Ga_5O_{12}$ (Dalmas de Réotier *et al.* 2003). In fact, we need to take into account the Heisenberg interaction for a complete modelisation of λ_Z. The simplest method consists in recalling a previous discussion which considers the fluctuations arising from the spin self-correlations. The intersite correlations are expected to matter only at low temperature. Referring to Eqs. 10.110 and 10.112 and setting $\Gamma^{zz}_{\rm ex} = 1/\tau_{\rm ex}$, we derive

$$\Gamma^{zz}_{\rm ex} = \sqrt{\frac{16\mathcal{J}^2 z_{\rm nn} J(J+1)}{3\pi\hbar^2}}. \tag{10.288}$$

The width $\Gamma^{zz}_{\rm ex}$ depends on the two-ion exchange interaction $\mathcal{J}$ that we have limited to the $z_{\rm nn}$ nearest neighbour ions. Therefore it reflects the environment of the ions under study, and it is temperature independent in the high-temperature limit we take. If needed, the influence of the temperature can be accounted for, working from Eq. F.14. The phonon-induced relaxation is easily taken into account in the expression for the total linewidth Γ^{zz}:

$$\Gamma^{zz} = \Gamma^{zz}_{\rm ex} + \Gamma^{zz}_{\rm ph}. \tag{10.289}$$

In the same way, for the electronic relaxation we have

$$\Gamma^{zz} = \Gamma^{zz}_{\rm ex} + \Gamma^{zz}_{\rm el}. \tag{10.290}$$

In this section we have studied the effect of CEF transitions on the temperature dependence of λ_Z. We have first made the approximation that the total angular momenta in the system behave independently. A simple phenomenological method to account for the self-correlations has just been given. The random-phase approximation exposed in Section 10.2.3.1 may be considered for a more complete description of the interactions. Beforehand the local correlation function $\Lambda^{\alpha\beta}(\omega)$ described here is required. Finally, we note that it is sometimes possible to model the spin dynamics in the ordered state of a compound using $\Lambda^{\alpha\beta}(\omega)$ computed with the presence of a molecular field and using the random-phase approximation to get $\Lambda^{\alpha\beta}(\mathbf{q}, \omega)$. This is explained in the book of Jensen and Mackintosh (1991).

10.3 Examples

We shall not attempt to review the published works on magnetic materials. Reports are available: see the references listed at the beginning of the chapter. We rather focus on the methodology. This explains the restricted number of examples given in this section. They are chosen so as to exemplify the variety of issues that the μSR techniques can address as far as magnetism is concerned. We successively consider the detection of magnetic phase transitions, the spontaneous local field at the muon site, the characterisation of magnetic domains in materials, and finally the study of spin dynamics.

We also note that many experimental studies of Knight shift in metals and semimetals have been performed. A theoretical background has been given in Section 5.3 and examples are shown in Sections 1.2.2, 9.4.1, 11.2.2, and 12.5. Many other examples are nicely reviewed by Schenck (1985) and there is no need to pursue this discussion here.

10.3.1 Detection of magnetic phase transitions

The μSR technique is popular in the condensed-matter community for its success in detecting magnetic phase transitions in compounds with small magnetic moments. In this section we examine two types of examples. Whereas for the first one, spontaneously precessing μSR signals are detected, pointing to the existence of a relatively long-range magnetic order, for the second type one observes only wide field distributions, a fingerprint of short-range magnetic order.

However, before discussing these examples, we show how the zero-field μSR technique can give information on the sharpness of a magnetic transition in a given sample through the thermal variation of the effective initial asymmetry. In a polycrystalline magnet, while $P_Z(t)$ is expected to be an exponential in a paramagnetic phase (Eq. 1.10) it is the sum of two terms in the ordered phase (see e.g. Eq. 1.5 for a simple model). In Fig. 10.12 we display the effective initial asymmetry for the hexagonal ferromagnet $GdNi_5$ and the tetragonal antiferromagnet UPt_2Si_2. The spectra have been recorded at a pulsed muon source, so that the spontaneous frequencies expected in the ordered states cannot be detected because of the limited time resolution at this type of muon source; see Section 2.2.2.1. The associated cosine signal is therefore averaged out to 0. Whereas for $GdNi_5$ the expected behaviour is observed, i.e. a sharp loss of about a factor 3 of the measured asymmetry at T_C, for UPt_2Si_2 the change at T_N is gradual and the ratio of the effective initial asymmetry in the paramagnetic and ordered state is smaller than 3. We conclude that the distribution of T_N values is relatively large in the UPt_2Si_2 sample. In addition, following the explanations at the beginning of Section 6.2, the crystallites of this sample are such that they are preferentially oriented with their spontaneous field parallel to the laboratory Z axis.

The distribution of $T_{\rm N}$ could be the signature of crystalline disorder: possibly some Pt and Si atoms are interchanged. Complementary measurements on a single-crystal sample of UPt_2Si_2 with $\mathbf{S}_\mu$ perpendicular to $\mathbf{B}_0$ (and therefore to the c axis) have shown that the effective initial asymmetry decreases to 0 around T_N in a temperature interval as large as $\sim$5 K as well. This result supports the conclusion deduced from the polycrystalline sample measurements. In summary, the measurement of the effective initial asymmetry offers a stringent test of the sample quality.

We now discuss the detection of long-range magnetic transitions, focusing first on a detailed presentation of the works on some organic compounds.

The search for purely organic molecular ferromagnets containing only light elements (carbon, hydrogen, oxygen, and nitrogen) is a subject of strong interest. The first such material to be found, the β crystal phase of *para*-nitrophenyl nitronyl nitroxide (*p*-NPNN, i.e. $C_{13}H_{16}N_3O_4$), was reported to have a Curie temperature $T_{\rm C} \sim 0.6$ K (Tamura *et al.* 1991). The unpaired spin is associated with the nitronyl nitroxide group (N–O group). The role of the rest of the molecule is to ensure the appropriate overlap of the correct orbitals on neighbouring molecules to produce three-dimensional ferromagnetism. A whole series of materials which incorporates this N–O group has been synthesised.

The first direct evidence of spontaneous magnetic order in *p*-NPNN was from μSR measurements (Le *et al.* 1993). It has subsequently been confirmed by zero-field neutron diffraction (Zheludev *et al.* 1994). In the left panel of Fig. 10.13 we present three

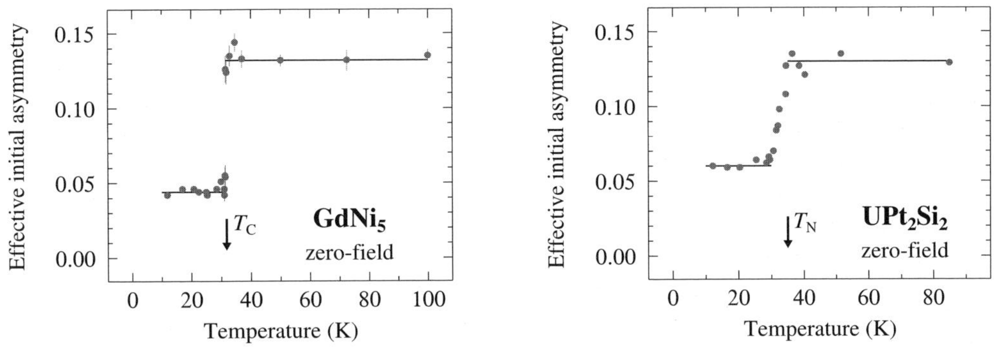

Fig. 10.12 Temperature dependence of the effective initial asymmetry $a_{\rm eff}$ measured for two polycrystalline samples. The notation $a_{\rm eff}$ is explained in Section 1.1.2. Whereas $a_{\rm eff}$ drops at T_C in a very small temperature range ($\sim$ 0.1 K) in the ferromagnet $GdNi_5$, it changes gradually near T_N (within $\sim$ 5 K) in the antiferromagnet UPt_2Si_2. The error bars in the second plot are smaller than the size of the symbols. The solid lines are guides to the eye. Reproduced with kind permission from Dalmas de Réotier and Yaouanc (1997).

zero-field μSR spectra. The mere observation of an oscillating signal at low temperature, i.e. of a spontaneous magnetic field, is a clear signature of the existence of rather long-range static magnetic correlations. Since a long-lived oscillation rather than an increase in damping is detected, *p*-NPNN orders with a well-defined magnetic structure. Measurements on single crystals indicate that $\mathbf{B}_0$ is nearly parallel to the b crystal axis, which is the easy axis (Le *et al.* 1993). The shapes of the spectra change drastically between 650 and 700 mK as seen in Fig. 10.13: the magnetic phase transition occurs between these two temperatures. The temperature dependence of B_0 is presented in the right panel. The solid line results from a fit to the function $B_0(T) \propto [1-(T/T_C)^\alpha]^\beta$. This compact formula makes it possible to discuss the spin wave ($T \ll T_C$) and the critical regimes: for $T \ll T_C$, $[B_0(0) - B_0(T)] \propto T^\alpha$ and $B_0(T) \propto (T - T_C)^\beta$ near T_C. The field $B_0(T)$ is well described in the whole ferromagnetic state by the compact formula with exponents $\alpha = 1.7\,(4)$ and $\beta = 0.36\,(5)$. The α value is consistent with the prediction of the Bloch $T^{3/2}$ law which attributes the decrease of the magnetisation, and therefore of B_0, to the thermal excitation of magnons. In three dimensions, for an isotropic antiferromagnet $\alpha = 2$ rather than $\alpha = \frac{3}{2}$ for an isotropic ferromagnet. Although indirect, this analysis indicates that the energy gap of the magnons is so small that they are still excited at this very low temperature. The uncertainty on the critical exponent β is so large that it is consistent with the theoretical predictions for the Ising, XY, and Heisenberg models as noted from Table 10.1. The weak magnetic anisotropy is accounted for by the dipolar interaction between the unpaired electron spins (Le *et al.* 1993). The small B_0 value found in the organic magnets when the temperature approaches zero is a strong indication that the magnetic moment carried by the unpaired electrons is small since it would be surprising that the dipolar field at the muon site accidentally cancels almost perfectly in all these compounds.

The transition temperature of the molecular crystals based on the nitronyl nitroxide

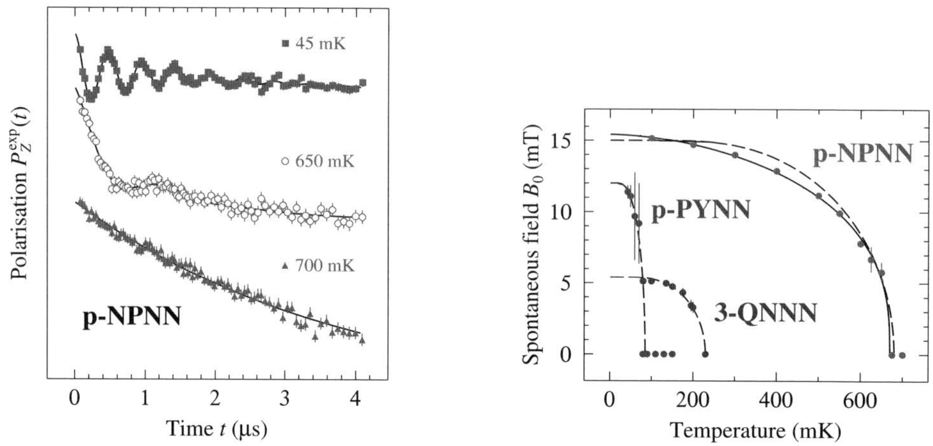

Fig. 10.13 Magnetic phase transition in organic compounds. (Left) Some zero-field spectra recorded for *p*-NPNN. The solid lines result from fits. (Right) Temperature dependence of the local spontaneous magnetic field at the muon site, B_0, in *p*-NPNN, *p*-PYNN, and 3-QNNN. The solid line is the result of a fit to the function $B_0(T) \propto [1 - (T/T_C)^\alpha]^\beta$, and the dashed lines correspond to the prediction of the molecular field model with spin $S = \frac{1}{2}$. Reproduced with kind permission from Dalmas de Réotier and Yaouanc (1997); the original data are from Blundell *et al.* (1994, 1995).

radical strongly depends on their chemical formula. This is nicely seen in the right panel of Fig. 10.13: $T_C \approx 90$ mK for *p*-PYNN, $T_C \approx 230$ mK for 3-QNNN, and $T_C \approx 670$ mK for *p*-NPNN.

The work on the molecular magnets — see also Lappas *et al.* (1995) for evidence of spontaneous ordering in a fullerene charge-tranfer salt — nicely shows the ease with which a magnetic phase transition temperature can be determined with the zero-field μSR technique. This capability has been used intensively to establish magnetic phase diagrams. As examples, we cite the phase diagram of a cuprate material by Weidinger *et al.* (1989) (see right panel of Fig. 1.12) and the determination of the iron and rare-earth long-range magnetic orders in a series of iron pnictides by Maeter *et al.* (2009).

Hoping to gain insight into the mechanism of superconductivity in high-T_c oxides, some experimentalists and theoreticians have turned their attention to systems as simple as chains and ladders of copper and oxygen atoms with the aim of applying their findings to the high-T_c superconductors where Cu–O planes play a significant role. A key prediction is that only ladders with an even number of legs have a singlet ground state separated from the triplet state by a large spin gap (Rice *et al.* 1993). However, the odd-leg systems are expected to reach a magnetically ordered ground state in the presence of inter-ladder interactions.

The $Sr_{n-1}Cu_{n+1}O_{2n}$ oxides are realisations of ladders. In Fig. 10.14 we present μSR spectra recorded for 2- and 3-leg ladders. The strong depolarisation of the zero-field spectra for the 3-leg system at low temperature shows that the ground state is

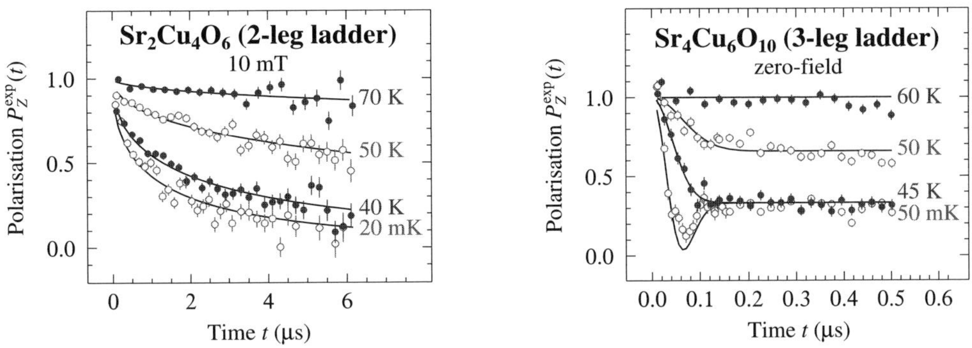

Fig. 10.14 Some spectra recorded for $Sr_2Cu_4O_6$ and $Sr_4Cu_6O_{10}$ as a function of temperature. The solid lines result from fits. Reproduced with kind permission from Dalmas de Réotier and Yaouanc (1997); the original data are from Kojima *et al.* (1995).

magnetic. Since no wiggles are detected, the disorder in the compound is important. By comparing the spectra recorded at 50 and 60 K, we infer that a magnetic transition occurs between these temperatures.

The magnetic behaviour of the 2- and 3-leg ladder systems differ remarkably as seen in Fig. 10.14. The polarisation functions for the 2-leg system are described with a square-root exponential function, appropriate for dilute fluctuating moments which may be associated with defects in the sample; see Section 6.4.1 and Eq. 6.90. Therefore no static magnetic ordering is detected.

In summary, the μSR measurements confirm the theoretical predictions that a 3-leg system has a magnetic ground state at low temperature, in contrast to a 2-leg system.

From a methodological point of view, it is worthwhile to note that the measurements described in this section were performed with powder samples and key information was extracted from the spectra without any knowledge of the muon localisation site(s).

10.3.2 Spontaneous local field at the muon site

When experimentalists begin to perform μSR experiments in magnetic materials they are usually pleased to observe a spontaneous muon spin precession. They immediately plot the spontaneous frequency versus temperature, with the idea in mind that the plot will give information on the order parameter of the phase transition. Below we provide examples showing that the physics may be more involved than sometimes assumed.

The spontaneous field in ferromagnets is often found to be collinear to the magnetisation vector **M** and it generally keeps a constant orientation relative to **M** when the temperature is changed; i.e. it is either parallel or antiparallel to **M**. This statement also holds for collinear antiferromagnets where the staggered magnetisation must be considered in place of the magnetisation. However, our interest in this section is in ferromagnets and we shall show that metallic cobalt provides a counterexample to the general rule. Figure 10.15 displays the temperature dependence of the spontaneous field measured for single crystals of metallic cobalt from 4 to 1100 K. It is deduced

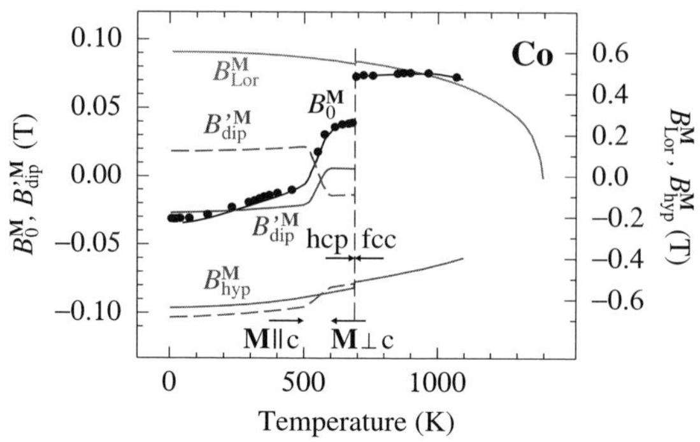

Fig. 10.15 Temperature dependence for metallic cobalt of the projection $B_0^{\mathbf{M}}$ along the local magnetisation vector of the spontaneous field (filled circles), as well as the projections of various fields contributing to $B_0^{\mathbf{M}}$, i.e. the hyperfine field $B_{\rm hyp}^{\mathbf{M}}$, the Lorentz field $B_{\rm Lor}^{\mathbf{M}}$, and the dipole field arising from moments inside the Lorentz sphere $B_{\rm dip}^{\prime \mathbf{M}}$. The solid and dashed curves for $B_{\rm dip}^{\prime \mathbf{M}}$ and $B_{\rm hyp}^{\mathbf{M}}$ refer to the octahedral and tetrahedral interstitial site assignments, respectively. The easy axis of magnetisation changes from being parallel to the c axis to a direction perpendicular to that axis in the temperature range 500 to 600 K marked by two arrows. A phase change from hexagonal-close-packed to face-centred-cubic marked by the vertical dashed line occurs at 690 K. Adapted from Graf *et al.* (1976).

from μSR spectra recorded in zero field. In order to grasp the full information from the experimental data, instead of plotting B_0, which is always positive by definition, here we plot the projection $B_0^{\mathbf{M}} \equiv \mathbf{B}_0 \cdot \mathbf{M}/M$ of $\mathbf{B}_0$ along the local magnetisation direction. While the absolute value of $B_0^{\mathbf{M}}$ is obtained from the spontaneous frequency, its sign is determined from the sign of $P_Y(t)$ at short times, knowing that the muon spin rotates clockwise when looking at $\mathbf{B}_0$ head on; see Section 3.1.1. This determination requires the measurements to be performed on a polarised sample, i.e. a sample with macroscopic magnetisation; see comment at the beginning of Section A.2.

As seen in the figure, the thermal behaviour of $B_0^{\mathbf{M}}$ is complicated. Three regimes can be distinguished. In the first one, from 4 to 450 K, $B_0^{\mathbf{M}}$ is negative at low temperature and its absolute value decreases smoothly from 4 to 450 K. In this temperature region, the easy axis of magnetisation is along the hexagonal c axis. In the second region extending from 550 to 680 K, $B_0^{\mathbf{M}}$ is positive and increases smoothly. The gap in the data points between 450 and 550 K corresponds to a temperature range where no μSR signal is detected. A sudden jump to high field occurs at 690 K, a temperature which corresponds to a structural phase transition from hexagonal-close-packed to face-centred-cubic. In the third region, above 690 K, only a weak temperature dependence of $B_0^{\mathbf{M}}$ is found.

As explained in Section 5.4, in zero field, the spontaneous field of an unmagnetised ferromagnetic sample is the sum of the hyperfine field, the Lorentz field and the dipole field arising from moments inside the Lorentz sphere (Eq. 5.81). This last field can be computed for a given muon site. The results are shown in the figure. The hyperfine field is then deduced by subtracting $\mathbf{B}_{\rm Lor}$ and $\mathbf{B}'_{\rm dip}$ from $\mathbf{B}_0$. Since there is no reason for

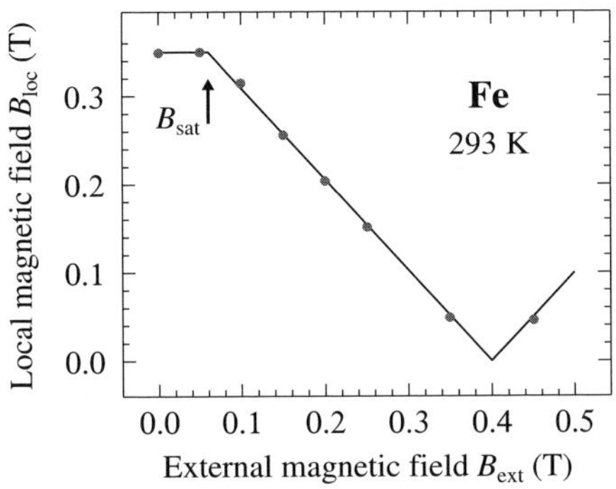

Fig. 10.16 The local field at the muon site versus the external field measured on a single crystal of iron at 293 K. The change of slope around $B_{ext} = 0.4$ T is due to the fact that $\mathbf{B}_0$ is antiparallel to $\mathbf{B}_{ext}$ when $B_{ext} < B_0$ and parallel at a higher external field. Adapted from Denison *et al.* (1979).

a sudden change in the thermal variation of B^M_{hyp} within the hexagonal-close-packed (hcp) phase, the muon site is inferred to be the octahedral site. Note that we have used the generic notation $\mathbf{B}_{hyp}$ for the hyperfine field. Here, in a metal, this is more specifically the field $\mathbf{B}_{RKKY}$, which was introduced in Section 5.1.3.

In conclusion, these results on metallic cobalt illustrate an ubiquitous feature of μSR: the dipole contribution to the local field at the muon is far from being negligible.

The external field dependence of the local field at the muon provides another way for the determination of the orientation of $\mathbf{B}_0$ with respect of the local domain magnetisation. In Fig. 10.16 we present B_{loc} versus B_{ext} for metallic iron. Note that we have not used here the notation B_0: the measured field is indeed not purely spontaneous since an external field is applied. The fact that B_{loc} is independent of B_{ext} for $B_{ext} < B_{sat}$, where B_{sat} is the saturation field, finds an explanation in the mechanism of magnetic shielding; see Section 5.4. When increasing B_{ext} the bulk magnetisation increases up to the saturation magnetisation, $\mathbf{M}_{sat}$, at $B_{ext} = B_{sat}$. Above saturation Eq. 5.79 becomes

$$\mathbf{B}_{loc} = \mathbf{B}_{ext} + \mathbf{B}'_{dip} + \mathbf{B}_{Lor} + \mathbf{B}_{sat} + \mathbf{B}_{hyp}, \tag{10.291}$$

because $\mathbf{B}_{dem} = -\mu_0 \boldsymbol{N} \mathbf{M}_{sat} \equiv \mathbf{B}_{sat}$. In iron, it is believed that the muon diffuses sufficiently fast that $\mathbf{B}'_{dip}$ averages out to 0 (Seeger 1978). Since $\mathbf{B}_{Lor}$, $\mathbf{B}_{sat}$, and $\mathbf{B}_{hyp}$ are field independent, Eq. 10.291 predicts $\mathbf{B}_{loc}$ to vary linearly with $\mathbf{B}_{ext}$. This variation is indeed experimentally observed for $B_{sat} < B_{ext} \lesssim 0.4$ T: see Fig. 10.16 which displays the modulus B_{loc} of $\mathbf{B}_{loc}$ versus B_{ext}. The fact that B_{loc} is found to decrease for $B_{ext} \gtrsim B_{sat}$ is a proof that the spontaneous field $\mathbf{B}_0$, i.e. the field at the muon site in the absence of external field, is antiparallel to $\mathbf{B}_{ext}$ and therefore to the magnetisation. In addition $\mathbf{B}_{hyp}$ can be deduced from the measurements since $\mathbf{B}_{sat}$ and $\mathbf{B}_{Lor}$ are known.

Values of spontaneous and hyperfine fields measured for five elemental ferromagnetic metals are listed in Table 10.3. In addition, we give the quantity $r_\mu H/(4\pi)$

Table 10.3 Component along the local magnetisation orientation of the spontaneous and hyperfine fields measured at low temperature for some elemental ferromagnetic metals. In addition, we list the parameter $r_\mu H/(4\pi)$ which is used to describe the spin–lattice relaxation rate for some models. The two values of $B_{\rm hyp}$ for dysprosium reflect the fact that the muon site is unknown for that metal. Data from Denison *et al.* (1979).

	Fe	Co	Ni	Gd	Dy
$B_0^{\rm M}$ (T)	−0.367 (10)	−0.0317 (10)	+0.148 (1)	+0.110 (5)	±1.23 (2)
$B_{\rm hyp}^{\rm M}$ (T)	−1.11 (2)	−0.61 (2)	−0.071 (1)	−0.698 (10)	−2.52 (10) or −0.07 (10)
$\frac{r_\mu H}{4\pi}$ (-)	−0.51 (1)	−0.338 (10)	−0.111 (2)	−0.728 (10)	−0.68 (4) or −0.02 (3)

which is proportional to the hyperfine field (see Section 5.2.4) and which appears in Section 10.2 for some expressions of λ_Z. It is remarkable that the hyperfine field is antiparallel to the bulk magnetisation for the first four metals of the table, and likely to be so for the last one, i.e. dysprosium. As already pointed out in Section 5.1.3 and discussed in some details by Denison *et al.* (1979), $B_{\rm hyp}$ does not reflect the intrinsic magnetisation density at the muon site; it is indeed strongly perturbed by the muon electric charge. The measured values of $B_{\rm hyp}$ in metallic iron, nickel, and cobalt have reasonably been reproduced using a finite cluster calculation and computing the electronic structure in the framework of the self-consistent local density theory (Ellis and Lindgren 1984). Remarkably, the screening radius of the muon charge is rather small: a calculation for iron shows that charge neutrality is attained within a radius about 20% larger than the Bohr radius.

The hyperfine field may not scale with the order parameter of a metallic magnetic compound. An example is given in Fig. 10.17 where we compare for iron $B_{\rm hyp}(T)$ with $M_{\rm sat}(T)$ (this is the order parameter for a ferromagnet). To quantify the deviation of $B_{\rm hyp}(T)$ from $M_{\rm sat}(T)$, it is convenient to consider the ratio

$$\frac{B_{\rm hyp}(T)}{B_{\rm hyp}(0)} = A(T)\frac{M_{\rm sat}(T)}{M_{\rm sat}(0)}. \tag{10.292}$$

In the right panel of the same figure we display $[A(T) - 1]$ for nickel, iron, and cobalt (hcp). The hyperfine field $B_{\rm hyp}(T)$ does not track $M_{\rm sat}(T)$ for these metals. This is attributed to the zero-point motion of the muon; see Section 10.1. Note that if the hyperfine field is much smaller than the dipole field at the muon site, $B_0(T)$ is expected to track the temperature dependence of the order parameter.

10.3.3 Investigation of magnetic domains

The μSR techniques have been intensively used to detect magnetic domains (or regions) embedded in a compound. These domains may be intrinsic; i.e. their presence does not

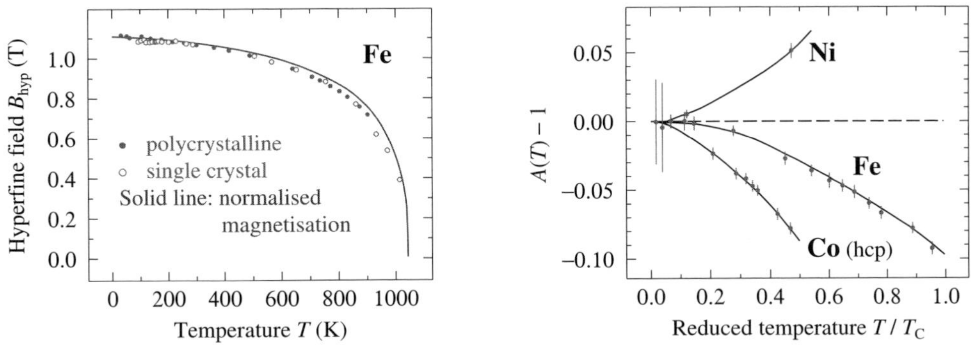

Fig. 10.17 The local field at the muon site in a few elemental ferromagnets. (Left) $B_{\rm hyp}(T)$ compared with normalised $M_{\rm sat}(T)$ for iron; adapted from Denison *et al.* (1979). (Right) $A(T) - 1$ versus the temperature for three elemental metallic ferromagnets. The parameter $A(T)$ is defined in Eq. 10.292. The lines are guides to the eye. Adapted from Nishida *et al.* (1978).

necessarily reflect the poor quality of a sample. Usually the polarisation function is modelled as a weighted sum of at least two components, one of them being associated with the magnetic domains and the other with the non-magnetic volume of the sample. Implicit here is the assumption that the volume occupied by the domain walls is negligible. In the simplest case, the magnetic domains are characterised in zero field either by wiggles or by a polarisation function of the Kubo–Toyabe type (see Eq. 6.17) with a large variance, reflecting the magnetic disorder. The second component may also be described by a Kubo–Toyabe function, this time with a small variance arising from the nuclear magnetic moments. The relative amplitude of each component gives an estimate of the relative volume of the sample associated with each of them.

An example is shown in Fig. 10.18. These data distinguish the volume (it does not have to be connected) in which electronic magnetic moments are in the quasi-static limit from the volume for which the fluctuation rate of the magnetic moments is fast. The polarisation function modelling these latter moments is the Kubo–Toyabe function with strong dynamics accounted for by the strong-collision model which is described in Section 6.3.2.

An alternative method for evidencing the coexistence of such domains in a compound is through transverse-field measurements in a weak field on the order of 10 mT. As in zero field, the response is again split into two components which now oscillate. The one related to the magnetic domains is more rapidly damped and the other exhibits wiggles on a much longer time range. It may occur that the field distribution arising from the magnetic region is so broad that the associated oscillations are damped out before the muon signal is observed. As in the case of zero-field measurements the ratio of the signal amplitudes gives a good estimate for the volume fraction of the different domains.

As another example of the ability of the zero-field μSR technique to probe the homogeneity of a sample, we mention the detection of a minority (3% in volume)

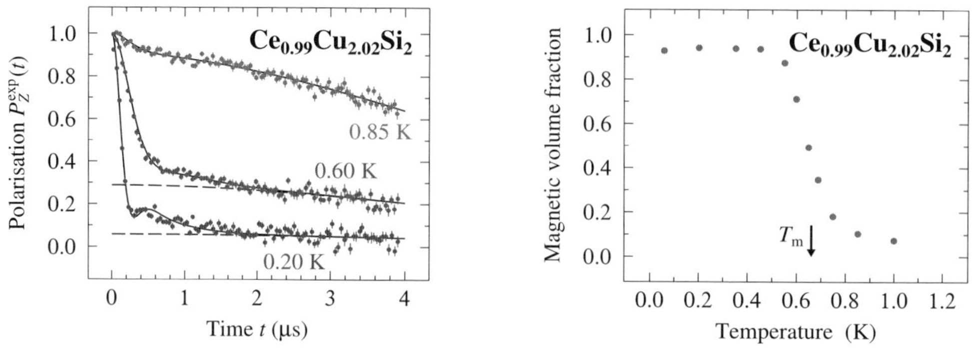

Fig. 10.18 Measurement of the magnetically ordered volume fraction for $Ce_{0.99}Cu_{2.02}Si_2$. (Left) Time dependence of the zero-field polarisation function at three temperatures. Note the pronounced two-component structure of the signal, which is composed of a slowly (dashed line) and a rapidly decaying component, corresponding to paramagnetic and magnetic domains, respectively. The curves are results of fits. (Right) Temperature dependence of the relative volume of the magnetic component. The arrow gives an indication of the temperature at which the broad transition to the magnetically ordered state is estimated to occur. Adapted from Feyerherm *et al.* (1997).

magnetic phase in a sample of $YBa_2Cu_3O_{6.6}$ used previously for neutron diffraction experiments (Sonier *et al.* 2009). This work suggests the reported extremely weak neutron magnetic Bragg reflections to only be a signature of a minority phase.

At this point we stress a key difference between the information obtained from neutron or X-ray scattering on one hand and μSR experiments on the other, when they are performed in samples consisting of fractions of magnetic and non-magnetic volumes. Basically, the intensity of a magnetic Bragg reflection is proportional to the product of the magnetic volume illuminated by the neutron or X-ray beam and the square of its (staggered) magnetisation. It is not possible to separate the two contributions to the intensity. On the other hand, the muons normally probe directly the order parameter (at least if $\mathbf{B}_{hyp}$ can be neglected) through the spontaneous field and the magnetic volume fraction through the related asymmetry or signal amplitude. In fact, it might be interesting to compare the scattering intensity to the product of the asymmetry and the square of the spontaneous field given by μSR. In a quite similar way, bulk magnetisation data measured with a magnetometer in collinear ferromagnets give access to the sum of the magnetic moments in the sample. This is therefore the product of the magnetisation in the magnetic regions times their volume. A comparison can then be performed with the product of the asymmetry and the modulus of the spontaneous field. This is further discussed, for example, by Brooks *et al.* (2005).

We have considered the case of phase separation in strongly correlated electronic systems. Other types of magnetic domains are expected in the intermediate state of superconductors and for diamagnets. Here we present an example of this latter case, that is, the detection of Condon domains in beryllium.

In general terms, we first note that the relation between induction, external field, and uniform magnetisation for a given material is given by the relation

$$B = B_{\text{ext}} - \mu_0(N^Z - 1)M, \tag{10.293}$$

where N^Z is a demagnetisation coefficient. It is assumed for simplicity that the sample is ellipsoidal and that the Z axis is parallel to one of the ellipsoid principal axes. Therefore, according to Eq. 5.37 (see also Fig. 5.2), if $\mathbf{B}_{\text{ext}}$ is applied along the revolution axis of an extremely oblate ellipsoid, i.e. in practice if $\mathbf{B}_{\text{ext}}$ is perpendicular to a plate-like sample with extremely small thickness relative to radius, Eq. 10.293 yields $B = B_{\text{ext}}$; i.e. the induction should follow the external field.

It is well known that the magnetic response of a metal at high field oscillates as a function of B_{ext}. This is the de Haas–van Alphen (dHvA) effect, which is used to study Fermi surfaces (Shoenberg 1984). Condon (1966) — see also the review of Gordon *et al.* (2003) — noted that if the oscillating amplitude is sufficiently large within some part of each dHvA oscillation period, i.e. for a given B_{ext} range, the conduction electron states in this range are thermodynamically unstable and cannot follow the electrodynamics relation $B = B_{\text{ext}}$. Therefore the electronic system should jump periodically over the forbidden intervals of B_{ext}. A compromise could be the splitting of the magnetic energy into alternating diamagnetic (with induction smaller than B_{ext}) and paramagnetic (with induction larger than B_{ext}) domains. With such a domain structure the relation $B = B_{\text{ext}}$ is fulfilled as an average over the sample. The so-called Condon domains are spectacular manifestations of the collective behaviour of electrons in quantised cyclotron orbits, i.e. Landau states.

The first direct observation of the Condon domain formation was carried out on silver by NMR measurements (Condon and Walstedt 1968). One had to wait another 28 years for a second report on the domain formation: using transverse-field μSR measurements, Solt *et al.* (1996) have reported the observation of domains in beryllium. Interestingly, the prediction of Condon concerned beryllium.

In Fig. 10.19 the measured local field(s) at the muon site is presented as a function of the external field in a small field region around $B_{\text{ext}} \simeq 2.739\,\text{T}$ at which the magnetisation is not homogeneous throughout the sample. Two fields are observed in the central region of the plot. As expected, within a given field range, B_{loc}, which is a measure of the induction, does not follow B_{ext}. The diamagnetic and paramagnetic signals are clearly seen, their respective populations changing smoothly as B_{ext} increases. Interestingly, the domains could not be visualised by NMR in beryllium because of the excessive line broadening produced by the electric field gradient. Since the muon does not have a quadrupole moment, it does not couple to this gradient. The Condon domains have been detected by a point-like probe such as the muon because the volume occupied by the domains is much larger than the volume of the domain walls: at $B_{\text{ext}} = 1\,\text{T}$, the domain and wall sizes are believed to be $\sim$30 and $\sim 1\,\mu$m, respectively. Note that the observation of the domains means that they are pinned (static) within the time scale of the muon experiment ($\sim 1\,\mu$s). For a detailed discussion of the Condon domains in beryllium as revealed by μSR, we refer to the paper of Solt and Egorov (2002).

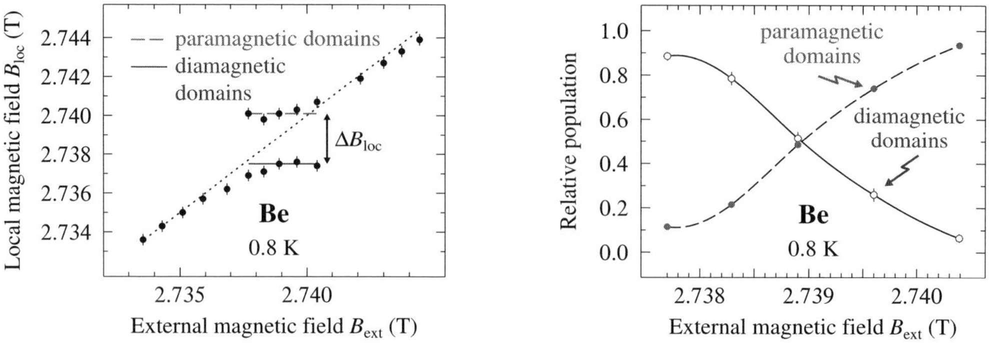

Fig. 10.19 Evidence for Condon domain formation in beryllium at 0.8 K. The external field is applied perpendicular to a plate-like sample. (Left) B_{loc} versus B_{ext}. The difference between the local fields in the paramagnetic and diamagnetic domains is clearly resolved. For reference the line of equation $B_{loc} = B_{ext}$ is plotted (dotted line). (Right) Relative populations of the paramagnetic and diamagnetic domains versus B_{ext}. The lines are guides to the eye. Reproduced with kind permission from Dalmas de Réotier and Yaouanc (1997); the original data are from Solt *et al.* (1996).

The splitting of the two local fields in the dHvA oscillating region, i.e. ΔB_{loc} in the left panel of Fig. 10.19, is the order parameter of a phase transition within the conduction electrons. It involves orbital currents rather than spin moments, as for a usual magnetic phase transition. The size of ΔB_{loc} gives information on the topology of the Fermi surface (Gordon *et al.* 1998).

10.3.4 Spin dynamics

A nice advantage of μSR compared to NMR is the possibility of easily performing measurements in zero field, thus avoiding the possible influence of the applied field. Together with the high sensitivity of the technique, this characteristic probably explains the large number of experimental μSR work devoted to the study of spin dynamics.

We first start by discussing four examples of spin dynamics studies in compounds exhibiting conventional second order magnetic phase transitions. Next a description of the spin dynamics observed above the freezing temperature for a spin-glass compound is given. It is followed by comments concerning the possible application of the methods developed for analysing spin glasses to other magnetic systems. Finally we highlight two typical examples of the investigation of the dynamics in compounds for which the other experimental techniques do not detect any signature of a magnetic phase transition. The interpretations of the data will be based on the material contained in Section 10.2.

As already noted in Section 10.3.1, $GdNi_5$ is an intermetallic hexagonal ferromagnet with a Curie temperature of $T_C \simeq 30.85$ K. Some μSR experimental data were already presented in Section 9.4.1. It is an axial magnet — with c axis as the easy axis — characterised by an extremely weak magnetic anisotropy (the anisotropy field is $B_{ani} = 0.21$ T at 5 K) arising from the dipole interaction between the Gd^{3+} magnetic moments

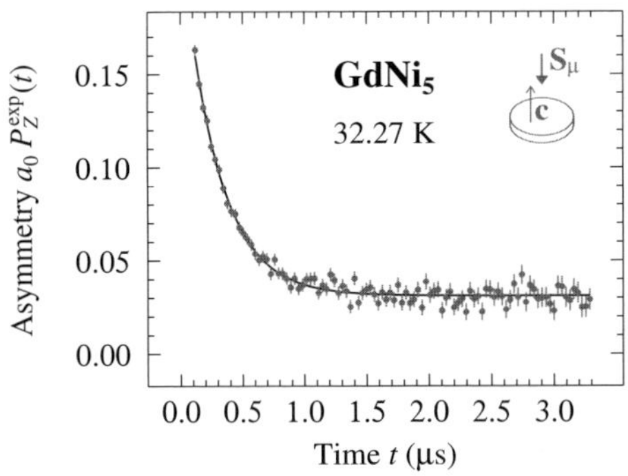

Fig. 10.20 A typical zero-field μSR spectrum recorded in the paramagnetic phase of a single crystal of $GdNi_5$. The initial muon beam polarisation is parallel to the c axis of the crystal structure. The measured asymmetry, recorded with a pulsed muon source, is the sum of two components, a time-dependent asymmetry proportional to the polarisation function of the compound under study, and a constant asymmetry due to the muon implanted in the environment of the compound: mainly sample holder and cryostat wall. A better design of the environment has subsequently led to an almost complete suppression of the second component: this is the so-called fly-past configuration discussed in Section 2.2.2.1. Reproduced with kind permission from Yaouanc *et al.* (1996).

which are only of spin origin. Zero-field μSR measurements were carried out between 5 K and room temperature on two single crystals which differ by the orientation of the c axis relative to the initial muon beam polarisation, $\mathbf{S}_\mu$, which is either parallel or perpendicular to that axis (Yaouanc *et al.* 1996). The spectra were recorded using a pulsed source where the expected coherent oscillations of $P_Z(t)$ below the Curie temperature cannot be observed; see beginning of Section 10.3.1. From the signal amplitudes observed for the two crystals, it was deduced that the spontaneous field is parallel to the c axis. All the spectra with a non vanishing asymmetry could be analysed assuming an exponential relaxation function. An example is displayed in Fig. 10.20.

The overall thermal and orientation behaviour of the muon spin–lattice relaxation rate λ_Z is presented in the left panel of Fig. 10.21.

Far above T_C, λ_Z is temperature independent, reflecting the vanishing of the pair correlations at high temperature. If the coupling of the Gd spin with the conduction electron spin were strong enough, we would observe a decrease of λ_Z which would be proportional to the inverse temperature. This is not the case experimentally, reflecting the relatively low density of states at the Fermi level (Mulders *et al.* 2003). The difference between the levels of relaxation reached for measurements performed with $\mathbf{S}_\mu$ parallel and perpendicular to the c axis is explained by the anisotropy of the coupling tensor between the muon spin and the Gd^{3+} spins.

Below T_C, the muon spin relaxation, which is only observed if $\mathbf{S}_\mu$ is parallel to the c axis, is driven by spin waves. As explained in Section 10.2.4.1, due to energy conservation, a relaxation process involving a single magnon is not possible. Instead,

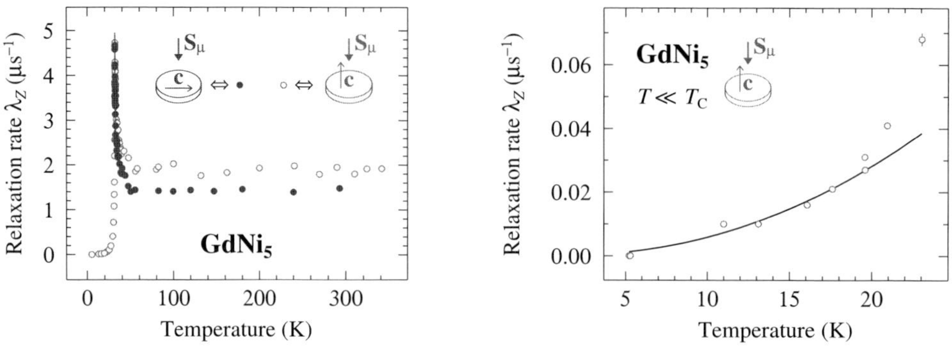

Fig. 10.21 The muon spin–lattice relaxation rate measured for the hexagonal ferromagnet $GdNi_5$. (Left) An overview of the zero-field longitudinal relaxation rate. The two sets of data refer to measurements performed on two crystals which differ in the orientation of $\mathbf{S}_\mu$ relative to the c axis. (Right) The relaxation rate in the ordered state. The solid line is the prediction for the relaxation induced by the Raman spin-wave process. Reproduced with kind permission from Yaouanc *et al.* (1996) and Dalmas de Réotier and Yaouanc (1997).

a two-magnon process (i.e. Raman process) is allowed. Accordingly, λ_Z follows the $T^2 \ln T$ law predicted for such a process, at least below about two-thirds of T_C. The interaction between magnons is indeed expected to be negligible in this temperature range. From a fit to Eq. 10.218 (see right panel of Fig. 10.21), a magnon stiffness constant of $D_{FM} = 3.2\,(1)\,\mathrm{meV\AA^2}$ is extracted. It is in the expected range.

The experimental investigation of critical spin dynamics at a second order magnetic phase transition enables the test of the hypothesis of dynamical scaling which is an extension of the static scaling theory for phase transitions. At the beginning of the eighties the situation for isotropic Heisenberg ferromagnets was frustrating since the available theory was not able to explain a wealth of data obtained by local spin probes and neutron diffraction techniques. It is only after a proper treatment of the dipole interaction between the magnetic ions, in the framework of a mode-coupling theory, that a consistent picture emerged; see the excellent review of Frey and Schwabl (1994). The analysis revealed the crucial importance of distinguishing the longitudinal from the transverse fluctuations (relative to the wave vector). The zero-field μSR technique has played a key role in this matter. The theoretical framework treating the muon spin relaxation has been detailed in Section 10.2.3.3 and we give now an example of it.

The relaxation rate measured by μSR near T_C for a dipolar Heisenberg ferromagnet such as $GdNi_5$ is expressed as a weighted sum of two functions accounting for the longitudinal and transverse fluctuations, respectively. The longitudinal fluctuations saturate as T_C is approached from above, whereas the transverse fluctuations diverge. It happens that the longitudinal fluctuations are better probed by μSR since λ_Z is dominated by these fluctuations in the temperature range experimentally accessible. A good proof is given by the plots of the left panel of Fig. 10.22. The experimental data are fitted to Eq. 10.190 with $a_{T_1}/(a_{T_1} + a_L) = 0.021$ and $a_{T_1} = 0$ for the full and dashed lines, respectively. In the right panel of the same figure is shown $\lambda_Z(T)$

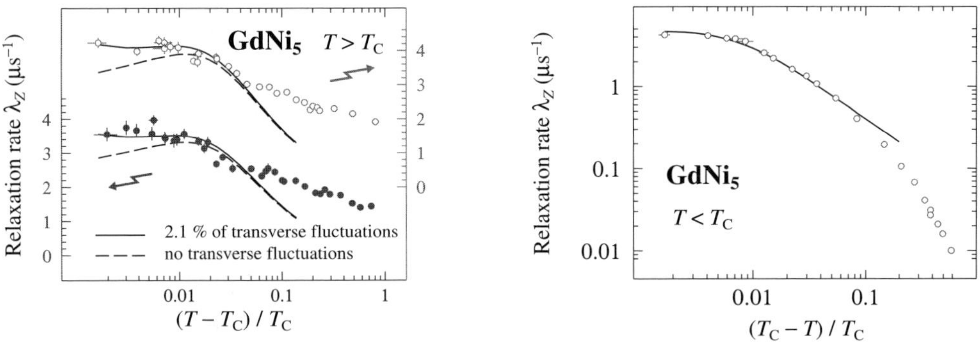

Fig. 10.22 The critical spin dynamics of $GdNi_5$ measured in zero field. (Left) λ_Z recorded in the paramagnetic phase given as a function of the relative temperature normalised to T_C and the orientation of $\mathbf{S}_\mu$ relative to the c axis (same symbol convention as in Fig. 10.21). The full lines are predictions of the mode-coupling theory for the paramagnetic critical behaviour of $\lambda_Z(T)$ in a dipolar Heisenberg ferromagnet. Only 2.1% of transverse fluctuations contribute to the measured relaxation. (Right) λ_Z in the ferromagnetic state versus the relative temperature normalised to T_C. The full line is the prediction for the critical paramagnetic fluctuations with a renormalised correlation length. For both plots the model used to fit the data derives from critical behaviour theory and consequently it is expected to break down outside the critical regime, which is indeed observed. Reproduced with kind permission from Yaouanc *et al.* (1996).

near T_C in the ferromagnetic state. The observed similarity between the paramagnetic and the ferromagnetic critical dynamics is remarkable. This is a strong support for the dynamical scaling theory of critical dynamics put forward by Halperin and Hohenberg (1967). Overall, these data provide one of the strongest experimental support for the mode-coupling description of critical dynamics for a dipolar Heisenberg ferromagnet.

The increase of λ_Z near the magnetic phase transition of $GdNi_5$ is relatively modest. In fact near T_C it saturates, in contrast to the expectation of slowing down of the magnetic fluctuations near a second order phase transition. Because of the dipole interaction, λ_Z, which is mostly driven by the longitudinal fluctuations, becomes temperature independent. This effect is not expected for an antiferromagnet. As an illustration we discuss the first detailed study of critical fluctuations by μSR techniques (de Renzi *et al.* 1984). It dealt with the rutile structure insulator MnF_2 which exhibits an antiferromagnetic phase transition at $T_N = 67.34$ K. The measurements were carried out at the beginning of the eighties at the European organization for nuclear research (CERN) where it was quite difficult to perform zero- or longitudinal-field measurements. The thermal behaviour of the transverse relaxation rate is displayed in Fig. 10.23. Relative to a ferromagnet such as $GdNi_5$, the variation of the rate for MnF_2 is larger near the ordering temperature. An analysis of the expression for λ_X predicts that $\lambda_X(\mathbf{B}_{\rm ext} \parallel a)/\lambda_X(\mathbf{B}_{\rm ext} \parallel c) = \frac{1}{2}$ near T_N. From dipole sums, one computes $\lambda_X(\mathbf{B}_{\rm ext} \parallel a)/\lambda_X(\mathbf{B}_{\rm ext} \parallel c) = 1.4$ far above T_N. This implies a crossover, which is effectively observed. The critical exponent is found to be $\varpi = 0.8$ (2). There are

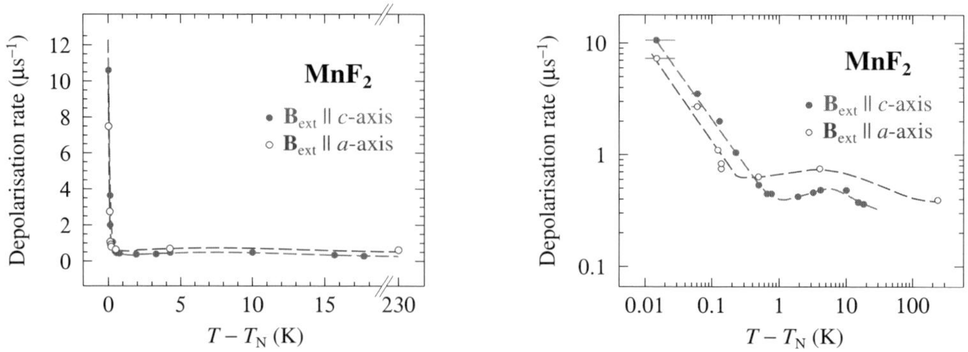

Fig. 10.23 Transverse depolarisation rate for MnF_2 versus temperature above the Néel temperature measured in an external field of 0.3 T. For $T - T_N \geq 1$ K, the data have been fitted to the product of an exponential and a Gaussian function. The former function accounts for the electronic magnetic moment fluctuations and the latter for the field distribution arising from the nuclear moments. The data which have been recorded for two orientations of $\mathbf{B}_{\text{ext}}$ relative to the crystal axes, display an anisotropy and an inversion around $T - T_N = 0.4$ K. The two panels show the same data with different scales: linear or logarithmic. Adapted from de Renzi *et al.* (1984).

two reasons for this large uncertainty on ϖ. An important fraction of the muons forms muonium centres and therefore does not probe the spin dynamics of MnF_2 (Uemura *et al.* 1986). In addition, the sample mass was about three orders of magnitude larger than today's samples. Therefore the authors probably had difficulties obtaining a homogeneous and stable sample temperature. The measured exponent is surprisingly consistent with the value expected for a three-dimensional Ising antiferromagnet; see Eq. 10.170 and Tables 10.1 and 10.2. Later on, results obtained from zero-field measurements confirmed the presence of a strong anisotropy in the spin dynamics of MnF_2 (Uemura *et al.* 1986).

Given the expected isotropic magnetic character of the Mn atom in MnF_2, the observed anisotropy of the spin dynamics reflects the anisotropy of the coupling of the muon spin to its environment. On the other hand, there are compounds for which a large anisotropy is expected because of the magnetic anisotropy of the magnetic ions. As an example, we refer to the work of Gubbens *et al.* (1994) for the ternary intermetallic compound $NdRh_2Si_2$ which crystallises in the tetragonal $ThCr_2Si_2$-type structure. This compound exhibits a single antiferromagnetic phase transition at $T_N \simeq 57$ K. At low temperature the Nd^{3+} magnetic moments are aligned along the c crystallographic direction: they form ferromagnetic ab planes which are antiferromagnetically stacked along the c axis. Single-crystal measurements show that λ_Z recorded for the initial muon beam polarisation $\mathbf{S}_\mu$ parallel to the c axis is augmented by less than 50% when approaching T_N from above. On the other hand, when $\mathbf{S}_\mu$ is perpendicular to the c axis λ_Z increases by about a factor of 170 in the accessible temperature range.

The application of an external pressure on a compound may lead to the discovery of exotic states of matter. Numerous examples are found in the fields of strongly

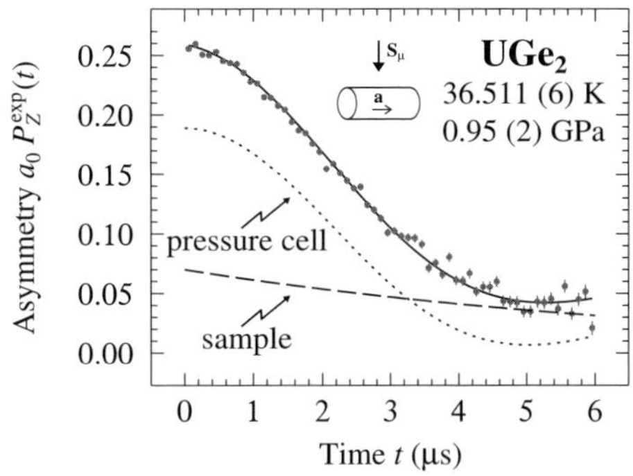

Fig. 10.24 UGe_2 spectrum recorded in zero field under a pressure of 0.95 (2) GPa at 36.511 (6) K. The corresponding reduced temperature is $\tau \equiv (T - T_C)/T_C = 0.0007\,(3)$. The sample is a single-crystalline cylinder whose a axis, i.e. the easy axis, is oriented perpendicular to the initial muon beam polarisation. The full line is the result of a fit to a sum of two components accounting for the response of the sample — exponential function, dashed line — and of the pressure cell — Kubo–Toyabe function, dotted line — respectively. Reproduced with kind permission from Sakarya *et al.* (2010).

correlated metals and superconductors. The interest is to probe such systems under pressure at any temperature down to well below 1 K. The number of such μSR measurements is still restricted. Until now, studies have been reported for applied pressures up to 1.5 GPa and down to 0.235 K. The sample under study is inserted in a pressure cell which itself is introduced in a cryostat. The wall thickness of the pressure cell being on the order of 1 cm, a high momentum muon beam is needed to ensure that a sizeable fraction of the muons is implanted into the sample; see Section 2.2.2.2. The contribution to the measured signal from muons stopped in the cell is nevertheless appreciable. Therefore the muon response of the cell should be properly taken into account in the fitting procedure. In Fig. 10.24 we give an example which concerns the intermetallic UGe_2. It is a ferromagnet with a Curie temperature $T_C \simeq 54\,$K at ambient pressure. Its main interest comes from its superconducting phase observed at low temperature under high pressure. It has been argued that the spin fluctuations could be involved in the formation of the Cooper pairs; hence, there is interest to investigate them. Although only about 30% of the measured asymmetry arises from UGe_2, it is still possible to characterise its critical spin dynamics at high pressure. Thanks to further developments, it is now possible to perform measurements at high pressure for which about half of the signal originates from muons implanted into the sample.

The intensive investigation of glass and spin-glass forming systems in the eighties has shown that the key to a comprehensive understanding of the transition lies in the dynamics (Fisher and Hertz 1991, Mydosh 1993). The spin self-correlation function is the most important quantity in the spin-glass systems since cross-correlations are small; i.e. one can neglect the wave vector dependence of the correlation function. Therefore for an isotropic spin-glass system one should only consider the correlation

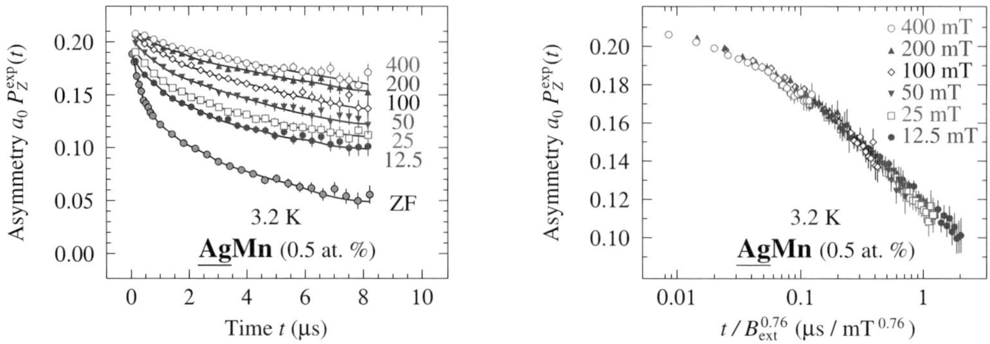

Fig. 10.25 Spin dynamics of the spin-glass AgMn (0.5 at. %). (Left) The dependence of the μSR spectra on the intensity of the longitudinal magnetic field expressed in units of milliteslas. The measurements have been performed at 3.2 K, i.e. above the spin-glass transition temperature ($T_{\rm g} = 2.95$ K). The solid lines are guides to the eye. (Right) The same spectra, but plotted as a function of the scaling variable $t/B_{\rm ext}^{0.76}$ and for the same values of $B_{\rm ext}$ except zero. Reproduced with kind permission from Dalmas de Réotier and Yaouanc (1997); the orginal data are from Keren *et al.* (1996).

function $\Lambda(t, \mathbf{q} = \mathbf{0})$, the functional form of which can be determined from the field dependence of $P_Z(t)$ as explained in Section 10.2.2: at sufficiently high field $P_Z(t, B_{\rm ext})$ is expected to follow the scaling law $P_Z(t/B_{\rm ext}^{\gamma})$ where the exponent γ characterises the functional form of $\Lambda(t, \mathbf{q} = \mathbf{0})$.

A good example is provided by Keren *et al.* (1996). The measurements were performed for the canonical Heisenberg spin-glass system AgMn (0.5 at. %). In the left panel of Fig. 10.25 is presented the field dependence of the spectra recorded at 3.2 K, i.e. just above the spin-glass transition temperature $T_{\rm g}$. In the right panel, the validity of the scaling law is demonstrated for $\gamma = 0.76$ (5) over more than two orders of magnitude in $t/B_{\rm ext}^{\gamma}$. According to Eq. 10.85, an instantaneous relaxation should occur as $B_{\rm ext} \rightarrow 0$ since then $P_Z(t) \rightarrow 0$ at any t. This is not experimentally observed. However, as $B_{\rm ext} \rightarrow 0$ the exponential term of the cutoff power law (Eq. 10.86) provides a cutoff. Therefore, only this law describes the field dependence of $P_Z(t)$. This function appears as an intrinsic, homogeneous feature of spin glasses (Pappas *et al.* 2003). This conclusion is achieved assuming no specific functional form for $P_Z(t)$.

A complementary approach is to test a functional form for $P_Z(t)$. In the same AgMn (0.5 at. %) alloy, it has been found that the stretched-exponential function given in Eq. 6.94 provides an adequate representation of the relaxation function at various temperatures around $T_{\rm g}$ when the system is cooled down in a longitudinal field of 12.5 mT. The stretched exponent $\beta_{\rm se}$ is temperature dependent and saturates to a value close to $\frac{1}{3}$ as the temperature is lowered towards $T_{\rm g}$. The stretched exponential form for $P_Z(t)$ is further discussed in Section 6.4.1.2.

It is important to note that a scaling law such as $P_Z(t, B_{\rm ext}) = P_Z(t/B_{\rm ext}^{\gamma})$ is not only a characteristic of spin glasses. For example, it has been also observed for the structurally ordered non-Fermi liquid $\rm YbRh_2Si_2$ for which the longitudinal polarisation function is an exponential function, a signature of homogeneous spin dynamics (Ishida *et al.* 2003).

Even for a non spin-glass system, it is quite common in the published literature to find fits of data recorded in a zero or longitudinal field to a stretched-exponential function. Usually it should simply be taken as a way of parametrising the data without any physical significance. It can even be misleading. For instance, the intrinsic exponential nature of the relaxation function for a single crystal with anisotropic response may lead to a stretched-exponential function when the powder average is performed (Pinkpank *et al.* 2000). Even worse, a compound with schematically two magnetic regions may sometimes be analysed with a stretched-exponential function, overshadowing the intrinsic inhomogeneity. This has been shown by combining μSR with neutron spin echo (Heffner *et al.* 2000) or NMR (Allodi *et al.* 2001) measurements.

Systems at the border line of exhibiting a magnetic phase transition are of great interest. The goal is to understand the reasons for which a given compound displays a magnetic phase transition while another compound with *a priori* similar properties does not. Related to this sensitivity of magnetism to the fine tuning of a physical parameter is the general problem of the relation between magnetic and superconducting ground states. Since the magnetic moments for the compounds of interest in this field are usually small, the sensitivity of μSR to weak moments is a definitive advantage.

In this respect, we first discuss results obtained for the ternary tetragonal strongly correlated system $CeRu_2Si_2$. This is a so-called heavy-fermion metal: its low-temperature specific heat $C_p = \gamma_{\rm Som} T$ is characterised by a particularly large Sommerfeld coefficient $\gamma_{\rm Som}$. This means that the electronic correlations are intense at low temperature. However, no magnetic phase transition has been detected by any experimental technique. Zero-field μSR experiments performed at low temperature indicated the occurrence of quasi-static magnetic correlations (Amato 1994). In Fig. 10.26 we present high statistic relaxation rate measurements recorded at a pulsed muon source which confirm the first indication of the build-up of magnetic correlations below about 1 K. Note that an extremely small magnetic field quenches λ_Z. Interestingly, the figure shows that a relaxation rate as small as $0.003\,\mu{\rm s}^{-1} = 3\,{\rm ms}^{-1}$ can be measured. The muon localisation site has experimentally been determined through Knight shift measurements. Thanks to this information and the data of Fig. 10.26, a rough estimate gives $\sim 10^{-3}\,\mu_{\rm B}/{\rm Ce}$ for the electronic moments at the origin of the increase of λ_Z below about 1 K (Amato 1994). A qualitative picture of the magnetic correlations in $CeRu_2Si_2$ emerges from these measurements. We first note that the additional relaxation measured below 1 K is of dynamical origin since $P_Z(t)$ is found to be an exponential function (Yaouanc *et al.* 1999). As inferred from Fig. 7.7, the measurements only probe quasi-static magnetic fluctuations. Since λ_Z is extremely small, in the heavy-fermion regime, i.e. below $\sim$15 K, the vast majority of the correlations are strongly inelastic. In other words, the spectral weight of the quasielastic correlations is extremely small. They are detected below 1 K because their correlation time happens to enter the range where they can be detected, acknowledging the extremely small magnetic moments at play. As suggested in Sections 10.2.3.1 and 10.2.3.3, the quasi-static nature of the detected moments goes along with their small size. The theoretical challenge remains to explain why there are no quasielastic magnetic fluctuations with

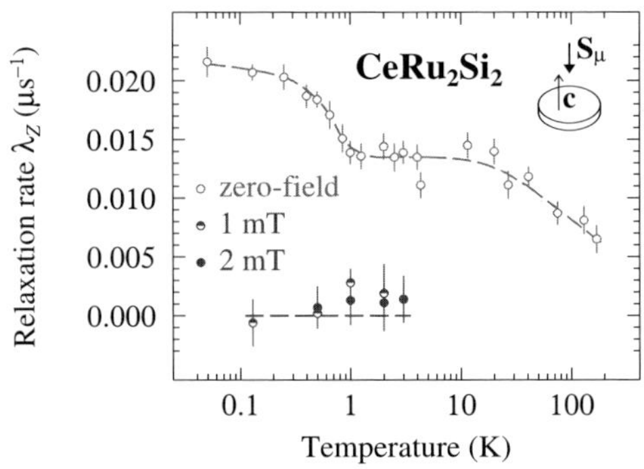

Fig. 10.26 Thermal dependence of λ_Z measured on crystals of the tetragonal heavy-fermion $CeRu_2Si_2$ with the c axis parallel to the initial muon beam polarisation. The spectra have been recorded either in zero field or in small longitudinal fields as indicated in the figure. The dashed lines are guides to the eye. Reproduced with kind permission from Yaouanc *et al.* (1999).

appreciable magnetic moments in the heavy-fermion regime of $CeRu_2Si_2$.

When no static magnetic correlations are found by any experimental techniques including μSR, the muon measurements can serve as establishing a stringent upper bound for the magnitude of frozen ordered or disordered electronic moments. We use here the example of plutonium as an illustration. Among the materials formed from a single chemical element, plutonium is certainly one of the most intriguing and for this reason it has been the subject of many fundamental theoretical and experimental studies alike. The $5f$ electronic shell of this element is partially filled and the interest mainly arises from the fact that these $5f$ states are at the borderline of localisation and delocalisation. Even more fascinating, six different crystalline allotropic phases are known for plutonium, among which the two most common ones are the α phase, which is considered as delocalised and the δ phase, which is believed to be partially localised. Theories have predicted a ground state with magnetic order in this latter phase. Zero-field μSR measurements performed in each phase lead to the detection of static fields of nuclear origin down to a temperature of 4 K (Heffner *et al.* 2006). Estimates by these authors set the upper bound of any static electronic moments to a value as low as $10^{-3}\,\mu_B$ per plutonium atom, consistent with expectation for the α phase but more surprising for the δ phase.

The plutonium work is remarkable from the experimental viewpoint. Although the sample had to be placed in a heavy container, as for NpO_2 to be discussed in the next section, the measurements were able to provide key results.

10.4 Complementarity of μSR and other techniques

The purpose of this section is to show, through selected examples, the complementary nature of the physical information extracted from measurements performed by μSR and other experimental techniques. For each system we single out the key information obtained from each experimental method. The interested reader can find the detailed arguments in the quoted references.

Uranium arsenide, which crystallises with the face-centred-cubic rocksalt structure, undergoes two first order magnetic phase transitions. Between about 125 and 62 K, it orders in the so-called type-I antiferromagnetic collinear ($\parallel c$ axis) structure made of alternating ferromagnetic sheets stacked along the c axis. This structure is described by a single propagation wave vector $\mathbf{k} = (001)$ in reciprocal lattice units. The transition at 125 K must be accompanied by a crystal symmetry breaking which slightly distorts the cubic unit cell into a tetragonal one with a lattice parameter ratio $c/a < 1$. Below 62 K, the magnetic structure is coined type IA and is more complex. It is a double-$\mathbf{k}$ structure — $\mathbf{k}_1 = (\frac{1}{2}00)$ and $\mathbf{k}_2 = (0\frac{1}{2}0)$ — and the moments no longer collinear lie along either the [110] or $[1\bar{1}0]$ direction. While a second lattice distortion to a tetragonal structure, this time with $c/a > 1$, was predicted by theory in the low-temperature phase, it was not experimentally observed until μSR experiments were performed by Asch *et al.* (1989). Spectra with full amplitude and no spontaneous muon spin precession are measured in the higher temperature ordered phase. This is the expected result if the muon localises in a site of high symmetry, namely the centre of the cubic cell. Indeed, at this site, the dipole and hyperfine contributions to the spontaneous field vanish by symmetry. In the low-temperature phase, beside the spin–lattice relaxation channel accounting for $\frac{1}{3}$ of the total amplitude, a signal consisting in three components with amplitudes in the ratio 1:2:1 is measured. The corresponding spontaneous fields are finite for two of them and vanish for the third one. This result can only be understood in the presence of a lattice distortion. In addition, because of the specific magnetic structure of UAs, it is found that the muon localisation site splits in fact into six different magnetic sites. A close inspection of these sites predicts the existence of the three spontaneous field values which are observed together with their experimental population. This tetragonal distortion in the low-temperature phase, which could not be detected with neutron or conventional X-ray scattering techniques, was eventually confirmed by McWhan *et al.* (1990) thanks to synchrotron radiation.

For half a century the low-temperature properties of NpO_2 have mystified theorists and experimentalists alike. Upon cooling from room temperature, a specific-heat peak is observed at $T_0 \simeq 25.5$ K. According to neutron diffraction, ^{237}Np Mössbauer spectroscopy, and static susceptibility, this thermodynamic phase transition does not correspond to a magnetic one. Even more puzzling, there is no evidence for a crystallographic phase transition or distortion at this temperature.

In this context, the zero-field μSR measurements which detected a magnetic transition at T_0 came as a real surprise. A consistent explanation came later on pointing to the presence of a magnetic transition involving higher order terms than the dipole ones in the Hamiltonian expansion (Santini and Amoretti 2000).

In Fig. 10.27 some μSR spectra of NpO_2 are shown. The appearance of a spontaneous muon spin precession signal is a safe indicator for the presence of a relatively long-range magnetic structure. The method used for recording these spectra was fairly standard. However, because neptunium is harmful from the chemical and radiation points of view, the sample environment was a bit special: an unusual large amount of powder sample (about 2 g) was doubly encapsulated in aluminium cylinders. Because

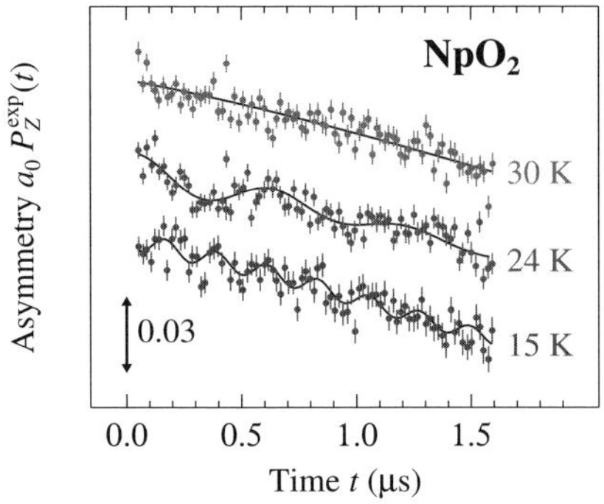

Fig. 10.27 Some zero-field μSR spectra recorded for NpO_2. Whereas oscillations are observed at 24 and 15 K, the measured asymmetry decays monotonously at 30 K; i.e. no oscillation is observed. These spectra establish unambiguously that a magnetic phase transition occurs between 24 and 30 K. For clarity, the spectra are presented with an asymmetry shift. The asymmetry scale is indicated in the graph. Adapted from Kopmann *et al.* (1998).

of this fairly massive encapsulation, the measurements had to be carried out at a decay muon beam, i.e. at a high momentum beam; see Section 2.2.2.2. Although a sizeable portion of muons were stopped in the sample environment, a significant piece of information was obtained from the spectra: the existence of a magnetic phase transition at T_0, probably of second order. In fact, the data analysis indicates that two precession frequencies are present in the signal, the amplitude of the lower frequency dominating. This lower frequency corresponds to $B_0 \simeq 51\,\mathrm{mT}$ at low temperature. A conventional interpretation would assume $\mathbf{B}_0$ to be built up from the contribution of the dipole magnetic moment of the Np atoms. However, this would be inconsistent with the neutron, Mössbauer, and susceptibility data. On the other hand, assuming that $\mathbf{B}_0$ arises from magnetic-octupole order instead of magnetic-dipole order, the whole set of data can be understood (Santini and Amoretti 2000). The assumed order lifts the degeneracy of the Np^{4+} Kramers doublet such that the ground state is a non-magnetic singlet state and thus cannot carry a dipole magnetic moment.

Resonant X-ray scattering experiments performed at the Np $M_{4,5}$ edges indicate the appearance of Bragg reflections, i.e. of a long-range order, below T_0 (Paixão *et al.* 2002). They correspond to the ordering of the Np electric quadrupoles induced by magnetic multipoles which drive the magnetic ordering. The primary order parameter of the phase transition is now suggested to be a rank-5 triakontadipole — or alternatively named dotriacontapole — (Santini *et al.* 2006). It is consistent with the observation of a spontaneous muon spin precession. To the best of our knowledge, NpO_2 is the first recognised example of a phase transition with a magnetic multipole as the primary order parameter.

A magnetic phase transition to long-range order takes place at low temperatures for most crystallographically ordered compounds containing a lattice of magnetic ions. Its classical signature is an anomaly in the magnetic specific heat at the phase transition temperature T_c. However, for some peculiar lattice structures, the long-range magnetic

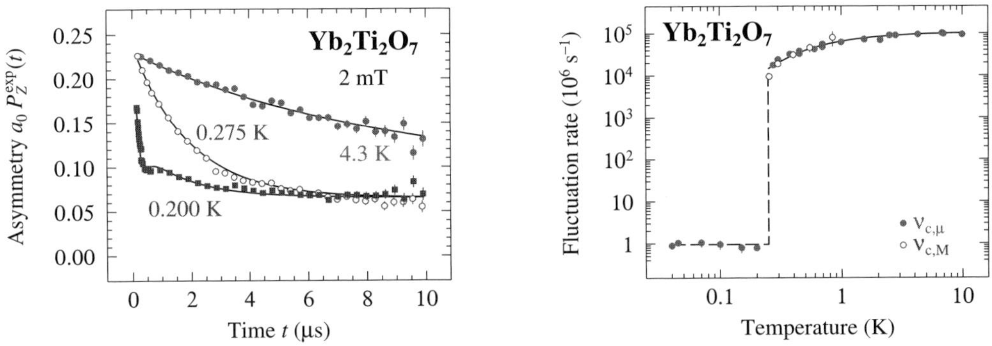

Fig. 10.28 (Left) Longitudinal-field μSR spectra recorded for $Yb_2Ti_2O_7$ in a field of 2 mT. This field only serves to quench the possible relaxation stemming from the nuclear moments of the compound. A marked change occurs on crossing the temperature ($T_0 \simeq 0.24$ K) of the specific-heat peak. Above this temperature the relaxation is exponential and below it displays a quasi-static Kubo–Toyabe shape, modified according to Eq. 6.101 to account for the correlations among the magnetic moments. The solid lines are results of fits. Reproduced with kind permission from Hodges *et al.* (2002). (Right) Estimate of the Yb^{3+} fluctuation rates as obtained from ^{170}Yb Mössbauer ($\nu_{c,M}$) and μSR ($\nu_{c,\mu}$) measurements. Note that we take the field correlation frequency ν_c to be equal to $\nu_{c,\mu}$. A first-order change in the fluctuation rate takes place at the thermodynamic phase transition, i.e. at T_0, and an appreciable rate is measured below T_0. This means that the observed short-range correlations are still dynamical in nature although quite slow in that temperature range. The full line is the result of a fit to an Arrhenius law with an activation energy of 0.5 K in temperature units, and the dashed line a guide to the eye. Reproduced with kind permission from Yaouanc *et al.* (2003).

ordering of the magnetic ions may be impeded by their geometric arrangement which gives rise to frustration of their magnetic interactions. We present results obtained for $Yb_2Ti_2O_7$ for which the geometrical frustration effectively prevents a long-range dipole order to occur, even though a specific heat peak is observed (Hodges *et al.* 2002).

$Yb_2Ti_2O_7$ crystallises in the pyrochlore crystal structure. In this structure the rare-earth ions are arranged on a motif of corner-sharing tetrahedra. An intense specific heat peak is detected at $T_0 \simeq 0.24$ K indicating a thermodynamic phase transition. The observation of a Zeeman splitting of the ^{170}Yb Mössbauer line at low temperature strongly suggests the transition to be magnetic. However, no magnetic Bragg reflections could be detected by neutron diffraction. This apparent conflict has been resolved by the longitudinal-field μSR technique. In the left panel of Fig. 10.28 are shown some spectra. A change in their shape is detected in the vicinity of T_0. The characteristic fluctuation time of the Yb^{3+} moments below that temperature is in fact found to be in the microsecond range, as revealed by the weakly damped tail of the 0.200 K spectrum in the left panel of Fig. 10.28. The Yb^{3+} fluctuation rate versus temperature is plotted in the right panel of the same figure. This long time scale explains why Mössbauer spectroscopy detects static magnetic correlations. Because the corre-

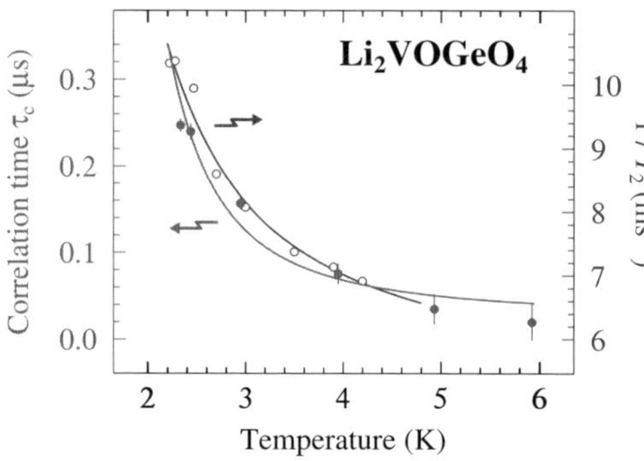

Fig. 10.29 Probing spin dynamics in Li_2VOGeO_4 by μSR and NMR techniques. We display the temperature dependence of the correlation time τ_c as measured by the longitudinal-field μSR technique and the ^{7}Li NMR transverse (or spin–spin) relaxation rate $1/T_2$. The latter data are shown with a 6 ms^{-1} offset which accounts for the temperature-independent nuclear dipole contribution to $1/T_2$. The shifted rate $1/T_2$ is expected to be proportional to τ_c. The solid lines are the predictions for activated behaviours. Adapted from Carretta *et al.* (2002).

lations are short range rather than long range, they escape detection by the neutron diffraction technique.

In conclusion, combining neutron diffraction, Mössbauer, and μSR spectroscopies, it is inferred that short-range quasi-static magnetic correlations build up below 0.24 K. The broad bump in the specific heat observed above around 2 K is probably an additional signature of these correlations.

Two-dimensional $S = \frac{1}{2}$ Heisenberg antiferromagnets have attracted a lot of interest, mainly after the discovery of high T_c superconductors. These systems are characterised by strong quantum fluctuations in view of the reduced dimensionality and low spin value. Here we focus on the model system Li_2VOGeO_4 for which the spins are located on a square lattice (Carretta *et al.* 2002). The fluctuations are believed to be enhanced by the frustration of the antiferromagnetic interactions: the next nearest neighbour interaction along the diagonal of the square competes with the nearest neighbour interaction along the side of the square. Below $T_N \simeq 2.1$ K, a collinear magnetic structure is formed. The spontaneous field at the muon site is described by a power law above $\sim$1.5 K with the exponent $\beta = 0.235$, consistent with a realisable two-dimensional XY system, i.e. a system with a finite size (Bramwell and Holdsworth 1994). In Fig. 10.29 we compare the temperature dependence of the paramagnetic dynamics as probed by μSR and NMR. They track each other. This definitely rules out the possibility of muon diffusion as an explanation for $\tau_c(T)$. The measured dynamics is anomalously slow if we refer to the scale of the exchange energy which corresponds to 5 K. In fact, the discussed dynamics is attributed to the motion of domain walls separating coexisting collinear domains. The spectral weight is schematically composed of two components: the one just discussed and a much faster conventional component, observed by longitudinal μSR measurements in fields above 10 mT and by the NMR spin–lattice relaxation rate $1/T_1$. The observation of the slow component by μSR and

NMR provides key information on the system.

The activated temperature dependence for the correlation time in Li_2VOGeO_4 is not a peculiarity of this compound. It is also found for the related oxide, Li_2VOSiO_4, and for the spin chain with Ising character (DMeFc)(TCNE). This latter compound is a molecular ferromagnet (Le *et al.* 2002). In fact, spin dynamics in layered and chain compounds are markedly different from the dynamics in three-dimensional systems. Useful references are given in the two mentioned μSR papers.

References

Abragam A. and Bleaney B. (1970) *Electron Paramagnetic Resonance of Transition Ions* (Clarendon Press, Oxford).

Akiyama R., Ikedo Y., Månsson M., Goto T., Sugiyama J., Andreica D., Amato A., Matan K., and Sato T.J. (2010) *Physical Review* **B 81**, 024404, 1–9.

Allodi G., Cestelli Guidi M., de Renzi R., Caneiro A., and Pinsard L. (2001) *Physical Review Letters* **87**, 127206, 1–4.

Amato A. (1997) *Reviews of Modern Physics* **69**, 1119–1180.

Amato A., Feyerherm R., Gygax F.N., Schenck A., Flouquet J., and Lejay P. (1994) *Physical Review* **B 50**, 619–622.

Anderson P.W. and Weiss P.R. (1953) *Reviews of Modern Physics* **25**, 269–276.

Asch L., Kalvius G.M., Kratzer A., Litterst F.J., Schenck A., Hitti B., Gygax F.N., Scott Ch., Mattenberger K., and Vogt O. (1989) *Europhysics Letters* **10**, 673–678.

Ashcroft N.W. and Mermin N.D. (2002) *Solid State Physics* (Holt, Rinehart and Winston, New York).

Balucani U., Lee M.H., and Tognetti V. (2003) *Physics Reports* **373**, 409–492.

Barbara B., Gignoux D., and Vettier C. (1988) *Lectures on Modern Magnetism* (Science Press, Beijing and Spinger-Verlag, Berlin).

Becker K.W., Fulde P., and Keller J. (1977) *Zeitschrift für Physik* **B 28**, 9–18.

Beeman D. and Pincus P. (1968) *Physical Review* **166**, 359–375.

Blinc R. (1981) *Physics Reports* **79**, 331–398.

Blundell S.J. (2001) *Magnetism in Condensed Matter* (Oxford University Press, Oxford).

Blundell S.J., Pattenden P.A., Valladares R.M., Pratt F.L., Sugano T., and Hayes W. (1994) *Solid State Communications* **92**, 569–572

Blundell S.J., Pattenden P.A., Pratt F.L., Valladares R.M., Sugano T., and Hayes W. (1995) *Europhysics Letters* **31**, 573–578.

Bramwell S.T. and Holdsworth P.C.W. (1994) *Physical Review* **B 49**, 8811–8814.

Brooks M.L., Blundell S.J., Lancaster T., Hayes W., Pratt F.L., Frampton P.P.C., and Battle P.D. (2005) *Physical Review* **B 71**, R220411, 1–4.

Budnick J.I., Chamberland B., Yang D.P., Nierdermayer Ch., Golnik A., Recknagel E., Rossmanith M., and Weidinger A. (1988) *Europhysics Letters* **5**, 651–656

Carretta P., Melzi R., Papinutto N., and Millet P. (2002) *Physical Review Letters* **88**, 047601, 1–4.

Chaikin P.M. and Lubensky T.C. (1995) *Principles of Condensed Matter Physics* (Cambridge University Press, Cambridge).

Chappert J. and Yaouanc A. (1986) in *Microscopic Methods in Metals*, edited by Gonser U., Topics in Current Physics, vol. 40 (Springer, Berlin).

Chiba M., Yoshitami A., Kikuchi H., Maegawa S., and Morimoto T. (1992) *Journal of the Physical Society of Japan* **61**, 1758–1771.

Collins M.F. (1989) *Magnetic Critical Scattering* (Oxford University Press, New York).

Condon J.H. (1966) *Physical Review* **145**, 526–535.

Condon J.H. and Walstedt R.E. (1968) *Physical Review Letters* **21**, 612–614.

Dalmas de Réotier P. and Yaouanc A. (1994) *Physical Review Letters* **72**, 290–293.

Dalmas de Réotier P. and Yaouanc A. (1995) *Physical Review* **B 52**, 9155–9158.

Dalmas de Réotier P. and Yaouanc A. (1997) *Journal of Physics: Condensed Matter* **9**, R9113–R9166.

Dalmas de Réotier P., Yaouanc A., and Frey E. (1994) *Physical Review* **B 50**, 3033–3036.

Dalmas de Réotier P., Gubbens P.C.M., and Yaouanc A. (2004) *Journal of Physics: Condensed Matter* **16**, S4687–S4705.

Dalmas de Réotier P., Yaouanc A., Gubbens P.C.M., Kaiser C.T., Baines C., and King P.J.C. (2003) *Physical Review Letters* **91**, 167201, 1–4.

Dalmas de Réotier P., Yaouanc A., Keller L., Cervellino A., Roessli B., Baines C., Forget A., Vaju C., Gubbens P.C.M., Amato A., and King P.J.C. (2006) *Physical Review Letters* **96**, 127202, 1–4.

Denison A.B., Graf H., Kündig W., and Meier P.F. (1979) *Helvetica Physica Acta* **52**, 460–517.

de Renzi R. (1999) in *Muon Science: Muons in Physics, Chemistry, and Materials*, edited by Lee S.L., Kilcoyne S.H., and Cywinski R. (Institute of Physics, Bristol).

de Renzi R., Guidi G., Podini P., Tedeschi R., Bucci C., and Cox S.F.J. (1984) *Physical Review* **B 30**, 197–204.

du Trémolet de la Lacheisserie E., Gignoux D., and Schlenker M. (2002) editors of *Magnetism, I. Fundamentals, II Materials & Applications* (Klumer Academic, Dordrecht).

Ellis D.E. and Lindgren B. (1984) *Hyperfine Interactions* **17-19**, 279–286.

Estreicher S. and Meier P.F. (1982) *Physical Review* **B 25**, 297–310.

Fawcett E. (1988) *Reviews of Modern Physics* **60**, 210–288.

Fazekas P. (1999) *Lectures Notes on Electron Correlation and Magnetism* (World Scientific, Singapore).

Feyerherm R., Amato A., Geibel C., Gygax F.N., Hellmann P., Heffner R.H., MacLaughlin D.E., Múller-Reisener R., Nieuwenhuys G.J., Schenck A., and Steglich F. (1997) *Physical Review* **B 56**, 699–710.

Fisher K.H. and Hertz J.A. (1991) *Spin Glasses* (Cambridge University Press, Cambridge)

Frey E. and Schwabl F. (1994) *Advances in Physics* **43**, 577–683.

Fulde P. and Loewenhaupt M. (1986) *Advances in Physics* **34**, 589–661.

Galli F., Nieuwenhuys G.J., MacLaughlin D.E., Heffner R.H., Amato A., Bernal O.O., and Mydosh J.A. (2002) *Physica*, **B 319**, 282–285.

Giamarchi T. and Le Doussal P. (1995) *Physical Review* **B 52**, 1242–1270.

Giamarchi T. and Le Doussal P. (1998) in *Spin Glasses and Random Fields*, edited by Young A.P. (World Scientific, Singapore).

Gibbs D., Moncton D.E., D'Amico K.L., Bohr J., and Grier B.H. (1985) *Physical Review Letters* **55**, 234–237.

Gist G.A. and Dodds S.A. (1984) *Physical Review* **B 30**, 2340–2344.

Gordon A., Vagner I.D., and Wyder P. (2003) *Advances in Physics* **52**, 385–454.

Gordon A., Itskovsky M.A., Vagner I.D., and Wyder P. (1998) *Physical Review Letters* **81**, 2787–2790.

Graf H., Kündig W., Patterson B.D., Reichart W., Roggwiller P., Camani M., Gygax F.N., Rüegg W., Schenck A., Schilling H., and Meier P.F. (1976) *Physical Review Letters* **37**, 1644–1647.

Gubbens P.C.M., Dalmas de Réotier P., Yaouanc A., Menovsky A.A., and Snel C.E. (1994) *Hyperfine Interactions* **85**, 245–250.

Halperin B.I. and Hohenberg P.C. (1967) *Physical Review Letters* **19**, 700–703.

Halperin B.I. and Hohenberg P.C. (1977) *Reviews of Modern Physics* **49**, 435–479.

Hartmann O., Karlsson E., Wäppling R., Chappert J., Yaouanc A., Asch L., and Kalvius G.M. (1986) *Journal of Physics F: Metal Physics* **16**, 1593–1615.

Heffner R.H. and Nagamine K. (2004), editors of a special issue on μSR of the *Journal of Physics: Condensed Matter* **16**, number 40.

Heffner R.H., Sonier J.E., MacLaughlin D.E., Nieuwenhuys G.J., Ehlers G., Mezei F., Cheong S.-W., Gardner J.S., and Röder H. (2000) *Physical Review Letters* **85**, 3285–3288.

Heffner R.H., Morris G.D., Fluss M.J., Chung B., McCall S., MacLaughlin D.E., Shu L., Ohishi K., Bauer E.D., Sarrao J.L., Higemoto W., and Ito T.U. (2006) *Physical Review* **B 73**, 094453, 1–5.

Henneberger S., Frey E., Maier P.G., Schwabl F., and Kalvius G.M. (1999) *Physical Review* **B 60**, 9630–9649.

Hodges J.A., Bonville P., Forget A., Yaouanc A., Dalmas de Réotier P., André G., Rams M., Królas K., Ritter C., Gubbens P.C.M., Kaiser C.T., King P.J.C., and Baines C. (2002) *Physical Review Letters* **88**, 077204, 1–4.

Hohenemser C., Rosov N., and Kleinhammes A. (1989) *Hyperfine Interactions* **49**, 267–324.

Holstein T. and Primakoff H. (1940) *Physical Review* **58**, 1098–1113.

Huberman T., Coldea R., Cowley R.A., Tennant D.A., Leheny R.L., Christianson R.J., and Frost C.D. (2005) *Physical Review* **B 72**, 014413, 1–9.

Ishida K., MacLaughlin D.E., Young Ben-Li, Okamoto K., Kawasaki Y., Kitaoka Y., Nieuwenhuys G.J., Heffner R.H., Bernal O.O., Higemoto W., Koda A., Kadono R., Trovalli O., Geibel C., and Steglich F. (2003) *Physical Review* **B 68**, 184401, 1–14.

Izyumov Yu.A. and Syromyatnikov V.N. (1990) *Phase Transitions and Crystal Symmetry* (Kluwer Academic, Dordrecht).

Izyumov Yu.A. and Chernoplekov N.A. (1994) *Neutron Spectroscopy* (Consultants Bureau, New York).

Jensen J. and Mackintosh A.R. (1991) *Rare Earth Magnetism, Structures and Excitations* (Clarendon Press, Oxford).

Kalvius G.M., Noakes D.R., and Hartmann O. (2001) in *Handbook on the Physics and Chemistry of Rare Earths*, vol. 32, edited by Gschneidner K.A., Eyring L., and Lander G.H. (Elsevier, Amsterdam).

Karlsson E.B. (1995) *Solid State Phenomena As Seen by Muons, Protons, and Excited Nuclei* (Clarendon Press, Oxford).

Kawamura H. (1998) *Journal of Physics: Condensed Matter* **10**, 4707–4754.

Keffer F. (1966) *Handbuch der Physik*, vol. 18/2 (Springer, Berlin).

Keren A. (2004) *Journal of Physics: Condensed Matter* **16**, S4603–S4618.

Keren A., Mendels Ph., Campbell I.A., and Lord J. (1996) *Physical Review Letters* **77**, 1386–1389.

Keren A., Gardner J.S., Ehlers G., Fukaya A., Segal E., and Uemura Y.J. (2004) *Physical Review Letters* **92**, 107204, 1–4.

Keren A., Shafir O., Shimshoni E., Marvaud V., Bachschmidt A., and Long J. (2007) *Physical Review Letters* **98**, 257204, 1–4.

Kittel C. (1963) *Quantum Theory of Solids* (Wiley, New York).

Kittel C. (1996) *Introduction to Solid State Physics*, seventh edition (Wiley, New York).

Kojima K., Keren A., Luke G.M., Nachumi B., Wu W.D., Uemura Y.J., Azuma M., and Takano M. (1995) *Physical Review Letters* **74**, 2812–2815.

Kopmann W., Litterst F.J., Klauss H.-H., Hillberg M., Wagener W., Kalvius G.M., Schreier E., Burghart F.J., Rebizant J., and Lander G.H. (1998) *Journal of Alloys and Compounds* **271-273**, 463–466.

Korringa J. (1950) *Physica* **16**, 601–610.

Koumoulis D., Panopoulos N., Reyes A., Fardis M., Pissas M., Douvalis A., Bakas T., Argyriou D. N., and Papavassiliou G. (2010) *Physical Review Letters* **104**, 077204 1–4.

Kübler J. (2000) *Theory of Itinerant Electron Magnetism* (Clarendon Press, Oxford).

Lappas A., Prassides K., Vavekis K., Arcon D., Blinc R., Cevc P., Amato A., Feyerherm R., Gygax F.N., and Schenck A. (1995) *Science* **267**, 1799–1802.

Le L.P., Keren A., Luke G.M., Wu W.D., Uemura Y.J., Tamura M., Ishikawa M., and Kinoshita M. (1993) *Chemical Physics Letters* **206**, 405–408.

Le L.P., Keren A., Larkin M.I., Luke G.M., Wu W.D., Uemura Y.J., Miller J. S., and Epstein A.J. (2002) *Physical Review* **B 65**, 024432, 1–5.

Lonzarich G.G. and Taillefer L. (1985) *Journal of Physics C: Solid State Physics* **18**, 4339–4371.

Lovesey S.W. (1986) *Theory of Neutron Scattering From Condensed Matter*, vol. 2 (Clarendon Press, Oxford).

McMillan W.L. (1976) *Physical Review* **B 14**, 1496–1502.

McWhan D.B., Vettier C., Isaacs E.D., Ice G.E., Siddons D.P., Hastings J.B., Peters C., and Vogt O. (1990) *Physical Review* **B 42**, 6007–6017.

Maeter H., Luetkens H., Pashkevich Yu.G, Kwadrin A., Khasanov R., Amato A., Gusev A.A., Lamonova K.V., Chervinskii D.A., Klingeler R., Hess C., Behr G., Büchner B., and Klauss H.-H. (2009) *Physical Review* **B 80**, 094524, 1–19.

Mattis D.C. (1965) *The Theory of Magnetism* (Harper & Row, New York).

Mazenko G.F. (2003) *Fluctuations, Order and Defects* (Wiley-Interscience, Hoboken, NJ).

Meier P. (1984) in *Muons and Pions in Materials Research*, edited by Chappert J. and Grynszpan R.I. (North-Holland, Amsterdam).

Mermin N.D. and Wagner H. (1966) *Physical Review Letters* **17**, 1133–1136.

Mezei F. (1984) *Journal of Magnetism and Magnetic Materials* **45**, 67–73.

Mila F. and Rice T.M. (1989) *Physical Review* **B 40**, 11382–11385.

Mishra S.G. (2001) in *Band-Ferromagnetism*, edited by Baberschke K., Donath M., and Nolting W. (Springer, Berlin).

Mitchell A.H. (1957) *Journal of Chemical Physics* **27**, 17–19.

Miyashita S. and Saito K. (2001) *Journal of the Physical Society of Japan* **70**, 3238–3246.

Mohn P. and Khmelevskyi S. (2001) in *Band-Ferromagnetism*, edited by Baberschke K., Donath M., and Nolting W. (Springer, Berlin).

Moriya T. (1956) *Progress of Theoretical Physics* **16**, 23–44; ibid. **16**, 641–657.

Moriya T. (1962) *Progress of Theoretical Physics* **28**, 371–400.

Moriya T. (1985) *Spin Fluctuations in Itinerant Electron Magnetism* (Springer Verlag, Berlin).

Moriya T. and Ueda K. (1974) *Solid State Communications* **15**, 169–172.

Mulders A.M., Kaiser C.T., Harker S.J., Gubbens P.C.M., Amato A., Gygax F.N., Schenck A., Dalmas de Réotier P., Yaouanc A., Buschow K.H.J., and Menovsky A.A. (2003) *Physical Review* **B 67**, 014303, 1–10.

Murata K.K. (1975) *Physical Review* **B 12**, 282–296.

Mydosh J.A. (1993) *Spin Glasses: an Experimental Introduction* (Taylor & Francis, London).

Nagamine K. (2003) *Introductory Muon Science* (Cambridge University Press, Cambridge).

Nattermann T. and Scheidl S. (2000) *Advances in Physics* **49**, 607–704.

Nishida N., Nagamine K., Hayano R.S., Yamazaki T., Fleming D.G., Duncan R.A., Brewer J.H., Ahktar A., and Yasuoka H. (1978) *Journal of the Physical Society of Japan* **44**, 1131–1141.

Ogielski A.T. (1985) *Physical Review* **B 32**, 7384–7398.

Ohira S., Tamura M., Kato R., Watanabe I., and Iwasaki M. (2004) *Physical Review* **B 70**, R220404, 1–4.

Orbach R. (1961) *Proceedings of the Physical Society of London* **A 264**, 458–483.

Overhauser A.W. (1960) *Journal of Physics and Chemistry of Solids* **13**, 71–80.

Paixão J.A., Detlefs C., Longfield M.J., Caciuffo R., Santini P., Bernhoeft N., Rebizant J., and Lander G.H. (2002) *Physical Review Letters* **89**, 187202, 1–4.

Pappas C., Mezei F., Ehlers G., and Campbell I.A. (2003) *Physical Review* **B 68**, 054431, 1–5.

Pelissetto A. and Vicari E. (2002) *Physics Reports* **368**, 549–727.

Pinkpank M., Amato A., Andreica D., Gygax F.N., Ott H.R., and Schenck A. (2000) *Physica* **B 289-290**, 295–298.

Rainford B.D. (1999) in *Muon Science: Muons in Physics, Chemistry, and Materials*, edited by Lee S.L., Kilcoyne S.H., and Cywinski R. (Institute of Physics, Bristol).

Ramirez A.P. (2001) in *Handbook of Magnetic Materials*, vol. 13, edited by Buschow K.H.J. (Elsevier, Amsterdam).

Redfield A.G. (1955) *Physical Review* **98**, 1787–1809.
Redfield A.G. (1957) *IBM Journal of Research and Development* **1**, 19–31.
Rice T.M., Gopalan S., and Sigrist M. (1993) *Europhysics Letters* **23**, 445–449.
Richards P.M. (1984) in *Magnetic Excitations and Fluctuations*, edited by Lovesey S.W., Balucani U., Borsa F., and Tognetti V. (Springer, Berlin).
Ross J.H., Wang Jr. Z., and Slichter C.P. (1986) *Physical Review Letters* **56**, 663–666.
Sakarya S., Gubbens P.C.M., Yaouanc A., Dalmas de Réotier P., Andreica D., Amato A., Zimmermann U., van Dijk N.H., Brück E., Huang Y., and Gortenmulder T. (2010) *Physical Review* **B 81**, 024429, 1-19.
Santini P. and Amoretti G. (2000) *Physical Review Letters* **85**, 2188–2191; ibid. **85**, 5481–5481.
Santini P., Carretta S., Magnani N., Amoretti G., and Caciuffo R. (2006) *Physical Review Letters* **97**, 207203, 1-4.
Schenck A. (1985) *Muon Spin Rotation Spectroscopy* (Adam Hilger, Bristol).
Schenck A. (1999) in *Muon Science: Muons in Physics, Chemistry, and Materials*, edited by Lee S.L., Kilcoyne S.H., and Cywinski R. (Institute of Physics, Bristol).
Schenck A. and Gygax F.N. (1995) in *Handbook of Magnetic Materials*, vol. 9, edited by Buschow K.H.J. (Elsevier, Amsterdam).
Schenck A., Andreica D., Gygax F.N., and Ott H.R. (2001) *Physical Review* **B 65**, 024444, 1–7.
Seeger A. (1978) in *Hydrogen in Metals*, edited by Alefeld G. and Völkl J., Springer Topics in Current Physics, vol. 28 (Springer, Berlin).
Shoenberg D. (1984) *Magnetic Oscillations in Metals* (Cambridge University Press, Cambridge).
Silbernagel B.G., Jaccarino V., Pincus P., and Wernick J.H. (1968) *Physical Review Letters* **20**, 1091–1094.
Sinha K.P. and Kumar N. (1980) *Interactions in Magnetically Ordered Solids* (Oxford University Press, Oxford).
Slichter C.P. (1996) *Principles of Magnetic Resonance*, third enlarged and updated edition (Springer, Berlin).
Smilga V.P. and Belousov Yu.M. (1994) *The Muon Method in Science*, Proceedings of the Lebedev Physics Institute, vol. 219 (Nova Science, Commack, NY).
Solt G. and Egorov V.S. (2002) *Physica* **B 318**, 231–250.
Solt G., Baines C., Egorov V.S., Herlach D., Krasnoperov E., and Zimmermann U. (1996) *Physical Review Letters* **76**, 2575–2578.
Sonier J.E., Pacradouni V., Sabok-Sayr S.A., Hardy W.N., Bonn D.A., Liang R., and Mook H.A. (2009) *Physical Review Letters* **103**, 167002, 1–4.
Squires G.L. (1997) *Introduction to the Theory of Thermal Neutron Scattering* (Dover, Mineola, NY).
Steer C.A. (2006) *Physical Review* **B 73**, 012408, 1–4.
Tamura M., Nakazawa Y., Shiomi D., Nozawa K., Hosokoshi Y., Ishikawa M., Takahashi M., and Kinoshita M. (1991) *Chemical Physics Letters* **186**, 401–404.
Turov E.A. and Petrov M.P. (1972) *Nuclear Magnetic Resonance in Ferro- and Antiferromagnets* (Wiley, New York).

Uemura Y.J. (1999) in *Muon Science: Muons in Physics, Chemistry, and Materials*, edited by Lee S.L., Kilcoyne S.H., and Cywinski R. (Institute of Physics, Bristol).

Uemura Y.J., Keitel R., Senba M., Kiefl R.F., Kreitzman S.R., Noakes D.R., Brewer J.H., Harshman D.R., Ansaldo E.J., Crowe K.M., Portis A.M., and Jaccarino V. (1986) *Hyperfine Interactions* **31**, 313–318.

van Kranendonk J. and Bloom M. (1956) *Physica* **22**, 545–560.

Weidinger A., Niedermayer Ch., Golnik A., Simon R., Recknagel E., Budnick J.I., Chamberland B., and Baines C. (1989) *Physical Review Letters* **62**, 102–105.

Yaouanc A. (2009) *http://arxiv.org/abs/0912.3739*, 1–11.

Yaouanc A. and Dalmas de Réotier P. (1991) *Journal of Physics: Condensed Matter* **3**, 6195–6201.

Yaouanc A., Dalmas de Réotier P., and Frey E. (1993) *Physical Review* **B 47**, 796–809; *Europhysics Letters* **21**, 93–98.

Yaouanc A., Dalmas de Réotier P., Gubbens P.C.M., Mulders A.M., Kayzel F.E., and Franse J.J.M. (1996) *Physical Review* **B 53**, 350–353.

Yaouanc A., Dalmas de Réotier P., Gubbens P.C.M., Kaiser C.T., Menovsky A.A., Mihalik M., and Cottrell S.P. (2002) *Physical Review Letters* **89**, 147001, 1–4.

Yaouanc A., Dalmas de Réotier P., Bonville P., Hodges J.A., Gubbens P.C.M., Kaiser C.T., and Sakarya S. (2003) *Physica* **B 326**, 456–459.

Yaouanc A., Dalmas de Réotier P., Gubbens P.C.M., Sakarya S., Lapertot G., Hillier A.D., and King P.J.C. (2005) *Journal of Physics: Condensed Matter* **17**, L129–L135.

Yaouanc A., Dalmas de Réotier P., Chapuis Y., Marin C., Lapertot G., Cervellino A., and Amato A. (2008) *Physical Review* **B 77**, 092403, 1-4.

Yaouanc A., Dalmas de Réotier P., Gubbens P.C.M., Kaiser C.T., Bonville P., Hodges J.A., Amato A., Schenck A., Lejay P., Menovsky A.A., and Mihalik M. (1999) *Physica* **B 259-261**, 126–127.

Yushankhai V.Yu. (1989) *Hyperfine Interactions* **50**, 775–779.

Zheludev A., Ressouche E., Schweizer J., Turek P., Wan M., and Wang H. (1994) *Solid State Communications* **90**, 233-235.

11 Investigation of superconductors

Started at the end of the 1980s, the rapid development of the μSR techniques owes much to the discovery of the high-T_c superconductors in 1986. As explained in Section 1.2.2, it was the zero-field technique which first established $YBa_2Cu_3O_{6.2}$ as a compound with a magnetic phase transition at low temperature. In the present chapter we shall discuss the investigation of superconductors, focusing on the electronic and magnetic properties peculiar to the superconducting states. However, one should not forget that a large number of compounds exhibit both superconducting and magnetic phase transitions. The experimental method for studying the latter type of phase transitions and the concepts involved in the data analysis and interpretation are identical to those used for a conventional magnetic material. Therefore we shall not treat that subject here: we refer to Chapter 10. Materials that are both ferromagnetic and superconducting might exhibit the interesting phenomenon of spontaneous vortices; i.e. vortices might form in the absence of external field due to the presence of the spontaneous magnetisation, but as yet a signature of them might have only been found in one superconductor (de Visser *et al.* 2009).

As already stressed for other themes in this book, we shall not review the μSR literature on superconductors. Instead we shall focus on the μSR methodology, using some examples to support our arguments.

Many other examples are discussed by Karlsson (1995), Amato (1997), Dalmas de Réotier and Yaouanc (1997), and Nagamine (2003). In addition, Lee (1999) and Uemura and Cywinski (1999) have written chapters dedicated to superconductivity as seen by μSR in the proceedings of a Summer School. The Institute of Physics (IOP) has published a large number of articles discussing results obtained for superconducting materials in a special issue of the *Journal of Physics: Condensed Matter* (Heffner and Nagamine 2004). As textbooks of general interest on superconductivity, we mention the monographs of de Gennes (1966), Abrikosov (1988), Poole *et al.* (1995), Tinkham (1996), Ketterson and Song (1999), Annett (2004), and Leggett (2006). In 2003 and 2004, Bennermann and Ketterson edited two books totaling more than a thousand pages which describe the compounds known to exhibit superconductivity at low temperatures and describe the state of understanding of their physics.

We shall provide in the first section a theoretical background for data analysis and interpretation, followed in the next section by examples for illustration, tackling two different types of problems: vortex matter, and the spin susceptibility and dynamics of superconducting fluids.

11.1 Theoretical background for data analysis

Here we provide tools for analysing μSR data recorded in a superconducting phase of a superconductor.

The description of the response of a superconducting material to an applied external magnetic field $\mathbf{B}_{\rm ext}$ depends on whether one is dealing with a so-called type I or a type II superconductor. A field applied to a type I superconductor does not penetrate into it, except for a thin layer. It is screened out by diamagnetic currents flowing close to the sample surface. This picture is valid if $B_{\rm ext}$ is less than the critical field B_c, i.e. if the sample is in the Meissner state. For $B_{\rm ext} > B_c$ the material is in the normal state. The number of type I superconductors is quite limited. For example, sufficiently pure aluminium metal belongs to this family. As a matter of fact, most of the superconductors are of type II. Their key characteristic is the existence of two critical magnetic fields, i.e. the lower and the upper critical fields denoted by B_{c1} and B_{c2}, respectively. This means that two superconducting phases appear instead of one. When $B_{\rm ext}$ is smaller than B_{c1}, a Meissner phase is formed as for a type I superconductor and $B_{\rm ext}$ penetrates into the material only within a thin layer. When $B_{c1} < B_{\rm ext} < B_{c2}$, the material is in the mixed or vortex phase, also called the Shubnikov phase. In this phase the magnetic field penetrates into the sample in the form of a regular array of flux tubes, each carrying a quantum of flux. In the simplest case, the axis of these tubes is parallel to $\mathbf{B}_{\rm ext}$. Within each unit cell of the array, there is a vortex of supercurrents concentrating the flux towards the vortex centre in which sits a core of quasi-normal material. This description of the mixed phase is only valid at the mean-field level. More detailed developments have shown that the periodic order of the vortex lattice is destroyed by crystalline disorder and thermal fluctuations; in general we deal with vortex matter.

μSR studies of vortex matter usually involve the transverse-field technique; i.e. the measurement gives access to the field distribution characterising this mixed phase. We shall first describe the field distribution in the mixed phase, assuming the mean-field approximation to be valid. This means that the effects of thermal fluctuations and disorder are neglected. In a second step, a theoretical framework for analysing data in real materials will be developed. Then, the case of a moving vortex lattice will be considered. Finally, the phenomenological parameters introduced to model vortex matter will be expressed in terms of microscopic properties of superconductors.

Some of the properties of a superconductor do not not depend on the characteristics of the vortex state. We shall deal with them in the second part of the current section.

11.1.1 Probing vortex matter

The monograph of Huebener (2001) is a well-known reference on flux-line structures in superconductors. Three general reviews are also available: Blatter *et al.* (1995), Brandt (1995), and Nattermann and Scheidl (2000). In addition, a short introduction to the analysis of transverse-field μSR data has been written by Lee (1999). Two extensive μSR reviews are also published (Sonier *et al.* 2000a, Sonier 2007). The book of Greer and Kossler (1995) presents a large number of field distributions computed within the London model.

As already mentionned, the field only penetrates within a surface layer of a specimen provided that $B_{\rm ext} < B_c$ or $B_{\rm ext} < B_{c1}$ for type I or II superconductors, respectively. In order to describe this Meissner state, London postulated a linear relationship between the supercurrent density $\mathbf{J}_{\rm s}$ and the vector potential $\mathbf{A}$: $\mathbf{J}_{\rm s} = -\mathbf{A}/(\mu_0\lambda_{\rm L}^2)$. Combining this equation with Ampere's law $\mu_0\mathbf{J}_{\rm s} = \mathbf{curl}\,\mathbf{B} = \mathbf{curl}\,\mathbf{curl}\,\mathbf{A}$ and Eq. C.9, it is concluded that the induction exponentially decreases from the surface towards the interior of a semi-infinite superconductor slab. The characteristic decay length is $\lambda_{\rm L}$, the so-called London penetration depth. London's assumption that $\mathbf{J}_{\rm s}$ at a given point only depends on $\mathbf{A}$ at that point is not always valid. Counterexamples are found e.g. in type I or in unconventional type II superconductors where effects of non-local electrodynamics are possible; see Section 1.2.3.

In the mixed phase of type II superconductors, i.e. for $B_{c1} < B_{\rm ext} < B_{c2}$, the decay of the field from a single vortex is represented, within the same London theory, by a modified Bessel function of the second kind. The characteristic decay length is again identified to $\lambda_{\rm L}$. Section 1.2.3 provides an example in which the equality of the field decay lengths in the Meissner and mixed states has experimentally been checked.

However, a condition is required for $\lambda_{\rm L}$ to be the characteristic length of the induction decay for a type I or type II superconductor in their Meissner or mixed phases. Only when $\lambda_{\rm L}$ is much larger than the Pippard–Bardeen–Cooper–Schrieffer coherence length ξ_0, and also when the mean free path for ordinary transport of the normal electrons is much longer than ξ_0, i.e. $\ell_{\rm mfp} \gg \xi_0$, is the decay length of the field expected to be equal to $\lambda_{\rm L}$.

The depth $\lambda_{\rm L}$ can be derived, for example, from microwave or low-energy muon beam measurements of field penetration, and from the analysis of data recorded in the mixed state by small angle neutron scattering (SANS), nuclear magnetic resonance (NMR), or surface muon beam techniques. At low temperatures $\lambda_{\rm L}$ can be as long as 600 nm for the heavy-fermion superconductor $\rm UPt_3$ (Yaouanc *et al.* 1998) and as short as 20 nm for niobium (Miyata *et al.* 2001). If $\lambda_{\rm L}$ is too large to be measured, only a bound on its value can be given, for instance $\lambda_{\rm L} > 1210$ nm in $\rm UBe_{13}$ (Dalmas de Réotier *et al.* 2000). For $\rm YBa_2Cu_3O_{6.95}$, experiments yield $\lambda_{\rm L} = 119\,(2)$ or $133\,(2)$ nm, depending on the assumed vortex lattice structure, triangular or square, respectively (Sonier *et al.* 2000a). More precisely, for $\rm UPt_3$ and $\rm YBa_2Cu_3O_{6.95}$ we refer to $\lambda_{\rm L}^{ab}$. By definition, $\lambda_{\rm L}^{ab} = (\lambda_{\rm L}^{a}\lambda_{\rm L}^{b})^{1/2}$ where $\lambda_{\rm L}^{\alpha}$ is the London penetration depth which describes the field screening by supercurrents flowing along the α axis. For the quoted examples, $\mathbf{B}_{\rm ext}$ is applied along the c axis of the hexagonal ($\rm UPt_3$) or orthorhombic ($\rm YBa_2Cu_3O_{6.95}$) crystal structures.

Here we are interested in modelling the field distribution in the mixed phase and, following common practice, the decay length of the field is taken to be $\lambda_{\rm L}$.

We recall that muons stop at interstitial sites and that the length scale between adjacent muon sites is given by the lattice parameter. Since the vortex lattice parameter is much larger than the atomic lattice parameter, the muons uniformly probe the induction field. We shall find that the width of the field distribution arising from the induction field in the mixed phase is much smaller than its mean. Therefore, according to the first line of Eq. 6.11, the transverse-field μSR technique measures the cosine Fourier transform of the distribution of the component of the induction field along the

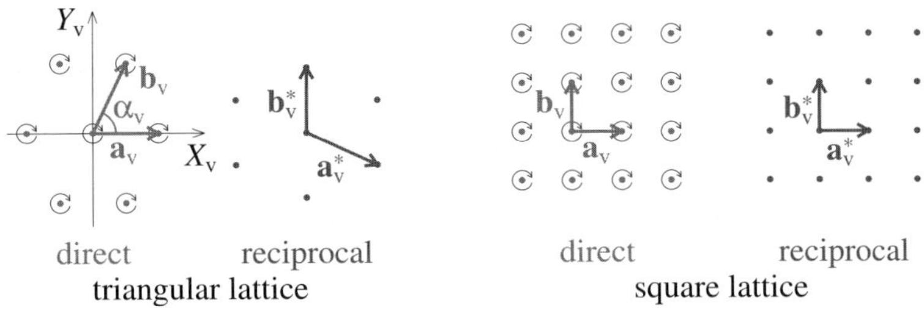

Fig. 11.1 Triangular and square vortex lattices. The primitive unit cells are defined by vectors $\{\mathbf{a}_v, \mathbf{b}_v\}$ and $\{\mathbf{a}_v^*, \mathbf{b}_v^*\}$ in the direct and reciprocal spaces, respectively. Each direct cell contains one current vortex which is symbolised by a circular arrow. The sense of rotation of the currents is clockwise, the external field being applied head-on, perpendicular to the vortex plane. For the triangular lattice the opening angle α_v has been set to 65°: the lattice is made of isosceles triangles. The X_v and Y_v axes are used from Eq. 11.10 onwards and for computations in Section 11.1.1.2. Assuming that the magnetic field is applied parallel to one of the principal axes of tensor $\mathbf{\Lambda}_{\rm sup}$ which is defined in Eq. 11.4, X_v and Y_v are the other two. One of the two possible configurations for a triangular FLL in the $\{X_v, Y_v\}$ reference frame has been chosen for the drawing (Kogan 1981).

Z axis. Hence, defining $\mathbf{r}$ as a two-dimensional vector running in a plane perpendicular to the flux tubes, we can identify the distribution of $B^Z(\mathbf{r})$ with $D_{c,\mathbf{B}_{\rm ext}}(B^Z)$, which is measured by the transverse-field μSR technique as explained in Section 6.2. The mean value of this distribution is denoted as $\overline{B^Z(\mathbf{r})}$, or $\overline{B^Z}$ in short.

Understanding the field distribution in the mixed phase requires modelling $B^Z(\mathbf{r})$. For that purpose, we shall assume as a first step the flux tubes to be arranged into a perfect two-dimensional lattice, the so-called flux-line lattice (FLL). That such a lattice exists was predicted a long time ago (Abrikosov 1957) and can be checked experimentally by imaging techniques such as decoration, scanning-tunnelling microscopy, and SANS. Note that only SANS provides imaging information on the bulk of the material. When $B^Z(\mathbf{r})$ is specified, it is then possible to compute the associated field distribution. A close-packed array of triangles, squares, or rhombuses is found experimentally for the FLL. In the triangular case, each vortex is surrounded by a hexagonal array of other vortices. The basic triangle is equilateral or more generally isosceles. In Fig. 11.1 are schematically drawn triangular and square vortex arrays in direct and reciprocal spaces. Because of the close-packed crystal structure and since a vortex carries one unit of quantum flux Φ_0 ($\Phi_0 = h/(2e) = 2.06783\times10^{-15}\,{\rm T\,m^2}$; see the table of numerical values at the beginning of the book), $N\Phi_0 = \overline{B^Z}\,S$, where N is the number of flux lines and S the cross-sectional area of the sample. Therefore, in terms of the area of the vortex lattice unit cell s_c, we derive

$$s_c = \frac{\Phi_0}{\overline{B^Z}}. \tag{11.1}$$

Let us denote the distance between two nearest neighbour vortices for the triangular and square lattices by $a_\triangle$ and $a_\square$, respectively. From the expression of s_c and the geometry of the vortex lattices, these distances can be computed:

$$a_\triangle = \sqrt{\frac{2}{\tan\alpha_\mathrm{v}}}\sqrt{\frac{\Phi_0}{\overline{B^Z}}} \text{ if } \alpha_\mathrm{v} \geq \frac{\pi}{3} \text{ and } a_\triangle = \sqrt{\frac{1}{2\sin\alpha_\mathrm{v}\cos\alpha_\mathrm{v}}}\sqrt{\frac{\Phi_0}{\overline{B^Z}}} \text{ otherwise,}$$

$$a_\square = \sqrt{\frac{\Phi_0}{\overline{B^Z}}}, \tag{11.2}$$

where α_v is the opening angle of a basic triangle of the triangular lattice. It follows for the commonly observed equilateral triangle based lattice for which $\alpha_\mathrm{v} = \pi/3$,

$$a_\triangle = \left(\frac{4}{3}\right)^{1/4}\sqrt{\frac{\Phi_0}{\overline{B^Z}}} = 1.07457\sqrt{\frac{\Phi_0}{\overline{B^Z}}}. \tag{11.3}$$

The magnetisation is negligible when B_ext is sufficiently intense, and therefore $\overline{B^Z} \simeq B_\mathrm{ext}$; see e.g. Eq. 10.293. Assuming $B_\mathrm{ext} = 1$ T, we evaluate $a_\triangle = 49\,\mathrm{nm}$ for a regular triangular FLL. This is usually smaller than λ_L. On the other hand, we compute $a_\triangle = 155\,\mathrm{nm}$ for $B_\mathrm{ext} = 0.1\,\mathrm{T}$. This lattice parameter is on the order of magnitude of λ_L for $YBa_2Cu_3O_{6.95}$. Therefore the supercurrents may interact and thus $\mathbf{B}(\mathbf{r})$ may not be a simple linear superposition of the induction from each vortex in the sample.

11.1.1.1 London model. The simplest description of the FLL follows from the inhomogeneous London equation for the induction. As usual in this textbook, $\mathbf{B}_\mathrm{ext}$ is applied along the Z axis of an orthogonal reference frame $\{X, Y, Z\}$. From the anisotropic London equation (Balatskiĭ *et al.* 1986),

$$\mathbf{B}(\mathbf{r}) + \mathbf{curl}\left[\boldsymbol{\Lambda}_\mathrm{sup}\,\mathbf{curl}\,\mathbf{B}(\mathbf{r})\right] = \Phi_0 \sum_i \delta\left(\mathbf{r} - \mathbf{i}\right)\hat{\mathbf{Z}}, \tag{11.4}$$

where $\hat{\mathbf{Z}}$ is the unit vector along the vortex tubes. The symbol $\delta\left(\mathbf{r}\right)$ denotes a two-dimensional Dirac delta function (see Section C.7.2) and $\mathbf{i}$ is the two-dimensional vector labelling the position of vortex i. The anisotropy of the magnetic response of the superconductor is contained in the tensor $\boldsymbol{\Lambda}_\mathrm{sup}$. In this description the flux tubes have no radial extension: they are rigid lines running along the Z direction.

Because of the translational invariance of the FLL, it is useful to express the London equation in reciprocal space, i.e. to consider the Fourier series of $\mathbf{B}(\mathbf{r})$ with Fourier coefficient $\mathbf{B}_\mathbf{K}$ where $\mathbf{K}$ is a vector of the reciprocal space of the vortex lattice; see Section E.2. It is sufficient to consider this lattice as a two-dimensional lattice since the flux lines are straight lines by hypothesis and therefore there is no induction variation along the Z axis. This means that $\mathbf{K}$ is a two-dimensional vector of components $(K^X,\ K^Y)$. Therefore we introduce the Fourier component

$$\mathbf{B}_\mathbf{K} = \frac{1}{s_c}\int_{s_c} \mathbf{B}(\mathbf{r})\exp\left(-i\mathbf{K}\cdot\mathbf{r}\right)\mathrm{d}^2\mathbf{r}, \tag{11.5}$$

of the periodic induction

$$\mathbf{B}(\mathbf{r}) = \sum_{\mathbf{K}} \mathbf{B}_{\mathbf{K}} \exp\left(i\mathbf{K}\cdot\mathbf{r}\right). \tag{11.6}$$

It is useful to write the London equation for the component $B^m_{\mathbf{K}}$ in the laboratory frame $\{X,Y,Z\}$. For this purpose, considering Eq. 11.5, we multiply each side of Eq. 11.4 by $\exp\left(-i\mathbf{K}\cdot\mathbf{r}\right)$ and integrate over a FLL unit cell. Using the fact that the curl of a vector can be expressed in terms of the Levi–Civita tensor $\epsilon^{\alpha\beta\gamma}$ (see Eq. C.6), we obtain

$$B^m_{\mathbf{K}} - \epsilon^{mnk}\, K^n\, \Lambda^{k\ell}_{\rm sup}\, \epsilon^{\ell pq} K^p\, B^q_{\mathbf{K}} = \overline{B^Z}\,\delta^{mZ}. \tag{11.7}$$

The summation over repeated indices is implicit.

The tensor $\mathbf{\Lambda}_{\rm sup}$ is diagonal in a reference frame attached to the crystal axes and adapted to its symmetry. As usual, we denote this frame as $\{x,y,z\}$; see Appendix D. In this frame, the elements of $\mathbf{\Lambda}_{\rm sup}$ can be written as

$$\Lambda^{k\ell}_{\rm sup} = \Lambda^{(1)}_{\rm sup}\delta^{k\ell} + \Lambda^{(2)}_{\rm sup}x^k x^\ell + \Lambda^{(3)}_{\rm sup}y^k y^\ell, \tag{11.8}$$

where $\{x^k\}$ ($\{y^k\}$), with k belonging to the $\{X,Y,Z\}$ set, are the coordinates of the unit vector $\hat{\mathbf{x}}$ ($\hat{\mathbf{y}}$) in the laboratory frame. To justify this form of $\Lambda^{k\ell}_{\rm sup}$, we note that when $\hat{\mathbf{X}} \parallel \hat{\mathbf{x}}$, $\hat{\mathbf{Y}} \parallel \hat{\mathbf{y}}$, and $\hat{\mathbf{Z}} \parallel \hat{\mathbf{z}}$, $\mathbf{\Lambda}_{\rm sup}$ is diagonal, as expected. In this case, the diagonal elements are therefore $\Lambda^X_{\rm sup} = (\lambda^X_{\rm L})^2 = \Lambda^{(1)}_{\rm sup} + \Lambda^{(2)}_{\rm sup}$, $\Lambda^Y_{\rm sup} = (\lambda^Y_{\rm L})^2 = \Lambda^{(1)}_{\rm sup} + \Lambda^{(3)}_{\rm sup}$, and $\Lambda^Z_{\rm sup} = (\lambda^Z_{\rm L})^2 = \Lambda^{(1)}_{\rm sup}$. For simplicity we write $\Lambda^k_{\rm sup}$ instead of $\Lambda^{kk}_{\rm sup}$. Plugging the decomposition of $\Lambda^{k\ell}_{\rm sup}$ in Eq. 11.7, using the Maxwell equation $\operatorname{div}\mathbf{B}(\mathbf{r}) = 0$ and Eq. C.3, we derive

$$\begin{aligned}(1 + \Lambda^{(1)}_{\rm sup}K^2)B^m_{\mathbf{K}} &+ \Lambda^{(2)}_{\rm sup}\,(\mathbf{K}\times\hat{\mathbf{x}})^m\,(\mathbf{K}\times\hat{\mathbf{x}})^q\,B^q_{\mathbf{K}} \\ &+ \Lambda^{(3)}_{\rm sup}\,(\mathbf{K}\times\hat{\mathbf{y}})^m\,(\mathbf{K}\times\hat{\mathbf{y}})^q\,B^q_{\mathbf{K}} = \overline{B^Z}\,\delta^{mZ},\end{aligned} \tag{11.9}$$

with the notation $K^2 = (K^X)^2 + (K^Y)^2$.

We now assume for simplicity that $\mathbf{B}_{\rm ext}$ is applied parallel to one of the three axes, i.e. x, y, or z. The Z axis is consequently one of them and we denote as $X_{\rm v}$ and $Y_{\rm v}$ the other two which define therefore the plane perpendicular to $\mathbf{B}_{\rm ext}$. Obviously there are two solutions for associating $X_{\rm v}$ and $Y_{\rm v}$ to these other two axes. The final result is of course independent of the chosen solution. We shall come back to this point below. With such notations, the expression for $B^Z_{\mathbf{K}}$ reads:

$$B^Z_{\mathbf{K}} = \frac{\Phi_0}{s_c}\frac{1}{1 + \left(\lambda^{X_{\rm v}}_{\rm L}K^{Y_{\rm v}}\right)^2 + \left(\lambda^{Y_{\rm v}}_{\rm L}K^{X_{\rm v}}\right)^2}. \tag{11.10}$$

In addition, there is no transverse-field components, i.e. $B^{X_{\rm v}}_{\mathbf{K}} = B^{Y_{\rm v}}_{\mathbf{K}} = 0$. If the three eigenvalues of $\mathbf{\Lambda}_{\rm sup}$ are degenerate and equal to $\lambda^2_{\rm L}$, the previous formula drastically simplifies:

$$B^Z_{\mathbf{K}} = \frac{\Phi_0}{s_c}\frac{1}{1 + \lambda^2_{\rm L}K^2}. \tag{11.11}$$

This is a well-known result; see e.g. de Gennes (1966). In most cases of interest $\lambda^2_{\rm L}K^2$ is numerically much larger than 1 even for the smallest reciprocal lattice vector. Therefore the first term in the denominator of Eq. 11.11, as well as of Eq. 11.10, can often be safely dropped.

The opening angle α_v of a triangular vortex lattice is related to the anisotropy of the $\mathbf{\Lambda}_\mathrm{sup}$ tensor. For the geometry pictured in Fig. 11.1, minimising the free energy, Kogan (1981) showed that

$$\tan \alpha_\mathrm{v} = \sqrt{3}\left(\lambda_\mathrm{L}^{X_\mathrm{v}}/\lambda_\mathrm{L}^{Y_\mathrm{v}}\right). \tag{11.12}$$

An ambiguity exists as to the choice of the X_v and Y_v axes in this formula. An additional piece of information is needed to select between the two possible values deduced for α_v from the penetration-depth ratio. SANS measurements, which probe the geometry of the flux-line reciprocal lattice, can indicate whether α_v is larger or smaller than 60° and consequently help alleviate this indetermination. An example is given in the literature (Yaouanc *et al.* 1998).

For the computation of $B_\mathbf{K}^Z$ we need to specify the geometry of the vortex lattice. We shall compute the components K^{X_v} and K^{Y_v} for a triangular lattice. The interested reader can extend the work for a square lattice. As usual, denoting the unit vectors parallel to the X_v and Y_v axes by $\hat{\mathbf{X}}_\mathrm{v}$ and $\hat{\mathbf{Y}}_\mathrm{v}$ and referring again to Fig. 11.1, we have

$$\mathbf{a}_\mathrm{v} = a_\mathrm{v}^x \hat{\mathbf{X}}_\mathrm{v}, \qquad \mathbf{b}_\mathrm{v} = b_\mathrm{v}^x \hat{\mathbf{X}}\mathrm{v} + b_\mathrm{v}^y \hat{\mathbf{Y}}_\mathrm{v}, \qquad \text{and} \qquad \mathbf{c}_\mathrm{v} = \hat{\mathbf{Z}}. \tag{11.13}$$

We note the relations:

$$b_\mathrm{v}^x = \frac{a_\mathrm{v}^x}{2}, \qquad t_\mathrm{K} \equiv \frac{\tan \alpha_\mathrm{v}}{2} = \frac{b_\mathrm{v}^y}{a_\mathrm{v}^x}, \qquad \text{and} \qquad s_c = a_\mathrm{v}^x b_\mathrm{v}^y. \tag{11.14}$$

Kogan's formula given in Eq. 11.12 can be rewritten in terms of the new parameter t_K:

$$t_\mathrm{K}^2 = \frac{3}{4}\left(\frac{\lambda_\mathrm{L}^{X_\mathrm{v}}}{\lambda_\mathrm{L}^{Y_\mathrm{v}}}\right)^2. \tag{11.15}$$

The two basic reciprocal lattice vectors can be expressed in terms of $\hat{\mathbf{X}}_\mathrm{v}$ and $\hat{\mathbf{Y}}_\mathrm{v}$:

$$\mathbf{a}_\mathrm{v}^* = \frac{2\pi}{s_c}\mathbf{b}_\mathrm{v} \times \mathbf{c}_\mathrm{v} = \frac{2\pi}{s_c}\left(b_\mathrm{v}^y \hat{\mathbf{X}}_\mathrm{v} - b_\mathrm{v}^x \hat{\mathbf{Y}}_\mathrm{v}\right), \quad \mathbf{b}_\mathrm{v}^* = \frac{2\pi}{s_c}\mathbf{c}_\mathrm{v} \times \mathbf{a}_\mathrm{v} = \frac{2\pi}{s_c} a_\mathrm{v}^x \hat{\mathbf{Y}}_\mathrm{v}. \tag{11.16}$$

With this material at hand, we can write the reciprocal lattice vector $\mathbf{K}_{pq}$ in the $\left\{\hat{\mathbf{X}}_\mathrm{v}, \hat{\mathbf{Y}}_\mathrm{v}\right\}$ reference frame:

$$\mathbf{K}_{pq} = p\,\mathbf{a}_\mathrm{v}^* + q\,\mathbf{b}_\mathrm{v}^* = \frac{2\pi}{s_c}\left[p\, b_\mathrm{v}^y \hat{\mathbf{X}}_\mathrm{v} + \left(q - \frac{p}{2}\right) a_\mathrm{v}^x \hat{\mathbf{Y}}_\mathrm{v}\right], \tag{11.17}$$

where p and q are integers. Hence,

$$K_{pq}^{X_\mathrm{v}} = 2\pi\, p \sqrt{\frac{t_\mathrm{K}}{s_c}} \qquad \text{and} \qquad K_{pq}^{Y_\mathrm{v}} = 2\pi\left(q - \frac{p}{2}\right)\sqrt{\frac{1}{s_c t_\mathrm{K}}}. \tag{11.18}$$

Dropping the term 1 in the denominator of Eq. 11.10, and using Eq. 11.15, the field Fourier components read

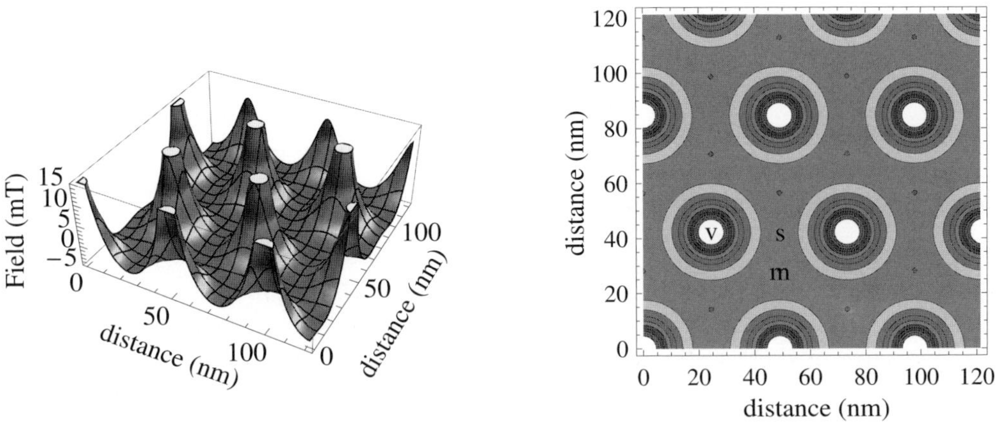

Fig. 11.2 Induction field in a three-dimensional representation (Left) and associated contour plot (Right) computed with the London model. We assume the flux-line lattice to crystallise in a equilateral triangular structure. We take the London penetration depth $\lambda_{\rm L} = 150\,{\rm nm}$ and the applied external field $B_{\rm ext} = 1\,{\rm T}$. For ease of presentation, only the deviation of the internal induction with respect to $B_{\rm ext}$ is represented. The observed anisotropy of the field profile around a given flux line, especially visible on the contour plot, is generic to flux-line lattices. The field distribution associated with this field map is displayed in Fig. 11.3 where the field scale is $\Phi_0/\lambda_{\rm L}^2 = 91.9\,{\rm mT}$. The letters m, v, and s in the contour plot show the locations of the minimum and maximum fields and saddle point, respectively. Note that the maximum field (located at the centre of the vortex) is infinite in the framework of the London model: the field profile has therefore been cut in the vicinity of each flux-line centre.

$$B^Z_{\mathbf{K}_{pq}} = B_{\rm L}\frac{1}{p^2 - pq + q^2}, \tag{11.19}$$

with

$$B_{\rm L} = \frac{\sqrt{3}}{8\pi^2}\frac{\Phi_0}{\lambda_{\rm L}^{Y_{\rm v}}\lambda_{\rm L}^{X_{\rm v}}}, \tag{11.20}$$

and $(p,q) \neq (0,0)$. This result is remarkable in two respects: the Fourier components do not depend on the $B_{\rm ext}$ value and on the opening angle of the triangular FLL. These properties are verified provided that $B_{\rm ext}$ is large enough that the first term in the denominator of Eq. 11.10 can be safely discarded.

Finally, setting $\mathbf{r}_{\alpha\beta} = \alpha\,\mathbf{a}_{\rm v} + \beta\,\mathbf{b}_{\rm v}$, we derive from Eq. 11.6

$$B^Z(\mathbf{r}_{\alpha\beta}) = \overline{B^Z} + B_{\rm L}\sum_{(p,q)\neq(0,0)}\frac{\exp\left[i2\pi\left(\alpha p + \beta q\right)\right]}{p^2 - pq + q^2}. \tag{11.21}$$

In Fig. 11.2, we present a typical induction field map and associated contour plot computed according to this formula.

Following Section C.7.1, the measured field distribution is mathematically given by

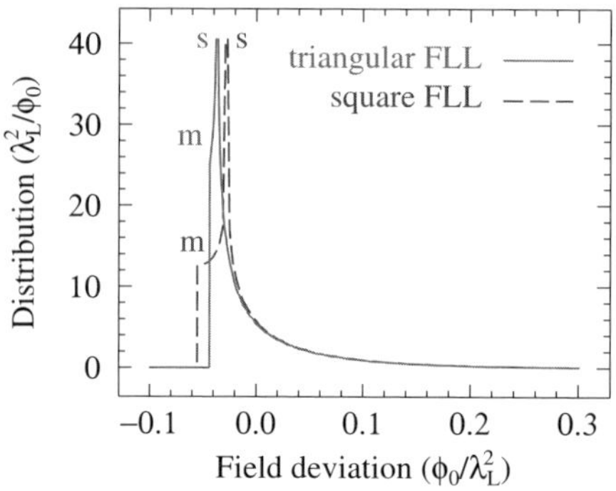

Fig. 11.3 Distinguishing triangular from square flux-line lattices looking at their field distribution. We present the field distributions computed from the London model for the two structures. The field deviation is defined as $B^Z - \overline{B^Z} \simeq B^Z - B_{\rm ext}$. The letters m and s show the location of the minimum field and saddle point where the distribution diverges. We recall that the maximum field is infinite within the London model.

$$D_{c,\mathbf{B}_{\rm ext}}\left(B^Z\right) = \int_{S_c} \delta\left[B^Z(\mathbf{r}) - B^Z\right] \frac{{\rm d}^2\mathbf{r}}{S_c}, \tag{11.22}$$

where the integral only extends over a single unit cell since $B^Z(\mathbf{r})$ is periodic. Figure 11.3 presents this distribution which is numerically obtained from an histogram of the field map presented in Fig. 11.2. Since $B^Z(\mathbf{r}_{\alpha\beta})$ in Eq. 11.21 only depends on α and β, the shape of the distribution is independent of $B_{\rm ext}$ and $\alpha_{\rm v}$.

The distribution is far from being symmetric and Gaussian-like. It presents van Hove singularities. They correspond to field regions for which the gradient of B^Z is zero. Here two van Hove singularities are observed. This is so when the magnetic response is assumed to be isotropic; i.e. the tensor $\mathbf{\Lambda}_{\rm sup}$ is scalar. The first singularity corresponds to the minimum field $B_{\rm m}$: indeed $D_{c,\mathbf{B}_{\rm ext}}\left(B^Z\right)$ vanishes for $B^Z < B_{\rm m}$. For the triangular FLL, the minimum field region lies at the centroid of the triangle formed by three first neighbour flux lines. The field step at the minimum field value observed in Fig. 11.3 is a feature common to all FLL field distributions and not restricted to the isotropic London model we have considered. The second singularity is the saddle point. In a triangular FLL, it is located in the middle of two flux-line neighbours. The nature of this singularity (i.e. logarithmic, power law, etc.) depends on the symmetry of $B^Z(\mathbf{r})$ at this point (Brandt and Seeger 1986). Note that a FLL field distribution may exhibit more than one saddle point. For example, let us assume a London tensor of uniaxial symmetry. As shown by Greer and Kossler (1995), $D_{c,\mathbf{B}_{\rm ext}}(B^Z)$ displays a single saddle point when the uniaxial axis is parallel to $\mathbf{B}_{\rm ext}$. However, when this axis is at an angle away from $\mathbf{B}_{\rm ext}$ such as 30°, two saddle point singularities appear. Finally, considering the high-field part of the distribution, it must be noted that there is no maximum field and no van Hove singularity: $D_{c,\mathbf{B}_{\rm ext}}\left(B^Z\right)$ tends to 0 only for an infinite field value. This reflects the fact that, within the London model, the field diverges at the vortex core. A more elaborate model that we shall present in Section 11.1.1.2 is required to evaluate the field variation in this region in a more realistic way.

Most of the available information on the geometry of the vortex lattice of superconductors has been derived from SANS measurements. However, transverse-field μSR can also yield information on the geometry. In Fig. 11.3 we compare $D_{\mathrm{c},\mathbf{B}_{\mathrm{ext}}}(B^Z)$ for regular triangular and square lattices. The distance between B_{m} and B_{s} are clearly different for the two crystal structures. This can be understood from the analytical results of Sidorenko *et al.* (1990). These authors performed the sum over $\mathbf{K}$ in the expression of $B^Z(\mathbf{r})$ given in Eq. 11.21 and derived formulae for the fields at the minimum and saddle points:

$$B_{\mathrm{m}}^{\triangle} = \overline{B^Z} - 0.79\frac{\Phi_0}{4\pi\lambda_{\mathrm{L}}^2}\ln 2 \quad \text{and} \quad B_{\mathrm{s}}^{\triangle} = \overline{B^Z} - \frac{2}{3}\frac{\Phi_0}{4\pi\lambda_{\mathrm{L}}^2}\ln 2,$$

$$B_{\mathrm{m}}^{\square} = \overline{B^Z} - \frac{\Phi_0}{4\pi\lambda_{\mathrm{L}}^2}\ln 2 \quad \text{and} \quad B_{\mathrm{s}}^{\square} = \overline{B^Z} - \frac{1}{2}\frac{\Phi_0}{4\pi\lambda_{\mathrm{L}}^2}\ln 2. \tag{11.23}$$

Note that these results are again obtained neglecting the term 1 in the denominator of Eq. 11.11. These equations predict $(B_{\mathrm{s}}^{\square} - B_{\mathrm{m}}^{\square}) > (B_{\mathrm{s}}^{\triangle} - B_{\mathrm{m}}^{\triangle})$, as observed in Fig. 11.3. As mentioned earlier, $D_{\mathrm{c},\mathbf{B}_{\mathrm{ext}}}(B^Z)$ is independent of the FLL opening angle α_{v}. Therefore the above formula which concerns triangular FLLs can be extended to anisotropic materials. For example, when the field is applied along the c axis of an orthorhombic crystal, λ_{L}^2 must be substituted by $\lambda_{\mathrm{L}}^a\lambda_{\mathrm{L}}^b$.

An important parameter characterising a distribution is its variance. From Eq. C.39 the variance of $D_{\mathrm{c},\mathbf{B}_{\mathrm{ext}}}\left(B^Z\right)$ is

$$\Delta_{Z,\mathrm{v}}^2 = \frac{1}{s_c}\int_{s_c}\left|B^Z(\mathbf{r})\right|^2 \mathrm{d}^2\mathbf{r} - \frac{1}{s_c^2}\left|\int_{s_c} B^Z(\mathbf{r})\mathrm{d}^2\mathbf{r}\right|^2 = \sum_{\mathbf{K}\neq\mathbf{0}} |B_{\mathbf{K}}^Z|^2. \tag{11.24}$$

This result has been derived with the help of the relation

$$\int_{s_c} \exp\left[i\left(\mathbf{K}-\mathbf{K}'\right)\cdot\mathbf{r}\right]\mathrm{d}^2\mathbf{r} = s_c\,\delta_{\mathbf{K},\mathbf{K}'}. \tag{11.25}$$

Using Eq. 11.19, the variance is given by

$$\Delta_{Z,\mathrm{v}}^2 = B_{\mathrm{L}}^2 \sum_{(p,q)\neq(0,0)} \frac{1}{(p^2 - pq + q^2)^2}. \tag{11.26}$$

A numerical evaluation of the double sum leads to the London standard deviation

$$\Delta_{Z,\mathrm{L}} = 0.06092\frac{\Phi_0}{\lambda_{\mathrm{L}}^{X_{\mathrm{v}}}\lambda_{\mathrm{L}}^{Y_{\mathrm{v}}}}. \tag{11.27}$$

For a square FLL the factor 0.06092 is replaced by 0.06219. The similarity of these two prefactors shows that $\Delta_{Z,\mathrm{L}}$ is not very sensitive to the FLL structure. A measurement of $\Delta_{Z,\mathrm{L}}$ therefore gives a good estimate of the penetration depth even if the FLL structure is unknown. Applying successively $\mathbf{B}_{\mathrm{ext}}$ along each of three axes $\{x, y, z\}$ which, we recall, are the principal axes of tensor $\mathbf{\Lambda}_{\mathrm{sup}}$ of an anisotropic superconductor, it is then possible to determine the three London penetration depths.

We have noted that $D_{c,\mathbf{B}_{ext}}(B^Z)$ is far from being symmetric. A parameter characterising this asymmetry is the skewness ϑ^Z which is defined in Eq. C.38. A numerical evaluation gives $\vartheta^Z = 1.44$ for a triangular lattice. This value is obviously independent of B_{ext} and α_v. The signature of the strong asymmetry is reflected in the clear departure of ϑ^Z from 0.

As indicated earlier, we have restricted ourselves to the case where $\mathbf{B}_{ext}$ is parallel to one of the axes $\{x, y, z\}$. The interested reader treats the general case. As a matter of fact, we note that the orientation dependence of the field distribution variance gives access to the penetration depth ratio through the formula

$$\Delta^2_{Z,v}(\theta_\alpha) = \Delta^2_{Z,v}(0)\left[\cos^2\theta_\alpha + \left(\lambda^\beta_L/\lambda^\alpha_L\right)^2 \sin^2\theta_\alpha\right], \tag{11.28}$$

where θ_α is the angle between the Z and the α axes, for rotations about the γ axis, i.e. the third axis. We recall that α, β, and γ stand for x, y, or z. For an example of data analysis with Eq. 11.28, we refer to the work of Ager *et al.* (2000).

11.1.1.2 Ginzburg–Landau model. A vortex in the mixed phase of a superconductor comprises a cylindrical core of a finite radius surrounded by supercurrents extending up to a length scale given by λ_L. The superconducting electron density in the centre of the core vanishes and the induction is as a corollary, maximum. In fact, the material in the core is quasi-normal. Electronic bound states are expected (Caroli *et al.* 1964), at least if there is no node in the superconducting gap. The London model, introduced in the previous section, assumes the core radius to be negligible since the source of induction is modelled by a delta function. This assumption is the origin, for example, of the poor description of the high-induction tail of the field distribution. In addition, of special concern to us, it cannot explain the observed intrinsic B_{ext} dependence of the amplitude of the Fourier components of the induction. Here we are going to compute the induction with a model which approximately accounts for the core size.

The computation of $B^Z(\mathbf{r})$ for arbitrary temperature and field is difficult, even numerically, and even for a conventional (isotropic) Bardeen–Cooper–Schrieffer (BCS) superconductor. Such a computation requires solving the BCS–Gor'kov equations. In this section we shall compute $B^Z(\mathbf{r})$ from the Ginzburg–Landau (GL) equations resulting from the minimisation of the GL free energy. We shall focus on the case of a conventional isotropic superconductor. The GL model has the advantage that it provides a reasonable description of the vortex cores of the FLL. A discussion of the validity of this approach and possible extensions will be presented at the end of this section. The GL model is based on a phenomenological expansion of the free energy in terms of the superconducting order parameter and its gradient. Obviously, the order parameter vanishes at the superconducting phase transition. Two length scales are introduced: λ_L already encountered and the GL coherence length ξ_{GL}. Minimisation of the free energy provides two coupled differential equations: one for the order parameter and one for the currents, that is, for the diamagnetic response of the superconductor. These two GL equations can be written in terms of a single parameter, the so-called GL parameter $\kappa_{GL} \equiv \lambda_L/\xi_{GL}$. Unfortunately they cannot be solved analytically, not even in the limit $\kappa_{GL} \to \infty$, which would be sufficient in most cases. Yaouanc *et al.* (1997)

have proposed an approximate analytical solution for $B^Z(\mathbf{K})$. Here we summarise their results.

The best analytical expression available for $B^Z(\mathbf{r})$ was obtained a long time ago by Clem (1975) for isotropic superconductors at low inductions, i.e. when the vortices are isolated. In practice a single vortex is considered. Clem uses a simple educated guess for the functional form of the order parameter of this isolated vortex located at the origin of the coordinates, namely

$$\Psi(\mathbf{r}) = \frac{r}{(r^2 + \xi_{\rm v}^2)^{1/2}} \exp(-i\varphi), \tag{11.29}$$

where φ is the polar angle of $\mathbf{r}$ in the plane perpendicular to $\mathbf{B}_{\rm ext}$. The order parameter vanishes at the centre of the vortex and its modulus tends to unity far away from it. The parameter $\xi_{\rm v}$ which enters the formula is determined by minimising the free energy: it is found that $\xi_{\rm v} = \sqrt{2}\xi_{\rm GL}$. In order to determine the field $\mathbf{b}$ associated with a single vortex, it is necessary to find the corresponding vector potential $\mathbf{a}$. Using the second GL equation, Ampere's law, and the equation $\mathbf{b} = \mathbf{curl}\,\mathbf{a}$, $\mathbf{a}$ is found to be the solution of a differential equation. This solution, which we do not reproduce here, involves a Bessel function. The field $\mathbf{b}$ is parallel to the Z axis, and when $\kappa_{\rm GL} \gg 1$ its Z component can be written as

$$b^Z(\mathbf{r}) = \frac{\Phi_0}{2\pi\lambda_{\rm L}^2} K_0 \left[\frac{\left(r^2 + 2\xi_{\rm GL}^2\right)^{1/2}}{\lambda_{\rm L}} \right]. \tag{11.30}$$

The field $b^Z(\mathbf{r})$ and the modulus of the order parameter $\Psi(\mathbf{r})$ are both shown in Fig. 11.4. The function $K_0(x)$ is a modified Bessel function of the second type which diverges as the $x = 0$ point is approached. On the other hand, $K_0(x) = \sqrt{\pi/(2x)}\exp(-x)$ when $x \gg 1$; see Eq. C.99. Note that the London model leads to the same expression of $b^Z(\mathbf{r})$ provided that $\xi_{\rm GL}$ is set to 0 in Eq. 11.30. The predictions of the London and GL models are therefore consistent far away from the vortex core. The main difference is in the vicinity of the core where the GL solution is reasonable, in contrast to the London solution which diverges.

The field $B^Z(\mathbf{r})$ is built up from the contribution of all the flux tubes in the FLL. Neglecting for the time being the possible effect of the interaction between the fields from the different flux lines, we write $B^Z(\mathbf{r})$ as a convolution product:

$$B^Z(\mathbf{r}) = \int_S b^Z(\mathbf{r} - \mathbf{r}')\varrho(\mathbf{r}')\mathrm{d}^2\mathbf{r}', \tag{11.31}$$

where the integration extends over the cross-sectional area of the sample and $\varrho(\mathbf{r})$ is the vortex density,

$$\varrho(\mathbf{r}) = \sum_{\mathbf{i}} \delta(\mathbf{r} - \mathbf{i}). \tag{11.32}$$

The set of vectors $\mathbf{i}$ spans the vector position of the flux tubes in the vortex lattice. Since $\varrho(\mathbf{r})$ is a periodic function, it can be expanded as a Fourier series:

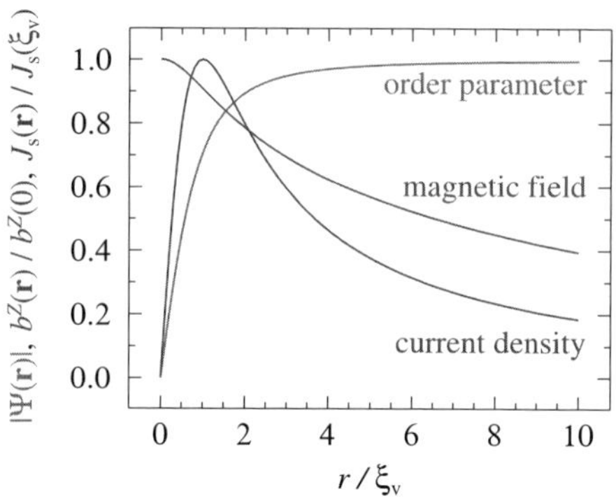

Fig. 11.4 Structure of an isolated vortex tube in a type II isotropic superconductor. We display the radial dependence of the modulus of the superconducting order parameter Ψ, of the field b^Z, and of the modulus of the supercurrent density J_s. We use Clem's solution of the Ginzburg–Landau model (Clem 1975). The field is normalised to its value at the vortex centre which is the origin of the distances. The current density which is obtained from Ampere's law is normalised to its maximum value, which is reached at $r = \xi_v$. The plot has been drawn for a London penetration depth $\lambda_L = 150\,\mathrm{nm}$ and $\xi_v = \sqrt{2}\xi_{GL} = 4\,\mathrm{nm}$, where ξ_{GL} is the Ginzburg–Landau coherence length. For these two parameters the computed field at the vortex core, which according to Eq. 11.30 depends only on these two parameters, is 0.0547 T.

$$\varrho(\mathbf{r}) = \frac{1}{s_c} \sum_{\mathbf{K}} \exp(i\mathbf{K} \cdot \mathbf{r}). \tag{11.33}$$

It is convenient to express $B^Z(\mathbf{r})$ as the Fourier transform of $B^Z(\mathbf{q})$ (Eq. C.52):

$$B^Z(\mathbf{r}) = \int_{\mathrm{RS}} B^Z(\mathbf{q}) \exp(i\mathbf{q} \cdot \mathbf{r}) \frac{\mathrm{d}^2\mathbf{q}}{(2\pi)^2} = \int_{\mathrm{RS}} b^Z(\mathbf{q}) \varrho(\mathbf{q}) \exp(i\mathbf{q} \cdot \mathbf{r}) \frac{\mathrm{d}^2\mathbf{q}}{(2\pi)^2}. \tag{11.34}$$

The integrals are performed over the two-dimensional reciprocal space. The second step in Eq. 11.34 follows because $B^Z(\mathbf{r})$ is written as a convolution product in Eq. 11.31. The Fourier transform $b^Z(\mathbf{q})$ is obtained from the expression of $b^Z(\mathbf{r})$ (Eq. 11.30) and the Fourier transform definition (Eq. C.51):

$$b^Z(\mathbf{q}) = \Phi_0 \frac{\tilde{q} K_1(\tilde{q})}{1 + \lambda_L^2 q^2} \quad \text{with} \quad \tilde{q} = \sqrt{2}\xi_{GL} \left(q^2 + \lambda_L^{-2}\right)^{1/2}. \tag{11.35}$$

In addition,

$$\varrho(\mathbf{q}) = \frac{(2\pi)^2}{s_c} \sum_{\mathbf{K}} \delta(\mathbf{q} - \mathbf{K}). \tag{11.36}$$

This last relation has been derived using Eq. 11.33 and the integral representation of the Dirac delta function; see Eq. C.48. Plugging Eqs. 11.35 and 11.36 into Eq. 11.34, we recover Eq. 11.6 for the component $B^Z(\mathbf{r})$ with the Fourier coefficient

$$B^Z_{\mathbf{K}} = \frac{\Phi_0}{s_c} \frac{\tilde{k} K_1(\tilde{k})}{1 + \lambda_L^2 K^2}, \quad \text{with} \quad \tilde{k} = \sqrt{2}\xi_{GL} \left(K^2 + \lambda_L^{-2}\right)^{1/2}. \tag{11.37}$$

To arrive at this expression we have used Eqs. C.82, C.101, and C.97. The asymptotic limits of the modified Bessel function $K_1(x)$ are such that $K_1(x) = 1/x$ when $x \ll 1$ and $K_1(x) = \sqrt{\pi/(2x)}\exp(-x)$ when $x \gg 1$; see Eq. C.99. Therefore, when the core size shrinks to 0, we recover the London expression for $B^Z_{\mathbf{K}}$ (Eq. 11.11).

Clem's approximate analytical theory for the dilute FLL was extended to large fields and to anisotropic superconductors by Hao *et al.* (1991). The analysis of these results leads to the following formula valid if $\kappa_{\rm GL} \gg 1$:

$$B^Z_{\mathbf{K}} = \frac{\Phi_0}{s_c}(1 - b^4_{\rm nor})\frac{vK_1(v)}{\left(\lambda^{X_{\rm v}}_{\rm L}K^{Y_{\rm v}}\right)^2 + \left(\lambda^{Y_{\rm v}}_{\rm L}K^{X_{\rm v}}\right)^2}. \tag{11.38}$$

Here $b_{\rm nor} = \overline{B^Z}/B_{c2}$ and $vK_1(v)$ is an anisotropic cutoff factor with

$$v^2 = 2\left[\left(\xi^{X_{\rm v}}_{\rm GL}K^{X_{\rm v}}\right)^2 + \left(\xi^{Y_{\rm v}}_{\rm GL}K^{Y_{\rm v}}\right)^2\right](1 + b^4_{\rm nor})\left[1 - 2b_{\rm nor}(1 - b_{\rm nor})^2\right]. \tag{11.39}$$

If needed, the normalised field variable $b_{\rm nor}$ can be written in terms of GL coherence lengths since

$$B_{c2} = \frac{\Phi_0}{2\pi\,\xi^{X_{\rm v}}_{\rm GL}\xi^{Y_{\rm v}}_{\rm GL}}, \tag{11.40}$$

when pairing is limited by the Lorentz force acting on the electrons (orbital limit). The shape of the vortex core cross section is an ellipse with semi-axes denoted by $\xi^\alpha_{\rm v}$ where

$$\xi^\alpha_{\rm v} = \xi^\alpha_{\rm GL}\sqrt{2}(1 + b^4_{\rm nor})^{1/2}\left[1 - 2b_{\rm nor}(1 - b_{\rm nor})^2\right]^{1/2}. \tag{11.41}$$

Note that the mean induction may be equated to $B_{\rm ext}$ when $2b_{\rm nor}\,\kappa^2_{\rm GL} \gg 1$.

Using the expressions of $K^{X_{\rm v}}_{pq}$ and $K^{Y_{\rm v}}_{pq}$ (Eq. 11.18) and Kogan's formula (Eq. 11.15), the form factor $B^Z_{\mathbf{K}_{pq}}$ factorises into

$$B^Z_{\mathbf{K}_{pq}} = B_{\rm L}\, b_{pq}(b_{\rm nor}), \tag{11.42}$$

where $B_{\rm L}$ is defined in Eq. 11.20 and $b_{pq}(b_{\rm nor})$ is a universal function:

$$b_{pq}(b_{\rm nor}) = (1 - b^4_{\rm nor})\frac{v_{pq}K_1(v_{pq})}{p^2 - pq + q^2}, \tag{11.43}$$

$$v_{pq} = \frac{2\sqrt{2\pi}}{3^{1/4}}b^{1/2}_{\rm nor}\left(1 + b^4_{\rm nor}\right)^{1/2}\left[1 - 2b_{\rm nor}\left(1 - b_{\rm nor}\right)^2\right]^{1/2}\left(p^2 - pq + q^2\right)^{1/2}.$$

As was found for the London model, the Fourier components do not depend on the opening angle of the vortex lattice, but now they do depend on $B_{\rm ext}$ since $\overline{B^Z} \simeq B_{\rm ext}$.

With this material at hand, we are in a position to numerically compute the field distribution $D_{{\rm c},\mathbf{B}_{\rm ext}}(B^Z)$. Examples are given in Fig. 11.5. Qualitatively, $D_{{\rm c},\mathbf{B}_{\rm ext}}(B^Z)$ computed with the analytical solution of the GL equations is similar to the distribution obtained from the London model only at low field (Fig. 11.3). As $b_{\rm nor}$, i.e. $B_{\rm ext}$, increases, $D_{{\rm c},\mathbf{B}_{\rm ext}}(B^Z)$ narrows. In addition, a pronounced cutoff effect is observed at large $b_{\rm nor}$.

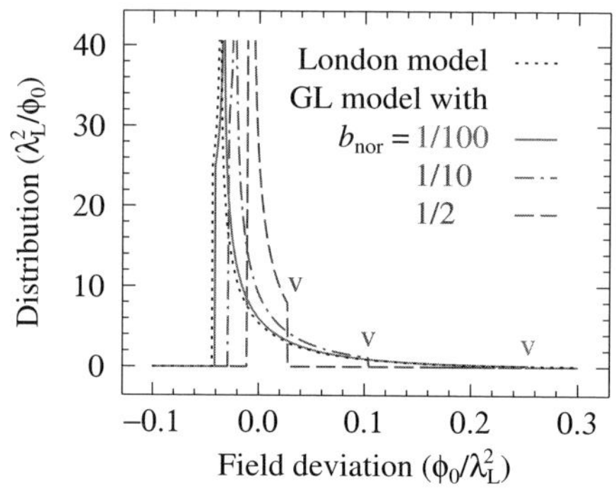

Fig. 11.5 Field distributions in a triangular FLL deduced from the approximate analytical solution of the GL equations presented in Section 11.1.1.2. As for Fig. 11.3, the field deviation is $B^Z - \overline{B^Z} \simeq B^Z - B_{\rm ext}$. Three curves are plotted for different values of $b_{\rm nor} = B_{\rm ext}/B_{c2}$. The chosen parameters are $\lambda_{\rm L} = 150\,{\rm nm}$ and $\xi_{\rm GL} = 10\,{\rm nm}$, i.e. $\kappa_{\rm GL} = 15$. The corresponding upper critical field computed from Eq. 11.40 is 3.29 T and $\Phi_0/\lambda_{\rm L}^2 = 91.9\,{\rm mT}$. For comparison, the result of the London model already shown in Fig. 11.3 is also presented. In addition to the van Hove singularities associated with the minimum field and the saddle point, which are not shown in this figure for the sake of clarity, a third one appears in the GL model. It is the signature of the maximum field in the vortex cores, $B_{\rm v}$, the position of which is indicated by the letter v in the plot.

The modulus $|B^Z_{\mathbf{K}_{pq}}|$ can be measured by the SANS technique. Under favourable experimental conditions, relatively high-order Bragg reflections are observed. The field distribution is not only sensitive to the modulus but also to the sign of $B^Z_{\mathbf{K}_{pq}}$. Within the present Clem–Hao model, all $B^Z_{\mathbf{K}_{pq}}$ are positive, unlike the result from the Abrikosov solution, valid only near B_{c2} — see e.g. Tinkham (1996). Models, like some two-component GL order parameter models, predict the sign of some $B^Z_{\mathbf{K}_{pq}}$ to be negative. The reconstruction of $B^Z(\mathbf{r})$ from SANS data, assuming the sign of the $B^Z_{\mathbf{K}_{pq}}$ components, can be tested against the field distribution measured by μSR; see Kealey *et al.* (2000).

The variance of $D_{{\rm c},\mathbf{B}_{\rm ext}}(B^Z)$ is readily obtained from Eqs. 11.24, 11.42, and 11.43. We find the standard deviation to separate into two factors,

$$\Delta_{Z,{\rm v}} = \Delta_{Z,{\rm L}} f_v(b_{\rm nor}), \tag{11.44}$$

where $\Delta_{Z,{\rm L}}$ defined in Eq. 11.27 is the London limit ($\xi^{X_{\rm v}}_{\rm GL}, \xi^{Y_{\rm v}}_{\rm GL} \rightarrow 0$) result. The universal function $f_v(b_{\rm nor})$ which accounts for the core size is such that

$$f_v^2(b_{\rm nor}) = 0.12968 \sum_{(p,q)\neq(0,0)} b^2_{pq}(b_{\rm nor}). \tag{11.45}$$

In Fig. 11.6 we display f_v as well as the skewness parameter ϑ^Z as a function of $b_{\rm nor}$ or $b^{1/2}_{\rm nor}$. The six $B^Z_{\mathbf{K}_{10}}$ equivalent terms dominate f_v (Yaouanc *et al.* 1997). Note the

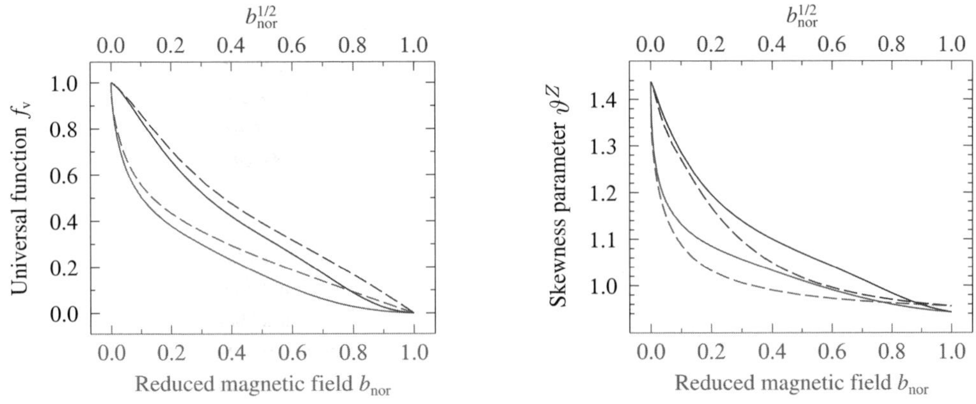

Fig. 11.6 The standard deviation of the FLL field distribution is proportional to the universal function f_v plotted in the left panel. By definition, $b_{\rm nor} = B_{\rm ext}/B_{c2}$. In the right panel is displayed the distribution skewness parameter ϑ^Z. The vortex lattice is assumed to be triangular. For both plots f_v and ϑ^Z are computed from the Fourier coefficients $B^Z_{\mathbf{K}_{pq}}$ obtained either from the analytical expression written in Eqs. 11.42 and 11.43 (full lines) or from a numerical solution of the GL equations — dashed lines; Brandt (1997, 2003). Note that f_v can alternatively be obtained analytically from Eq. 11.45 without resorting to the preliminary computation of the distribution. To set the scales, in the London model, f_v and ϑ^Z are equal to 1 and 1.44, respectively, irrespective of the $b_{\rm nor}$ value. In addition ϑ^Z would vanish if the distribution were symmetric. For a given panel, the lower (higher) set of curves is plotted against $b_{\rm nor}$ ($b_{\rm nor}^{1/2}$). The left panel shows that the curves $f_{\rm v}(b_{\rm nor}^{1/2})$ are not too far away from a straight line.

strong field dependence of f_v and ϑ^Z, in particular in low fields, a region of most experimental interest since the experiments are usually restricted to $b_{\rm nor} < \frac{1}{3}$. The strong asymmetry of $D_{c,\mathbf{B}_{\rm ext}}(B^Z)$, typical of a FLL, is expected to be observed only at low field as indicated by $\vartheta^Z(b_{\rm nor})$. The functions f_v and ϑ^Z are computed under the hypothesis that the orbital limit for the upper critical field applies (Eq. 11.40). However, experiments show that the coherence lengths at low temperature and field may not reliably be estimated with this relation (Sonier *et al.* 2000a). Hence, these two functions should be computed using the measured upper critical field or coherence length.

It is useful to get an estimate for $\Delta_{Z,\rm L}$. Taking $\lambda_{\rm L}^{X_{\rm v}} = \lambda_{\rm L}^{Y_{\rm v}} = \lambda_{\rm L} = 150\,\text{nm}$, we compute $\Delta_{Z,\rm L} = 5.60\,\text{mT}$. Such a value is easily measured.

Since $\Delta_{Z,\rm v} \propto \lambda_{\rm L}^{-2}$, the relative experimental uncertainty on the London length is only half of the relative uncertainty on $\Delta_{Z,\rm v}$. This property is experimentally interesting since it suggests a determination of the London length from μSR data with a reasonable precision to be possible even if the precision on the standard deviation is somewhat restricted.

Two basic lengths characterising a superconducting material, namely $\lambda_{\rm L}$ and $\xi_{\rm GL}$, can *a priori* be deduced from μSR and SANS data through fits. It would be of interest to compare values obtained from these two bulk techniques. Interestingly, it is also possible to estimate the core size using $B^Z(\mathbf{r})$, which is obtained from the fitted field distribution, with the help of Ampere's law $\mu_0 \mathbf{J}_\mathrm{s}(\mathbf{r}) = \mathbf{curl}\,\mathbf{B}(\mathbf{r})$, where $\mathbf{J}_s(\mathbf{r})$ is the supercurrent vector. In terms of current and field components, we get

$$\mu_0 J_s^X(\mathbf{r}) = \frac{\partial B^Z(\mathbf{r})}{\partial Y} \quad \text{and} \quad \mu_0 J_s^Y(\mathbf{r}) = -\frac{\partial B^Z(\mathbf{r})}{\partial Y}. \tag{11.46}$$

The core size is defined to be the distance measured from the core centre, where $|\mathbf{J}_s(\mathbf{r})|$ reaches its maximum value (Sonier *et al.* 2000a). It can be compared with $\xi_\mathrm{v} = \sqrt{2}\xi_\mathrm{GL}$; see Fig. 11.4 for a single vortex, i.e. when the interaction between vortices is neglected.

It is certainly a good idea to investigate the vortex lattice of a superconductor using a single-crystal sample. However, a crystal is not always available. Fortunately, it has been shown that, from a measurement on a powder sample, it is possible to deduce the smaller penetration depth of a superconductor of uniaxial symmetry, $\lambda_\mathrm{L}^\mathrm{min}$, when the penetration-depth anisotropy is greater than about 5. Denoting $\lambda_\mathrm{L}^\mathrm{powder}$ the penetration depth estimated from the variance of the powder distribution, Barford and Gunn (1988) have shown that $\lambda_\mathrm{L}^\mathrm{min} \simeq 0.81\lambda_\mathrm{L}^\mathrm{powder}$.

To explain the observed trend of the transformation of the vortex lattice geometry from triangular to square as the field is increased for (at least some) tetragonal materials with $\mathbf{B}_\mathrm{ext}$ applied along the c axis, the London model has been generalised. Two approaches have been used. Starting from a microscopic description, nonlocal electrodynamic terms contributing to $B_\mathbf{K}^Z$ have been added (Kogan *et al.* 1997). Similar terms have also been inferred based only on symmetry arguments (Affleck *et al.* 1997). For both methods it is found that

$$B_\mathbf{K}^Z = \frac{\Phi_0}{s_c} \frac{1}{1 + \lambda_\mathrm{L}^2 K^2 + \rho^2 \lambda_\mathrm{L}^2 \left(K^{X_\mathrm{v}} K^{Y_\mathrm{v}}\right)^2}, \tag{11.47}$$

where ρ is a parameter weighting the fourfold anisotropy of $B_\mathbf{K}^Z$. We have neglected the term proportional to K^4 given by Kogan and collaborators since its weight is negligible relative to the weight of the $(K^{X_\mathrm{v}})^2(K^{Y_\mathrm{v}})^2$ term. The latter biquadratic contribution is expected to be appreciable for low-κ_GL materials, and also for materials with nodes in their gap. As an obvious extension of Eqs. 11.38 and 11.39, we propose

$$B_\mathbf{K}^Z = \frac{\Phi_0}{s_c}(1 - b_\mathrm{nor}^4) \frac{v K_1(v)}{\lambda_\mathrm{L}^2 K^2 + \rho^2 \lambda_\mathrm{L}^2 \left(K^{X_\mathrm{v}} K^{Y_\mathrm{v}}\right)^2}, \tag{11.48}$$

with

$$v^2 = 2\left[\xi_\mathrm{GL}^2 K^2 + \rho^2 \xi_\mathrm{GL}^2 \left(K^{X_\mathrm{v}} K^{Y_\mathrm{v}}\right)^2\right] \left(1 + b_\mathrm{nor}^4\right) \left[1 - 2b_\mathrm{nor}(1 - b_\mathrm{nor})^2\right], \tag{11.49}$$

since the spatial symmetries of the core and supercurrent orbits are the same; see for example Figure 9.1 of Tinkham (1996).

Until now we have discussed $D_{\mathrm{c},\mathbf{B}_\mathrm{ext}}(B^Z)$ derived from an approximate analytical solution of the original GL model which deals with a conventional isotropic superconductor. In addition, a simple extension for a superconductor of tetragonal symmetry has been proposed. The approximate solution provides a fair description for most cases of physical interests, i.e. those with a large GL parameter (Yaouanc *et al.* 1997). This limit can be overcome by numerically integrating the GL equations (Brandt 1997, 2003). The numerical method is efficient enough to be used in data analysis. Figure 11.6 compares $f_{\rm v}$ and ϑ^Z computed from the approximate and numerical solutions of the GL model.

We have only discussed conventional superconductors. Based on symmetry arguments, a free energy of the GL type can be written down for any unconventional superconductor. If it is not too complex, it should be possible to derive an expression for $B^Z_{\mathbf{K}}$. For instance, evidence for superconductors with multiple gaps has been presented and a Ginzburg–Landau expression for the free energy of such superconductors has become available (Zhitomirsky and Dao 2004). It would be of interest to derive a $B^Z_{\mathbf{K}}$ expression for these compounds.

The GL approach is certainly the first step to take when analysing data. However, $D_{{\rm c},\mathbf{B}_{\rm ext}}(B^Z)$ obtained from the GL model may not be reliable for all experimental conditions. Recalling that the GL free energy is written in terms of an expansion of the order parameter and its gradient, the computed $D_{{\rm c},\mathbf{B}_{\rm ext}}(B^Z)$ is expected to be valid close to the phase transition. In addition, the GL result is also typically a good approximation for impure superconductors, i.e. with a mean free path $\ell_{\rm mfp}$ much smaller than ξ_0 (for a summary, see Brandt 1995). However, the high-$T_{\rm c}$ superconductors are believed to be in the clean limit, i.e. $\ell_{\rm mfp} \gg \xi_0$.

Although the GL approach cannot be justified for most cases of physical interest, it is usually believed to give a good qualitative description of $D_{{\rm c},\mathbf{B}_{\rm ext}}(B^Z)$. With this framework, the two lengths introduced to model the FLL, i.e. $\xi_{\rm GL}$ and $\lambda_{\rm L}$, are considered as effective parameters expected to depend on field intensity and temperature, reflecting the limitations of the GL description of $B^Z_{\mathbf{K}}$. For instance, the BCS–Gor'kov equations in the quasi-classical limit, i.e. the Eilenberger equations (for a discussion, see e.g. Ketterson and Song 1999), have numerically been solved for a conventional superconductor and their results have been compared with those of the associated GL model (Laiho *et al.* 2007). This analysis seems to explain the thermal and field dependences of the two effective parameters observed for some superconductors. The mentioned successful interpretation of the two effective parameters suggests the following procedure to be used: one fits the field distribution deduced from a simple GL model to the experimental data and then compares the extracted parameters with the predictions obtained from the solution of the Eilenberger equations.

However, this methodology is not of general validity. As an example, the distribution recorded for niobium at low temperature and high field exhibits a high-field tail which vanishes linearly with the field; see Fig. 1.15 and Section 1.2.3. This behaviour, which cannot be explained by the GL model, is a signature of the electron diffraction by the FLL of this clean superconductor. This effect plays a role because $\ell_{\rm mfp}$ is far longer than the distance between the vortex cores and the thermal disorder is negligible. Only the microscopic Eilenberger equations capture this physics, as shown by Delrieu (1972, 1974). Interestingly, corresponding to this exotic $D_{{\rm c},\mathbf{B}_{\rm ext}}(B^Z)$, an interchange of the lattice positions for the saddle point and the minimum of the field is predicted.

To conclude this discussion on the application of a microscopic model, even if the symmetry of the superconducting parameter is known, the microscopic description of $D_{c,\mathbf{B}_{ext}}(B^Z)$ remains a difficult task. In addition to being able to numerically solve the Eilenberger equations, one needs to know the material parameters entering into the theory (for example, a description of the Fermi surface is required) with sufficient accuracy.

11.1.1.3 Effect of disorder on a field distribution. According to our discussion until now, $D_{c,\mathbf{B}_{ext}}(B^Z)$ in a type II superconductor should exhibit van Hove singularities at the maximum, minimum, and saddle points of $B^Z(\mathbf{r})$. Although usually observed experimentally, these singularities are always broadened.

One reason for this smearing stems from the nuclear magnetic moments of the specimen if any. Their effect is taken into account by convoluting the FLL distribution computed in the previous section with the field distribution from the nuclear magnetic moments which is well approximated by a Gaussian distribution; see Section 7.2. The convolution product is justified since the two distributions are independent. The second origin of smearing is technical. A transverse-field μSR experiment directly gives access to the transverse-field polarisation function. As explained at the end of Section 2.3 and detailed in Section A.2, a Fourier transform of the polarisation function is often helpful in order to get a flavour of $D_{c,\mathbf{B}_{ext}}(B^Z)$ and to compare it with a theoretically motivated distribution. The experimental field distribution can alternatively be obtained from the maximum entropy technique. In both cases an apodisation of the experimental data, which broadens the obtained distribution, is required. However, we recall that the visualisation of the experimental field distribution is only performed for convenience. It helps to catch the physics at play. Since the final analysis should be performed in the time domain, the apodisation has no effect on the results derived from a quantitative examination of the data.

In addition to the nuclear and apodisation effects, which will not be discussed further, one must account in a real material for the interaction of the flux tubes with inhomogeneities and defects which act as pinning centres. In addition to smearing the singularities of $D_{c,\mathbf{B}_{ext}}(B^Z)$, these perturbations reduce or increase the width relative to that of an ideal distribution. Below, we discuss the influence of pinning centres on $D_{c,\mathbf{B}_{ext}}(B^Z)$.

We first summarise the present knowledge on vortex matter. For reviews, see Nattermann and Scheidl (2000) and Giamarchi and Bhattacharya (2002). A FLL lattice is established at low temperature and low field in a superconducting material even if randomly distributed weak point-like pinning centres are present. Although the system is disordered by the flux-line pinning, a quasi-long-range positional order is established as proven by the observation of Bragg reflections by SANS (see e.g. Cubitt *et al.* 1993 or Klein *et al.* 2001). Such a vortex phase is the so-called Bragg-glass phase (Giamarchi and Le Doussal 1995). It is a glass: it has no extended defects such as dislocations and its dynamical magnetic properties are similar to those of a glass. A key feature of this phase is the predicted power-law decay of the translational order correlation function at long scale. Two indirect experimental proofs have been presented (Klein *et al.* 2001, Laver *et al.* 2008). As B_{ext} is increased at low temperature, the interaction energy between the vortices and the defects increases (because the mean distance between the vortices and the defects decreases) up to the point where the quasi-long-range order disappears (no Bragg peaks) but strong short-ranged correlations still persist. The exact nature of the high-field phase is still under debate. We denote it as the vortex-glass phase for simplicity, acknowledging that more than one phase may exist. On the other hand, when $B_{\rm ext}$ is still modest but the temperature is sufficiently high, it is definitively established that the FLL melts. As expected, melting occurs as a first order phase transition. The described phase diagram is supposed to be universal, although it is only with the high-$T_{\rm c}$ superconductors that its richness has been recognised.

The different phases of vortex matter are thought to result from the competition between the elastic and pinning energies of the vortex array. Point-like as well as extended pinning centres are encountered. In addition, pinning can be introduced by irradiation. The Bragg-glass phase is in the elastic limit of the vortex matter. On the other hand, it is possible that the plastic limit applies for the vortex-glass phase.

Relative to the ideal distribution, a broadening is generally observed in the Bragg-glass phase. Numerical simulations show that the shape of the distribution near $B_{\rm m}$ depends drastically on the disorder. Its effect is milder near $B_{\rm s}$ and $B_{\rm v}$ (Brandt 1988). The location of these three fields is shown in Figs. 11.3 and 11.5. The disorder is conventionally accounted for by the convolution of the ideal distribution with a Gaussian function. While this method works quite well numerically for the Bragg-glass phase, it is not satisfactory because the involved physics is not clear. In addition, it completely fails to account for the shape of the measured distribution in the vortex glass and above the melting transition. Brandt (1991) — see also Menon *et al.* (1999) — has argued that highly flexible vortex cores affected by pinning and thermal fluctuations threading the superconducting layers of some high-$T_{\rm c}$ compounds may reduce the width of the distribution. In contrast, the fluctuations of the positions of the flux tubes increase the width. In this latter case the basic physical object is a stiff flux tube, whereas a highly flexible tube is better modelled as a stack of pancakes.

Here we report on a development performed with the aim of building a theoretical framework able to microscopically account for the effects of pinning-induced disorder on $D_{{\rm c},{\bf B}_{\rm ext}}(B^Z)$ (Dalmas de Réotier *et al.* 2004). It directly provides the variance of the field distribution arising from a disturbed FLL. This subject is still not fully developed and should therefore be subject to much improvement in the future.

We shall first consider the FLL to be soft. Thereafter we shall deal with a stiff FLL.

11.1.1.3.1 Soft flux-line lattice. Our previous discussion indicates that we should not always consider vortex matter as a system of rigid tubes. Rather we need to account for the soft nature of the tubes: we must tackle a three-dimensional system. Hence, instead of the $B^Z({\bf r})$ expression given in Eq. 11.34, we have

$$B^Z({\bf r}) = \int_{\rm RS} b^Z({\bf q})\varrho({\bf q}) \exp(i{\bf q}\cdot{\bf r}) \frac{{\rm d}^3{\bf q}}{(2\pi)^3}, \tag{11.50}$$

where ${\bf r} = {\bf r}_\perp + r_Z\hat{\bf Z}$ and ${\bf q} = {\bf q}_\perp + q_Z\hat{\bf Z}$. We picture the pancakes to be localised on a three-dimensional array. Following the methodology of Section 10.1.3, we note that they do not sit at the ideal positions $\{{\bf i}\}$ but at the effective positions $\{{\bf i} + \tilde{\bf u}({\bf i})\}$

which deviate from the equilibrium positions by the displacement vectors $\{\tilde{\mathbf{u}}(\mathbf{i})\}$. The pancake density is therefore written as

$$\varrho(\mathbf{r}) = \sum_{\mathbf{i}} \delta\left[\mathbf{r} - \mathbf{i} - \tilde{\mathbf{u}}(\mathbf{i})\right]. \tag{11.51}$$

Assuming the displacement field to be smooth,

$$\varrho(\mathbf{r}) \simeq \frac{1}{v_{\mathrm{c}}} \sum_{\mathbf{K}} \exp\left\{i\mathbf{K} \cdot \left[\mathbf{r} - \mathbf{u}(\mathbf{r})\right]\right\}, \tag{11.52}$$

where v_{c} is the volume of the three-dimensional array unit cell and $\mathbf{u}(\mathbf{r})$ the displacement of a pancake as a function of its real position. Here $\mathbf{K}$ which denotes a reciprocal lattice vector is a three-dimensional vector. Using the expression of the density in terms of its Fourier components, we derive

$$B^Z(\mathbf{r}) = \sum_{\mathbf{K}} \int_{\mathrm{RS}} \mathcal{B}^Z(\mathbf{q}) \exp(i\mathbf{q} \cdot \mathbf{r}) \int_V \exp\left[-i\left(\mathbf{q} - \mathbf{K}\right) \cdot \mathbf{r}'\right] \exp\left[-i\mathbf{K} \cdot \mathbf{u}(\mathbf{r}')\right] \mathrm{d}^3\mathbf{r}' \frac{\mathrm{d}^3\mathbf{q}}{(2\pi)^3}. \tag{11.53}$$

We have defined

$$\mathcal{B}^Z(\mathbf{q}) = b^Z(\mathbf{q})/v_{\mathrm{c}}. \tag{11.54}$$

We note that $B^Z(\mathbf{q})$ and $b^Z(\mathbf{q})$ are Fourier transforms. This is not the case for $\mathcal{B}^Z(\mathbf{q})$.

The computation of $B^Z(\mathbf{r})$ requires the determination of $b^Z(\mathbf{q})$ for a single pancake in each of the layers. The Z axis is perpendicular to the layers. According to Clem (1991), for a pancake located at the origin of the coordinates

$$b^Z(\mathbf{r}) = \frac{\Phi_0 d_{\mathrm{v}}}{4\pi\lambda_{\mathrm{L}}^2 r} \exp\left(-\frac{r}{\lambda_{\mathrm{L}}}\right), \tag{11.55}$$

where d_{v} is the interlayer spacing. This result is derived within the London approximation; i.e. the vortex core has been neglected. The Fourier transform of this field component is easily obtained:

$$b^Z(\mathbf{q}) = \int_{\mathrm{S}} b^Z(\mathbf{r}) \exp\left(-i\mathbf{q} \cdot \mathbf{r}\right) \mathrm{d}^3\mathbf{r} = \Phi_0 d_{\mathrm{v}} \frac{1}{1 + \lambda_{\mathrm{L}}^2 q^2}. \tag{11.56}$$

As was done for a flux line in the previous section, we need to account for the core of the pancake through a cutoff function. A natural extension of the result given in Eq. 11.35 leads to

$$b^Z(\mathbf{q}) = \Phi_0 d_{\mathrm{v}} \frac{\tilde{q} K_1(\tilde{q})}{1 + \lambda_{\mathrm{L}}^2 q^2} \quad \text{with} \quad \tilde{q} = \sqrt{2}\xi_{\mathrm{GL}} \left(q^2 + \lambda_{\mathrm{L}}^{-2}\right)^{1/2}. \tag{11.57}$$

At first sight, the $b^Z(\mathbf{q})$ formula for a pancake and a flux tube are similar; compare Eqs. 11.57 and 11.35. Here, however, $\mathbf{q}$ is a three-dimensional vector whereas it is a two-dimensional vector for a flux tube. In addition, while there is no in-plane field component for a flux tube, such a component exists for a pancake. However, we do not need to bother with it since only the Z component of the field matters in the assumed high-transverse-field approximation, which is here justified; see Section 6.2.1. With the help of Eq. C.100 it is simple matter to derive the $b^Z(\mathbf{r})$ expression for a single pancake with a proper description of the core:

$$b^Z(\mathbf{r}) = \int_{\rm RS} b^Z(\mathbf{q}) \exp(i\mathbf{q}\cdot\mathbf{r}) \frac{\mathrm{d}^3\mathbf{q}}{(2\pi)^3}$$

$$= \frac{\Phi_0 d_v}{4\pi\lambda_{\rm L}^2} \frac{1}{\left(r^2 + 2\xi_{\rm GL}^2\right)^{1/2}} \exp\left(-\frac{\left(r^2+2\xi_{\rm GL}^2\right)^{1/2}}{\lambda_{\rm L}}\right). \qquad (11.58)$$

In contrast to the result given in Eq. 11.55, $b^Z(\mathbf{r})$ is now finite at the core centre.

As for an ideal FLL discussed in the previous section, we need to extend the solution for a single pancake to an array of pancakes. Guided by the expression of $B^Z_{\mathbf{K}}$ written in Eq. 11.38, we propose

$$\mathcal{B}^Z(\mathbf{q}) = \frac{\Phi_0 d_{\rm v}}{v_{\rm c}}(1-b_{\rm nor}^4)\frac{vK_1(v)}{1+\lambda_{\rm L}^2 q^2}. \qquad (11.59)$$

with

$$v^2 = 2\xi_{\rm GL}^2 q^2 (1+b_{\rm nor}^4)\left[1 - 2b_{\rm nor}(1-b_{\rm nor})^2\right]. \qquad (11.60)$$

Since q is now a continuous variable, $\lambda_{\rm L}^2 q^2$ may not be large compared with 1. Therefore in the present case we need to keep the two terms in the denominator of Eq. 11.59. We have assumed for simplicity the superconductor to be isotropic. If needed, the anisotropy can easily be included.

To get a feeling of the effects of pinning-induced disorder on $D_{\rm c,\mathbf{B}_{ext}}(B^Z)$, we now study the variance of this field distribution which is expressed as a function of the displacement field. As in our discussion in Section 10.1.3 about the significance of long-range magnetic order in the context of μSR, we need to remember two points. Only the thermal average of a physical quantity can be measured and, because we are dealing with a disordered system, an ensemble average must be performed. As in Section 10.1.3, we shall assume the thermal average already performed and, as usual, we shall denote the ensemble average of A by $\overline{A}$. The generalisation of Eq. 11.24 to a disordered system is given by the expression

$$\Delta_{Z,\rm v}^2 = \frac{1}{V}\overline{\int_V \left|B^Z(\mathbf{r})\right|^2 \mathrm{d}^3\mathbf{r}} - \frac{1}{V^2}\overline{\left|\int_V B^Z(\mathbf{r})\mathrm{d}^3\mathbf{r}\right|^2}. \qquad (11.61)$$

Note that unlike the expression of the variance for a magnetic structure considered in Section 10.1.3 (Eq. 10.42), the variance for a vortex lattice requires performing two integrations over the sample volume. Expressing $B^Z(\mathbf{r})$ as a function of $\mathcal{B}^Z(\mathbf{q})$ (see Eq. 11.53) in the first term on the right-hand side of the formula for $\Delta^2_{Z,\rm v}$, we derive for this term

$$\frac{1}{V}\sum_{\mathbf{K},\mathbf{K}'}\int_{\rm RS}\left|\mathcal{B}^Z(\mathbf{q})\right|^2 \int_V\int_V \exp\left[i\left(\mathbf{q}-\mathbf{K}'\right)\cdot\mathbf{r}\right]\exp\left[i\left(\mathbf{K}-\mathbf{K}'\right)\cdot\mathbf{r}'\right] \qquad (11.62)$$

$$\times \overline{\exp\left[-i\mathbf{K}\cdot\mathbf{u}\left(\mathbf{r}'\right)\right]\exp\left[i\mathbf{K}'\cdot\mathbf{u}\left(\mathbf{r}'+\mathbf{r}\right)\right]}\mathrm{d}^3\mathbf{r}\mathrm{d}^3\mathbf{r}'\frac{\mathrm{d}^3\mathbf{q}}{(2\pi)^3}.$$

We encounter the correlation function $\overline{\exp\left[-i\mathbf{K}\cdot\mathbf{u}\left(\mathbf{r}'\right)\right]\exp\left[i\mathbf{K}'\cdot\mathbf{u}\left(\mathbf{r}'+\mathbf{r}\right)\right]}$. As for any such function, it should not depend on the origin of the coordinates. It is also negligible unless $\mathbf{K} = \mathbf{K}'$. This leads us to define the translational order correlation function

$$\mathcal{C}_{\mathbf{K}}(\mathbf{r}) = \overline{\exp\left\{i\mathbf{K}\cdot\left[\mathbf{u}\left(\mathbf{r}\right) - \mathbf{u}\left(\mathbf{0}\right)\right]\right\}}. \tag{11.63}$$

The integration over $\mathbf{r}'$ in Eq. 11.62 is now easily performed. This results in a drastically simplified formula for the first term on the right-hand side of Eq. 11.61:

$$\int_{\mathrm{RS}} \left|\mathcal{B}^Z(\mathbf{q})\right|^2 \sum_{\mathbf{K}} \int_V \exp\left[i\left(\mathbf{q}-\mathbf{K}\right)\cdot\mathbf{r}\right]\mathcal{C}_{\mathbf{K}}(\mathbf{r})\mathrm{d}^3\mathbf{r}\frac{\mathrm{d}^3\mathbf{q}}{(2\pi)^3}. \tag{11.64}$$

Following the same methodology, it is not difficult to realise that the second term on the right-hand side of Eq. 11.61 is simply the $\mathbf{K} = \mathbf{0}$ term of the above sum. Therefore, we finally get

$$\Delta^2_{Z,\mathrm{v}} = \int_{\mathrm{RS}} \left|\mathcal{B}^Z(\mathbf{q})\right|^2 \sum_{\mathbf{K}\neq\mathbf{0}} \int_V \exp\left[i\left(\mathbf{q}-\mathbf{K}\right)\cdot\mathbf{r}\right]\mathcal{C}_{\mathbf{K}}(\mathbf{r})\mathrm{d}^3\mathbf{r}\frac{\mathrm{d}^3\mathbf{q}}{(2\pi)^3}. \tag{11.65}$$

This formula accounts for the expected interference between the different terms contributing to the variance. The structure of Eq. 11.65 is strikingly reminiscent of Eq. 10.47, which accounts for the variance of the field distribution arising from a non-perfect magnetic structure of a magnet. A first difference is yet to be noted: while the integration volume for the integral over $\mathbf{q}$ is the Brillouin zone in the magnet case, it is the entire reciprocal space as far as the FLL is concerned. This is because the field in the FLL was obtained from the sum of the contribution of the individual vortices, unlike the magnet case for which the periodic nature of the field was recognised from the very beginning. It would be of interest to attempt a reformulation of the non-perfectly ordered FLL variance introducing the FLL Brillouin zone concept. A second difference is the following: while the variance in Eq. 11.65 chacterises the actual field distribution, only the broadening due to disorder is accounted for in Eq. 10.47.

To get a physical insight into our result, we investigate asymptotic limits. Let us first assume a disorder-free FLL. This means that we neglect any displacement of the pancake positions. Referring to the definition of $\mathcal{C}_{\mathbf{K}}(\mathbf{r})$, we conclude that $\mathcal{C}_{\mathbf{K}}(\mathbf{r}) = 1$. Hence, according to Eq. 11.65,

$$\Delta^2_{Z,\mathrm{v}} = \int_{\mathrm{RS}} \left|\mathcal{B}^Z(\mathbf{q})\right|^2 \sum_{\mathbf{K}\neq\mathbf{0}} \int_V \exp\left[i\left(\mathbf{q}-\mathbf{K}\right)\cdot\mathbf{r}\right]\mathrm{d}^3\mathbf{r}\frac{\mathrm{d}^3\mathbf{q}}{(2\pi)^3} = \sum_{\mathbf{K}\neq\mathbf{0}} \left|\mathcal{B}^Z_{\mathbf{K}}\right|^2. \tag{11.66}$$

This is obviously the expected result for an ideal FLL; see Eq. 11.24. Although Eq. 11.65 has been derived with the assumption of a smooth displacement from the position of the ideal vortices, it is still instructive to examine the strongly disordered vortex matter limit. For this purpose, we define

$$\mathcal{A} = \sum_{\mathbf{K}\neq\mathbf{0}} \int_V \exp\left[i\left(\mathbf{q}-\mathbf{K}\right)\cdot\mathbf{r}\right] \mathcal{C}_{\mathbf{K}}(\mathbf{r})\mathrm{d}^3\mathbf{r}. \tag{11.67}$$

In the strong-disorder limit the correlation function is proportional to a Dirac delta function. Then $\mathcal{A}$ becomes independent of $\mathbf{q}$ and $\mathbf{K}$; i.e. $\mathcal{A}$ is a constant with $\mathcal{A} = v_{\mathrm{c}}$. The evaluation of the integral $\int_{\mathrm{RS}} \left|\mathcal{B}^Z(\mathbf{q})\right|^2 v_{\mathrm{c}}\mathrm{d}^3\mathbf{q}/(2\pi)^3$ can analytically be performed only in the London limit, with the result

$$\Delta^2_{Z,\mathrm{v}} = \frac{\overline{B^Z}\,\Phi_0\,d_{\mathrm{v}}}{8\pi\lambda_{\mathrm{L}}^3}, \tag{11.68}$$

which follows from Eqs. 11.54 and 11.56. This is the field variance for a system of vortices with no spatial correlations. It can directly be compared with the variance for a disorder-free FLL in the London limit; see Eq. 11.27 with $\lambda_{\mathrm{L}}^{X_{\mathrm{v}}} = \lambda_{\mathrm{L}}^{Y_{\mathrm{v}}} = \lambda_{\mathrm{L}}$. It is remarkable that the variance for the disordered system is linear in $\overline{B^Z}$, whereas it is field independent for a disorder-free FLL. Assuming $\lambda_{\mathrm{L}} = 150\,\mathrm{nm}$, $d_{\mathrm{v}} = 1\,\mathrm{nm}$, and $B_{\mathrm{ext}} \simeq \overline{B^Z} = 10\,\mathrm{mT}$, we compute $\Delta_{Z,\mathrm{v}} = 0.494\,\mathrm{mT}$, i.e. more than an order of magnitude smaller than the low-field limit for an ideal FLL (see after Eq. 11.45, where we compute $\Delta_{Z,\mathrm{L}} = 5.60\,\mathrm{mT}$ when $\lambda_{\mathrm{L}} = 150\,\mathrm{nm}$). The narrowing of the field distribution at low field is experimentally observed in layered materials with strong two-dimensional character and was first explained by Brandt (1991). A qualitative picture of the narrowing is also offered by this author: when the positions of the pancakes are randomly distributed, the field contrast is poor, and therefore the field width is small.

11.1.1.3.2 Stiff flux-line lattice. Having derived two well-known asymptotic limits of Eq. 11.65, we increase further our insight by considering a superconducting material with parallel stiff flux tubes presenting a positional disorder. Because of the two-dimensional character of that system, it is more appropriate to start with an expression for the variance in two dimensions. Obvious modifications of Eqs. 11.61 and 11.65 lead to

$$\begin{aligned}\Delta^2_{Z,\mathrm{v}} &= \frac{1}{S}\overline{\int_S \left|B^Z(\mathbf{r}_\perp)\right|^2 \mathrm{d}^2\mathbf{r}_\perp} - \frac{1}{S^2}\overline{\left|\int_S B^Z(\mathbf{r}_\perp)\mathrm{d}^2\mathbf{r}_\perp\right|^2} \\ &= \int_{\mathrm{RS}} \left|\mathcal{B}^Z(\mathbf{q}_\perp)\right|^2 \sum_{\mathbf{K}_\perp\neq\mathbf{0}} \int_S \exp\left[i\left(\mathbf{q}_\perp-\mathbf{K}_\perp\right)\cdot\mathbf{r}_\perp\right] \mathcal{C}_{\mathbf{K}_\perp}(\mathbf{r}_\perp)\mathrm{d}^2\mathbf{r}_\perp \frac{\mathrm{d}^2\mathbf{q}_\perp}{(2\pi)^2},\end{aligned} \tag{11.69}$$

where S denotes the sample cross section in the $\mathbf{B}_{\mathrm{ext}}$ direction. The translational order correlation function is now given by the formula

$$\mathcal{C}_{\mathbf{K}_\perp}(\mathbf{r}_\perp) = \overline{\exp\left\{i\mathbf{K}_\perp\cdot\left[\mathbf{u}_\perp\left(\mathbf{r}_\perp\right) - \mathbf{u}_\perp\left(\mathbf{0}\right)\right]\right\}}. \tag{11.70}$$

Explicitly we have

$$\mathcal{B}^Z(\mathbf{q}_\perp) \simeq \frac{\Phi_0}{s_c}(1-b_{\rm nor}^4)\frac{vK_1(v)}{1+\lambda_{\rm L}^2 q_\perp^2}, \tag{11.71}$$

with

$$v^2 = 2\xi_{\rm GL}^2 q_\perp^2 (1+b_{\rm nor}^4)\left[1-2b_{\rm nor}(1-b_{\rm nor})^2\right]. \tag{11.72}$$

The $\mathbf{K}_\perp$, $\mathbf{q}_\perp$, $\mathbf{u}_\perp$, and $\mathbf{r}_\perp$ vectors are perpendicular to the vortex tubes. The superconductor is assumed to be isotropic for simplicity. These equations describe the field variance for a vortex lattice with rigid flux tubes. If these tubes are too soft, the three-dimensional version of the previous four equations is required, and therefore the dependence of the correlation function on the coordinates along the field direction must be taken into account. For the Bragg-glass phase, however, this dependence is expected to be negligible relative to the in-plane dependence. Here we focus on the physics of this phase.

As for a three-dimensional system and with the same reservation, when the disorder is strong it is convenient to rewrite Eq. 11.69 introducing

$$\mathcal{A}_\perp = \sum_{\mathbf{K}_\perp \neq \mathbf{0}} \int_S \exp\left[i\left(\mathbf{q}_\perp - \mathbf{K}_\perp\right)\cdot \mathbf{r}_\perp\right] \mathcal{C}_{\mathbf{K}_\perp}(\mathbf{r}_\perp){\rm d}^2\mathbf{r}_\perp. \tag{11.73}$$

We take $\mathcal{A}_\perp$ to be a constant with $\mathcal{A}_\perp = s_{\rm c}$. From the London limit of $b^Z(\mathbf{q})$ which we derive from Eq. 11.35 and the relation $K_1(x) \simeq 1/x$ when x is small (see Section C.10.3) we get

$$\Delta_{Z,{\rm v}}^2 = \frac{\overline{B^Z}\,\Phi_0}{4\pi\lambda_{\rm L}^2}. \tag{11.74}$$

This is the field variance for a strongly disordered vortex system with rigid vortex tubes parallel to the Z axis. The variance is again linearly dependent of $\overline{B^Z}$. With the same parameters as above, we compute $\Delta_{Z,{\rm v}} = 8.55\,{\rm mT}$, i.e. about 50% larger than for the disorder-free FLL; see text below Eq. 11.45. The qualitative difference between materials with highly flexible and stiff flux tubes was first recognised by Brandt (1991). The skewness parameter is found to be given by

$$\vartheta^Z = 1.92108\left(\frac{\overline{B^Z}}{\Phi_0}\right)^{1/6}\lambda_{\rm L}^{1/3}. \tag{11.75}$$

Numerically we get $\vartheta^Z = 1.33$ again with the same parameters ($\overline{B^Z} = 10\,{\rm mT}$ and $\lambda_{\rm L} = 150\,{\rm nm}$). This is to be compared to $\vartheta^Z = 1.44$, which is the low-field limit for the disorder-free vortex lattice; see right panel of Fig. 11.6. As expected, the disorder has therefore slightly decreased the field asymmetry.

As already pointed out, the basic hypothesis required to derive the formula for the variance in terms of the translational correlation function has been to assume the validity of the decomposition of the vortex density in terms of Fourier components. Therefore it cannot be used to describe phases such as the so-called vortex glass because of the expected proliferation of dislocations for that phase. On the other hand, it should be well suited to account for the field variance in the so-called Bragg-glass phase. Strictly speaking, our derivations of the two expressions for the variance in the limit of strong disorder are not valid because of the large density of dislocations expected when the disorder is strong. However, the predictions derived from them should be qualitatively correct.

The Bragg-glass model introduces two length scales: $R_a^\perp$ and R_a^Z (Giamarchi and Le Doussal 1995). They are, respectively, the translational correlation lengths in the plane perpendicular to $\mathbf{B}_{\rm ext}$ and in the direction parallel to that field. The weaker the disorder is, the longer they are. Since in the Bragg-glass regime $R_a^\perp$ is much shorter than R_a^Z, the in-plane disorder is stronger than the out-of-plane disorder. Thus, as a first approximation, we can assume the flux tubes to be stiff and thus Eqs. 11.69 through 11.72 are relevant. Whereas a transverse-field μSR experiment should give access to $R_a^\perp$, a SANS measurement yields an estimate of R_a^Z (Klein *et al.* 2001). When performed with the appropriate experimental geometry, the SANS technique has also the capability of measuring $R_a^\perp$ (Laver *et al.* 2008). Combining transverse-field μSR and SANS experiments on a given sample, detailed experimental information of the Bragg-glass phase should be obtained. This has not yet been achieved.

To summarise our discussion, we present results of the computation of the standard deviation using Eqs. 11.69 through 11.72. To proceed further, we need a model for the correlation function. Although Giamarchi and Le Doussal (1995 and 1998) have shown that it algebraically decays at long distance in the Bragg-glass phase, for the sake of simplicity and ease of convergence we shall use here a conventional exponential decay, $\mathcal{C}_{\mathbf{K}_\perp}(\mathbf{r}_\perp) = \exp(-r_\perp/R_a^\perp)$. Figure 11.7 presents examples of normalised $\Delta_{Z,{\rm v}}$ versus normalised $R_a^\perp$ for different values of $B_{\rm ext}$. The plot shows that the disorder influences the standard deviation of the field distribution even when $R_a^\perp$ is quite large. This is consistent with experimental results since the width of the measured distributions is usually found to be substantially larger than the theoretical prediction for a disorder-free vortex lattice. The field distribution broadening shown in Fig. 11.7 should be considered as preliminary because of our simplifying hypothesis of an exponential correlation function. Anyhow, the methodology is promising since one can account for the field distribution broadening due to disorder in terms of a correlation function, and a correlation length if it can be defined. The effect of disorder can be related to the root-mean-square of the vortex displacement which scales as $\lambda_{\rm L}^2$, if the factor 1 in the denominator of the expression of $B_{\mathbf{K}}^Z$ (see Eq. 11.37) can be neglected (Riseman *et al.* 1995). This suggests $R_a^\perp$ to strongly depend on $\lambda_{\rm L}$. The study of this dependence is left to the reader.

A measurement gives access to the whole distribution and not only to its variance. Therefore an analytical study of the effect of quenched disorder on $D_{{\rm c},\mathbf{B}_{\rm ext}}(B^Z)$ is needed. This has not yet been performed. The only method currently available for accounting for disorder is the convolution technique mentioned at the beginning of the section. It provides an efficient phenomenological description for the pinning-induced disorder of the Bragg-glass phase. However, unsuspected interesting physics may be missed.

We have just addressed the effect of static disorder on a measured distribution. In the next part we compute $P_X(t)$ for a moving perfect vortex lattice.

11.1.1.4 Effect of a moving vortex lattice on the μSR signal. Until now the FLL has been assumed to be static. It is well known that when a transport current is

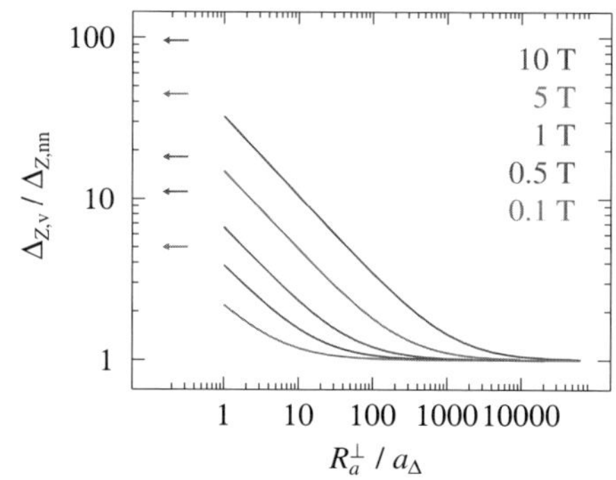

Fig. 11.7 Standard deviation $\Delta_{Z,\mathrm{v}}$ of the field distribution in a weakly disordered stiff vortex lattice as a function of the in-plane correlation length $R_a^\perp$ assuming an exponential correlation function. The sum over $\mathbf{K}_\perp$ in Eq. 11.69 is restricted to the first six $\mathbf{K}_\perp$ vectors of a triangular flux-line lattice. The normalisation factor $\Delta_{Z,\mathrm{nn}}$ is defined as $\sum'_{\mathbf{K}_\perp \neq \mathbf{0}} \left| B^Z_{\mathbf{K}_\perp} \right|^2$ where the sum runs over the same $\mathbf{K}_\perp$ vectors. The length $R_a^\perp$ is given in units of the triangular lattice parameter $a_\triangle$. We have set $B_{c2} = 18.5\,\mathrm{T}$, $\lambda_\mathrm{L} = 126\,\mathrm{nm}$, and $\xi_\mathrm{GL} = 1.8\,\mathrm{nm}$. These are approximately the high-field material parameters measured at 3.8 K in V_3Si for which we neglect the triangular-to-square FLL transition observed when the field is ramped up (Sonier *et al.* 2004). The different curves from top to bottom correspond to different external field magnitudes indicated in the same order. The arrows, again in the same order, show the respective limits in the case of extreme disorder. These limits are obtained from Eq. 11.69 with $\mathcal{B}^Z(\mathbf{q}_\perp)$ given by Eq. 11.71 and $\mathcal{A}_\perp = s_c$; $\mathcal{A}_\perp$ defined in Eq. 11.73.

applied to a type II superconductor in its mixed state, the FLL can break free from pinning and move in the direction of the Lorentz force. A moving vortex lattice is an example of a nonequilibrium system. It has been established to be more ordered in this state than in the pinned state. The first study of a moving FLL by a local probe was performed by Delrieu (1973). Charalambous *et al.* (2006) have reported a detailed study by transverse-field μSR measurements. In this section we shall explain how $P_X(t)$ provides information on a FLL moving as a whole with velocity $\mathbf{v}$ relative to the laboratory reference frame. The muons will be taken as static. The geometry of the experiment is pictured in Fig. 11.8.

Here, for simplicity, we study a moving perfect vortex lattice. The extension to a disordered vortex lattice can be performed with the material of the previous part, in particular Eq. 11.53. As usual we work in the high-transverse-field limit and therefore we only need to consider the component of the induction along the Z direction; see Section 6.2.1. Generalising Eq. 11.6, it can be expressed as a Fourier series:

$$B^Z(\mathbf{r}, t) = \sum_{\mathbf{K}} B^Z_{\mathbf{K}} \exp\left[i\mathbf{K} \cdot (\mathbf{r} - \mathbf{v}t)\right]. \tag{11.76}$$

As explained in Section 6.2.1, a transverse-field experiment provides information on the decay of the muon polarisation resulting from the dephasing of the individual muon spin precession. Referring to Section 11.1.1, we know that the muons uniformly probe

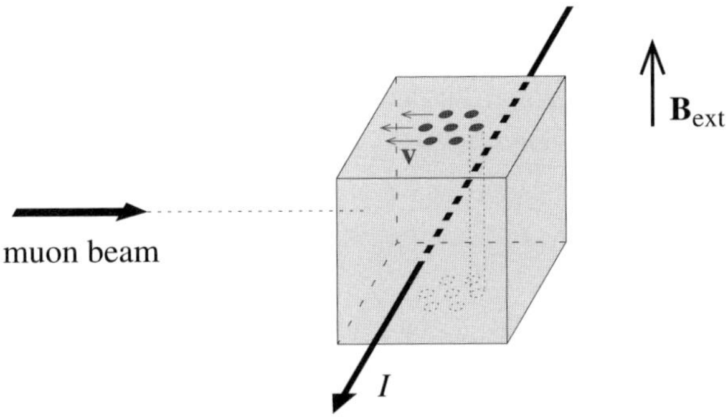

Fig. 11.8 Schematic diagram of the experimental setup used to study the motion of a vortex lattice. A current I flows through the sample which is submitted to an external field. The flux lines move with velocity $\mathbf{v}$ under the action of the Lorentz force. Here we assume a standard transverse-field geometry. It is in use when $B_{\rm ext}$ is not too large; see Section 2.2.2.3.

the induction field. Hence, resorting to the complex polarisation function $P_{X+iY}(t) \equiv P_X(t) + iP_Y(t)$ (see Section A.2), the measured polarisation is written as an average over a unit cell of the FLL:

$$P_{X+iY}(t) = \overline{P_{X+iY}(\mathbf{r},t)} = \int_{s_c} P_{X+iY}(\mathbf{r},t)\frac{\mathrm{d}^2\mathbf{r}}{s_c}, \tag{11.77}$$

where

$$P_{X+iY}(\mathbf{r},t) = \exp\left[i\int_0^t \gamma_\mu B^Z(\mathbf{r},t')\mathrm{d}t'\right]. \tag{11.78}$$

Working out these last two equations the polarisation functions $P_X(t)$ and $P_Y(t)$ can be computed. Here, we shall only consider the implications of Eq. 11.78. Two cases need to be distinguished: either $\mathbf{K}\cdot\mathbf{v}$ vanishes for some set of FLL reciprocal vectors, or $\mathbf{K}\cdot\mathbf{v}$ is finite for all $\mathbf{K}$, except of course $\mathbf{K} = 0$.

We first consider that the velocity vector is perpendicular to no reciprocal lattice vector. Substituting Eq. 11.76 into Eq. 11.78, we get

$$P_{X+iY}(\mathbf{r},t) = \exp\left(i\gamma_\mu\overline{B^Z}t\right)\prod_{\mathbf{K}\neq 0}\exp\left\{-\frac{\gamma_\mu B^Z_{\mathbf{K}}}{\mathbf{K}\cdot\mathbf{v}}\exp(i\mathbf{K}\cdot\mathbf{r})[\exp(-i\mathbf{K}\cdot\mathbf{v}t)-1]\right\}. \tag{11.79}$$

In the high-velocity limit, the product over $\mathbf{K}\neq 0$ is equal to 1 and

$$P_{X+iY}(t) = \exp\left(i\gamma_\mu\overline{B^Z}t\right). \tag{11.80}$$

We recover the extreme motional narrowing limit since the oscillation is no longer damped. The field distribution as seen by the muons therefore corresponds to a Dirac delta function centred at $\overline{B^Z}$.

We now examine a FLL moving in the $\hat{\mathbf{X}}_\mathrm{v}$ direction, i.e. $\mathbf{v} = v\hat{\mathbf{X}}_\mathrm{v}$; see Fig. 11.8 for a schematic diagram of the experimental setup and Fig. 11.1 for the definition of $\hat{\mathbf{X}}_\mathrm{v}$. Considering a given reciprocal lattice vector $\mathbf{K}_{pq}$, according to Eq. 11.17, $\mathbf{K}_{pq} \cdot \mathbf{v}$ vanishes only if p does. It is convenient at this stage to write $P_{X+iY}(\mathbf{r}, t)$ in terms of $\mathbf{K}_{pq}$ and $B^Z_{\mathbf{K}_{pq}}$. Using Eqs. 11.17 and 11.18, we obtain

$$\begin{aligned} P_{X+iY}(\mathbf{r}, t) = {} & \exp\left(i\gamma_\mu \overline{B^Z} t\right) \\ & \times \prod_{q\neq 0} \exp\left[i\gamma_\mu B^Z_{\mathbf{K}_{0,q}} t \cos\left(K^{Y_\mathrm{v}}_{0,q} r^{Y_\mathrm{v}}\right)\right] \exp\left[-\gamma_\mu B^Z_{\mathbf{K}_{0,q}} t \sin\left(K^{Y_\mathrm{v}}_{0,q} r^{Y_\mathrm{v}}\right)\right] \\ & \times \prod_{p\neq 0,q} \left\{ \exp\left\{ -\frac{\gamma_\mu B^Z_{\mathbf{K}_{pq}}}{K^{X_\mathrm{v}}_{pq} v} \left[\cos\left(\mathbf{K}_{pq}\cdot\mathbf{r} - K^{X_\mathrm{v}}_{pq} vt\right) - \cos\left(\mathbf{K}_{pq}\cdot\mathbf{r}\right)\right] \right\} \right. \\ & \left. \times \exp\left\{ -i\frac{\gamma_\mu B^Z_{\mathbf{K}_{pq}}}{K^{X_\mathrm{v}}_{pq} v} \left[\sin\left(\mathbf{K}_{pq}\cdot\mathbf{r} - K^{X_\mathrm{v}}_{pq} vt\right) - \sin\left(\mathbf{K}_{pq}\cdot\mathbf{r}\right)\right] \right\} \right\}, \end{aligned} \tag{11.81}$$

where we have introduced $r^{Y_\mathrm{v}} = \mathbf{r} \cdot \hat{\mathbf{Y}}_\mathrm{v}$. Besides the factor oscillating with an angular frequency $\gamma_\mu \overline{B^Z}$ in the first line of the previous equation, we have products of real and purely imaginary exponential functions of time. The former factors correspond to the damping of the oscillations described by the latter. Let us consider one of the oscillating factors linear in time:

$$P_q(\mathbf{r}, t) = \exp\left[i\gamma_\mu B^Z_{\mathbf{K}_{0,q}} t \cos\left(K^{Y_\mathrm{v}}_{0,q} r^{Y_\mathrm{v}}\right)\right]. \tag{11.82}$$

The argument of the exponential function in $P_q(\mathbf{r}, t)$ contains a cosine term. This suggests the use of Eq. C.94 after setting $\lambda = i \exp\left(iK^{Y_\mathrm{v}}_{0,q} r^{Y_\mathrm{v}}\right)$. We get

$$P_q(\mathbf{r}, t) = \sum_{n=-\infty}^{+\infty} i^n \exp\left(inK^{Y_\mathrm{v}}_{0,q} r^{Y_\mathrm{v}}\right) J_n\left(\gamma_\mu B^Z_{\mathbf{K}_{0,q}} t\right). \tag{11.83}$$

The fact that the Bessel functions $J_n\left(\gamma_\mu \mathbf{B}^Z_{\mathbf{K}_{0,q}} t\right)$ are present in the expression for $P_q(\mathbf{r}, t)$ is remarkable. Referring to Section 10.1.1, the presence of the J_0 term in Eq. 11.83 is reminiscent of the polarisation function observed in magnets with an incommensurate structure; see Eq. 10.8. The analogy between the two physical situations is even stronger in the limit of a large FLL velocity since in this case only J_0 survives in Eq. 11.83. Indeed, at large velocity, the product appearing in the last two lines of Eq. 11.81 is equal to 1 and therefore $P_{X+iY}(t)$ is obtained from an average of the remaining factors over a unit cell of the FLL (see Eq.11.77). When $\mathbf{r}$ spans the FLL unit cell, the angle $\alpha = K^{Y_\mathrm{v}}_{0,q} r^{Y_\mathrm{v}}$ spans the interval $[0, 2q\pi[$. This suggests that only the term $n = 0$ of the $P_q(\mathbf{r}, t)$ factors contribute to $P_{X+iY}(t)$. This means that, beside a shift equal to $\overline{B^Z} \simeq B_\mathrm{ext}$, the field distributions probed by static muons in a fast-moving FLL and in an incommensurate magnet are similar.

To conclude this part, we mention that a comparison of the observed and predicted field distributions enables a direct estimate of the FLL velocity. A nice example is given by Charalambous *et al.* (2006).

11.1.1.5 Superconducting length scales. Four types of length scales have been introduced: R_a, ξ_0, $\xi_{\rm GL}$, and $\lambda_{\rm L}$.

The positional correlation lengths $R_a^\perp$ and R_a^Z parametrise the pinning-induced disorder perpendicular and along the applied magnetic field for a material in its Bragg-glass phase. They should not be confused with the two Larkin–Ovchinikov pinning lengths directly related to the current above which the flux tubes start to move; see e.g. Tinkham (1996). Since they have been introduced more recently, the thermal and field behaviours of $R_a^\perp$ and R_a^Z have not yet been experimentally established. However, because the effective density of defects increases with the field, we expect $R_a^\perp$ and R_a^Z to shrink at large fields.

The other three types of length scales are related to microscopic properties of the superconducting ground state. They can be influenced by impurity scattering. For instance, if the mean free path $\ell_{\rm mfp}$ is not very large compared with ξ_0, the length which is deduced from the measured field standard deviation and Eq. 11.44 is not the London penetration depth. It is rather an effective penetration depth $\lambda_{\rm eff}$ which is linked to $\lambda_{\rm L}$ through the phenomenological relation

$$\lambda_{\rm eff} = \lambda_{\rm L} \left(1 + \frac{\xi_0}{\ell_{\rm mfp}}\right)^{1/2}; \tag{11.84}$$

see for example Tinkham (1996). However, the quality of the available materials is usually sufficiently good for the effect of impurity scattering to be neglected. Thereafter, we shall take this assumption for granted.

The Pippard–BCS coherence length ξ_0 can be expressed in terms of the Fermi velocity $v_{\rm F}$ and the energy gap at zero temperature Δ_0:

$$\xi_0 = \frac{\hbar v_{\rm F}}{\pi \Delta_0}. \tag{11.85}$$

This is a microscopic parameter related to the possible nonlocal character of the electrodynamics.

The last two length scales, the GL coherence length $\xi_{\rm GL}$ and London penetration depth $\lambda_{\rm L}$, can *a priori* be extracted from transverse-field μSR experiments as explained earlier in this chapter. They provide microscopic information on a superconducting phase. In addition, $\xi_{\rm GL}$ is related to B_{c2} (see Eq. 11.40) and $\lambda_{\rm L}$ mainly to B_{c1} through the well-known relationship (see for example de Gennes (1966))

$$B_{c1} = \frac{\Phi_0}{4\pi \lambda_{\rm L}^{X_{\rm v}} \lambda_{\rm L}^{Y_{\rm v}}} \ln \kappa_{\rm GL}. \tag{11.86}$$

The dependence of B_{c1} on $\kappa_{\rm GL} = \lambda_{\rm L}/\xi_{\rm GL}$ is mild since it appears in a logarithm function. In some superconducting materials such as the high-$T_{\rm c}$ superconductors, it is impossible to measure B_{c2} directly because of the presence of the liquid vortex phase at high temperature. It is also difficult to safely estimate B_{c1} from bulk magnetisation measurements. Using Eq. 11.86 and Eq. 11.40 when it applies, the transverse-field μSR technique is an experimental method for indirectly estimating values of both B_{c1} and B_{c2} from measurements of $\lambda_{\rm L}$ and $\xi_{\rm GL}$.

We first consider $\xi_{\rm GL}$. As first noted by Caroli *et al.* (1964), a single vortex core can be seen as a cylinder of normal metal. The density of states per volume unit associated with the bound states of an array of noninteracting vortices is written as (Fetter and Hohenberg 1969)

$$N_b(E_{\rm F}) = N(E_{\rm F})\pi\xi_0 \frac{\pi I}{\left.\frac{\partial\Psi(r)}{\partial r}\right|_{r=0}} \frac{\overline{B^Z}}{\Phi_0}, \tag{11.87}$$

where $N(E_{\rm F})$ is the density (per spin) of normal electron states at the Fermi level, $\partial\Psi(r)/\partial r|_{r=0}$ the derivative of the order parameter at the vortex centre, and I a constant of the order of unity. Using Clem's variational solution of the GL equations (Eq. 11.29), since $\partial\Psi(r)/\partial r|_{r=0} = 1/\xi_{\rm v} \simeq 1/(\sqrt{2}\xi_{\rm GL})$, $N_b(E_{\rm F})$ is found to be proportional to ξ_0 and $\xi_{\rm GL}$. Unlike ξ_0, $\xi_{\rm GL}$ depends on the temperature. A classical method for estimating the density of states of a system in solid state physics is through the measurement of its specific heat. It is well known that the electronic contribution to the specific heat of a metal is proportional to the temperature and to the density of states. In the case of noninteracting vortex cores, from Eq. 11.87 we derive for the specific heat of the cores,

$$C_{\rm core}\,(T, B_{\rm ext}) = \frac{2}{3}\pi^2 N_b(E_{\rm F}) k_B^2 T = \sqrt{2}\pi^2 \xi_0 I \xi_{\rm GL} \frac{\overline{B^Z}}{\Phi_0} \gamma_{\rm Som} T, \tag{11.88}$$

where $\gamma_{\rm Som}T = (2/3)\pi^2 N(E_{\rm F}) k_B^2 T$ is the normal-state electronic specific heat. This formula has been used to interpret the field dependence of specific-heat data recorded in the mixed phase of a conventional superconductor in terms of the field dependence of $\xi_{\rm GL}$ measured by the transverse-field μSR technique; see Section 11.2.1.

We now turn our attention to the London penetration depth. The physics is described by the two tensors $\boldsymbol{\rho}_{\rm sup}$ and $\boldsymbol{m}_{\rm car}$ that we shall describe below. These tensors are assumed to be diagonal in the same principal axes, and we shall write relations in terms of components along each of these axes. For simplicity we shall use a single superscript for the tensor components instead of two identical. The key physical quantity is not $\lambda_{\rm L}^{\alpha}(T)$ but rather

$$[\lambda_{\rm L}^{\alpha}(T)]^{-2} = [\lambda_{\rm L}^{\alpha}(0)]^{-2} \rho_{\rm sup}^{\alpha}(T). \tag{11.89}$$

The parameter $\rho_{\rm sup}^{\alpha}(T)$ is the normalised superfluid density, i.e. a pure number with $\rho_{\rm sup}^{\alpha}(T) = n_{\rm sup}^{\alpha}(T)/n_{\rm car}^{\alpha}$, where $n_{\rm sup}^{\alpha}(T)$ is the volume density of the carriers flowing along the α direction with $n_{\rm sup}^{\alpha}(T=0) = n_{\rm car}^{\alpha}$. Denoting $m_{\rm car}^{\alpha}$ as the effective mass of the same carriers, we have

$$\frac{1}{[\lambda_{\rm L}^{\alpha}(0)]^2} = \frac{\mu_0 e^2 n_{\rm car}^{\alpha}}{m_{\rm car}^{\alpha}}. \tag{11.90}$$

In simple cases the two aforementioned tensors are scalar and the carriers are the conduction electrons which all superconduct at zero temperature.

We have previously derived the variance of $D_{{\rm c},\mathbf{B}_{\rm ext}}(B^Z)$ to be proportional to $(\lambda_{\rm L}^{X_{\rm v}}\lambda_{\rm L}^{Y_{\rm v}})^{-1}$ for an ideal FLL; see Eqs. 11.44 and 11.27. Hence, from measurements at low temperature of the variance for different appropriate orientations of $\mathbf{B}_{\rm ext}$ it is possible to estimate the ratio of the concentration of the superconducting carriers over their effective mass for currents flowing along the different axes.

By definition $\rho^{\alpha}_{\rm sup}(T = 0) = 1$. It decreases as the sample is warmed towards its superconducting critical temperature, at which it vanishes. The tensor $\boldsymbol{\rho}_{\rm sup}$ is scalar for an isotropic material. If, in addition, the order parameter does not have nodes, the single non-zero element of $\boldsymbol{\rho}_{\rm sup}$ displays an activated thermal behaviour with a temperature-dependent activation energy. The superfluid density is then almost independent of temperature up to relatively high temperatures. However, for most of the superconductors of modern interest, the gap presents points and possibly lines of nodes. As a result, $\boldsymbol{\rho}_{\rm sup}$ is usually both anisotropic and strongly temperature dependent even at low temperature. In order to express the temperature dependence of $\boldsymbol{\rho}_{\rm sup}$, it is convenient to define the paramagnetic tensor $\boldsymbol{\rho}_{\rm par}$ which accounts for the depletion of the normalised superfluid density by thermal paramagnetic excitations as the superconductor is heated:

$$\boldsymbol{\rho}_{\rm sup} = \boldsymbol{I} - \boldsymbol{\rho}_{\rm par}. \tag{11.91}$$

It is found at low temperatures that

$$\rho^{\alpha}_{\rm par}(T) = c_{\alpha} T^{n_{\alpha}}. \tag{11.92}$$

The exponent n_{α} depends not only on the dimension of the manifold of gap nodes (i.e. points, lines, etc.) but also on the shape of the gap in the neighbourhood of its zeros. If a high density of states is available in the α direction, the normal quasiparticles are easily excited and $[\lambda^{\alpha}_{\rm L}(T)]^{-2}$ decreases rapidly when warming the sample. Therefore the study of the anisotropy and temperature dependence of $(\lambda^{\alpha}_{\rm L})^{-2}$ is a valuable method for gathering information on the characteristics of the nodes of a superconductor. The value of the exponent strongly depends on the $\mathbf{B}_{\rm ext}$ orientation relative to the crystal axes (Gross *et al.* 1986, Groß-Alltag *et al.* 1991, Mineev and Samokhin 1999). For example, depending on the orientation, n_{α} takes values equal to 1 or 3 for the so-called polar state characterised by an equatorial line of nodes, and 2 or 4 for the axial state with point nodes at the poles. Note that not only the exponent n_{α} is of interest; the value of the factor c_{α} carries information on the Fermi surface (Gross *et al.* 1986, Groß-Alltag *et al.* 1991). It should also be remarked that impurities might change the behaviour of $[\lambda^{\alpha}_{\rm L}(T)]^{-2}$ at low temperature, such that any conclusion concerning the symmetry of the order parameter may be erroneous. This was the case for $YBa_2Cu_3O_{6.95}$ for many years after its discovery. In fact, the comparison of $\rho^{\alpha}_{\rm par}(T)$ at low temperature for a pure and a doped compound provides information on the superconducting gap. Additional experiments, such as angle-resolved photoemission spectroscopy (ARPES) or Josephson effect, are also helpful for obtaining information in this respect.

Finally we point out that the presence of nodes in the gap is not necessarily a signature of the unconventional nature of a superconductor, i.e. with a gap function of symmetry lower than that of the group symmetry of the crystal. Gap nodes may also exist for conventional (or s-wave) pairing; see e.g. Annett (2004).

The neutron FLL cross section usually becomes extremely small at high temperature because of its proportionality to $(\lambda^{\alpha}_{\rm L})^{-4}$. On the other hand, it may be possible to follow $[\lambda^{\alpha}_{\rm L}(T)]^{-2}$ up to $T_{\rm c}$ with the transverse-field μSR method; see for example the discussion in Section 11.2.1. Therefore we need an expression for $[\lambda_{\rm L}^{\alpha}(T)]^{-2}$ valid over the whole temperature range, and not only in the asymptotic low-temperature limit. Referring to the literature (Gross *et al.* 1986), the $\boldsymbol{\rho}_{\rm par}$ tensor components in its principal axes are given by

$$\rho_{\rm par}^{\alpha}(T) = \frac{1}{n_{\rm car}^{\alpha}}\frac{\hbar^2}{m_{\rm car}^{\alpha}}\int_{\rm FV}(k^{\alpha})^2\frac{-\partial n_{\rm FD}(\beta E_{\mathbf{k}})}{\partial E_{\mathbf{k}}}\frac{{\rm d}^3\mathbf{k}}{(2\pi)^3}, \tag{11.93}$$

where the integral is performed over the volume enclosed by the Fermi surface and

$$E_{\mathbf{k}} = \sqrt{\xi^2(\mathbf{k}) + \Delta_{\rm sup}^2(\mathbf{k})}. \tag{11.94}$$

In Eq. 11.93, $n_{\rm FD}(\beta E_{\mathbf{k}})$ is the Fermi–Dirac function (Eq. C.30) with the origin shifted by $E_{\rm F}$ (Tinkham 1996): $n_{\rm FD}(\beta E_{\mathbf{k}}) = [\exp(\beta E_{\mathbf{k}}) + 1]^{-1}$. The function $\xi(\mathbf{k})$ is the dispersion relation of the quasiparticles and $\Delta_{\rm sup}(\mathbf{k})$, the energy gap function, which is usually written in the product form

$$\Delta_{\rm sup}(\mathbf{k}) = \Delta_0(T)f(\theta,\varphi), \tag{11.95}$$

where θ and φ denote the polar and azimuthal angles of $\mathbf{k}$, respectively. The energy gap function written in the previous equation is a generalisation of the original BCS isotropic gap for which $f(\theta,\varphi)$ is a constant equal to 1. An interpolation formula is available for $\Delta_0(T)$ in the weak coupling approximation (Gross *et al.* 1986):

$$\Delta_0(T) = \Delta_0 \tanh\left(a\sqrt{\frac{T_{\rm c}}{T} - 1}\right), \tag{11.96}$$

where the parameters Δ_0 and a are independent of temperature. An order of magnitude for a is 1. In general terms $f(\theta,\varphi)$ is a weighted sum of spherical harmonics. In favourable cases, the analysis of $[\lambda_{\rm L}^{\alpha}(T)]^{-2}$ helps to determine which harmonics contribute significantly to $\Delta_{\rm sup}(\mathbf{k})$ and thus to obtain information on the symmetry of the superconducting order parameter. This symmetry is directly reflected by the polar and azimuthal dependences of $\Delta_{\rm sup}(\mathbf{k})$. An example is found in the literature for the heavy-fermion superconductor UPt_3 (Yaouanc *et al.* 1998).

Let us consider a spherical Fermi surface of radius $k_{\rm F}$. The electron density, i.e. the carrier density, is equal to

$$n_{\rm car} = \int_{\rm FV}\frac{{\rm d}^3\mathbf{k}}{(2\pi)^3} = \frac{1}{(2\pi)^3}\frac{4\pi}{3}k_{\rm F}^3 = \frac{k_{\rm F}^3}{6\pi^2}. \tag{11.97}$$

For simplicity we have suppressed the superscript α of $n_{\rm car}$. Assuming the order parameter to be isotropic, i.e. $\Delta_{\rm sup}(\mathbf{k}) = \Delta_0(T)$ is independent on the orientation of $\mathbf{k}$, the tensor $\boldsymbol{\rho}_{\rm par}$ is scalar as well as the effective mass tensor. From Eq. 11.93, we compute one of its elements

$$\rho_{\rm par}(T) = \frac{1}{24\pi^2}\frac{\hbar^2}{n_{\rm car}m_{\rm car}}\frac{1}{k_{\rm B}T}\int_0^{\infty}k^4\cosh^{-2}\left(\frac{\sqrt{\xi^2(k)+\Delta_0^2(T)}}{2k_{\rm B}T}\right){\rm d}k, \tag{11.98}$$

where we have taken the quasiparticle dispersion relation $\xi(\mathbf{k})$ to depend only on the modulus of $\mathbf{k}$. Again we have dropped the superscript α in $\rho_{\rm par}(T)$ and $m_{\rm car}$. Using the usual form $\xi(k) = \hbar^2 k^2/(2m_{\rm car}) - E_{\rm F}$, we can proceed with the change of variable $k \longrightarrow \xi$ in the integral to arrive at

$$\rho_{\rm par}(T) = \frac{1}{24\pi^2} \frac{(2m_{\rm car})^{3/2}}{n_{\rm car}\hbar^3} \frac{1}{k_{\rm B}T} \int_{-E_{\rm F}}^{\infty} (\xi + E_{\rm F})^{3/2} \cosh^{-2}\left(\frac{\sqrt{\xi^2 + \Delta_0^2(T)}}{2k_{\rm B}T}\right) {\rm d}\xi. \tag{11.99}$$

The inverse square hyperbolic cosine factor, which results from the derivative of the Fermi-Dirac stastics is close to 1 when its argument is very close to 0 and vanishes otherwise. Since $\Delta_0(T) \ll E_{\rm F}$, ξ is small compared to $E_{\rm F}$. It is then justified to replace the factor $\xi + E_{\rm F}$ in the integrand with $E_{\rm F}$. The lower bound of the integral can also be set to $-\infty$. Acknowledging the relation $E_{\rm F} = \hbar^2 k_{\rm F}^2/(2m_{\rm car})$ and using Eq. 11.97,

$$\rho_{\rm par}(T) = \frac{1}{4k_{\rm B}T} \int_{-\infty}^{\infty} \cosh^{-2}\left(\frac{\sqrt{\xi^2 + \Delta_0^2(T)}}{2k_{\rm B}T}\right) {\rm d}\xi. \tag{11.100}$$

It is simple matter to check that $\rho_{\rm par}(T_{\rm c}) = 1$ as it should be.

More involved cases can be considered. Here we investigate the normalised superfluid density when $\xi(\mathbf{k})$ depends on the orientation of the wave vector $\mathbf{k}$. We assume the isoenergetic surfaces to be ellipsoids with a revolution axis taken as the c axis. The centre of the ellipsoid is located at the centre of the reciprocal space. For this simple case two functions are relevant: $\rho^c_{\rm par}$ and $\rho^\perp_{\rm par}$. They stand for the normalised density of thermal paramagnetic excitations propagating along the c axis and in the plane perpendicular to it.

To proceed further we need to explicitly write the dispersion relation. In Cartesian coordinates

$$\begin{aligned} \xi\left(k^x, k^y, k^z\right) &= \frac{\hbar^2}{2m^x_{\rm car}} \left(k^x\right)^2 + \frac{\hbar^2}{2m^y_{\rm car}} \left(k^y\right)^2 + \frac{\hbar^2}{2m^z_{\rm car}} \left(k^z\right)^2 - E_{\rm F} \\ &= \frac{\hbar^2}{2m_{\rm car}c_{\rm ani}} \left(k^x\right)^2 + \frac{\hbar^2}{2m_{\rm car}c_{\rm ani}} \left(k^y\right)^2 + \frac{\hbar^2 c^2_{\rm ani}}{2m_{\rm car}} \left(k^z\right)^2 - E_{\rm F}, \end{aligned} \tag{11.101}$$

where $c_{\rm ani}$ characterises the mass tensor anisotropy: $m^x_{\rm car} = m^y_{\rm car} = c^3_{\rm ani} m^z_{\rm car} = c_{\rm ani} m_{\rm car}$. Taking advantage of the axial symmetry of the Fermi surface, we write in spherical coordinates $\xi\left(k, \theta, \varphi\right) = \xi\left(k, \theta\right)$, where

$$\xi\left(k, \theta\right) = \frac{\hbar^2 k^2}{2m_{\rm car}} \left[\frac{1}{c_{\rm ani}} + \cos^2\theta \left(c^2_{\rm ani} - \frac{1}{c_{\rm ani}}\right)\right] - E_{\rm F}. \tag{11.102}$$

With the definition adopted for the anisotropy parameter $c_{\rm ani}$, the volume enclosed by the Fermi surface is independent of it and Eq. 11.97 applies. With this material at hand and referring to the literature (Gross *et al.* 1986, Groß-Alltag *et al.* 1991),

$$\begin{aligned} \rho^c_{\rm par} &= \frac{3}{8} \frac{\hbar^2}{m_{\rm car} k_{\rm F}^3 k_{\rm B}T} \int_0^{\pi} \cos^2\theta \sin\theta {\rm d}\theta \int_0^{\infty} \frac{k^4}{\cosh^2\left(\dfrac{E_{\mathbf{k}}}{2k_{\rm B}T}\right)} {\rm d}k, \\ \rho^\perp_{\rm par} &= \frac{3}{16} \frac{\hbar^2}{m_{\rm car} k_{\rm F}^3 k_{\rm B}T} \int_0^{\pi} \sin^2\theta \sin\theta {\rm d}\theta \int_0^{\infty} \frac{k^4}{\cosh^2\left(\dfrac{E_{\mathbf{k}}}{2k_{\rm B}T}\right)} {\rm d}k. \end{aligned} \tag{11.103}$$

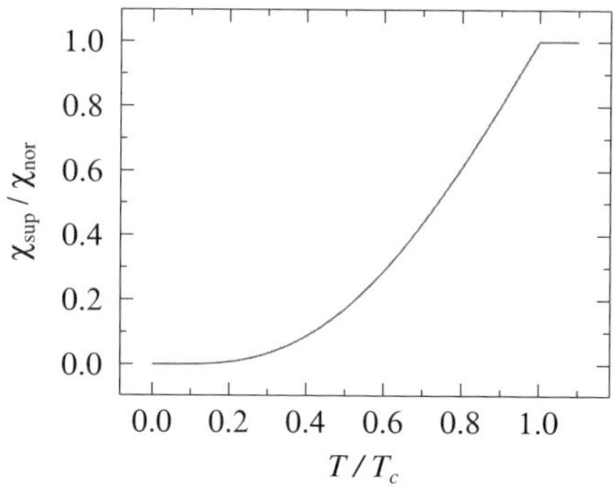

Fig. 11.9 Spin susceptibility in the superconducting state normalised to the spin susceptibility in the normal state versus T/T_c as computed from the BCS theory by Yosida (1958).

11.1.2 Some superconducting fluid properties

As discussed in Section 10.2.3.2, the conduction electron density at the muon site in a non-magnetic metal induces a relaxation, usually extremely weak, of the muon spin. According to the Korringa relation, to the relaxation in a longitudinal-field experiment corresponds a frequency shift, i.e. a Knight shift, in a transverse-field experiment. This relation can also be seen as a consequence of the Kramers–Kronig relation. Quite naturally, we also expect to observe a relaxation and a Knight shift due to a superconducting fluid. These effects are well known from NMR. The theoretical framework is well developed, at least for conventional superconductors; see the review of MacLaughlin (1976). Since the same physical ideas are applicable to μSR, we shall only briefly discuss the response of a superconducting fluid to an external field.

The experimental determination of the Knight shift gives the possibility of measuring the conduction electron susceptibility for a metal if the Fermi contact interaction can be computed; see Eq. 5.78. However, this computation is usually not possible. On the other hand, one can follow the susceptibility as a metal passes to the superconducting state. At first sight the spin susceptibility at low temperature in the superconducting state must be equal to 0, at least in a usual superconductor. Indeed, in the ground state the electrons are bound into singlet Cooper pairs and the excited states are separated from the ground state by an energy gap. Hence, a small magnetic field is not capable of generating magnetisation; i.e. the susceptibility is zero. The ratio $\chi_{\rm sup}/\chi_{\rm nor}$, where $\chi_{\rm sup}$ and $\chi_{\rm nor}$ are, respectively, the spin susceptibilities in the superconducting and normal states, can be computed from BCS theory. Figure 11.9 presents the result.

However, it is well known from NMR that the spin susceptibility deep in the superconducting state is usually finite and can reach a value closed to its value in the normal state, even for a high-quality material with singlet Cooper pairing. This is attributed to the spin–orbit interaction of the conduction electrons. This interaction leads to a ground state wavefunction which contains an admixture of both spin directions. In view of this, polarisation of the conduction electrons in a weak field is possible. The BCS theory is based on plane-wave states with fixed spin. In the case of an appreciable spin–orbit coupling, these states are no longer good one-electron functions. Using a simple extension of the BCS plane-wave approximation, Anderson (1959) has computed the ratio $\chi_{\rm sup}/\chi_{\rm nor}$. Abrikosov and Gor'kov (1962) have performed the same calculation, taking into account the effect of spin-reversing scattering on the energy gap and on the density of states for quasiparticle excitations. However, they had to assume $\ell_{\rm SO} \gg \ell_{\rm mfp}$, where $\ell_{\rm SO}$ and $\ell_{\rm mfp}$ are the mean free paths for spin–orbit scattering and ordinary transport, respectively. Simple expressions for the susceptibility ratio are derived at $T = 0$ in two limiting cases:

$$\frac{\chi_{\rm sup}(T=0)}{\chi_{\rm nor}} \simeq 1 - \varepsilon_p \frac{\ell_{\rm SO}}{\xi_0} \quad \text{when} \quad \ell_{\rm SO} \ll \xi_0,$$
$$\frac{\chi_{\rm sup}(T=0)}{\chi_{\rm nor}} \simeq \varepsilon_b \frac{\xi_0}{\ell_{\rm SO}} \quad \text{when} \quad \ell_{\rm SO} \gg \xi_0. \tag{11.104}$$

According to Anderson $\varepsilon_p = 2/\pi$, whereas the estimate of Abrikosov and Gor'kov is $\varepsilon_p = \frac{3}{4}$. These two values are quite similar. On the other hand, the result of Anderson and of Abrikosov and Gor'kov are clearly different for ε_b: $\varepsilon_b = \pi/6$ and $\varepsilon_b = \pi^2/6$, respectively.

Examples from transverse-field μSR experiments of the spin susceptibility in the superconducting state of materials are discussed in Section 11.2.2. Here, we point out that the spin susceptibility may drop when going through the superconducting temperature, even when the Cooper pairs are in a triplet state. It is sufficient for the pairs to be in a state where the electrons have opposite spin projections. Finally, we note that according to Eq. 11.104 the measurement of the Knight shift potentially provides information on $\xi_0/\ell_{\rm SO}$.

In Section 10.2.3.2 we have argued that the conduction electron spin density at the muon site relaxes the muon spin (or any nuclear spin) through the hyperfine interaction. A key characteristics of a superconductor is the presence of an energy gap below the superconducting temperature. Because of the gap, deep in the superconducting state, the nuclear spin is not expected to be relaxed. This is commonly observed by NMR. It is also seen by μSR, but it is not so common simply because the rate λ_Z is usually already small above $T_{\rm c}$. We would expect the relaxation to increase smoothly up to $T_{\rm c}$ as the superconductor is warmed up. However, in contrast to this expectation, Hebel and Slichter (1959) observed a peak just below $T_{\rm c}$ for the ^{27}Al spin–lattice relaxation rate in aluminium. They realised that the simple argument neglects completely the coherence of the BCS ground state wavefunction. The quasiparticle excitations which relax the spin of the nuclear probe are not those of free particles. They are still correlated by the phase correlations between states of opposite momentum and spin which is a fundamental property of the ground state. Hebel and Slichter derived an expression for the ratio of the spin–lattice relaxation rate in the superconducting and normal states:

$$\frac{\lambda_{Z,\rm sup}}{\lambda_{Z,\rm nor}} = \frac{2}{k_{\rm B}T} \int_{\Delta_{\rm sup}}^{\infty} \left[N_{\rm sup}^2(E) + M_{\rm sup}^2(E)\right] n_{\rm FD}(E) \left[1 - n_{\rm FD}(E)\right] {\rm d}E. \tag{11.105}$$

Here, the energy E is again measured relative to $E_{\rm F}$ and $N_{\rm sup}(E)$ and $M_{\rm sup}(E)$ are normalised densities of states. The singular form of the BCS density of states is not realistic because it neglects the possible gap anisotropy, the finite lifetime of the electronic excitations due to the phonon–electron interaction and the electronic correlations. A rough but simple method to account for these effects is through a broadening function $B(E' - E)$ such that

$$N_{\rm sup}(E) = \int B\left(E' - E\right) \frac{E'}{\left(E'^2 - \Delta_{\rm sup}^2\right)^{1/2}} {\rm d}E',$$

$$M_{\rm sup}(E) = \int B\left(E' - E\right) \frac{\Delta_{\rm sup}}{\left(E'^2 - \Delta_{\rm sup}^2\right)^{1/2}} {\rm d}E'. \tag{11.106}$$

If we take $B(E' - E) = \delta(E' - E)$, the simple BCS formulae for $N_{\rm sup}(E)$ and $M_{\rm sup}(E)$ are found. Better methods to account for the deviation from the BCS model than the convolution technique are possible; see e.g. MacLaughlin (1976) and MacFarlane *et al.* (1998).

11.2 Examples

We shall first discuss the physics of vortex matter as probed by the transverse-field μSR technique. We shall restrict ourselves to five examples, our purpose being to display the unique possibilities of μSR techniques. Additional examples are shown in Section 1.2.3. We shall complete this section by presenting the results of some measurements which provide direct information on the properties of the superconducting fluid.

11.2.1 Vortex matter

Before starting our discussion on the vortex matter, we like to point out that μSR can measure the field penetration depths, even with powder samples. This is nicely exemplified by the measurements performed to establish the so-called "Uemura line"; see Section 1.2.3.

To be of some technical use, a superconductor must pin the flux lines. It is quite simple to determine whether the flux lines are pinned by μSR measurements. An example is given in Fig. 11.10. Obviously, using the method described in the caption of this figure, it is possible to shift significantly the background contribution to the distribution (sample holder, cryostat walls, etc.) from that of the superconductor itself. This can be useful for a quantitative analysis of a spectrum. Conversely, pinning can affect the measured field distribution, which in turn may not reflect the intrinsic properties of the superconducting fluid. This is discussed in Section 11.1.1.3.

As a second example, we discuss the investigation of the vortex state in the highly anisotropic superconductor $Bi_{2.15}Sr_{1.85}CaCu_2O_{8+\delta}$. We focus on the μSR response at low temperature. In this system the vortices are best described as layered systems of two-dimensional interacting pancake vortices. The disorder is expected to increase with $B_{\rm ext}$; see Section 11.1.1.3. At low temperature and for $B_{\rm ext}$ less than the crossover field $B_{\rm cr}$, a lattice of extended flux lines is observed. Yet, at low temperature and for $B_{\rm ext} > B_{\rm cr}$, the field distribution narrows as first reported by Harshman *et al.* (1991). This surprising result has been explained by Brandt; see Section 11.1.1.3. It occurs principally through the truncation of the high-field tail of the distribution, which may

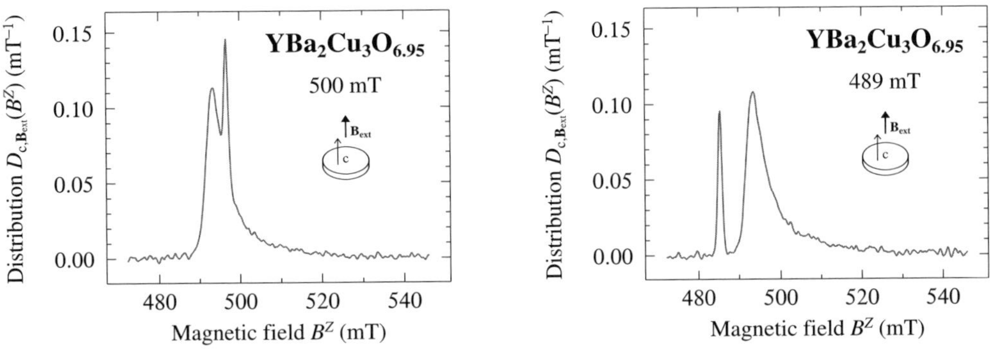

Fig. 11.10 Typical field distributions recorded in $YBa_2Cu_3O_{6.95}$ after cooling the sample in a field of 500 mT. While the distribution of the left panel has been recorded at 5.4 K with this field value, the distribution of the right panel was measured at 6 K after decreasing the field to 489 mT. Because of the high critical temperature of this compound, the difference in temperature for the two measurements is not expected to be significant. The sharp lines just below 500 mT in the left panel and below 490 mT for the other panel represent the background contribution, i.e. muons which are not stopped in the sample. The other part of the distribution arises from the superconductor and is not affected by the field decrease. These measurements prove the vortices to be pinned. Reproduced with kind permission from Dalmas de Réotier and Yaouanc (1997); the original data are from Sonier *et al.* (1994).

be understood as follows. The tail arises from regions close to the centres of vortices; see Fig. 11.2. The onset of disorder along the vortex line washes out this tail since the pancake cores in neighbouring layers no longer coincide. This is in contrast to the effect of disorder on a system of rigid flux lines, where the width of the distribution increases with disorder. The cutoff of the high-field tail is nicely observed in the field distribution and is quantitatively demonstrated by the field dependence of the skewness parameter as shown in Fig. 11.11.

The study of the FLL of some superconductors has resulted in remarkable achievements; see the reviews of Sonier *et al.* (2000a) and Sonier (2007). The thermal variation of the normalised superfluid density of $YBa_2Cu_3O_{6.95}$ at low temperature has been found to decrease linearly with the temperature as the superconductor is warmed; i.e. the exponent $n_\perp$ in Eq. 11.92 is equal to 1. This is the expected behaviour for a d-wave superconductor. A series of experiments performed by Sonier and collaborators has shown that the two basic superconducting lengths, that is, the London penetration depth and the GL coherence length, are field dependent at low field, even for a s-wave superconductor with no gap nodes such as $NbSe_2$. This was first met with skepticism until it was shown that the well-known anomalous field dependence of the specific heat can be explained if the field dependence of the GL coherence length is taken into account (Sonier *et al.* 1999). The field dependence exhibits a downward curvature, which is attributed to vortex interactions that affect the vortex density of states. The experimental data have been interpreted within the theoretical framework outlined in Section 11.1.1.5.

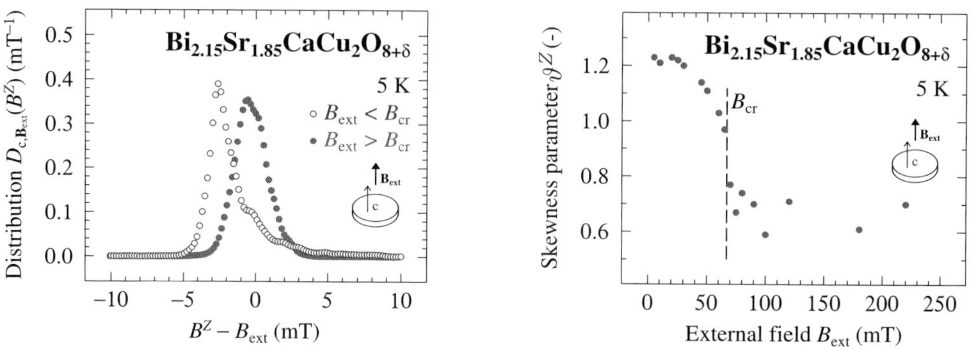

Fig. 11.11 The vortex state in the highly anisotropic superconductor $Bi_{2.15}Sr_{1.85}CaCu_2O_{8+\delta}$ as the intensity of the external magnetic field is increased. (Left) The field distributions measured for external fields of 30 and 100 mT, respectively, i.e. below and above the crossover field B_{cr}. The measurements were both performed after cooling in a field from above T_c down to 5 K, i.e. far below T_c. (Right) The field dependence of the skewness parameter, which allows one to quantify the change in the shape of the distribution. Note the sharpness of the crossover at $B_{cr} = 67\,(2)$ mT, as indicated by the dashed line. Reproduced with kind permission from Dalmas de Réotier and Yaouanc (1997); the original data are from Aegerter *et al.* (1996) and Aegerter (1997).

When analysing a μSR spectrum recorded on a superconductor at low temperature, one approach consists of fitting the data assuming a field distribution obtained from the GL model of Section 11.1.1.2. This method models the whole distribution, and not only its variance. Therefore, it is *a priori* excellent. However, if the field distribution predicted from the GL model is too far from the actual one, the results of the analysis could be unreliable. Rather than trying to analyse the whole field distribution, one may only focus on the variance and skewness parameters since they can be extracted from the spectra without theoretical input. Figure 11.12 presents the results of this approach for the high-T_c superconductor $(BiPb)_2(SrLa)_2CuO_{6+\delta}$. Khasanov *et al.* (2008) have selected this compound because (i) angular-resolved photoemission data are available and (ii) only a single gap is expected to be important for the portion of the Fermi surface probed by the measurements. As seen from the left panel, the *d*-wave symmetry model explains the data. According to the right panel, the skewness parameter is field independent, which suggests that the geometry of the flux line is not modified by the applied field. In this study a successful description of the temperature dependence of the variance is also obtained assuming an angular gap symmetry as derived from angular-resolved photoemission experiments. The whole analysis is based on the hypothesis that the FLL disorder, which must yield an appreciable contribution to the measured variance, does not influence the results of the data analysis. The field independence of the skewness parameter for $B_{ext} \leq 600$ mT at low temperature supports this hypothesis; see right panel of Fig. 11.12. Yet, the available skewness data display a non-negligible temperature dependence above $T_c/2$. Obviously more work is needed to establish the reliability of the data analysis. However, Khasanov and collaborators

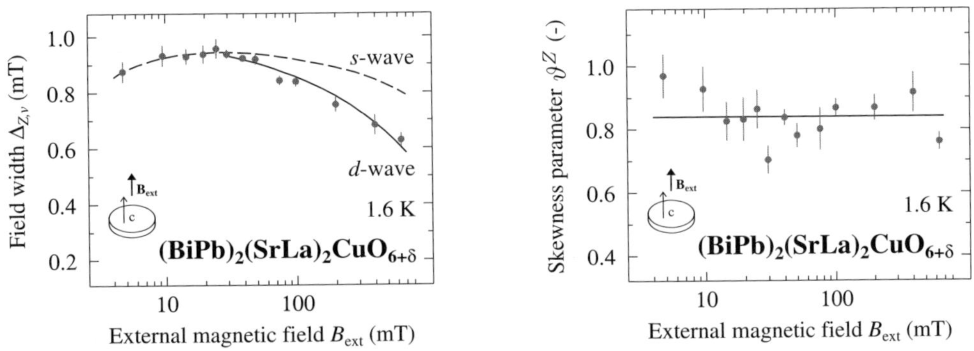

Fig. 11.12 Field dependence of the standard deviation and skewness of the field distribution measured in $(BiPb)_2(SrLa)_2CuO_{6+\delta}$ at 1.6 K. The magnetic field is applied along the c axis. The contribution from the nuclear magnetic moments to the measured variance has been subtracted to get the variance originating from superconductivity. (Left) The dashed line corresponds to the prediction of the GL model for an s-wave superconductor. The solid line displays the expected behaviour for a d-wave superconductor. The fit is only performed for $B_{ext} > 40$ mT because the available theory is only valid deep into the mixed phase. The straight solid line in the right panel results from a fit to a constant value. Adapted from Khasanov *et al.* (2008).

nicely show the advantage of combining the analysis of μSR and other experimental techniques data.

The intermetallic UPt_3 is the only stoichiometric compound for which two superconducting phases in zero field are established beyond doubt. They occur around 0.5 K and are separated by about 50 mK; for a review, see Joynt and Taillefer (2002). The London penetration depth is particularly long in the low-temperature phase. Hence the neutron scattering cross section is small. This limits neutron measurements to low temperatures. On the other hand, the μSR measurements can be extended in the whole low-temperature superconducting phase. The analysis of the second moment of the field distribution is reported in the literature (Yaouanc *et al.* 1998). The opening angle of the vortex lattice at low temperature is deduced from the μSR data and found to be in agreement with the SANS result. As shown in Fig. 11.13, the measurement of this angle is extended up to almost the lower critical temperature, in contrast to the SANS determination which only concerns the low-temperature part of the phase diagram. Information on the symmetry of the order parameter is inferred from the temperature and orientation dependence of the normalised superfluid density.

11.2.2 Superconducting fluid

The observation by NMR of a maximum in the temperature dependence of the ratio $\lambda_{Z,\text{sup}}/\lambda_{Z,\text{nor}}$ just below the superconducting temperature, the so-called Hebel–Slichter coherence peak, is nowadays taken as a reliable signature of a conventional s-wave superconductivity. This effect has not been seen by longitudinal-field μSR measurements for inorganic superconductors, simply because the relaxation is too small even in the

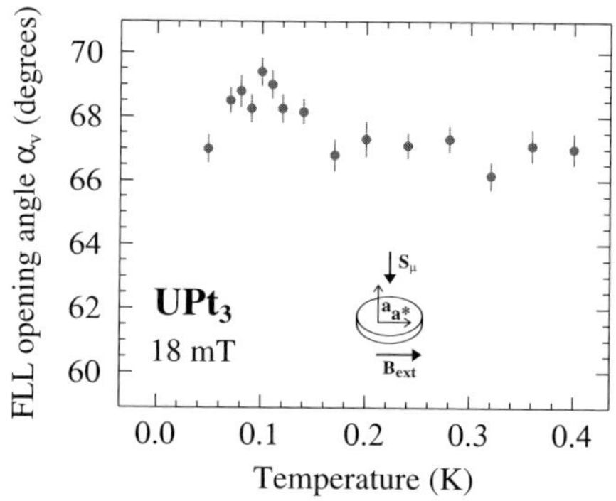

Fig. 11.13 Temperature dependence of the opening angle α_v of the UPt_3 FLL with $\mathbf{B}_{ext}$ applied in the basal plane of the hexagonal crystal structure; see Fig. 11.1 for the definition of α_v. While its temperature dependence is mild, α_v clearly deviates from the value 60° found in isotropic superconductors. Adapted from Yaouanc *et al.* (1998), where the angle is defined as the complementary angle to α_v.

normal state. This is not the case for the alkali-fulleride superconductors. A small fraction of the implanted muons forms endohedral muonium, i.e. a muonium centre located inside the C_{60} cage. This entity is denoted as Mu@C_{60}. The relaxation rate for the complex Mu@C_{60} in a longitudinal field exhibits a Korringa-like temperature dependence in the normal state, i.e. a typical Fermi-liquid signature, and a small Hebel–Slichter coherence peak followed by an activation law thermal dependence well below T_c. An example is presented in Fig. 11.14. The theoretical framework for the interpretation of the Hebel–Slichter coherence peak is given in Section 11.1.2.

As argued in Section 11.1.2, the measurement of the spin susceptibility of the

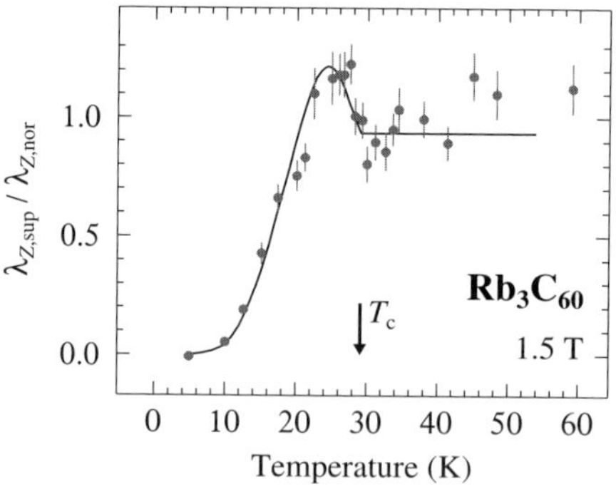

Fig. 11.14 Relaxation rate $\lambda_{Z,sup}$ in units of its value above T_c versus temperature for Mu@C_{60} in Rb_3C_{60}. A longitudinal field of 1.5 T is applied. The solid line is the result of the fit to the Hebel–Slichter theory with an account of the broadening of the superconducting density of states arising from the finite lifetime of the quasiparticle excitations. Adapted from MacFarlane *et al.* (1998).

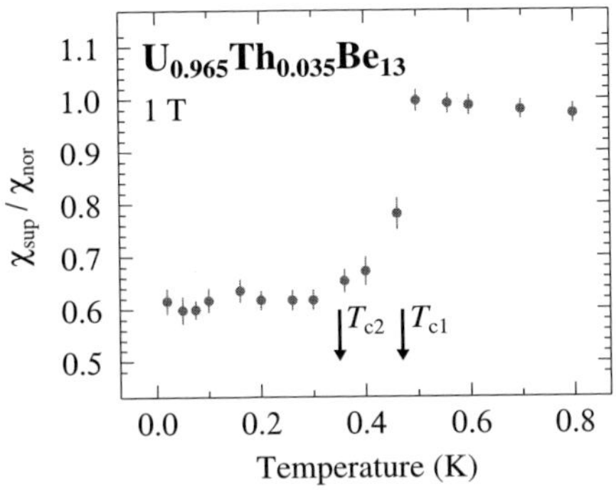

Fig. 11.15 Probing the spin state of the Cooper pairs in the heavy-fermion superconductor $U_{0.965}Th_{0.035}Be_{13}$. This alloy has two superconducting temperatures at $T_{c1} = 0.47\,(5)$ K and $T_{c2} = 0.37\,(2)$ K. We present the temperature dependence of the ratio $\chi_{\rm sup}/\chi_{\rm nor}$, where $\chi_{\rm sup}$ and $\chi_{\rm nor}$ are the spin susceptibilities in the superconducting and normal states, respectively. A transverse field of 1 T is applied. The drop of the susceptibility ratio below T_{c1} is a strong indication for singlet Cooper pairs between T_{c1} and T_{c2}. Adapted from Sonier *et al.* (2000b).

conduction electrons in a superconducting phase of a material is a method for obtaining information on the spin state of the Cooper pairs. A nice example is furnished by Feyerherm *et al.* (1994). The spin susceptibility is found to be isotropic in the strongly anisotropic UPd_2Al_3 superconductor. UPd_2Al_3 is therefore a dual system, with strongly anisotropic localised $5f$ electrons and roughly isotropic conduction electrons. Since a reduction of the susceptibility is observed below the superconducting temperature, the Cooper pairs are in a substate of opposite spin projection. The spin susceptibility has been found to be field dependent, which has been attributed to the breaking of the Cooper pairs in this compound. In Fig. 11.15 we show the result of the measurement of the normalised spin susceptibility for $U_{0.965}Th_{0.035}Be_{13}$. This is an alloy with two superconducting phases. Whereas the normalised susceptibility decreases between the higher and lower superconducting temperatures, no change is detected below the lower superconducting temperature.

References

Abrikosov A.A. (1957) *Soviet Physics JETP* **5**, 1174–1182.

Abrikosov A.A. (1988) *Fundamentals of the Theory of Metals* (North-Holland, Amsterdam).

Abrikosov A.A. and Gor'kov L.P. (1962) *Soviet Physics JETP* **15**, 752–757.

Aegerter C.M. (1997) private communication.

Aegerter C.M., Lee S.L., Keller H., Forgan E.M., and Lloyd S.H. (1996) *Physical Review* **B 54**, R15661–R15664.

Affleck I., Franz M., and Amin M.H.S. (1997) *Physical Review* **B 55**, R704–R707.

Ager C., Ogrin F.Y., Lee S.L., Aegerter C.M., Romer S., Keller H., Savić I.M., Lloyd S.H., Johnson S.J., Forgan E.M., Riseman T., Kealey P.G., Tajima S., and Rykov A. (2000) *Physical Review* **B 62**, 3528–3533.

Amato A. (1997) *Reviews of Modern Physics* **69**, 1119–1180.
Anderson P.W. (1959) *Physical Review Letters* **3**, 325–326.
Annett J.F. (2004) *Superconductivity, Superluids and Condensates* (Oxford University Press, Oxford).
Balatskiĭ A.V., Burlachkov L.I., and Gor'kov L.P. (1986) *Soviet Physics JETP* **63**, 866–871.
Barford W. and Gunn J.M.F. (1988) *Physica* **C 156**, 515–522.
Bennemann K.H. and Ketterson J.B. (2003, 2004) editors of *The Physics of Superconductors*, vol. 1 and vol. 2 (Springer-Verlag, Berlin).
Blatter G., Feigel'man M.V., Geshkenbein V.B., Larkin A.I., and Vinokur V.M. (1995) *Reviews of Modern Physics* **66**, 1125–1388.
Brandt E.H. (1988) *Journal of Low Temperature Physics* **73**, 355–390.
Brandt E.H. (1991) *Physical Review Letters* **66**, 3213–3216.
Brandt E.H. (1995) *Reports on Progress in Physics* **58**, 1465-1594.
Brandt E.H. (1997) *Physical Review Letters* **78**, 2208–2211.
Brandt E.H. (2003) *Physical Review* **B 68**, 054506, 1–11.
Brandt E.H. and Seeger A. (1986) *Advances in Physics* **35**, 189–236.
Caroli C., de Gennes P.G., and Matricon J. (1964) *Physics Letters* **9**, 307–309.
Charalambous D., Forgan E.M., Ramos S., Brown S.P., Lycett R.J., Ucko D.H., Drew A.J., Lee S.L., Fort D., Amato A., and Zimmerman U. (2006) *Physical Review* **B 73**, 104514, 1–9.
Clem J.R. (1975) *Journal of Low Temperature Physics* **18**, 427–434.
Clem J.R. (1991) *Physical Review* **B 43**, 7837–7846.
Cubitt R., Forgan E.M., Yang G., Lee S.L., Paul D.McK., Mook H.A., Yethiraj M., Kes P.H., Li T.W., Menovsky A.A., Tarnawski Z., and Mortensen K. (1993) *Nature* **365**, 407–411.
Dalmas de Réotier P. and Yaouanc A. (1997) *Journal of Physics: Condensed Matter* **9**, R9113–R9166.
Dalmas de Réotier P., Gubbens P.C.M., and Yaouanc A. (2004) *Journal of Physics: Condensed Matter* **16**, S4687–S4705.
Dalmas de Réotier P., Yaouanc A., Heffner R.H., Smith J.L., Gubbens P.C.M., and Kaiser C.T. (2000) *Physical Review* **B 70**, 6377–6380.
de Gennes P.G. (1966) *Superconductivity of Metals and Alloys* (Benjamin, New York; reissued in 1989 by Addison-Wesley, Reading).
Delrieu J.M. (1972) *Journal of Low Temperature Physics* **6**, 197–219.
Delrieu J.-M. (1973) *Journal of Physics F: Metal Physics* **3**, 893–906.
Delrieu J.M. (1974) PhD Thesis, Université Paris-Sud.
de Visser A., Huy N.T., Gasparini A., de Nijs D.E., Andreica D., Baines C., and Amato A. (2009) *Physical Review Letters* **102**, 167003, 1–4.
Fetter A.L. and Hohenberg P. (1969) in *Superconductivity*, vol. 2, 817–923, edited by Parks R.D. (Marcel Dekker, New York).
Feyerherm R., Amato A., Gygax F.N., Schenck A., Geibel C., Steglich F., Sato N., and Komatsubara T. (1994) *Physical Review Letters* **73**, 1849–1852.
Giamarchi T. and Le Doussal P. (1995) *Physical Review* **B 52**, 1242–1270.

Giamarchi T. and Le Doussal P. (1998) in *Spin Glasses and Random Fields*, edited by Young A.P. (World Scientific, Singapore).

Giamarchi T. and Bhattacharya S. (2002) in *High Magnetic Fields: Applications in Condensed Matter Physics and Spectroscopy*, edited by Berthier C., Levy L.P., and Martinez G. (Springer-Verlag, Berlin).

Greer A.J. and Kossler W.J. (1995) *Low Magnetic Fields in Anisotropic Superconductors* (Springer, Berlin).

Gross F., Chandrasekhar B.S., Einzel D., Andres K., Hirschfeld P.J., Ott H.R., Beuers J., Fisk Z., and Smith J.L. (1986) *Zeitschrift für Physik* **B 64**, 175–188.

Groß-Alltag F., Chandrasekhar B.S., Einzel D., Hirschfeld P.J., and Andres K. (1991) *Zeitschrift für Physik* **B 82**, 243–255.

Hao Z., Clem J.R., McElfresh W., Civale L., Malozemoff A.P., and Holtzberg F. (1991) *Physical Review* **B 43**, 2844–2852.

Harshman D.R., Kleiman R.N., Inui M., Espinosa G.P., Mitzi D.B., Kapitulnik A., Pfiz T., and Williams D.Li. (1991) *Physical Review Letters* **67**, 3152–3155.

Hebel L.C. and Slichter C.P. (1959) *Physical Review* **113**, 1504–1519.

Heffner R.H. and Nagamine K. (2004), editors of a special issue on μSR of the *Journal of Physics: Condensed Matter* **16**, number 40.

Huebener R.P. (2001) *Magnetic Flux Structures in Superconductors* (Springer-Verlag, Berlin).

Joynt R. and Taillefer L. (2002) *Reviews of Modern Physics* **74**, 235–294.

Karlsson E.B. (1995) *Solid State Phenomena As Seen by Muons, Protons, and Excited Nuclei* (Clarendon Press, Oxford).

Kealey P.G., Riseman T.M., Forgan E.M., Galvin L.M., MacKenzie A.P., Lee S.L., Paul D.McK., Cubitt R., Agterberg D.F., Heeb R., Mao Z.Q., and Maneo Y. (2000) *Physical Review Letters* **84**, 6094–6097.

Ketterson J.B. and Song S.N. (1999) *Superconductivity* (Cambridge University Press, Cambridge).

Khasanov R., Kondo T., Strässle S., Heron D.O.G., Kaminski A., Keller H., Lee S.L., and Tsunehiro T. (2008) *Physical Review Letters* **101**, 227002, 1–4.

Klein T., Joumard I., Blanchard S., Marcus J., Cubitt R., Giamarchi T., and Le Doussal P. (2001) *Nature* **413**, 404–406.

Kogan V.G. (1981) *Physics Letters* **A 85**, 298–300.

Kogan V.G, Bullock M., Harmon B., Miranovic P., Dobrosavljevic-Grujic Lj., Gammel P.L., and Bishop D.J. (1997) *Physical Review* **B 55**, R8693–R8696.

Laiho R., Safonchik M., and Traito K.B. (2007) *Physical Review* **B 75**, 174524, 1-6.

Laver M., Forgan E.M., Abrahamsen A.B., Bowell C., Geue Th., and Cubitt R. (2008) *Physical Review Letters* **100**, 107001, 1-4.

Lee S. (1999) in *Muon Science: Muons in Physics, Chemistry, and Materials*, edited by Lee S.L., Kilcoyne S.H., and Cywinski R. (Institute of Physics, Bristol).

Leggett A.J. (2006) *Quantum Liquids: Bose Condensation and Cooper Pairing in Condensed-Matter Systems* (Oxford University Press, Oxford).

MacFarlane W.A., Kiefl R.F., Dunsiger S., Sonier J.E., Chakhalian J., Fischer J.E., Yildirim T., and Chow K.H. (1998) *Physical Review* **B 58**, 1004–1024.

MacLaughlin D.E. (1976) in *Solid State Physics*, vol. 31, 1–69, edited by Ehrenreich H., Seitz F., and Turnbull D. (Academic Press, New York).
Menon G.I., Dasgupta C., and Ramakrishnan T.V. (1999) *Physical Review* **B 60**, 7607–7622.
Mineev V.P. and Samokhin K.V. (1999) *Introduction to Unconventional Superconductivity* (Gordon and Breach, Amsterdam).
Miyata S., Osamura K., and Suzuki J. (2001) *Journal of the Physical Society of Japan* **70**, 2098–2101.
Nagamine K. (2003) *Introductory Muon Science* (Cambridge University Press, Cambridge).
Nattermann T. and Scheidl S. (2000) *Advances in Physics* **49**, 607–704.
Poole Jr. C.P., Farack H.A., and Creswick R.J. (1995) *Superconductivity* (Academic Press, San Diego).
Riseman T.M., Brewer J.H., Chow K.H., Hardy W.N., Kiefl R.F., Kreitzman S.R., Liang R., MacFarlane W.A., Mendels P., Morris G.D., Rammer J., Schneider J.W., Niedermayer C., and Lee S.L. (1995) *Physical Review* **B 52**, 10569–10580.
Sidorenko A.D., Smilga V.P., and Fesenko V.I. (1990) *Physica* **C 166**, 167–176.
Sonier J.E. (2007) *Reports on Progress in Physics* **70**, 1717–1755.
Sonier J.E., Brewer J.H., and Kiefl R.F. (2000a) *Reviews of Modern Physics* **72**, 769–811.
Sonier J.E., Hundley M.F., Thompson J.D., and Brill J.W. (1999) *Physical Review Letters* **82**, 4914–4917.
Sonier J.E., Callaghan F.D., Miller R.I., Boaknin E., Taillefer L., Kiefl R.F., Brewer J.H., Poon K.F., and Brewer J.D. (2004) *Physical Review Letters* **93**, 017002, 1–4.
Sonier J.E., Heffner R.H., MacLaughlin D.E., Nieuwenhuys G.J., Bernal O., Movshovich R., Pagliuso P.G., Cooley J., Smith J.L., and Thompson J.D. (2000b) *Physical Review Letters* **85**, 2821–2824.
Sonier J.E., Kiefl R.F., Brewer J.H., Bonn D.A., Carolan J.F., Chow K.H., Dosanjh P., Hardy W.N., Liang Ruixing, MacFarlane W.A., Mendels P., Morris G.D., Riseman T.M., and Schneider J.W. (1994) *Physical Review Letters* **72**, 744–747.
Tinkham M. (1996) *Introduction to Superconductivity* (McGraw-Hill, New York).
Uemura Y.J. and Cywinski R. (1999) in *Muon Science: Muons in Physics, Chemistry, and Materials*, edited by Lee S.L., Kilcoyne S.H., and Cywinski R. (Institute of Physics, Bristol).
Yaouanc A., Dalmas de Réotier P., and Brandt E.H. (1997) *Physical Review* **B 55**, 11107–11110.
Yaouanc A., Dalmas de Réotier P., Huxley A., Flouquet J., Bonville P., Gubbens P.C.M., and Mulders A.M. (1998) *Journal of Physics: Condensed Matter* **10**, 9791–9798.
Yosida K. (1958) *Physical Review* **110**, 769–770.
Zhitomirsky M.E. and Dao V.-H. (2004) *Physical Review* **B 69**, 054508, 1-11.

12 Muonium centres in materials

Energetic spin-polarised positive muons are injected into the sample of interest for all the μSR techniques. Towards the end of their thermalisation track, it is believed that they neutralise by capturing electrons from their environment, forming muonium atoms. However, in metals they eventually lose their bound electrons since only diamagnetic muons are detected. Inversely, there is a large probability of detecting muonium states, the so-called muonium centres, in gases, liquids, semiconductors, dielectrics, and solids formed with molecules.

Many studies on muonium centres have been performed for semiconductors. This is justified because of the possible connection with the presence of hydrogen in these materials. The behaviour of hydrogen has been studied extensively since the early 1980s following the discovery that hydrogen can passivate shallow donors and acceptors in silicon. However, the available information on isolated hydrogen is relatively scarce. For example, the first investigation of an hydrogen centre in silicon by electron spin resonance has only been reported in 1987 — Gorelkinskii and Nevinnyi (1987, 1991); for a review see Gorelkinskii (1999). This explains the great interest for muonium centres in semiconductors. In this context muonium is considered as an isotope of hydrogen. In fact, based on the available muonium and hydrogen information, Lichti (1999) argues for the relevance of the study of muonium centres in semiconductors as models for hydrogen centres. Hydrogen is usually incorporated during the process of sample preparation, whereas muonium is observed after the implantation of energetic muons into the sample, facilitating observation of the isolated defects.

The interests in muonium research are on the properties of the muonium centres themselves and on their diffusion properties. Here we are concerned with the first type of works, the pure diffusion studies having been discussed in Chapter 9. It is important to remember that most muonium studies of matter are performed out of thermodynamical equilibrium. During the time interval spanned by a few muon lifetimes, a muonium atom created in a given electronic state may jump to a site of lower potential energy where its electronic state is different from the initial one. Therefore metastable muonium centres are expected to be found. They can be studied provided their lifetime is comparable to that of the muon.

As usual, our purpose is not to review the available literature on the applications of the μSR techniques to the study of muonium centres in materials. Rather, we have selected studies which can serve as examples. For more information, the reader is referred to the articles of Cox (1999) and Roduner (1999), as well as the short review of Blundell (1999) and the book of Nagamine (2003). Works such as the books of Walker (1983) and Smilga and Belousov (1994), as well as the article of Reid and Roduner

(2000), deserve special attention since μSR studies in gases and liquids are reported there. In the present book these studies are only mentioned shortly in Section 12.6. The Institute of Physics (IOP) has published in a special issue of the *Journal of Physics: Condensed Matter* six articles discussing muonium centres in materials (Heffner and Nagamine 2004). Concerning the study of semiconductors, we point out the review of Patterson (1988). It has been followed by updates of Chow *et al.* (1998), Lichti (1999), and Cox (2009).

Among the semiconductors silicon is of great technological interest. This is the reason for the large number of μSR studies devoted to this material. In addition, these works provide typical examples of the study of muonium centres in semiconductors. We describe these centres in the first section. Information on specific muonium states are presented in the subsequent sections: shallow donor and quasi-static neutral states. Then we discuss the muonium formation and charge states. Finally, the field of muoniated free radicals is introduced.

Rather than discussing the mechanisms of muonium formation first, we have chosen to start with the presentation of well-established results. However, before that we need to specify the muonium states encountered in materials. A muonium atom in vacuum is a bound state of a muon with an electron: it is denoted as Mu; see Chapter 8. It is a neutral atom. A muonium centre in a material is specified by its electric charge state and interstitial location. For example, a neutral muonium centre at or very near a tetrahedral interstitial site is labelled Mu^0_T. Until now three charge states have been recognised: Mu^0, Mu^+, and Mu^-. The second species Mu^+ results from the ionisation of the neutral muonium atom. The capture of an electron by Mu^0 leads to the hydride-like ion Mu^-. Whereas Mu^0 is paramagnetic, Mu^+ is diamagnetic. The species Mu^- is also found to be diamagnetic.

12.1 Inferring muonium centres from silicon studies

In Section 1.2.4 a transverse-field μSR frequency spectrum that was recorded for a *p*-type doped silicon sample at an intermediate field intensity is shown. Two pairs of transitions are observed. For these experimental conditions an isotropic muonium centre is characterised by a pair of transitions as explained in Section 8.2.2. Comparing the splitting for the pair detected at high frequency in Si and SiO_2 (quartz), it is deduced that the hyperfine interaction is reduced by almost a factor 2 in Si relative to SiO_2. Indeed, for an isotropic muonium centre the splitting of a pair is inversely proportional to the magnitude of the hyperfine interaction; see Eqs. 8.25 and 8.27, which are valid in the limit of small B_ext. Assuming an atomic-like picture for muonium (see Section 8.1), we infer the high-frequency pair for silicon to correspond to a muonium centre with a radius slightly larger than in vacuum. The hyperfine interaction in SiO_2 is indeed close to the vacuum value. A word of caution is in order here: thinking of a muonium centre with the atomic picture is convenient for visualisation but puts aside that the muonium and the adjacent atomic wavefunctions hybridise, leading to a molecular orbital system; see Fig. 12.2 below.

At this stage, digressing from our inference of the physical properties of muonium centres, it is of interest to relate the magnitude of the hyperfine interaction to a parameter associated with the material in which the centre is observed. Figure 12.1

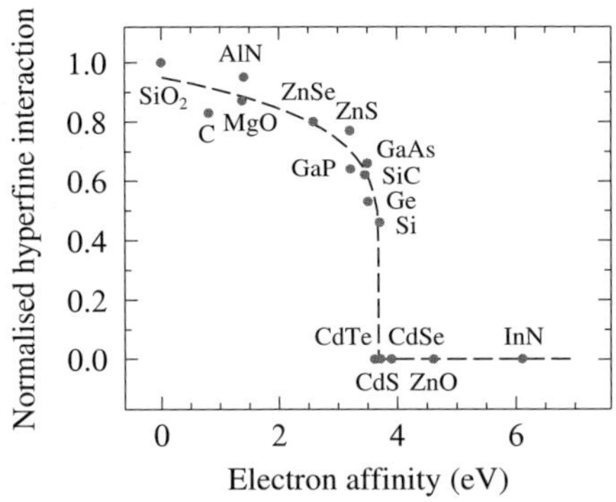

Fig. 12.1 Magnitude of the hyperfine interaction $\omega_{\rm iso}$, normalised to the free muonium value $\omega_0 \equiv 2\pi\nu_0$, of isotropic muonium centres in a series of semiconductors plotted against the electron affinity of the host material. For the definitions of ω_0 and $\omega_{\rm iso}$, see Eqs. 8.6 and 8.7, respectively. To the left of the transition found around an affinity of 3.5 eV, the hyperfine interaction varies from approximately the free muonium value in SiO_2 to about a half of it in Si. To the right of the transition, the ratio $\omega_{\rm iso}/\omega_0$ is around 10^{-4} for all the semiconductors shown here. The electron affinity values result from a band-offset diagram computed by Van de Walle and Neugebauer (2003). The dashed line is a guide to the eye. Adapted from Cox (2003).

shows the relevant parameter to be the electron affinity of the host material, i.e. the depth of the conduction band minimum below the vacuum level. Following the simple atomic picture the magnitude of the hyperfine interaction can be taken as a measure of the dilation of the electron wavefunction. Hence, Fig. 12.1 indicates that above a critical electron affinity of about 3.5 eV, the wavefunction delocalises almost completely into conduction-band states. The detection of these shallow donor states is presented in Section 12.2.

Back to the identification of the muonium centres, the most surprising result of Fig. 1.19 is the observation of a pair of transitions at low frequency. The corresponding anomalous muonium state has been further investigated experimentally by Patterson *et al.* (1978). The field and orientation dependence of the frequencies are both explained by an effective dipole-like Hamiltonian such that of Eq. 8.38, with the crystal axis [111] as the symmetry axis. In addition, the hyperfine couplings are quite small. Similar anomalous muonium centres have also been found in many other semiconductors (Chow *et al.* 1998).

It is interesting to note that this anomalous centre was quite unanticipated. It prompted experimental and theoretical works which eventually led to two important results (Gorelkinskii and Nevinnyi 1991, Cox and Lichti 1997). The first is the discovery by electron spin resonance (ESR) in silicon of the hydrogen analogue of the anomalous muonium, that is, $H^0_{\rm BC}$ — the notation $H^0_{\rm BC}$ will become clear below. The second result is that *ab initio* calculations seem to show this centre to be actually the ground state of interstitial hydrogen in silicon; we refer to Table 4 of the review by Cox (2009) for a list of references.

The question now is how to infer the possible muonium sites associated with these centres. Of course the symmetry of the muonium state must reflect the symmetry of the site. We recall that silicon crystallises in the diamond structure. The isotropic muonium centre is located at the tetrahedral interstitial site of this structure. The symmetry at this site implies the existence of four threefold axes, along to each of the cube diagonals. This means that the hyperfine tensor $\boldsymbol{A}$, as defined in Section 8.3, should be invariant under rotations around each of the threefold axes through an angle $\pm 2\pi/3$. This leads to the conclusion that $\boldsymbol{A}$ is scalar. It is then consistent to associate the isotropic muonium centre to this tetrahedral site, hence the label Mu_T^0 for this centre. This result has been accepted for a long time.

The assignment of the anomalous muonium centre, which was originally denoted by Mu^*, has been much more controversial. Its site is now believed to be located half-way between two neighbouring atoms on a [111] or symmetry equivalent axis. It is therefore at a bond centre. We shall present below the experiments which unambiguously led to this conclusion. The muonium centre at this position is then naturally denoted by $\mathrm{Mu}_\mathrm{BC}^0$. Taking the z axis along the bond direction, and accounting for the point symmetry at this site, $\boldsymbol{A}$ must be invariant under $\pm 2\pi/3$ rotations around z. This implies that $\boldsymbol{A}$ is uniaxial; i.e. its principal values along the axes x and y perpendicular to z are equal. Referring to Section 8.3, the hyperfine Hamiltonian in the bond centre is therefore dipole-like. The left part of Fig. 12.2 shows the location of the bond centre and illustrates how the anisotropy of this Hamiltonian can be so large.

These results for $\mathrm{Mu}_\mathrm{BC}^0$ raise questions. How has the $\mathrm{Mu}_\mathrm{BC}^0$ assigment been obtained? What are the lattice distortions introduced by the muonium centre? How should the phenomenologically introduced Hamiltonian describing the measured spectra be interpreted? While the field- and orientation-dependent precession frequencies observed with transverse-field μSR accurately determine the muonium hyperfine tensor, it is difficult with standard μSR to precisely measure the corresponding nuclear hyperfine coupling constants. These constants contain important information on the electronic wavefunction of the muonium centre reflecting its overlap with neighbouring nuclei.

An experimental method which provides key informations regarding the previous questions is the avoided-level-crossing (ALC) spectroscopy, the principles of which are explained in Sections 4.2 and 8.4.2. Together with conventional μSR experiments, these ALC experiments were a crucial part in the assignment of the anomalous muonium centre to $\mathrm{Mu}_\mathrm{BC}^0$. We briefly report on this work.

A combined analysis of extremely high statistic time-differential spectra (8×10^8 muon decay events) and ALC spectra has enabled Kiefl *et al.* (1988) to directly establish the anomalous muonium centre in intrinsic silicon to be located at the bond centre between two Si atoms on the [111] symmetry axis. It should be realised that these measurements are made particularly difficult by the low isotopic abundance of ^{29}Si (4.7%; see Appendix G), which is the only non-spinless natural isotope of silicon. The time-differential spectra nicely exhibit the influence of the ^{29}Si hyperfine interaction on the anomalous muonium centre. The ALC spectroscopy (for an example, see right panel of Fig. 12.2) provides accurate values for the nuclear hyperfine constants of the two equivalent Si neighbour sites, including their sign; see Table 12.1. The detailed

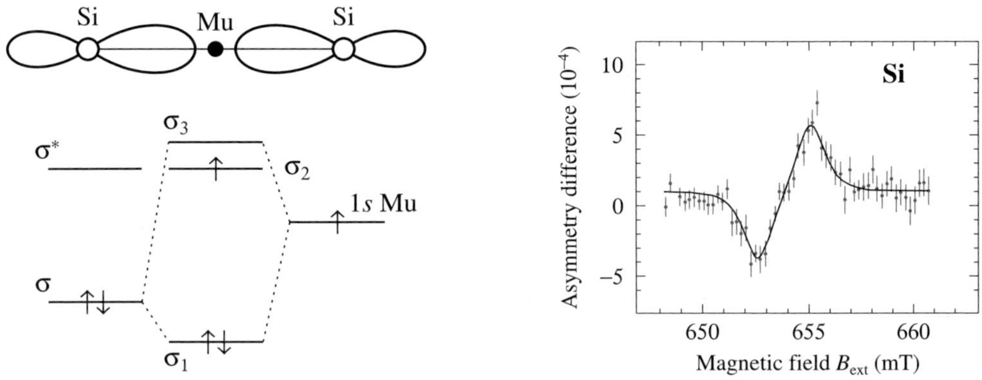

Fig. 12.2 (Left) The top drawing shows the bond-centre model for anomalous muonium centre in silicon. This centre is located half-way between two Si atoms on a threefold axis ([111] axis or symmetry equivalent) which is horizontal on the schematic drawing. The lower part shows the singly occupied electronic orbital derived from the Si–Si antibonding orbital, i.e. the σ^* orbital (adapted from Cox and Symons 1986). This panel sketches only the bond-centre site and the relevant electronic structure. It explains how the electron can be so close to the muon without overlapping onto it, i.e. how the dipole anisotropy can be so strong while the overall contact interaction is so low relative to the case of the muonium at the tetragonal site: the muon lies at or near a node of the unpaired electron wavefunction. (Right) Here we show an avoided-level-crossing (ALC) resonance with nearest neighbour ^{29}Si nuclei presented as an asymmetry difference spectrum. This panel is adapted from Kiefl *et al.* (1988).

analysis indicates the spin density at these sites to be large, positive, and mostly p-like.

In Table 12.1 we also compare the hyperfine parameters for $\mathrm{Mu}^0_{\mathrm{BC}}$ and $\mathrm{H}^0_{\mathrm{BC}}$ and for their nearest neighbour nuclei deduced from μSR and ESR experiments. Their absolute values are remarkably consistent.

As an invitation to further reading, we point out that a muonium ALC study has been performed for the CuCl semiconductor (Schneider *et al.* 1990). It shows that a wealth of information can be extracted from such a study.

The assignments of muonium centres to interstitial sites have been performed assuming them to be static. However, it is known that $\mathrm{Mu}^0_{\mathrm{T}}$ is highly mobile even at low temperature, whereas $\mathrm{Mu}^0_{\mathrm{BC}}$ is immobile, at least at low temperature. This comes about because $\mathrm{Mu}^0_{\mathrm{T}}$ induces little local distortion of the tetrahedral cage, whereas the intrinsic bond is stretched by about 40% to accommodate $\mathrm{Mu}^0_{\mathrm{BC}}$. So it is just possible for $\mathrm{Mu}^0_{\mathrm{T}}$ that the instantaneous site is of lower symmetry than assumed.

12.2 Shallow muonium states

For a long time, muonium centres have been of interest to researchers involved in the study of elemental semiconductors as well as the so-called III–V compounds. These latter compounds are formed from an element of the boron column of the periodic table and an element of the nitrogen column. They are characterised by muonium frequencies in the range of at least hundreds of megahertz. This means that they

Table 12.1 Hyperfine parameters for Mu^0_{BC} and H^0_{BC} centres in silicon and for their nearest neighbour nuclei from μSR and ESR experiments. The proton word is between quotation marks because, for ease of comparison, the related parameters have been rescaled by the quotient of the muon and proton gyromagnetic ratios, the value of which is about 3.183. The parameters given for Mu^0_{BC} and H^0_{BC} are those entering the Hamiltonian of Eq. 8.38. As a reference, the parameter $\omega_{iso}/(2\pi)$ (Eq. 8.4) for Mu^0_T is 2006 (2) MHz (Patterson 1988). The signs for the μSR data are deduced from measurements, although using hypotheses. The signs for the ESR data have not been determined. Because there is no risk of confusion, we denote the hyperfine parameters for the muon, proton, and ^{29}Si with the same symbols. Relative to the discussion in Section 8.4.2, the hyperfine interaction for ^{29}Si nucleus is here dipole-like rather than isotropic. Adapted from Chow *et al.* (1998) with the exception of the data in the first line of the table which are due to Ivanter *et al.* (2003).

Centre	Nucleus	$\omega_{\parallel}/(2\pi)$ (MHz)	$\omega_{\perp}/(2\pi)$ (MHz)
Mu^0_{BC}	Muon	−16.52 (6)	−92.58 (2)
	^{29}Si	−137.5 (1)	−73.96 (5)
H^0_{BC}	"proton"	19.7 (3.0)	99.9 (3.0)
	^{29}Si	140.1 (1.0)	72.9 (1.0)

have tightly localised electron wavefunctions. The corresponding energy levels of such muonium centres are then deep in the energy gap. Interest has subsequently grown to investigate II–VI compounds — made of elements of the zinc and oxygen families respectively — for which muonium centres with hyperfine frequencies in the megahertz and even smaller have been detected. Here we discuss in some details the example of ZnO which is one of the most promising semiconductors for the next generation of electronic and optoelectrical devices. A topical review on shallow centres has been published (Cox 2003).

Zinc oxide, a wide-band-gap semiconductor which crystallises in the wurtzite structure, was studied in the 1950s. It was found that hydrogen introduced as an impurity in it acts as a donor. For applications it is crucial to control the bulk electronic conductivity of the compound. But it is notoriously difficult to obtain intrinsic ZnO, which always displays a strong n-type conductivity. The origin of this unintentional carrier doping was controversial. A first-principles investigation, based on density functional theory, strongly suggests that hydrogen acts as a source of conductivity and behaves as a shallow donor (Van de Walle 2000). This theoretical prediction, entirely consistent with the old hydrogen data derived from bulk measurements, requires microscopic support. The μSR results obtained by Cox *et al.* (2001b) and Shimomura *et al.* (2002) offer such a support. Here, we report on the latter work performed on single crystals.

Some angular and temperature dependences of the Fourier transforms of the spectra obtained by the maximum entropy method are shown in Fig. 12.3; this method as applied to μSR is described in Section A.2.2. At low temperature two pairs of satellite lines are seen with their position symmetrically situated about the central line corresponding to the precession of diamagnetic muons. The splitting of these satellites was

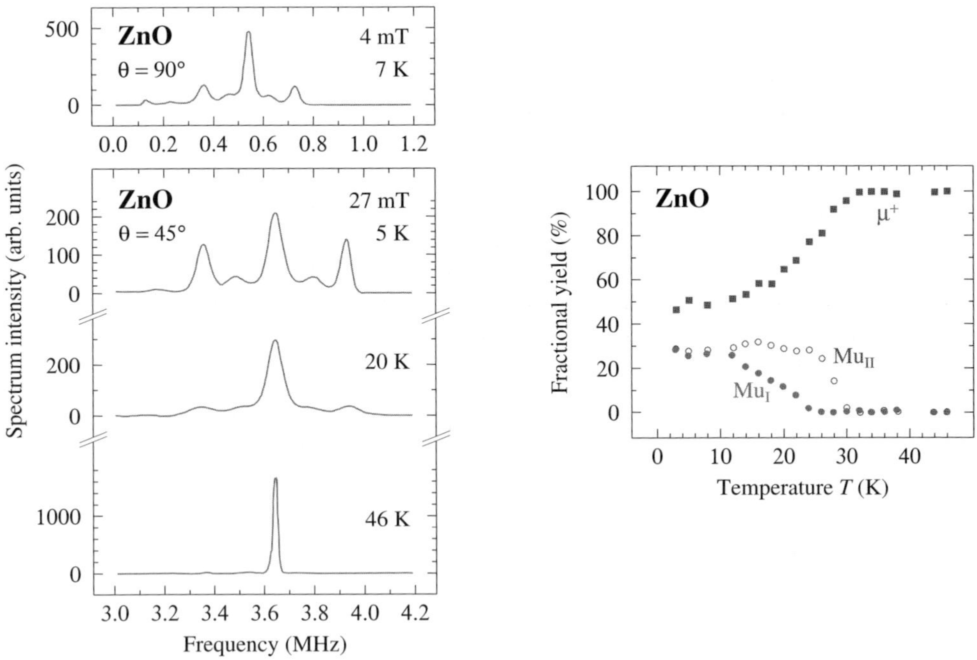

Fig. 12.3 Transverse-field measurements on a single-crystalline wafer of ZnO. (Left) Frequency spectra obtained by the maximum entropy method. For the top panel the sample was at 7 K and a field of 4 mT was applied perpendicular to the hexagonal [0001] axis. For the other plots a field of 27 mT, 45° tilted to the [0001] axis was applied. (Right) Temperature dependence of the fractional yields of $\mathrm{Mu_I}$ (filled circles), $\mathrm{Mu_{II}}$ (empty circles), and the diamagnetic state in the form of a free muon μ^+ (filled squares). Adapted from Shimomura *et al.* (2002).

found to be independent of the applied field intensity. This situation is reminiscent of the high-field limit for the transverse-field muonium polarisation function; see Section 8.2.2. Moreover, qualitatively, the structure of the frequency spectrum does not depend on the orientation of the field relative to the crystal axis as noticed from a comparison of the spectra taken at 7 and 5 K. Recalling the results of Section 8.3.2.1 and the fact that the predicted two high-frequency transitions are too weak to be seen, it is deduced that two muonium centres $\mathrm{Mu_I}$ and $\mathrm{Mu_{II}}$, with anisotropic hyperfine interactions, exist in ZnO. Fits of time spectra yield the parameter controlling the isotropic part of the Hamiltonian (see Eq. 8.40): $\omega_{\mathrm{iso}}/(2\pi) = 491$ and 293 kHz for the two centres, respectively. These values are extremely small, about 10^{-4} of those for free muonium. Assuming hydrogen-like defects, they correspond to Bohr radii a_{Mu} of about $21 a_{\mathrm{Mu},0} \simeq 1.1$ nm and $25 a_{\mathrm{Mu},0} \simeq 1.3$ nm, respectively (see Eq. 8.5). We recall that $a_{\mathrm{Mu},0}$ is the free muonium ground state radius. The Bohr radius for such a defect can also be estimated from the known dielectric constant and electron effective mass. It is computed to be 1.3 nm, in good accord with the estimates from the μSR transverse-field measurements. These large radii justify the name of shallow states given to these two muonium centres. This is in clear contrast to the usual deep muonium centres characterised by radii on the order of $a_{\mathrm{Mu},0}$. The right panel of Fig. 12.3 shows the percentage of detected $\mathrm{Mu_I}$, $\mathrm{Mu_{II}}$ and diamagnetic muon as a function of the temperature. The two muonium atoms are ionised to a diamagnetic state above $\sim$40 K. Activation energies of 3 and 25 meV are derived from these data for $\mathrm{Mu_I}$ and $\mathrm{Mu_{II}}$, respectively. Assuming $E_d = 2E_{\mathrm{ac}}$, where E_d is the level energy and E_{ac} the activation energy of a muonium centre, level energies of 6 and 50 meV are deduced. The latter is fairly consistent with the estimate of 66 meV for an hydrogenic defect and the value of 61 meV deduced for hydrogen from the analysis of Hall effect measurements (Look *et al.* 1998). In fact this analysis suggests a second hydrogen centre with $E_d =$ 31 meV. This is clearly at variance with the energy of 6 meV derived from μSR. While much work is still needed, both experimentally and theoretically, the available μSR results of Shimomura *et al.* (2002) as well as those of Cox *et al.* (2001b) provide a strong support to the theoretical analysis of Van de Walle (2000), which suggests that hydrogen is the primary origin of unintentional n-type conductivity in ZnO.

12.3 Quasi-static neutral muonium states

Here we introduce a nice experimental result which shows that the concepts developed for diamagnetic muons can sometimes be employed directly to interpret muonium data.

In zero field we expect to observe the Kubo–Toyabe shape for the muonium relaxation if the value of the muonium angular frequency ω_{24} is outside the detection band of the spectrometer and if the field at the muonium centre is quasi-static in the μSR time window. This latter assumption is just opposite to the fast fluctuation hypothesis made for the derivation of the Wangsness–Bloch equations in Section 8.4.3. It must be realised that the field distributions probed by muonium centres and bare muons are quite different. This is so because the precession frequency in low field is more than hundred times larger for muonium than for muon (see Eq. 8.29), implying that the time scale of the Kubo–Toyabe function (see e.g. Fig. 1.4) is squeezed by the same ratio. Therefore, fields produced by nuclear dipoles which are nearest neighbours to a muonium centre might even be too large to be resolved.

In Fig. 12.4 we present the measured asymmetry for solid krypton at 20.3 K. The width of the field is not too large, and therefore measurable, because only 11.5% of the natural krypton nuclei posses a nonzero magnetic moment: this is the ^{83}Kr isotope; see Table G.1. A large fraction of the muonium centres can therefore have spinless nuclei as first neighbours. The solid line is the result of a fit assuming a small mobility of muonium which is accounted for by the strong-collision model; see Section 6.3.1. The typical Kubo–Toyabe feature of a minimum of the asymmetry followed by its partial recovery at long time is observed. The interpretation of the zero-field spectrum is confirmed by the strong influence of a weak longitudinal field as shown in Fig. 12.4. Relative to a typical work with diamagnetic muons, the asymmetry is much reduced. This is not surprising for two reasons. (i) Only half of the muonium asymmetry can display the Kubo–Toyabe shape. The other half is related to the angular frequency ω_{24} as explained in Section 8.2.2. (ii) Muonium atoms which have at least one ^{83}Kr

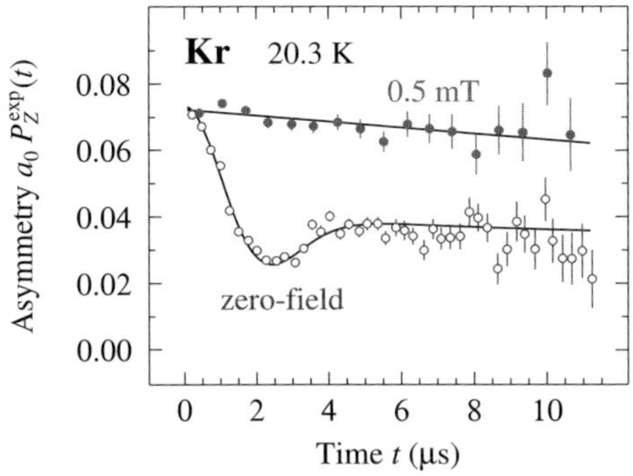

Fig. 12.4 Evolution of the muon polarisation measured for solid krypton at 20.3 K. The Kubo–Toyabe function with slow dynamics is observed in zero field. The depolarisation is quenched by a weak longitudinal magnetic field of 0.5 mT. This is the first observation of a Kubo–Toyabe type response for a paramagnetic centre. Adapted from Storchak *et al.* (1996).

nucleus neighbour experience strong nuclear hyperfine interactions and are rapidly and completely depolarised at short times. In addition, it must be noted that a fraction of the implanted muons does not form a muonium centre.

12.4 Muonium state formation

Muonium centres have been detected in semiconductors, dielectrics, and molecular systems but not in metals (with the exception of metallic alkali-metal doped fullerides). It is believed that, although formed at the early stage of implantation, a muonium state in a metal loses its bound electron due to the strong screening of the Coulomb interaction by the conduction electrons.

Up to the beginning of the 1990s, most of the information on muonium state formation was derived from numerical simulation with theoretical input (Patterson 1988). It is only more recently that detailed information on the muonium state formation mechanisms in condensed-matter systems has experimentally been obtained.

During the last stage of its thermalisation in the material into which it is implanted a positive muon can pick up an electron to form a muonium atom. Generally it will subsequently be stripped of this electron, and bind again with another one. This process occurs several times until the muon is fully thermalised. If the last step leaves atomic muonium in its neutral charge state, it is said to have been formed promptly. If the muon thermalises as a bare particle, there is a high probability that some of the radiolysis electrons generated during thermalisation can reach it before it decays. This process leads to delayed muonium formation. An essential feature for it to occur is that the electrons start out spatially separated from the muon; this property is the key for the experimental differentiation between delayed and prompt muonium formation. This differentiation can be achieved through the application of an external electric field. A relatively weak field can sometimes overcome the muon–electron Coulomb attraction and thus reduce the probability for delayed muonium state formation. Low-energy muon experiments support the picture given above for the formation

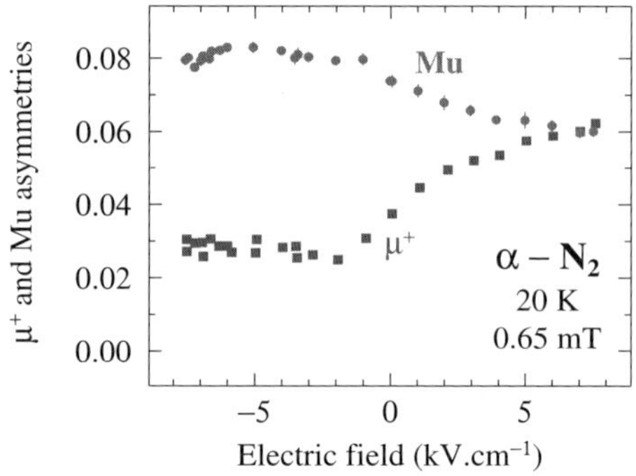

Fig. 12.5 Electric field dependence of the transverse magnetic field signal amplitude measured for $\alpha - \mathrm{N}_2$ at 20 K with $B_{\mathrm{ext}} = 0.65\,\mathrm{mT}$. Solid nitrogen under its own vapour pressure crystallises in two solid phases. The low-temperature $\alpha - \mathrm{N}_2$ phase is face-centred cubic for temperatures up to 35.6 K. The circles and squares refer to the asymmetries for the muonium and diamagnetic components, respectively. Adapted from Storchak *et al.* (1999).

of this state since it strongly depends on the availability of excess (radiolysis) electrons (Prokscha *et al.* 2007).

As mentioned in Section 1.2.4, it was experimentally shown in 1992 (for superfluid helium) that a muonium state may arise from the binding of a radiolytic electron to a muon. This result was deduced from experiments performed with a sample submitted to an electric field. Two years later the first study of muonium formation in a solid was published (Storchak *et al.* 1994), again through the study of the effect of an applied electric field on the signal amplitudes. An example of typical results is shown in Fig. 12.5. Both asymmetries do indeed change with the field. On the average, we expect the bare muon to thermalise downstream from the radiolytic electrons of the muon ionisation track. By definition, a positive sign for the electric field corresponds to a field applied parallel to the initial muon momentum direction. Thus, a positive field pulls the muon and radiolytic electrons apart, giving rise to an increase of the diamagnetic component as observed experimentally. The decrease of the muonium component corroborates this observation. Note that the muonium asymmetry decreases by about half as much as the diamagnetic asymmetry increases. This is because the asymmetry of the muonium signal represents only half of the muonium ensemble when the transitions between singlet and triplet states cannot be observed by the apparatus in use, as explained in Section 8.2.2. For a negative electric field, the muon and radiolytic electrons are pulled towards each other, increasing the probability for muonium formation. This is consistent with the data of Fig. 12.5. The average initial muon–electron distance is estimated to be about 50 nm. It is also possible to determine the characteristic time for the electron transport to the bare muon from the magnetic field dependence of the muonium asymmetry measured in transverse-field geometry. In fact, during the μSR time window two states compete, the muonium and diamagnetic states, and as time evolves, diamagnetic muons transform into muonium centres. The two-state model of Section 6.4.2 is directly applicable to get the polarisation function. Physically, because muonium states are formed at different times, the phase coherence among them is lost. The higher the applied magnetic field, the stronger the dephasing. This means that the muonium asymmetry is reduced as the field is increased. The rate of its reduction is controlled by the electron transport rate. Hence, it is possible to determine this rate and compare the μSR results with the traditional macroscopic time of flight technique. This latter technique consists of measuring the mobility of electrons submitted to an electric field through the transient current detected on electrodes. The carriers are excited by an energetic electron beam pulse.

Measurements under high electric fields are performed using alternating electric fields to avoid accumulation of charges near the sample surface that would screen the electric field inside the sample (Eshchenko *et al.* 1999). The time scale of this screening depends on the sample–electrode connection and the radiolytic electron generation rate. Typical switching periods of up to 5 s have been used.

The work reported above for solid nitrogen convincingly shows that muonium states may be formed *via* transport of radiolytic electrons to thermalised muons; that is, a positive field prevents an initially distant electron from reaching an immobile bare muon, whereas a negative field favours the approach of the electron to the muon site. With this knowledge the next obvious task is to determine the influence of these electrons on the formation of the two muonium centres commonly encountered in semiconductors. This has been achieved for GaAs; see Eshchenko *et al.* (2002).

We briefly discuss this work. Since $\mathrm{Mu}_{\mathrm{T}}^{0}$ is a tightly bound state, fields of atomic scale strength, i.e. much higher than $10\,\mathrm{kV\,cm^{-1}}$, would be needed to affect this muonium centre. Therefore the response of the diamagnetic component asymmetry to the field is only related to the $\mathrm{Mu}_{\mathrm{BC}}^{0}$ centres. This has experimentally been checked. The response is found roughly symmetric relative to the point taken without field, in clear contrast to the results obtained for dielectrics; see for example Fig. 12.5. Therefore a strong field prevents muonium formation in either direction. This means that a mechanism different from the one operating in dielectrics drives the field response. Because of the large dielectric constant and the very small effective electron mass, Eshchenko and collaborators argue that $\mathrm{Mu}_{\mathrm{BC}}^{0}$ must have an excited or weakly bound precursor that can be field-ionised, in which case the cascade to the ground state is prevented.

To conclude this section, we mention that the possible muonium formation mechanisms are still hotly debated. In particular, there is no agreement on the mechanism(s) leading to the formation of the so-called shallow states. We stress to the reader that shallow and weakly bound excited states should not be confused. In addition, they are not necessarily related. The latter states may be formed as intermediate states, while the former ones can only be found in final states.

12.5 Different species of muonium centres

Experimentally, a measured μSR signal is taken as arising from muons in a diamagnetic state if the full initial asymmetry expected for this state is measured, or if the typical muonium frequencies are not observed. However, it should be realised that the physics may in fact be quite involved.

As a first example, we consider graphite. At first sight, a conventional diamagnetic signal is observed. In fact, this semimetal is commonly used to calibrate the asymmetry associated to a μSR spectrometer. The diamagnetic signal is surprising since the conduction electron density ($\sim 3 \times 10^{18}\,\mathrm{cm}^{-3}$) is at least four orders of magnitude smaller than in a simple metal. At this density, one would expect muonium states to be formed. However, they are not detected. Experimentally, the spin–lattice relaxation rate and the Knight shift are found to be anomalously large compared to those of simple metals and both have unusual temperature dependence; see Section 5.3 for an introduction to Knight shifts. Investigators (Cox *et al.* 2001a, Chakhalian *et al.* 2002) believe that a kind of muonium state is formed but with the electron spin density predominantly located on the neighbouring carbon atoms, much as for a radical.

As a second example, we discuss heavily doped n-type silicon, that is, silicon with carriers concentration from the $\sim 10^{17}$ to the $\sim 10^{19}\,\mathrm{cm}^{-3}$ range; see Chow *et al.* (2000). Rapid spin exchange between the free carriers and the muonium-bound electron averages out the muonium hyperfine fields, thereby eliminating the characteristic frequency pattern. Consequently, $\mathrm{Mu}^0_{\mathrm{BC}}$ centres are not directly observed, but a large frequency shift is measured in transverse-field experiments, establishing the paramagnetic nature of the $\mathrm{Mu}^0_{\mathrm{BC}}$ centres.

The strong spin exchange between the muonium electron and the free carriers randomises the spin orientation of the muonium electron, causing a rapid muon spin depolarisation if the spin dynamics is slow. However, it must be noted that longitudinal-field measurements do not suffer from this depolarisation and may provide a means of studying this type of muonium centre. Indeed, when scanning the external field, a value may be found for which the field at the muonium site is fully transverse for half of the muonium atoms as explained in Section 8.4.3.2. This yields an increase in the relaxation rate λ_Z. For that to happen, the anisotropy of the muonium hyperfine tensor must be strong enough and the spin dynamics should not be too fast. The field value for which the resonance occurs is accessible to experiments only when the muonium hyperfine coupling is small enough. Figure 12.6 provides an example of this resonance for $\mathrm{Mu}^0_{\mathrm{BC}}$ in moderately doped n-type silicon and this is a unmistakable signature of the spin-exchange relaxation mechanism. The phenomenological model used to interpret the data was introduced by Nosov and Yakovleva (see Section 8.4.3). This model accounts for the spin-exchange relaxation of the muonium electron which is induced by the scattering of free carriers. This relaxation is characterised by the rate ν_{NY}. From the fit value of ν_{NY} the cross section for spin-exchange scattering can be estimated. Remarkably, this resonant-like spin relaxation provides a method for the determinaton of the hyperfine parameters under conditions where a transverse-field μSR signal is difficult, if not impossible, to observe.

Note that the spin-exchange mechanism evidenced for n-type silicon could be of importance for graphite as well.

The study of the field dependence of the spin–lattice relaxation rate in semiconductors may reveal the fate of the different muonium states. We refer, for example, to the work at several high temperatures for intrinsic silicon (Chow *et al.* 1993). The analysis of the data was performed with a two-state model corresponding to alternating charge states Mu^0 and Mu^+; see Section 6.4.2. From the value which is found for the hyperfine coupling, it is deduced that the neutral state is $\mathrm{Mu}^0_{\mathrm{T}}$, suggesting that the neutral muonium spends a significant fraction of time at or near the tetragonal

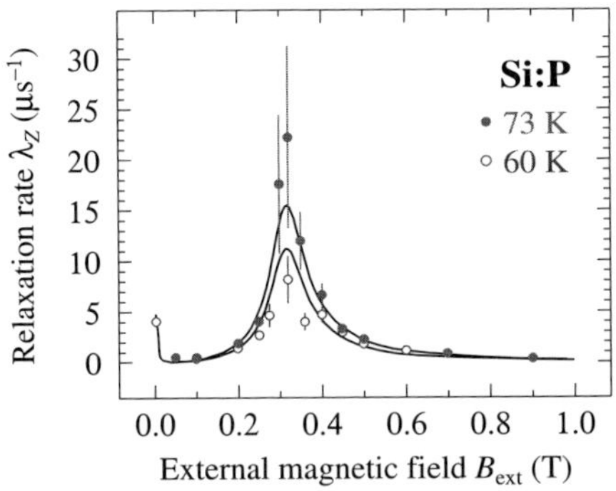

Fig. 12.6 Field dependence of the muon spin–lattice relaxation rate at 60 and 73 K in a Si:P single crystal with a carrier concentration of 10^{14} cm^{-3}. The solid lines are results from fits of the Nosov–Yakovleva theory to the data. The spin-exchange relaxation rate ν_{NY} is found to be 11.2 (5) and 15.4 (5) μs^{-1} for the data at 60 and 73 K, respectively. Adapted from Chow *et al.* (1994).

site rather than at the bond centre. Conversely, the ionised state could be associated to the bond centre Mu^+_{BC}, which would imply a change of site while an electron is lost or captured.

Heavily doped p-type semiconductors are of special interest from the μSR point of view. Indeed, the large concentration of holes implies that the Fermi level is close to the valence band, and thus the formation of the positively charged state Mu^+ is expected to be preferred (Herring *et al.* 2001, Van de Walle and Neugebauer 2003). This expectation is confirmed by Mansour *et al.* (2008), who have performed μSR measurements on a silicon sample doped with boron at a concentration of 10^{18} cm^{-3} and who have found that essentially all the signal is due to a nonparamagnetic state. These experiments also show that, at low temperature, Mu^+ is located at the bond centre. In addition, the investigation performed between 200 and ~600 K nicely illustrate the possibility of studying the diffusion and trapping of Mu^+ in silicon, mostly as it was carried out for the bare muon in niobium doped with nitrogen; see Section 1.2.1. Two other examples of muonium centre diffusion studies can be found in Section 9.4.2.

Four states have been recognised for muonium in semiconductors: Mu^0_T, Mu^0_{BC}, Mu^+_{BC}, and Mu^-_T. The transitions between these states involve charge transitions (such as $Mu^0_{BC} \longrightarrow Mu^+_{BC} + e^-$), site transition ($Mu^0_T \longrightarrow Mu^0_{BC}$), and combined charge/site transitions (such as $Mu^+_{BC} + e^- \longrightarrow Mu^0_T$). Since there is no phase coherence between these states, the transverse-field technique is not well adapted for studying the transitions. On the other hand, while the radio-frequency technique (RF-μSR; see Section 4.1) yields information similar to that of the transverse-field technique with regard to muonium state identification, it has the definite advantage that it does not require any phase coherence. In addition, the presence of a RF field implies that only the muonium centre which satisfies the resonance condition is detected. A typical application is presented by Hitti *et al.* (1999). In this respect, working at a pulsed muon source is a definitive advantage as nicely shown by Cottrell *et al.* (2003).

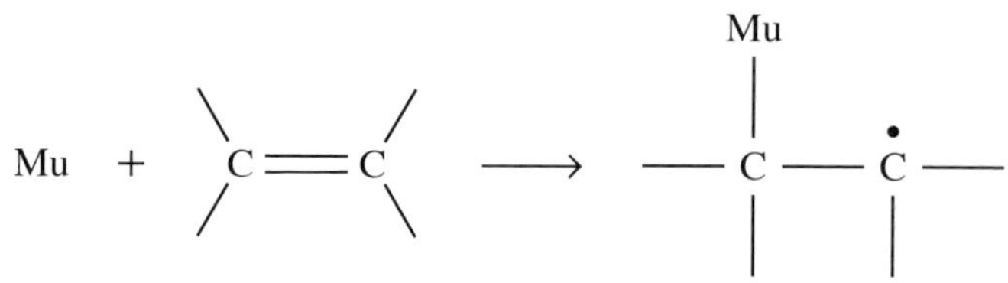

Fig. 12.7 Chemical reaction of a muonium atom with an unsaturated carbon bond. Here the unpaired electron is localised on the carbon adjacent to the muoniated one. The reaction leaves the muon as a polarised spin label chemically bound to a free radical.

Finally, we point out that muonium centres in semiconductors may undergo fast spin relaxation under illumination due to a cyclic charge-exchange reaction with photo-induced carriers (Kadono *et al.* 1994).

12.6 Muoniated free radicals

When injecting muons into molecules and molecular compounds, there is usually a large probability for the formation of free radicals by muonium addition to unsaturated organic compounds (Roduner 1988). This reaction is illustrated in Fig. 12.7. If localised, the unpaired electron usually resides in an orbital centred on the carbon nucleus next to the muoniated carbon. If delocalised through an extended π-bond, it contributes to a spin density spread over more distant centres. In both cases, the magnitude of the hyperfine interaction is reduced compared to that for a free muonium atom.

The physics and chemistry of free radicals in matter is the subject of this section. In general, due to the presence of protons, the radicals must be regarded as multispin systems, with complex hyperfine spectra (Section 8.4.1).

Some molecular solids exhibit magnetic or superconducting phase transitions at low temperature. The book edited by Veciana in 2001 provides a useful introduction to the field of magnetism in molecular materials. Here, we shall not deal with the magnetic and superconducting properties since they are covered in Chapters 10 and 11, respectively. Instead we shall focus on properties related to chemical physics.

The chemical analogy between the proton and muon, and the concept of muonium as a light isotope of hydrogen were important insights, recognized from the outset. As well as the specific application to defects in semiconductors that we have already seen, exploitation of these analogies has given rise to the large field of muonium chemistry, pursued in solids, liquids, and gases.

A popular topic, beyond the scope of this book, is the comparison of muonium reaction rates with those of atomic hydrogen. That is, kinetic isotope effects, chiefly in the fluid phases, and lately also in superfluid phases. As an example, we mention the study of the kinetics of muonium addition to acetone in sub- and supercritical water (Brodovitch *et al.* 2009). We note, however, the equally important isotope effects in the molecular conformation and dynamics of the reaction products. The great majority of the studies concerns organic radicals, of which many hundreds are now known.

As a guide to the literature, we refer to the works of McCollum *et al.* (2009) and McKenzie *et al.* (2006), and references therein. The muonium-substituted or muoniated isotopomers (or isotopic isomers) are easily recognised by comparison with their normal or protiated counterparts, usually already known to ESR spectroscopy. In a few instances, however, the μSR observation came first and prompted ESR searches. We have seen how identification of Mu^0_{BC} led to the search for H^0_{BC} in silicon. Likewise the muonium adduct of fullerene, MuC_{60}, was discovered before the monohydride HC_{60} (by then sought eagerly by ESR chemists and astrophysicists alike) and assisted its subsequent assignment of its ESR spectrum. In these cases, comparison of the ESR and μSR hyperfine parameters shows the importance of zero-point energy in the chemical bond to light atoms. This is already important on going from deuterium to hydrogen, but the effects are much enhanced (so easier to interpret) on going from hydrogen to muonium. Temperature dependences of hyperfine constants reveal the energy barriers or separations involved in soft librational modes. In a few instances, the corresponding states of protium are still unknown to ESR, e.g. its defect centres in diamond, or the endohedral trapped atom in fullerenes denoted $Mu@C_{60}$; see Section 11.2.2. Thus $Mu@C_{60}$ and MuC_{60} are seen coexisting in μSR spectra, but $H@C_{60}$ has yet to be prepared, whether chemically or by proton bombardment.

The ALC technique has proved of great importance to a variety of aspects. Fullerenes are again a case in point. The $\Delta M = 0$ resonances are used for maping out the spin density around the carbon skeleton, i.e. the distribution of the unpaired electron in its singly occupied molecular orbital. The $\Delta M = 1$ resonances are supremely sensitive to rotation or reorientation and demonstrate a surprising difference in behaviour for C_{60} and C_{70} in their plastic phases. The work of Roduner *et al.* (1995) is of special interest in this respect. An introduction to the physics of the resonances is given in Section 8.4.2.

The muon provides a convenient spin label for the fullerene study, not interfering too much with the topic of investigation. The same is true for monitoring secondary chemical reactions for many of the other organic radicals. The ALC technique is invariably displayed in integral counting mode (see Section 4.2.3), but spin relaxation in the time domain provides a complementary means of probing temperature-dependent molecular dynamics. And just as the study of muonium defect centres in semiconductors was extended to their interaction with charge carriers, whether through spin exchange or charge exchange, so in conducting polymers and synthetic metals the form of the depolarisation function and its dependence on magnetic field and temperature inform on the mobility of solitons, polarons, or other spin carriers, charged or neutral. An example is detailed below; see Section 12.6.2.

Before discussing in some detail two physical-chemistry studies of muoniated free radical centers, we recall that a work on the muoniated cyclohexadienyl radical is reported in Section 1.2.4.

We shall first highlight a work carried out for crystal structures of industrial interest, the zeolites, and then consider polymers.

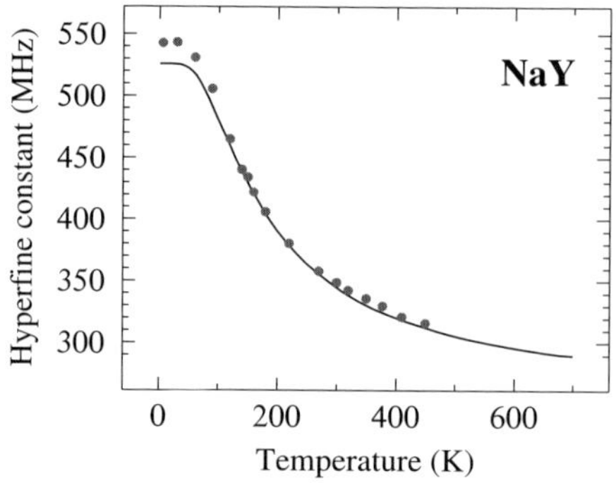

Fig. 12.8 Comparison of the experimental and theoretical values for the temperature dependence of the isotropic hyperfine constant $\omega_{\text{iso}}/(2\pi)$ for the Mu-ethyl free radical in NaY faujasite at a loading of one molecule per supercage. Adapted from Bridges *et al.* (2007).

12.6.1 Zeolites

Zeolites are important industrial catalysts but microscopic information on reaction mechanisms is scarce, particularly the reactions involving neutral H-adduct free radicals, which might be important intermediates in petrochemical pathways. As an example, we have chosen to single out the study of the Mu-ethyl (C_2H_4Mu) radical in faujasites by Bridges *et al.* (2007). Of the many possible zeolite frameworks, the synthetic faujasites have large internal pore volumes with "supercages" of about 12 Å diameter, accessed by "windows" of ~7.5 Å in diameter. There are eight of these supercages per unit cell, within which a variety of guest organic molecules can be accommodated, up to saturation loadings of five to six molecules per supercage. Transverse-field and ALC experiments have been performed. In Fig. 12.8 is displayed the isotropic hyperfine constant in NaY zeolite versus temperature at a loading of one molecule per supercage. The temperature dependence mainly reflects the geometric structure around the radical. The remarkable point is the ability of the *ab initio* calculation (density functional theory with a given geometrical model) to reproduce the measured hyperfine constant. The maximum deviation between the experimental data and the result of the theory, which is observed at low temperature, is only 5%.

12.6.2 Polymers

A free radical muonium centre formed in fullerenes and zeolites acts as a label. Here we discuss muonium centres found in polymers. Of much interest is the possibility for such a centre to induce a mobile defect travelling along the polymer chain. Polaronic motion in doped conducting polymers can also be measured using NMR and ESR (Mizoguchi 1995). However, for these techniques to work, the carrier density needs to be sufficiently large. This is a stringent restriction. In contrast to transport experiments, which provide results inevitably dominated by the transport process with the smallest correlation time, μSR measurements give information on the intrinsic transport processes governing the mobility of the electronic defect.

Here we focus on the investigation of the emeraldine base form of polyaniline for which a muon-induced electronic defect has been unravelled. Its dynamics has been studied by the longitudinal-field technique for fields up to 0.4 T (Pratt *et al.* 1997). The data have been analysed with the Risch and Kehr relaxation function (Section 8.4.3.4). At 6 K the expected linear dependence of $\lambda_{\rm chain}$ as a function of $1/B_{\rm ext}$ is observed; see Eq. 8.85. This is a characteristic of a fast one-dimensional intrachain defect diffusion. It enables the measurement of the transfer rate of the defect along the chain, $\nu_\parallel$. However, at 300 K the $1/B_{\rm ext}$ regime holds only down to 30 mT or so. This has been taken by Pratt and collaborators as the signature of the importance of the three-dimensional diffusion regime at high temperature. Since Risch and Kehr did not consider the interchain diffusion, $\lambda_{\rm chain}(B_{\rm ext})$ has phenomenologically been analysed with the form $\lambda_{\rm chain}(B_{\rm ext}) = \lambda_0/(1 + B_{\rm ext}/B_{\rm cross})$. The rate of transitions between adjacent chains, $\nu_\perp$, has been taken as given by the relation $\nu_\perp = g_{\rm e}\mu_{\rm B}B_{\rm cross}/\hbar$. As shown in the original work, information on the transport properties of the defect is obtained through the measurements of $\nu_\parallel(T)$ and $\nu_\perp(T)$.

It is interesting to understand the origin of the muon-induced defect in polyaniline. For this purpose it is useful to consider the hydrogenation of that polymer. According to the molecular orbital model of Duke *et al.* (1987), hydrogenation of the emeraldine form of polyaniline corresponds to adding a single electron to the lowest unoccupied molecular orbital (LUMO). Following rapid electronic and structural relaxation of the surrounding molecule, a negative polaron is formed. μSR provides a way of studying this defect which moves away from its initial site, diffusing through the system.

References

Blundell S.J. (1999) *Contemporary Physics* **40**, 175–192.

Bridges M.D., Arseneau D.J., Fleming D.G., and Ghandi K. (2007) *Journal of Physical Chemistry* **C 111**, 9779–9793.

Brodovitch J.-C., McCollum B.M., and Percival P.W. (2009) *Physica* **B 404**, 950–952.

Chakhalian J.A., Kiefl R.F., Dunsiger S.R., Macfarlane W.A., Miller R., Sonier J.E., and Fischer J.E. (2002) *Physical Review* **B 66**, 155107, 1–8.

Chow K.H., Hitti B., and Kiefl R.F. (1998) in *Identification of Defects in Semiconductors*, edited by Stavola M. (Academic Press, New York).

Chow K.H., Kiefl R.F., Hitti B., Estle T.L., and Lichti R.L. (2000) *Physical Review Letters* **84**, 2251–2254.

Chow K.H., Kiefl R.F., Schneider J.W., Hitti B., Estle T.L., Lichti R.L., Schwab C., Duvarney R.C., Kreitzman S.R., MacFarlane W.A., and Senba M. (1993) *Physical Review* **B 47**, R16004–R16007.

Chow K.H., Lichti R.L., Kiefl R.F., Dunsiger S., Estle T.L., Hitti B., Kadono R., MacFarlane W.A., Schneider J.W., Schumann D., and Shelley M. (1994) *Physical Review* **B 50**, R8918–R8921.

Cottrell S.P., Johnson C., Cox S.F.J., and Scott C.A. (2003) *Physica* **B 326**, 248–251.

Cox S.F.J. (1999) in *Muon Science: Muons in Physics, Chemistry, and Materials*, edited by Lee S.L., Kilcoyne S.H., and Cywinski R. (Institute of Physics, Bristol).

Cox S.F.J. (2003) *Journal of Physics: Condensed Matter* **15**, R1727–R1780.

Cox S.F.J. (2009) *Reports on Progress in Physics* **72**, 116501 1–130.

Cox S.F.J. and Symons M.C.R. (1986) *Chemical Physics Letters* **126**, 516–525.

Cox S.F.J. and Lichti R.L. (1997) *Journal of Alloys and Compounds* **253-254**, 414–419.

Cox S.F.J., Cottrell S.P., Charlton M., Donnelly P.A., Ewels C., Heggie M., and Hourahine B. (2001a) *Journal of Physics: Condensed Matter* **13**, 2169–2175.

Cox S.F.J., Davis E.A., Cottrell S.P., King P.J.C., Lord J.S., Gil J.M., Alberto H.V., Vilão R.C., Piroto Duarte J., Ayres de Campos N., Weidinger A., Lichti R.L., and Irvine S.J.C. (2001b) *Physical Review Letters* **86**, 2601–2604.

Duke C.B., Paton A., Conwell E.M., Salaneck W.R., and Lundström I. (1987) *Journal of Chemical Physics* **86**, 3414–3423.

Eshchenko D.G., Storchak V.G., and Morris G.D. (1999) *Physics Letters* **A 264**, 226–231.

Eshchenko D.G., Storchak V.G., Brewer J.H., and Lichti R.L. (2002) *Physical Review Letters* **89**, 226601, 1–4.

Gorelkinskii Y.V. (1999) in *Hydrogen in Semiconductors II*, edited by Nickel N.H. (Academic Press, New York).

Gorelkinskii Y.V. and Nevinnyi N.N. (1987) *Soviet Technical Physics Letters* **13**, 45–46.

Gorelkinskii Y.V. and Nevinnyi N.N. (1991) *Physica* **B 170**, 155–167.

Heffner R.H. and Nagamine K. (2004), editors of a special issue on μSR of the *Journal of Physics: Condensed Matter* **16**, number 40.

Herring C., Johnson N.M., and Van de Walle C.G. (2001) *Physical Review* **B 64**, 125209, 1–27.

Hitti B., Kreitzman S.R., Estle T.L., Bates E.S., Dawdy M.R., Head T.L., and Lichti R.L. (1999) *Physical Review* **B 59**, 4918–4924.

Ivanter I.G., Krasnoperov E.P., Nikol'sky B.N., Ponomarev A.N., Nezhivoy A.N., Zimmermann U., Duginov V.N., and Gritsaj K.I. (2003) *Journal of Physics: Condensed Matter* **15**, 7419–7422.

Kadono R., Matsushita A., Macrae R.M., Nishiyama K., and Nagamine K. (1994) *Physical Review Letters* **73**, 2724–2727.

Kiefl R.F., Celio M., Estle T.L., Kreitzman S.R., Luke G.M., Riseman T.M., and Ansaldo E.J. (1988) *Physical Review Letters* **60**, 224–226.

Lichti R.L. (1999) in *Hydrogen in Semiconductors II*, edited by Nickel N.H. (Academic Press, New York).

Look D.C., Reynolds D.C., Sizelove J.R., Jones R.L., Litton C.W., Cantwell G., and Harsch W.C. (1998) *Solid State Comunications* **105**, 399–401.

McCollum B.M., Brodovitch J.-C., Clyburne J.A.C., Percival P.W., and West R. (2009) *Physica* **B 404**, 940–942.

McKenzie I., Brodovitch J.-C., Ghandi K., and Percival P.W. (2006) *Physica* **B 374-375**, 299–302.

Mansour A.I., Salman Z., Chow K.H., Fan I., King P.J.C., Hitti B., Jung J., and Cottrell S.P. (2008) *Physical Review Letters* **100**, 257602, 1–4.

Mizoguchi K. (1995) *Japanese Journal of Applied Physics* **34**, 1–19.

Nagamine K. (2003) *Introductory Muon Science* (Cambridge University Press, Cambridge).

Patterson B.D. (1988) *Reviews of Modern Physics* **60**, 69–159.

Patterson B.D., Hinterman A., Kündig W., Meier P.F., Waldner F., Graf H., Recknagel E., Weidinger A., and Wichert Th. (1978) *Physical Review Letters* **40**, 1347–1350.

Pratt F.L., Blundell S.J., Hayes W., Nagamine K., Ishida K., and Monkman A.P. (1997) *Physical Review Letters* **79**, 2855–2858.

Prokscha T., Morenzoni E., Eshchenko D.G., Garifianov N., Glückler H., Khasanov R., Luetkens H., and Suter A. (2007) *Physical Review Letters* **98**, 227401, 1–4.

Reid I.D. and Roduner E. (2000) in *Encyclopedia of Spectroscopy and Spectrometry*, edited by Tranter G.E. and Holmes J.L. (Academic Press, San Diego)

Roduner E. (1988) *The Positive Muon As a Probe in Free Radical Chemistry*, edited by Berthier G., Lecture Notes in Chemistry, vol. 49 (Springer, Berlin).

Roduner E. (1999) in *Muon Science: Muons in Physics, Chemistry, and Materials*, edited by Lee S., Kilcoyne S.H., and Cywinski R. (Institute of Physics, Bristol).

Roduner E., Prassides K., Macrae R.M., Thomas I.M., Niedermayer C., Binninger U., Bernhard C., Hofer A., and Reid I.D. (1995) *Chemical Physics* **192**, 231–237.

Schneider J.W., Celio M., Keller H., Kündig W., Odermatt W., Pümpin B., Savíc I.M., Simmler H., Estle T.L., Schwab C., Kiefl R.F., and Renker D. (1990) *Physical Review* **B 41**, R7254–R7257.

Shimomura K., Nishiyama K., and Kadono R. (2002) *Physical Review Letters* **89**, 255505, 1–4.

Smilga V.P. and Belousov Yu.M. (1994) *The Muon Method in Science*, Proceedings of the Lebedev Physics Institute, vol. 219 (Nova Science, Commack, NY).

Storchak V., Brewer J.H., and Morris G.D. (1994) *Physics Letters* **A 193**, 199–205.

Storchak V., Brewer J.H., Morris G.D., Arseneau D.J., and Senba M. (1999) *Physical Review* **B 59**, 10559–10572.

Storchak V., Cox S.F.J., Cottrell S.P., Brewer J.H., Morris G.D., and Prokof'ev N.V. (1996) *Physical Review* **B 53**, 662–667.

Veciana J. (2001) editor of *π-Electron Magnetism: From Molecules to Magnetic Materials, Structure and Bonding*, vol. 100 (Springer-Verlag, Berlin).

Walker D.C. (1983) *Muon and Muonium Chemistry* (Cambridge University Press, Cambridge).

Van de Walle C.G. (2000) *Physical Review Letters* **85**, 1012–1015.

Van de Walle C.G. and Neugebauer J. (2003) *Nature* **423**, 626–628.

13 Epilogue: trends for the future

In this epilogue we shall focus on technical developments which could be achieved within the next few years and finally mention opportunities for new muon sources.

The future of any experimental technique rests in part on the opportunities for technical and operational developments. In the case of μSR, three types of improvements can be recognised: (i) a significant extension of the performance and variety of sample environment equipment, (ii) a development of spectrometers with notably improved performance, which is within reach at pulsed sources, and (iii) a provision of better access to the instruments for scientists with no μSR background.

The variety of sample environment still suffers from limitations. While a wide range of equipment for cooling the sample down to very low temperature is available to experimentalists — ^{4}He and ^{3}He cryostats as well as dilution and closed cycle refrigerators — the possibility of performing measurements at high temperatures is more restricted. At the time of writing the highest magnetic field which can be applied is 8 T, and this is only possible at CMMS and for temperatures above about 2 K. Magnetic fields of 15 T are nowadays common at neutron diffractometers and spectrometers, and such fields would be highly desirable for μSR studies. However, it is quite a difficult technical challenge to perform μSR experiments in this range of fields, especially in transverse fields. The main issues lie in the dephasing of the muon spins in the magnet stray fields, in the detector time resolution, and in the spiraling radius of the positron trajectories. In fact, spectrometers with applied fields up to 10 and 5 T equipped with a dilution fridge are currently being built at SμS and ISIS, respectively, offering an interesting field to temperature ratio. Experiments under relatively high pressure are also feasible. At SμS such studies under good hydrostatic conditions are even possible down to 0.3 K. Nonetheless, the maximum pressure available for muon experiments is somewhat restricted: up to $\sim$1.5 GPa. Within a couple of years, pressure cells with a maximum applied pressure of $\sim$3 GPa should be available, with the possibility of working in a dilution refrigerator. Even though 3 GPa is a small pressure compared to what is available at synchrotron radiation facilities, it is reasonable compared with inelastic neutron scattering experiments.

Although less frequently used, other types of sample environments are sometimes available at μSR facilities, for example illumination of sample or application of an electric field to it. Some spectrometers are also equipped with an alternating current magnetic susceptibility apparatus which *in situ* enables the detection of a magnetic or superconducting transition in the sample under investigation by μSR. This is of particular interest for studies performed in the vicinity of phase transitions.

The time range of a spectrum extends up $\sim$10 and $\sim$15 μs at a pseudo-continuous and pulsed source, respectively. With the μSR spectrometers currently available, routine measurements are performed within about one hour with a sample of ~50 mg at a pseudo-continuous source. A sample heavier by an order of magnitude is required at a pulsed source. As mentioned at the beginning of Chapter 1, measurements at CMMS with a sample mass as small as 3.7 mg have even been reported. Studies with such small samples are quite sparse. Such a small sample is yet still large relative to acceptable sizes for a third generation synchrotron radiation facility. However, such a mass is usually too small for a neutron scattering measurement.

Unless an unforeseen technical advance occurs, no major progress on the spectrometer front is envisaged at the pseudo-continuous sources. On the other hand, since the count rate at pulsed sources is not limited by the number of incoming muons but rather by dead-time effects in detectors, an upgrade of the current spectrometers is possible through detector segmentation. For example, a threefold increase in data rate is expected in 2010 at one of the spectrometers of ISIS, and a spectrometer with $\sim 10^3$ positron counters is being designed for J-PARC. In the future, keeping a data acquisition time to one hour, it should be possible to perform a measurement with good statistics up to $\sim 15\,\mu$s with samples smaller than now using the fly-past technique (see Section 2.2.2.1). In addition, using a kicker magnet or an electrostatic device, one may dream of slicing the muon pulses, such that the detection of muon spin precession in a field of 0.5 T becomes possible. The feasibility of this exciting opportunity has yet to be demonstrated. Finally, as explained in Section 4.1, pulsed sources are well suited for RF-μSR measurements. To mature, this technique requires extension of the available experimental possibilities.

The investigations of surfaces and interfaces by the low-energy muon and β-detected NMR techniques are still in their infancy. Thanks to the recently available specialised spectrometers, the number of studies is bound to increase.

Originally the development of the μSR techniques owed much to nuclear physicists who turned to solid state physics. Nowadays, a typical visiting scientist at a μSR instrument is using a range of experimental techniques for investigating a given field in condensed-matter physics or chemistry. To perform successful μSR studies, he/she relies on well-trained experts organised in small teams at an instrument. In fact, as for neutron scattering and synchrotron radiation, most of the instruments tend to belong to facilities. This has been the trend in Europe for the past one or two decades and it is now under way at CMMS and J-PARC. Much effort is still needed, in particular in training new users and providing them with the possibility of working on their own.

In order to progress, established experimental techniques need an outlook. Current examples are the high-intensity neutron spallation sources or the X-ray sources based on free electron lasers. As explained in the introduction of Chapter 1 a new muon source has just been built in Japan, at J-PARC, using a high-performance proton beam. A μSR facility has been constructed there. A similar facility could possibly be built at a future high-intensity pulsed neutron spallation source in Europe. However, a completely different route to new muon sources is actively pursued. Thanks to advances in accelerator technology (Clery 2010), a high-intensity, stand-alone muon source dedicated to and fully optimised for μSR studies could be built. The proponents of this route suggest that new proton accelerators with a duty cycle approaching the

μSR requirement optimum can be designed at reasonable cost. Such accelerators could feed spectrometers similar to those available at today's pulsed sources, increasing the counting rate by two orders of magnitude, and in addition have the potentiality for a steady state operation matching that of the best current pseudo-continuous beam sources (Cywinski *et al.* 2009).

References

Clery D. (2010) *Science* **327**, 142–143.

Cywinski R., Bungau A.E., Poole M.W., Smith S., Dalmas de Réotier P., Barlow R., Edgecock R., King P.J.C., Lord J.S., Pratt F.L., Clausen K.N., and Shiroka T. (2009) *Physica* **B** **404**, 1024–1027.

Appendices

Appendix A
Handling experimental data

In this appendix we explain how to extract an asymmetry spectrum from a recorded positron count and how to visualise the magnetic field distribution at the muon site from a measured oscillating asymmetry spectrum. For the latter purpose, two methods are presented: Fourier transform and maximum entropy. In addition, the method for displaying an oscillating spectrum in a rotating reference frame is given.

A.1 From positron counts to time asymmetry spectra

It must be realised that, while the positron counts $N_\pm(t_n)$ is the measured quantity, in most experiments the physical quantities of interest are the so-called asymmetry spectra $a_0\eta_\mu P_\alpha^{\rm exp}(t_n)$ with $\alpha = X$ or Z depending on the experimental setup. We use the notations defined in Section 2.3. The relationships between these quantities have been spelt out in Eqs. 2.12 and 2.14 and are of quite general validity. In order to obtain the product $a_0\eta_\mu P_\alpha^{\rm exp}(t_n)$ from $N_\pm(t_n)$, one needs to determine the scaling constant N_0 and the background parameters $b_{\rm elec,\pm}$. This can be achieved through a fit. However, this procedure is not satisfactory because the quantity $a_0\eta_\mu P_\alpha^{\rm exp}(t_n)$ may not be unambiguously derived. This can be shown in the following way. Let us assume for simplicity that we deal with a longitudinal setup, $b_{\rm elec,\pm}$ being negligible, and that $P_Z^{\rm exp}(t_n) = 1$ in the time range of interest. Inspecting Eq. 2.14 it can be observed that for a given set of say $N_+(t_n)$ data, the following two extreme solutions cannot be distinguished: (i) N_0 is equal to a number that we denote as A and $\eta_\mu = 1$, and (ii) $N_0 = A(1 + a_0)$ and $\eta_\mu = 0$.

A way out of this issue is to combine the positron counts of forward and backward detectors located in opposite orientations relative to the sample. The asymmetry spectrum $a_0\eta_\mu P_\alpha^{\rm exp}(t_n)$ is then directly obtained through the simple operation described in Eq. 2.10 for the transverse setup. The same operation also applies to the longitudinal setup. No ambiguity exists on $a_0\eta_\mu P_\alpha^{\rm exp}(t_n)$ since no parameter determination is required. In practice, however, because of the experimental conditions, the symmetry requirements needed for Eqs. 2.12 and 2.14 to apply are not fulfilled. These two equations must then be rewritten in the generic form

$$N_\pm(t_n) = N_{0\pm}\{\exp(-t_n/\tau_\mu)\,[1 \pm a_0\eta_\mu P_\alpha^{\rm exp}(t_n)] + b_{\rm elec,\pm}\}. \tag{A.1}$$

Indeed, the parameters N_{0+} and N_{0-} generally differ for different reasons, for example the unequivalent solid angles subtended by each detector as seen from the sample or their difference in efficiency. An asymmetric sample environment (including the sample thickness itself) which absorbs the low-energy positrons differently for the two detectors can also be at the origin of an unbalance in the detected counts. If the ratio $\alpha_{\rm bal} \equiv N_{0+}/N_{0-}$ is known it is a simple matter to get $a_0 \eta_\mu P_\alpha^{\rm exp}(t_n)$ from $N_\pm(t_n)$:

$$a_0 \eta_\mu P_\alpha^{\rm exp}(t_n) = \frac{N_+(t_n) - \alpha_{\rm bal} N_-(t_n)}{N_+(t_n) + \alpha_{\rm bal} N_-(t_n)}. \tag{A.2}$$

Equation A.2 has been derived assuming $b_{\rm elec,\pm} = 0$. We shall discuss later on the case $b_{\rm elec,\pm} \neq 0$. In principle the balance parameter $\alpha_{\rm bal}$ could be calculated, but it is far better to experimentally determine it for each series of measurements on a given sample. In fact, the knowledge of $a_0 \eta_\mu P_\alpha^{\rm exp}(t_n)$ for given experimental conditions (e.g. a given temperature) is enough for the determination of $\alpha_{\rm bal}$, through the use of Eq. A.2. In practice the parameter $\alpha_{\rm bal}$ is often determined from a measurement in a weak transverse field (a few milliteslas). In such an experiment, in systems such as magnets at temperatures well above their ordering point, or superconductors above their critical temperature, one expects the polarisation function to be a damped cosine. The amplitude at $t_{\rm offset}$, the earliest time at which data are available, gives $a_0 \eta_\mu$ since $t_{\rm offset}$ is sufficiently short that $\cos(\gamma_\mu B_{\rm ext} t_{\rm offset}) \simeq 1$. If there is no missing fraction, $\eta_\mu = 1$. A very precise estimate for $\alpha_{\rm bal}$ is achieved with a couple of millions of positron counts, i.e. after a few minutes counting at current muon spectrometers. This estimate is precise even if the shape of the damping function is not accurately known. The obtained $\alpha_{\rm bal}$ value is then used for all the subsequent measurements in the series.

Note that an incorrect determination of $\alpha_{\rm bal}$ will result in first order in a shift of the asymmetry spectrum, at least at short enough times, as shown below. Let us call $\alpha'_{\rm bal}$ the incorrectly estimated value for the balance parameter and $c_{\alpha_{\rm bal}}$ the parameter which links it to the true $\alpha_{\rm bal}$ through the relation $\alpha'_{\rm bal} = \alpha_{\rm bal}(1 - c_{\alpha_{\rm bal}})/(1 + c_{\alpha_{\rm bal}})$. Then from Eq. A.2, it can be derived that

$$\frac{N_+(t_n) - \alpha'_{\rm bal} N_-(t_n)}{N_+(t_n) + \alpha'_{\rm bal} N_-(t_n)} = \frac{c_{\alpha_{\rm bal}} + a_0 \eta_\mu P_\alpha^{\rm exp}(t_n)}{1 + c_{\alpha_{\rm bal}} a_0 \eta_\mu P_\alpha^{\rm exp}(t_n)}, \tag{A.3}$$

which can be approximated to $c_{\alpha_{\rm bal}} + a_0 \eta_\mu P_{X,Z}^{\rm exp}(t_n)$ if $c_{\alpha_{\rm bal}} \ll 1$. In the same limit $\alpha'_{\rm bal} \simeq \alpha_{\rm bal}(1 - 2c_{\alpha_{\rm bal}})$. This means that if the balance parameter is, say, 2% off, the asymmetry spectrum is shifted by -0.01. This remark can be used in an iterative procedure to get the proper value for $\alpha_{\rm bal}$. It works as follows. A trial value is first set for this parameter. The ratio $[N_+(t_n) - \alpha_{\rm bal} N_-(t_n)]/[N_+(t_n) + \alpha_{\rm bal} N_-(t_n)]$ of forward and backward data measured in a weak transverse field is fitted to a damped cosine function plus a constant. From the value of this constant, a new value for $\alpha_{\rm bal}$ is estimated according to the rule just described and the procedure is iterated until convergence is reached. In practice this is when the shift is 0 within error bars; i.e. it is less than 10^{-3} or 10^{-4} in absolute value.

So far, we have considered the case where $b_{\rm elec,\pm}$ are negligible. If it is not, Eq. A.2 can still be used provided that $N_\pm(t_n)$ are replaced with $N_\pm(t_n) - N_{0\pm} b_{\rm elec,\pm}$. This raises the problem of the determination of $N_{0\pm} b_{\rm elec,\pm}$. A first possible way is to fit $N_\pm(t_n)$ with Eq. A.1. This yields satisfactory results in the case of transverse-field measurements, i.e. when the polarisation function is purely oscillating, or in the case of zero- or longitudinal-field experiments when the polarisation decays rapidly to 0. Conversely, if the decay is very slow, inspection of Eq. A.1 shows that $N_{0\pm} b_{\rm elec,\pm}$

cannot be obtained from a fit of $N_\pm(t_n)$ if $\eta_\mu P_\alpha^{\rm exp}(t_n)$ is not known. An alternative solution for an estimate of $N_{0\pm}b_{\rm elec,\pm}$ is an inspection of $N_\pm(t_n)$ at time channels for which there are no muons in the sample, i.e. at times long enough that $N_{0\pm}\exp(-t_n\tau_\mu)$ is negligible or at "negative" times, that is to say, before the muons are implanted in the sample. This latter possibility depends on the electronic settings used for the data acquisition.

In any case, much care must be exercised when considering an asymmetry spectrum for times t_n satisfying the relation $\exp(-t_n/\tau_\mu) \lesssim b_{\rm elec,\pm}$. Indeed in this time range, its shape crucially depends on the value chosen for $b_{\rm elec,\pm}$ to compute it from $N_\pm(t_n)$. Let us estimate the typical time t_c — defined as $\exp(-t_c/\tau_\mu) = b_{\rm elec,\pm}$ — after which an asymmetry spectrum might be distorted due to the presence of the $b_{\rm elec,\pm}$ terms, for different kinds of experimental setups. The magnitude of $b_{\rm elec,\pm}$ at state-of-the-art spectrometers on surface muon beams at a pseudo-continuous source is 1%. This corresponds to $t_c \simeq 4.6\tau_\mu$ or $10\,\mu$s. Using a device which extracts one muon from the beam upon request (see Section 2.2.2.1), $b_{\rm elec,\pm}$ can be reduced by two orders of magnitude. This extends t_c roughly to $9\tau_\mu$ or $20\,\mu$s. However, it must be noted that the counterpart of using such a device is a small distortion of the asymmetry spectrum at early times ($t \leq 0.5\ \mu$s). Spectrometers at pulsed muon sources are characterised by an extremely low background which is even difficult to measure, $b_{\rm elec,\pm} \lesssim 10^{-5}$, resulting in no distortion in the experimental time range. Pulsed muon sources suffer, however, from a few specific drawbacks compared to pseudo-continuous sources. These are briefly described below.

Pulsed muon sources have a few additional specificities compared to pseudo-continuous sources that it is worth mentioning here. We recall that at a pulsed muon beam source, a bunch of muons is implanted at a time taken as 0 (see Section 2.2.2.1). This means that a positron detector might be hit by two or more particles in a very short time interval, especially shortly after the instant of muon implantation, i.e. when the decay positrons are many. If this time interval is too short, the probability of pile up in the positron detection increases, and events, i.e. positron counts, might be missed. The effect of dead time in the electronic circuitry is similar. If the average number of events counted by the detector, per time bin of width Δt and per muon bunch, is much less than 1, the observed positron count $N_\pm^{\rm pu}(t_n)$ with the pile-up effect is related to the ideal count through the relation

$$N_\pm^{\rm pu}(t_n) = n_p\{1 - \exp[-N_\pm(t_n)/n_p]\}, \tag{A.4}$$

where n_p is the number of muon bunches which have been taken to record the spectrum. This formula follows from the Poisson distribution of particles resulting from a radioactive decay. It can be used to compute $N_\pm(t_n)$ from $N_\pm^{\rm pu}(t_n)$ and correct for the distortion observed at short times at pulsed sources. The interested reader will find more details in the article by Kadono *et al.* (1989). Anyhow, if the initial shape of the signal is important, it is wise to run with a muon beam intensity sufficiently small for the distortion to be negligible.

Another specificity of pulsed muon sources is the fact that the muons of a single bunch are not implanted in the sample exactly at the same time. As indicated in Section 2.2.2.1, a typical time distribution of the muons is a Gaussian function of full-width at half-maximum equal to 80 ns. This time is not negligible compared to the characteristic times in a μSR experiment. For instance, in transverse-field experiments, the spins of earlier muons will have substantially precessed while the later ones are still being implanted. Numerically, in a field of 46 mT, the muon spin precesses by half a turn in 80 ns. The result of such an effect is a strong reduction of the amplitude of the oscillating signal as the field is increased. In a similar way, in zero- or longitudinal-field experiments performed on strongly depolarising compounds, the polarisation of earlier muons is lost before the later ones are implanted, resulting again in a decrease of the observed amplitude. Quantitatively, the observed count $N_{\pm}^{\rm fw}(t_n)$ with the finite width of the muon bunch is related to the ideal count through the convolution product

$$N_{\pm}^{\rm fw}(t_n) = \int_{-\infty}^{+\infty} D(\tau) N_{\pm}(t_n - \tau)\mathrm{d}\tau, \tag{A.5}$$

where $D(t)$ is the time distribution function of the muons in the bunch. When $D(t)$ is a Gaussian function of 80 ns full-width at half-maximum, the effect of the convolution is to divide by 2 the amplitude of a signal precessing at a 6-MHz frequency.

A.2 Oscillating spectra

We have so far detailed processing procedures of interest for data recorded with the two geometries generally used in μSR, namely the longitudinal- or zero-field geometry on one hand and the transverse-field geometry on the other; see Section 1.1.1. When the transverse geometry is used, e.g. for the investigation of the frequency shift and therefore the local susceptibility of a compound or the study of the field distribution in the mixed phase of a type II superconductor, a large external field $\mathbf{B}_{\rm ext}$ is applied. The experimentalist is basically interested in the field distribution at the muon site, which is generally discrete and continuous for the two types of measurements that we have mentioned, respectively. We recall that sufficiently large transverse-field measurements probe basically the distribution of the field component parallel to the external field; see Section 6.2.1. As shown at the beginning of this section, provided this distribution is a symmetric function, the resulting $P_X(t)$ function is the product of an oscillating function, $\cos(\gamma_\mu B_{\rm ext} t)$, by the cosine Fourier transform of the field distribution. The presence of the cosine term makes it difficult to directly get a flavour of the actual shape of the damping function or to inspect the quality of a fit of a spectrum to a model. A naive way out of this problem would simply consist in dividing $P_X(t)$ by $\cos(\gamma_\mu B_{\rm ext} t)$. Obviously, this method does not give satisfactory results when $\cos(\gamma_\mu B_{\rm ext} t)$ is close to 0.

Before we proceed further, the reader will have noticed that we have now replaced the sampling times t_n with the continuous variable t. In the same way the polarisation functions have been denoted as $P_\alpha(t)$ instead of $P_\alpha^{\rm exp}(t)$. The reason is that the procedures detailed here can be applied similarly to experimental and simulated spectra. For the sake of simple notation we have also dropped the product $a_0\eta_\mu$.

The proper solution to the situation described above is to consider a rotating reference frame; see Riseman and Brewer (1990). This method requires not only the measurement of $P_X(t)$ but also $P_Y(t)$. We then introduce the complex polarisation function $P_{X+iY}(t) \equiv P_X(t) + iP_Y(t)$, and a frame $\{\hat{\mathbf{X}}', \hat{\mathbf{Y}}', \hat{\mathbf{Z}}'\}$ rotating around $\hat{\mathbf{Z}}$ at angular frequency $\omega_{\rm rf}$. The relation between the frame vectors $\hat{\mathbf{X}}', \hat{\mathbf{Y}}', \hat{\mathbf{Z}}'$ and $\hat{\mathbf{X}}, \hat{\mathbf{Y}}, \hat{\mathbf{Z}}$ is given in Eq. 4.3. The polarisation functions $P_{X'}(t)$ and $P_{Y'}(t)$ expressed in the rotating frame are then

$$P_{X'}(t) = \mathcal{R}e\left\{\exp(i\omega_{\rm rf}t)P_{X+iY}(t)\right\} \text{ and } P_{Y'}(t) = \mathcal{I}m\left\{\exp(i\omega_{\rm rf}t)P_{X+iY}(t)\right\}, \quad \text{(A.6)}$$

respectively. Note that the sign of the argument of the exponential term has been chosen positive since the muon spin rotates clockwise in the XY plane when the field is oriented parallel to the Z axis (see Section 3.1.1.1). Now, which $\omega_{\rm rf}$ value do we select? A first possibility is to take a $\omega_{\rm rf}$ value slightly lower than $\gamma_\mu B_{\rm ext}$. With this choice, wiggles are still present in the rotating reference frame spectra, informing the observer that he/she is dealing with transverse-field measurements, while the reduced precession frequency provides a good feeling of how the wiggles are damped. The alternative choice is to set $\omega_{\rm rf} = \omega_\mu$. In the physical context where $P_X(t)$ and $P_Y(t)$ contain $\cos(\omega_\mu t)$ and $\sin(\omega_\mu t)$ factors, respectively, this second choice enables a direct visualisation of the envelope function. Obviously the rotating-reference-frame technique can be used to generate a nice display of spontaneous oscillations, reflecting the spontaneous magnetic field at the muon site in the ordered state of a magnetic compound. However, difficulties may appear if more than one frequency is present.

Using the rotating reference frame has an additional benefit when the fitting model is very demanding in terms of computation time. Since the number of wiggles can be drastically reduced, it is possible to perform a massive binning of the data. Consequently, when expressed in the rotating frame, the model function needs to be computed at much fewer time values.

It is important to note that the information contained in $P_Y(t)$ cannot always be predicted from $P_X(t)$. As an illustration, let us consider the two following static distributions for which the X and Y components of $\mathbf{B}_{\rm loc}$ vanish. While in the first case, all the muons are submitted to the same field $\mathbf{B}_0$ parallel to $\hat{\mathbf{Z}}$, in the second, the field distribution is equipopulated among the $\pm\mathbf{B}_0$ values. The associated distributions $D_{\rm c}(B^Z_{\rm loc})$ read $\delta(B^Z_{\rm loc} - B_0)$ and $[\delta(B^Z_{\rm loc} - B_0) + \delta(B^Z_{\rm loc} + B_0)]/2$, respectively. The computation of $P_X(t)$ (Eq. 6.5) and $P_Y(t)$ (Eq. 3.14) leads to $P_X(t) = \cos(\gamma_\mu B_0)$ in both situations while $P_Y(t) = -\sin(\gamma_\mu B_0 t)$ in the former case and $P_Y(t) = 0$ in the latter. When performing transverse-field experiments it is therefore important to consider both $P_X(t)$ and $P_Y(t)$, and, e.g. in the case of the study of the mixed phase of type II superconductors, to fit both functions to a field distribution model.

Another use of the complex polarisation function is found in the analysis of data recorded with positron detectors which are not only located along the $\pm\mathbf{X}$ and $\pm\mathbf{Y}$ directions. This situation is encountered e.g. at pulsed muon sources, where the optimisation of the counting rate leads to the design of spectrometers with massively segmented detectors, each of them covering a relatively small solid angle. Let us denote by $P_\varphi(t)$ the asymmetry spectrum recorded by the detector located in the XY plane at angle φ from $\hat{\mathbf{X}}$ and let us assume a second detector to be present at angle $\varphi + \pi/2$ with associated spectrum $P_{\varphi+\pi/2}(t)$. As before we construct the complex polarisation function $P_{\varphi+i(\varphi+\pi/2)}(t) = P_\varphi(t) + iP_{\varphi+\pi/2}(t)$. Obviously, the corresponding polarisation functions rotated in the $\{X, Y\}$ frame are $\mathcal{R}e\{\exp(i\varphi)P_{\varphi+i(\varphi+\pi/2)}(t)\}$

and $\mathcal{I}m\{\exp(i\varphi)P_{\varphi+i(\varphi+\pi/2)}(t)\}$ for $P_X(t)$ and $P_Y(t)$, respectively. The procedure can be repeated for any pair of detectors separated by a $\pi/2$ angle: the results are added together to get the global polarisation functions $P_X(t)$ and $P_Y(t)$. Of course, when detectors located in opposite direction (say at angles φ and $\varphi+\pi$) are available, it is a good idea to compute $P_\varphi(t)$ from the counts in the respective detectors following Eq. A.2.

When oscillations are present in an asymmetry spectrum it is often convenient to get its Fourier spectrum. This is of great help in the choice of the model function to be used in order to fit the data. Two methods are generally applied in the context of μSR: the direct Fourier transform, which is the subject of the next section, and the maximum entropy method, described in the subsequent section. Since this discussion is more specifically intended to measured data, we return to the experimental polarisation functions measured at discrete times $P_\alpha^{\exp}(t_n)$ and reintroduce the product $a_0\eta_\mu$.

A.2.1 Fourier transform method

We first recall the general definition of the discrete Fourier transform. Let us assume that we have a set of discrete data $\{m_n\}$ with $n=\{0,1,\ldots,N_{\rm s}-1\}$ where the number of samples $N_{\rm s}$ is taken to be even. We define the discrete Fourier transform of this set by

$$M_p = \frac{1}{\sqrt{N_{\rm s}}}\sum_{n=0}^{N_{\rm s}-1} m_n \exp\left(i\frac{2\pi\, n\, p}{N_{\rm s}}\right), \tag{A.7}$$

with $p=\{0,1,\ldots,N_{\rm s}-1\}$. According to the Fourier theorem, m_n can be computed from the set $\{M_p\}$:

$$m_n = \frac{1}{\sqrt{N_{\rm s}}}\sum_{p=0}^{N_{\rm s}-1} M_p \exp\left(-i\frac{2\pi\, n\, p}{N_{\rm s}}\right). \tag{A.8}$$

We now apply the Fourier transform technique to $\{m_n\}$ where we have $m_n = a_0\eta_\mu P_\alpha^{\exp}(t_n)$. We like to compute the weight of the field distribution at channel p for a detector with the geometrical angle φ. This weight denoted by A_p^φ is obtained by a discrete cosine Fourier transform using the relation

$$A_p^\varphi = \frac{1}{\sqrt{N_{\rm s}}}\sum_{n=0}^{N_{\rm s}-1} m_n \cos\left(\omega_p\, t_n + \varphi\right) = \frac{1}{\sqrt{N_{\rm s}}}\sum_{n=0}^{N_{\rm s}-1} m_n \cos\left(\frac{2\pi\, pn}{N_{\rm s}} + \frac{2\pi\, p t_{\rm offset}}{N_{\rm s}\Delta t} + \varphi\right). \tag{A.9}$$

We have set $t_n = n\Delta t + t_{\rm offset}$ and $\omega_p = p[2\pi/(N_{\rm s}\Delta t)]$ where Δt is the time resolution of the spectrometer; see Section 2.3. From Eqs. A.7 and A.9 and since m_n is real for any n, we derive

$$A_p^\varphi = \cos\left(\frac{2\pi\, p t_{\rm offset}}{N_{\rm s}\Delta t} + \varphi\right)\mathcal{R}e\{M_p\} - \sin\left(\frac{2\pi\, p t_{\rm offset}}{N_{\rm s}\Delta t} + \varphi\right)\mathcal{I}m\{M_p\}. \tag{A.10}$$

Hence, we need to compute M_p. The direct application of Eq. A.7 would require $N_{\rm s}^2$ complex number multiplications. The fast Fourier transform, or FFT, is an algorithm for performing the Fourier transform much more efficiently than the straightforward calculation — see for example, James *et al.* (1999), Thijssen (1999), or Press *et al.* (2007). It uses the fact that in determining the Fourier transform, the factor $\exp[i(2\pi\, n\, p)/N_s]$ assumes the same value for different combinations of n and p. The number N_s is taken as $N_s = 2^\gamma$ where γ is an integer. Then the number of multiplications is about $N_s \log_2 N_s$ ($\log_2 N_s = \ln N_s/\ln 2$) instead of N_s^2 for the direct evaluation of the Fourier transform. In the case where the number of available channels is not a power of 2, the asymmetry spectrum can be padded with zeros so as to reach such a number of channels. Alternatively, more sophisticated algorithms also exist which are still rapid and do not suffer from the FFT limitation as far as the number of samples is concerned.

We have here described the Fourier transform of a single asymmetry spectrum. It can be a good idea to consider complex asymmetry data obtained from two detectors separated by a $\pi/2$ angle, following the procedure described earlier in Section A.2. We leave to the interested reader the task of adapting Eqs. A.9 and A.10.

In the context of Section A.2.2.2 — see below — the discrete Fourier transform of an assumed field distribution is needed. Obviously, the amplitude of the field, M_p, is real for any channel p. From the method used for the computation of A_p^φ, we derive the weight, a_n^φ, at channel n, i.e. at time $t_n = n\Delta t + t_{\rm offset}$, for a detector at angle φ:

$$a_n^\varphi = \cos\varphi\, \mathcal{R}e\{m_{n+\ell}\} + \sin\varphi\, \mathcal{I}m\{m_{n+\ell}\}. \tag{A.11}$$

Here we have supposed $t_{\rm offset} = \ell\Delta t$ where ℓ is an integer. Obviously, m_n is periodical because of the equality $m_{n+N_s} = m_n$ which follows from Eq. A.8. As previously, m_n should be computed with the FFT algorithm.

A.2.2 Maximum entropy method

A.2.2.1 The entropy in information theory. Entropy is an old concept. Its first introduction into thermodynamics by Clausius dates back to the middle of the nineteenth century. Later on, Boltzmann derived the entropy expression in terms of statistical mechanics, i.e. the well-known Boltzmann relation,

$$S = k_{\rm B} \ln \Omega, \tag{A.12}$$

where $k_{\rm B}$ is the Boltzmann constant and Ω is the number of microstates for the particular macrostate. The entropy in thermodynamics or statistical mechanics is a measure of the degree of disorder existing in a system; see any standard textbooks on thermodynamics and statistical mechanics, e.g. Zemansky (1968).

Entropy is also a modern concept important in the theory of information which was introduced by Shannon. In the common case, we have N_e propositions labelled by $n = 0, 1, \ldots, N_e - 1$ and we assign to them a set of probabilities $\{p_n\}$ which encode our knowledge relevant to choosing one rather than another. The amount of uncertainty of the information source is represented by the so-called information-theory entropy $S_{\rm I}$. A possible form is (Buck and Macaulay 1991, Wu 1997)

$$S_{\rm I} = -\sum_{n=0}^{N_e-1} p_n \ln\left(\frac{p_n}{w_n}\right), \tag{A.13}$$

where w_n is the so-called prior estimate of p_n and obviously $\sum_{n=0}^{N_e-1} p_n = 1$.

The significance of Eq. A.13 can be understood with a simple example such as a system of N_e particles identical but distinguishable.

We first apply the tools of statistical mechanics. Suppose we have a set of energy levels $\{\epsilon_n\}$, each of them with a degeneracy g_n. The number of quantum states corresponding to a given macrostate $\{n_n\}$ is

$$\Omega\left(\{n_n\}\right) = \frac{(N_e - 1)!}{\prod_{n=0}^{N_e-1} n_n!} \prod_{n=0}^{N_e-1} g_n^{n_n}, \tag{A.14}$$

where n_n is the number of particles with energy ϵ_n. Under the constraints $\sum_{n=0}^{N_e-1} n_n = N_e$ and $\sum_{n=0}^{N_e-1} n_n \epsilon_n = E$ where E is the energy of the macrostate, maximising the expression of $\ln \Omega$ leads to the well-known result

$$n_n = \frac{N_e g_n}{Z} \exp\left(-\lambda \epsilon_n\right), \tag{A.15}$$

where Z is the partition function, $Z = \sum_{n=0}^{N_e-1} g_n \exp\left(-\lambda \epsilon_n\right)$, and λ a Lagrange coefficient. Thermodynamics makes it possible to identify λ with $1/(k_B T)$.

We analyse now the same system from the information viewpoint. Coefficient p_n in Eq. A.13 is then interpreted as the probability that a particle is found to be in energy level ϵ_n and w_n is taken as g_n. Maximising S_I subject to the constraints $\sum_{n=0}^{N_e-1} p_n = 1$ and $\sum_{n=0}^{N_e-1} p_n \epsilon_n = E/N_e$ gives (see e.g. Wu (1997))

$$p_n = \frac{g_n}{Z} \exp\left(-\lambda \epsilon_n\right). \tag{A.16}$$

Since $p_n = n_n/N_e$, the method of information theory recovers the result of statistical mechanics.

The previous discussion suggests that maximising S_I, subject to constraints reflecting the model under consideration, is a method for assigning values to the set of probabilities $\{p_n\}$.

A.2.2.2 The maximum entropy method in data analysis. In the so-called inverse problem, one is required to construct from a limited and noisy set of data a functional form for it. The maximum entropy method (MEM) is a scheme for the construction with takes the form of a density distribution. It is therefore adapted for determining the static field distribution behind the μSR polarisation function. Its first use in the field of μSR has been by Rainford and Daniel (1994) and its value for the study of flux-line lattices in superconductors has been strongly advocated — see e.g. Riseman and Forgan (2000). It is still a method under development, in particular for the analysis of μSR data. The reader should therefore be aware that improvements of the method might be proposed in the future.

The strategy — see e.g Buck and Macaulay (1991), Daniell (1991) — is to build a distribution (in the frequency domain) agreeing with the known data (in time domain) according to some criterion. The key point is the requirement that the proper distribution maximises S_I, i.e. is the most probable among all those which agree with observation.

We first recall the definition of the chi-square, χ^2:

$$\chi^2 = \sum_{n=0}^{N_e-1} \frac{(d_n - m_n)^2}{\sigma_n^2}. \tag{A.17}$$

Here $d_n = a_0 \eta_\mu P_\alpha^{\rm exp}(t_n)$ is the measured polarisation function at channel n and σ_n the related standard deviation. The image in time domain at channel n of the distribution we are looking for is denoted by $m_n = m_n(M_0, M_1, \ldots, M_{N_e-1})$, where M_n is the spectral weight for the precession frequency $\nu_{\mu,n}$. The computation results, as explained below, from an iterative procedure. An unbiased choice for the starting spectral weight is a uniform one. The image number m_n is computed by discrete Fourier transform from M_n; see Section A.2.1. In contrast to the usual Fourier transform method for which the calculation is performed on the experimental spectrum and the value of d_n must be set to 0 when n is large, we now Fourier-transform a test distribution and compare the result with the measured spectrum by maximising the multivariate function $F(M_0, M_1, \ldots, M_{N_e-1})$. This justifies the name of inverse problem given to this procedure. Thus,

$$F(M_0, M_1, \ldots, M_{N_e-1}) = -\alpha \sum_{n=0}^{N_e-1} M_n \ln\left(\frac{M_n}{w_n}\right) + \frac{1}{2}\chi^2(M_0, M_1, \ldots, M_{N_e-1}). \tag{A.18}$$

The value of α is an arbitrary real number. It is usually chosen to make χ^2 equal to N_e, although other possibilities have been proposed (Daniell 1991). So far, the choice for w_n has been $w_n = 1$ for all n.

The problem of finding numerically the maximum of $F(M_0, M_1, \ldots, M_{N_e-1})$ defined in Eq. A.18 is not trivial because of the large number of variables. Fortunately, routines are available, see e.g Skilling and Bryan (1984). Since the computation is iterative, it is slow in comparison to the direct fast Fourier transform of the polarisation data. However, relative to that latter method, it presents two advantages. First, the field distribution is built-in positive. Second, since a well-defined test distribution is used, its Fourier transform can be evaluated without the need of truncation. In the conventional method the truncation, which reflects the practical finite time range of an experiment, introduces spurious effects.

A.2.3 Apodisation

No matter the method employed to extract a field distribution from a measured spectrum, the direct discrete Fourier transform or maximum entropy method, the computed distribution displays ringing phenomena, i.e. oscillations in the distribution of purely numerical origin. They reflect the fact that the experimental data result from particle counts and are therefore affected by statistical uncertainties. Because of the muon decay, these uncertainties are more prominent at long time. The usual method used to damp these artefacts is the so-called apodisation, which is a classical means of signal processing. Prior to the Fourier transform computation, one multiplies the asymmetry spectrum by a window or apodisation function. In the case of μSR, Gaussian or exponential functions with characteristic decay times on the order of the muon lifetime are well adapted. Hence the data at long time, which are less accurately measured, have a lower weight in the Fourier transform. While smoothing the distribution, this technique presents a disadvantage: it broadens the frequency (or field) distribution. Indeed, a multiplication in the time domain results in a convolution in frequency. This is not too serious a problem since the quantitative analysis of the data is in any case performed with the recorded spectrum (in time domain), the computed field distribution being only used for presentation and physical discussion of the results.

References

Buck B. and Macaulay V.A. (1991) in *Maximum Entropy in Action*, edited by Buck B. and Macaulay V.A. (Clarendon Press, Oxford).

Daniell G.J. (1991) in *Maximum Entropy in Action*, edited by Buck B. and Macaulay V.A. (Clarendon Press, Oxford).

James G., Burley D., Dyke P., Searl J., Steele N., and Wright J. (1999) *Advanced Modern Engineering Mathematics*, second edition (Addison-Wesley, Harlow).

Kadono R., Imazato J., Matsuzaki T., Nishiyama K., Nagamine K., Yamazaki T., Richter D., and Welter J.-M. (1989) *Physical Review* **B 39**, 23–41

Press W.H., Teukolsky S.A., Vetterling W.T., and Flannery B.P. (2007) *Numerical Recipes: The Art of Scientific Computing*, third edition (Cambridge University Press, Cambridge).

Rainford B.D. and Daniell G.J. (1994) *Hyperfine Interactions* **87**, 1129–1134.

Riseman T.M. and Brewer J.H. (1990) *Hyperfine Interactions* **65**, 1107–1112.

Riseman T.M. and Forgan E.M. (2000) *Physica* **B 289-290**, 718–721.

Skilling J. and Bryan R.K. (1984) *Monthly Notices of the Royal Astronomical Society* **211**, 111–124; information is available at the following internet address: http://www.maxent.co.uk/.

Thijssen J.M. (1999) *Computational Physics* (Clarendon Press, Oxford).

Wu N. (1997) *The Maximum Entropy Method*, Springer Series in Information Science, vol. 32 (Springer, Berlin).

Zemansky M.W. (1968) *Heat and Thermodynamics* (McGraw-Hill, New York).

Appendix B
Polarisation functions: a summary

Here we provide in Table B.1 a short summary of widely used polarisation functions, which is to serve as a guide to the reader who is new to the μSR techniques. Description of experimental data may require considering a weighted sum of the product of some polarisation functions.

Table B.1 For each polarisation function we give its physical context and recall the section(s) where it is described. The index provides another way of finding the relevant information about the functions.

Name or origin of $P_X(t)$ or $P_Z(t)$	Physical context	Sections
Abragam	Dyamical transverse field distribution. The static (strongly dynamical) case corresponds to the Gaussian (exponential) function. Might be also used in zero field when the dynamics is not too slow.	6.3.1, 6.3.2.3
Bessel	Spontaneous field from a simple static incommensurate amplitude-modulated magnetic structure.	10.1.1.2
Cosine	Single static transverse field.	6.2
Discrete stochastic field	Spontaneous field flipping between two opposite values.	6.3.2.6
Dynamical Kubo–Toyabe	Extension to dynamical fields of the Kubo–Toyabe function. Its limit in the case of extremely fast dynamics is the exponential function.	6.3.1
Exponential	Motional-narrowing limit relaxation.	6.3.2.3
Gaussian	Static Gaussian field distribution in transverse field.	6.2.1

Table B.1 *continued*

Name or origin of $P_X(t)$ or $P_Z(t)$	Physical context	Sections
Koptev–Tarasov	Magnetically ordered phase in a powder sample.	6.2.3
Kubo–Toyabe	Static Gaussian isotropic field distribution in zero field, or by extension in longitudinal field.	6.2.1
Noakes–Kalvius	System with short-range magnetic correlations.	6.4.1.3
Risch–Kehr	Intermittent hyperfine interaction in one-dimensional systems.	8.4.3
Squared-Lorentzian distribution	System with diluted magnetic impurities in zero field.	6.4.1.1, 6.4.1.2
Stretched exponential	System with distributed relaxation.	6.4.1.2
Two-state model	Two clearly differentiated muon environments. A given muon might alternatively probe each of them.	6.4.2

Appendix C
Mathematical concepts and formulae

The purpose of this appendix is to remind the reader of some mathematical concepts and formulae frequently used in various chapters.

C.1 Tensors

We consider a reference frame defined by a set of three orthogonal unit vectors $\{\hat{\mathbf{X}}, \hat{\mathbf{Y}}, \hat{\mathbf{Z}}\}$.

A set specified by 3^n components forms a tensor of rank n if they follow the appropriate transformation rule under change of reference frame. For example, for a tensor of the first rank ($n = 1$) the transformation rules are the normal vector transformations. Let us consider a tensor $\boldsymbol{G}$ that we shall denote as $\mathbf{G}$ since it is also a vector. Its components G^α in reference frame $\{\hat{\mathbf{X}}, \hat{\mathbf{Y}}, \hat{\mathbf{Z}}\}$ transform into components G'^α in reference frame $\{\hat{\mathbf{X}}', \hat{\mathbf{Y}}', \hat{\mathbf{Z}}'\}$. The transformation rule reads

$$G^{\alpha'} = \sum_\beta l^{\alpha'\beta}\, G^\beta. \tag{C.1}$$

The direction cosines $l^{\alpha'\beta}$ do not form a tensor of rank two but are the elements of a matrix since they relate the components of $\mathbf{G}$ in two different reference frames; see e.g. Nye (1985) or Chapter 7 of Dove (2003). The tensor $\mathbf{G} = G^X\hat{\mathbf{X}} + G^Y\hat{\mathbf{Y}} + G^Z\hat{\mathbf{Z}}$ is simply a vector where G^α is equal to the scalar product of $\mathbf{G}$ by $\hat{\boldsymbol{\alpha}}$, i.e. $G^\alpha = \mathbf{G}\cdot\hat{\boldsymbol{\alpha}}$.

The physics described in this monograph requires the consideration of a restricted class of tensors formed by the tensor product of tensors of lower rank. For example, let us consider the tensor of rank one $\mathbf{g}$ with $\sum_\alpha (g^\alpha)^2 = 1$. The tensor product of $\mathbf{g}$ by itself is the tensor of rank two with components $P_{\mathrm{L}}^{\alpha\beta}(\mathbf{g}) = g^\alpha g^\beta$.

Let us define two tensors of rank one (vectors) $\mathbf{E}$ and $\mathbf{F}$ such that $\mathbf{E} = \boldsymbol{P}_{\mathrm{L}}(\mathbf{g})\,\mathbf{F}$. Since

$$E^\alpha = \sum_\beta P_{\mathrm{L}}^{\alpha\beta}(\mathbf{g})\,F^\beta = \sum_\beta g^\alpha\, g^\beta\, F^\beta = g^\alpha\,(\mathbf{g}\cdot\mathbf{F})\,, \tag{C.2}$$

we have the remarkable result $\mathbf{E} = (\mathbf{g}\cdot\mathbf{F})\,\mathbf{g}$. Thus $\boldsymbol{P}_{\mathrm{L}}(\mathbf{g})\,\mathbf{F}$ is a vector pointing along the direction of $\mathbf{g}$ with a magnitude equal to $\mathbf{g}\cdot\mathbf{F}$. Therefore $\boldsymbol{P}_{\mathrm{L}}(\mathbf{g})$ acts on a vector as a longitudinal projector.

We use the antisymmetric unit pseudotensor (also called the unit axial tensor or Levi–Civita tensor) with components $\epsilon^{\alpha\beta\gamma}$. By definition, $\epsilon^{\alpha\beta\gamma}$ is equal to 0 if two or more of the superscripts are equal, 1 if the set $\{\alpha\,\beta\,\gamma\}$ is deduced from $\{X\,Y\,Z\}$ by an even permutation, and -1 otherwise. Calculations sometimes require the evaluation of products of pseudotensor elements. The following identity is useful for this purpose:

$$\sum_{\nu} \epsilon^{\alpha\beta\nu} \epsilon^{\lambda\mu\nu} = \delta_{\alpha,\lambda}\delta_{\beta,\mu} - \delta_{\alpha,\mu}\delta_{\beta,\lambda}, \tag{C.3}$$

where $\delta_{\alpha,\beta}$ is the Kronecker symbol, i.e. $\delta_{\alpha,\beta} = 1$ if $\alpha = \beta$ and 0 otherwise.

As expected for a vectorial space of three dimensions, in addition to the longitudinal projector, two transverse projection operators can be built. These projectors select two independent vectors perpendicular to $\mathbf{g}$. One of them is $\boldsymbol{P}_{\mathrm{T}_1}(\mathbf{g}) = \boldsymbol{I} - \boldsymbol{P}_{\mathrm{L}}(\mathbf{g})$, where $\boldsymbol{I}$ is the unit tensor with components $\delta_{\alpha,\beta}$. Note that $\boldsymbol{P}_{\mathrm{L}}(\mathbf{g})$ and $\boldsymbol{P}_{\mathrm{T}_1}(\mathbf{g})$ are symmetric tensors since $P_{\mathrm{L}}^{\alpha\beta}(\mathbf{g}) = P_{\mathrm{L}}^{\beta\alpha}(\mathbf{g})$ and $P_{\mathrm{T}_1}^{\alpha\beta}(\mathbf{g}) = P_{\mathrm{T}_1}^{\beta\alpha}(\mathbf{g})$. In addition, $\boldsymbol{P}_{\mathrm{L}}(\mathbf{g})$ and $\boldsymbol{P}_{\mathrm{T}_1}(\mathbf{g})$ satisfy the relations

$$\sum_{\beta} P_{\mathrm{L}}^{\alpha\beta}(\mathbf{g}) P_{\mathrm{T}_1}^{\beta\gamma}(\mathbf{g}) = 0 \quad \text{and} \quad \sum_{\beta} P_{\mathrm{L}}^{\alpha\beta}(\mathbf{g}) P_{L}^{\beta\gamma}(\mathbf{g}) = P_{\mathrm{L}}^{\alpha\gamma}(\mathbf{g}). \tag{C.4}$$

The second transverse projector (tensor) $\boldsymbol{P}_{\mathrm{T}_2}(\mathbf{g})$ is defined as $P_{\mathrm{T}_2}^{\alpha\beta}(\mathbf{g}) = \sum_{\gamma} \epsilon^{\alpha\gamma\beta}\, g^{\gamma}$. It is antisymmetric, i.e. $P_{\mathrm{T}_2}^{\alpha\beta}(\mathbf{g}) = -P_{\mathrm{T}_2}^{\beta\alpha}(\mathbf{g})$. It is interesting to consider the effect of $\boldsymbol{P}_{\mathrm{T}_2}(\mathbf{g})$ on a tensor of rank one such as $\mathbf{F}$. Defining $\mathbf{H} = \boldsymbol{P}_{\mathrm{T}_2}(\mathbf{g})\,\mathbf{F}$, we derive

$$H^{\alpha} = \sum_{\beta\gamma} \epsilon^{\alpha\gamma\beta}\, g^{\gamma}\, F^{\beta} = (\mathbf{g} \times \mathbf{F})^{\alpha}. \tag{C.5}$$

This means that $\mathbf{H}$ is the vector product of $\mathbf{g}$ by $\mathbf{F}$.

As a by-product of the previous relation, we note that the curl of a vector field $\mathbf{F}$ is a vector, the components of which can be written in a compact form with the Levi–Civita tensor and the notation $\nabla^{\beta} = \partial/\partial r^{\beta}$:

$$(\mathbf{curl}\,\mathbf{F})^{\alpha} = \sum_{\beta\gamma} \epsilon^{\alpha\beta\gamma}\, \nabla^{\beta}\, F^{\gamma}. \tag{C.6}$$

The divergence of a vector field $\mathbf{F}$ is also of interest; it is a scalar defined as

$$\operatorname{div}\mathbf{F} = \sum_{\alpha} \nabla^{\alpha} F^{\alpha}. \tag{C.7}$$

The gradient of the scalar field f is the vector of components

$$(\mathbf{grad}\, f)^{\alpha} = \nabla^{\alpha} f. \tag{C.8}$$

The Laplacian of the scalar field f is $\Delta f = \operatorname{div}\mathbf{grad}\, f$, and the Laplacian of the vectorial field $\mathbf{F}$ is the vector $\Delta\mathbf{F}$ whose components are the respective Laplacians of the components of $\mathbf{F}$. The following relation between these differential operators holds (see e.g. Chow 2000):

$$\mathbf{curl}\,\mathbf{curl}\,\mathbf{F} = \mathbf{grad}\operatorname{div}\mathbf{F} - \Delta\mathbf{F}. \tag{C.9}$$

C.2 Operators

Let us consider two operators A and B. The following very useful cycling rule is obeyed,

$$\mathrm{Tr}\{AB\} = \mathrm{Tr}\{BA\}, \tag{C.10}$$

where $\mathrm{Tr}\{A\}$ stands for the trace of operator A.

We now give three useful relations which involve the exponential function of operators. This function is quite common in quantum mechanics. Let us consider the commutator of two operators A and B, which is by definition the operator $C = [A, B] = AB - BA$. If both A and B commute with C, then

$$\exp(A + B) = \exp(A)\,\exp(B)\,\exp(-C/2) = \exp(B)\,\exp(A)\,\exp(C/2). \tag{C.11}$$

A proof is given by Slichter — Appendix A in Slichter (1996).

Writing $\exp(A)$ as a Maclaurin series, it can be shown that

$$B^{-1}\exp(A)B = \exp(B^{-1}AB), \tag{C.12}$$

where B^{-1} is the inverse operator of B; i.e. $B^{-1}B = BB^{-1} = \mathcal{I}$ where $\mathcal{I}$ is the unit operator.

The following relation is derived by expanding the exponential operators in power:

$$\exp(iBt)\,A\exp(-iBt) = A - i[A, B]\,t - \frac{1}{2}[[A, B], B]\,t^2 + \mathcal{O}(t^3). \tag{C.13}$$

C.3 Spin operators

Let us consider the spin vector operator $\mathbf{J}$. Its Cartesian components verify the commutation rule

$$\left[J^\alpha, J^\beta\right] = i\sum_\gamma \epsilon^{\alpha\beta\gamma}\,J^\gamma. \tag{C.14}$$

It is convenient to introduce the raising and lowering spin operators, J^+ and J^- respectively, defined by the equations

$$J^+ = J^X + iJ^Y \quad \text{and} \quad J^- = J^X - iJ^Y. \tag{C.15}$$

We may express J^X and J^Y in terms of J^+ and J^- by solving Eq. C.15, getting

$$J^X = \frac{1}{2}\left(J^+ + J^-\right) \quad \text{and} \quad J^Y = \frac{1}{2\,i}\left(J^+ - J^-\right). \tag{C.16}$$

The following relations are satisfied: $(J^+)^\dagger = J^-$ and $(J^-)^\dagger = J^+$, where for example $(J^+)^\dagger$ is the Hermitian conjugate of J^+. The matrix elements of any spin operator can be computed using Eq. C.16 and the relations

$$\begin{aligned} J^+|J, m\rangle &= \sqrt{J(J+1) - m(m+1)}\,|J, m+1\rangle, \\ J^-|J, m\rangle &= \sqrt{J(J+1) - m(m-1)}\,|J, m-1\rangle, \\ J^Z|J, m\rangle &= m|J, m\rangle. \end{aligned} \tag{C.17}$$

The vector $|J, m\rangle$ is an eigenvector of the J^Z operator. We recall that $J^2|J, m\rangle = J(J+1)|J, m\rangle$.

C.4 Rotation operators

In quantum mechanics, the operator associated with a rotation of angle ϱ about the u axis is $R_u(\varrho) = \exp(-i\,\varrho\,J^u)$. The effect of the rotation on an operator Q is described by the product operator $R_u(\varrho)\,Q\,R_u^{-1}(\varrho)$ where $R_u(\varrho)R_u^{-1}(\varrho) = R_u^{-1}(\varrho)R_u(\varrho) = \mathcal{I}$. For example, the following two identities are derived:

$$\exp(-i\,\varrho\,J^Z)\,J^{\pm}\,\exp(+i\,\varrho\,J^Z) = \exp(\mp i\,\varrho)\,J^{\pm}. \tag{C.18}$$

More information on the rotation operators can be found in standard textbooks on quantum mechanics such as, for example, Rose (1957) or Messiah (1965).

Any rotation in a three-dimensional vectorial space can be decomposed as the product of up to three basic rotations — see for example Messiah (1965), exercising care with respect to misprints. Each of them is specified by an Euler angle and the rotation axis about which the rotation is performed. Refering to Fig. D.1 we denote these angles by φ, θ, and ψ. Gathering the information above, the result of the rotation, defined by these Euler angles, on an operator Q is the operator $\tilde{Q}$

$$\tilde{Q} = R(\varphi,\theta,\psi)QR^{-1}(\varphi,\theta,\psi), \tag{C.19}$$

where $R(\varphi,\theta,\psi)$ is the operator for the rotation:

$$R(\varphi,\theta,\psi) = R_Z(\varphi)R_Y(\theta)R_Z(\psi). \tag{C.20}$$

Considering a spin $J = \frac{1}{2}$, the effect of the rotation on the basis states $\{|\frac{1}{2}\rangle, |-\frac{1}{2}\rangle\}$ is described by (see e.g. Rose (1957))

$$R(\varphi,\theta,\psi)\,|m\rangle = \sum_{m'} D^{(1/2)}(\varphi,\theta,\psi)^{m'm}\,|m'\rangle\,, \tag{C.21}$$

where

$$D^{(1/2)}(\varphi,\theta,\psi) = \begin{pmatrix} \exp\left(-i\dfrac{\varphi+\psi}{2}\right)\cos\left(\dfrac{\theta}{2}\right) & -\exp\left(i\dfrac{\psi-\varphi}{2}\right)\sin\left(\dfrac{\theta}{2}\right) \\ \exp\left(i\dfrac{\varphi-\psi}{2}\right)\sin\left(\dfrac{\theta}{2}\right) & \exp\left(i\dfrac{\varphi+\psi}{2}\right)\cos\left(\dfrac{\theta}{2}\right) \end{pmatrix}. \tag{C.22}$$

The entity $D^{(1/2)}(\varphi,\theta,\psi)$ is a representation for a spin $\frac{1}{2}$ within the framework of group theory as applied to symmetry in physics; see Wigner (1959) or Bradley and Cracknell (1972). Note that the opposite operator $-D^{(1/2)}(\varphi,\theta,\psi)$ defines the same rotation. This reflects the 2 to 1 homomorphism of $SU(2)$ into $SO(3)$. This has no consequences in the context of this monograph.

C.5 Pauli operators

The spin vector operator $\mathbf{J}$ for a spin $J = \frac{1}{2}$ has remarkable mathematical properties. They can be expressed in a compact formula with the Pauli vector operator defined by $\boldsymbol{\sigma} = 2\,\mathbf{J}$; for instance

$$\sigma^{\alpha}\,\sigma^{\beta} = \delta_{\alpha\,\beta}\mathcal{I} + i\sum_{\gamma}\epsilon^{\alpha\,\beta\,\gamma}\,\sigma^{\gamma}, \tag{C.23}$$

where σ^{α} is the Cartesian component α of operator $\boldsymbol{\sigma}$. This result shows that the square of any component of the Pauli operator is the unit operator $\mathcal{I}$, the trace of which is 2, and the product of two different components of the Pauli operator is proportional to the third component of the Pauli operator.

As done for the spin operators, it is convenient to introduce the raising and lowering operators, σ^{+} and σ^{-} respectively, defined by the equations

$$\sigma^{+} = \sigma^{X} + i\sigma^{Y} \qquad \text{and} \qquad \sigma^{-} = \sigma^{X} - i\sigma^{Y}. \tag{C.24}$$

We may express σ^{X} and σ^{Y} in terms of σ^{+} and σ^{-} by solving Eq. C.24, getting

$$\sigma^{X} = \frac{1}{2}\left(\sigma^{+} + \sigma^{-}\right) \qquad \text{and} \qquad \sigma^{Y} = \frac{1}{2\,i}\left(\sigma^{+} - \sigma^{-}\right). \tag{C.25}$$

The Pauli operators are practical because they follow the rule given in Eq. C.23. Explicitly,

$$\begin{aligned} &\sigma^{Z}\sigma^{+} = \sigma^{+} = -\sigma^{+}\sigma^{Z}, \qquad \sigma^{-}\sigma^{Z} = \sigma^{-} = -\sigma^{Z}\sigma^{-}, \\ &\left(\sigma^{+}\right)^{2} = \left(\sigma^{-}\right)^{2} = 0, \qquad \sigma^{+}\sigma^{-} = 2\left(\mathcal{I} + \sigma^{Z}\right), \qquad \sigma^{-}\sigma^{+} = 2\left(\mathcal{I} - \sigma^{Z}\right). \end{aligned} \tag{C.26}$$

If the quantisation axis is along the Z axis, the three Pauli operators are represented by 2×2 traceless matrices

$$\sigma^{X} = \begin{pmatrix} 0 & 1 \\ 1 & 0 \end{pmatrix}, \qquad \sigma^{Y} = \begin{pmatrix} 0 & -i \\ i & 0 \end{pmatrix}, \qquad \sigma^{Z} = \begin{pmatrix} 1 & 0 \\ 0 & -1 \end{pmatrix}. \tag{C.27}$$

Note that $\sigma^{+}|-\rangle = 2\,|+\rangle$ and $\sigma^{-}|+\rangle = 2\,|-\rangle$, where $|+\rangle$ and $|-\rangle$ are the eigenvectors of σ^{Z} associated with the eigenvalues 1 and -1, respectively.

The operator $R^{\sigma}_{\alpha}(\rho)$ which describes the rotation of angle ϱ about the α axis for a Pauli operator has the simple expression

$$R^{\sigma}_{\alpha}(\rho) = \exp(-i\rho\sigma^{\alpha}/2) = \mathcal{I}\cos\left(\rho/2\right) - i\sigma^{\alpha}\sin\left(\rho/2\right). \tag{C.28}$$

Note that the argument of the cosine and sine functions is half of the angle of rotation.

The following two identities are then easily derived:

$$\exp(-i\,\rho\,\sigma^{Z})\,\sigma^{\pm}\,\exp(+i\,\rho\,\sigma^{Z}) = \exp(\mp i\,2\rho)\,\sigma^{\pm}. \tag{C.29}$$

Obviously, the same two relations can be obtained directly from Eq. C.18.

C.6 Statistical thermodynamics

In this monograph, we use statistical functions such as the well-known Fermi–Dirac and Planck distributions — see e.g. Kittel and Kroemer (1980).

The probability for a fermion to occupy a state of energy E at temperature T is given by the Fermi–Dirac statistics:

$$n_{\rm FD}\left(\frac{E}{k_{\rm B}T}\right) = \frac{1}{\exp\left(\frac{E-E_{\rm F}}{k_{\rm B}T}\right)+1}, \tag{C.30}$$

where $E_{\rm F}$ is the Fermi energy.

The Planck distribution function valid for excitations such as phonons and magnons reads

$$n_{\rm P}\left(\frac{E}{k_{\rm B}T}\right) = \frac{1}{\exp\left(\frac{E}{k_{\rm B}T}\right)-1}. \tag{C.31}$$

The Bose–Einstein distribution function is an extension of the previous relation when the chemical potential matters.

C.7 Distributions

The Dirac delta distribution, or simply the Dirac delta function, is commonly used to take into account the conservation of energy and momentum. It also allows one to write an expression for a distribution in a transparent way. Below we detail the case of the field distributions used in this monograph. Then we list some properties of the Dirac delta distribution.

C.7.1 Field distributions

From the expression of the field at each point of space, $\mathbf{B}(\mathbf{r})$, the vector field distribution $D_{\rm v}(\mathbf{B})$ can be computed with the formula

$$D_{\rm v}(\mathbf{B}) = \frac{\int \delta\left[\mathbf{B}(\mathbf{r})-\mathbf{B}\right] {\rm d}^d\mathbf{r}}{\int {\rm d}^d\mathbf{r}}, \tag{C.32}$$

where $\delta(\mathbf{B})$ is the Dirac delta function and d the dimension of space. In practice, for the distributions of this monograph d equals 2 or 3. In certain instances distributions derived from $D_{\rm v}(\mathbf{B})$ are used. These are the distribution of field component α,

$$D_{\rm c}(B^\alpha) = \frac{\int \delta\left[B^\alpha(\mathbf{r})-B^\alpha\right] {\rm d}^d\mathbf{r}}{\int {\rm d}^d\mathbf{r}}, \tag{C.33}$$

and the distribution of the modulus B of $\mathbf{B}$,

$$D_{\rm m}(B) = \frac{\int \delta\left[B(\mathbf{r})-B\right] {\rm d}^d\mathbf{r}}{\int {\rm d}^d\mathbf{r}}. \tag{C.34}$$

The relation between the latter distribution and $D_v(\mathbf{B})$ is particularly useful. In the three-dimensional case and in spherical coordinates, the relation reads

$$D_m(B) = \int D_v(B, \theta, \varphi) B^2 \sin\theta d\theta d\varphi. \tag{C.35}$$

In the particular case where $D_v(B, \theta, \varphi)$ is independent of θ and φ, i.e. when it is isotropic, we have

$$D_m(B) = 4\pi B^2 D_v(B). \tag{C.36}$$

A usual way of characterising a distribution is through its moments. The mth moment of a field component distribution is defined by

$$\overline{(B^\alpha)^m} = \int (B^\alpha)^m D_c(B^\alpha) dB^\alpha, \tag{C.37}$$

where m is a positive integer. The integration is performed over all possible B^α values. The first moment $\overline{B^\alpha}$ is simply the field component average. The second moment $\overline{(B^\alpha)^2}$ is of frequent use. If $D_c(B^\alpha_{loc})$ is an even function ($D_c(B^\alpha) = D_c(-B^\alpha)$), we have the remarkable result $\overline{(B^\alpha)^{2m+1}} = 0$. It is sometimes of interest to chacterise the symmetry of a distribution with respect to its average value: we refer for example to the field distribution characterising a flux-line lattice of a superconductor. The skewness parameter ϑ^α makes it possible to quantify the lack of symmetry. It is defined by

$$\vartheta^\alpha = \frac{\left[\overline{\left(B^\alpha - \overline{B^\alpha}\right)^3}\right]^{1/3}}{\left[\overline{\left(B^\alpha - \overline{B^\alpha}\right)^2}\right]^{1/2}}. \tag{C.38}$$

For a symmetric distribution, $\vartheta^\alpha = 0$. The previous equation defines the skewness for component α of the field. It is of course possible to define the skewness of, for example, the distribution of the field modulus through the substitution $B^\alpha \to B$ in Eq. C.38.

When considering distributions one parameter is ubiquitous: it is the variance or dispersion defined as

$$\Delta_\alpha^2 = \overline{\left(B^\alpha - \overline{B^\alpha}\right)^2} = \overline{(B^\alpha)^2} - \overline{B^\alpha}^2. \tag{C.39}$$

It is the square of the standard deviation Δ_α, which is also called the root-mean-square (rms) of the distribution of field component α.

C.7.2 The Dirac delta distribution

As indicated at the beginning of this section, the Dirac delta function is used in more general contexts than the field distributions and it is necessary to detail some of its properties. Let us first consider the one-dimension case. By definition,

$$\int_a^b f(x)\delta(x - x_0) dx = f(x_0), \tag{C.40}$$

for any function $f(x)$, provided the interval $]a, b[$ contains the point x_0.

Below, we list some properties of the Dirac delta function in one dimension:

$$\delta(x) = \delta(-x), \qquad \delta(a\,x) = \frac{\delta(x)}{|a|} \quad \text{for} \quad a \neq 0, \tag{C.41}$$

$$\delta(x) = \frac{1}{2\pi}\int_{-\infty}^{+\infty} \exp(iqx)\mathrm{d}q, \tag{C.42}$$

$$\delta[g(x)] = \sum_n \frac{1}{|g'(x_n)|}\delta(x - x_n). \tag{C.43}$$

The variable x_n must be such that $g(x_n) = 0$ and $g'(x_n) \equiv \mathrm{d}g/\mathrm{d}x|_{x=x_n} \neq 0$.

A useful representation of the Dirac function is

$$\delta(x) = \frac{1}{\pi}\lim_{a\to\infty}\frac{1-\cos(ax)}{ax^2}. \tag{C.44}$$

In spaces of dimension $d > 1$, the Dirac function $\delta(\mathbf{r} - \mathbf{r}_0)$ is defined as a natural extension of the one-dimension case with the relation

$$\int F(\mathbf{r})\delta(\mathbf{r} - \mathbf{r}_0)\mathrm{d}^d\mathbf{r} = F(\mathbf{r}_0), \tag{C.45}$$

where the integration is performed over a domain including the point $\mathbf{r}_0$ of coordinates $\left(x_0^{(1)}, x_0^{(2)}, \cdots, x_0^{(d)}\right)$. The Dirac function fulfills the properties

$$\begin{aligned}\delta(\mathbf{r} - \mathbf{r}_0) &= \delta\left(x^{(1)} - x_0^{(1)}\right)\delta\left(x^{(2)} - x_0^{(2)}\right)\cdots\delta\left(x^{(d)} - x_0^{(d)}\right) \\ &= \frac{1}{(2\pi)^d}\int \exp[i\mathbf{q}\cdot(\mathbf{r} - \mathbf{r}_0)]\mathrm{d}^d\mathbf{q}.\end{aligned} \tag{C.46}$$

In spherical coordinates ($d = 3$), denoting as $(r_0, \theta_0, \varphi_0)$ the coordinates of $\mathbf{r}_0$, we have

$$\begin{aligned}\delta(\mathbf{r} - \mathbf{r}_0) &= \frac{1}{r^2\sin^2\theta}\delta(r - r_0)\delta(\theta - \theta_0)\delta(\varphi - \varphi_0) \\ &= \frac{1}{r^2}\delta(r - r_0)\delta(\cos\theta - \cos\theta_0)\delta(\varphi - \varphi_0).\end{aligned} \tag{C.47}$$

From Eq. C.46,

$$\delta(\mathbf{q} - \mathbf{q}_0) = \frac{1}{(2\pi)^d}\int \exp[i\mathbf{r}\cdot(\mathbf{q} - \mathbf{q}_0)]\mathrm{d}^d\mathbf{r}. \tag{C.48}$$

C.8 Linear transforms

C.8.1 Time and space Fourier transforms

We define the time Fourier transform of a function $f(\tau)$ as follows:

$$f(\omega) = \int_{-\infty}^{\infty} \exp(i\omega\tau)\, f(\tau)\,\mathrm{d}\tau. \tag{C.49}$$

As it is well known, the following complementary relation holds:

$$f(\tau) = \int_{-\infty}^{\infty} \exp(-i\omega\tau)\, f(\omega)\, \frac{\mathrm{d}\omega}{2\pi}. \tag{C.50}$$

The space Fourier transform of a function $f(\mathbf{r})$ is written as:

$$f(\mathbf{q}) = \int_{\mathrm{DS}} \exp(-i\mathbf{q}\cdot\mathbf{r})\, f(\mathbf{r})\, \mathrm{d}^d\mathbf{r}. \tag{C.51}$$

The inverse relation is given by

$$f(\mathbf{r}) = \int_{\mathrm{RS}} \exp(i\mathbf{q}\cdot\mathbf{r})\, f(\mathbf{q})\, \frac{\mathrm{d}^d\mathbf{q}}{(2\pi)^d}. \tag{C.52}$$

The integration volumes of these integrals are the whole direct (DS) and reciprocal (RS) spaces, respectively.

In some instances, the Fourier power is used. It is defined as $|f(\omega)|^2$ and $|f(\mathbf{q})|^2$ in the time and space domains, respectively.

C.8.2 Laplace transform

For any positive real s, the Laplace transform $f(s)$ of function $f(t)$ is by definition

$$f(s) = \int_{0}^{+\infty} f(t) \exp(-st)\mathrm{d}t. \tag{C.53}$$

C.9 Density of states

The spin–lattice and spin–spin relaxation rates can sometimes be expressed in terms of a weighted sum of correlation functions integrated over the first Brillouin zone. It can be convenient to switch from an integration over this Brillouin zone to an integration over energy. This is performed by introducing the density of excitations responsible for the muon spin relaxation. Here we describe the relation between the two types of integration. But we first need to recall two concepts.

A key characteristic of an excitation is its dispersion relation, i.e. the relationship between its energy E and its wave vector $\mathbf{q}$. Concerning this matter, only three-dimensional spaces are considered in this monograph. We therefore set the space dimension to $d = 3$. Assuming the energy to only depend on $q = |\mathbf{q}|$, we write

$$E = f\,(q)\,. \tag{C.54}$$

We shall assume that the reciprocal function $f^{-1}(E)$ can be defined:

$$q = f^{-1}\,(E)\,. \tag{C.55}$$

We now introduce the density of states. The number of excitations with wave vectors comprised between the shells of radius q and $q + \mathrm{d}q$ is equal to $V\mathrm{d}^3\mathbf{q}/(2\pi)^3$, where V is the volume of the sample. By definition, this is also given by $Vg(E)\mathrm{d}E$, where $g(E)$ is the density of excitation modes per volume unit of the sample; see e.g. Kittel (1996) or Ashcroft and Mermin (2002). Hence,

$$g(E) = \frac{1}{2\pi^2} \frac{q^2}{\left.\frac{\mathrm{d}f}{\mathrm{d}q}\right|_q} = \frac{1}{2\pi^2} \frac{\left[f^{-1}(E)\right]^2}{\left.\frac{\mathrm{d}f}{\mathrm{d}q}\right|_{f^{-1}(E)}} = \frac{1}{2\pi^2} \left[f^{-1}(E)\right]^2 \left.\frac{\mathrm{d}f^{-1}}{\mathrm{d}E}\right|_E . \tag{C.56}$$

We have introduced the relation

$$\left.\frac{\mathrm{d}f^{-1}}{\mathrm{d}E}\right|_E = \frac{1}{\left.\frac{\mathrm{d}f}{\mathrm{d}q}\right|_{f^{-1}(E)}} . \tag{C.57}$$

The density $g(E)$ can also formally be expressed in terms of a δ function:

$$g(E) = \int_{v_c^*} \delta\left(E - E(q)\right) \frac{\mathrm{d}^3\mathbf{q}}{(2\pi)^3} = \int_{v_c^*} \delta\left(E - f(q)\right) \frac{\mathrm{d}^3\mathbf{q}}{(2\pi)^3} . \tag{C.58}$$

It can be shown that formulae C.56 and C.58 for $g(E)$ are equivalent using Eq. C.43.

The dispersion relation and the density of states of the excitations being defined, we now look for an expression of the integral over the Brillouin zone, $\int_{v_c^*} A(q)\mathrm{d}^3\mathbf{q}$, in terms of the density of excitation modes. First we change the variables of integration from Cartesian to spherical $(q_x, q_y, q_z) \rightarrow (E, \theta, \varphi)$ with $q = f^{-1}(E)$, and get

$$\int_{v_c^*} A(q)\mathrm{d}^3\mathbf{q} = \int A\left[f^{-1}(E)\right] \frac{\partial\left(q_x, q_y, q_z\right)}{\partial\left(E, \theta, \varphi\right)} \mathrm{d}E\mathrm{d}\theta\mathrm{d}\varphi, \tag{C.59}$$

where $\partial(q_x, q_y, q_z)/\partial(E, \theta, \varphi,)$ is the Jacobian of (q_x, q_y, q_z) with respect to the variables (E, θ, φ); see e.g. Chow (2000). Since

$$\frac{\partial\left(q_x, q_y, q_z\right)}{\partial\left(E, \theta, \varphi\right)} = \left[f^{-1}(E)\right]^2 \left.\frac{\mathrm{d}f^{-1}}{\mathrm{d}E}\right|_E \sin\theta, \tag{C.60}$$

using Eq. C.56 we derive

$$\frac{\partial\left(q_x, q_y, q_z\right)}{\partial\left(E, \theta, \varphi,\right)} = 2\pi^2 g(E) \sin\theta, \tag{C.61}$$

and therefore

$$\int_{v_c^*} A(q)\mathrm{d}^3\mathbf{q} = (2\pi)^3 \int A\left[f^{-1}(E)\right] g(E)\mathrm{d}E. \tag{C.62}$$

C.10 Some integrals and other formulae

Here we give some integrals as well as mathematical functions properties. A good reference is Gradshteyn and Ryzhik (1994, 2000). The parameters x, ν, d and q are real quantities, while a, b, c, and p are strictly positive real numbers, n is a natural integer and λ a complex number. Note that some of the formulae given in this section might be of wider generality. For instance, although expressed in terms of real variables, they sometimes apply also for complex variables. For the sake of simplicity, we restrict ourselves to our direct need. We first list some definite integrals of elementary functions.

C.10.1 Definite integrals

$$\int_0^{+\infty} \exp\left(-p^2x^2\right)\cos(qx)\mathrm{d}x = \frac{\sqrt{\pi}}{2p}\exp\left(-\frac{q^2}{4p^2}\right). \tag{C.63}$$

$$\int_0^{+\infty} x^2\exp\left(-p^2x^2\right)\cos(qx)\mathrm{d}x = \sqrt{\pi}\frac{2p^2-q^2}{8p^5}\exp\left(-\frac{q^2}{4p^2}\right). \tag{C.64}$$

$$\int_0^{+\infty} x\exp\left(-p^2x^2\right)\sin(ax)\mathrm{d}x = \frac{a\sqrt{\pi}}{4p^3}\exp\left(-\frac{a^2}{4p^2}\right). \tag{C.65}$$

$$\int_0^{+\infty} x^3\exp\left(-p^2x^2\right)\sin(ax)\mathrm{d}x = \sqrt{\pi}\frac{6ap^2-a^3}{16p^7}\exp\left(-\frac{a^2}{4p^2}\right). \tag{C.66}$$

$$\int_0^{+\infty} \exp\left(-ax^2-\frac{b}{x^2}\right)\mathrm{d}x = \frac{1}{2}\sqrt{\frac{\pi}{a}}\exp\left(-2\sqrt{ab}\right). \tag{C.67}$$

$$\int_0^{+\infty} x^2\exp\left(-ax^2-\frac{b}{x^2}\right)\mathrm{d}x = \frac{1}{4}\sqrt{\frac{\pi}{a^3}}\left(1+2\sqrt{ab}\right)\exp\left(-2\sqrt{ab}\right). \tag{C.68}$$

$$\int_0^{+\infty} \frac{\cos(ax)}{b^2+x^2}\mathrm{d}x = \frac{\pi}{2b}\exp(-ab). \tag{C.69}$$

$$\int_0^{+\infty} \frac{\cos(ax)}{(b^2+x^2)^2}\mathrm{d}x = \frac{\pi}{4b^3}(1+ab)\exp(-ab). \tag{C.70}$$

$$\int_0^{+\infty} \frac{x^2\cos(ax)}{\left(b^2+x^2\right)^2}\mathrm{d}x = \frac{\pi}{4b}(1-ab)\exp(-ab). \tag{C.71}$$

$$\int_0^{+\infty} \frac{x\sin(ax)}{b^2+x^2}\mathrm{d}x = \frac{\pi}{2}\exp(-ab). \tag{C.72}$$

$$\int_{-\infty}^{+\infty} \frac{b+dx}{p+2qx+x^2}\cos(ax)\mathrm{d}x \tag{C.73}$$

$$= \pi\left[\frac{b-dq}{\sqrt{p-q^2}}\cos(aq)+d\sin(aq)\right]\exp\left(-a\sqrt{p-q^2}\right), \quad p>q^2.$$

$$\int_{-\infty}^{+\infty} \frac{b+dx}{p+2qx+x^2}\sin(ax)\mathrm{d}x \tag{C.74}$$

$$= \pi \left[\frac{dq - b}{\sqrt{p - q^2}} \sin(aq) + d\cos(aq) \right] \exp\left(-a\sqrt{p - q^2}\right), \quad p > q^2.$$

$$\int_0^{+\infty} x \exp(-px) \cos(2x^2 - px) \mathrm{d}x = \frac{p\sqrt{\pi}}{8} \exp\left(\frac{-p^2}{4}\right). \tag{C.75}$$

$$\int_0^{+\infty} \frac{1 - \cos x}{x^2} \mathrm{d}x = \frac{\pi}{2}. \tag{C.76}$$

$$\int_0^{+\infty} x \exp(-bx) \sin(ax) \mathrm{d}x = \frac{2ab}{(a^2 + b^2)^2}. \tag{C.77}$$

$$\int_0^{+\infty} x^{p-1} \exp(-bx) \cos(ax) \mathrm{d}x = \frac{(p-1)!}{(a^2 + b^2)^{p/2}} \cos\left[p \arctan\left(\frac{a}{b}\right)\right]. \tag{C.78}$$

$$\int_0^{+\infty} \frac{\mathrm{d}x}{ax^2 + (1 + a)x + 1} = \frac{\ln(a)}{a - 1}. \tag{C.79}$$

$$\int_0^{1} (1 - x^2)^{a-1} \mathrm{d}x = \frac{(-1/2)!(a-1)!}{2(a - 1/2)!}. \tag{C.80}$$

$$\int_0^{b} (b^2 - x^2)^{\nu - 1/2} \cos(ax) \mathrm{d}x = \frac{\sqrt{\pi}}{2} \left(\nu - \frac{1}{2}\right)! \left(\frac{2b}{a}\right)^{\nu} J_\nu(ab), \tag{C.81}$$

for $\nu > -\frac{1}{2}$. Here $J_\nu(x)$ denotes a Bessel function of the first kind, an expression of which is given in Eq. C.93.

$$\int_0^{\pi} \exp(ia\cos x) \cos(nx) \mathrm{d}x = i^n \pi J_n(a). \tag{C.82}$$

C.10.2 Other integrals

The following integral is used in Chapter 10:

$$I = \int_{-\infty}^{+\infty} \frac{1}{(x + p)^2 + b^2} \frac{1 - \cos(ax)}{x^2} \mathrm{d}x. \tag{C.83}$$

To compute I, the integration of $f(z) = [1 - \exp(iaz)]/\{z^2[(z + p)^2 + b^2]\}$ over the contour shown in Fig. C.1 is performed using the residue theorem — see, for example, Chow (2000). The only pole of $f(z)$ within this contour is at $-p + ib$. Introducing the change of variable $z = r\exp(i\theta)$, the integrals of $f(z)$ over the semicircles γ and Γ are evaluated. Taking the limits $R \to \infty$ and $r \to 0$, we get

$$I = \frac{\pi a}{p^2 + b^2} + \frac{\pi}{b} \frac{1}{(p^2 + b^2)^2} \left\{ (p^2 - b^2) \left[1 - \exp(-ab)\cos(ap)\right] - 2bp \exp(-ab) \sin(ap) \right\}, \tag{C.84}$$

since the integral of $f(z)$ on semicircle Γ vanishes. As expected, the limit of I/a when $a \to \infty$ is consistent with the direct computation from Eq. C.83 using the representation of the Dirac δ function given in Eq. C.44.

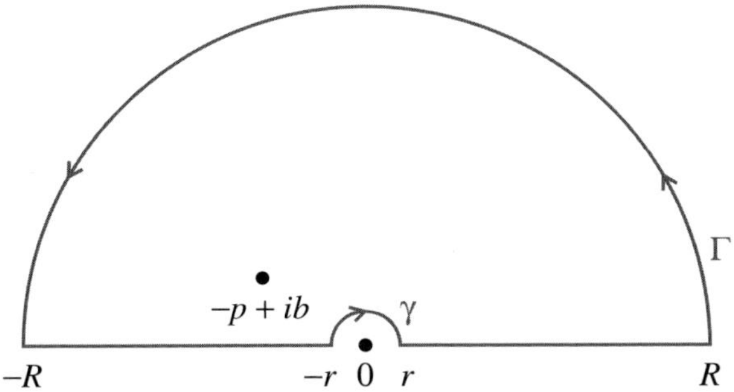

Fig. C.1 Complex plane contour used for the computation of integral I in Eq. C.83. The semicircles Γ and γ are centred on the origin and their radii are R and r, respectively.

The following double integral is encountered when investigating the effect of dynamics on the polarisation functions:

$$J = \int_0^t \mathrm{d}t' \int_0^{t'} f(t' - t'')\,\mathrm{d}t''. \tag{C.85}$$

We are going to express J in terms of a single integral. Let us first define the function g such that

$$\int_0^{t'} f(t' - t'')\,\mathrm{d}t'' = \int_0^{t'} f(\tau)\,\mathrm{d}\tau = g(t'). \tag{C.86}$$

With this definition we derive

$$J = \int_0^t g(t')\mathrm{d}t' = \int_0^t (t-\tau) f(\tau)\mathrm{d}\tau, \tag{C.87}$$

where the second step has been achieved by integrating by parts. Finally we get

$$J = \int_0^t \mathrm{d}t' \int_0^{t'} f(t' - t'')\,\mathrm{d}t'' = \int_0^t (t-\tau) f(\tau)\mathrm{d}\tau. \tag{C.88}$$

A second double integral appears in the study of the polarisation functions. It differs from the previous only by a different upper bound for the second integral:

$$K = \int_0^t \mathrm{d}t' \int_0^{t} f(t' - t'')\,\mathrm{d}t''. \tag{C.89}$$

We first note that

$$K = \int_0^t \mathrm{d}t' \int_0^{t'} f(t' - t'')\,\mathrm{d}t'' + \int_0^t \mathrm{d}t' \int_{t'}^{t} f(t' - t'')\,\mathrm{d}t''. \tag{C.90}$$

If $f(t)$ is an even function, it can be shown that the second term on the right-hand side of Eq. C.90 is equal to the first. The proof is obtained after several changes of variables in the double integral. Hence, from Eq. C.89, we get

$$\int_0^t \mathrm{d}t' \int_0^t f\left(t' - t''\right) \mathrm{d}t'' = 2 \int_0^t (t - \tau) f(\tau) \mathrm{d}\tau. \tag{C.91}$$

The following relation, which holds for a well-behaved f function and which is obtained after two integrations by parts, is useful when dealing with Laplace transforms:

$$\int_0^\infty \exp(-st) \int_0^t (t - \tau) f(\tau) \mathrm{d}\tau \, \mathrm{d}t = \frac{1}{s^2} \int_0^\infty \exp(-st) f(t) \mathrm{d}t. \tag{C.92}$$

C.10.3 A few functions and their properties

Here we list functions and a few of their properties of interest in this monograph.

The Bessel function of the first kind $J_\nu(x)$ satisfies the relation

$$J_\nu(x) = \left(\frac{x}{2}\right)^\nu \sum_{k=0}^{\infty} \frac{(-1)^k}{k!(\nu + k)!} \left(\frac{x}{2}\right)^{2k}. \tag{C.93}$$

When ν is an integer ($\nu = n$),

$$\sum_{n=-\infty}^{+\infty} J_n(x) \lambda^n = \exp\left[\frac{x}{2}\left(\lambda - \frac{1}{\lambda}\right)\right]; \tag{C.94}$$

see Jeffreys and Jeffreys (1962). The first term of its asymptotic expansion at large x reads

$$J_\nu(x \to \infty) = \sqrt{\frac{2}{\pi x}} \cos\left(x - \frac{\pi}{2}\nu - \frac{\pi}{4}\right). \tag{C.95}$$

Two spherical Bessel functions of the first kind are used in Chapter 6. They can be expressed with trigonometric functions:

$$j_0(x) = \frac{\sin x}{x}, \quad \text{and} \quad j_1(x) = \frac{\sin x}{x^2} - \frac{\cos x}{x}. \tag{C.96}$$

The function $K_\nu(x)$ denotes a modified Bessel function of the second kind. When ν is an integer, this Bessel function satisfies the relation

$$K_{-n}(x) = K_n(x). \tag{C.97}$$

An integral representation for $K_\nu(x)$ is

$$K_\nu(x) = \frac{\sqrt{\pi}}{(\nu - 1/2)!} \left(\frac{x}{2}\right)^\nu \int_1^{+\infty} \exp(-xt)(t^2 - 1)^{\nu - 1/2} \mathrm{d}t, \tag{C.98}$$

where $x > 0$ and $\nu > -\frac{1}{2}$. The series representation of $K_\nu(x)$ is complicated. We simply note that $K_0(x)$ diverges for $x \to 0$ and the first term of the expansion of

$K_1(x)$ in the same limit is equal to $1/x$. The leading term of the asymptotic large x expansion of $K_\nu(x)$ is

$$K_\nu(x \to \infty) = \sqrt{\frac{\pi}{2x}} \exp(-x). \tag{C.99}$$

The following relation holds for $K_1(x)$:

$$\int_0^{+\infty} x \frac{K_1\left[a\left(x^2+b^2\right)^{1/2}\right]}{\left(x^2+b^2\right)^{1/2}} \sin(cx)\mathrm{d}x = \frac{\pi}{2} \frac{c}{a\left(a^2+c^2\right)^{1/2}} \exp\left[-b\left(a^2+c^2\right)^{1/2}\right]. \tag{C.100}$$

An equation linking Bessel functions of the first and second kinds reads:

$$\int_0^{+\infty} J_0(ax) K_0\left(b\sqrt{x^2+c^2}\right) x\mathrm{d}x = \frac{c}{\sqrt{a^2+b^2}} K_{-1}\left(c\sqrt{a^2+b^2}\right). \tag{C.101}$$

A useful value:

$$\left(-\frac{1}{2}\right)! = \sqrt{\pi}. \tag{C.102}$$

An integral representation of the Bernoulli numbers is

$$B_{2n} = (-1)^{n-1} \pi^{-2n} \int_0^{+\infty} \frac{x^{2n}}{\sinh^2 x} \mathrm{d}x. \tag{C.103}$$

Some useful Bernoulli numbers:

$$B_2 = \frac{1}{6} \quad \text{and} \quad B_4 = -\frac{1}{30}. \tag{C.104}$$

The error function is defined as

$$\mathrm{erf}(x) = \frac{2}{\sqrt{\pi}} \int_0^x \exp(-z^2)\mathrm{d}z. \tag{C.105}$$

In Chapter 6, an expansion is needed for $\mathrm{erf}(x)$ in the limit $x \to \infty$. For that purpose we use the so-called complementary error function $\mathrm{erfc}(x)$, whose expansion is

$$\mathrm{erfc}(x) = 1 - \mathrm{erf}(x) = \frac{\exp(-x^2)}{x\sqrt{\pi}} \left[1 - \frac{1}{2x^2} + \frac{3}{4x^4} + \mathcal{O}\left(\frac{1}{x^6}\right)\right]. \tag{C.106}$$

The definition of the Gamma function is

$$\Gamma(p) = \int_0^{+\infty} x^{p-1} \exp(-x)\mathrm{d}x. \tag{C.107}$$

It is related to the factorial function through the relation

$$\Gamma(p) = (p-1)!. \tag{C.108}$$

C.10.4 Trotter and miscellaneous formulae

Trotter has established the following formula for N_{T} matrices A_i of finite dimension (Trotter 1959):

$$\exp\left(\sum_{i=1}^{N_{\mathrm{T}}} A_i\right) = \lim_{\ell\to\infty}\left[\prod_{i=1}^{N_{\mathrm{T}}} \exp\left(\frac{A_i}{\ell}\right)\right]^{\ell}. \tag{C.109}$$

We recall that only if the A_i matrices commute the following formula is valid:

$$\exp\left(\sum_{i=1}^{N_{\mathrm{T}}} A_i\right) = \prod_{i=1}^{N_{\mathrm{T}}} \exp\left(A_i\right). \tag{C.110}$$

Finally, for I positive integer or half integer:

$$\sum_{m=-I}^{I} m^2 = \frac{1}{3} I(I+1)(2I+1). \tag{C.111}$$

References

Ashcroft N.W. and Mermin N.D. (2002) *Solid State Physics* (Holt, Rinehart and Winston, New York).

Bradley C.J. and Cracknell A.P. (1972) *The Mathematical Theory of Symmetry in Solids. Representation Theory for Point Groups and Space Groups* (Clarendon Press, Oxford).

Chow T.L. (2000) *Mathematical Methods for Physicists* (Cambrige University Press, Cambridge).

Dove M. T. (2003) *Structure and Dynamics* (Oxford University Press, Oxford).

Gradshteyn I. S. and Ryzhik I.M. (1994) *Table of Integrals, Series and Products,* fifth edition (Academic Press).

Gradshteyn I. S. and Ryzhik I.M. (2000) *Table of Integrals, Series and Products,* sixth edition (Academic Press); see errata at website http://www.mathtable.com/gr/.

Jeffreys H. and Jeffreys B. (1962) *Methods of Mathematical Physics* (Cambridge University Press, Cambridge).

Kittel C. (1996) *Introduction to Solid State Physics,* seventh edition (Wiley, New York).

Kittel C. and Kroemer H. (1980) *Thermal Physics* (Freeman, San Francisco).

Messiah A. (1965) *Quantum Mechanics*, third reprint, vol. 2 (North-Holland, Amsterdam).

Nye J.F. (1985) *Physical Properties of Crystals* (Clarendon Press, Oxford).

Rose M.E. (1957) *Elementary Theory of Angular Momentum* (Wiley, New York).

Slichter C.P. (1996) *Principles of Magnetic Resonance,* third enlarged and updated edition (Springer, Berlin).

Trotter H.F. (1959) *Proceedings of the American Mathematical Society* **10**, 545–551.

Wigner E.P. (1959) *Group Theory and Its Application to the Quantum Mechanics of Atomic Spectra* (Academic Press, New York).

Appendix D
Laboratory and crystal reference frames

The physical quantities measured by μSR have been usually derived in the main text using the coordinates in the laboratory reference frame. Here they are expressed in terms of the coordinates of a frame attached to the crystal. Their symmetry properties are more transparent in this representation. We assume these two frames to be orthogonal.

We first provide a mathematical background and then proceed by deriving formulae for various quantities. Finally, examples are given of the use of symmetry arguments for a drastic simplification of formulae.

D.1 Mathematical background

We recall that we only consider periodic crystals in this text; see Appendix. E. We first define the two reference frames of interest.

For simplicity we suppose the compound under study to have only one type of localised magnetic moment (electronic or nuclear) and the muon to occupy only one type of interstitial site. We consider two orthonormal reference frames: the laboratory reference frame, $\{X, Y, Z\}$, and a reference frame attached to the crystal axes, $\{x, y, z\}$. The latter frame should be adapted to the symmetry of the compound; i.e. its unit vectors should be chosen to be parallel to crystal axes of high symmetry. For crystal symmetries such that these axes do not form an orthogonal frame it is sometimes possible to build a new reference frame with orthogonal axes, see e.g. Section E.1. The frame $\{X, Y, Z\}$ is defined in the $\{x, y, z\}$ frame through the Euler angles φ, θ, and ψ; see Fig. D.1. Note that θ and φ are also the polar and azimuthal angles of the Z axis in the $\{x, y, z\}$ frame. We recall that Z is the direction of the external field for a transverse-field measurement and of the initial muon beam polarisation for a zero- or longitudinal-field measurement.

We now list the properties of the associated Legendre functions, P_ℓ^n, which we shall use thereafter. The P_ℓ^0 functions obey the relation

$$P_\ell^0(\cos\theta_i) = \sum_{m=-\ell}^{\ell} \frac{(\ell - |m|)!}{(\ell + |m|)!} P_\ell^{|m|}(\cos\Theta_i) P_\ell^{|m|}(\cos\theta) \exp[i\, m\, (\Phi_i - \varphi)], \qquad \text{(D.1)}$$

where $\ell = 0, 1, 2, \ldots$. We use the notations of Messiah (1965). The angles Θ_i and Φ_i are the polar and azimuthal angles of the radius vector $\mathbf{r}_i$ in the $\{x, y, z\}$ frame and

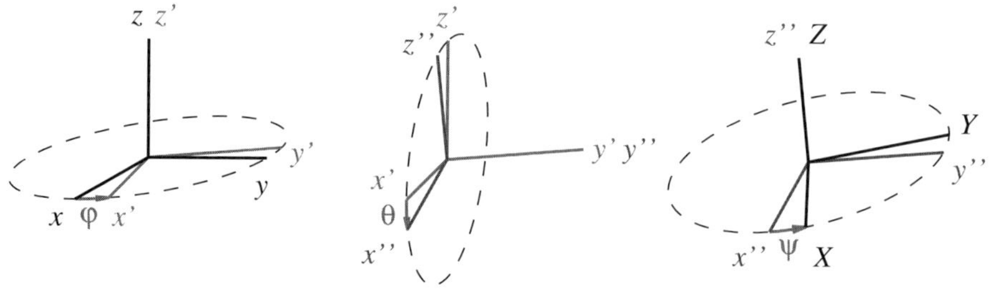

Fig. D.1 The Euler angles φ, θ, and ψ defining the reference frame $\{X, Y, Z\}$ relative to $\{x, y, z\}$. The figure shows the subsequent rotations and the intermediate reference frames: after rotation of φ about z, $\{x, y, z\}$ is transformed into $\{x', y', z'\}$ where $z' \equiv z$. This latter frame is then rotated about y' by θ to yield $\{x'', y'', z''\}$ with $y'' \equiv y'$. Finally $\{X, Y, Z\}$ is obtained from $\{x'', y'', z''\}$ after rotation of ψ about $z'' \equiv Z$.

θ_i the polar angle of the same vector in the $\{X, Y, Z\}$ frame. Equation D.1 offers the possibility of expressing an associated Legendre function written with the laboratory coordinate θ_i as a function of coordinates in the crystal reference frame and of θ and φ. We list the associated Legendre functions of interest to us. The three functions of second order are

$$P_2^0(\cos\theta) = \frac{1}{2}(3\cos^2\theta - 1), \quad P_2^1(\cos\theta) = 3\cos\theta\sin\theta, \quad P_2^2(\cos\theta) = 3\sin^2\theta. \qquad \text{(D.2)}$$

The five functions of fourth order read

$$\begin{aligned} P_4^0(\cos\theta) &= \frac{1}{8}(35\cos^4\theta - 30\cos^2\theta + 3), \quad P_4^1(\cos\theta) = \frac{5}{2}\cos\theta\sin\theta(7\cos^2\theta - 3), \\ P_4^2(\cos\theta) &= \frac{15}{2}\sin^2\theta(7\cos^2\theta - 1), \quad P_4^3(\cos\theta) = 105\cos\theta\sin^3\theta, \\ P_4^4(\cos\theta) &= 105\sin^4\theta. \end{aligned} \qquad \text{(D.3)}$$

The associated Legendre functions form a basis on which a function of θ_i can be developed. In particular,

$$[P_2^0(\cos\theta_i)]^2 = \frac{1}{5} + \frac{2}{7}P_2^0(\cos\theta_i) + \frac{18}{35}P_4^0(\cos\theta_i). \qquad \text{(D.4)}$$

D.2 Components of the local field

In this section the components of the local field in the laboratory frame, denoted by $B^{\Gamma}_{\text{loc,l}}$, are written as a function of the components $B^{\alpha}_{\text{loc,c}}$ in the crystal frame. The superscripts Γ and α stand for one of the Cartesian axes. We have

$$B^{\Gamma}_{\text{loc,l}} = \sum_{\alpha} R_{\Gamma\alpha}(\varphi, \theta, \psi) B^{\alpha}_{\text{loc,c}}. \qquad \text{(D.5)}$$

The R matrix is

$$\begin{pmatrix} \cos\psi\cos\varphi\cos\theta - \sin\psi\sin\varphi & \cos\psi\sin\varphi\cos\theta + \sin\psi\cos\varphi & -\cos\psi\sin\theta \\ -\sin\psi\cos\varphi\cos\theta - \cos\psi\sin\varphi & \cos\psi\cos\varphi - \sin\psi\sin\varphi\cos\theta & \sin\psi\sin\theta \\ \cos\varphi\sin\theta & \sin\varphi\sin\theta & \cos\theta \end{pmatrix}. \tag{D.6}$$

For instance, $R_{Zx} = \cos\varphi\sin\theta$.

Since R is a unitary matrix, i.e. $RR^\dagger = R^\dagger R = \mathcal{I}$, where $R^\dagger$ is the Hermitian conjugate of R, it is easy to express $B^{\alpha}_{\rm loc,c}$ in terms of $B^{\Gamma}_{\rm loc,l}$ using the matrix elements listed above. Explicitly, $B^{\alpha}_{\rm loc,c} = \sum_\Gamma R^{-1}_{\alpha\Gamma}(\varphi,\theta,\psi) B^{\Gamma}_{\rm loc,l}$, with $R^{-1}_{\alpha\Gamma}(\varphi,\theta,\psi) = R_{\Gamma\alpha}(\varphi,\theta,\psi)$.

D.3 Longitudinal and transverse relaxation rates

Here we give formulae for the exponential relaxation rates as a function of the spin correlation tensor described in a frame attached to the crystal axes of the material and of θ and φ.

Within the extreme motional narrowing limit, the zero- or longitudinal-field relaxation function has an exponential form as well as the envelope of the transverse-field polarisation function; see Section 7.1. In that section we have shown that the associated relaxation rates can be written as a sum of field correlation functions at the muon site expressed in the laboratory reference frame. However, it is physically better suited to give the rates in terms of the coupling tensor $\boldsymbol{G}(\mathbf{q})$ and the spin correlation tensor $\boldsymbol{\Lambda}_{dd'}(\mathbf{q},\omega_\mu)$ written in a frame adapted to the symmetry of the crystal structure. For this purpose, we use the results of Sections 7.1.4.3 and D.2 to derive:

$$\lambda_Z = \frac{\mathcal{D}}{2}\sum_{\beta,\alpha} L^{\beta\alpha}(\theta,\varphi)\int_{v_c^*}\sum_{\gamma,\gamma'}\sum_{d,d'} G_d^{\alpha\gamma}(\mathbf{q}) G_{d'}^{\gamma'\beta}(-\mathbf{q})\Lambda_{dd'}^{\gamma\gamma'}(\mathbf{q},\omega_\mu)\frac{\mathrm{d}^3\mathbf{q}}{(2\pi)^3}. \tag{D.7}$$

We have defined $\mathcal{D} = (\mu_0/4\pi)^2\gamma_\mu^2 (g\mu_B)^2/v_{\rm c}$ where $v_{\rm c}$ is the volume of the unit cell. Here is the only section of the current appendix where we consider the possibility of having several magnetic ions in the unit cell. The fact that the symmetry axes of the magnet do not necessarily coincide with the laboratory frame axes is accounted for by the symmetric matrix $L(\theta,\varphi)$ which is written as

$$L(\theta,\varphi) = \begin{pmatrix} \cos^2\varphi\cos^2\theta + \sin^2\varphi & -\cos\varphi\sin\varphi\sin^2\theta & -\cos\varphi\cos\theta\sin\theta \\ -\cos\varphi\sin\varphi\sin^2\theta & \sin^2\varphi\cos^2\theta + \cos^2\varphi & -\sin\varphi\cos\theta\sin\theta \\ -\cos\varphi\cos\theta\sin\theta & -\sin\varphi\cos\theta\sin\theta & \sin^2\theta \end{pmatrix}. \tag{D.8}$$

For the transverse relaxation rate, we get

$$\lambda_X = \frac{\lambda_Z}{2} + \frac{\mathcal{D}}{2}\sum_{\beta,\alpha} T^{\beta\alpha}(\theta,\varphi)\int_{v_c^*}\sum_{\gamma,\gamma'}\sum_{d,d'} G_d^{\alpha\gamma}(\mathbf{q}) G_{d'}^{\gamma'\beta}(-\mathbf{q})\Lambda_{dd'}^{\gamma\gamma'}(\mathbf{q},\omega=0)\frac{\mathrm{d}^3\mathbf{q}}{(2\pi)^3}. \tag{D.9}$$

We have introduced the symmetric matrix $T(\theta,\varphi)$:

$$T(\theta,\varphi) = \begin{pmatrix} \cos^2\varphi\sin^2\theta & \cos\varphi\sin\varphi\sin^2\theta & \cos\varphi\cos\theta\sin\theta \\ \cos\varphi\sin\varphi\sin^2\theta & \sin^2\varphi\sin^2\theta & \sin\varphi\cos\theta\sin\theta \\ \cos\varphi\cos\theta\sin\theta & \sin\varphi\cos\theta\sin\theta & \cos^2\theta \end{pmatrix}. \tag{D.10}$$

D.4 Field variances

Here we use the transformation properties of associated Legendre functions to derive formulae for various variances in terms of the orientation of the Z axis in the crystal frame.

Let us first discuss the variance of the field distribution in the case of a zero-field measurement. If the electric-field gradients acting on the sources of depolarisation are negligible, from Eqs. 6.151 and D.1 we derive

$$\Delta^2_{\mathrm{G,ZF}} = \left(\frac{\mu_0}{4\pi}\right)^2 \sum_{i=1}^{N} \frac{\gamma_i^2 \hbar^2}{r_i^6} \frac{I_i\,(I_i+1)}{3} [2 - P_2^0(\cos\theta_i)], = 2\mathcal{A}_0 - \mathcal{A}_2(\theta,\varphi). \tag{D.11}$$

We recall that $\Delta^2_{\mathrm{G,ZF}}$ is the variance of each component of the assumed isotropic Gaussian-field distribution. We have introduced the constant $\mathcal{A}_0$,

$$\mathcal{A}_0 = \left(\frac{\mu_0}{4\pi}\right)^2 \sum_{i=1}^{N} \frac{\gamma_i^2 \hbar^2}{r_i^6} \frac{I_i\,(I_i+1)}{3}, \tag{D.12}$$

and the function $\mathcal{A}_{2p}(\theta,\varphi)$ (p is a strictly positive integer):

$$\mathcal{A}_{2p}(\theta,\varphi) = \mathcal{C}_{2p}^0 P_{2p}^0(\cos\theta) + \sum_{m=1}^{2p} P_{2p}^m(\cos\theta)\left[\mathcal{C}_{2p}^m \cos(m\varphi) + \mathcal{S}_{2p}^m \sin(m\varphi)\right]. \tag{D.13}$$

The coefficients of the expansion of $\mathcal{A}_{2p}(\theta,\varphi)$ are defined as

$$\mathcal{C}_{2p}^0 = \left(\frac{\mu_0}{4\pi}\right)^2 \sum_{i=1}^{N} \frac{\gamma_i^2 \hbar^2}{r_i^6} \frac{I_i\,(I_i+1)}{3} P_{2p}^0(\cos\Theta_i), \tag{D.14}$$

and for $0 < m \leq 2p$,

$$\begin{aligned} \mathcal{C}_{2p}^m &= \left(\frac{\mu_0}{4\pi}\right)^2 \frac{2(2p-m)!}{(2p+m)!} \sum_{i=1}^{N} \frac{\gamma_i^2 \hbar^2}{r_i^6} \frac{I_i\,(I_i+1)}{3} P_{2p}^m(\cos\Theta_i)\cos(m\,\Phi_i), \\ \mathcal{S}_{2p}^m &= \left(\frac{\mu_0}{4\pi}\right)^2 \frac{2(2p-m)!}{(2p+m)!} \sum_{i=1}^{N} \frac{\gamma_i^2 \hbar^2}{r_i^6} \frac{I_i\,(I_i+1)}{3} P_{2p}^m(\cos\Theta_i)\sin(m\,\Phi_i). \end{aligned} \tag{D.15}$$

Using the same method, $\Delta^2_{\mathrm{G,ZF}}$ can be rewritten if the splitting of the nuclear levels is dominated by the effect of the electric-field gradients. For integral spins, from Eq. 6.161 we derive

$$\Delta^2_{\mathrm{G,ZF}} = \frac{4}{3}\left[\mathcal{A}_0 - \mathcal{A}_2(\theta,\varphi)\right]. \tag{D.16}$$

For half-integral spins, we get from Eq. 6.162

$$\Delta^2_{\mathrm{G,ZF}} = \frac{4}{3}\left[\mathcal{A}_0 + \mathcal{A}_0' - \mathcal{A}_2(\theta,\varphi) + \frac{1}{2}\mathcal{A}_2'(\theta,\varphi)\right]. \tag{D.17}$$

We have introduced the additional constant

$$\mathcal{A}'_0 = \left(\frac{\mu_0}{4\pi}\right)^2 \sum_{i=1}^{N} \frac{\gamma_i^2 \hbar^2}{r_i^6} \frac{(2I_i+1)}{16}, \tag{D.18}$$

and the function

$$\mathcal{A}'_{2p}(\theta,\varphi) = \mathcal{C}'^{0}_{2p} P^0_{2p}(\cos\theta) + \sum_{m=1}^{2p} P^m_{2p}(\cos\theta)\left[\mathcal{C}'^{m}_{2p}\cos(m\varphi) + \mathcal{S}'^{m}_{2p}\sin(m\varphi)\right], \tag{D.19}$$

with

$$\mathcal{C}'^{0}_{2p} = \left(\frac{\mu_0}{4\pi}\right)^2 \sum_{i=1}^{N} \frac{\gamma_i^2 \hbar^2}{r_i^6} \frac{(2I_i+1)}{16} P^0_{2p}(\cos\Theta_i), \tag{D.20}$$

and for $0 < m \leq 2p$,

$$\mathcal{C}'^{m}_{2p} = \left(\frac{\mu_0}{4\pi}\right)^2 \frac{2(2-m)!}{(2+m)!} \sum_{i=1}^{N} \frac{\gamma_i^2 \hbar^2}{r_i^6} \frac{(2I_i+1)}{16} P^m_{2p}(\cos\Theta_i)\cos(m\,\Phi_i),$$

$$\mathcal{S}'^{m}_{2p} = \left(\frac{\mu_0}{4\pi}\right)^2 \frac{2(2-m)!}{(2+m)!} \sum_{i=1}^{N} \frac{\gamma_i^2 \hbar^2}{r_i^6} \frac{(2I_i+1)}{16} P^m_{2p}(\cos\Theta_i)\sin(m\,\Phi_i). \tag{D.21}$$

We now deal with $\Delta^2_{\mathrm{G,TF}}$ in the zero-electric-field-gradient approximation. From Eqs. 6.144 and D.4,

$$\begin{aligned}\Delta^2_{\mathrm{G,TF}} &= 4\left(\frac{\mu_0}{4\pi}\right)^2 \sum_{i=1}^{N} \frac{\gamma_i^2 \hbar^2}{r_i^6} \frac{I_i\,(I_i+1)}{3}\left[\frac{1}{5} + \frac{2}{7}P^0_2(\cos\theta_i) + \frac{18}{35}P^0_4(\cos\theta_i)\right] \\ &= 4\left[\frac{1}{5}\mathcal{A}_0 + \frac{2}{7}\mathcal{A}_2(\theta,\varphi) + \frac{18}{35}\mathcal{A}_4(\theta,\varphi)\right].\end{aligned} \tag{D.22}$$

The different variances depend only on the orientation of the Z axis relative to the crystal axes and coefficients which can be computed for a given muon localisation site. However, even without evaluating these coefficients, the symmetry at the muon site dictates which of them must vanish. This is discussed in Section D.7.

D.5 Lorentz field

Here, we first discuss the dipole field arising from magnetic moments inside the Lorentz sphere, $\mathbf{B}'_{\mathrm{dip}}$, estimated with the direct lattice sum technique. According to Eq. 5.21,

$$B'^{\alpha}_{\mathrm{dip,l}} = \frac{\mu_0}{4\pi}\frac{1}{v_c}\sum_{\beta}\sum_{i=1}^{N_{\mathrm{L}}} D^{\alpha\beta}_{\mathbf{r}_i,\mathrm{l}} m^{\beta}_{i,\mathrm{l}}, \tag{D.23}$$

where $\boldsymbol{D}_{\mathbf{r}_i}$ is the dipole field tensor associated to site $\mathbf{r}_i$. The sum is restricted to the Lorentz sphere; i.e. it extends up to N_{L} rather than N. The quantity N_{L} is the number of dipoles in the sphere contributing to the field. All the vector and tensor components of Eq. D.23 are written in terms of the coordinates of the laboratory reference frame.

Our purpose is to derive an expression for the components of the field in the crystal frame, $B'^{\alpha}_{\rm dip,c}$, with respect to its components in the laboratory frame $B'^{\Gamma}_{\rm dip,l}$. From Section D.2 and Eq. D.23, we have

$$\begin{aligned} B'^{\alpha}_{\rm dip,c} &= \sum_{\Gamma} R^{-1}_{\alpha\Gamma} B'^{\Gamma}_{\rm dip,l} = \frac{\mu_0}{4\pi}\frac{1}{v_c}\sum_{\Gamma,\beta}\sum_{i=1}^{N_{\rm L}} R^{-1}_{\alpha\Gamma} D^{\Gamma\beta}_{\mathbf{r}_i,{\rm l}} m^{\beta}_{i,{\rm l}} \\ &= \frac{\mu_0}{4\pi}\frac{1}{v_c}\sum_{\Gamma,\beta,\gamma}\sum_{i=1}^{N_{\rm L}} R^{-1}_{\alpha\Gamma} D^{\Gamma\beta}_{\mathbf{r}_i,{\rm l}} R_{\beta\gamma} m^{\gamma}_{i,{\rm c}}. \end{aligned} \tag{D.24}$$

This means that the components of $\mathbf{B}'_{\rm dip}$ in the crystal frame can be obtained from the components of the magnetic moments in the same frame using a relation similar to Eq. D.23 in which the components of $\boldsymbol{D}_{\mathbf{r}_i}$ are replaced with those of $R^{-1}\boldsymbol{D}_{\mathbf{r}_i}R$.

D.6 Knight shift

In this section we express the dipole and hyperfine contributions to the Knight shift in terms of the susceptibility and coupling tensors written in the crystal reference frame. We first analyse the contribution from the dipoles inside the Lorentz sphere, $K'_{\rm dip}$.

If the susceptibility is isotropic, i.e. $\chi^{\beta Z} = \chi_{\rm iso}\,\delta_{\beta,Z}$, one gets from Eqs. 5.75 and 5.23

$$K'_{\rm dip} = \frac{\chi_{\rm iso}}{4\pi}\sum_{i=1}^{N_{\rm L}} D^{ZZ}_{\mathbf{r}_i} = \frac{\chi_{\rm iso} v_c}{2\pi}\sum_{i=1}^{N_{\rm L}}\frac{1}{r_i^3}P_2^0(\cos\theta_i) = \mathcal{A}''_2(\theta,\varphi)\,\chi_{\rm iso}. \tag{D.25}$$

We have defined $\mathcal{A}''_{2p}(\theta,\varphi)$:

$$\mathcal{A}''_{2p}(\theta,\varphi) = \mathcal{C}''^{0}_{2p} P^0_{2p}(\cos\theta) + \sum_{m=1}^{2p} P^m_{2p}(\cos\theta)\left[\mathcal{C}''^{m}_{2p}\cos(m\varphi) + \mathcal{S}''^{m}_{2p}\sin(m\varphi)\right]. \tag{D.26}$$

We obtain for the coefficients of the expansion of $\mathcal{A}''_{2p}(\theta,\varphi)$:

$$\mathcal{C}''^{0}_{2p} = \sum_{i=1}^{N_{\rm L}}\frac{v_{\rm c}}{2\pi}\frac{1}{r_i^3}P^0_{2p}(\cos\Theta_i), \tag{D.27}$$

and for $0 < m \le 2p$,

$$\begin{aligned} \mathcal{C}''^{m}_{2p} &= \frac{v_{\rm c}(2-m)!}{\pi(2+m)!}\sum_{i=1}^{N_{\rm L}}\frac{1}{r_i^3}P^m_{2p}(\cos\Theta_i)\cos(m\,\Phi_i), \\ \mathcal{S}''^{m}_{2p} &= \frac{v_{\rm c}(2-m)!}{\pi(2+m)!}\sum_{i=1}^{N_{\rm L}}\frac{1}{r_i^3}P^m_{2p}(\cos\Theta_i)\sin(m\,\Phi_i). \end{aligned} \tag{D.28}$$

However, the susceptibility may not be isotropic. In that case the complete formula for the Knight shift should be evaluated, i.e. from Eq. 5.75,

$$K'_{\rm dip} = \frac{1}{4\pi} \sum_{\beta} \left(\sum_{i=1}^{N_{\rm L}} D^{Z\beta}_{\mathbf{r}_i} \right) \chi^{\beta Z}. \tag{D.29}$$

We need to compute the matrix elements $\sum_{i=1}^{N_{\rm L}} D^{Z\beta}_{\mathbf{r}_i}$ and $\chi^{\beta Z}$ when β is one element of the $\{X, Y, Z\}$ set. These quantities are written in terms of the coordinates of the laboratory reference frame. The components $\chi_{\rm l}^{\alpha\beta}$ and $D^{\alpha\beta}_{\mathbf{r}_i,{\rm l}}$ of the susceptibility and dipole tensors, respectively, in this latter frame are given by the relations $\chi_l^{\alpha\beta} = \sum_{\gamma\gamma'} R_{\alpha\gamma} \chi_{\rm c}^{\gamma\gamma'} R^{-1}_{\gamma'\beta}$ and $D^{\alpha\beta}_{\mathbf{r}_i,{\rm l}} = \sum_{\gamma\gamma'} R_{\alpha\gamma} D^{\gamma\gamma'}_{\mathbf{r}_i,{\rm c}} R^{-1}_{\gamma'\beta}$, where $\chi_{\rm c}^{\gamma\gamma'}$ and $D^{\gamma\gamma'}_{\mathbf{r}_i,{\rm c}}$ are the components of the susceptibility and dipole tensors in the crystal frame, respectively.

The susceptibility tensor is a symmetric tensor of rank two. We take it to be diagonal in the crystal frame, $\chi_{\rm c}^{\alpha\beta} = \delta_{\alpha,\beta} \chi_{\rm c}^{\alpha\alpha} \equiv \delta_{\alpha,\beta} \chi_{\rm c}^{\alpha}$. We refer to the textbook of Nye (1985) for a detailed discussion of the symmetry properties of tensors of rank two for the seven possible crystal systems. We introduce the tensor $\mathfrak{D} = [1/(4\pi)] \sum_{i=1}^{N_{\rm L}} \boldsymbol{D}_{\mathbf{r}_i}$. *A priori* this tensor is not diagonal in the crystal frame. We derive

$$\begin{aligned} K'_{\rm dip} = & \frac{1}{3} \left(\mathfrak{D}_{\rm c}^{xx} \chi_{\rm c}^{x} + \mathfrak{D}_{\rm c}^{yy} \chi_{\rm c}^{y} + \mathfrak{D}_{\rm c}^{zz} \chi_{\rm c}^{z} \right) \\ & + \frac{2}{3} \left[\mathfrak{D}_{\rm c}^{zz} \chi_{\rm c}^{z} - \frac{1}{2} \left(\mathfrak{D}_{\rm c}^{xx} \chi_{\rm c}^{x} + \mathfrak{D}_{\rm c}^{yy} \chi_{\rm c}^{y} \right) \right] P_2^0(\cos\theta) \\ & + \frac{1}{3} \mathfrak{D}_{\rm c}^{xz} \left(\chi_{\rm c}^{x} + \chi_{\rm c}^{z} \right) P_2^1(\cos\theta) \cos\varphi \\ & + \frac{1}{3} \mathfrak{D}_{\rm c}^{yz} \left(\chi_{\rm c}^{y} + \chi_{\rm c}^{z} \right) P_2^1(\cos\theta) \sin\varphi \\ & + \frac{1}{6} \left(\mathfrak{D}_{\rm c}^{xx} \chi_{\rm c}^{x} - \mathfrak{D}_{\rm c}^{yy} \chi_{\rm c}^{y} \right) P_2^2(\cos\theta) \cos(2\varphi) \\ & + \frac{1}{6} \mathfrak{D}_{\rm c}^{xy} \left(\chi_{\rm c}^{x} + \chi_{\rm c}^{y} \right) P_2^2(\cos\theta) \sin(2\varphi). \end{aligned} \tag{D.30}$$

This formula has been given by Schenck (1999). The contribution $K'_{\rm dip}$ from the dipoles inside the Lorentz sphere is the sum of an angle-independent term and a linear combination of associated Legendre functions of second order multiplied by trigonometric functions. If the susceptibility is isotropic, the constant term vanishes (the trace of $\mathfrak{D}$ is zero) and we recover the result given in Eq. D.25. If the susceptibility tensor is axial with $\chi_{\rm c}^{x} = \chi_{\rm c}^{y} \equiv \chi_{\rm c}^{\perp}$ and $\chi_{\rm c}^{z} \equiv \chi_{\rm c}^{\parallel}$, and the tensor $\mathfrak{D}$ is diagonal in the crystal frame, with its two components perpendicular to the z axis equal, the previous formula simplifies considerably:

$$K'_{\rm dip} = \frac{1}{3} \mathfrak{D}_{\rm c}^{zz} \left[(\chi_{\rm c}^{\parallel} - \chi_{\rm c}^{\perp}) + (2\chi_{\rm c}^{\parallel} + \chi_{\rm c}^{\perp}) P_2^0(\cos\theta) \right]. \tag{D.31}$$

The results for $K'_{\rm dip}$ have been obtained only using the fact that $\mathfrak{D}$ is a tensor. Therefore the same analysis can be carried out for $K_{\rm hyp}$ with the definition

$\mathfrak{H} = [1/(4\pi)]\sum_{i=1} H_{\mathbf{r}_i}$. Obviously it leads to a formula such as Eq. D.30 with the components $\mathfrak{H}_{\rm c}^{\alpha\beta}$ substituted for $\mathfrak{D}_{\rm c}^{\alpha\beta}$. Assuming $\mathfrak{H}$ to be diagonal in the crystal reference frame and $\mathfrak{H}_{\rm c}^{xx} = \mathfrak{H}_{\rm c}^{yy} \equiv \mathfrak{H}_{\rm c}^{\perp}$ (the trace of $\mathfrak{H}$ has no reason to be zero),

$$K_{\rm hyp} = \frac{1}{3}\left(\mathfrak{H}_{\rm c}^{zz}\chi_{\rm c}^{\parallel} + 2\mathfrak{H}_{\rm c}^{\perp}\chi_{\rm c}^{\perp}\right) + \frac{2}{3}\left(\mathfrak{H}_{\rm c}^{zz}\chi_{\rm c}^{\parallel} - \mathfrak{H}_{\rm c}^{\perp}\chi_{\rm c}^{\perp}\right) P_2^0(\cos\theta). \tag{D.32}$$

Sometimes the Knight shift is recorded in a polycrystalline sample. Thus, it is of practical interest to perform the 4π steradian solid angle average of the expression of $K'_{\rm dip}$ given in Eq. D.30. When this average is performed, a single term remains in $K'_{\rm dip}$: this is the first term of the right-hand side of this equation. This means that the dipole field contributes nothing to the shift for an isotropic powder sample. On the other hand, the hyperfine term does not have to vanish for such a sample, as is easily deduced from Eq. D.32.

D.7 Implications of the muon site symmetry

All the formulae derived above may seem rather complicated. However, they display remarkable simple symmetry properties which make them convenient when trying to infer properties of the muon–sample system such as the muon site.

As examples of the information which can be extracted from symmetry arguments, in this section we shall discuss variances, Lorentz fields, dipole fields, and Knight shifts for selected muon site symmetries.

As a first example we consider the angular dependence of the variance of the field distribution for a zero-field measurement when the electric-field gradient acting on the sources of depolarisation is negligible. Let us first consider the case where the muon sits in the octahedral site of a face-centred-cubic lattice; see Fig. 9.1 for a sketch of this site. This lattice is described by space group $Fm\bar{3}m$. The international tables for crystallography (Hahn 1996) tell us that the symmetry at the octahedral site is $m\bar{3}m$. This means in particular that the muon environment is symmetric with respect to the (100), (110), and (101) crystallographic planes. Considering the first mirror, one expects the responses measured with the Z axis successively oriented in the (001) plane at arbitrary angles φ and $\pi - \varphi$ to be the same, or in other words $\Delta^2_{\rm G,ZF}(\theta = \pi/2, \varphi) = \Delta^2_{\rm G,ZF}(\theta = \pi/2, \pi - \varphi)$. The presence of the second mirror implies $\Delta^2_{\rm G,ZF}(\theta = \pi/2, \varphi) = \Delta^2_{\rm G,ZF}(\theta = \pi/2, 3\pi/2 - \varphi)$. Since the last two equations must be fulfilled for any angle φ, referring to Eqs. D.11, D.12, D.13, and D.15, this means that $\mathcal{C}_2^2 = \mathcal{S}_2^2 = 0$. Because of the presence of the two previous mirror planes, the above relations concerning $\Delta^2_{\rm G,ZF}$ must also be fulfilled at any angle θ. Choosing an arbitrary value for θ different from 0, $\pi/2$, and π, one concludes that $\mathcal{C}_2^1 = \mathcal{S}_2^1 = 0$. Now, the (101) mirror plane dictates for instance that $\Delta^2_{\rm G,ZF}(\theta, \varphi = 0) = \Delta^2_{\rm G,ZF}(\pi/2 + \theta, \varphi = \pi)$. We end up with $\mathcal{C}_2^0 = 0$. Therefore $\mathcal{A}_2(\theta, \varphi)$ vanishes: the variance of the field distribution is finally $\Delta^2_{\rm G,ZF} = 2\mathcal{A}_0$, irrespective of the crystal orientation in the muon beam. In a very similar way one can show that for the same $m\bar{3}m$ site symmetry, $\mathcal{C}_2^{\prime 2} = \mathcal{S}_2^{\prime 2} = \mathcal{C}_2^{\prime 1} = \mathcal{S}_2^{\prime 1} = \mathcal{C}_2^{\prime 0} = 0$. Therefore the invariance of $\Delta^2_{\rm G,ZF}$ upon changes in the direction of the Z axis also holds in the case of dominating electric-field gradient effects.

For a muon sitting in a crystallographic site having a fourfold symmetry axis parallel to the [001] direction, we expect $\Delta^2_{\rm G,ZF}(\theta, \varphi) = \Delta^2_{\rm G,ZF}(\theta, \varphi + \pi/2)$. This enforces

$\mathcal{C}_2^2 = \mathcal{S}_2^2 = \mathcal{C}_2^1 = \mathcal{S}_2^1 = 0$ and $\mathcal{C}_2'^2 = \mathcal{S}_2'^2 = \mathcal{C}_2'^1 = \mathcal{S}_2'^1 = 0$. In summary, for such a site $\Delta^2_{\rm G,ZF}$ is equal to a constant plus a term proportional to $P_2^0(\cos\theta)$. This type of analysis can be pursued for any symmetry.

As another example of the efficiency of symmetry considerations, we examine the dipole field arising from magnetic moments inside the Lorentz sphere, $\mathbf{B}'_{\rm dip}$. Even restricted to the Lorentz sphere (Section 5.2.3), the summation over the dipoles still involves a large number of dipoles. However, symmetry arguments can drastically reduce the amount of numerical work. For example, let us discuss the case of a muon surrounded by dipole moments of equal magnitude and direction. From Section D.5

$$B'^{\alpha}_{\rm dip,c} = \frac{\mu_0}{4\pi} \sum_{\Gamma} \left(\sum_{i=1}^{N_{\rm L}} \frac{1}{v_{\rm c}} D^{\alpha\Gamma}_{\mathbf{r}_i,{\rm c}} \right) m^{\Gamma}_{\rm c}. \tag{D.33}$$

Let us assume again a muon localised at a site with $m\bar{3}m$ symmetry. Lorentz showed that for such a case $\mathbf{B}'_{\rm dip} = 0$ (Seeger 1979). Because this result is obtained no matter the direction of the magnetic moments, the tensor $\sum_{i=1}^{N_{\rm L}} \boldsymbol{D}_{\mathbf{r}_i}$ is the zero tensor. Group theory can also be used to infer properties of the latter tensor as a rank two tensor subject to point symmetry. For the $m\bar{3}m$ cubic point symmetry one expects a scalar tensor. Since we know that its trace is zero, this is necessarily the nil tensor. Other simple examples of the power of symmetry for the analysis of the dipole field are given by Seeger (1979).

Symmetry properties are also useful for deriving information on the tensor $\boldsymbol{C}(\mathbf{q})$ which is used for the evaluation of the dipole field using the reciprocal space method; see Section 5.2.4. The relaxation rates are also given in terms of that tensor. In this case it is already written in the coordinates of the crystal frame; see Section D.3. If expressed in the laboratory frame as for the dipole field (see Section 5.2.4), it is a simple matter to get it in the crystal frame through the transformation already discussed in Section D.5. Thereafter we assume $\boldsymbol{C}(\mathbf{q})$ to be written in the crystal frame.

We discuss the case of a tetrahedral or an octahedral interstitial site of face-centred- or body-centred-cubic crystals, fcc and bcc, respectively. Here the reference frame with its axes parallel to the cube edges is the proper frame to use. We restrict ourselves to the tensor at $\mathbf{q} = \mathbf{0}$. In these cubic axes, $C^{\alpha\beta}(\mathbf{q} = \mathbf{0})$ is diagonal. We recall that the trace of $\boldsymbol{C}(\mathbf{q} = 0)$ is equal to 1; see Eq. 5.58. In the case of a fcc crystal, symmetry dictates the tensor to be scalar and therefore $C^{\alpha\beta}(\mathbf{q} = \mathbf{0}) = \frac{1}{3}\delta^{\alpha\beta}$ for a muon in either a tetrahedral or octahedral site.

The situation is not so simple for a bcc crystal (space group $Im\bar{3}m$). Let us consider a tetrahedral interstitial site with the tetragonal axis taken along the fourfold axis. We have

$$C^{xx}(\mathbf{q} = \mathbf{0}) = C^{yy}(\mathbf{q} = \mathbf{0}) \equiv C_1, \qquad C^{zz}(\mathbf{q} = \mathbf{0}) = 1 - 2C_1 \equiv C_2. \tag{D.34}$$

As the tetragonal axis can be parallel to one of the three $\langle 100 \rangle$ axis, there are three types of $\boldsymbol{C}(\mathbf{q} = \mathbf{0})$ tensors for a given type of interstitial site. The diagonal elements are equal to (C_1, C_1, C_2), (C_1, C_2, C_1), or (C_2, C_1, C_1). The same results apply for the octahedral sites. As a consequence, we have here an illustration of the following rule:

to one muon crystallographic site may correspond more than one magnetic site — see e.g. Section 10.1.1.

Part of the interest of using the Fourier method for the evaluation of $\mathbf{B}'_{\rm dip}$, instead of performing directly the crystal sum, appears when dealing with a magnetic structure characterised by a propagation wave vector $\mathbf{k}$ different from 0. Whereas if the direct space sum technique is used, one first must numerically simulate the magnetic structure and then perform the crystal sum, with the Fourier technique the only non-trivial work is to evaluate $\boldsymbol{D}(\mathbf{k})$; see Eq. 5.54 for the definition of this tensor. Symmetry considerations are still useful although no general rule exists for the trace of $\boldsymbol{D}(\mathbf{k})$: one must examine the symmetry properties of the reciprocal space around the point $\mathbf{k}$.

Finally we consider the contribution to the Knight shift of the dipoles inside the Lorentz sphere. If we assume an isotropic susceptibility, this contribution is given by Eq. D.25. We look for the influence of the muon site symmetry on it. In a manner very similar to the one used for the variance of the field distribution (see the beginning of the current section), one concludes that for a muon site of symmetry $m\bar{3}m$, all the coefficients of the expansion given in Eq. D.26 vanish. Therefore the Knight shift arising from the dipoles inside the Lorentz sphere vanishes for a muon site of cubic symmetry if the susceptibility is isotropic.

D.8 Implications of the Hamiltonian symmetry

Here we show the implications of the symmetry of a Hamiltonian on the correlations probed. We shall focus on a Hamiltonian describing the evolution of the spin operators of a single ion. It must be invariant under the operations of the group formed by the local symmetry operators at the ion position. As an example, let us assume that one of the symmetry operations of the group is a rotation of angle $\varrho = 2\pi/n$ about the z axis. The z axis is therefore a n-fold symmetry axis. Mathematically this invariance means

$$[\mathcal{H}, R_z(\varrho)] = 0, \tag{D.35}$$

where $\mathcal{H}$ is the Hamiltonian and $R_z(\varrho) = \exp(-i\varrho J^z)$ stands for the rotation operator about the z axis of angle ϱ. Using Eq. D.35, we derive, for example,

$$\left\langle J^-(t) J^z \right\rangle = \exp(i\varrho) \left\langle J^-(t) J^z \right\rangle . \tag{D.36}$$

This relation is obtained from the usual definition of the correlation function and the expression of $J^-(t)$. Explicitly,

$$\begin{aligned} \left\langle J^-(t) J^z \right\rangle &= \frac{\mathrm{Tr}\{\exp(-\beta\mathcal{H}) J^-(t) J^z\}}{\mathrm{Tr}\{\exp(-\beta\mathcal{H})\}} \\ &= \frac{\mathrm{Tr}\{\exp(-\beta\mathcal{H}) \exp(i\mathcal{H}t/\hbar) J^- \exp(-i\mathcal{H}t/\hbar) J^z\}}{\mathrm{Tr}\{\exp(-\beta\mathcal{H})\}} . \end{aligned} \tag{D.37}$$

One must also insert the unit operator $\mathcal{I} = R_z^{-1}(\varrho) R_z(\varrho)$ on both sides of J^- in Eq. D.37 and use Eqs. C.18 and C.10. Equation D.36 implies that $\langle J^-(t) J^z \rangle$ vanishes. Proceeding with this methodology, it is found that only three correlations are of interest for $n \geq 3$. Therefore if the z axis is at least a threefold rotation axis, the only three symmetrised correlation functions (see Appendix. F) to be considered are $\Lambda^{+-}(\mathbf{q}, \omega)$, $\Lambda^{-+}(\mathbf{q}, \omega)$, and $\Lambda^{zz}(\mathbf{q}, \omega)$. Since it can be shown that $\Lambda^{-+}(\mathbf{q}, \omega) = \Lambda^{+-}(\mathbf{q}, -\omega)$, only two of them are actually independent. In the basis $\{+, -, z\}$ we consider here, the correlation tensor is therefore not diagonal. It is usually helpful to diagonalise it. This is achieved through a rotation about the z axis.

The reader will easily adapt the above discussion to symmetry elements of a point group other than the rotations considered here.

References

Hahn T. (1996) *International Tables for Crystallography*, vol. A: Space-group symmetry, fourth revised edition (Kluwer Academic, Dordrecht).

Messiah A. (1965) *Quantum Mechanics*, third reprint, vols. 1 and 2 (North-Holland, Amsterdam).

Nye J. F. (1985) *Physical Properties of Crystals* (Clarendon Press, Oxford).

Schenck A. (1999) in *Muon Science: Muons in Physics, Chemistry, and Materials*, edited by Lee S.L., Kilcoyne S.H., and Cywinski R. (Institute of Physics, Bristol).

Seeger A. (1978) in *Hydrogen in Metals,* edited by Alefeld G. and Völkl J., Springer Topics in Current Physics, vol. 28 (Springer, Berlin).

Appendix E
Translational symmetry and Fourier transform

This appendix is an introduction of the application of the space Fourier transform method for the description of periodic crystals. More information can be found in any introductory text on solid state physics, e.g. Ziman (1972), Kittel (1996), or Ashcroft and Mermin (2002). The book of Dove (2003) contains a clear physical presentation of the crystal symmetries and their general implications. We shall first recall some fundamental properties of crystals and then summarise the space Fourier technique intensively used in some chapters (Chapters 5, 7, 10 and 11).

The discovery in 1984 of quasicrystals, for which long-range orientational order is observed without translational symmetry by Shechtman *et al.* (1984), sparked a scientific revolution which led to questioning the definition of a crystal. Nowadays one distinguishes the so-called periodic crystals — the most common ones — and the aperiodic crystals such as quasicrystals and incommensurately modulated crystals: see e.g. the discussion of Steurer (2007) and subsequent contributions in the same issue of *Zeitschrift für Kristallographie*. Quasicrystals do not have the translational symmetry of crystals, but they do have a subtle long-range order. Along any symmetry axis of a quasicrystal, the placement of atoms can be expressed as a sum of two or more periodic functions, but the wavelength ratio for any two of those functions is irrational, and as a result, no single periodic function can represent the atomic positions. The atom's quasiperiodicity — indeed, "quasicrystal" is short for quasiperiodic crystal — implies long-range correlations and, as with conventional crystals, allows quasicrystals to display sharp, structure-revealing diffraction patterns (Blau 2009). In this monograph, we only consider the former class of crystals, i.e. periodic crystals, with the exception of Section 10.1.1 where we discuss incommensurate magnetic structures for which the field at the muon site is aperiodic.

E.1 Lattices and unit cells

A periodic structure, such as a crystal or an arrangement for flux lines, is constructed by regularly repeating a basic structural unit known as the unit cell. The points at which the unit cells are located form a lattice. We write for any vector $\mathbf{i}$ of the lattice

$$\mathbf{i} = m\mathbf{a} + n\mathbf{b} + p\mathbf{c}, \tag{E.1}$$

where $\mathbf{a}$, $\mathbf{b}$, and $\mathbf{c}$ are the basis lattice vectors, and m, n, and p are integers. The vectors $\mathbf{a}$, $\mathbf{b}$, and $\mathbf{c}$ define a unit cell. It is important to note that the choice of the

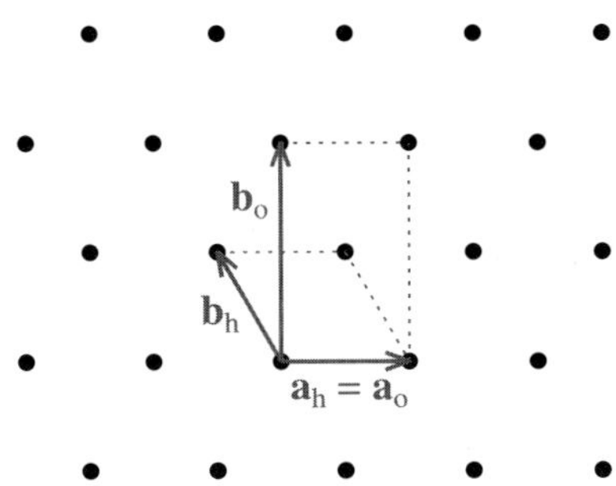

Fig. E.1 Projection in the plane perpendicular to $\mathbf{c}_h$ of the hexagonal lattice characterised by the basis vectors $\mathbf{a}_h$ and $\mathbf{b}_h$. The so-called orthohexagonal basis vectors $\mathbf{a}_o = \mathbf{a}_h$ and $\mathbf{b}_o = \mathbf{a}_h + 2\mathbf{b}_h$ are represented. Together with $\mathbf{c}_o = \mathbf{c}_h$, they form an orthogonal set. Unit cells of both hexagonal and orthohexagonal lattices are shown with dotted lines.

basis lattice vectors is arbitrary to a large extent. However, we can always choose these vectors such that the volume they define is minimum. This is known as the primitive unit cell and is defined by the primitive lattice vectors. It follows that the primitive cell contains just a single lattice point. It is not compulsory to work with such a cell. Other choices may be more practical in specific cases.

A given lattice has characteristic symmetries which not only include translations but also rotations, mirror and inversion symmetries and rotoinversions. A lattice in real space is known as a direct lattice to distinguish it from its associated reciprocal lattice defined in the next section.

For the complete description of a crystal structure we need to associate a basis with each lattice point. The basis describes the different atomic positions in a unit cell. Figure 5.3 represents an example of a crystal with two atoms (shown as + and ×) in the basis. A given atom in a crystal may be thought of as belonging to a basis which is itself associated with a particular unit cell. Thus, in general, an angular momentum carried by an atom will be denoted by $\mathbf{J}_{i+d}$. The vector $\mathbf{d}$ defines the position of atom d relative to $\mathbf{i}$.

For convenience, we shall sometimes describe crystal structures in terms of orthogonal lattice structures. For example, let us consider metallic gadolinium which crystallises in the hexagonal-close-packed (hcp) structure. The relations between the orthohexagonal and hexagonal basis vectors are $\mathbf{a}_o = \mathbf{a}_h$, $\mathbf{b}_o = \mathbf{a}_h + 2\mathbf{b}_h$, and $\mathbf{c}_o = \mathbf{c}_h$. Although gadolinium is described by four identical Bravais lattices in the orthohexagonal structure instead of only two in the hexagonal structure (see Fig. E.1), the advantage of using the former is that the formulae are easier to write down.

E.2 Reciprocal space and lattice Fourier transform

The set of vectors $\{\mathbf{a}^*, \mathbf{b}^*, \mathbf{c}^*\}$, with

$$\mathbf{a}^* = 2\pi \frac{\mathbf{b} \times \mathbf{c}}{\mathbf{a} \cdot (\mathbf{b} \times \mathbf{c})}, \quad \mathbf{b}^* = 2\pi \frac{\mathbf{c} \times \mathbf{a}}{\mathbf{a} \cdot (\mathbf{b} \times \mathbf{c})}, \quad \mathbf{c}^* = 2\pi \frac{\mathbf{a} \times \mathbf{b}}{\mathbf{a} \cdot (\mathbf{b} \times \mathbf{c})}, \tag{E.2}$$

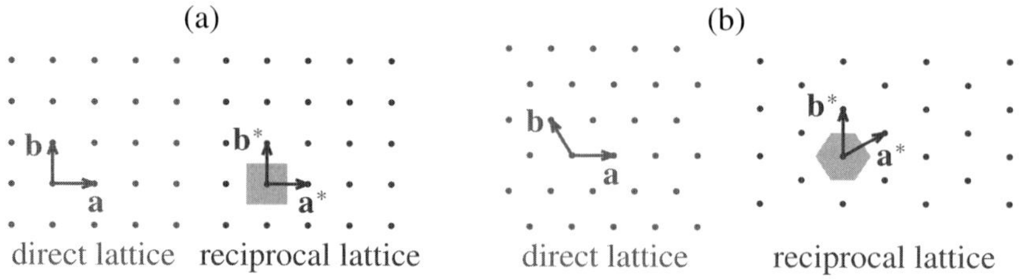

Fig. E.2 Comparison of direct and reciprocal lattices in two dimensions for two cases: (a) square lattice and (b) triangular lattice. The first Brillouin zone for each of the two reciprocal lattices is indicated by the shaded area.

defines the so-called reciprocal lattice first introduced by Gibbs. The mixed product $\mathbf{a}\cdot(\mathbf{b}\times\mathbf{c})$ is equal to the unit cell volume v_c. We use the solid state physicist convention which explicitly displays the factor 2π in the definition of the reciprocal basis vectors. In Fig. E.2 we compare direct and reciprocal lattices for two lattices in two dimensions. The third vector $\mathbf{c}$ is chosen to be normal to the plane spanned by $\mathbf{a}$ and $\mathbf{b}$. For a square lattice the reciprocal lattice is also made of squares with a lattice spacing $2\pi/a$; see left panel of Fig. E.2. If the axes are not orthogonal as is the case for the triangular lattice, the basis vectors in real and reciprocal space are not parallel to each other as indicated in the right panel of Fig. E.2.

Any vector $\mathbf{K}$ in the reciprocal lattice can be generated from the three basis vectors of the reciprocal lattice as

$$\mathbf{K} = h\mathbf{a}^* + k\mathbf{b}^* + \ell\mathbf{c}^*, \tag{E.3}$$

where h, k, and ℓ are integers. To see the significance of $\mathbf{K}$, we note for the scalar product $\mathbf{K}\cdot\mathbf{i} = 2\pi(hm + kn + \ell p) = 2\pi u$, where u is an integer. It follows that $\exp[i\mathbf{K}\cdot(\mathbf{i}+\mathbf{d})] = \exp(i\mathbf{K}\cdot\mathbf{d})$, i.e. $\exp[i\mathbf{K}\cdot(\mathbf{i}+\mathbf{d})]$ does not depend on $\mathbf{i}$.

The interest in the reciprocal lattice stems from the fact that a periodic physical quantity such as a density can be expressed as a Fourier series:

$$\eta(\mathbf{r}) = \sum_{\mathbf{K}} \eta_{\mathbf{K}} \exp(i\mathbf{K}\cdot\mathbf{r}), \tag{E.4}$$

with

$$\eta_{\mathbf{K}} = \frac{1}{v_c}\int_{v_c} \eta(\mathbf{r}) \exp(-i\mathbf{K}\cdot\mathbf{r})\, \mathrm{d}^3\mathbf{r}. \tag{E.5}$$

Note that the weight factors $\eta_{\mathbf{K}}$ are only defined at the lattice points of the reciprocal lattice which we suppose perfect. In contrast, $\eta(\mathbf{r})$ is a continuous function. Substituting the scalar functions $\eta(\mathbf{r})$ and $\eta_{\mathbf{K}}$ by vector functions, relations similar to Eqs. E.4 and E.5 can be written.

The so-called first Brillouin zone, which we simply call the Brillouin zone in this textbook, is of special importance for crystalline solids. It is the smallest volume entirely enclosed by planes that are the perpendicular bisectors of the reciprocal lattice vectors drawn from the origin. The volume of the Brillouin zone is $v_c^* = (2\pi)^3/v_c$. Two examples are given in Fig. E.2. A fundamental property of a crystal excitation is that its wave vector can always be taken to belong to the Brillouin zone.

The following orthogonality relation is obeyed:

$$\sum_{\mathbf{q}} \exp[i\mathbf{q}\cdot(\mathbf{i}-\mathbf{i}')] = n_c\,\delta_{\mathbf{i},\mathbf{i}'}, \tag{E.6}$$

where the sum extends over n_c vectors of the Brillouin zone. We recall that n_c is also the number of unit cells of the crystal of volume $V = n_c\,v_c$, and $\delta_{\mathbf{i},\mathbf{i}'}$ is a Kronecker symbol which is equal to 1 if $\mathbf{i} = \mathbf{i}'$ and 0 otherwise. This relation is the analog for discrete sums of the delta function integral representation $\int_{-\infty}^{\infty} \exp(i\mathbf{q}\cdot\mathbf{x})\,[\mathrm{d}^3\mathbf{q}/(2\pi)^3] = \delta(\mathbf{x})$.

A sibling relation holds for sums over the direct lattice vectors:

$$\sum_{\mathbf{i}} \exp[i(\mathbf{q}\cdot\mathbf{i})] = n_c\,\delta_{\mathbf{q},\mathbf{K}}, \tag{E.7}$$

for any $\mathbf{q}$ vector in the reciprocal space. If $\mathbf{q}$ is in the Brillouin zone, the only reciprocal lattice it can match is $\mathbf{K} = \mathbf{0}$. In addition, for any $\mathbf{q}$ and $\mathbf{q}'$ vectors in the Brillouin zone,

$$\sum_{\mathbf{i}} \exp[i(\mathbf{q}-\mathbf{q}')\cdot\mathbf{i}] = n_c\,\delta_{\mathbf{q},\mathbf{q}'}, \tag{E.8}$$

since the only reciprocal lattice vector spanned by $\mathbf{q}-\mathbf{q}'$ is the origin.

As an illustration of the use of the Brillouin zone concept, let us define the $\mathbf{q}$ component in the Brillouin zone of the angular momentum, $\mathbf{J}_{d,\mathbf{q}}$, such as

$$\mathbf{J}_{d,\mathbf{q}} = \frac{1}{\sqrt{n_c}} \sum_{i=1}^{n_c} \mathbf{J}_{i+d} \exp[-i\mathbf{q}\cdot(\mathbf{i}+\mathbf{d})], \tag{E.9}$$

and consider the lattice Fourier transform $(1/\sqrt{n_c})\sum_{\mathbf{q}} \mathbf{J}_{d,\mathbf{q}} \exp[i\mathbf{q}\cdot(\mathbf{i}+\mathbf{d})]$. We have

$$\begin{aligned}\frac{1}{\sqrt{n_c}} \sum_{\mathbf{q}} \mathbf{J}_{d,\mathbf{q}} \exp[i\mathbf{q}\cdot(\mathbf{i}+\mathbf{d})] &= \frac{1}{\sqrt{n_c}} \sum_{\mathbf{q}} \exp[i\mathbf{q}\cdot(\mathbf{i}+\mathbf{d})] \\ &\quad \times \left\{ \frac{1}{\sqrt{n_c}} \sum_{i'=1}^{n_c} \mathbf{J}_{i'+d} \exp[-i\mathbf{q}\cdot(\mathbf{i}'+\mathbf{d})] \right\} \\ &= \frac{1}{n_c} \sum_{i'=1}^{n_c} \mathbf{J}_{i'+d} \sum_{\mathbf{q}} \exp[i\mathbf{q}\cdot(\mathbf{i}-\mathbf{i}')] = \mathbf{J}_{i+d}.\end{aligned} \tag{E.10}$$

The last step requires Eq. E.6. Therefore we have derived

$$\mathbf{J}_{i+d} = \frac{1}{\sqrt{n_c}} \sum_{\mathbf{q}} \mathbf{J}_{d,\mathbf{q}} \exp[i\mathbf{q}\cdot(\mathbf{i}+\mathbf{d})], \tag{E.11}$$

with the definition of $\mathbf{J}_{d,\mathbf{q}}$ given in Eq. E.9. Since n_c is large, it is convenient for the evaluation of the sum over $\mathbf{q}$, to consider $\mathbf{q}$ as a continuous variable and therefore to replace the sum by an integral with the result

$$\mathbf{J}_{i+d} = \frac{V}{\sqrt{n_c}} \int_{v_c^*} \mathbf{J}_d(\mathbf{q}) \exp[i\mathbf{q}\cdot(\mathbf{i}+\mathbf{d})] \frac{\mathrm{d}^3\mathbf{q}}{(2\pi)^3}$$
$$= \sqrt{n_c}\, v_c \int_{v_c^*} \mathbf{J}_d(\mathbf{q}) \exp[i\mathbf{q}\cdot(\mathbf{i}+\mathbf{d})] \frac{\mathrm{d}^3\mathbf{q}}{(2\pi)^3}, \quad \text{(E.12)}$$

since $V = n_c v_c$. The integral extends over the Brillouin zone. We write $\mathbf{J}_d(\mathbf{q})$ instead of $\mathbf{J}_{d,\mathbf{q}}$ if $\mathbf{q}$ is a continuous variable.

The operators $J^+_{d,\mathbf{q}} \equiv J^X_{d,\mathbf{q}} + iJ^Y_{d,\mathbf{q}}$ and $J^-_{d,\mathbf{q}} \equiv J^X_{d,\mathbf{q}} - iJ^Y_{d,\mathbf{q}}$ are often used. Their Hermitian conjugates satisfy the relations

$$(J^+_{d,\mathbf{q}})^\dagger = J^-_{d,-\mathbf{q}} \quad \text{and} \quad (J^-_{d,\mathbf{q}})^\dagger = J^+_{d,-\mathbf{q}}. \quad \text{(E.13)}$$

Finally we note the useful following relation which directly derives from Eq. E.9:

$$\mathbf{J}_{d,\mathbf{q}+\mathbf{K}} = \mathbf{J}_{d,\mathbf{q}} \exp(-i\mathbf{K}\cdot\mathbf{d}). \quad \text{(E.14)}$$

References

Ashcroft N.W. and Mermin N.D. (2002) *Solid State Physics* (Holt, Rinehart and Winston, New York).

Blau S.K. (2009) *Physics Today*, August, 14.

Dove M.T. (2003) *Structure and Dynamics* (Oxford University Press, Oxford).

Kittel C. (1996) *Introduction to Solid State Physics*, seventh edition (Wiley, New York).

Shechtman D., Blech I., Gratias D., and Cahn J.W. (1984) *Physical Review Letters* **53**, 1951–1953.

Steurer W. (2007) *Zeitschrift für Kristallographie* **222**, 308-309.

Ziman J.M. (1972) *Principles of the Theory of Solids*, second edition (Cambrige University Press, Cambridge).

Appendix F
Correlation functions and linear response theory

The objective of this appendix is to express a symmetrised spin-correlation function in terms of different well-known functions derived from the linear response theory: generalised magnetic susceptibility, static susceptibility and spectral-weight, and Green functions. Some of their properties are reproduced here. No basic derivations are given. They can be found, for example, in the textbooks of Marshall and Lovesey (1971), Lovesey (1987), Jensen and Mackintosh (1991), Chaikin and Lubensky (1995), and White (2007). This appendix has mainly been written using the work of Jensen and Mackintosh since we use their definition of the time Fourier transform.

F.1 Generalised susceptibility

The bulk static susceptibility is the dimensionless ratio $\chi_0^{\alpha\beta} = M^\alpha/H^\beta$, where M^α is the magnetisation of the compound along the α axis when it is submitted to a field $\mathbf{H}$ along the β direction. Here, the α and β indices refer to the crystal reference frame. A proper choice of the crystal axes is dictated by the crystal symmetries (see Section D.8). Note that for an experimental determination of $\chi_0^{\alpha\beta}$, if the susceptibility and the demagnetisation coefficient in the β direction are appreciable, the raw measurement results must be corrected; see beginning of Section 5.3.

It is meaningful to consider a generalised susceptibility which depends on space and time because the magnetic response of a sample is expected to depend on the position at which it is measured and a time-dependent field may be used. According to Kubo,

$$\chi^{\alpha\beta}(\mathbf{q},\omega) = \frac{i}{\hbar}\frac{\mu_0 g^2 \mu_{\rm B}^2}{v_{\rm c}} \lim_{\epsilon\to 0+} \int_0^\infty \langle[\delta J^\alpha(\mathbf{q},t), \delta J^\beta(-\mathbf{q})]\rangle \exp[i(\omega + i\epsilon)t]{\rm d}t. \qquad \text{(F.1)}$$

We recall that we consider periodic crystals in this textbook; see Appendix. E. The susceptibility $\chi^{\alpha\beta}(\mathbf{q},\omega)$ is also dimensionless. For definitiveness, we have chosen to work with $\mathbf{q}$ as a continuous variable. The quantity $\langle[A,B]\rangle$ stands for the thermal average of the commutator of operators A and B. Note that the lower bound of the integral is 0 and not $-\infty$. The susceptibility $\chi^{\alpha\beta}(\mathbf{q},\omega)$ is the boundary value on the real axis of the analytical function $\chi^{\alpha\beta}(\mathbf{q},z)$, which is defined in the complex plane. The parameter g is the spectroscopic splitting factor, e.g. the Landé factor for a free rare-earth ion (see Appendix H), $\mu_{\rm B}$ the electronic Bohr magneton, and $v_{\rm c}$ the volume of the unit cell. For simplicity, we have assumed one magnetic ion per unit cell. The susceptibility $\chi^{\alpha\beta}(\mathbf{q},\omega)$ is written in terms of the fluctuation part of the components of the total angular momentum, such as $\delta J^{\alpha}(\mathbf{q},t)$. Here, we implicitly consider ions with a partially filled f-shell in the $L-S$ coupling scheme. In the case of $3d$ ions, one may substitute the total angular momentum $\mathbf{J}$ by the spin angular momentum $\mathbf{S}$; see Appendix H. In practical work α and β in Eq. F.1 can be replaced with $\{+,-,z\}$.

It is always possible to write

$$\chi^{\alpha\beta}(\mathbf{q},\omega) = \mathcal{R}e\{\chi^{\alpha\beta}(\mathbf{q},\omega)\} + i\mathcal{I}m\{\chi^{\alpha\beta}(\mathbf{q},\omega)\}. \tag{F.2}$$

The real and imaginary parts of the susceptibility are connected by integral relations known as the Kramers–Kronig relations. One of these reads

$$\mathcal{R}e\{\chi^{\alpha\beta}(\mathbf{q},\omega)\} = \frac{1}{\pi}\mathcal{P}\int_{-\infty}^{\infty}\frac{\mathcal{I}m\{\chi^{\alpha\beta}(\mathbf{q},\omega')\}}{\omega'-\omega}\mathrm{d}\omega', \tag{F.3}$$

where $\mathcal{P}\int f(x)\mathrm{d}x$ stands for the Cauchy principal value of the integral of $f(x)$. Note that $\mathcal{R}e\{\chi^{\alpha\beta}(\mathbf{q},\omega=0)\}$ is in fact the static wave vector-dependent susceptibility, i.e.

$$\mathcal{R}e\{\chi^{\alpha\beta}(\mathbf{q},\omega=0)\} = \chi^{\alpha\beta}(\mathbf{q}). \tag{F.4}$$

Obviously, with our conventions $\chi^{\alpha\beta}(\mathbf{q})$ is dimensionless, consistent with the fact that $\chi_0^{\alpha\beta} = \chi^{\alpha\beta}(\mathbf{q}=\mathbf{0})$.

The susceptibility $\chi^{\alpha\beta}(\mathbf{q},\omega)$ is characterised by its singularities. Let us assume the singularities to be only poles, as it is usually the case. They correspond to peaks in $\mathcal{I}m\{\chi^{\alpha\beta}(\mathbf{q},\omega)\}$ which may be observed by neutron scattering techniques. In order to be more explicit, let us consider here a simple one-pole approximation in the vicinity of $\omega=0$:

$$\left[\chi^{\alpha\beta}(\mathbf{q},\omega)\right]^{-1} = \left[\chi^{\alpha\beta}(\mathbf{q})\right]^{-1}\left[1 - i\,\omega/\Gamma^{\alpha\beta}(\mathbf{q})\right]. \tag{F.5}$$

From this formula we derive

$$\mathcal{R}e\{\chi^{\alpha\beta}(\mathbf{q},\omega)\} = \chi^{\alpha\beta}(\mathbf{q})\frac{\left[\Gamma^{\alpha\beta}(\mathbf{q})\right]^2}{\left[\Gamma^{\alpha\beta}(\mathbf{q})\right]^2+\omega^2}, \tag{F.6}$$

and

$$\frac{\mathcal{I}m\{\chi^{\alpha\beta}(\mathbf{q},\omega)\}}{\omega} = \chi^{\alpha\beta}(\mathbf{q})\frac{\Gamma^{\alpha\beta}(\mathbf{q})}{\left[\Gamma^{\alpha\beta}(\mathbf{q})\right]^2+\omega^2}. \tag{F.7}$$

The ratio $\mathcal{I}m\{\chi^{\alpha\beta}(\mathbf{q},\omega)\}/\omega$ has a Lorentzian dependence relative to ω. A two-pole approximation is discussed in the next section.

F.2 Correlation, static susceptibility, and spectral-weight functions

It is convenient to introduce the correlation tensor with components $\Lambda^{\alpha\beta}(\mathbf{q},\omega)$:

$$\Lambda^{\alpha\beta}(\mathbf{q},\omega) = \int_{-\infty}^{\infty}\langle\{\delta J^{\alpha}(\mathbf{q},t)\delta J^{\beta}(-\mathbf{q})\}\rangle\exp(i\omega t)\,\mathrm{d}t. \tag{F.8}$$

The symbol $\langle\{AB\}\rangle$ stands for the thermal average of the symmetrised correlation function of operators A and B: $2\langle\{AB\}\rangle = \langle AB\rangle + \langle BA\rangle$. A non-symmetrised correlation function is sometimes used:

$$\tilde{\Omega}^{\alpha\beta}(\mathbf{q},\omega) = \int_{-\infty}^{\infty} \langle \delta J^{\alpha}(\mathbf{q},t)\delta J^{\beta}(-\mathbf{q})\rangle \exp(i\omega t)\,\mathrm{d}t. \tag{F.9}$$

It fulfills the detailed balance relation

$$\tilde{\Omega}^{\alpha\beta}(\mathbf{q},\omega) = \exp\left(\frac{\hbar\omega}{k_{\mathrm{B}}T}\right)\tilde{\Omega}^{\beta\alpha}(-\mathbf{q},-\omega). \tag{F.10}$$

A theorem, known as the fluctuation–dissipation (Nyquist's) theorem, links the thermal fluctuations, as described by a symmetrised correlation function, to the generalised susceptibility:

$$\Lambda^{\alpha\beta}(\mathbf{q},\omega) = \frac{\hbar v_{\mathrm{c}}}{\mu_0 g^2\mu_{\mathrm{B}}^2}\coth\left(\frac{\hbar\omega}{2k_{\mathrm{B}}T}\right)\mathcal{I}m\{\chi^{\alpha\beta}(\mathbf{q},\omega)\}. \tag{F.11}$$

Because the muon spin usually only probes magnetic excitations at small energy transfer $\hbar\omega$, we shall consider the theorem in the limit $\hbar\omega \ll k_{\mathrm{B}}T$. We then get

$$\Lambda^{\alpha\beta}(\mathbf{q},\omega) = \frac{2k_{\mathrm{B}}Tv_{\mathrm{c}}}{\mu_0 g^2\mu_{\mathrm{B}}^2}\frac{\mathcal{I}m\{\chi^{\alpha\beta}(\mathbf{q},\omega)\}}{\omega}. \tag{F.12}$$

It occurs in many cases that the energy dependence of $\mathcal{I}m\{\chi^{\alpha\beta}(\mathbf{q},\omega)\}/\omega$ can be factorised. It is then of interest to introduce the spectral-weight function, $F^{\alpha\beta}(\mathbf{q},\omega)$, which, by definition, accounts for the whole energy dependence of $\mathcal{I}m\{\chi^{\alpha\beta}(\mathbf{q},\omega)\}/\omega$. It is in general an even function of ω, and is normalised such that

$$\int_{-\infty}^{\infty} F^{\alpha\beta}(\mathbf{q},\omega)\mathrm{d}\omega = 1. \tag{F.13}$$

With our normalisation of the time Fourier transform (Eq. C.50), $F^{\alpha\beta}(\mathbf{q},t=0) = 1/(2\pi)$. Although it may be disturbing that $F^{\alpha\beta}(\mathbf{q},t=0)\neq 1$, obviously the results for the μSR polarisation functions do not depend on the definition of the Fourier transform and the chosen normalisation for the spectral-weight function. It is useful to derive an expression of the nth order moment of ω with respect to $F^{\alpha\beta}(\mathbf{q},t=0)$. Limiting ourselves to the second order moment, we have

$$\int_{-\infty}^{\infty} \omega^2 F^{\alpha\beta}(\mathbf{q},\omega)\mathrm{d}\omega = -\frac{\mu_0 g^2\mu_{\mathrm{B}}^2}{\hbar^2 v_{\mathrm{c}}\chi^{\alpha\beta}(\mathbf{q})}\left\langle\left[\left[\mathcal{H},\delta J^{\alpha}(\mathbf{q})\right],\delta J^{\beta}(-\mathbf{q})\right]\right\rangle, \tag{F.14}$$

where the Hamiltonian $\mathcal{H}$ drives the dynamics of $\mathbf{J}$. This relation is simple enough that the second moment can explicitly be evaluated in specific cases.

Using Eqs. F.13, F.12, F.3, and F.4, we derive

$$\Lambda^{\alpha\beta}(\mathbf{q},\omega) = \frac{2\pi v_{\mathrm{c}}}{\mu_0 g^2\mu_{\mathrm{B}}^2}k_{\mathrm{B}}T\chi^{\alpha\beta}(\mathbf{q})F^{\alpha\beta}(\mathbf{q},\omega). \tag{F.15}$$

The function $F^{\alpha\beta}(\mathbf{q},\omega)$ is convenient for the description of magnetic fluctuations because these are quasielastic excitations, i.e. centred at $\omega = 0$, and their energy width is relatively small; see the discussion at the beginning of Section 10.2. For example, assuming a Lorentzian energy dependence, i.e.

$$F^{\alpha\beta}(\mathbf{q},\omega) = \frac{1}{\pi}\frac{\Gamma^{\alpha\beta}(\mathbf{q})}{[\Gamma^{\alpha\beta}(\mathbf{q})]^2+\omega^2}, \tag{F.16}$$

we obtain

$$\Lambda^{\alpha\beta}(\mathbf{q},\omega) = \frac{2v_c}{\mu_0 g^2\mu_B^2}\, k_B T \chi^{\alpha\beta}(\mathbf{q})\frac{\Gamma^{\alpha\beta}(\mathbf{q})}{[\Gamma^{\alpha\beta}(\mathbf{q})]^2+\omega^2}. \tag{F.17}$$

One of the main advantages of the μSR techniques is the possibility of performing truly zero-field magnetic measurements. As explained in Chapter 10, this means that we have $\omega_\mu = 0$ and therefore can set $\omega = 0$ in Eq. F.17. We then get the important result

$$\Lambda^{\alpha\beta}(\mathbf{q},\omega=0) = \frac{2v_c}{\mu_0 g^2\mu_B^2}\, k_B T\frac{\chi^{\alpha\beta}(\mathbf{q})}{\Gamma^{\alpha\beta}(\mathbf{q})}. \tag{F.18}$$

The time-integrated correlation function of mode $\mathbf{q}$ is therefore proportional to the temperature and the static wave vector-dependent susceptibility of mode $\mathbf{q}$ and inversely proportional to the relaxation rate of that mode. This rate is the half-width at half-maximum of the quasielastic excitation.

If one is interested in the description of low-energy excitations, it may be more appropriate to use the spectral-weight function of the damped-harmonic-oscillator (see, for example, Reich *et al.* (2003)):

$$F^{\alpha\beta}(\mathbf{q},\omega) = \frac{2}{\pi}\frac{\Gamma^{\alpha\beta}(\mathbf{q})\left[\Omega^{\alpha\beta}(\mathbf{q})\right]^2}{\left\{\omega^2-\left[\Omega^{\alpha\beta}(\mathbf{q})\right]^2\right\}^2+4\omega^2\left[\Gamma^{\alpha\beta}(\mathbf{q})\right]^2}. \tag{F.19}$$

This function fulfills the normalisation requirement of Eq. F.13. This is easily shown using the decomposition in partial fractions — see Eq. 10.63 for the case $\Gamma^{\alpha\beta}(\mathbf{q}) < \Omega^{\alpha\beta}(\mathbf{q})$. This simple shape for $F^{\alpha\beta}(\mathbf{q},\omega)$ is very convenient for investigating the influence of low-energy excitations on the spin–lattice relaxation function as shown in Section 10.2.1.

The instantaneous correlation function is in fact proportional to the static wave vector-dependent susceptibility since, from Eqs. F.15, F.13, and C.50, we derive

$$\Lambda^{\alpha\beta}(\mathbf{q},t=0) = \frac{k_B T v_c}{\mu_0 g^2\mu_B^2}\,\chi^{\alpha\beta}(\mathbf{q}). \tag{F.20}$$

Therefore

$$\left\langle\left\{\delta J^\alpha(\mathbf{q})\delta J^\beta(-\mathbf{q})\right\}\right\rangle = \frac{k_B T v_c}{\mu_0 g^2\mu_B^2}\,\chi^{\alpha\beta}(\mathbf{q}), \tag{F.21}$$

or

$$\left\langle\left\{\delta J^\alpha_{\mathbf{q}}\delta J^\beta_{-\mathbf{q}}\right\}\right\rangle = \frac{k_B T v_c}{\mu_0 g^2\mu_B^2}\,\chi^{\alpha\beta}_{\mathbf{q}}, \tag{F.22}$$

assuming $\mathbf{q}$ to be a discrete variable. It is to be noted that the Curie law relating the homogeneous susceptibility $\chi_0 = \chi(\mathbf{q}=0)$ to the temperature in the high-temperature limit follows from Eq. F.20.

We now give a physical insight in the consequences of Eq. F.20. We recall the relation $\delta m_i^\alpha = -g\mu_{\rm B}\delta J_i^\alpha$. Since, according to Eq. E.9,

$$\delta m_{\mathbf{q}}^\alpha = \frac{1}{\sqrt{n_c}} \sum_{i=1}^{n_c} \delta m_i^\alpha \exp\left(-i\mathbf{q}\cdot\mathbf{i}\right), \tag{F.23}$$

we derive

$$\sum_{\mathbf{q}} \left\langle \{\delta m_{\mathbf{q}}^\alpha \delta m_{-\mathbf{q}}^\alpha\}\right\rangle = \sum_{i=1}^{n_c} \left\langle (\delta m_i^\alpha)^2 \right\rangle = n_c \overline{\left\langle (\delta m^\alpha)^2 \right\rangle}. \tag{F.24}$$

The first step is reached because of Eq. E.6 and then we have defined $\overline{\left\langle (\delta m^\alpha)^2 \right\rangle}$ as the average of $\left\langle (\delta m_i^\alpha)^2 \right\rangle$. We recall that $\overline{A}$ stands for the ensemble average of A and $\langle B \rangle$ the thermal average of B. From this result and Eq. F.22 we get

$$\frac{\overline{\langle (\delta m^\alpha)^2 \rangle}}{v_0^2} = \frac{k_{\rm B}T}{\mu_0} \int_{v_c^*} \chi^{\alpha\alpha}(\mathbf{q}) \frac{{\rm d}^3\mathbf{q}}{(2\pi)^3}. \tag{F.25}$$

The integral extends over the first Brillouin zone volume. The volume normalisation v_0 is the volume per magnetic ion, which is here the volume per unit cell since one atom per unit cell is assumed. This sum rule provides a relation between the α component of the fluctuating local magnetic moment, viewed as a random variable with a variance $\overline{\langle (\delta m^\alpha)^2 \rangle}$, and the integral over the Brillouin zone of the static wave vector-dependent susceptibility. However, since we use the small ω (or high-temperature) limit of the fluctuation–dissipation theorem the integral over ω that has been performed to obtain Eq. F.20 is actually underestimated. The integration over $\mathbf{q}$ in Eq. F.25 must therefore be limited to modes such that $\hbar\Gamma^{\alpha\beta}(\mathbf{q}) \lesssim k_{\rm B}T$. This is further discussed in Section 10.2.3.2 where we describe itinerant long-wavelength modes (small $\mathbf{q}$ modes).

The advantage of expressing $\Lambda^{\alpha\beta}(\mathbf{q},\omega)$ in terms of a generalised susceptibility is clearly seen when, for example, the random-phase approximation can be used to model the spin dynamics, as in Section 10.2.3.1. To describe experimental data it may be convenient to write $\Lambda^{\alpha\beta}(\mathbf{q},\omega)$ in terms of the product of a static susceptibility and a spectral-weight function. This enables the extraction of meaningful parameters from the data. However, a microscopic description of the physics involved requires, most of the time, the computation of Green functions.

F.3 Green functions

Here, we provide a short introduction to Green functions, mainly with the purpose to relate $\Lambda^{\alpha\beta}(\mathbf{q},\omega)$ to these functions. We shall limit ourselves to the case of zero temperature. Among the Green functions that can be defined, the most common is the so-called retarded Green function,

$$G_{\rm R}^{\alpha\beta}(\mathbf{q},t) = -\frac{i}{\hbar}\frac{\mu_0 g^2\mu_{\rm B}^2}{v_{\rm c}}\theta_{\rm H}(t)\left\langle [\delta J^\alpha(\mathbf{q},t), \delta J^\beta(-\mathbf{q})]\right\rangle_{T=0}, \tag{F.26}$$

where the Heaviside step function has been introduced: $\theta_{\rm H}(t) = 1$ when $t > 0$, and $\theta_{\rm H}(t) = 0$ otherwise. The notation $\langle\cdots\rangle_{T=0}$ means that we are dealing with the expectation value of the physical quantity of interest, the system being in the ground state. In order to grasp the physical significance of this Green function, let us compute $G_{\mathrm{R}}^{+-}(\mathbf{q}, t)$ for a ferromagnet in the harmonic approximation. According to Eq. E.9,

$$\delta J^+(\mathbf{q}, t) = \frac{1}{\sqrt{n_\mathrm{c}}} \sum_i \delta J_i^+(t) \exp(-i\mathbf{q} \cdot \mathbf{i}). \tag{F.27}$$

Since we focus on the harmonic approximation, we have from Eq. 10.192

$$\delta J_i^+(t) = \sqrt{2J} a_i^\dagger(t), \tag{F.28}$$

because $\langle J^+ \rangle_{T=0} = 0$ in the considered harmonic approximation. Inverting the second equation of Eq. 10.195, we finally derive

$$\delta J^+(\mathbf{q}, t) = \sqrt{2J} a^\dagger(-\mathbf{q}, t) = \sqrt{2J} \exp\left[i\omega(-\mathbf{q})t\right] a^\dagger(-\mathbf{q}). \tag{F.29}$$

The last step follows from Eq. 10.206. Using Eq. E.13, we obtain

$$\delta J^-(-\mathbf{q}) = \sqrt{2J} a(-\mathbf{q}). \tag{F.30}$$

From this material, and accounting for the magnon commutation relation derived from Eqs. 10.195 and 10.193, namely $[a(\mathbf{q}), a^\dagger(\mathbf{q}')] = \delta_{\mathbf{q},\mathbf{q}'}$, we obtain

$$G_{\mathrm{R}}^{+-}(\mathbf{q}, t) = \frac{i}{\hbar} \frac{2J\mu_0 g^2 \mu_\mathrm{B}^2}{v_\mathrm{c}} \exp\left[i\omega(-\mathbf{q})t\right], \tag{F.31}$$

if $t > 0$ and zero otherwise. The Green function is found to be proportional to a plane wave for $t > 0$.

In general terms, a boson field is described by the two Green functions

$$G_{11}(\mathbf{q}, t) = -i\theta_\mathrm{H}(t) \left\langle a^\dagger(\mathbf{q}, t) a(\mathbf{q}) \right\rangle_{T=0}, \quad G_{12}(\mathbf{q}, t) = -i\theta_\mathrm{H}(t) \left\langle a(\mathbf{q}, t) a(-\mathbf{q}) \right\rangle_{T=0}. \tag{F.32}$$

These are the so-called normal and anomalous Green functions, respectively — see, for example, Fetter and Walecka (1971). We use a conventional notation for these functions. Although $G_{12}(\mathbf{q}, t) = 0$ in the harmonic approximation, it may acquire a finite value if non-linear terms of the Holstein–Primakoff transformation are kept.

However, it is not the time-dependent Green functions which are of interest, but their Fourier transforms. According to the definition of such a time Fourier transform (see Eq. C.49), we get the following simple result for the retarded Green function,

$$G_{\mathrm{R}}^{\alpha\beta}(\mathbf{q}, \omega) = \lim_{\epsilon \to 0^+} \int_{-\infty}^{\infty} G_{\mathrm{R}}^{\alpha\beta}(\mathbf{q}, t) \exp[i(\omega + i\epsilon)t] \, \mathrm{d}t = -\chi^{\alpha\beta}(\mathbf{q}, \omega), \tag{F.33}$$

which follows from Eq. F.1. At first look, it does not seem interesting to work with $G_{\mathrm{R}}^{\alpha\beta}(\mathbf{q}, \omega)$ since it is equal to the related generalised susceptibility except for the sign. However, equations of motion can be written for Green functions. Within approximations, they can be solved. Moreover, Dyson equations can also be established for them. This opens up the possibility of computing Green functions in perturbation. In fact the Green function method is very powerful. Once the Green functions at finite temperature have been calculated, it is possible to directly evaluate the correlation functions at any temperature. It is therefore far simpler than the alternative method consisting in the diagonalisation of the Hamiltonian, which usually requires some approximations, and the subsequent calculation of the thermal average of the quantities of interest. We refer to the specialised literature for more information.

References

Chaikin P.M. and Lubensky T.C. (1995) *Principles of Condensed Matter Physics* (Cambridge University Press, Cambridge).

Fetter A.L. and Walecka J.D. (1971) *Quantum Theory of Many-Particle Systems* (McGraw-Hill, New York).

Jensen J. and Mackintosh A.R. (1991) *Rare Earth Magnetism, Structures and Excitations* (Clarendon Press, Oxford).

Lovesey S.W. (1987) *Theory of Neutron Scattering From Condensed Matter*, vol. 1 (Clarendon Press, Oxford).

Marshall W. and Lovesey S.W. (1971) *Theory of Thermal Neutron Scattering* (Clarendon Press, Oxford).

Reich C., Böni P., Roessli B., and Rastelli E. (2003) *Physical Review Letters* **91**, 157203, 1–4.

White R.M. (2007) *Quantum Theory of Magnetism*, third edition (Springer, Berlin).

Appendix G
Nuclear data relevant to μSR

Here we give nuclear data relevant to the context of this monograph. These are the natural abundance and spin, magnetic moment, gyromagnetic ratio, and quadrupolar moment of the naturally occurring isotopes in their ground state.

We first recall the definitions of the basic interaction energies in terms of the listed parameters.

For a given nucleus sitting at site i, the Zeeman Hamiltonian reads

$$\mathcal{H}_{\mathrm{Z},i} = -\hbar\omega_i \, I_i^Z, \tag{G.1}$$

assuming the magnetic field acting on the nuclear spin to be parallel to the Z axis. Obviously, I_i^Z is the projection of the nuclear spin operator along this axis. The gyromagnetic ratio is the constant of proportionality between the nuclear angular frequency, ω_i, and the Z component of the magnetic field acting on the nuclear spin, $B^Z_{\mathrm{loc},i}$: $\gamma_i = \omega_i / B^Z_{\mathrm{loc},i}$. The Zeeman energy can be written in terms of the nuclear magnetic moment, $m_{\mathrm{n},i}$, with

$$m_{\mathrm{n},i} = \gamma_i \, \hbar I_i. \tag{G.2}$$

It is common practice to give $m_{\mathrm{n},i}$ in units of the nuclear magneton, $\mu_{\mathrm{n}} = 5.05078 \times 10^{-27}\,\mathrm{J\,T^{-1}}$; see the numerical values of fundamental constants tabulated at the begining of the book.

The interaction energy between a nuclear charge density and a potential due to external (to the nucleus) sources can be expanded in a Taylor series about the site of the nucleus. Beyond the lowest orders which are not of interest to us, the interaction Hamiltonian is relatively simple if written in the principle axes x, y, and z of the electric-field gradient tensor:

$$\mathcal{H}_{\mathrm{Q}_i} = \frac{e^2 q_i Q_i}{4I_i\,(2I_i - 1)} \left\{ \left[3\,(I_i^z)^2 - (I_i)^2 \right] + \eta_i \left[(I_i^x)^2 - (I_i^y)^2 \right] \right\}, \tag{G.3}$$

where Q_i is the quadrupole moment of the nucleus at site i. The parameters η_i and q_i are the so-called electric-field gradient asymmetry parameter and the electric-field gradient at the nuclear site i, respectively, and e is the proton electric charge.

Table G.1 For each chemical element, we list the non-spinless natural isotopes with their natural abundance, spin, nuclear magnetic moment in units of μ_n, gyromagnetic ratio in megaradians per second and per tesla, and quadrupolar moment in barns (1 barn $\equiv 10^{-28}\,m^2$). Dash signs are shown when the data are irrelevant: for instance, the chemical element is not found in a natural state (e.g. technetium, promethium, polonium), it has no natural non-spinless isotope (cerium, thorium) or the quadrupole moment is not defined for nuclei with a spin equal to 0 or $\frac{1}{2}$. The data are taken from Harris *et al.* (2001), with the exception of those for protactinium, which are from http://www.webelements.com/protactinium/nmr.html as of February 2010. For the sake of simplicity, we have retained at most six significant figures in the gyromagnetic ratios even if more digits are sometimes available. The magnetic moments are calculated from the gyromagnetic ratio, spin values, and Eq. G.2. Again, at most six significant figures are reported. Note that the definition of the nuclear magnetic moment adopted by Harris *et al.* (2001) differs from Eq. G.2.

Chemical element	Isotope	Natural abundance	Spin	Magnetic moment	Gyromagn. ratio	Quadrup. moment
$_1$H	^{1}H	0.9999	$\frac{1}{2}$	2.79285	267.522	—
	^{2}H	$1.15\,10^{-4}$	1	0.857438	41.0663	0.002860
$_2$He	^{3}He	$1.37\,10^{-4}$	$\frac{1}{2}$	−2.12762	−203.802	—
$_3$Li	^{6}Li	0.0759	1	0.822056	39.3717	$-8.08\,10^{-4}$
	^{7}Li	0.9241	$\frac{3}{2}$	3.25646	103.977	−0.0401
$_4$Be	^{9}Be	1	$\frac{3}{2}$	−1.17749	−37.5967	0.05288
$_5$B	^{10}B	0.199	3	1.80064	28.7468	0.08459
	^{11}B	0.801	$\frac{3}{2}$	2.68865	85.8470	0.04059
$_6$C	^{13}C	0.0107	$\frac{1}{2}$	0.702412	67.2828	—
$_7$N	^{14}N	0.99632	1	0.403761	19.3378	0.02044
	^{15}N	0.00368	$\frac{1}{2}$	−0.283189	−27.12618	—
$_8$O	^{17}O	$3.8\,10^{-4}$	$\frac{5}{2}$	−1.89380	−36.2808	−0.02558
$_9$F	^{19}F	1	$\frac{1}{2}$	2.62887	251.815	—
$_{10}$Ne	^{21}Ne	0.0027	$\frac{3}{2}$	−0.661797	−21.1308	0.10155
$_{11}$Na	^{23}Na	1	$\frac{3}{2}$	2.21766	70.8085	0.104
$_{12}$Mg	^{25}Mg	0.1000	$\frac{5}{2}$	−0.855464	−16.3887	0.1994
$_{13}$Al	^{27}Al	1	$\frac{5}{2}$	3.64150	69.7627	0.1466
$_{14}$Si	^{29}Si	0.046832	$\frac{1}{2}$	−0.55529	−53.190	—
$_{15}$P	^{31}P	1	$\frac{1}{2}$	1.13160	108.394	—
$_{16}$S	^{33}S	0.0076	$\frac{3}{2}$	0.643821	20.5568	−0.0678
$_{17}$Cl	^{35}Cl	0.7578	$\frac{3}{2}$	0.821874	26.2420	−0.08165
	^{37}Cl	0.2422	$\frac{3}{2}$	0.684123	21.8437	−0.06435
$_{18}$Ar	—	—	—	—	—	—

Table G.1 *continued*

Chemical element	Isotope	Natural abundance	Spin	Magnetic moment	Gyromagn. ratio	Quadrup. moment
$_{19}$K	^{39}K	0.932581	$\frac{3}{2}$	0.391507	12.5006	0.0585
	^{40}K	$1.17\,10^{-4}$	4	−1.29810	−15.5429	−0.073
	^{41}K	0.067302	$\frac{3}{2}$	0.214870	6.86068	0.0711
$_{20}$Ca	^{43}Ca	0.00135	$\frac{7}{2}$	−1.31764	−18.0307	−0.0408
$_{21}$Sc	^{45}Sc	1	$\frac{7}{2}$	4.75648	65.0880	−0.220
$_{22}$Ti	^{47}Ti	0.0744	$\frac{5}{2}$	−0.78846	−15.105	0.302
	^{49}Ti	0.0541	$\frac{7}{2}$	−1.10417	−15.1095	0.247
$_{23}$V	^{50}V	0.00250	6	3.34569	26.7065	0.210
	^{51}V	0.99750	$\frac{7}{2}$	5.14870	70.4551	−0.052
$_{24}$Cr	^{53}Cr	0.09501	$\frac{3}{2}$	−0.47455	−15.152	−0.150
$_{25}$Mn	^{55}Mn	1	$\frac{5}{2}$	3.46872	66.4525	0.330
$_{26}$Fe	^{57}Fe	0.02119	$\frac{1}{2}$	0.0906230	8.68062	—
$_{27}$Co	^{59}Co	1	$\frac{7}{2}$	4.627	63.32	0.420
$_{28}$Ni	^{61}Ni	0.011399	$\frac{3}{2}$	−0.75003	−23.948	0.162
$_{29}$Cu	^{63}Cu	0.6917	$\frac{3}{2}$	2.22734	71.1179	−0.220
	^{65}Cu	0.3083	$\frac{3}{2}$	2.38161	76.0435	−0.204
$_{30}$Zn	^{67}Zn	0.0410	$\frac{5}{2}$	0.875205	16.7669	0.150
$_{31}$Ga	^{69}Ga	0.60108	$\frac{3}{2}$	2.01659	64.3885	0.171
	^{71}Ga	0.39892	$\frac{3}{2}$	2.56226	81.8117	0.107
$_{32}$Ge	^{73}Ge	0.0773	$\frac{9}{2}$	−0.879467	−9.36030	−0.196
$_{33}$As	^{75}As	1	$\frac{3}{2}$	1.43947	45.9616	0.314
$_{34}$Se	^{77}Se	0.0763	$\frac{1}{2}$	0.535074	51.2539	—
$_{35}$Br	^{79}Br	0.5069	$\frac{3}{2}$	2.10640	67.2562	0.313
	^{81}Br	0.4931	$\frac{3}{2}$	2.27056	72.4978	0.262
$_{36}$Kr	^{83}Kr	0.1149	$\frac{9}{2}$	0.970671	−10.3310	0.259
$_{37}$Rb	^{85}Rb	0.7217	$\frac{5}{2}$	1.35335	25.9270	0.276
	^{87}Rb	0.2783	$\frac{3}{2}$	2.75182	87.8640	0.1335
$_{38}$Sr	^{87}Sr	0.0700	$\frac{9}{2}$	−1.09360	−11.6394	0.335
$_{39}$Y	^{89}Y	1	$\frac{1}{2}$	−0.137415	−13.1628	—
$_{40}$Zr	^{91}Zr	0.1122	$\frac{5}{2}$	−1.30362	−24.9743	−0.176
$_{41}$Nb	^{93}Nb	1	$\frac{9}{2}$	6.1705	65.674	−0.320
$_{42}$Mo	^{95}Mo	0.1592	$\frac{5}{2}$	−0.9140	−17.51	−0.022
	^{97}Mo	0.0955	$\frac{5}{2}$	−0.9333	−17.88	0.255

Table G.1 *continued*

Chemical element	Isotope	Natural abundance	Spin	Magnetic moment	Gyromagn. ratio	Quadrup. moment
$_{43}$Tc	—	—	—	—	—	—
$_{44}$Ru	^{99}Ru	0.1276	$\frac{5}{2}$	−0.6415	−12.29	0.079
	^{101}Ru	0.1706	$\frac{5}{2}$	−0.7188	−13.77	0.457
$_{45}$Rh	^{103}Rh	1	$\frac{1}{2}$	−0.08840	−8.468	—
$_{46}$Pd	^{105}Pd	0.2233	$\frac{5}{2}$	−0.642	−12.3	0.66
$_{47}$Ag	^{107}Ag	0.51839	$\frac{1}{2}$	−0.113680	−10.8892	—
	^{109}Ag	0.48161	$\frac{1}{2}$	−0.130691	−12.5186	—
$_{48}$Cd	^{111}Cd	0.1280	$\frac{1}{2}$	−0.594886	−56.9831	—
	^{113}Cd	0.1222	$\frac{1}{2}$	−0.622301	−59.6092	—
$_{49}$In	^{113}In	0.0429	$\frac{9}{2}$	5.5289	58.845	0.799
	^{115}In	0.9571	$\frac{9}{2}$	5.5408	58.972	0.810
$_{50}$Sn	^{115}Sn	0.0034	$\frac{1}{2}$	−0.91883	−88.013	—
	^{117}Sn	0.0768	$\frac{1}{2}$	−1.00104	−95.8879	—
	^{119}Sn	0.0859	$\frac{1}{2}$	−1.04728	−100.317	—
$_{51}$Sb	^{121}Sb	0.5721	$\frac{5}{2}$	3.3634	64.435	−0.36
	^{123}Sb	0.4279	$\frac{7}{2}$	2.5498	34.892	−0.490
$_{52}$Te	^{123}Te	0.0089	$\frac{1}{2}$	−0.736947	−70.5910	—
	^{125}Te	0.0707	$\frac{1}{2}$	−0.888505	−85.1084	—
$_{53}$I	^{127}I	1	$\frac{5}{2}$	2.81327	53.8957	−0.710
$_{54}$Xe	^{129}Xe	0.2644	$\frac{1}{2}$	−0.777976	−74.5210	—
	^{131}Xe	0.2118	$\frac{3}{2}$	0.691862	22.0908	−0.114
$_{55}$Cs	^{133}Cs	1	$\frac{7}{2}$	2.58202	35.3325	−0.00343
$_{56}$Ba	^{135}Ba	0.06592	$\frac{3}{2}$	0.837941	26.7550	0.160
	^{137}Ba	0.11232	$\frac{3}{2}$	0.937364	29.9295	0.245
$_{57}$La	^{138}La	0.00090	5	3.71364	35.5724	0.450
	^{139}La	0.99910	$\frac{7}{2}$	2.78304	38.0833	0.200
$_{58}$Ce	—	—	—	—	—	—
$_{59}$Pr	^{141}Pr	1	$\frac{5}{2}$	4.1754	81.907	−0.0589
$_{60}$Nd	^{143}Nd	0.122	$\frac{7}{2}$	−1.065	−14.57	−0.630
	^{145}Nd	0.083	$\frac{7}{2}$	−0.656	−8.98	−0.330
$_{61}$Pm	—	—	—	—	—	—
$_{62}$Sm	^{147}Sm	0.1499	$\frac{7}{2}$	−0.8148	−11.15	−0.259
	^{149}Sm	0.1382	$\frac{7}{2}$	−0.6717	−9.192	0.074

Table G.1 *continued*

Chemical element	Isotope	Natural abundance	Spin	Magnetic moment	Gyromagn. ratio	Quadrup. moment
$_{63}$Eu	^{151}Eu	0.4781	$\frac{5}{2}$	3.4717	66.510	0.903
	^{153}Eu	0.5219	$\frac{5}{2}$	1.5330	29.369	2.412
$_{64}$Gd	^{155}Gd	0.1480	$\frac{3}{2}$	−0.25723	−8.2132	1.270
	^{157}Gd	0.1565	$\frac{3}{2}$	−0.33727	−10.769	1.350
$_{65}$Tb	^{159}Tb	1	$\frac{3}{2}$	2.014	64.31	1.432
$_{66}$Dy	^{161}Dy	0.1891	$\frac{5}{2}$	−0.4803	−9.201	2.507
	^{163}Dy	0.2490	$\frac{5}{2}$	0.6728	12.89	2.648
$_{67}$Ho	^{165}Ho	1	$\frac{7}{2}$	4.173	57.10	3.580
$_{68}$Er	^{167}Er	0.2293	$\frac{7}{2}$	−0.56385	−0.77157	3.565
$_{69}$Tm	^{169}Tm	1	$\frac{1}{2}$	−0.2316	−22.18	—
$_{70}$Yb	^{171}Yb	0.1428	$\frac{1}{2}$	0.49367	47.288	—
	^{173}Yb	0.1613	$\frac{5}{2}$	−0.67988	−13.025	2.800
$_{71}$Lu	^{175}Lu	0.9741	$\frac{7}{2}$	2.2327	30.552	3.490
	^{176}Lu	0.0259	7	3.1692	21.684	4.970
$_{72}$Hf	^{177}Hf	0.1860	$\frac{7}{2}$	0.7936	10.86	3.365
	^{179}Hf	0.1362	$\frac{9}{2}$	−0.6409	−6.821	3.793
$_{73}$Ta	^{181}Ta	0.99988	$\frac{7}{2}$	2.3705	32.438	3.170
$_{74}$W	^{183}W	0.1431	$\frac{1}{2}$	0.117785	11.2824	—
$_{75}$Re	^{185}Re	0.3740	$\frac{5}{2}$	3.1871	61.057	2.180
	^{187}Re	0.6260	$\frac{5}{2}$	3.2197	61.682	2.070
$_{76}$Os	^{187}Os	0.0196	$\frac{1}{2}$	0.0646519	6.19289	—
	^{189}Os	0.1615	$\frac{3}{2}$	0.659933	21.0713	0.856
$_{77}$Ir	^{191}Ir	0.373	$\frac{3}{2}$	0.1507	4.812	0.816
	^{193}Ir	0.627	$\frac{3}{2}$	0.1637	5.227	0.751
$_{78}$Pt	^{195}Pt	0.33832	$\frac{1}{2}$	0.60952	58.385	—
$_{79}$Au	^{197}Au	1	$\frac{3}{2}$	0.148158	4.73060	0.547
$_{80}$Hg	^{199}Hg	0.1687	$\frac{1}{2}$	0.505885	48.4579	—
	^{201}Hg	0.1318	$\frac{3}{2}$	−0.560226	−17.8877	0.386
$_{81}$Tl	^{203}Tl	0.29524	$\frac{1}{2}$	1.62226	155.393	—
	^{205}Tl	0.70476	$\frac{1}{2}$	1.63821	156.922	—
$_{82}$Pb	^{207}Pb	0.221	$\frac{1}{2}$	0.582582	55.8046	—
$_{83}$Bi	^{209}Bi	1	$\frac{9}{2}$	4.1106	43.750	−0.516
$_{84}$Po	—	—	—	—	—	—

Table G.1 *continued*

Chemical element	Isotope	Natural abundance	Spin	Magnetic moment	Gyromagn. ratio	Quadrup. moment
$_{85}$At	—	—	—	—	—	—
$_{86}$Rn	—	—	—	—	—	—
$_{87}$Fr	—	—	—	—	—	—
$_{88}$Ra	—	—	—	—	—	—
$_{89}$Ac	—	—	—	—	—	—
$_{90}$Th	—	—	—	—	—	—
$_{91}$Pa	^{231}Pa	1	$\frac{3}{2}$	1.01	32.1	−1.72
$_{92}$U	^{235}U	0.007200	$\frac{7}{2}$	−0.38	−5.2	4.936

References

Harris R.K., Becker E.D., Cabral de Menezes S.M., Goodfellow R., and Granger P. (2001) *Pure and Applied Chemistry* **73**, 1795–1818.

Appendix H
f and $3d$ electronic shell data

In this appendix we provide some information on the electronic ground state of ions with partially filled f or d shells. An introduction has been written by Kittel (1996). For a detailed discussion we refer to the textbook of Abragam and Bleaney (1970) and the references therein. Useful information can also be found in the book of Jensen and Mackintosh (1991).

In the simplest case, a definite electronic valence can be ascribed to an ion sitting in a compound. We shall restrict our present discussion to this physical situation. A number of interactions determine the electronic ground state. A given electron in an ion is Coulomb attracted by the ion nucleus. This attraction is modified by the repulsive field from the other electrons. This results in electronic levels grouped into configurations. The residual electrostatic interaction of the electrons, not represented by the central field which describes the configuration, gives rise, in the $L-S$ coupling scheme, to terms characterised by different L and S built from the same configuration. A set of L and S values, the total orbital and spin angular momenta, respectively, defines a term. For example, the ground state term of an atom with a single electron in the d shell is determined by Hund's first two rules and is labelled as ^{2}D. The letter D means that $L = 2$ and the number 2 corresponds to the spin degeneracy: in our case $S = \frac{1}{2}$ since $2S + 1 = 2$. The energy levels of a term are split by the spin–orbit interaction taken as a perturbation. Finally the magnetic spin–spin interaction must be taken into account. This Russell–Saunders description is valid for free $3d$ transition ions and rare-earth ions ($4f$). It is usually only a rough approximation for an element of the actinide ($5f$) group because the spin–orbit and residual electrostatic interactions are there of about the same strength.

However, we are interested in magnetic ions embedded in compounds. Hence, the additional ligand field (or crystal field) must be taken into account.

In the case of f electrons, the crystal-field interaction is a perturbation relative to the Coulomb and spin–orbit interactions. One must first deal with the spin–orbit interaction. This leads to a splitting of the ground-state term into a set of multiplets. Each of these multiplets is specified by a total angular momentum J which arises from a combination of L and S. The ground state multiplet is determined by Hund's third rule: $J = |L - S|$ for a f shell which is less than half-filled, $J = |L + S|$ for a shell more than half-filled, and $J = S$ in the half-filled case. This ground state describes fairly well the physics of rare-earth ions except the Eu^{3+} and Sm^{3+} ions, for which the influence of the higher lying multiplets cannot be neglected. In Table H.1 are listed the J, as well as the Landé factor g and the product $g\,J$ values for the f shell elements

Table H.1 For each f count (i.e. for each given number of f electrons), we list examples of f ions, the total angular momentum in $\hbar$ units, the Landé factor, and the product of the Landé factor with the total angular momentum in $\hbar$ units. These data are relative to the ground state multiplet.

f count	Examples	J ($\hbar$)	g (-)	gJ ($\hbar$)
1	Ce^{3+}	$\frac{5}{2}$	$\frac{6}{7}$	$\frac{15}{7}$
2	Pr^{3+}, U^{4+}, Pu^{6+}	4	$\frac{4}{5}$	$\frac{16}{5}$
3	Nd^{3+}, U^{3+}, Np^{4+}	$\frac{9}{2}$	$\frac{8}{11}$	$\frac{36}{11}$
4	U^{2+}, Np^{3+}, Pu^{4+}	4	$\frac{3}{5}$	$\frac{12}{5}$
5	Sm^{3+}	$\frac{5}{2}$	$\frac{2}{7}$	$\frac{5}{7}$
6	Eu^{3+}	0	0	0
7	Gd^{3+}, Eu^{2+}, Tb^{4+}	$\frac{7}{2}$	2	7
8	Tb^{3+}	6	$\frac{3}{2}$	9
9	Dy^{3+}	$\frac{15}{2}$	$\frac{4}{3}$	10
10	Ho^{3+}	8	$\frac{5}{4}$	10
11	Er^{3+}	$\frac{15}{2}$	$\frac{6}{5}$	9
12	Tm^{3+}	6	$\frac{7}{6}$	7
13	Yb^{3+}	$\frac{7}{2}$	$\frac{8}{7}$	4

in their ground state. The crystal-electric-field Hamiltonian for a single ion is nicely written in operator form

$$\mathcal{H}_{\mathrm{CEF}} = \sum_{\ell} \sum_{m=-\ell}^{m=+\ell} B_\ell^m \mathcal{O}_\ell^m, \tag{H.1}$$

where B_ℓ^m are parameters which, in practice, must be experimentally determined and $\mathcal{O}_\ell^m$ Stevens' operators. This Hamiltonian is normally written in the reference frame $\{x, y, z\}$ which is adapted to the point symmetry at the f ion site. The symmetry elements dictate which ones of the terms vanish in Eq. H.1. The higher the symmetry, the fewer number of terms. Stevens' operators are written in terms of polynomials of components of $\mathbf{J}$. Lists of them are given, for example, by Abragam and Bleaney (1970) and Jensen and Mackintosh (1991). There is a key result, the Kramers theorem, which should be remembered: in zero field the degeneracies of the energy levels of an half-integral J ion are even. There is no rule for integral J values: if a degeneracy is even, this is accidental. In the literature, an ion with an half-integral J value is commonly referrred to as a Kramers ion, while if J has an integer value the ion is said to be non-Kramers.

The description of the electronic structure of $3d$ ions is more complicated than in the f ion case, the essential reason being that, depending on the environment of the d ion, the crystal field can be stronger or weaker than the residual electrostatic interaction. A constant feature is anyhow that the crystal field is always relatively strong. This is so because the $3d$ shell is the outermost one in $3d$ ions, while the $4f$ shell

Table H.2 For each d count, we give examples of $3d$ ions and specify the ground state (GS) term of the free ion. Then information on the ground state is given for an ion in the weakly bound and strongly bound cases. The total spin is given in $\hbar$ units as well as the orbital degeneracy in an octahedral ligand field. For the strongly bound state, the interaction energy of the electron spins is assumed to be smaller than the crystal-field energy. For the weakly bound case, the tetrahedral symmetry is also considered. Adapted from Table 7.2 of Abragam and Bleaney (1970).

d count	Examples	GS term free ion	Weakly bound S ($\hbar$)	Weakly bound Deg. Octa.	Weakly bound Deg. Tetra.	Strongly bound S ($\hbar$)	Strongly bound Deg. Octa.
1	V^{4+}, Ti^{3+}	^2D	$\frac{1}{2}$	3	2	$\frac{1}{2}$	3
2	Cr^{4+}, V^{3+}	^3F	1	3	1	1	3
3	Mn^{4+}, Cr^{3+}, V^{2+}	^4F	$\frac{3}{2}$	1	3	$\frac{3}{2}$	1
4	Mn^{3+}, Cr^{2+}	^5D	2	2	3	1	3
5	Fe^{3+}, Mn^{2+}, Cr^{+}	^6S	$\frac{5}{2}$	1	1	$\frac{1}{2}$	3
6	Co^{3+}, Fe^{2+}, Mn^{+}	^5D	2	3	2	0	1
7	Ni^{3+}, Co^{2+}, Fe^{+}	^4F	$\frac{3}{2}$	3	1	$\frac{1}{2}$	2
8	Cu^{3+}, Ni^{2+}, Co^{+}	^3F	1	1	3	1	1
9	Cu^{2+}, Ni^{+}	^2D	$\frac{1}{2}$	2	3	$\frac{1}{2}$	2

lies deep inside the rare-earth ions. In Table H.2 we summarise some characteristics of the ground state in limiting cases. General characteristics of the table and examples of applications are given below.

The strong interaction of an ion with the crystal field has a major effect: the coupling of **L** and **S** is largely broken up, so that the states are specified by their S values.

Let us suppose the crystal field to be a perturbation relative to the Coulomb field. This is the weakly bound case. Then it is justified to determine the electronic ground state starting from a term such as ^2D, for example. The crystal field splits the energy levels of the ground state term. Let us pick up two examples. From Table H.2, we learn that the ^2D term describes ions with d counts equal to 1 and 9. Obviously $S = \frac{1}{2}$ for these two types of ions. The nature of the ground state is determined by the point symmetry at the ion site but also by the sign of the parameters of the crystal-field Hamiltonian, as the inversion in the values of the degeneracies shows. Physically, this is easily understood since an ion with 9 electrons can be regarded as having a full electronic shell with a hole in it and is equivalent to an ion with one electron in its d shell.

The weakly bound model explains a lot of experimental results. However, the crystal field is not always a small perturbation (Abragam and Bleaney 1970). An alternative is the strongly bound model for which the five orbital states are split into a lower triplet and an upper doublet. This is, of course, consistent with the splitting between the D and F terms of the free ion model. The nomenclature t_2 and e is used for the triplet and doublet orbitals, respectively. Obviously the crystal-field energy is minimised by placing as many electrons as possible in the lower t_2 states, but we must take into account the magnetic coupling between the electron spins as expressed by first Hund's rule. We can consider this coupling smaller or larger than the crystal field. In Table H.2 the crystal field of octahedral symmetry is assumed to be stronger than the magnetic coupling. It is important to note in the table that orbital singlets (i.e. with degeneracy equal to 1) are possible. Isotropic magnetic properties are expected in such cases. However, breakdown of the octahedral crystal-field symmetry may lead to magnetic properties with small anisotropy.

While the two models give consistent predictions for d counts equal to 1, 2, 3, 8, and 9, the differences are large when the count is equal to 4, 5, 6, or 7. The reader may find surprising that the degeneracy is equal to 1 for the half-filled shell case when the crystal field is weak. This simply reflects the presence of residual interactions as explained by Abragam and Bleaney (1970).

References

Abragam A. and Bleaney B. (1970) *Electron Paramagnetic Resonance of Transition Ions* (Clarendon Press, Oxford).

Jensen J. and Mackintosh A.R. (1991) *Rare Earth Magnetism, Structures and Excitations* (Clarendon Press, Oxford).

Kittel C. (1996) *Introduction to Solid State Physics*, seventh edition (Wiley, New York).

Index